CONTINENTAL MARGIN SEDIMENTATION

Other publications of the International Association of Sedimentologists

SPECIAL PUBLICATION NUMBER 37 OF THE INTERNATIONAL ASSOCIATION OF SEDIMENTOLOGISTS

Continental Margin Sedimentation: From Sediment Transport to Sequence Stratigraphy

EDITED BY

Charles A. Nittrouer, James A. Austin, Michael E. Field, Joseph H. Kravitz, James P.M. Syvitski and Patricia L. Wiberg

SERIES EDITOR

Ian Jarvis
School of Earth Sciences & Geography
Centre for Earth & Environmental Science Research
Kingston University
Penrhyn Road
Kingston upon Thames KT1 2EE
UK

Blackwell Publishing

© 2007 International Association of Sedimentologists
and published for them by
Blackwell Publishing Ltd

BLACKWELL PUBLISHING
350 Main Street, Malden, MA 02148-5020, USA
9600 Garsington Road, Oxford OX4 2DQ, UK
550 Swanston Street, Carlton, Victoria 3053, Australia

First published 2007 by Blackwell Publishing Ltd

1   2007

Library of Congress Cataloging-in-Publication Data
Continental margin sedimentation : from sediment transport to sequence
stratigraphy / edited by C.A. Nittrouer . . . [*et al.*].
     p. cm. – (Special publication number 37 of the International Association of
Sedimentologists)
   Includes bibliographical references and index.
   ISBN 978-1-4051-6934-9 (hardback : alk. paper)
   1. Sedimentation and deposition.   2. Sedimentology.   3. Sequence stratigraphy.
   4. Sediment transport.   5. Continental margins.   I. Nittrouer, Charles A.

   QE571.C57 2007
   551.3′53–dc22

                                                                    2007007166

A catalogue record for this title is available from the British Library.

Set in 10.5/12.5pt Palatino
by Graphicraft Limited, Hong Kong
Printed and bound in Singapore
by Markono Print Media Pte Ltd

The publisher's policy is to use permanent paper from mills that operate a sustainable
forestry policy, and which has been manufactured from pulp processed using acid-free
and elementary chlorine-free practices. Furthermore, the publisher ensures that the
text paper and cover board used have met acceptable environmental accreditation
standards.

For further information on
Blackwell Publishing, visit our website:
www.blackwellpublishing.com

Contents

Preface

'If we don't learn from history, we're doomed to relive it'. Unlike human history, most of the events that form the record of Earth history are out of our control. However, we may still learn from them and prepare ourselves for future environmental events (e.g. storm surges, sea-level rise). Understanding continental-margin sedimentation is important for many reasons, as diverse as finding natural resources and maintaining safe navigation. In addition, the stratigraphy that results from margin sedimentation provides an extremely rich record of Earth history – including the natural processes and, more recently, the human impacts operating both on land and in the sea. Unfortunately, we cannot learn from this record until we can read it. Large portions of the following text have this purpose, and collectively provide a unique contribution to the continuing legacy of studies to unravel the secrets of margin stratigraphy.

GOALS AND ORGANIZATION

This volume is an outgrowth of the STRATAFORM programme (STRATA FORmation on Margins) funded by the US Office of Naval Research (ONR). Consequently, the goals and organization of the volume reflect those of STRATAFORM. In that programme, we set out to integrate across three major domains in our geological and geophysical examination of continental-margin sedimentation: environments, from inner shelves to distal slopes; processes, from discrete events to the long-term preserved stratigraphy; and techniques, from observations to modelling. Pieces of this integrated approach have been undertaken previously, but STRATAFORM broke new ground in its holistic investigation across such a complex matrix.

Construction of this volume has followed a similar pattern, and has experienced the same challenges. First, continental-margin sedimentation is an extremely broad field and we have had to define workable boundaries, so the scope of the volume is tractable. Future investigators and funding agencies are offered this result as

a blue print for studies of margin sedimentation in other environments. Second, participants have had to think beyond their individual disciplinary specialities, so integration of results could be balanced and fair. This has not always been easy, but the consensus of the group has made it happen (and ONR programme manager, Dr Joseph Kravitz, was persuasive).

Finally, the actual mechanics of merging many people and their diverse contributions has probably been the toughest challenge of all. Rather than creating a 'project volume' with a pot-pourri of loosely related papers, we have envisioned a written document that is comprehensive and presents continua of ideas across the spectra of the research. For independent-minded scientists experienced in writing research papers in their areas of speciality, a contiguous blend of summary papers with finite boundaries and required contents is a challenge. However, we succeeded, and the results are presented in the papers that follow.

THANKS

There are many people to thank for the scientific research, operations, leadership and support that have carried the STRATAFORM programme from its inception through the completion of this volume. The research was undertaken first, and we are indebted to the legions of investigators, students and technicians at participating institutions who were involved in STRATAFORM cruises, experiments, and programming. The ONR was the funding agency, and we appreciate its commitment to this extended research effort. Among the ranks of ONR managers the greatest supporter is honoured below.

The authors created the text and the editors helped make it better. Great thanks go to the lead authors, who stuck to the task long after the programme funding ended. A diverse group of reviewers provided constructive advice, and included people outside and inside STRATAFORM, as well as the editors. Each of these receives our appreciation, and they are listed below.

Bob Aller
Mead Allison
Carl Amos
James Austin
Sam Bentley
Jeff Borgeld
Dave Cacchione
Dick Faas
Mike Field
Roger Flood
Carl Friedrichs
Jim Gardner
Rocky Geyer
John Goff
Steve Goodbred
Courtney Harris
John Jaeger
Gail Kineke

Joe Kravitz
Steve Kuehl
Lonnie Leithold
Tim Milligan
Dave Mohrig
Beth Mullenbach
Alan Niedoroda
Chuck Nittrouer
Andrea Ogston
Dan Orange
Chris Paola
Harry Roberts
Rudy Slingerland
James Syvitski
Peter Traykovski
Gert Jan Weltje
Pat Wiberg
Don Wright

CHUCK NITTROUER, JAMIE AUSTIN, MIKE FIELD, JAMES SYVITSKI and PAT WIBERG (as co-editors)

On behalf of all scientists involved in this volume and in STRATAFORM

Dr Joseph Kravitz along the bank of the Eel River, near its mouth. (Photograph courtesy of Rob Wheatcroft.)

We appreciate the efforts of Ian Jarvis and his assistant Stella Bignold at the editorial office of IAS special publications (in Kingston University) and the efforts of personnel at Blackwell Publishing, who helped us to produce the volume we envisioned.

DEDICATION

STRATAFORM would not have been possible without the stalwart support of Dr Joseph Kravitz, and we express our recognition and appreciation by dedicating this volume to him.

Joe is Pennsylvania born and raised. Educated at Syracuse and George Washington Universities, he has worked his way through life with drive and determination. He managed a number of programmes over the years at ONR and NOAA, and his last was his most ambitious in terms of scientific goals and scope.

Joe provided stern, but caring leadership. He nurtured investigators in a way that allowed them to employ their best creative talents. Any successes that came from STRATAFORM were made possible by Joe. It was a special period in the professional lives of all those involved. The individuals and the science benefited from the leadership and vision he brought to the programme. Good deeds deserve recognition, and this volume is our gift, and our thanks, to Joe Kravitz.

Writing a Rosetta stone: insights into continental-margin sedimentary processes and strata

CHARLES A. NITTROUER*, JAMES A. AUSTIN JR†, MICHAEL E. FIELD‡,
JOSEPH H. KRAVITZ§, JAMES P.M. SYVITSKI¶ and PATRICIA L. WIBERG**

*School of Oceanography and Department of Earth and Space Sciences, University of Washington, Seattle, WA 98195, USA
(Email: nittroue@ocean.washington.edu)
†Institute for Geophysics, John A. and Katherine G. Jackson School of Geosciences, University of Texas, Austin, TX 78759, USA
‡Pacific Science Centre, US Geological Survey, Santa Cruz, CA 95060, USA
§Department of Earth and Environmental Sciences, George Washington University, Washington, DC 20052, USA
¶Institute for Arctic and Alpine Research, University of Colorado, Boulder, CO 80309, USA
**Department of Environmental Sciences, University of Virginia, Charlottesville, VA 22904, USA

ABSTRACT

Continental margins are valuable for many reasons, including the rich record of Earth history that they contain. A comprehensive understanding about the fate of fluvial sediment requires knowledge that transcends time-scales ranging from particle transport to deep burial. Insights are presented for margins in general, with a focus on a tectonically active margin (northern California) and a passive margin (New Jersey). Formation of continental-margin strata begins with sediment delivery to the seabed. Physical and biological reworking alters this sediment before it is preserved by burial, and has an impact upon its dispersal to more distal locations. The seabed develops strength as it consolidates, but failure can occur and lead to sediment redistribution through high-concentration gravity flows. Processes ranging from sediment delivery to gravity flows create morphological features that give shape to continental-margin surfaces. With burial, these surfaces may become seismic reflectors, which are observed in the subsurface as stratigraphy and are used to interpret the history of formative processes. Observations document sedimentary processes and strata on a particular margin, but numerical models and laboratory experimentation are necessary to provide a quantitative basis for extrapolation of these processes and strata in time and space.

Keywords Continental margin, continental shelf, continental slope, sedimentation, stratigraphy.

INTRODUCTION

The history of processes influencing the Earth is recorded in many ways. The sedimentary strata forming around the fringes of the ocean contain an especially rich record of Earth history, because they are impacted by a complex array of factors within the atmosphere (e.g. climate), the lithosphere (e.g. mountain building) and the biosphere (e.g. carbon fluxes).

Events that occur in coastal oceans and adjacent land surfaces have great impacts on humans, because most people live near the sea and depend on the bountiful resources formed or found there. Landslides, river floods, storm surges and tsunamis are examples of processes that can have sudden and catastrophic consequences for coastal regions. Other important processes have characteristic time-scales that are longer and the processes are somewhat more predictable; e.g. sea-level rise or fall, crustal uplift or subsidence, sediment accumulation or erosion. The confluence of terrestrial and marine processes occurs in the physiographical region known as the **continental margin**, extending

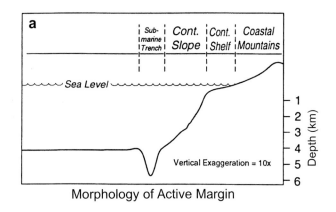

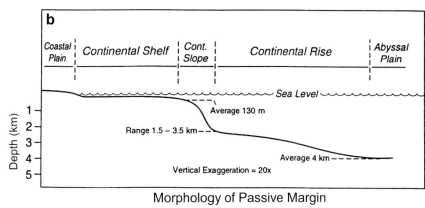

Fig. 1 Morphology of continental margins. (a) Typical morphology for a tectonically active continental margin, where oceanic and continental plates collide and subduction occurs. (b) A passive margin, where the continental and oceanic crust moves in concert. Significant distinctions include the presence of a coastal mountain range, narrow and steep continental shelf, and submarine trench (which can be filled with sediment) for the active margin. The passive margin is characterized by a coastal plain, broad continental shelf, and continental rise. (From Brink *et al.*, 1992.)

from coastal plains and coastal mountain ranges, across shorelines, to shallow continental shelves, and steeper and deeper continental slopes and rises (Fig. 1).

The interplay of terrestrial and marine processes on continental margins creates a complex mixture of stratigraphic signals in the sediments that accumulate there. This region of Earth, however, has the largest sediment accumulation rates, which create the potential for resolving diverse signals imparted over a range of time-scales (e.g. signals of river floods, and of sea-level change). Not only are the continental margins diverse and complex, but they are also very energetic. Waves, tides and currents are strong here, and provide the means to erase as well as form sedimentary records. Continental-margin stratigraphy represents a great archive of Earth history, but the challenges of reading it are also great, and require a fundamental understanding (a Rosetta stone) for translating stratigraphic character into a record of sedimentary processes.

The goal of this introductory paper is to distill the knowledge presented in the following papers of this volume, and integrate the recent insights that have been developed regarding sedimentary processes on continental margins, their impacts on strata formation, and how the preserved strata can be used to unravel Earth history. In contrast to the following papers that isolate topics, this paper highlights the linkages that come from a multi-dimensional perspective of margins. This is a summary of continental-margin sedimentation: from sediment transport to sequence stratigraphy.

THE BOUNDARY CONDITIONS

The full range of topics relevant to continental-margin sedimentation is extensive. In high latitudes, present or past glacial processes and sediments have a strong impact on sedimentation. In some low-latitude settings, biogenic carbonate sediments and their unique mechanisms of formation (e.g. coral reefs) dominate sedimentation. However, from polar to tropical environments, rivers can be the overwhelming sediment source for strata formation on continental margins. Margins affected by fluvial sediment, therefore, are the focus of this discussion.

Rivers add to the complexity of continental-margin processes through their discharge of freshwater and solutes. Rivers are also the dominant suppliers of particulate material from land to sea (globally ~85–95% is fluvial sediment; Milliman & Meade, 1983; Syvitski *et al.*, 2003). The largest rivers create extensive deposits near their mouths (e.g. Amazon, Ganges–Brahmaputra, Mississippi), but the combined discharges of moderate and small rivers (especially from coastal mountain ranges) dominate global sediment supply (Milliman & Meade, 1983; Milliman & Syvitski, 1992) and, therefore, are important to the creation of continental-margin stratigraphy.

Fluvial sedimentation on tectonically **active** and **passive** margins (Fig. 1) can now be examined over time-scales ranging from wave periods of seconds, to the stratigraphy formed and preserved over 10^7 years. Studies can span this broad range of time-scales with new rigour because numerous instruments (e.g. acoustic sensors for particle transport) and techniques (e.g. short-lived radioisotopes for seabed dynamics) have been developed recently to provide insights into important sedimentary processes. Similarly, significant advances have been made in seismic tools (e.g. CHIRP reflection profiling, multibeam swath mapping) that allow better resolution of stratigraphic surfaces. Recent advances in numerical modelling and laboratory simulations provide the opportunity quantitatively to span the temporal gap between processes operating over seconds and stratigraphy developed over millions of years.

The continental shelf and slope are the primary targets of this discussion because they are among the most dynamic environments on Earth, and record a wealth of information about environmental processes. At the boundary between land and ocean, they are impacted by energetic events characteristic of both regions (e.g. river floods, storm waves). On longer time-scales as sea level rises and falls, shelves are flooded and exposed, and slopes switch from sediment starvation to become recipients of all fluvial sediment. The boundaries between subaerial and submarine settings (i.e. the **shoreline**) and between shelf and slope (i.e. the **shelf break**) represent two dominant environmental and physiographical transitions on Earth. The transfers of sediment across these boundaries are also of special interest, because the particles on each side experience much different processes and therefore different fates. For example, on active margins, sediment crossing the shelf break can be subducted, but sediment remaining on the shelf cannot.

In this paper, fluvial sediment supply is taken as a source function on the landward side, without extensive discussion about the myriad processes occurring on land. On the seaward side, the evaluation of sedimentary processes and their effects on the formation and preservation of strata stops short of the continental rise, and the submarine fans formed there. The goal is a general understanding of sedimentary processes and stratigraphy on the continental shelf and slope, and the complex interrelationships are highlighted through two common study areas.

THE COMMON THREADS

The discussions within this paper cascade from short to long time-scales, from surficial layers of the seabed to those buried deeply within, and from shallow to deep water. Continuity in discussions is provided through examples from two diverse continental margins, which have been studied intensely throughout the STRATAFORM programme (STRATA FORmation on Margins; Nittrouer, 1999). The continental margin of northern California, near the Eel River (between Cape Mendocino and Trinidad Head; Fig. 2), is undergoing active tectonic motions and experiencing a range of associated sedimentary processes. In contrast, the margin of New Jersey (Fig. 3) is moving passively in concert with the adjacent continental and oceanic crust, and a distinctly different history of sedimentary processes is recorded.

Eel River (California) continental margin

The Eel basin is typical for rivers draining tectonically active continental margins. It is small (~9000 km²), mountainous (reaching elevations > 2000 m), and composed of intensely deformed and easily erodible sedimentary rocks (Franciscan mélange and other marine deposits). These conditions lead to frequent subaerial landslides, especially because the high elevations cause orographic effects that intensify

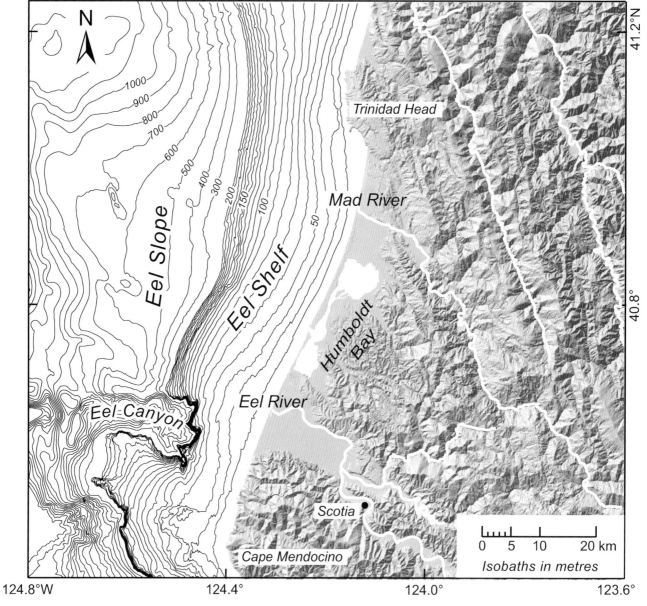

Fig. 2 The study area for the Eel margin, stretching from Cape Mendocino to Trinidad Head. The Eel River supplies an order of magnitude more sediment (~2 × 10^7 t yr^{-1}) than the Mad River. Below the town of Scotia (location of the lowermost river gauge), the river mouth has a small delta plain and most Eel River sediment escapes to the ocean. The shelf break is in a water depth of ~150 m, and is indented by Eel Canyon west of the river mouth. (Modified from Sommerfield *et al.*, this volume.)

rainfall from winter storm systems moving eastward off the Pacific. The annual **sediment yield** (mass discharge per basin area) is large (~2000 t km^{-2}), and although interannual discharge is highly variable, the mean value of sediment supplied to the ocean is estimated to be ~2 × 10^7 t yr^{-1} (Brown & Ritter, 1971; Wheatcroft *et al.*, 1997; Sommerfield & Nittrouer,

1999; Syvitski & Morehead, 1999). The grain size of the combined bedload and suspended load is relatively coarse (~25% sand; Brown & Ritter, 1971), due to the mountainous terrain and short length of the river (~200 km). Its size and orientation (generally parallel to the coastline) cause the entire basin to receive precipitation simultaneously during

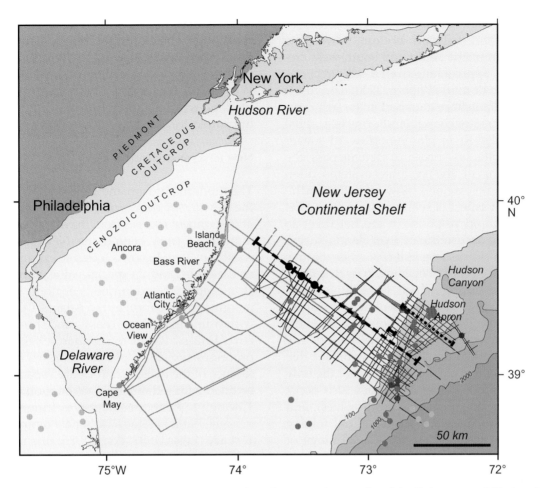

Fig. 3 The study area for the New Jersey margin, stretching between the mouths of the Delaware and Hudson Rivers. Most sediment is trapped in the estuaries at the river mouths and behind the New Jersey coastal barriers. The importance of the New Jersey margin is found in the underlying stratigraphy, which is a classic representation of passive-margin evolution. Some of the data used in this volume were collected at locations shown by the dots (drill sites) and lines (seismic profiles). Isobaths are metres. The shelf break is at ~100 m, and is indented by multiple submarine canyons including Hudson Canyon. (Modified from Mountain *et al.*, this volume.)

storms, and therefore the river discharge increases rapidly.

For the Eel River, major rainfall events commonly lead to episodic floods of the basin. Fluvial sediment discharge increases exponentially with water discharge (Syvitski *et al.*, 2000), and large floods dominate intra-annual and interannual variability of sediment transport. The mouth of the Eel River has no estuary and a very small delta plain (Fig. 2), so periods of sediment transport in the river become periods of sediment supply to the ocean. Most supply occurs during the winter (~90%; Brown & Ritter, 1971), and, for the past ~50 yr, decadal floods during the winter have had a significant impact on the river geomorphology and ocean sedimentation. The largest flood during this period was in 1964 and, more recently, a couplet of significant floods occurred in 1995 and 1997 (Wheatcroft & Borgeld, 2000).

Low-pressure cyclonic systems move eastward from the Pacific Ocean toward the west coast of North America. Commonly there is an asymmetry, such that the steepest pressure gradients are associated with the leading edges of the systems. Therefore, initial winds are strong, from the south or south-west, and Coriolis and frictional forces cause **Ekman transport** of surface water eastward toward the coast. Water elevations rise there, creating a seaward-sloping water surface that produces northward **barotropic** flow of shelf water. The eastward component of surface flows also causes **downwelling** and seaward bottom flows. The strong winds from the south and south-west create large waves approaching from those directions (as high as 10 m or more; Wiberg, 2000), and result in northward **alongshore transport** in the surf zone. This transport creates coastal landforms (e.g. spits) that direct the Eel River plume northward (Geyer et al., 2000). As the low-pressure systems pass, the trailing portions of the cyclonic systems often cause winds to reverse and blow from the north.

An important aspect of sedimentation on the Eel margin is the rapid response of the Eel River to rainfall, and the occurrence of river floods during energetic ocean storms (see Hill et al., this volume, pp. 49–99). These types of events can be described as **wet storms**, during which large fluvial discharges reach the ocean when sediment transport processes are strong. The river plume, coastal current, and wind waves during these periods are important dynamical processes for sediment dispersal on the Eel margin, but they are not the only processes. Energetic ocean conditions also occur without river floods (e.g. large swell waves), and these are described as **dry storms**. Tidal forcing is important on the Eel margin. A tidal range of ~2 m causes current speeds ~50 cm s^{-1} oriented primarily alongshelf. The tidal prism flowing in and out of Humboldt Bay (Fig. 2) influences shelf circulation near its mouth (Geyer et al., 2000). In addition, tidal forcing in deeper water initiates internal waves that maintain suspended sediment near and below the shelf break (McPhee-Shaw et al., 2004).

Sediment from the Eel River and the adjacent Mad River (~10% of the Eel discharge) is supplied to a relatively narrow continental shelf surface (~20 km wide) constrained by promontories: Cape Mendocino to the south and Trinidad Head to the north (Fig. 2). The shelf break is at ~150 m water depth and Eel Canyon incises the shelf surface just west of the river mouth. The morphological elements of the surface (e.g. narrow and steep shelf) and subsurface (e.g. structural folds and faults) are largely the result of tectonic activity. The present Eel margin is part of the larger Eel River Basin (Clarke, 1987, 1992; Orange, 1999), which became a forearc basin in the Miocene and accumulated > 3000 m of marine sediment by the middle Pleistocene (~1 Ma). At that time, the northward migration of the Mendocino Triple Junction and subduction associated with the Gorda Plate initiated modern tectonic conditions. The Gorda and North American plates are converging at ~3 cm yr^{-1} (DeMets et al., 1990), and create localized uplift and subsidence with a WNW–ESE orientation. This is the tectonic framework on which Eel margin sedimentation has been imprinted for the past million years.

New Jersey continental margin

The modern Hudson and Delaware Rivers bracket the New Jersey continental margin (Fig. 3), but very little sediment escapes from the estuaries at the river mouths or from behind the New Jersey barrier coastline. New Jersey is a classic example of a passive margin, and its special value comes from the stratigraphic record buried beneath its surface. The margin began to form as the Atlantic Ocean opened with rifting in the Late Triassic and spreading in the Early Jurassic (Grow & Sheridan, 1988). A range of processes typical of passive margins caused subsidence of the margin, and created space that could be filled with sediment (i.e. **accommodation space**). Through the Cretaceous, it was fringed by a barrier reef, but it became a carbonate ramp in the early Tertiary (Jansa, 1981; Poag, 1985) due to continued subsidence and sediment starvation.

Sediment accumulation rates dramatically increased (to ~10–100 m Myr^{-1}) in the late Oligocene and early Miocene, due to tectonic activity in the source area that increased fluvial sediment supply to the margin (Poag, 1985; Poag & Sevon, 1989). The resulting stratigraphic record has been examined by many seismic and drilling investigations (Mountain et al., this volume, pp. 381–458). Cycles of sea-level fluctuation are recorded by repetitive sequences of strata: a basal layer of **glauconite** sand (an authigenic mineral indicating negligible sedimentation) overlain by silt, which coarsens upward into quartz sand (Owens & Gohn, 1985; Sugarman & Miller, 1997). These sequences reflect sea-level rise, followed by seaward migration of shelf and nearshore sedimentary environments. During the Miocene, most of the sediment accumulation resulted from migration on the shelf of morphological structures known as **clinoforms** (Greenlee et al., 1992). These have a shallow, gently dipping **topset** region of upward growth and, farther offshore, a steeper **foreset** region of seaward growth (see below). The extent of sea-level fluctuations during the Miocene is controversial, but probably was subdued (20–30 m;

Kominz *et al.*, 1998; Miller *et al.*, 1998) relative to fluctuations that followed (> 100 m) in the Pleistocene.

Glacial erosion in the source area was largely responsible for supplying sediment to the marine environment during the Pleistocene. Earlier sedimentation had built a wide shelf with a gentle gradient, but margin subsidence had slowed and was producing little new accommodation space on the inner shelf. During lowered sea level, glacial outwash streams incised the shelf and icebergs even scraped the surface (Duncan & Goff, 2001; Fulthorpe & Austin, 2004). Generally, sediment accumulation was displaced seaward to the outer shelf and upper slope, dramatically changing the sedimentation regime (Greenlee *et al.*, 1988, 1992; Mountain *et al.*, this volume, pp. 381–458). Clinoforms were active there, and the inflection in their bathymetric gradient became the shelf break. Sedimentation on the continental slope increased significantly, which caused seaward growth of the shelf break to its present position > 100 km from shore. The slope also grew seaward, but the influx of sediment initiated localized erosional processes. Miocene **submarine canyons** and smaller erosional features (**gullies**) were buried or reactivated by the substantial sediment supply to the relatively steep slope (Mountain, 1987; Pratson *et al.*, 1994). The long history of the New Jersey margin provides an opportunity to observe how a diverse range of sedimentary processes impacts the preserved strata on a passive margin.

SEDIMENT DELIVERY

Detailed aspects of sediment delivery on continental margins have been addressed in this volume by Hill *et al.* (pp. 49–99) and Syvitski *et al.* (pp. 459–529).

General considerations

The first step in the formation of continental-margin strata is sediment delivery. The timing and content of fluvial discharge depend on many factors, such as basin character, weather, glaciation and groundwater flow (Beschta, 1987), which can be observed and modelled. Commonly, a **rating curve** is developed to relate sediment flux to river discharge (Cohn, 1995; Syvitski *et al.*, 2000). The observations needed to generate a rating curve are confounded by difficulty in making measurements over a range of flow conditions – especially during large flood events, which are important periods because much sediment is transported (Wheatcroft *et al.*, 1997). Other difficulties are imposed by changes in the curve that occur when the river basin is altered naturally (e.g. landslides) or unnaturally (e.g. land use). Asymmetry in sediment discharge is commonly associated with rise and fall of river stage, and can cause a **hysteresis** whereby different sediment fluxes occur for the same discharge (Brown & Ritter, 1971; Meade *et al.*, 1990). Over longer time-scales of climatic and sea-level changes, adjustments to the snow pack and basin size have an impact upon the timing and amount of discharge (Mulder & Syvitski, 1996). Fluctuations in regional precipitation patterns also can modify the shape of the river hydrograph and the dominance of sustained flows or episodic floods, which are conditions that affect sediment transport substantially. For example, strengthening of the monsoonal regime in the early Holocene caused the Ganges–Brahmaputra system to have more than twice its present sediment load (Goodbred & Kuehl, 2000).

Rivers supply a range of grain sizes to the ocean. Sediment in suspension (mostly silt and clay, i.e. < 64 μm) generally represents ~90% of the discharge, and the remainder is bedload (almost entirely sand; Meade, 1996). Early recognition of patterns for modern sediment distribution on continental margins provided suggestions about delivery mechanisms to the seabed. Commonly, sand is concentrated on the inner shelf, and silt and clay are found farther seaward. Potential mechanisms for dispersal of the fine sediment are:

1 a land source with high concentrations of mud that diffuse seaward through wave and tidal reworking (Swift, 1970);
2 erosion of nearshore fluvial sediment by physical processes that intensify toward shore, and advection by currents to deeper, quiescent settings (McCave, 1972);
3 resuspension of sediment in concentrations turbid enough to flow seaward under the influence of gravity (Moore, 1969).

All three mechanisms (and others) are possible, with one or another dominating under particular conditions.

The first step in sediment delivery is for particles to leave the river plume. Sand settles rapidly and reaches the seabed near the river mouth. Silts and clays sink from surface plumes within a few kilometres of the river mouth (Drake, 1976). Individual silt and clay particles settle too slowly to explain this latter observation; they must form larger aggregates that sink rapidly. One possible mechanism is **biogenic aggregation** (Drake, 1976) into faecal pellets by filter-feeding organisms, but this cannot explain broad spatial distribution of particle settling, especially in turbid plumes. Most fine particles have surface charges which, in freshwater, cause the development of large, repulsive ion clouds. In brackish water with salinities of a few parts per thousand, the ion clouds compress and allow **van der Waals' forces** of attraction to dominate, forming larger aggregates that settle rapidly. When

this process occurs inorganically (e.g. glacial meltwater), it is referred to as **coagulation**. If organic molecules help bridge the gap between particles, which is common in middle and low latitudes, the aggregation process is known as **flocculation**. In addition to the mechanism of aggregation, the length of time for aggregation, the suspended-sediment concentration and the turbulence of the environment are likely to control size and settling velocity (McCave, 1984; Hill, 1992; Milligan & Hill, 1998). Despite these complexities, aggregate settling velocities are generally ~1 mm s⁻¹ (ten Brinke, 1994; Hill *et al.*, 1998).

The character of the river plume has a strong impact on the delivery of particles to the seabed. Most plumes are **hypopycnal** with densities less than the ambient seawater. They flow and spread at the surface (Fig. 4), controlled by local winds,

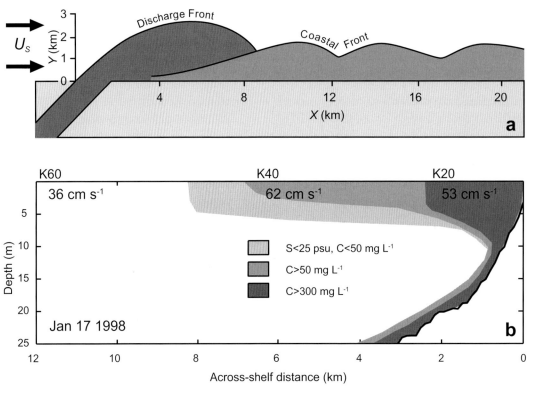

Fig. 4 Hypopycnal plumes. (a) A schematic map view for discharge of a general hypopycnal plume, with a river mouth at an angle to the shoreline. U_s is the velocity of an ambient current directed northward in the northern hemisphere. The combination of ambient current, Coriolis force, and mouth orientation causes the plume to flow to the right, creating a coastal current. (From Hill *et al.*, this volume; modified from Garvine, 1987.) (b) Cross-section (facing northward) of Eel plume on the continental shelf north of the river mouth during a period of northward winds (S = salinity; C = suspended-sediment concentration). The low-salinity and turbid river water extends offshore as a hypopycnal plume flowing northward; velocity measured 2 m below water surface shown in cm s⁻¹. Northward winds also produce downwelling against the coast and the seaward flow of bottom water with suspended sediment. (From Hill *et al.*, this volume; modified from Geyer *et al.*, 2000.)

currents, Coriolis force and the relative significance of inertial and buoyancy forces (Wright, 1977). The path of surface plumes (e.g. direction, speed) has an impact upon the trajectory of settling particles. Under special conditions, rivers can enter water bodies with similar densities, forming **homopycnal** plumes that spread throughout the water column as turbulent jets. If the density of the river plume is greater than the ambient seawater, it forms a **hyperpycnal** plume that sinks and moves near the bottom. Of special importance to this paper are conditions (e.g. floods) where freshwater has extremely high suspended-sediment concentrations (> 40 g L^{-1}) that cause the excess density. These plumes move as gravity-driven sediment flows deflected by Coriolis force and physical oceanographic conditions (e.g. currents), but primarily they follow the steepest bathymetric gradient. Although uncommon (Mulder & Syvitski, 1995), some rivers, especially those with mountainous drainage basins, can reach hyperpycnal conditions and transport massive amounts of sediment across continental margins.

During highstands of sea level, as at present, the processes of sediment delivery tend to be focused in shallow water. For fluvial systems where or when freshwater discharge is relatively weak, aggregation begins within **estuaries** at river mouths (or even within the rivers themselves) and sediment can be trapped there. This is particularly true for low-gradient rivers emptying onto passive margins, such as the Hudson and Delaware rivers. If river plumes with substantial sediment concentrations extend onto the shelves, sedimentation can occur there, and follow the mechanisms described previously in this section. Most active margins have coastal mountain ranges, steep river channels, small or no estuaries and narrow continental shelves (Fig. 1). Under these conditions, plumes can reach the continental slope. Hypopycnal plumes form surface **nepheloid layers** (diffuse clouds of turbid water), which are carried by the local currents and dissipate as suspended sediment settles onto the slope (known as **hemipelagic** sedimentation). Hyperpycnal plumes move down the steepest portions of the slope (commonly submarine canyons), and can accelerate to erode the seabed and refuel their excess density, thus becoming one of several means to create **turbidity currents**. Today, some submarine canyons extend into the mouths of rivers (e.g. Sepik River, Congo River) and gravity-driven

sediment flows (e.g. hyperpycnal plumes, turbidity currents) usually dominate sediment transport (Kineke *et al.*, 2000; Khripounoff *et al.*, 2003). During lower stands of sea level, such situations were common.

Delivery of Eel margin sediment

Initial northward winds and currents, a northward-pointed river mouth and the Coriolis force cause the early stages of Eel River flood discharges (associated with winter storms) to be directed northward. The **radius of curvature** defines the turning distance of the plume at the river mouth. This radius is controlled by plume speed and the Coriolis force (Garvine, 1987), and is ~10 km near the Eel mouth. The plume turns into a northward-flowing coastal current (Fig. 4) that is restricted to regions < 40 m deep and is moving at ~50 cm s^{-1} (maximum 130 cm s^{-1}; Geyer *et al.*, 2000). Suspended silts and clays, which dominate the discharge, aggregate (mean floc size 230 µm; Curran *et al.*, 2002) and are largely removed from the surface plume within 10 km of the river mouth (Hill *et al.*, 2000). The correlation of discharge events and oceanic storm conditions guarantees turbulence within the coastal current. This turbulence results from wind-driven downwelling that destroys water-density stratification, and from a storm-wave surf zone that extends seaward to as far as 15-m water depth (Curran *et al.*, 2002). The intense turbulence within the surf zone keeps fine sediments suspended, providing a mechanism to resupply the coastal current. As the coastal current moves northward, it experiences some seaward transport due to **Ekman veering** in the bottom boundary layer (Smith & Long, 1976; Drake & Cacchione, 1985). When winds reverse, northward transport is slowed and the plume broadens seaward (Geyer *et al.*, 2000). For periods of low river discharge, correlation with meteorological events is not evident, and variable winds preclude a net direction of sediment transport. In some years, southward transport of shelf sediment can be significant (Ogston & Sternberg, 1999; Ogston *et al.*, 2004).

During coupled discharge and storm events, wave activity has a significant control on aggregate properties observed along the shelf, due to continual injection of particles from the surf zone into the coastal current (Curran *et al.*, 2002). Beyond the surf zone (> 15 m depth), a shelf frontal zone (Fig. 4) concentrates suspended sediment on the inner shelf (Ogston *et al.*, 2000); here, wave activity can stimulate across-shelf sediment transport. Although waves provide little net direction for sediment transport, they can create high-concentration (> 10 g L^{-1}) **fluid muds** in the wave boundary layer (< 10 cm thick) that produce gravity-driven sediment flows moving seaward at 10–30 cm s^{-1} (Traykovski *et al.*, 2000). The signature of these flows occurs within the current boundary layer (lowermost several metres of water column) where velocity normally decreases logarithmically toward the seabed. When concentrations of suspended sediment are very large, velocity increases near the bed within the wave boundary layer (5–10 cm above seabed).

These wave-supported sediment gravity flows transport much sediment mass as they move across shelf. As near-bed wave activity decreases seaward, the gradient of the shelf seabed is not sufficient to allow continued flow, and the sediment stops moving (Wright *et al.*, 2001). Within the resulting flood deposits are fine laminae (centimetre-scale **sedimentary structures**) that record pulses of sediment flux (Wheatcroft & Borgeld, 2000). The location of the gravity-flow deposits generally coincides with the convergence of sediment transport from shelf currents (Wright *et al.*, 1999; Ogston *et al.*, 2000), and together these processes create a locus of sediment deposition on the Eel shelf between 50-m and 70-m water depth and ~10–30 km north of the river mouth (Fig. 5).

Not all sediment discharged to the Eel continental shelf reaches the seabed; much (> 50%) continues to the continental slope. Turbid water in the bottom boundary layer of the shelf can detach near the shelf break and move seaward along an isopycnal surface within the water column as an **intermediate nepheloid layer** (INL). These layers are maintained, in part, by internal waves (McPhee-Shaw *et al.*, 2004). Eel sediment is broadcast across the slope, and rapid delivery is confirmed by the

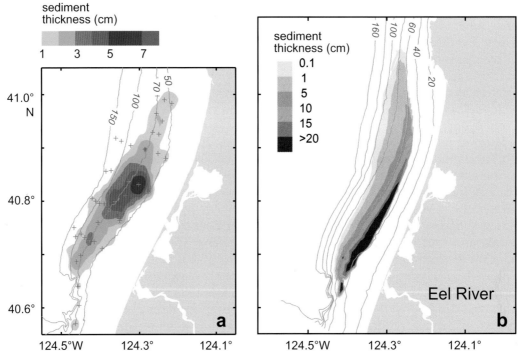

Fig. 5 Shelf sediments resulting from the 1997 flood of the Eel River. (a) Isopach map of the 1997 flood deposit. The thickest portion is found in ~70-m water depth and ~15–25 km north of the Eel River mouth. This compares well with the pattern of the 1995 flood deposit shown in Fig. 10. (From Hill *et al.*, this volume; based on Wheatcroft & Borgeld, 2000.) (b) Predicted thickness of a deposit resulting from wave-supported sediment gravity flows during the 1997 flood event (porosity assumed to be 0.75). Thicknesses are greater and extend farther north than those observed in (a), but the predicted pattern has many similarities to the flood deposit, including its shape and the location of the landward and seaward boundaries. (From Hill *et al.*, this volume; based on Scully *et al.*, 2003.)

presence of the short-lived radioisotope [7]Be (half-life 53 days) in sediment traps (Walsh & Nittrouer, 1999). This same isotope is found in the seabed of the open slope (Sommerfield *et al.*, 1999) and Eel Canyon (Mullenbach & Nittrouer, 2000), probably from input through intermediate nepheloid layers and other mechanisms. In the head of Eel Canyon, inverted velocity profiles (increasing near the seabed) similar to gravity-driven sediment flows on the shelf are observed (Puig *et al.*, 2003, 2004). Modelling studies indicate that substantial amounts of Eel sediment discharge are likely to be carried into the canyon by these flows (Scully *et al.*, 2003). During major flood periods (e.g. 1995, 1997), the river may become hyperpycnal, and bottom plumes may carry large fractions of the Eel discharge directly to the Canyon or the open slope north of the Canyon (Fig. 2; Imran & Syvitski, 2000). Therefore, a range of mechanisms associated with the Eel plume deliver sediment to the continental slope during the present highstand of sea level.

Modelling studies indicate that during the **Last Glacial Maximum** (LGM), the Eel basin was wetter and colder, and storm frequency was greater (Morehead *et al.*, 2001). These differences would have caused approximately a doubling of the water and sediment discharge (Syvitski & Morehead, 1999). Most discharge from the modern Eel River occurs with winter rains. For the LGM, increase in precipitation would have caused a more sustained discharge as snow pack melted during the spring and summer. Rains on low-elevation snow also would have caused more intense floods than today. These differences in water and sediment discharges and in the intra-annual variability of discharges distinguish modern and past conditions for sediment delivery to the Eel margin.

SEDIMENT ALTERATION

Processes affecting the preservation of the sediment record during deposition and the early stages of burial have been examined in this volume by Wheatcroft *et al.* (pp. 101–155).

General considerations

Sediment delivered to the seabed is altered in many ways before being preserved by burial. Especially important changes are those that alter the dynamical properties of the seabed thereby impacting lateral transfer of sediment across margins (e.g. alteration of particle-size distribution, bottom roughness, porosity), and those that cause vertical displacements of seabed particles thereby affecting stratigraphic signatures (e.g. alteration of sedimentary structures, acoustic properties). These alterations occur primarily over time-scales of days to years and over vertical length scales of millimetres to decimetres. Deposition of new particles applies a downward force (i.e. weight) to the underlying sediment. Physical processes erode and deposit particles, rearranging them based on hydrodynamic character. Macrobenthic organisms displace particles in the seabed through a wide assortment of activities, including ingestion and defaecation. Chemical processes also alter sediment after delivery to the seabed, but usually have less direct impact on transport and stratigraphy than the other processes (for summaries of chemical alteration see: Aller, 2004; McKee *et al.*, 2004).

Consolidation (also known as **compaction**) decreases **porosity**, as new overburden reduces pore space and displaces pore fluid. Initial changes occur near the surface of the seabed, such that a relatively uniform porosity is approached within a few centimetres (Fig. 6a), although consolidation continues much deeper in the seabed as overburden increases. Porosity profiles impact many properties in the seabed (e.g. bulk density, acoustic signature), and also influence sedimentary processes; high-porosity surface layers are easily eroded by weak shear stresses. Porosity profiles indicate whether the weight of overlying sediment is supported by a particle framework or by pore fluids, conditions that may ultimately determine the distribution of stresses and whether the seabed will fail. For all of these reasons, understanding the consolidation rate of natural sediment is important, as is understanding the factors affecting that rate (e.g. permeability, bioturbation). In general, sands consolidate quickly toward a minimum porosity of ~0.35 (fractional volume of pore space) and muds consolidate more slowly toward minimum values (Been & Sills, 1981; Wheatcroft, 2002). However, fluctuations in sedimentation complicate consolidation history of the seabed. Erosion of the seabed exposes sediment that is **overconsolidated** (Skempton, 1970) relative to

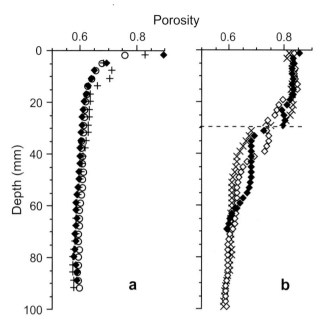

Fig. 6 Sediment porosity profiles on the Eel Shelf. (a) Replicate porosity profiles at a mid-shelf station (70-m water depth) six months before the 1997 flood of the Eel River. Relatively uniform porosity is reached within ~30 mm of seabed surface. (b) Replicate porosity profiles at the same station as (a), but 2 weeks after the 1997 flood. A uniform layer of higher porosity is observed within the upper ~30 mm, which is the thickness of the flood deposit at this location, as documented by X-radiography and radiochemistry. (From Wheatcroft *et al.*, this volume.)

what is expected at the surface. Rapid deposition of thick flood layers places sediment below the surface that is **underconsolidated** (Skempton, 1970) relative to what is expected at that depth in the seabed. Variable grain sizes and biological effects further complicate consolidation, and make modelling and prediction of porosity profiles more difficult.

Physical reworking adds and subtracts particles from locations on the seabed, often removing fine particles (i.e. **winnowing**) and coarsening (i.e. **armouring**) the surface. The fine particles (silts and clays) possess interparticle forces of attraction and, where these sediments deposit, the seabed develops **cohesion**. With consolidation, cohesive forces increase, and the fluid velocities needed for resuspension also increase. Armouring inhibits resuspension

by developing a coarse surface layer, and cohesion causes an abrupt decrease in erodibility just below the surface of muddy deposits.

The extent of physical reworking depends on the strength of operative processes (e.g. surface waves, coastal currents) as well as seabed properties (e.g. grain size, porosity). Under special conditions (e.g. equatorial settings with sustained trade winds, shallow tide-dominated coastal areas), reworking can be relatively continuous (Nittrouer *et al.*, 1995). However, most continental margins are dominated by cyclonic storms, which cause episodic physical reworking that is largely the result of surface waves (Komar *et al.*, 1972; Drake & Cacchione, 1985). Waves impact the seabed in water depths less than about half their wavelength, and large waves can rework bottom sediments to depths of 100–200 m in extreme events (Komar *et al.*, 1972). The near-bed wave orbital velocities increase toward shore, and are additionally dependent on wave height and period (Komar & Miller, 1975; Madsen, 1994; Harris & Wiberg, 2001). A velocity of ~14 cm s^{-1} has been observed as the critical value needed for resuspension of muddy shelf deposits by waves (Wiberg *et al.*, 1994, 2002) but this value is influenced by many factors, including grain size and consolidation state.

In non-cohesive sandy sediment, the **active layer** of moving sediment can be a few centimetres thick but, where bedforms develop and migrate, it is comparable to their height (~5–10 cm). In cohesive muddy sediment, the active layer is dependent on the thickness of high-porosity surficial sediment. Erosion and redeposition of sediment create a **graded** deposit (i.e. fining upward) within the active layer (Reineck & Singh, 1972; Nittrouer & Sternberg, 1981). Subsequent to deposition, benthic organisms alter the seabed through a range of activities. Ingestion of particles and formation of faecal pellets change the effective grain size of sediment. Together with formation of mounds and burrows, these processes increase seabed roughness (Jumars & Nowell, 1984) and alter porosity, all of which influence sediment transport. The mucous that glues animal faecal pellets is similar to organic substances produced by microalgae on seabed surfaces, and adhesive coatings from both sources tend to bind the seabed and reduce physical reworking. Feeding, locomotion and dwelling construction

Fig. 7 Wave energy on the Eel margin. (a) Spatial variation of wave characteristics measured by buoys (NOAA National Buoy Center) along the northern California coast (north of San Francisco). Buoy 46022 is located near the Eel River, and has the most energetic wave climate, as shown by the return period for a given significant wave height. (b) The probability is shown of exceeding various near-bed orbital velocities (U_b) at different water depths across the Eel shelf. For the mid-shelf deposits (~50–70 m water depth), a velocity of ~15 cm s^{-1} is likely to erode the surface sediment. (From Wheatcroft et al., this volume; modified from Wiberg, 2000.)

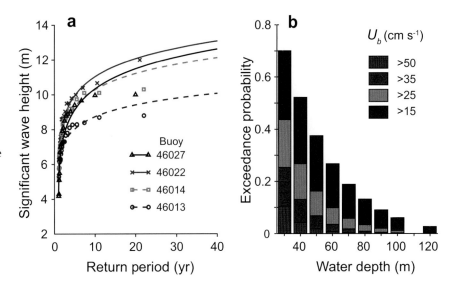

are processes by which benthic organisms stir sediment within the seabed (i.e. **bioturbation**), destroying physical sedimentary structures and creating biological structures. These processes occur within a region known as the **surface mixing layer**, which is ~5–20 cm thick.

Alteration of Eel margin sediment

The Eel margin is an instructive place to investigate seabed alteration, because the relevant processes operate intensely and cause the seabed to be dynamic. Floods of the river create thick layers of high-porosity sediment of variable grain size on the continental shelf. Energetic oceanic storms cause reworking of that sediment. An abundant and well-adapted benthic community rapidly mixes the seabed.

Beyond the inner-shelf sands (> 60 m depth), steady-state porosity profiles asymptotically approach values of 0.6–0.7 several centimetres below the seabed surface (Fig. 6a; Wheatcroft & Borgeld, 2000). The floods in 1995 and 1997 added significant perturbations, creating layers of uniform porosity many centimetres thick (up to ~8 cm) with values of 0.8–0.9 (Fig. 6b). The consolidation rate of this sediment had an important control on the erodibility of the seabed. The upper centimetre returned to steady-state porosities within months (< 4) and

made the seabed resistant to erosion, even though a couple of years were needed for deeper flood sediments to reach the lower values (Wheatcroft et al., this volume, pp. 101–155). Concurrent bioturbation imposed significant spatial variability on these general observations.

The Eel margin has the greatest wave energy along the northern California coast (north of San Francisco), with waves reaching heights > 10 m (Fig. 7a; Wiberg, 2000). The inner-shelf region (< 50 m depth) experiences relatively long durations when the near-bed wave orbital velocities exceed the critical value (totalling > 40% of the time; Fig. 7b). These events are sufficient to winnow most mud (silt and clay), and create a seabed dominated by sand. Farther seaward, mud becomes a substantial portion of the seabed (> 50%) and adds cohesion as a relevant property. Despite the energetic wave regime experienced by the Eel margin, the thickness of the active layer is surprisingly small. For a strong wave event estimated to have a 10-yr recurrence interval (December, 1995), erosion occurred to ~2 cm within the seabed at 50-m water depth (Wiberg, 2000). The estimated thickness increases to 5 cm for a 100-yr storm and to 10 cm for a 1000-yr storm. For most storms, however, the active layer is millimetres thick, especially in water depths > 50 m. In addition to redeposition of local sediment, some areas

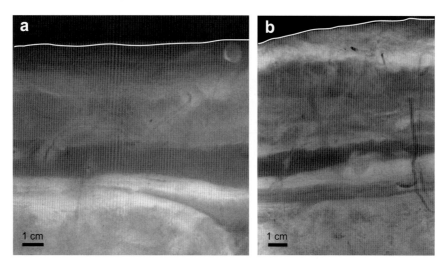

Fig. 8 X-radiograph negatives of sediment cores collected from the mid-shelf about (a) 15 km and (b) 25 km north of the Eel River mouth, illustrating various biogenic structures. (a) Collected from ~70-m water depth during February 1995, showing the January 1995 flood deposit. The burrow extending from middle left to upper right is most likely to be an escape structure of the bivalve mollusc at the sediment–water interface (upper right). (b) Collected from ~60-m water depth during July 1996. The physical sedimentary structures near the base of the radiograph are coarse silt and clay layers in the 1995 flood deposit. Bioturbation has partially destroyed the records of the flood and has imparted a general mottling to the sediment. In addition, animals have created discrete burrows that extend tens of centimetres into the seabed. In 18 months following the 1995 flood event, new sediment was added to the seabed above the flood deposit, a process that favoured preservation of the deposit. (X-radiographs are courtesy of R.W. Wheatcroft, Oregon State University; see also Wheatcroft et al., this volume.)

can experience a convergence of sediment flux during dry storms (e.g. transfer from inner-shelf to mid-shelf depths) adding millimetres to 1 cm of sediment (Zhang et al., 1999; Harris & Wiberg, 2002). The resulting storm deposits are graded, but bioturbation destroys them within weeks (Fig. 8; Harris & Wiberg, 1997; Bentley & Nittrouer, 2003; Wheatcroft & Drake, 2003).

Thicknesses of event deposits are greater during wet storms, due to the influx of new river sediment (Fig. 5). These deposits have relatively high clay contents (Drake, 1999), and can be easily identified by their physical sedimentary structures (Wheatcroft & Borgeld, 2000), radiochemical signatures (presence of ^{7}Be and low level of ^{210}Pb; Sommerfield et al., 1999) and terrestrial carbon composition (Leithold & Hope, 1999). Subsequent to the formation of clay-rich flood deposits, the seabed coarsens by the addition of silts and fine sands from the inner shelf (Drake, 1999). In addition, animal bioturbation gradually destroys physical sedimentary structures and creates discrete biogenic structures (Fig. 8).

Polychaete worms dominate macrofauna on the Eel margin, and most of the abundant species are sub-surface-deposit feeders (Bentley & Nittrouer, 2003; Wheatcroft, 2006). They produce many small burrows (millimetres diameter) within the upper 3–5 cm and build a few larger burrows (1–10 cm diameter, some with reinforced lining) extending down as much as 15–20 cm (Fig. 8). On the Eel shelf, the dominance of subsurface-deposit versus surface-deposit feeders minimizes the importance of faecal pelletization at the seabed surface (Drake, 1999). Biogenic seabed roughness is important seaward of ~60 m depth (Cutter & Diaz, 2000), but monitoring observations in these deeper shelf locations (Ogston et al., 2004) demonstrate significant temporal variability as storm events form ripples, even on substrates of silt and clay.

Subsurface bioturbation can be quantified from seabed profiles (upper 4–8 cm) of the short-lived radioisotope ^{234}Th (half-life 24 days; Aller & Cochran, 1976; Wheatcroft & Martin, 1996). The **bio-diffusion coefficient** is moderately high (3 cm^2 yr^{-1}

to > 100 cm^2 yr^{-1}, mean 20–30 cm^2 yr^{-1}; Bentley & Nittrouer, 2003; Wheatcroft, 2006) on the Eel margin. It reveals substantial small-scale variability over tens of metres, but also demonstrates a decrease between the shelf and the deeper continental slope (water depth > 500 m; Wheatcroft *et al.*, this volume, pp. 101–155). Most interesting is the temporal variability in bioturbation. Organism abundance shows an increase during summer and autumn, and a decrease in winter due to annual cycles of recruitment and growth (Wheatcroft, 2006). Although the extreme flood of January 1997 caused a subsequent drop in abundance, the mortality that year was comparable with other winters without major floods, and was consistent with weak seasonal changes in biodiffusive mixing intensity (slight increases in autumn). Winter is normally a period of low numbers of benthic organisms and low bioturbation activity in the seabed. Therefore, the Eel margin benthic community is well adapted to seasonal cycles in storm reworking and flood deposition.

The dominance of the subsurface-deposit feeders controls the preservation of sedimentary signals on the Eel margin. Important factors are thickness of event signals (storm reworking, flood deposits), thickness of the surface mixing layer, intensity of bioturbation (biodiffusion coefficient) and the sediment accumulation rate. Knowledge of these terms allows evaluation of the **transit time** for a signal to pass through the surface mixing layer, and the **dissipation time** for destruction of the signal (Wheatcroft, 1990). For the Eel shelf, the transit time is 9–65 yr and the dissipation time is ~2 yr; therefore, most signals are destroyed before they can be preserved (Wheatcroft & Drake, 2003). This is particularly true for physical sedimentary structures, which are lost due to particle mixing with overlying and underlying sediment. Event layers > 5 cm thick can be preserved, but those < 3 cm cannot. Other event signals (e.g. increased clay content, decreased ^{210}Pb activity, increased terrestrial carbon) are smeared vertically, but are still recognizable in preserved strata (Sommerfield & Nittrouer, 1999; Blair *et al.*, 2003; Wheatcroft & Drake, 2003). The timing of subsequent events can have an impact on preservation. For example, emplacement of the 1997 flood deposit effectively decreased the transit time for the 1995 flood deposit and allowed its partial preservation. Without such benefit, the 1997

flood deposit was destroyed in 2.5 yr (Wheatcroft & Drake, 2003).

SEDIMENT DISPERSAL SYSTEM

The dispersal of sediment on continental margins has been reviewed in this volume by Sommerfield *et al.* (pp. 157–212).

General considerations

Fluvial sediment is delivered to the seabed, where it undergoes alteration that influences its burial or transport to more distal locations. The integrated result over decades and centuries (i.e. longer than the transit time through the surface mixing layer) is a sedimentary deposit stretching along a succession of hydraulically contiguous sedimentary environments. This succession of environments is a **sediment dispersal system** (Sommerfield *et al.*, this volume, pp. 157–212) and the marine portion is just part of a longer system stretching from terrestrial sources. The expansion of time-scales brings new factors into the consideration of margin sedimentation. The slowing of **eustatic** (i.e. global) sea-level rise ~5000 yr ago (from ~5 mm yr^{-1} to ~2 mm yr^{-1}) has allowed some rivers to fill their estuaries, to extend sediment dispersal systems to the continental shelf and slope, and to form deposits with significant morphological expression (e.g. subaerial and subaqueous **deltas**). As such deposits build toward ambient sea level, they consume the space available for sediment accumulation (i.e. accommodation space). On active margins, vertical **tectonic motions** cause subsidence and uplift that adds or subtracts space for further sedimentation. Changes in accommodation space can put the seafloor into or out of energetic environments reworked by physical processes (e.g. surface waves), and can lead to displacement of sedimentation along a dispersal system.

The increased time-scale also brings climatic variability into consideration. Fluctuations in global precipitation patterns have caused periods, lasting from many decades to centuries during the late Holocene, when North America was wet and flood-prone (Ely *et al.*, 1993; Knox, 2000). On shorter time-scales, ENSO (El Niño–Southern Oscillation) and PDO (Pacific Decadal Oscillation) events have impacted fluvial discharge (Inman & Jenkins, 1999; Farnsworth & Milliman, 2003). Land use by humans has compounded the climatic impacts, both increasing sediment discharge (farming, logging) and decreasing discharge (damming, diverting). Global sediment budgets indicate there has been an anthropogenic increase in fluvial transport (from 14 to 16 Gt yr^{-1}). They also suggest ~30% trapping of this sediment landward of the coast, so that the net discharge to the ocean is ~10% less than natural levels (12.6 Gt yr^{-1}; Syvitski *et al.*, 2005). Global budgets mean nothing to individual rivers, where the scales of perturbations, the mechanisms associated with sediment routing and the storage capacity of the basin determine the impact of perturbations (Walling, 1999). Generally, these factors lead to anthropogenic impacts being greatest (and commonly most conspicuous) on rivers of small to moderate size.

The diversity and intensity of processes operating on continental margins creates the rich record of events preserved in the deposits of sediment dispersal systems. However, these same processes cause erosion and time gaps (i.e. **hiatuses**) in the record over a range of scales (e.g. storm erosion, sea-level change). In this regard, the metric for quantitatively evaluating sedimentation is the

mass flux into the seabed, averaged over some time-scale. Ephemeral placement on the seabed is **deposition**, but the sediment is subsequently impacted by **erosion**. The integrated sum of deposition and erosion through time is **accumulation**. The relevant time-scales for deposition rate and accumulation rate can be set for any processes of interest (McKee *et al.*, 1983). As described for this discussion of sediment dispersal systems, deposition refers to sediment placement over days/months and accumulation is the net growth of the seabed over decades/centuries. The distinction is important, because mass flux into the seabed is inversely related to the time-scale of integration (e.g. Fig. 9; Sadler, 1981; Sommerfield, 2006), as the result of more and of more severe hiatuses impacting strata formation over progressively longer time-scales.

Fortunately, a range of natural and artificial **radioisotopes** is found in terrestrial and marine environments, and they can serve as chronometers tagged to sediment particles. The large surface area (per gram of sediment) and the surface charges of silt and clay particles allow them to adsorb large concentrations of particle-reactive chemical components, including radioisotopes. Analytical techniques typically limit sedimentological use of radioisotopes to a time-scale < 4–5 half-lives. Of

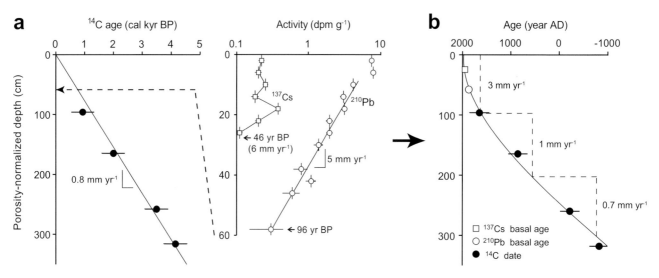

Fig. 9 Accumulation rates of Eel shelf sediments. (a) The two profiles show ^{14}C, ^{210}Pb and ^{137}Cs for the same site on the Eel shelf (95 m water-depth), and allow calculation of accumulation rates integrated over time-scales of ~3000 yr, ~100 yr and ~50 yr, respectively (data points have been adjusted vertically to a uniform bulk porosity). (b) Composite profile illustrating vertical changes in ages and accumulation rates within the seabed. The accumulation rates are greater for the uppermost sediment column, because it retains a record that is more complete than the underlying strata. (From Sommerfield *et al.*, this volume.)

special relevance here (Sommerfield *et al.*, this volume, pp. 157–212), ^{234}Th (half-life 24 days) and ^{7}Be (half-life 53 days) have primary sources, respectively, in ocean water (from decay of dissolved ^{238}U) and in terrestrial soils (from cosmogenic fallout). Lead-210 (half-life 22 yr) has several potential sources but, in ocean water, primarily comes from decay of ^{238}U-series radioisotopes. Lead-210 accumulation rates are commonly verified by profiles of ^{137}Cs (Fig. 9), a bomb-produced radioisotope. Caesium-137 was globally distributed by transport through the atmosphere and by subsequent fallout, and it first reached continental-margin sediments in ~1954. On the long end of these discussions, ^{14}C (half-life 5730 yr) ages are recorded in organic C (e.g. wood fragments) and inorganic $CaCO_3$ (e.g. shell fragments).

By using radiochemical tools with different half-lives, a composite understanding can be obtained for continental-margin sedimentation over a range of time-scales. For example, the dichotomy between deposition and accumulation rates can be related to processes and patterns of sediment dispersal. The Yangtze River undergoes flooding during the quiescent summer months, and rapidly deposits much sediment on the continental shelf near its mouth. However, longer-term accumulation rates indicate that winter storms remove and transport > 50% of this sediment to distal portions of the dispersal system (McKee *et al.*, 1983; DeMaster *et al.*, 1985). In contrast, the Amazon River has peak discharge during intervals of seasonally intense tradewinds and waves, and most of its sediment discharge (> 50%) is immediately displaced along the dispersal system > 200 km from the river mouth, to shelf areas where it deposits and near where it ultimately accumulates (Kuehl *et al.*, 1986, 1996).

Eel margin sediment dispersal system

The Eel basin has experienced multiple decades of sustained wet, dry and variable conditions during the past 100 yr (Sommerfield *et al.*, this volume, pp. 157–212). El Niño–Southern Oscillation events can bring unusual precipitation, but the location of the basin between latitudinal weather bands precludes a clear repetitive signal (e.g. El Niño brought the driest year in 1977, and the wettest year in 1983). The second half of the 1900s was a

period with increased logging in the Eel basin, and, together with enhanced precipitation (Sommerfield *et al.*, 2002), this land use significantly increased sediment yield (by 23–45%). Other forms of human interaction (e.g. damming) were minimal, so the increased sediment flux was transferred to the ocean.

In addition to the storm-related physical oceanographic processes near the Eel mouth that have been described above, regional circulation influences distal portions of the dispersal system. Seaward of the shelf break, the **California Current** flows southward (Hickey, 1979, 1998) and, on the shelf, the **Davidson Current** flows northward during the autumn and winter (Strub *et al.*, 1987). The local promontories (Cape Mendocino, Trinidad Head) can deflect these currents (Pullen & Allen, 2000), leading to the seaward transport of water and suspended sediment and to the development of eddies (Washburn *et al.*, 1993; Walsh & Nittrouer, 1999). Other morphological features on the Eel margin influence the fate of water and sediment, especially Eel Canyon, which forms a chasm across the southern boundary. More subtle across-margin ridges (**anticlines**) and depressions (**synclines**) are moving up and down at rates of millimetres per year (averaged over millennia; Orange, 1999).

Sediment deposition on the Eel margin is clearly demonstrated by the distribution patterns associated with the 1995 and 1997 flood events, which discharged ~24 × 10^6 t and ~29 × 10^6 t of sediment, respectively (Wheatcroft & Borgeld, 2000). Both events formed elliptical deposits on the middle shelf north of the Eel mouth (Figs 5 & 10), representing 20–30% and 15–30% of the mass discharged, respectively. The similarity of the two deposits suggests that the mechanisms of emplacement operated in a repetitive manner. The remainder of the sediment could deposit landward, northward, southward or seaward of these deposits. The inner-shelf sands contain some intermixed mud, and Humboldt Bay might receive some sediment through tidal exchange and estuarine circulation. The Davidson and California Currents could move some surface plumes of sediment beyond the confines of the Eel margin (e.g. Mertes & Warrick, 2001). However, the bulk of sediment is thought to be transported seaward of the Eel shelf by a combination of hyperpycnal flows, storm-induced fluid muds and intermediate nepheloid layers.

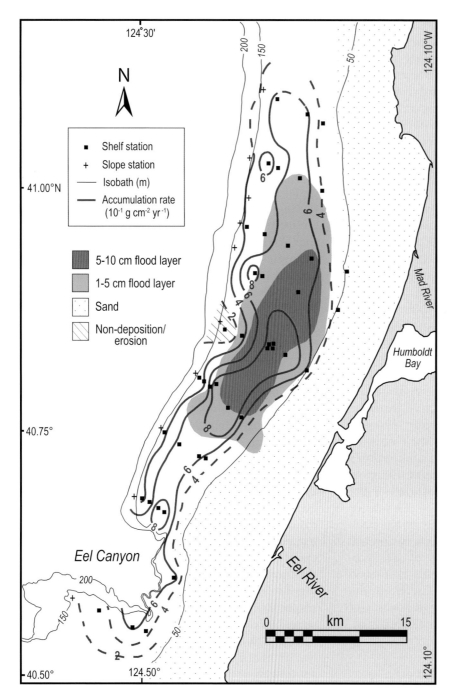

Fig. 10 Contour map of ^{210}Pb accumulation rates (red isopach lines) on the Eel shelf, superimposed on the thickness of the January 1995 flood deposit (green shaded areas). They coincide well, and indicate maximum values in ~50–70 m water depth and ~10–30 km north of the river mouth. Approximately 20–30% of sediment discharged by the 1995 flood remained on the shelf, and this fraction is similar to that retained over a 100-yr time-scale. (Modified from Sommerfield & Nittrouer, 1999.)

On time-scales of decades and centuries, the fate of silt and clay from the Eel River shows a similar pattern: ~10% is buried with inner-shelf sands (< 60-m water depth; Crockett & Nittrouer, 2004), ~20% accumulates on the middle and outer shelf (Fig. 10; Sommerfield & Nittrouer, 1999), and the remainder is exported to deeper water. Accumula-tion on the upper slope (150–800 m) accounts for ~20% of the Eel sediment discharge (Alexander & Simoneau, 1999) and Eel Canyon is the alternative pathway on the slope (Mullenbach & Nittrouer, 2000, 2006). These observations demonstrate that the Eel shelf traps less than a third of modern sediment discharge. They also highlight the importance of Eel Canyon for dispersing sediment seaward; as much as 50% of the river discharge could be moving into and through the canyon. Both the escape of sediment from the shelf and the large flux through Eel Canyon are occurring during the present highstand of sea level, and probably reflect sedimentation typical of narrow, tectonically active continental margins.

The accumulated strata reveal interesting sedimentary trends along the dispersal system. Grain size decreases progressively with distance from the Eel mouth, both northward and seaward. Fining continues across the slope, but includes an anomalously coarse zone below the shelf break (250–350 m water depth; Alexander & Simoneau, 1999), possibly due to winnowing by shoaling internal waves (Cacchione *et al.*, 2002). Organic carbon shows a progressive increase in the marine component relative to the terrestrial component with distance from the Eel River (Blair *et al.*, 2003). Temporal changes are also observed for the past ~4000 yr (Sommerfield *et al.*, this volume, pp. 157–212). The accumulating sediment has progressively become finer (less sand, more silt) as the dispersal system has evolved since sea-level rise slowed. For the past 200 yr the upward fining trend has been accelerated by human impacts on land use. The magnitude and frequency of flood events also have increased, imposing the sedimentary characteristics of those events: high sediment flux, increased clay content, much terrestrial carbon. The changes have been particularly acute during the past 50 yr (Sommerfield *et al.*, 2002), and reflect the combined effects of land use (clear cutting, road building) and climatic increases in precipitation intensity.

There is a distinct similarity of the shelf patterns in flood deposition and accumulation rates over decades/centuries (see Figs 5 & 10). This is due to the correlation of river discharge and energetic oceanic conditions. Most sediment is immediately transported to a stable location for accumulation (i.e. where it will not be eroded by strong boundary shear stresses), rather than being temporarily deposited and subsequently transported. In this regard, the Eel margin more closely approximates the conditions of shelf deposition/accumulation near the mouth of the Amazon River than those near the Yangtze River. However, the accumulation pattern over decades/centuries also matches well with thicknesses of late Holocene strata (see below),

which are related to tectonic features on the shelf (Orange, 1999; Burger *et al.*, 2002; Spinelli & Field, 2003). Millenial accumulation rates are < 1 mm yr^{-1} over anticlines, and reach 6 mm yr^{-1} in synclines. The similarity of accumulation patterns suggests that tectonic activity on the margin impacts sedimentation on scales as short as decades (as detailed in Sommerfield *et al.*, this volume, pp. 157–212). Likely candidates for the operative mechanisms are gravity flows, which are common on the Eel margin and respond to subtle gradients of the seabed.

SEABED FAILURE

The processes and products of seabed failure on continental margins have been addressed in this volume by Lee *et al.* (pp. 213–274) and Syvitski *et al.* (pp. 459–529).

General considerations

The dispersal of sediment to sites of accumulation is a continuing process; new sediment buries old sediment, causing consolidation and development of strength to resist subsequent shear forces. However, in some cases, forces exerted on the seabed are stronger than the strength developed, and the seabed fails. The resulting **mass movement** is driven by body forces (i.e. gravity) rather than by fluid stresses exerted on the seabed surface. In this way, mass movement differs from sediment erosion and transport. Some famous failures have occurred in the past 100 yr, including the 1929 Grand Banks (Heezen & Ewing, 1952), 1964 Alaska (Coulter & Migliaccio, 1966; Lemke, 1967) and 1998 Papua New Guinea (Tappin *et al.*, 1999; Geist, 2000) landslides; all were triggered by earthquakes and all initiated tsunamis. Failures can be triggered by other processes, including large waves associated with storms, such as Hurricane Camille in 1969 (Sterling & Strohbeck, 1973; Bea *et al.*, 1983) and more recent hurricanes in the Gulf of Mexico. Large landslides have also occurred in the geological past leaving scars and deposits as evidence, such as the Storrega landslides (Bryn *et al.*, 2003) during the Pleistocene and Holocene (most recently ~8200 yr ago). These removed a large piece of the Norwegian continental margin (~3000 km^3) and displaced it over a region stretching ~800 km.

The largest landslides on Earth are found in the ocean.

Underwater **landslides** move sediment with a range of speed and internal deformation of the deposit (Varnes, 1958). Subclasses of movement include **creep,** when the movement is slow, and **slumps,** when sediment blocks rotate along a curved failure surface. **Liquefaction** occurs when loosely packed particles temporarily lose contact with each other, and the weight of the deposit becomes supported by pore fluids. All styles of failure can lead to disintegration of the sediment deposits, and development of gravity flows (e.g. debris flows, turbidity currents).

Failures and landslides are prevalent in environments of the continental margin where thick deposits of soft sediment accumulate. Fjords can receive large amounts of rock flour (with limited cohesion) that rapidly accumulate on steep gradients (some > 5°). Fjord sediments are commonly organic rich and produce methane gas. Subsequent earthquakes or even very low tides can initiate failure (Syvitski & Farrow, 1983; Prior *et al.*, 1986). Deltas are also loci of rapid accumulation, and despite gentle gradients (usually < 2°) can fail in response to earthquakes or storms (Coleman *et al.*, 1980; Field *et al.*, 1982). Continental slopes are extensive and steep (> 4°) regions with a propensity for failure, which is accentuated during lowstands of sea level when fluvial and glacial sediment discharge occurs directly at the top of the slope. Gas and gas hydrates, which commonly form on continental slopes, can be responsible for failures (Field & Barber, 1993), especially with sealevel fall that reduces hydrostatic pressure on the seabed and causes dissociation of hydrates (Kayen & Lee, 1991). Submarine canyons are regions of preferential sediment accumulation, and failures near their heads can lead to gravity flows that supply sediment to submarine fans at the bases of the canyons (Hampton, 1972; Booth *et al.*, 1993). Especially important are failures triggered by earthquakes on active margins that cause turbidity currents to transport much sediment long distances (e.g. Goldfinger *et al.*, 2003). During the present highstand of sea level, continental slopes are generally below the depth of surface-wave influence, but the heads of submarine canyons are in shallower water and can be impacted by energetic waves (Puig *et al.*, 2004). From observations in a range of sedimentary environments, the factors recognized to influence failures are sediment accumulation rates, bathymetric gradients, seismicity, storm waves and gas.

Failures and landslides occur when and where driving stresses exceed shear resistance. Bathymetry is important because it defines the gravity-induced stresses. Earthquakes cause cyclic accelerations in addition to gravity (Lee & Edwards, 1986). Similarly, large storm waves produce alternating pressures that create stresses superimposed on those from gravity (Henkel, 1970). In opposition to the applied stresses is the **shear strength** of the seabed, which is defined as the limit of stress before failure. The shear strength of sediment increases as it is buried by subsequent accumulation and as the seabed consolidates. The **factor of safety** for the seabed is the shear strength divided by the shear stress. In addition to large stresses, the factor of safety can be reduced by a loss of shear strength. A common mechanism is the development of excess pore pressures, due to (i) the inability to remove pore fluids during consolidation (e.g. under high accumulation rates; Coleman & Garrison, 1977), (ii) the development of gas bubbles (e.g. from the decay of organic matter or the dissociation of hydrates; Kayen & Lee, 1991) and (iii) the infusion of additional water (e.g. by groundwater seepage). Earthquakes and storm waves apply stresses cyclically, which destroys particle fabric (i.e. grain-to-grain contact), causes liquefaction (Seed, 1968), and increases pore pressures. Human activity can cause failure as well, commonly from construction at or near the shoreline that destabilizes the seabed. Sometimes, the resulting landslides even stimulate tsunamis, e.g. during 1979 in Nice, France (Seed *et al.*, 1988) and during 1994 in Skagway, Alaska (Rabinovich *et al.*, 1999).

Whether by increased stresses, reduced strength or both, marine sediments can fail. After failure, they create landslide deposits or disintegrate into fluid flows (Hampton *et al.*, 1996), depending on the bulk density (i.e. porosity) of the sediment (Poulos *et al.*, 1985; Lee *et al.*, 1991). A critical threshold separating these two fates (i.e. slide deposit from fluid flow) can be defined for each sediment type. If seabed conditions have densities below this threshold (**contractive sediment**), then excess pore pressures will develop after failure, and the sediment will flow. Densities above this threshold (**dilatant sediment**) will cause the sediment to strengthen after failure, and it will not flow. The transition to flow can be facilitated during failure by other factors, including large amounts of energy exerted and water added.

Eel margin failure

The Eel margin exhibits conditions conducive to seabed failure: rapid sediment accumulation; steep bathymetric gradients; intense seismicity; energetic storm waves; and plentiful gas. The largest feature on the Eel margin (~90 km²) with the outward appearance of failure is known as the 'Humboldt Slide' (Field *et al.*, 1980), but its origin is controversial. It is found in an amphitheatre-like depression between 220-m and 650-m water depth (Fig. 11), just north of Eel Canyon. The upper portion of the amphitheatre (above 380-m water depth) is overconsolidated, consistent with removal of ~15 m of sediment (Lee *et al.*, 1981). Analysis of the sediment indicates that the density state would preclude transition to a flow, and that the failed sediment would create a deformed slide deposit at its base (Lee *et al.*, 1991). Recent multibeam surveys (Fig. 11; Goff *et al.*, 1999) and high-resolution seismic profiling (Fig. 12a; Gardner *et al.*, 1999) demonstrate that the lower portion of the deposit is a crenulated surface with ridges and swales, similar to subaerial landslide deposits. The profiles suggest that the greatest failure was in the middle of the 'slide', and that it underwent a small amount of downslope translation with shallow rotation, creating gentle compression folds at its base. Deeper profiles show older deposits of similar character and imply a long history of such events (Field *et al.*, 1980). The multibeam surveys also document many pockmarks, interpreted as evidence of gas escape from the seabed (Yun *et al.*, 1999), and many erosive gullies on the rim of the feature (above 380-m water depth, see Fig. 11).

These gullies possibly reflect processes related to an alternative origin: erosive gravity flows on the rim and depositional 'sediment waves' at the base (Fig. 11). Such processes and morphology have

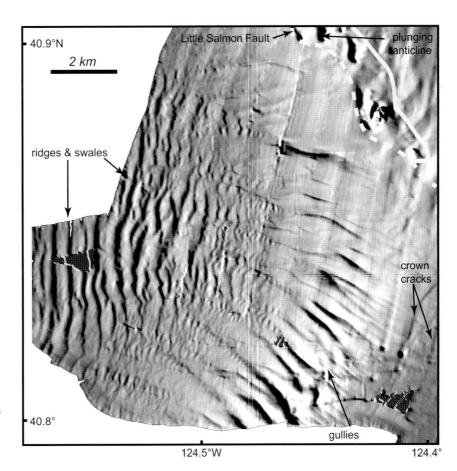

Fig. 11 Multibeam bathymetry of 'Humboldt Slide' on the Eel continental slope immediately north of Eel Canyon and south of the Little Salmon Anticline. Deep gullies are found on the upper rim, and ridges and swales are at the base. (Modified from Gardner *et al.*, 1999.)

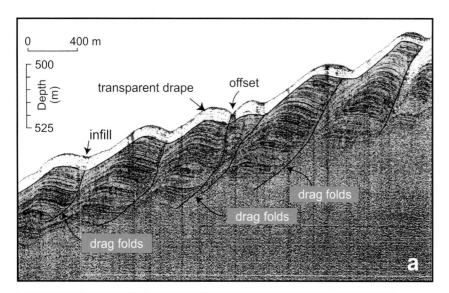

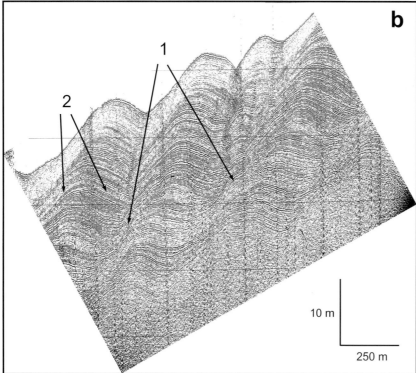

Fig. 12 Seismic profiles from the main body of the 'Humboldt Slide'. (a) Profile with an interpretation of folded and back-rotated slide blocks. Shear surfaces are shown by black lines, with drag folds near bases. (From Lee *et al.*, this volume; modified from Gardner *et al.*, 1999.) (b) Profile showing evidence of a 'sediment wave' origin: 1, internal reflectors can be followed between 'waves'; 2, beds on the upslope side are thicker than those on the downslope side. (From Lee *et al.*, this volume; modified from Lee *et al.*, 2002.)

been interpreted for similar features elsewhere in the world, and are extrapolated to the 'Humboldt Slide' (Lee *et al.*, 2002). Sediment transported as hyperpycnal flows directly from Eel River and sediment reconstituted with seawater to create fluid-mud flows move across the shelf break to the steeper slope. There, they undergo a transition into turbidity currents that erode the seabed. These currents create the gullies and expose overconsolidated sediment. Near the base, where the bathymetric gradient is more gentle, the turbidity currents deposit their load in the form of 'sediment waves', which explains the origin of the ridges and swales (Fig. 11). These contain internal reflectors that can be traced from one 'wave' to the next (Fig. 12b), which should not be the case for landslide deposits. The reflectors also show asymmetry, with greatest sediment thicknesses on the upslope side of the 'wave' (Fig. 12b) due to preferential sedimentation, a process observed for bedforms in other slope areas with active sediment waves.

Evidence remains on both sides of the argument whether 'Humboldt Slide' is truly the result of a slide or of sediment transport, but critical consideration of the controversy is important, because similar features abound on continental margins: off the west coast of Africa (Wynn *et al.*, 2000); off British Columbia (Bornhold & Prior, 1990); and in the Adriatic Sea (Correggiari *et al.*, 2001).

South of the 'Humboldt Slide', liquefaction has been invoked as an active mechanism for generating gravity flows of fluid mud in the head of Eel Canyon (Puig *et al.*, 2004), which is located at ~90-m water depth. These flows were observed soon after the beginning of dry storms with waves capable of reaching that depth. The speed of response, lack of river floods and measured wave impacts all suggest liquefaction due to wave-induced loading, a mechanism that was documented to occur many times during the winter (Puig *et al.*, 2004) and to provide seabed deposits widely observed through the canyon head (Mullenbach *et al.*, 2004; Drexler *et al.*, 2006).

Other than the controversial 'Humboldt Slide' and the liquefaction in the Eel Canyon head, the Eel margin is relatively devoid of evidence for failure, considering the environmental conditions that make it a prime candidate. Other factors have been invoked to explain the apparent stability of the seabed. Laboratory consolidation experiments suggest that the intense bioturbation on the Eel margin might be responsible for strengthening the seabed (Lee *et al.*, this volume, pp. 213–274). The sediment repackaging processes associated with faecal-pellet production can create sediment with greater bulk density than a rapidly deposited and unbioturbated seabed. This process is dependent on the ambient benthic community, because some other areas experience increased porosity from bioturbation (Bokuniewicz *et al.*, 1975; de Deckere *et al.*, 2001). Another possible explanation for the apparent stability of the Eel margin seabed is strengthening by seismic activity. Studies have shown that repeated seismic events below the threshold of failure can increase seabed density and strength (Boulanger, 2000). Laboratory experiments reveal

the mechanism to be a development of excess pore pressures during earthquakes, which led to the subsequent drainage of pore fluids. The result is a seabed with properties of overconsolidation.

GRAVITY FLOWS

The properties and significance of sediment gravity flows on continental margins have been reviewed in this volume by Parsons *et al.* (pp. 275–337) and Syvitski *et al.* (pp. 459–529).

General considerations

Many seabed failures (contractive sediment) produce **gravity flows**, which continue down slope due to suspended-sediment concentrations sufficient to exceed the density of the surrounding fluid. In other cases, dense concentrations are injected from rivers or are formed in shallow water, generating hyperpycnal and fluid-mud flows, respectively. The fundamental importance of gravity flows is their ability to transport quickly large masses of sediment across isobaths: across sedimentary regimes (e.g. from inner-shelf sand to mid-shelf mud), across the shelf break, and across the continental slope to build submarine fans on the continental rise. Turbidity currents were studied relatively early in the history of marine sedimentology, because they were generated by the 1929 Grand Banks earthquake and failure. These turbidity currents left multiple records of their occurrence: deposits known as **turbidites**, and the sequential destruction of telegraph cables lying along the continental slope and rise. Turbidity currents are **autosuspending** (or **ignitive**), which means that they erode the seabed (replacing sediment left behind as turbidites), refuelling the currents and allowing them to travel long distances (Parker, 1982; Pantin, 2001). Distinctive strata deposited by turbidity currents have been recognized since the 1800s as the **flysch** deposits defined in Europe, but the formative mechanisms were not linked to them until much later (Kuenen & Migliorini, 1950). Subsequently, the repetitive signatures of turbidites were documented to be a series of distinctive sedimentary structures, which are formed by bedload and suspended load during successive phases of waning flow (Bouma, 1962).

Although turbidity currents occur in the modern ocean (e.g. Congo Canyon, Khripounoff *et al.*, 2003; Monterey Canyon, Xu *et al.*, 2004), their unpredictable and energetic nature makes them difficult to observe directly. As a consequence, most of our understanding about the mechanics of these flows comes from laboratory simulations and numerical modelling. The operation of turbidity currents is strongly influenced by turbulent mixing at several boundaries:

1 the bottom boundary, where frictional drag on the seabed causes erosion and constrains deposition;
2 the top boundary, where the current interacts with the overlying fluid and loses sediment due to mixing associated with shear;
3 the leading edge (**front**), which has the most intense mixing. This mixing slows the front, so it moves with a velocity less than the body. Three-dimensional

instabilities occur at the front (Parsons, 1998), creating vortices that counter rotate in a plane perpendicular to the direction of current flow. The front develops an irregular shape with lobes and clefts (Hartel *et al.*, 2000).

The centre of a turbidity current is intensely erosional (García & Parker, 1993), but the periphery is depositional (Fig. 13a), creating constructional features on the sides that contain the flow (i.e. **levees**). The resulting channels have many morphological similarities with subaerial river channels (e.g. meanders, point bars, crevasse splays; Hagen *et al.*, 1994; Peakall *et al.*, 2000). However, the fluid surrounding the turbidity current is seawater of nearly equal density, not low-density air, and this distinction causes differences in operational processes and in details of the resulting morphological features. A good example is levee construction (Peakall

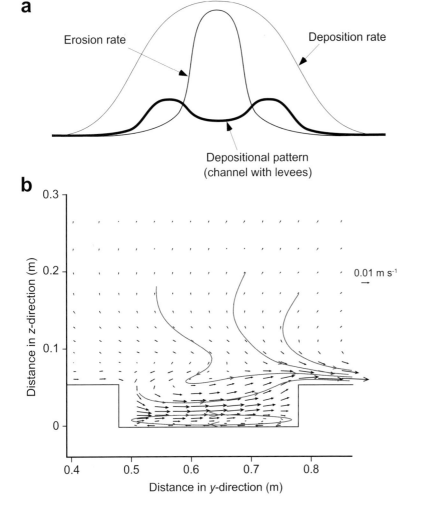

Fig. 13 (a) Diagram showing the relationships of deposition and erosion rates distributed across a turbidity current. The net result is a depositional pattern with levees and channel that confine the flow. (From Parsons *et al.*, this volume.) (b) From numerical modelling, secondary circulation (arrows) at a meander bend of a submarine channel experiencing a turbidity current. A suppressed circulation cell develops near the bed and substantial flow leaves the channel (known as flow stripping). (From Parsons *et al.*, this volume; modified from Kassem & Imran, 2004.)

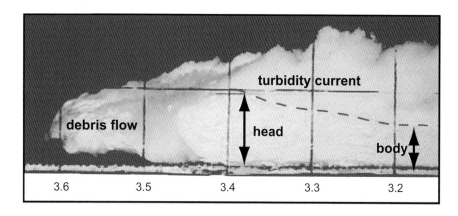

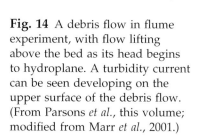

Fig. 14 A debris flow in flume experiment, with flow lifting above the bed as its head begins to hydroplane. A turbidity current can be seen developing on the upper surface of the debris flow. (From Parsons *et al.*, this volume; modified from Marr *et al.*, 2001.)

et al., 2000), which occurs due to **overspill** when turbidity-current water and suspended-sediment extend onto the flanks of the channel (Hiscott *et al.*, 1997). One cause of overspill is Coriolis force, which has a more significant impact in submarine channels than river channels (Klaucke *et al.*, 1998), and leads to dramatic asymmetry in levee heights (i.e. in direction of flow, the right levee is higher for channels in the northern hemisphere). **Super-elevation** of the turbidity-current surface occurs due to centrifugal forces on the outside of meander bends. Although modest in rivers (few centimetres), superelevation in submarine channels (Hay, 1987; Imran *et al.*, 1999) is comparable to channel depth (i.e. flow thickness doubles). As levees build up-ward due to overspill from Coriolis and centrifugal forces, so does the channel bed, creating channel systems that are perched well above the surround-ing seafloor (Flood *et al.*, 1991). Secondary circulation in these channels is much more complex (Fig. 13b) than in river channels (e.g. helical circulation), largely as a result of the ease with which turbid-ity currents can extend above the levees and flow independently (known as **flow stripping**; Piper & Normark, 1983).

Debris flows are gravity flows with greater sediment concentrations than turbidity currents, and in which turbulent motions are limited to the heads of the flows (Marr *et al.*, 2001). On land, debris flows have devastating impacts on communities in mountainous areas (Costa, 1984), and, in the ocean, they transport much sediment and create abundant strata (Elverhøi *et al.*, 1997). In contrast to turbid-ity currents, the dominant factors controlling the behaviour of debris flows come from pore pressures and grain-to-grain interactions. Some clay is neces-sary to reduce permeability and maintain excess

pore pressures (Marr *et al.*, 2001). The mechanics of operation are similar on land and underwater, and are characterized by flow as non-Newtonian fluids. Differences between subaerial and submarine debris flows are again due to the density difference between air and seawater. Very strong pressures are created by a submarine debris flow as it accelerates down slope and displaces water. If the water is not displaced quickly enough, the debris flow will separate from the seafloor (Fig. 14) and begin to **hydroplane** on a thin layer of lubricating water (Mohrig *et al.*, 1998). Consequently, the head regions of submarine flows move faster than those of sub-aerial flows, and the submarine flows cover more distance (known as **runout**). The head also moves faster than the body of a submarine debris flow (in contrast to turbidity currents), commonly causing the head to separate and achieve much greater runout than the body of the flow (Prior *et al.*, 1984; Nissen *et al.*, 1999).

The generation and deposition of different types of gravity flows can be interrelated. Turbidity cur-rents are commonly produced due to shear on the front and upper surface of a debris flow (Fig. 14). This process depends on how readily the debris-flow slurry breaks into pieces and becomes turbu-lent (Marr *et al.*, 2001), either by entrainment of fluid (most efficient) or by grain-by-grain erosion. After motion of a debris flow terminates, the associated turbidity current continues. Although only ~1% of the debris-flow sediment is needed to form a turbidity current, that current forms a flow about six times thicker (Mohrig *et al.*, 1998). When all motion has ended, the debris-flow deposit is sur-rounded by finer-grained turbidites on top and in front, reflecting another mechanism for creating turbidity currents.

8.2 Eel margin gravity flows

The modern Eel margin contains a broad range of gravity flows. Hyperpycnal plumes (> 40 g L^{-1} with freshwater) are thought to form during large, decadal flood events (Mulder & Syvitski, 1995), and possibly occurred during the 1995 and 1997 events (Imran & Syvitski, 2000). It is also possible that gravitational instability in turbid hypopycnal plumes (> 380 mg L^{-1}) from the Eel River (McCool & Parsons, 2004) caused a transfer of sediment to the bottom boundary layer by the process of **convective sedimentation** (Parsons *et al.*, 2001). This is one of several mechanisms that could create fluid-mud concentrations (> 10 g L^{-1} with salt water). Such concentrations are not reached by convergent estuarine flows on the Eel shelf (as described for the Amazon shelf by Kineke *et al.*, 1996), although a frontal zone might behave in a similar manner (Ogston *et al.*, 2000). The best-documented mechanism for reaching fluid-mud concentrations on the Eel shelf is by sediment resuspension within the wave boundary layer (< 10 cm thick; Ogston *et al.*, 2000; Traykovski *et al.*, 2000). Such thin layers are difficult to investigate, and require the use of downward-directed acoustic tools or the placement of optical sensors close to the seabed. Therefore, wave-supported fluid-mud flows are likely to be more common on other shelves than recognized in past studies.

Gravity flows generated by wave resuspension of seabed sediment differ from other types (e.g. turbidity currents, debris flows), because they commonly flow on very gentle seafloor gradients (continental shelves) and need continual infusion of energy from surface waves. The theory of sediment transport in the wave boundary layer (Grant & Madsen, 1979) has undergone some recent revision: e.g. with new information about the involvement of seabed permeability (Hsu & Hanes, 2004) and about the turbulence structure (Lamb *et al.*, 2004). However, a major factor affecting the mechanics of sediment transport remains the strong stratification at the top of the wave boundary layer. Below this, the large suspended-sediment concentrations are able to suppress further turbulence and limit seabed resuspension (Traykovski *et al.*, 2000; Wright *et al.*, 2001; as modified from Trowbridge & Kineke, 1994). A well-mixed boundary layer develops with a sharp turbidity (and density)

gradient at its surface boundary (known as a **lutocline**). At 60-m water depth on the Eel shelf, the wave boundary layer can be 5–10 cm thick and have concentrations in the order of 100 g L^{-1}, which abruptly decrease to < 1 g L^{-1} above the lutocline (Traykovski *et al.*, 2000; Wright *et al.*, 2001). Sediment within the layer flows down slope at velocities > 10 cm s^{-1} (as much as ~60 cm s^{-1}) and continues seaward as long as sufficient energy is supplied from surface gravity waves. In Eel Canyon, liquefaction from cyclic wave action initiates failure that starts wave-supported gravity flows moving down canyon (~15 cm s^{-1}; Puig *et al.*, 2003, 2004).

The difference in the Eel River hydrograph between present and LGM conditions has had an impact on the operation of gravity flows (Parsons *et al.*, this volume, pp. 275–337). Today, sediment supply is reduced and characterized by episodic winter floods/storms, deposits of which can be partially preserved. During the LGM, more sediment was discharged and generally in a more sustained manner (due to spring/summer snow melt). However, when flood events occurred, they were typically more severe, due to the combination of rainfall and melting of low-elevation snow. Consequently, modelling studies (Morehead *et al.*, 2001) suggest that the net impact of higher accumulation rates and more severe floods during the LGM would be better preservation of gravity-flow deposits in the stratigraphic record. During the LGM, sea level was lower and these flows travelled down the continental slope, both through channels entering Eel Canyon and gullies to the north.

MARGIN MORPHOLOGY

The morphology of continental margins has been considered in this volume by Pratson *et al.* (pp. 339–380) and Syvitski *et al.* (pp. 459–529).

General considerations

Sedimentary processes from flocculation to debris flows create deposits that change bathymetry and give shape to continental margins: in other words, they produce a **seascape**. The seascape is a stratal surface that is buried and viewed later in the subsurface as a record of the formative processes. However, seascapes are not inert features that only respond to sedimentary processes, they also directly influence those processes. For example, topographic highs can put the seabed within the realm of wave reworking, and steep seabed gradients may initiate failures and gravity flows.

An understanding of the seascape for continental margins requires the inclusion of processes operating over longer time-scales (10^4–10^6 yr) than those discussed previously. When continents rift to form new ocean basins, the initial margins have a stair-step shape, due to the abrupt descent of several kilometres from the surfaces of thick, low-density continental crust down to the surfaces of thin, dense oceanic crust. The margin then moves passively as both continental and oceanic crust is carried away from the mid-ocean ridge, which is the situation with the New Jersey margin (Fig. 1b). Eventually, global forces within the Earth change and different plates, one with continental crust and another with oceanic crust, move in opposition. This creates a tectonically active margin, where the oceanic plate is **subducted** beneath the continental plate, as with the Eel margin. The operative processes and resulting morphologies (Fig. 1) are very different for passive and active margins, and depend primarily on the mechanisms forming and filling accommodation space (Van Wagoner *et al.*, 1988).

Sea-level fluctuation is another process relevant to margin seascapes, and for a particular location depends on global (**eustatic**) sea level and on vertical motions of the margin. These fluctuations have many impacts, including forced migration of the shoreline and other morphological features. Disequilibria result as erosive mechanisms associated with fluvial and marine processes begin to operate on new surfaces. The waxing and waning of coastal plains and continental shelves demonstrate the intimate relationship of seascapes and landscapes as sea level fluctuates, and as the important boundaries of shoreline and shelf break migrate and even merge. Sea-level rise causes landward migration (i.e. **transgression**) of the **shoreface**, which is the region from the low-tide shoreline to ~10-m water depth, where reworking by surface waves is most intense. This migration destroys shoreline landscapes (e.g. beaches, aeolian dunes, tidal flats, marshes) creating a **ravinement surface** on the continental shelf. During sea-level fall, the shoreline migrates seaward (i.e. **regression**)

and these same landscapes are stranded on the newly formed coastal plain. The shelf seascape progressively dwindles as the shoreline approaches the shelf break. The new proximity of fluvial sediment supply dramatically impacts the seascape of the continental slope, causing both seaward growth and landward erosion (e.g. incisions by submarine canyons and gullies).

Glaciers and rivers are the primary suppliers of sediment to fill margin accommodation space. Temperate glaciers have especially large sediment yields (Milliman & Syvitski, 1992). During the LGM (and other glacial periods), the influence of glacial supply expanded toward lower latitudes, as demonstrated by sediment remnants on margins (e.g. moraines, jökulhlaup deposits; Shor & McClennen, 1988; Uchupi *et al.*, 2001). However, during all periods, fluvial sediment supply has dominated the global flux of terrestrial sediment to margins, supplying > 85% today (Milliman & Meade, 1983; Milliman & Syvitski, 1992). The pattern of fluvial discharge can be as a single, major river (e.g. Amazon, Ganges–Brahmaputra, Mississippi) within a large contiguous dispersal system, or as a collection of small rivers with coalescing dispersal systems (Jaeger & Nittrouer, 2000). The former are known as **point sources** and are usually limited to passive margins; the latter are **line sources** and are common on active margins. Substantial discharge of sediment associated with both geometries of sediment supply can create significant morphological features on continental shelves, including subaerial and subaqueous deltas.

Subaerial deltas (e.g. Mississippi and Nile Deltas) are well known because many people live on or near them, and the dynamic processes affecting morphology are easily recognized. One of the most dramatic processes results from consolidation of deltaic deposits, which leads to abrupt changes in the locus of deltaic sedimentation (i.e. **lobe switching**) associated with the formation and filling of accommodation space (Penland *et al.*, 1988). Subaqueous deltas are not as easily observed, but can represent the dominant sink for fluvial sediment (e.g. Amazon River; Nittrouer *et al.*, 1986). They are the result of energetic processes (waves, tides, currents) inhibiting sediment accumulation in shallow water (Walsh *et al.*, 2004), and of shelf width/depth providing sufficient accommodation space for accumulation in deeper water. Surface waves are especially important physical processes reworking the inner shelf. Their influence is greatest on the shoreface and continues to dominate seaward to a depth (known as **wave base**) dependent on wave properties (e.g. wavelength of storm waves) and seabed character (e.g. grain size). Fine-grained sediments are delivered farther seaward (e.g. by wave- or tide-driven diffusion, Ekman veering, downwelling bottom currents, or fluid-mud flows) and accumulate there. Silts and clays represent the bulk of fluvial supply, so accumulation rates can be great and subaqueous deltas can form. Shelf stratigraphy largely results from the interplay and relative motion of three important morphological features (Fig. 15a): the shoreline and possible subaerial delta; the subaqueous delta; and the shelf break (Swenson *et al.*, 2005).

The continental slope receives sediment shed from land and from landward portions of the continental margin, creating the ocean's largest repository for sediment mass (> 40% of marine sediment; Kennett, 1982). A complex mixture of sediment deposition, failure and mass movement impacts the morphology of the open slope (Adams & Schlager, 2000). This would predict a bathymetric gradient similar to the angles of repose for saturated sand, silt or clay, which are all > 10° (Allen, 1985). However, the observed gradient is ~4° and, therefore, other mechanisms must also be operating. Earthquakes and rapid sediment accumulation cause excess pore pressures, and could lead to reduced gradients (Pratson & Haxby, 1996; O'Grady *et al.*, 2000). Groundwater flowing out of continental slopes due to tectonic squeezing or differential loading can destabilize the seabed and reduce gradients (Iverson & Major, 1986; Orange & Breen, 1992). Internal waves triggered by tides or storms create a bore that propagates up slope, inhibiting deposition and creating nepheloid layers (Cacchione *et al.*, 2002; Cacchione & Pratson, 2004). This mechanism

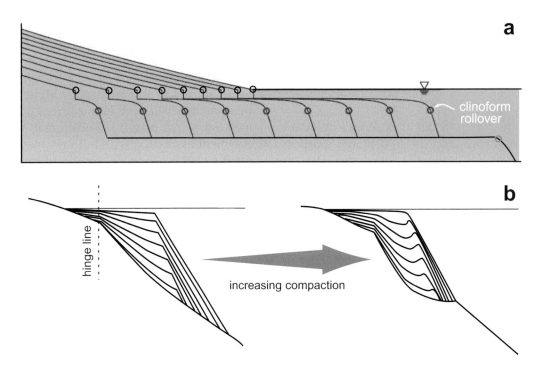

Fig. 15 Stratal geometries on continental margins. (a) Model representation of the shoreline (black circles), clinoform rollover (red circles) and shelf break (green circle), which are critical boundaries on continental margins. The stratigraphy produced by shelf sedimentation (especially on passive margins) is largely the result of the interplay and relative motions among these three important morphological features. (From Pratson *et al.*, this volume; modified from Swenson *et al.*, 2005.) (b) Model representation of shelf and slope strata impacted by isostatic subsidence (left) and sediment compaction (right) on a passive margin. Both of these processes operate where sediment accumulates, causing a feedback that creates the thickest sediment deposits, commonly near the shelf break. (From Pratson *et al.*, this volume; modified from Reynolds *et al.*, 1991.)

might constrain the gradient of the slope to the angle of internal-wave propagation (~4°), and is attractive because it provides a global process for explaining the bathymetric gradient of continental slopes. Turbidity currents are another possible mechanism, because they become erosional (ignitive) for seabeds steeper than ~4°, and become depositional for more gentle seabeds (Kostic *et al.*, 2002). A range of mechanisms is available, therefore, for controlling the bathymetric gradient of continental slopes, and it is not surprising that global observations converge to a relatively uniform value (e.g. Pratson & Haxby, 1996).

Although continental slopes undergo significant sediment accumulation, the Earth's most dramatic erosional features also occur there. These are submarine canyons, some of which dwarf the largest subaerial canyons (Normark & Carlson, 2003). However, a spectrum of sizes and shapes is found on continental slopes, probably related to their age and formative processes. Some smaller canyons (short, narrow, little vertical incision) have U-shaped cross-sections, with heads that do not reach the shelf break (Twichell & Roberts, 1982). Failures and subsequent landslides, with or without gravity flows, are probable mechanisms of formation (Farre *et al.*, 1983). In some cases, the regular spacing of these canyons suggests that groundwater seepage may play a role in their formation (Orange *et al.*, 1994). The largest canyons are V-shaped and incise the shelf break. Regardless of the mechanisms initially forming them, gravity flows (and associated scour) are the mechanisms ultimately causing the large magnitude of these canyons. Sedimentary deposits within the submarine fans at their termini indicate that turbidity currents are a more dominant transport process than debris flows (Ericson *et al.*, 1961). Excavation of the large canyons is tied to lowstands of sea level, when rivers delivered sediment directly to their heads (Emery & Uchupi, 1972). In addition to being dependent on river discharge, these canyons have morphologies similar to rivers (Shepard, 1977; Pratson, 1993): multiple tributaries at their heads; V-shaped cross-sections; concave-up longitudinal profiles; and division into distributaries at their bases (on submarine fans). The history of submarine-canyon development is made complex by periods of variable activity, and by multiple processes impacting their evolution (Shepard, 1981; Goodwin & Prior, 1989).

New Jersey margin morphology

On passive margins, **thermal subsidence** is an important process controlling accommodation space and sedimentation, as the plates cool with time and with spreading distance from the mid-ocean ridge (Parsons & Sclater, 1977). The rate of subsidence decreases landward to a negligible level at the **hinge line**, which is near the landward edge of the initial rift. Where subsidence is rapid, sediment accumulation leads to upward **aggradation** of the shelf and, where slow, seaward **progradation** of the slope is dominant (Reynolds *et al.*, 1991). Although thermal subsidence accounts for ~40% of accommodation space generated on the New Jersey margin (Steckler *et al.*, 1988), other processes are also operating. As sediment accumulates, the additional weight causes **isostatic subsidence** and the margin sinks into the plastic portions of the Earth's upper mantle (Watts & Ryan, 1976). Sediment consolidation is another important process, as thick deposits of sand and mud have their porosities reduced from high values at the seabed surface to 0.3 and 0.2, respectively, by burial to depths of many kilometres (Sclater & Christie, 1980; Bahr *et al.*, 2001). Isostatic subsidence and sediment consolidation operate where sediment accumulates (Fig. 15b), causing a feedback that creates the thickest sediment deposits, commonly beneath the shelf break (Reynolds *et al.*, 1991). Other processes can cause localized impacts on accommodation space. Extensional forces associated with rifting create **normal faults** that allow the underlying structure of the margin to break into segments that move downward and seaward. Similarly, in areas of rapid sediment accumulation (Suppe, 1985), **growth faults** form as thickening sedimentary deposits slowly rotate downward and seaward (Emery & Uchupi, 1984). However, New Jersey strata discussed in this paper of Cretaceous age and younger show relatively little evidence of such faulting (Poag, 1985).

The passive origin of the New Jersey margin has created the breadth and depth needed for substantial accumulation on its continental shelf. Major components of the stratigraphy underlying the New Jersey margin are those of deltaic sequences, with the typical morphology of **clinoforms**. Gently dipping topset strata change gradient across the **rollover point** to steeper-dipping foreset

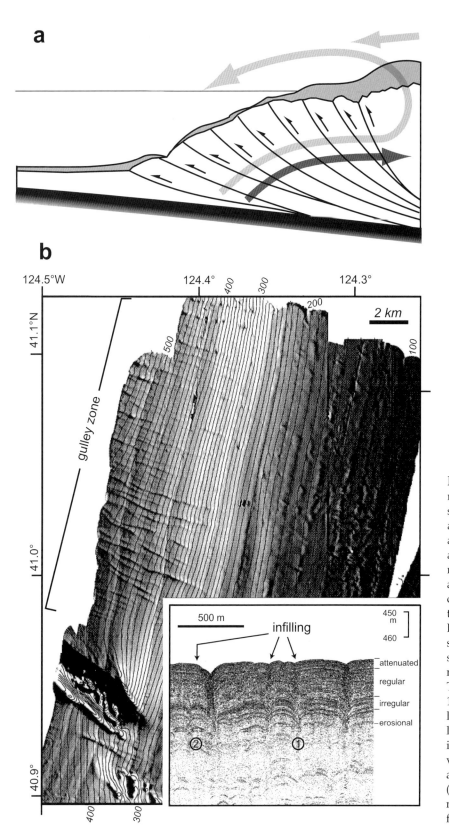

Fig. 16 (a) On an active margin, sediments scraped off the subducting oceanic plate create an accretionary prism, causing shelf and slope deposits to be uplifted and recycled by erosion. Uplift also reduces shelf accommodation space and displaces sedimentation to the continental slope. (From Pratson *et al.*, this volume; modified from Kulm & Fowler, 1974.) (b) Multibeam survey showing one mode of Eel slope sedimentation: aggradational gullies north of the Little Salmon Anticline. They are ~100 m wide and only 1–2 m deep in this figure, but extend long distances (> 10 km) and become larger farther down slope. The inset is a seismic profile proving (1) the vertical continuity of some gullies and (2) the termination of others. (From Pratson *et al.*, this volume; modified from Field *et al.*, 1999 and from Spinelli & Field, 2001.)

strata, and at the base become nearly flat **bottom-set** strata. The foreset region is comparable to the **delta front** of a subaerial delta, and the bottomset is similar to the **prodelta** region. A fundamental difference is that the rollover point is the shoreline for a subaerial delta, and is below sea level (usually 20–50 m) for a subaqueous delta. The latter is probably the dominant origin for the clinoforms found within the New Jersey margin. When energy expenditure inhibited sediment accumulation on the topset, it was displaced seaward, creating relatively steep foreset strata and an **oblique** clinoform shape (Mitchum *et al.*, 1977a; Pirmez *et al.*, 1998). When transport processes caused sediment accumulation to be distributed more broadly, the resulting clinoform had a gentler morphology and a **sigmoidal** shape. These clinoform features were created during periods of plentiful sediment supply, and were responsible for aggrading and prograding the New Jersey margin.

Eel margin morphology

On active margins, tectonic uplift associated with plate motions controls accommodation space and sedimentation. Sediments that have accumulated on the subducting oceanic plate are scraped off, folded and amalgamated into margin deposits known as **accretionary prisms**. Shelf and slope strata are rotated upward (Fig. 16a), ultimately providing mountainous source rock that can be eroded and returned to the margin (Kulm & Fowler, 1974; Kulm *et al.*, 1975), thus recycling sediment (like the Eel margin). Uplift also reduces shelf accommodation space, and the recycled sediment predominantly accumulates on the continental slope (also like the Eel margin).

Eel Canyon is an important conduit for sediment transfer from shelf to slope, but much sediment also crosses the shelf break to the open slope farther north. Regardless of the origin for the 'Humboldt Slide', the gullies found on its upper rim (Fig. 11) clearly indicate the paths for some sediment transfer. Similar features are found farther north (Fig. 16b), where the gullies start small near the shelf break and enlarge with distance down slope (Field *et al.*, 1999). Several different geometries of these features can be identified on continental slopes. **Rills** are ~5–300 m wide and ~1–40 m deep, extending downslope many kilometres (commonly > 10 km)

and converging with each other at low angles (Pratson *et al.*, 1994). **Dendritic gullies** are relatively short (< 5 km) and intersect submarine canyons at high angles, creating dendritic erosion patterns (Farre *et al.*, 1983). Although many rills and gullies are erosional, those on the northern Eel slope (Fig. 16b) are building upward (i.e. they are **aggradational gullies**), and grow by differential sediment accumulation associated with turbidity currents (Spinelli & Field, 2001). Sediment accumulation on the banks of adjacent gullies overlaps and grows upward together with the gully channels (similar to mechanisms shown in Fig. 13a). In contrast to a submarine canyon, gullies represent a line source of sediment to a continental slope and rise.

MARGIN STRATIGRAPHY

The startigraphy of continental margins has been reviewed in this volume by Mountain *et al.* (pp. 381–458) and Syvitski *et al.* (pp. 459–529).

General considerations

Sedimentary laminae (millimetres to centimetres thick) are stacked to create beds (metres to tens of metres thick), and beds are stacked to create sequences of strata (tens to hundreds of metres thick) that become the continental-margin sediment record. The sequences record repetitive cycles of sedimentation driven by processes that operate over large spatial scales (e.g. eustatic sea-level fluctuations, tectonic motions, sediment dispersal mechanisms). The time-scales of sequences that build margins vary with the inherent scales of the controlling processes, but are all > 10^4 yr. Examination of long-term stratigraphy is important for a number of reasons:

1 documenting the history of the Earth;
2 evaluating the operation of natural processes that fluctuate on long time-scales (parts of which may not be operative today);
3 validating predictive and diagnostic models relating sedimentary processes and strata.

The sedimentary record of continental margins is divided by abrupt breaks (**unconformities**) caused by large-scale erosion or non-deposition associated both with migration of the shoreline and reduction of sediment supply. Eustatic sea-level change has been suggested as a cause of stratal breaks that are globally synchronous over 10^6 yr (Vail *et al.*, 1977; Posamentier *et al.*, 1988) and even shorter time-scales (Haq *et al.*, 1987). This is the foundation of **sequence stratigraphy**. However, other processes (e.g. glacial rebound, tectonic motions, even deltaic lobe switching) cause similar transgressions and regressions on a local basis (Christie-Blick, 1991; Miall, 1991) and can complicate interpretation of the sedimentary record. Regardless of cause, margin stratigraphy is segmented on many time-scales and genetically related sediment deposits (e.g. from shelf clinoforms) are bounded by unconformities (Mitchum *et al.*, 1977a).

Recognition of stratigraphic relationships on continental margins is enabled by seismic tools (Payton, 1977; Hamilton, 1980), which detect the physical contrasts in sediments (e.g. changes in bulk density and sound velocity) that cause acoustic reflections. For adequate identification of seismic reflectors, the sedimentary contrasts must be vertically abrupt (scale of metres) and laterally persistent (scale of hundreds of metres). Direct sampling of strata (e.g. rotary drilling) gives access to sediment for age estimation (e.g. by biostratigraphy, magneto-geochronology and isotopic dating; Berggren *et al.*, 1995; Lowrie & Kent, 2004) and allows evaluation of a wide range of physical and chemical properties (by downhole measurements using electrical, nuclear and acoustic tools; Goldberg, 1997). Together, seismic profiling and seabed drilling provide the data necessary to document the geometric and temporal relationships of strata, which also can be adjusted (known as **backstripping**) for post-depositional changes that affect spatial relationships (e.g. isostatic subsidence, sediment consolidation; Steckler *et al.*, 1988, 1999). The angular terminations of strata and unconformities allow for development of insights about the processes creating sequences (e.g. sea-level fluctuations, tectonic motions, sediment supply, ocean currents; Karner & Driscoll, 1997). Among the common geometric relationships are (Fig. 17; Mitchum *et al.*, 1977b):

1 an unconformity defining the top of a sequence (**toplap**);
2 strata building landward against a basal unconformity (**onlap**);

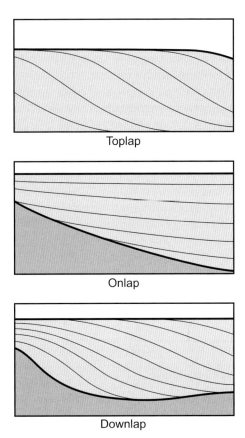

Fig. 17 Common geometric relationships between seismic reflectors (thin lines) and bounding unconformities (bold lines). Toplap, an unconformity defining the top of sequence; onlap, strata building landward against a basal unconformity; downlap, strata building seaward across an underlying unconformity. (From Mountain *et al.*, this volume pp. 381–458; modified from Mitchum *et al.*, 1977b.)

3 strata building seaward across an underlying unconformity (**downlap**).

Dating provides information that can be used in many ways (e.g. calculation of accumulation rates), and can determine whether the age of a seismic reflector varies spatially (i.e. time transgressive). With rare exceptions, all strata above a sequence boundary are younger than all strata below it (Vail *et al.*, 1977).

A common mechanism for creation of a sequence boundary is subaerial exposure (Van Wagoner *et al.*, 1988), which causes erosion into the underlying sequence. Another potential stratigraphic break is associated with sea-level rise and the resulting transgression, which creates the maximum flooding surface onto which subsequent sedimentation builds (e.g. prograding clinoforms). Intrasequence stratigraphic geometry commonly reveals mud sitting abruptly above sand, followed upward by gradual coarsening back to sand (representing seaward migration of the shoreline). The coarsening-upward strata are known as **parasequences** (Van Waggoner *et al.*, 1988), and are stacked to create larger-scale sequences. During the various stages of sea level (high, falling, low, rising), different types of parasequences are distributed across a continental margin to create unique patterns known as **systems tracts** (Posamentier *et al.*, 1988).

New Jersey margin stratigraphy

During the Miocene and Pleistocene, sediment accumulation reached rates of ~10–100 m Myr^{-1}. The Miocene phase of sedimentation contained drainage systems from the ancestral Delaware/ Susquehanna and Hudson Rivers that flowed southeastward (Poag & Sevon, 1989; Poag & Ward, 1993). They created at least eight sequences of well-preserved clinoforms (Fig. 18; Greenlee *et al.*, 1992), which compare favourably with eustatic sea-level fluctuations (Sugarman *et al.*, 1997). Sequences along the northern coastal plain and shelf of New Jersey are thicker than those along southern New

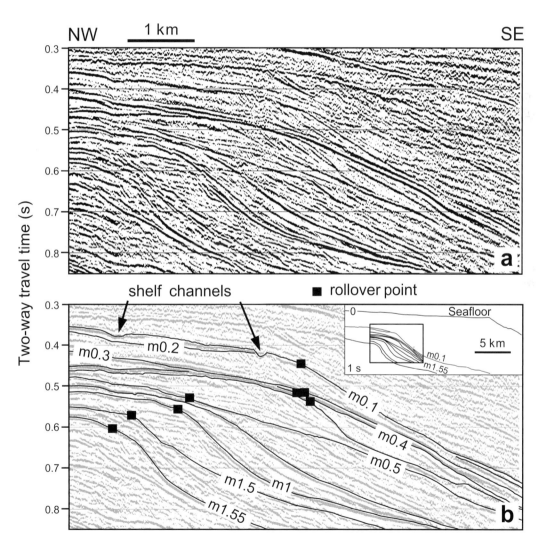

Fig. 18 Across-shelf seismic profile for Miocene sequences underlying the New Jersey margin. (a) Original profile. (b) Interpreted profile. These are prograding clinoforms, and rollover points are shown by black squares. Several channels can be seen incising topset surfaces. (From Mountain *et al.*, this volume; after Fulthorpe *et al.*, 1999.)

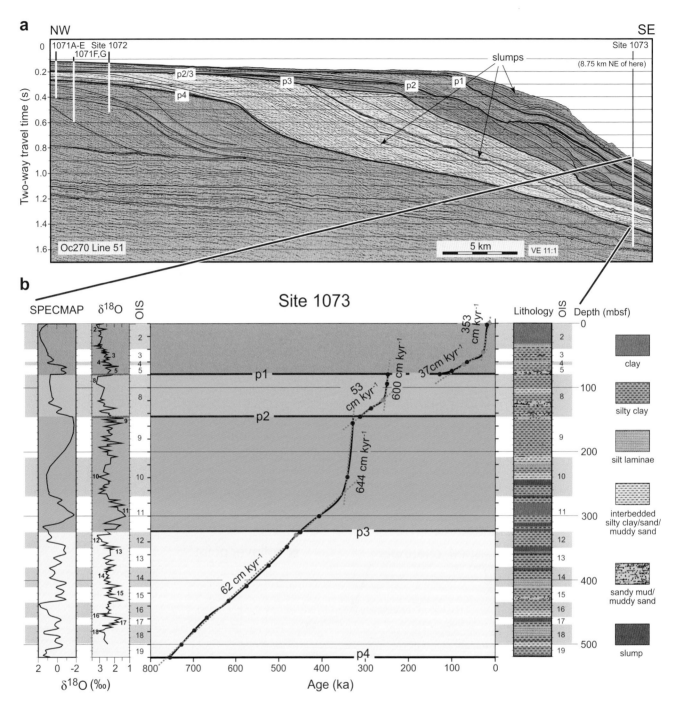

Fig. 19 Sediment profiles on the New Jersey margin. (a) Interpreted seismic profile crossing several Ocean Drilling Program (ODP) drill sites, including site 1073. Four Pleistocene sequences are identified by different colours, each representing clinoforms that have prograded across the outer shelf and upper slope. (From Mountain *et al.*, this volume.) (b) Age profiles for Site 1073, with defining isotopic data (on left; see Mountain *et al.*, this volume, for description of SPECMAP chronology), and lithological information and oxygen isotope stage (OIS; right). The top three Pleistocene sequences show an upward increase in accumulation rate. (From Mountain *et al.*, this volume; modified from McHugh & Olson, 2002.)

Jersey, presumably due to differences in accommodation space and fluvial sediment supply. A few clinoforms have foresets with small incisions but the surfaces are generally unscarred (Fulthorpe *et al.*, 2000), suggesting that sediment was delivered to the foresets as non-channelized flows. In contrast, there are well-developed valleys on the topsets, whose origins are linked to incisions during lowstand of sea level (Fulthorpe *et al.*, 1999). The contrast between topset and foreset incisions indicates that the rollover depth was about the same as the lower extent of sea-level fluctuations, ~20–30 m (Kominz *et al.*, 1998; Miller *et al.*, 1998). This magnitude of fluctuation compares well with numerical modelling based on backstripping of stratal surfaces (Steckler *et al.*, 1999).

During the Pleistocene, sedimentation extended the shelf width by tens of kilometres (Mountain *et al.*, this volume, pp. 381–458). The Hudson Apron is located on the upper slope south-west of Hudson Canyon and reveals four well-developed sedimentary sequences (Fig. 19a), whose bounding surfaces extend conformably across the shelf break and demonstrate the relationships of shelf and slope strata. The bathymetric gradient for the base of each sequence becomes progressively steeper (from ~1:60 to ~1:20) as they move seaward through time. A number of chaotic reflectors are observed within the sequences, suggesting failure and mass movement (i.e. slumps). Three of the four sequences have accumulation rates increasing upward by an order of magnitude (Fig. 19b). Eustatic sea level does not seem to be the only cause for clinoform evolution during this time, but, rather, it also involves processes associated with glacial advance/retreat, isostatic response and glacial sediment discharge. In areas where continental margins are impacted by glaciation, sequence stratigraphy is particularly complicated.

The uppermost strata on the New Jersey shelf reveal bathymetric features related to the most recent transgression. The seafloor was eroded (3–10 m) and modified into sand ridges as the shoreface moved landward, and subsequently waves and currents on the shelf have continued to rework the ridges (Goff *et al.*, 2005). Beneath these surface strata (0–20 m thick) lies evidence for multiple fluvial drainage systems, which were cut by glacial meltwater pulses flowing across the pre-existing coastal plain (Nordfjord *et al.*, 2005). Some of these pulses were probably associated with catastrophic breaks of glacial-lake dams (~12–19 ka; Uchupi *et al.*, 2001), which disrupted stratigraphy by creating erosional blocks of old, layered deposits and stranding them within buried channels now containing acoustically transparent sediments (Fulthorpe & Austin, 2004). A diverse assortment of incision geometries provides evidence for across-shelf flows and sediment supply to the New Jersey continental slope during sea-level lowstands in both the Miocene and the Pleistocene.

The New Jersey slope reveals Miocene gullies and canyons, many of which have been partially or completely filled (Miller *et al.*, 1987; Mountain, 1987), especially on the upper slope. Pleistocene processes excavated some of the old canyons, and many remained open and unfilled on the lower slope (Pratson *et al.*, 1994). The sedimentary deposits filling the canyons reveal a consistent pattern of coarse basal material, followed upward by fine-grained turbidites, and finally hemipelagic deposits (May *et al.*, 1983; Mountain *et al.*, 1996). The general expectation is for canyon filling to occur during sea-level rise (Posamentier *et al.*, 1988) and that may be the case for the burial of Miocene canyons. However, many canyons active during the LGM on the New Jersey slope are still exposed today (and elsewhere in the world; Fig. 3), so the process of filling during transgression is not guaranteed to be complete.

Eel margin stratigraphy

The modern tectonic conditions that started in the middle Pleistocene on the Eel margin created a widespread unconformity, and since then 13 more have formed (Fig. 20; Burger *et al.*, 2002). During that time, ~1 km of sediment has accumulated (Clarke, 1987), although tectonic structures cause significant local variability in thickness. The sequences bounded by the unconformities are not characterized by prograding clinoforms (as observed on the New Jersey margin), but are dominated by fill of relatively small, segmented basins. Early sources of sediment (> 500 ka) were from a northern river and the depocentre was located west of Trinidad Head. The locus of greatest sediment preservation shifted southward, and accumulation rates have generally decreased for the past 500 kyr (Burger *et al.*, 2002). During that

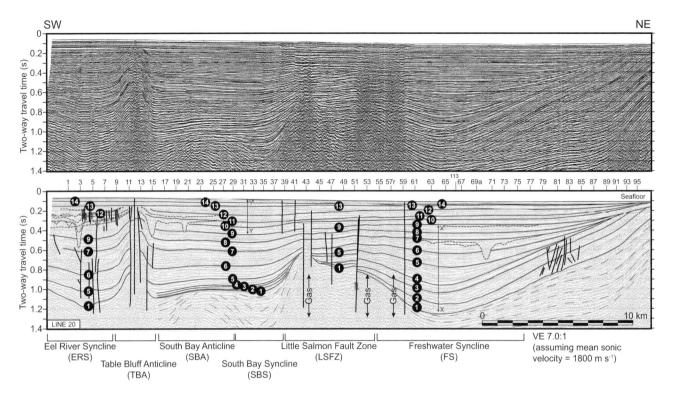

Fig. 20 Along-shelf seismic profile for Pleistocene sequences (past ~500 kyr) underlying the Eel margin. (a) Original profile. (b) Interpreted profile. The strata are segmented into basins by anticlines (labelled) and faults (near-vertical bold lines). Regional unconformities are shown by near-horizontal lines (14 total). Gas disturbs the seismic record in several areas. (From Mountain *et al.*, this volume; after Burger *et al.*, 2002.)

period, ancestral versions of both the Eel and Mad Rivers became significant sediment sources with recognizable basins of accumulation. For the past 360 kyr, Eel Canyon has been active and has provided a conduit for substantial sediment transport beyond the shelf break. Modern sediment dispersal processes (i.e. dominant input from Eel River, minor input from Mad River; Sommerfield *et al.*, this volume, pp. 157–212) have shaped the margin for the past ~43 kyr (Burger *et al.*, 2002).

Table Bluff Anticline is one of the dominant structural highs (Fig. 20) oriented across the shelf (Clarke, 1987), and has separated Eel and Mad River sedimentation at times in the past. Shelf channels on the surface of the anticline had a south-westward gradient, heading toward Eel Canyon (Burger *et al.*, 2002). The overall dendritic pattern of the channels is similar to channels on the New Jersey shelf (Duncan *et al.*, 2000), which have been interpreted as fluvial systems operating during lower stands of sea level (Austin *et al.*, 1996). Ten stratigraphic surfaces stacked vertically are

identified by incised channels, and represent the past ~500 kyr (Burger *et al.*, 2001). There are too many surfaces to be the result of only eustatic sea-level lowering, so tectonic uplift must be responsible for some of the erosional surfaces.

On the Eel slope, the Little Salmon Anticline (which is breached at its crest) is another cross-margin structural high that separates sedimentation regimes (Figs 11 and 16b). Eel Canyon and the large gullies on the rim of 'Humboldt Slide' occur to the south, and smaller, aggradational gullies are observed to the north. During lowstands of sea level, the Little Salmon Anticline separated the discharges of the Eel and Mad rivers, and the Mad probably had a major impact on the northern gullies at these times (Burger *et al.*, 2001). Biogenic and thermogenic gas are common on this portion of the Eel slope (Kvenvolden & Field, 1981; Lorenson *et al.*, 1998). Gas commonly impacts seismic observations (e.g. Figure 20) and is probably responsible for pockmarks observed by multibeam data (Yun *et al.*, 1999). Some of the pockmarks are found in linear orientations near the gullies on the slope, but no formative relationship has been established (Field *et al.*, 1999).

Modelling studies of the Eel margin (Mountain *et al.*, this volume, pp. 381–458) indicate a number of fundamental differences in strata formation relative to the New Jersey margin. Tectonic subsidence on the Eel margin (~2 mm yr^{-1}; Orange, 1999) has a significant impact on the observed sequences, because it leads to seaward divergence of shelf reflectors, not prograding clinoforms, as with New Jersey. Accommodation space on the Eel margin has been sufficient to balance sediment supply, and this results in a stationary shelf break, not seaward progradation, as with New Jersey. Other differences include the dominance of segmented depocentres on the Eel margin and the potential for erosional unconformities due to periods of uplift. The relative importance of operational processes leads to distinct differences in the stratigraphic imprints of these passive and active margins.

CONCLUSIONS

This paper integrates recent knowledge about continental-margin sedimentation across several domains, as highlighted in Table 1. The challenges and benefits are briefly summarized here.

1 *Shelves and slopes.* As with the topset and foreset on a clinoform, the linkage between the upward aggrading and seaward prograding segments of a margin (shelf and slope, respectively) depends on the operative processes. At the crossroads of land and ocean, continental margins exhibit a wealth of potential processes, and there is a nearly infinite range of combinations and magnitudes for those processes. However, the dual representation of stratigraphy in response to both shelf and slope processes provides twice the perspective for interpreting Earth history.

2 *Processes and stratigraphy.* The goal is to use a fundamental understanding of short-term processes transporting and accumulating sediment to interpret preserved stratigraphic signatures better. In return, stratigraphy documents the relative importance of sedimentary processes, not necessarily according to their spatial and temporal dominance in modern environments, but according to their impact on preserved signatures. This allows investigations to focus on understanding the stratigraphically important processes, which differ with the time-scales of interest.

3 *Observations and modelling.* Observations are important for describing a particular continental margin but, by themselves, provide limited means to extrapolate

Table 1 Examples of the multidimensional evaluation of continental-margin sedimentation, as presented by this paper and described in detail in this volume

Location	Focus	Method	Topic
Shelf	Process	Observation	Wave-supported sediment gravity flows (Hill *et al.*, this volume; Parsons *et al.*, this volume)
Shelf	Process	Modelling	Sediment delivery during the Last Glacial Maximum (Syvitski *et al.*, this volume)
Shelf	Stratigraphy	Observation	Flood-deposit signature, preservation and distribution (Wheatcroft *et al.*, this volume; Sommerfield *et al.*, this volume)
Shelf	Stratigraphy	Modelling	Backstripping of New Jersey shelf (Mountain *et al.*, this volume; Syvitski *et al.*, this volume)
Slope	Process	Observation	Gullies formed by turbidity currents (Pratson *et al.*, this volume; Mountain *et al.*, this volume)
Slope	Process	Modelling	Hydroplaning debris flows (Parsons *et al.*, this volume)
Slope	Stratigraphy	Observation	'Humboldt Slide' (Lee *et al.*, this volume)
Slope	Stratigraphy	Modelling	Submarine-canyon evolution (Pratson *et al.*, this volume; Syvitski *et al.*, this volume)

in time and space. That is the potential of numerical modelling and laboratory experimentation. Predictive models are valuable for developing a fundamental understanding of complex systems, but inverse models (that are diagnostic) provide special benefits for reading the stratigraphic record. Observational studies then give validation to the models. Knowledge is best attained by using observation and modelling in tandem.

The benefits of integrated studies of continental margins are limited by boundary conditions, such as those introduced at the beginning of this paper. Not all details of Earth history are recorded in fluvial deposits on shelves and slopes. Future integrated studies should expand to high-latitude settings, investigating systems dominated by both temperate glaciation and by the full range of polar processes. Similarly, low-latitude carbonate sedimentation has some unique attributes, and requires focused attention. The time-scales of the present paper have stopped short of the deep stratigraphy on margins and of rock records on land. Integration of margin studies is valuable for the time-scales of this paper, and should be even more important when extended to the full spectrum of margin stratigraphy. Terrestrial and coastal environments are largely untouched in this paper, and the same is true for submarine fans, continental rises and abyssal plains, all of which are intimately related to continental-margin sedimentation. Source-to-sink programmes examining sediment dispersal systems from mountain tops to deep ocean floors have been initiated in several places around the world, and this trend should be continued and encouraged.

ACKNOWLEDGEMENTS

The authors express great thanks to the participants in STRATAFORM and the authors of papers in this volume for contributing the material summarized here. We especially thank Paul Hill, Rob Wheatcroft, Chris Sommerfield, Homa Lee, Lincoln Pratson and Greg Mountain, who provided valuable comments that helped improve the paper. Tina Drexler kindly prepared the figures. The Office of Naval Research is responsible for supporting the research efforts that produced this paper, and we appreciate their commitment over many years. Other agencies also sponsored parts of the work summarized, and all deserve credit for support of investigations into continental-margin sedimentation.

REFERENCES

Adams, E.W. and Schlager, W. (2000) Basic types of submarine slope curvature. *J. Sediment. Res.*, **70**, 814–828.

Alexander, C.R. and Simoneau, A.M. (1999) Spatial variability in sedimentary processes on the Eel continental slope. *Mar. Geol.*, **154**, 243–254.

Allen, J.R.L. (1985) *Principles of Physical Sedimentology*. Allen & Unwin, Boston, 272 pp.

Aller, R.C. (2004) Conceptual models of early diagenetic processes: the muddy seafloor as an unsteady, batch reactor. *J. Mar. Res.*, **62**, 815–835.

Aller, R.C. and Cochran, J.K. (1976) ^{234}Th/^{238}U disequilibrium in near-shore sediment, particle reworking and diagenetic time scales. *Earth Planet. Sci. Lett.*, **29**, 37–50.

Austin, J.A., Fulthorpe, C.S., Mountain, G.S., Orange, D.L. and Field, M.E. (1996) Continental-margin seismic stratigraphy: Assessing the preservation potential of heterogeneous geological processes operating on continental shelves and slopes. *Oceanography*, **9**, 173–177.

Bahr, D.B., Hutton, E.W.H., Syvitski, J.P.M. and Pratson, L.F. (2001) Exponential approximations to compacted sediment porosity profiles. *Comput. Geosci.*, **27**, 691–700.

Bea, R.G., Wright, S.G., Sircar, P. and Niedoroda, A.W. (1983) Wave-induced slides in South Pass Block 70, Mississippi Delta. *J. Geotech. Eng.*, **109**, 619–644.

Been, K. and Sills, G.C. (1981) Self-weight consolidation of soft soils: an experimental and theoretical study. *Géotechnique*, **31**, 519–535.

Bentley, S.J. and Nittrouer, C.A. (2003) Emplacement, modification and preservation of event strata on a flood-dominated continental shelf, Eel shelf, northern California. *Cont. Shelf Res.*, **23**, 1465–1493.

Berggren, W.A., Kent, D.V., Swisher, C.C. and Aubry, M.-P. (1995) A revised Cenozoic geochronology and chronostratigraphy. In: *Geochronology, Time Scales and Global Stratigraphic Correlation* (Ed. W.A. Berggren), pp. 129–212. Special Publication 54, Society of Economic Paleontologists and Mineralogists, Tulsa, OK.

Beschta, R.L. (1987) Conceptual models of sediment transport in streams. In: *Sediment Transport in Gravel-bed Rivers* (Eds C.R. Thorne, J.C. Bathurst, J.C. and R.D. Hey), pp. 387–419. John Wiley & Sons, Chichester.

Blair, N.E., Leithold, E.L., Ford, S.T., *et al.* (2003) The persistence of memory: the fate of ancient sedimentary organic carbon in a modern sedimentary system. *Geochim. Cosmochim. Acta*, **67**, 63–73.

Bokuniewicz, H.J., Gordon, R. and Rhoads, D.C. (1975) Mechanical properties of the sediment–water interface. *Mar. Geol.*, **18**, 263–278.

Booth, J.S., O'Leary, D.W., Popenoe, P. and Danforth, W.W. (1993) U.S. Atlantic contintental slope landslides: their distribution, general attributes and implications. In: *Submarine Landslides: Selected Studies in the U.S. EEZ* (Eds W.C. Schwab, H.J. Lee and D.C. Twichell). *U.S. Geol. Surv. Bull.*, **2002**, 14–39.

Bornhold, B.D and Prior, D.B. (1990) Morphology and sedimentary processes on the subaqueous Noeick River delta, British Columbia, Canada. In: *Coarse-grained Deltas* (Eds A. Colella and D.B. Prior), pp. 169–184. Special Publication 10, International Association of Sedimentologists. Blackwell Scientific Publications, Oxford.

Boulanger, E. (2000) *Comportement cyclique des sédiments de la marge continentale de la rivière Eel: une explication possible pour le peu de glissements sous-marins superficiels dans cette region.* MSc thesis, Department Geology and Geological Engineering, Laval University, Quebec.

Bouma, A.H. (1962) *Sedimentology of Some Flysch Deposits: a Graphic Approach to Facies Interpretation.* Elsevier, Amsterdam.

Brink, K.H., Bane, J.M., Church, T.M., *et al.* (1992) *Coastal Ocean Processes: a Science Prospectus.* Technical Report WHOI–92–18, Woods Hole Oceanographic Institute, 88 pp.

Brown, W.M. and Ritter, J.R. (1971) Sediment transport and turbidity in the Eel River basin, California. *U.S. Geol. Surv. Wat. Supply Pap.*, **1986**, 70 pp.

Bryn, P., Solheim, A., Berg, K., *et al.* (2003) The Storegga slide complex: repeated large scale sliding in response to climatic cyclicity. In: *Submarine Mass Movements and their Consequences* (Eds J. Locat and J. Mienert), pp. 215–222. Kluwer, Amsterdam.

Burger, R.L., Fulthorpe, C.S. and Austin, J.A., Jr. (2001) Late Pleistocene channel incisions in the southern Eel River Basin, northern California: implications for tectonic vs. eustatic influences on shelf sedimentation patterns. *Mar. Geol.*, **177**, 317–330.

Burger, R.L., Fulthorpe, C.S., Austin Jr, J.A. and Gulick, S.P.S. (2002) Lower Pleistocene to present structural deformation and sequence stratigraphy of the continental shelf, offshore Eel River Basin, northern California. *Mar. Geol.*, **185**, 249–281.

Cacchione, D.A. and Pratson, L.F. (2004). Internal tides and the continental slope. *Am. Sci.*, **92**, 130–137.

Cacchione, D.A., Pratson, L.F. and Ogston, A.S. (2002) The shaping of continental slopes by internal tides. *Science*, **296**, 724–727.

Christie-Blick, N. (1991) Onlap, offlap and the origin of unconformity-bounded depositional sequences. *Mar. Geol.*, **97**, 35–56.

Clarke, S.H. (1987) Late Cenozoic geology and structure of the onshore-offshore Eel River Basin, northern California. In: *Tectonics, Sedimentation and Evolution of the Eel River and Other Coastal Basins of Northern California* (Eds H. Schymiczek and R. Suchland), pp. 31–40. Miscellaneous Publication 37, San Joaquin Geological Society.

Clarke, S.H. (1992) Geology of the Eel River Basin and adjacent region: implications for Late Cenozoic tectonics of the southern Cascadia subduction zone and Mendocino Triple Junction. *Bull. Am. Assoc. Petrol. Geol.*, **76**, 199–224.

Cohn, T.A. (1995) Recent advances in statistical methods for the estimation of sediment and nutrient transport in rivers. *Rev. Geophys.*, **33** (IUGG Suppl.), 1117–1123.

Coleman, J.M. and Garrison, L.E. (1977) Geological aspects of marine slope stability, northwestern Gulf of Mexico. *Mar. Geol.*, **2**, 9–44.

Coleman, J.M., Prior, D.B. and Garrison, L.E. (1980) Subaqueous sediment instabilities in the offshore Mississippi River delta. *U.S. Bur. Land Man. Open-file Rep.*, **80–01**, 60 pp.

Correggiari, A., Trincardi, F., Langone, L. and Roveri, M. (2001) Styles of failure in late Holocene highstand prodelta wedges on the Adriatic shelf. *J. Sediment. Res.*, **71**, 218–236.

Costa, J.E. (1984) Physical geomorphology of debris flows. In: *Developments and Applications of Geomorphology* (Eds J.E. Costa and P.J. Fleisher), pp. 269–317. Springer-Verlag, Berlin.

Coulter, H.W. and Migliaccio, R.R. (1966) Effects of the earthquake of March 27, 1964 at Valdez, Alaska. *U.S. Geol. Surv. Prof. Pap.*, **542-C**.

Crockett, J.S. and Nittrouer, C.A. (2004) The sandy inner shelf as a repository for muddy sediment: an example from northern California. *Cont. Shelf Res.*, **24**, 55–73.

Curran, K.J., Hill, P.S. and Milligan, T.G. (2002) Fine-grained suspended sediment dynamics in the Eel River flood plume. *Cont. Shelf Res.*, **22**, 2537–2550.

Cutter, G.R. and Diaz, R.J. (2000) Biological alteration of physically structured flood deposits on the Eel margin, northern California. *Cont. Shelf Res.*, **20**, 235–253.

De Deckere, E.M., Tolhurst, T.J. and de Brouwer, J.F.C. (2001) Destabilization of cohesive intertidal sediments by infauna. *Estuarine Coastal Shelf Sci.*, **53**, 665–669.

DeMaster, D.J., McKee, B.A., Nittrouer, C.A., Qian, J. and Cheng, G. (1985) Rates of sediment accumulation and particle reworking based on radiochemical measurements from continental shelf deposits in the East China Sea. *Cont. Shelf Res.*, **4**, 143–158.

DeMets, C., Gordon, R.G., Argus, D.F. and Stein, S. (1990) Current plate motions. *Geophys. J. Int.*, **101**, 425–478.

Drake, D.E. (1976) Suspended sediment transport and mud deposition on continental shelves. In: *Marine Sediment Transport and Environmental Management* (Eds D.J. Stanley and D.J.P. Swift), pp. 127–158. John Wiley & Sons, New York.

Drake, D.E. (1999) Temporal and spatial variability of the sediment grain-size distribution on the Eel shelf: the flood layer of 1995. *Mar. Geol.*, **154**, 169–182.

Drake, D.E. and Cacchione, D.A. (1985) Seasonal variation in sediment transport on the Russian River shelf, California. *Cont. Shelf Res.*, **4**, 495–514.

Drexler, T.M., Nittrouer, C.A. and Mullenbach, B.L. (2006) Impact of local morphology on sedimentation in a submarine canyon, ROV studies in Eel Canyon, northern California, U.S.A. *J. Sediment. Res.*, **76**, 839–853.

Duncan, C.S. and Goff, J.A. (2001) Relict iceberg keel marks on the New Jersey outer shelf, southern Hudson Apron. *Geology*, **29**(5), 411–414.

Duncan, C.S., Goff, J.A., Austin, J.A., Jr. and Fulthorpe, C.S. (2000) Tracking the last sea-level cycle: seafloor morphology and shallow stratigraphy of the latest Quaternary New Jersey middle continental shelf. *Mar. Geol.*, **170**, 395–421.

Elverhøi, A., Norem, H., Andersen, E.S., *et al.* (1997) On the origin and flow behavior of submarine slides on deep-sea fans along the Norwegian Barents Sea continental margin. *Geo-Mar. Lett.*, **17**, 119–125.

Ely, L.L., Enzel, Y., Baker, V.R. and Cayan, D.R. (1993) A 5000-year record of extreme floods and climate change in the southwestern United States. *Science*, **262**, 410–412.

Emery, K.O. and Uchupi, E. (1972) *Western North Atlantic Ocean; Topography, Rocks, Structure, Water, Life and Sediments.* Memoir 17, American Association of Petroleum Geologists, Tulsa, OK, 532 pp.

Emery, K.O. and Uchupi, E. (1984). *The Geology of the Atlantic Ocean.* Springer-Verlag, New York, 1050 pp.

Ericson, D.B., Ewing, M., Wollin, G. and Heezen, B.C. (1961) Atlantic deep-sea sediment cores. *Geol. Soc. Am. Bull.*, **72**, 193–285.

Farnsworth, K.L. and Milliman, J.D. (2003) Effects of climatic and anthropogenic change on small mountainous rivers: the Salinas River example. *Global Planet. Change*, **39**, 53–64.

Farre, J.A., McGregor, B.A., Ryan, W.B.F. and Robb, J.M. (1983). Breaching the shelf break; passage from youthful to mature phase in submarine canyon evolution. In: *The Shelf Break; Critical Interface on Continental Margins* (Eds D.J. Stanley and G.T. Moore), pp. 25–39. Special Publication 33, Society of Economic Paleontologists and Mineralogists, Tulsa, OK.

Field, M.E. and Barber, J.H., Jr. (1993) A submarine landslide associated with shallow sea-floor gas and gas hydrates off northern California. In: *Submarine Landslides: Selected Studies in the U.S. EEZ* (Eds W.C. Schwab, H.J. Lee and D.C. Twichell). *U.S. Geol. Surv. Bull.*, **2002**, 151–157.

Field, M.E., Clarke, S.H., Jr. and White, M.E. (1980) Geology and geologic hazards of offshore Eel River Basin, northern California continental margin. *U.S. Geol. Surv. Open-File Rep.*, **80–1080**.

Field, M.E., Gardner, J.V., Jennings, A.E. and Edwards, B.D. (1982) Earthquake-induced sediment failures on a 0.250 slope, Klamath River delta, California. *Geology*, **10**, 542–546.

Field, M.E., Gardner, J.V. and Prior, D.B. (1999) Geometry and significance of stacked gullies on the northern California slope. *Mar. Geol.*, **154**, 271–286.

Flood, R.D., Manley, P.L., Kowsmann, R.O., Appi, C.J. and Pirmez, C. (1991) Seismic facies and late Quaternary growth of Amazon submarine fan. In: *Seismic Facies and Sedimentary Processes of Modern and Ancient Submarine Fans* (Eds P. Weimer and M.H. Link), pp. 415–433. Spring-Verlag, New York.

Fulthorpe, C.S. and Austin, J.A., Jr. (2004) Shallowly buried, enigmatic seismic stratigraphy on the New Jersey outer shelf: evidence for latest Pleistocene catastrophic erosion? *Geology*, **32**, 1013–1016.

Fulthorpe, C.S., Austin, J.A., Jr. and Mountain, G.S. (1999) Buried fluvial channels off New Jersey: did sea-level lowstands expose the entire shelf during the Miocene? *Geology*, **27**, 203–206.

Fulthorpe, C.S., Austin, J.A., Jr. and Mountain, G.S. (2000) Morphology and distribution of Miocene slope incisions off New Jersey: are they diagnostic of sequence boundaries? *Geol. Soc. Am. Bull.*, **112**, 817–828.

García, M. and Parker, G. (1993) Experiments on the entrainment of sediment into suspension by a dense bottom current. *J. Geophys. Res.*, 98, 4793–4807.

Gardner, J.V., Prior, D.B. and Field, M.E. (1999) Humboldt slide – a large shear-dominated retrogressive slope failure. *Mar. Geol.*, 154, 323–338.

Garvine, R.W. (1987) Estuary plumes and fronts in shelf waters: a layer model. *J. Phys. Ocean.*, **17**, 1877–1896.

Geist, E.L. (2000) Origin of the 17 July 1998 Papua New Guinea tsunami; earthquake or landslide. *Seism. Res. Lett.*, **71**, 344–351.

Geyer, W.R., Hill, P.S., Milligan, T.G. and Traykovski, P. (2000) The structure of the Eel River plume during floods. *Cont. Shelf Res.*, **20**, 2067–2093.

Goff, J.A., Orange, D.L., Mayer, L.A. and Hughes Clarke, J.E. (1999) Detailed investigation of continental shelf morphology using a high-resolution swath sonar survey: the Eel margin, northern California. *Mar. Geol.*, **154**, 255–270.

Goff, J.A., Austin, J.A., Jr., Gulick, S., *et al.* (2005) Recent and modern erosion on the New Jersey outer shelf. *Mar. Geol.*, **216**, 275–296.

Goldberg, D. (1997) The role of downhole measurements in marine geology and geophysics. *Rev. Geophys.*, **35**, 315–342.

Goldfinger, C., Nelson, C.H. and Johnson, J.E. (2003) Holocene earthquake records from the Cascadia subduction zone and northern San Andreas fault based on precise dating of offshore turbidites. *Ann. Rev. Earth Planet. Sci.*, **31**, 555–577.

Goodbred, S.L. and Kuehl, S.A. (2000) Enormous Ganges–Brahmaputra sediment discharge during strengthened early Holocene monsoon. *Geology*, **28**, 1083–1086.

Goodwin, R.H. and Prior, D.B. (1989) Geometry and depositional sequences of the Mississippi Canyon, Gulf of Mexico. *J. Sediment. Petrol.*, **59**, 318–329.

Grant, W.D. and Madsen, O.S. (1979) Combined wave and current interaction with a rough bottom. *J. Geophys. Res.*, **84**, 1797–1808.

Greenlee, S.M., Schroeder, F.W. and Vail, P.R. (1988) Seismic stratigraphic and geohistory analysis of Tertiary strata from the continental shelf off New Jersey – calculation of eustatic fluctuations from stratigraphic data. In: *Atlantic Continental Margin: U.S.* (Eds R.E. Sheridan and J.A. Grow), pp. 437–444. DNAG Series, Geological Society of America, Boulder, CO.

Greenlee, S.M., Devlin, W.J., Miller, K.G., Mountain, G.S. and Flemings, P.B. (1992) Integrated sequence stratigraphy of Neogene deposits, New Jersey continental shelf and slope: comparison with the Exxon model. *Geol. Soc. Am. Bull.*, **104**, 1403–1411.

Grow, J.A. and Sheridan, R.E. (1988) U.S. Atlantic continental margin; a typical Atlantic-type or passive continental margin. In: *Atlantic Continental Margin: U.S.* (Eds R.E. Sheridan and J.A. Grow), pp. 1–8. DNAG Series, Geological Society of America, Boulder, CO.

Hagen, R.A., Bergersen, D., Moberly, R. and Coulbourn, W.T. (1994) Morphology of a large meandering submarine canyon system the Peru–Chile forearc. *Mar. Geol.*, **119**, 7–38.

Hamilton, E.L. (1980) Geoacoustic modeling of the seafloor. *J. Acoust. Soc. Am.*, **68**, 1313–1340.

Hampton, M. (1972) The role of subaqueous debris flow in generating turbidity currents. *J. Sediment. Petrol.*, **42**, 775–993.

Hampton, M.A., Lee, H.J. and Locat, J. (1996) Submarine landslides. *Rev. Geophys.*, **34**, 33–59.

Haq, B.U., Hardenbol, J. and Vail, P.R. (1987) Chronology of fluctuating sea levels since the Triassic (250 million years ago to Present). *Science*, **235**, 1156–1167.

Harris, C.K. and Wiberg, P.L. (1997) Approaches to quantifying long-term continental shelf sediment transport with an example from the northern California STRESS mid-shelf site. *Cont. Shelf Res.*, **17**, 1389–1418.

Harris, C.K. and Wiberg, P.L. (2001) A two-dimensional, time-dependent model of suspended sediment transport and bed reworking for continental shelves. *Comput. Geosci.*, **27**, 675–690.

Harris, C.K. and Wiberg, P.L. (2002) Across-shelf sediment transport, interactions between suspended sediment and bed sediment. *J. Geophys. Res.*, **107**(C1), doi: 10.1029/2000JC000634.

Hartel, C., Carlsson, F. and Thunblom, M. (2000) Analysis and direct numerical simulation of the flow at a gravity-current head, Part 2. The lobe-and-cleft instability. *J. Fluid Mech.*, **418**, 213–229.

Hay, A.E. (1987) Turbidity currents and submarine channel formation in Rupert Inlet, British Columbia, Part 2: The roles of continuous and surge type flow. *J. Geophys. Res.*, **92**, 2883–2900.

Heezen, B.C. and Ewing, M. (1952) Turbidity currents and submarine slumps and the 1929 Grand Banks Earthquake. *Am. J. Sci.*, **250**, 849–873.

Henkel, D.J. (1970) The role of waves causing submarine landslides. *Géotechnique*, **20**, 75–80.

Hickey, B.M. (1979) The California Current system-hypotheses and facts. *Progr. Ocean.*, **8**, 191–279.

Hickey, B.M. (1998) Coastal oceanography of western North America from the tip of Baja California to Vancouver Island. In: *The Sea*, Vol. 11, *The Global Coastal Ocean: Regional Studies and Syntheses* (Eds A.R. Robinson and K.H. Brink), pp. 345–393. John Wiley & Sons, New York.

Hill, P.S. (1992) Reconciling aggregation theory with observed vertical fluxes following phytoplankton blooms. *J. Geophys. Res.*, **97**, 2295–2308.

Hill, P.S., Syvitski, J.P., Cowan, E.A. and Powell, R.D. (1998) *In situ* observations of floc settling velocities in Glacier Bay, Alaska. *Mar. Geol.*, **145**, 85–94.

Hill, P.S., Milligan, T.G. and Geyer, W.R. (2000) Controls on effective settling velocity in the Eel River flood plume. *Cont. Shelf Res.*, **20**, 2095–2111.

Hiscott, R.N., Hall, F.R. and Pirmez, C. (1997) Turbidity-current overspill from the Amazon channel: Texture of the silt/sand load, pale flow from anisotropy of magnetic susceptibility and implications for flow processes. *Proc. ODP Sci. Res.*, **155**, 53–78.

Hsu, T.J. and Hanes, D.M. (2004) Effects of wave shape on sheet flow sediment transport. *J. Geophys. Res.*, **109**, Art. C05025.

Imran, J. and Syvitski, J.P.M. (2000) Impact of extreme river events on coastal oceans. *Oceanography*, **13**, 85–92.

Imran, J., Parker, G. and Pirmez, C. (1999) A nonlinear model of flow in meandering submarine and subaerial channels. *J. Fluid Mech.*, **400**, 295–331.

Inman, D.L. and Jenkins, S.A. (1999) Climate change and the episodicity of sediment flux of small California rivers. *J. Geol.*, **107**, 251–270.

Iverson, R.M. and Major, J.J. (1986) Groundwater seepage vectors and the potential for hillslope failure and debris flow mobilization. *Water Resour. Res.*, **22**, 1543–1548.

Jaeger, J.M. and Nittrouer, C.A. (2000). The formation of point- and multiple-source deposits on continental shelves. In: *Coastal Ocean Processes (CoOP): Transport and Transformation Processes over Continental Shelves with Substantial Freshwater Inflows* (Eds S. Henrichs, N. Bond, R. Garvine, G. Kineke and S. Lohrenz) pp. 78–89. Technical Report TS-237-00, Center for Environmental Science, University of Maryland, Cambridge, MD.

Jansa, L.F. (1981) Mesozoic carbonate platforms and banks of the eastern North American margin. *Mar. Geol.*, **44**, 97–117.

Jumars, P.A. and Nowell, A.R.M. (1984) Effects of benthos on sediment transport, difficulties with functional grouping. *Cont. Shelf Res.*, **3**, 115–130.

Karner, G.D. and Driscoll, N.W. (1997) Three-dimensional interplay of advective and diffusive processes in the generation of sequence boundaries. *J. Geol. Soc. London*, **154**, 443–449.

Kassem, A. and Imran, J. (2004) Three-dimensional modeling of density current. II. Flow in sinuous confined and unconfined channels. J. *Hydraul. Res.*, **42**, 591–602.

Kayen, R.E. and Lee, H.J. (1991) Pleistocene slope instability of gas hydrate-laden sediment on the Beaufort Sea margin. *Mar. Geotech.*, **10**, 125–141.

Kennett, J.P. (1982) *Marine Geology*. Prentice-Hall, Englewood Cliffs, NJ, 813 pp.

Khripounoff, A., Vangriesheim, A., Babonneau, N., *et al.* (2003) Direct observation of intense turbidity current activity in the Zaire submarine valley at 4000 m water depth. *Mar. Geol.*, **194**, 151–158.

Kineke, G.C., Sternberg, R.W., Trowbridge, J.H. and Geyer, W.R. (1996) Fluid-mud processes on the Amazon continental shelf. *Cont. Shelf Res.*, **16**, 667–696.

Kineke, G.C., Woolfe, K.J., Kuehl, S.A., *et al.* (2000) Sediment export from the Sepik River, Papua New Guinea: evidence for a divergent sediment plume. *Cont. Shelf Res.*, **20**, 2239–2266.

Klaucke, I., Hesse, R. and Ryan, W.B.F. (1998) Morphology and structure of a distal submarine trunk channel: The north-west Atlantic Mid-Ocean Channel between lat 53°N and 44°30′N. *Geol. Soc. Am. Bull.*, **110**, 22–34.

Knox, J.C. (2000) Sensitivity of modern and Holocene floods to climate change. *Quat. Sci. Rev.*, **19**, 439–457.

Komar, P.D. and Miller, M.C. (1975) On the comparison between the threshold of sediment motion under waves and unidirectional currents with a discussion of the practical evaluation of the threshold. *J. Sediment. Petrol.*, **45**, 362–367.

Komar, P.D., Neudeck, R.H. and Kulm, L.D. (1972) Observations and significance of deep-water oscillatory ripple marks on the Oregon continental shelf. In: *Shelf Sediment Transport: Process and Pattern* (Eds D.J.P. Swift, D.B. Duane and O.H. Pilkey), pp. 601–624. Dowden, Hutchinson and Ross, Stroudsburg, PA.

Kominz, M.A., Miller, K.G. and Browning, J.V. (1998) Long-term and short-term global Cenozoic sea-level estimates. *Geology*, **26**, 311–314.

Kostic, S., Parker, G. and Marr, J.G. (2002) Role of turbidity currents in setting the foreset slope of clinoforms prograding into standing fresh water. *J. Sediment. Res.*, **72**, 353–362.

Kuehl, S.A., DeMaster, D.J. and Nittrouer, C.A. (1986) Nature of sediment accumulation on the Amazon continental shelf, *Cont. Shelf Res.*, **6**, 209–225.

Kuehl, S.A., Nittrouer, C.A., Allison, M.A., *et al.* (1996) Sediment deposition, accumulation and seabed dynamics in an energetic, fine-grained coastal environment. *Cont. Shelf Res.*, **16**, 787–816.

Kuenen, Ph.H. and Migliorini, C.I. (1950) Turbidity currents as a cause of graded bedding. *J. Geol.*, **58**, 91–127.

Kulm, L.D. and Fowler, G.A. (1974) Oregon continental margin structure and stratigraphy; a test of the imbricate thrust model. In: *The Geology of Continental Margins* (Eds C.A. Burk and C.L. Drake), pp. 261–233. Springer-Verlag, New York.

Kulm, L.D., Roush, R.C., Harlett, J.C., Neudeck, R.H., Chambers, D.M. and Runge, E.J. (1975) Oregon continental shelf sedimentation; interrelationships of facies distribution and sedimentary processes. *J. Geol.*, **83**, 145–175.

Kvenvolden, K.A. and Field, M.E. (1981) Thermogenic hydrocarbons in unconsolidated sediments of the Eel River Basin, offshore northern California. *Bull. Am. Assoc. Petrol. Geol.*, **65**, 1642–1646.

Lamb, M.P., D'Asaro, E. and Parsons, J.D. (2004) Turbulent structure of high-density suspensions formed under waves. *J. Geophys. Res.*, **109**, Art. C12026.

Lee, H.J. and Edwards, B.D. (1986) Regional method to assess offshore slope stability. *ASCE J. Geotech. Eng.*, **112**, 489–509.

Lee, H.J., Edwards, B.D. and Field, M.E. (1981) Geotechnical analysis of a submarine slump, Eureka,

California. *Proceedings of the Offshore Technology Conference, Houston*, pp. 53–59.

Lee, H.J., Schwab, W.C., Edwards, B.D. and Kayen, R.E. (1991) Quantitative controls on submarine slope failure morphology. *Mar. Geotech.*, **10**, 143–158.

Lee, H.J., Syvitski, J.P.M., Parker, G., *et al.* (2002) Distinguishing sediment waves from slope failure deposits: field examples, including the 'Humboldt Slide' and modeling results. *Mar. Geol.*, **192**, 79–104.

Leithold, E.L. and Hope, R.S. (1999) Deposition and modification of a flood layer on the northern California shelf: lessons from and about the fate of terrestrial particulate organic carbon. *Mar. Geol.*, **154**, 183–195.

Lemke, R.W. (1967) Effects of the earthquake of March 27, 1964, at Seward, Alaska. *U.S. Geol. Surv. Prof. Pap.*, **542-E**.

Lorenson, T.D., McLaughlin, R.J., Kvenvolden, K.A., *et al.* (1998) Comparison of offshore and onshore gas occurrences, Eel River basin, Northern California. *U.S. Geol. Surv. Open-File Rep.*, **98–781**.

Lowrie, W. and Kent, D.V. (2004) Geomagnetic polarity timescale and reversal frequency regimes. In: *Timescales of the Internal Geomagnetic Field* (Eds J.E.T. Channell, D.V. Kent, W. Lowrie and J. Meert), pp. 287–298. Geophysical Monograph 145, American Geophysical Union, Washington, DC.

Madsen, O.S. (1994) Spectral wave-current bottom boundary layer flows. In: *Coastal Engineering 1994, Proceedings, 24th International Conference*, pp. 384–397. Coastal Engineering Research Council/American Society of Civil Engineers, Kobe, Japan.

Marr, J.G., Harff, P.A., Shanmugam, G. and Parker, G. (2001) Experiments on subaqueous sandy gravity flows: the role of clay and water content in flow dynamics and depositional structures. *Geol. Soc. Am. Bull.*, **113**, 1377–1386.

May, J.A., Warme, J.E. and Slater, R.A. (1983) Role of submarine canyons on shelfbreak erosion and sedimentation: modern and ancient examples. In: *The Shelfbreak; Critical Interface on Continental Margins* (Eds D.J. Stanley and G.T. Moore), pp. 315–332. Special Publication 33, Society of Economic Paleontologists and Mineralogists, Tulsa, OK.

McCave, I.N. (1972) Transport and escape of fine-grained sediment from shelf areas. In: *Shelf Sediment Transport: Process and Pattern* (Eds D.J.P. Swift, D.B. Duane and O.H. Pilkey), pp. 225–248. Dowden, Hutchinson and Ross, Stroudsburg, PA.

McCave, I.N. (1984) Size spectra and aggregation of suspended particles in the deep ocean. *Deep-sea Res.*, **31**, 329–352.

McCool, W.W. and Parsons, J.D. (2004) Sedimentation from buoyant fine-grained suspensions. *Cont. Shelf Res.*, **24**, 1129–1142.

McHugh, C.M.G. and Olson, H.C. (2002) Pleistocene chronology of continental margin sedimentation: New insights into traditional models, New Jersey. *Mar. Geol.*, **186**, 389–411.

McKee, B.A., Nittrouer, C.A. and DeMaster, D.J. (1983) The concepts of sediment deposition and accumulation applied to the continental shelf near the mouth of the Yangtze River. *Geology*, **11**, 631–633.

McKee, B.A., Aller, R.C., Allison, M.A., Bianchi, T.A. and Kineke, G.C. (2004) Transport and transformation of dissolved and particulate materials on continental margins influenced by major rivers: benthic boundary layer and seabed processes. *Cont. Shelf Res.*, **24**, 899–926.

McPhee-Shaw, E.E., Sternberg, R.W., Mullenbach, B. and Ogston, A.S. (2004) Observations of intermediate nepheloid layers on the northern California continental margin. *Cont. Shelf Res.*, **24**, 693–720.

Meade, R.H. (1996) River-sediment inputs to major deltas. In: *Sea-level Rise and Coastal Subsidence* (Eds J.D. Milliman and B.U. Haq), pp. 63–85. Kluwer Academic Publishers, Dordrecht.

Meade, R.H., Yuzyk, T.R. and Day, T.J. (1990) Movement and storage of sediment in rivers of the United States and Canada. In: *Geology of North America: Surface Water Hydrology* (Eds M.G. Wolman and H.C. Riggs), pp. 255–280. Geological Society of America, Boulder, CO.

Mertes, L.A.K. and Warrick, J.A. (2001) Measuring flood output from 110 coastal watersheds in California with field measurements and SeaWiFS. *Geology*, **29**, 659–662.

Miall, A.D. (1991) Stratigraphic sequences and their chronostratigraphic correlation. *J. Sediment. Petrol.*, **61**, 497–505.

Miller, K.G., Melillo, A.J., Mountain, G.S., Farre, J.A. and Poag, C.W. (1987) Middle to late Miocene canyon cutting on the New Jersey continental slope: biostratigraphic and seismic stratigraphic evidence. *Geology*, **15**, 509–512.

Miller, K.G., Mountain, G.S., Browning, J.S., *et al.* (1998) Cenozoic global sea level, sequences and the New Jersey transect: results from coastal plain and continental slope drilling. *Rev. Geophys.*, **36**, 569–601.

Milligan, T.G. and Hill, P.S. (1998) A laboratory assessment of the relative importance of turbulence, particle composition and concentration in limiting maximal floc size. *J. Sea Res.*, **39**, 227–241.

Milliman, J.D. and Meade, R.H. (1983) World-wide delivery of river sediment to the oceans. *J. Geol.*, **91**, 1–21.

Milliman, J.D. and Syvitski, J.P.M. (1992) Geomorphic/tectonic control of sediment discharge to the ocean:

the importance of small mountainous rivers. *J. Geol.*, **100**, 525–544.

Mitchum, R.M., Vail, P.R. and Thompson, S. (1977a) Seismic stratigraphy and global changes of sea level; Part 2, The depositional sequence as a basic unit for stratigraphic analysis. In: *Seismic Stratigraphy; Applications to Hydrocarbon Exploration* (Ed. C.E. Payton), pp. 53–62. Memoir 26, American Association of Petroleum Geologists, Tulsa, OK.

Mitchum, R.M., Vail, P.R. and Sangree, J.B. (1977b) Seismic stratigraphy and global changes of sea level; Part 6, Stratigraphic interpretation of seismic reflection patterns in depositional sequences. In: *Seismic Stratigraphy; Applications to Hydrocarbon Exploration* (Ed. C.E. Payton), pp. 117–133. Memoir 26, American Association of Petroleum Geologists, Tulsa, OK.

Mohrig, D., Whipple, K.X., Hondzo, M., Ellis, C. and Parker, G. (1998) Hydroplaning of subaqueous debris flows. *Geol. Soc. Am. Bull.*, **110**, 387–394.

Moore, D.G. (1969) *Reflection Profiling Studies of the California Continental Borderland: Structure and Quaternary Turbidite Basins.* Special Paper 107, Geological Society of America, Boulder, CO, 142 pp.

Morehead, M.D., Syvitski, J.P.M. and Hutton, E.W.H. (2001) The link between abrupt climate change and basin stratigraphy: A numerical approach. *Global Planet. Change*, **28**, 115–135.

Mountain, G.S. (1987) Cenozoic margin construction and destruction offshore New Jersey. In: *Timing and Depositional History of Eustatic Sequences: Constraints on Seismic Stratigraphy* (Eds C. Ross and D. Haman), pp. 57–83. Special Publication 24, Cushman Foundation for Foraminiferal Research.

Mountain, G.S., Damuth, J.E., McHugh, C.M.G., Lorenzo, J.M. and Fulthorpe, C.S. (1996) Origin, reburial and significance of a middle Miocene canyon, New Jersey continental slope. *Proc. ODP Sci. Results*, **150**, 283–292.

Mulder, T. and Syvitski, J.P.M. (1995) Turbidity currents generated at river mouths during exceptional discharges to the world oceans. *J. Geol.*, **103**, 285–299.

Mulder, T. and Syvitski J.P.M. (1996) Climatic and morphologic relationships of rivers: Implications of sea level fluctuations on river loads. *J. Geol.*, **104**, 509–523.

Mullenbach, B.L. and Nittrouer, C.A. (2000) Rapid deposition of fluvial sediment in the Eel Canyon, northern California. *Cont. Shelf Res.*, **20**, 2191–2212.

Mullenbach, B.L. and Nittrouer, C.A. (2006) Decadal record of sediment export to the deep sea via Eel Canyon. *Cont. Shelf Res.*, **26**, 2157–2177.

Mullenbach, B.L., Nittrouer, C.A., Puig, P. and Orange, D.L. (2004) Sediment deposition in a modern submarine canyon: Eel Canyon, northern California. *Mar. Geol.*, **211**, 101–119.

Nissen, S.E., Haskell, N.L, Steiner, C.T. and Coterill, K.L. (1999) Debris flow outrunner blocks, glide tracks and pressure ridges identified on the Nigerian continental slope using 3-D seismic coherency. *Leading Edge*, **18**, 595–599.

Nittrouer, C.A. (1999) STRATAFORM: overview of its design and synthesis of its results. *Mar. Geol.*, **154**, 3–12.

Nittrouer, C. and Sternberg, R.W. (1981) The formation of sedimentary strata in an allochthonous shelf environment: application to the Washington continental shelf. *Mar. Geol.*, **42**, 201–232.

Nittrouer, C.A., Kuehl, S.A., DeMaster, D.J. and Kowsmann, R.O. (1986) The deltaic nature of Amazon shelf sedimentation. *Geol. Soc. Am. Bull.*, **97**, 444–458.

Nittrouer, C.A., Kuehl, S.A., Sternberg, R.W., Figueiredo, A.G. and Faria, L.E.C. (1995) An introduction to the geological significance of sediment transport and accumulation on the Amazon continental shelf. *Mar. Geol.*, **125**, 177–192.

Nordfjord, S., Goff, J.A., Austin, J.A., Jr. and Sommerfield, C.K. (2005) Seismic geomorphology of buried channel systems on the New Jersey outer shelf: assessing past environmental conditions. *Mar. Geol.*, **214**, 339–364.

Normark, W.R. and Carlson, P.R. (2003) Giant submarine canyons: is size any clue to their importance in the rock record? In: *Extreme Depositional Environments: Mega End Members in Geologic Time* (Eds M.A. Chan and A.W. Archer), pp. 1–15. Special Paper 370, Geological Society of America, Boulder, CO.

O'Grady, D.B., Syvitski, J.P.M., Pratson, L.F. and Sarg, J.F. (2000) Categorizing the morphologic variability of siliciclastic passive continental margins. *Geology*, **28**, 207–210.

Ogston, A.S. and Sternberg, R.W. (1999) Sediment-transport events on the northern California continental shelf. *Mar. Geol.*, **154**, 69–82.

Ogston, A.S., Cacchione, D.A., Sternberg, R.W. and Kineke, G.C. (2000) Observations of storm and river flood-driven sediment transport on the northern California continental shelf. *Cont. Shelf Res.*, **20**, 2141–2162.

Ogston, A.S., Guerra, J.V. and Sternberg, R.W. (2004) Interannual variability of nearbed sediment flux on the Eel River shelf, northern California. *Cont. Shelf Res.*, **24**, 117–136.

Orange, D.L. (1999) Tectonics, sedimentation and erosion in northern California: submarine geomorphology and sediment preservation potential as a result of three competing forces. *Mar. Geol.*, **154**, 369–382.

Orange, D.L. and Breen, N.A. (1992). The effects of fluid escape on accretionary wedges; 2, Seepage force, slope failure, headless submarine canyons and vents. *J. Geophys. Res.*, **97**, 9277–9295.

Orange, D.L., Anderson, R.S. and Breen, N.A. (1994) Regular canyon spacing in the submarine environment; the link between hydrology and geomorphology. *GSA Today*, **4**, 29, 36–39.

Owens, J.P. and Gohn, G.S. (1985) Depositional history of the Cretaceous series in the U.S. coastal plain: stratigraphy, paleoenvironments and tectonic controls of sedimentation. In: *Geologic Evolution of the United States Atlantic Margin* (Ed. C.W. Poag), pp. 25–86. Van Nostrand Reinhold, New York.

Pantin, H.M. (2001) Experimental evidence for autosuspension. In: *Particulate Gravity Currents* (Eds W.D. McCaffrey, B.C. Kneller and J. Peakall), pp. 189–205, Special Publication 31, International Association of Sedimentologists. Blackwell Science, Oxford.

Parker, G. (1982) Conditions for the ignition of catastrophically erosive turbidity currents. *Mar. Geol.*, **46**, 307–327.

Parsons, B. and Sclater, J.G. (1977) An analysis of the variation of ocean floor bathymetry and heat flow with age. *J. Geophys. Res.*, **82**, 803–827.

Parsons, J.D. (1998) *Mixing mechanisms in density intrusions*. Unpublished PhD thesis, University of Illinois, Urbana-Champaign.

Parsons, J.D., Bush, J.W.M. and Syvitski, J.P.M. (2001) Hyperpycnal plumes with small sediment concentrations. *Sedimentology*, **48**, 465–478.

Payton, C.E. (Ed.) (1977) *Seismic Stratigraphy – Applications to Hydrocarbon Exploration*. Memoir 26, American Association of Petroleum Geologists, Tulsa, OK, 516 pp.

Peakall, J., McCaffrey, B. and Kneller, B. (2000) A process model for the evolution and architecture of sinuous submarine channels. *J. Sediment. Res.*, **70**, 434–448.

Penland, S., Boyd, R. and Suter, J.R. (1988) Transgressive depositional systems of the Mississippi delta plain: a model for barrier shoreline and shelf sand development. *J. Sediment. Petrol.*, **58**, 932–949.

Piper, D.J.W. and Normark, W.R. (1983) Turbidite depositional patterns and flow characteristics, Navy submarine fan, California Borderland. *Sedimentology*, **30**, 681–694.

Pirmez, C., Pratson, L.F. and Steckler, M.S. (1998) Clinoform development by advection-diffusion of suspended sediment; modeling and comparison to natural systems. *J. Geophys. Res.*, **103**, 24,141–24,157.

Poag, C.W. (1985) Depositional history and stratigraphic reference section for central Baltimore Canyon trough. In: *Geologic Evolution of the United States Atlantic Margin* (Ed. C.W. Poag), pp. 217–263. Van Nostrand Reinhold, New York.

Poag, C.W. and Sevon, W.D. (1989) A record of Appalachian denudation in postrift Mesozoic and Cenozoic sedimentary deposits of the U.S. middle Atlantic continental margin. *Geomorphology*, **2**, 119–157.

Poag, C.W. and Ward, L.W. (1993) Allostratigraphy of the U.S. middle Atlantic continental margin – characteristics, distribution and depositional history of principal unconformity-bounded Upper Cretaceous and Cenozoic sedimentary units. *U.S. Geol. Surv. Prof. Pap.*, **1542**.

Posamentier, H.W., Jervey, M.T. and Vail, P.R. (1988) Eustatic controls on clastic deposition I: conceptual framework. In: *Sea Level Changes: an Integrated Approach* (Eds C.K. Wilgus, B.S. Hastings, C.G. St.C. Kendall, *et al.*), pp. 125–154. Special Publication 42, Society of Economic Paleontologists and Mineralogists, Tulsa, OK.

Poulos, S.G., Castro, G. and France, J.W. (1985) Liquefaction evaluation procedure. *J. Geotech. Eng.*, **111**, 772–791.

Pratson, L.F. (1993) *Morphologic studies of submarine sediment drainage*. PhD dissertation, Columbia University, New York.

Pratson, L.F. and Haxby, W.F. (1996) What is the slope of the U.S. continental slope? *Geology*, **24**, 3–6.

Pratson, L.F., Ryan, W.B.F., Mountain, G.S. and Twichell, D.C. (1994) Submarine canyon initiation by downslope-eroding sediment flows; evidence in late Cenozoic strata on the New Jersey continental slope. *Geol. Soc. Am. Bull.*, **106**, 395–412.

Prior, D.B., Bornhold, B.D. and Johns, M.W. (1984) Depositional characteristics of a submarine debris flow. *J. Geol.*, **92**, 707–727.

Prior, D.B., Bornhold, B.D. and Johns, M.W. (1986) Active sand transport along a fjord-bottom channel, Bute Inlet, British Columbia. *Geology*, **14**, 581–584.

Puig, P., Ogston, A.S., Mullenbach, B.L., Nittrouer, C.A. and Sternberg, R.W. (2003) Shelf-to-canyon sediment-transport processes on the Eel continental margin (northern California). *Mar. Geol.*, **193**, 129–149.

Puig, P., Ogston, A.S., Mullenbach, B.L., *et al.* (2004) Storm-induced sediment-gravity flows at the head of the Eel submarine canyon. *J. Geophys. Res.*, **109**, C03019, doi:10.1029/2003JC001918.

Pullen, J.D. and Allen, J.S. (2000) Modeling studies of the coastal circulation off Northern California: shelf response to a major Eel River flood event. *Cont. Shelf Res.*, **20**, 2213–2238.

Rabinovich, A.B., Thomson, R.E., Kulikov, E.A., *et al.* (1999) The landslide-generated tsunami November 3, 1994, in Skagway. *Geophys. Res. Lett.*, **26**, 3009–3012.

Reineck, H.-E. and Singh, I.B. (1972) Genesis of laminated sand and graded rhythmites in storm-sand layers of shelf mud. *Sedimentology*, **18**, 123–128.

Reynolds, D.J., Steckler, M.S. and Coakley, B.J. (1991) The role of the sediment load in sequence stratigraphy; the influence of flexural isostasy and compaction. *J. Geophys. Res.*, **96**, 6931–6949.

Sadler, P.M. (1981) Sediment accumulation rates and the completeness of stratigraphic sections. *J. Geol.*, **89**, 569–584.

Sclater, J.G. and Christie, P.A.F. (1980) Continental stretching; an explanation of the post-Mid-Cretaceous subsidence of the central North Sea basin. *J. Geophys. Res.*, **85**, 3711–3739.

Scully, M.E., Friedrichs, C.T. and Wright, L.D. (2003) Numerical modeling of gravity-driven sediment transport and deposition on an energetic continental shelf: Eel River, northern California. *J. Geophys. Res.*, **108** (C4), paper 17, 1–14.

Seed, H.B. (1968) Landslides during earthquakes due to soil liquefaction. *ASCE J. Soil Mech. Found. Div.*, **94**, 1055–1122.

Seed, H.B., Seed, R.B., Schlosser, F., Blondeau, F. and Juran, I. (1988) *The Landslide at the Port of Nice on October 16, 1979*. Report UCB/EERC–88/10, Earthquake Engineering Research Center, University of California, Berkeley, 68 pp.

Shepard, F.P. (1977) *Geological Oceanography: Evolution of Coasts, Continental Margins and the Deep-Sea Floor*. Crane, Russak and Co., New York, 214 pp.

Shepard, F.P. (1981) Submarine canyons: multiple causes and long-time persistence. *Bull. Am. Assoc. Petrol. Geol.*, **65**, 1062–1077.

Shor, A.N. and McClennen, C.E. (1988) Marine physiography of the U.S. Atlantic margin. In: *The Atlantic Continental Margin; U.S.* (Eds R.E. Sheridan and J.A. Grow), pp. 9–18. The Geology of North America, I–2. Geological Society of America, Boulder, CO.

Skempton, A.W. (1970) The consolidation of clays by gravitational compaction. *Q. J. Geol. Soc. London*, **125**, 373–411.

Smith, J.D. and Long, C.E. (1976) The effects of turning in the bottom boundary layer on continental shelf sediment transport. *Soc. Roy. Sci. Liege Mem.*, **10**, 369–396.

Sommerfield, C.K. (2006) On sediment accumulation rates and stratigraphic completeness: Lessons from Holocene ocean margins. *Cont. Shelf Res.*, **26**, 2225–2240.

Sommerfield, C.K. and Nittrouer, C.A. (1999) Modern accumulation rates and a sediment budget for the Eel shelf: a flood-dominated depositional environment. *Mar. Geol.*, **154**, 227–241.

Sommerfield, C.K., Nittrouer, C.A. and Alexander, C.R. (1999) [7]Be as a tracer of flood sedimentation on the northern California continental margin. *Cont. Shelf Res.*, **19**, 335–361.

Sommerfield, C.K., Drake, D.E. and Wheatcroft, R.A. (2002) Shelf record of climatic changes in flood magnitude and frequency, north-coastal California. *Geology*, **30**, 395–398.

Spinelli, G.A. and Field, M.E. (2001). Evolution of continental slope gullies on the Northern California margin. *J. Sediment. Res.*, **71**, 237–245.

Spinelli, G.A. and Field, M.E. (2003) Controls of tectonics and sediment source locations on along-strike variations in transgressive deposits on the northern California margin. *Mar. Geol.*, **197**, 35–47.

Steckler, M.S., Watts, A.B. and Thorne, J.A. (1988) Subsidence and basin modeling at the U.S. Atlantic passive margin. In: *The Atlantic Continental Margin; U.S.* (Eds R.E. Sheridan and J.A. Grow), pp. 399–416. The Geology of North America, I–2. Geological Society of America, Boulder, CO.

Steckler, M.S., Mountain, G.S., Miller, K.G. and Christie-Blick, N. (1999) Reconstruction of Tertiary progradation and clinoform development on the New Jersey passive margin by 2-D backstripping. *Mar. Geol.*, **154**, 399–420.

Sterling, G.H. and Strohbeck, G.E. (1973) The failure of South Pass 70B Platform in Hurricane Camille. *Proceedings of the Offshore Technology Conference*, Vol. 1, 123–150.

Strub, P.T., Allen, J.S., Huyer, A. and Smith, R.L. (1987) Seasonal cycles of currents, temperatures, winds and sea level over the Pacific continental shelf: 35N to 48N. *J. Geophys. Res.*, **92**, 1507–1526.

Sugarman, P.J. and Miller, K.G. (1997) Correlation of Miocene sequences and hydrogeologic units, New Jersey Coastal Plain. *Sediment. Geol.*, **108**, 3–18.

Sugarman, P., McCartan, L., Miller, K., *et al.* (1997) Strontium-isotopic correlation of Oligocene to Miocene sequences, New Jersey and Florida. *Proc. ODP Sci. Results*, **150X**.

Suppe, J. (1985) *Principles of Structural Geology*. Prentice-Hall, Englewood Cliffs, NJ, 537 pp.

Swenson, J.B., Paola, C., Pratson, L., Voller, V.R. and Murray, A.B. (2005) Fluvial and marine controls on combined subaerial and subaqueous delta progradation: morphodynamic modeling of compound clinoform development. *J. Geophys. Res.*, **110**, 2013–2029.

Swift, D.J.P. (1970) Quaternary shelves and the return to grade. *Mar. Geol.*, **8**, 5–30.

Syvitski, J.P.M. and Farrow, G.E. (1983) Structures and processes in bayhead deltas: Knight and Bute Inlet, British Columbia. *Sediment. Geol.*, **36**, 217–244.

Syvitski, J.P.M. and Morehead, M.D. (1999) Estimating river-sediment discharge to the ocean: application to the Eel Margin, northern California. *Mar. Geol.*, **154**, 13–28.

Syvitski, J.P.M., Morehead, M.D., Bahr, D. and Mulder, T. (2000) Estimating fluvial sediment transport: the rating parameters. *Water Resour. Res.*, **366**, 2747–2760.

Syvitski, J.P.M., Peckham, S.D., Hilberman, R.D. and Mulder, T. (2003) Predicting the terrestrial flux of sediment to the global ocean: a planetary perspective. *Sediment. Geol.*, **162**, 5–24.

Syvitski, J.P.M., Vörösmarty, C.J., Kettner, A.J. and Green, P. (2005) Impact of humans on the flux of terrestrial sediment to the global coastal ocean. *Science*, **308**, 376–380.

Tappin, D.R., Matsumoto, T. and Shipboard Scientists (1999) Offshore surveys identify sediment slump as likely cause of devastating Papua New Guinea Tsunami 1998. *Trans. Am. Geophys. Union*, **80**, 329, 334, 340.

Ten Brinke, W.B.M. (1994) Settling velocities of mud aggregates in the Oosterchelde tidal basin (the Netherlands), determined by a submersible video system. *Estuarine Coastal Shelf Sci.*, **39**, 549–564.

Traykovski, P., Geyer, W.R., Irish, J.D. and Lynch, J.F. (2000) The role of wave-induced density-driven fluid mud flows for cross-shelf transport on the Eel River continental shelf. *Cont. Shelf Res.*, **20**, 2113–2140.

Trowbridge, J.H. and Kineke, G.C. (1994) Structure and dynamics of fluid muds on the Amazon continental-shelf. *J. Geophys. Res.*, **99**, 865–874.

Twichell, D.C. and Roberts, D.G. (1982) Morphology, distribution and development of submarine canyons on the United States Atlantic continental slope between Hudson and Baltimore canyons. *Geology*, **10**, 408–412.

Uchupi, E., Driscoll, N., Ballard, R.D. and Bolmer, S.T. (2001) Drainage of late Wisconsin glacial lakes and the morphology and late Quaternary stratigraphy of the New Jersey–southern New England continental shelf and slope. *Mar. Geol.*, **172**, 117–145.

Vail, P.R., Mitchum, R.M. and Thompson, S. (1977) Seismic stratigraphy and global changes of sea level, Part 4: Global cycles of relative changes of sea level. In: *Seismic Stratigraphy – Applications to Hydrocarbon Exploration* (Ed. C.E. Payton), pp. 83–98. Memoir 26, American Association of Petroleum Geologists, Tulsa, OK.

Van Wagoner, J.C., Posamentier, H.W., Mitchum, R.M., et al. (1988) An overview of the fundamentals of sequence stratigraphy and key definitions. In: *Sea-level Changes: an Integrated Approach* (Eds C.K. Wilgus, B.S. Hastings, C.A. Ross, et al.), pp. 39–45. Special Publication 42, Society of Economic Paleontologists and Mineralogists, Tulsa, OK.

Varnes, D.J. (1958) Landslide types and processes, In: *Landslides and Engineering Practice* (Ed. E.D. Eckel),

pp. 20–47. Special Report 29, Highway Research Board, Washington, DC.

Walling, D.E. (1999) Linking land use, erosion and sediment yields in river basins. *Hydrobiologia*, **410**, 223–240.

Walsh, J.P. and Nittrouer, C.A. (1999) Observations of sediment flux to the Eel continental slope, northern California. *Mar. Geol.*, **154**, 55–68.

Walsh, J.P., Nittrouer, C.A., Palinkas, C.M., et al. (2004) Clinoform mechanics in the Gulf of Papua. *Cont. Shelf Res.*, **24**, 2487–2510.

Washburn, L., Swenson, M.S., Largier, J.L., Kosro, P.M. and Ramp, S.R. (1993) Cross-shelf sediment transport by an anticyclonic eddy off northern California. *Science*, **261**, 1560–1564.

Watts, A.B. and Ryan, W.B.F. (1976) Flexure of the lithosphere and continental margin basins. *Tectonophysics*, **36**, 25–44.

Wheatcroft, R.A. (1990) Preservation potential of sedimentary event layers. *Geology*, **18**, 843–845.

Wheatcroft, R.A. (2002) *In situ* measurements of near-surface porosity in shallow-marine sands. *J. Ocean. Eng.*, **27**, 561–570.

Wheatcroft, R.A. (2006) Time-series measurements of macrobenthos abundance and sediment bioturbation intensity on a flood-dominated shelf. *Progr. Ocean*, **71**, 88–122.

Wheatcroft, R.A. and Borgeld, J.C. (2000) Oceanic flood deposits on the northern California shelf: large-scale distribution and small-scale physical properties. *Cont. Shelf Res.*, **20**, 2163–2190.

Wheatcroft, R.A. and Drake, D.E. (2003) Post-depositional alteration and preservation of sedimentary event layers on continental margins, I. Role of episodic sedimentation. *Mar. Geol.*, **199**, 123–137.

Wheatcroft, R.A. and Martin, W.R. (1996) Spatial variation in short-term (^{234}Th) sediment bioturbation intensity along an organic-carbon gradient. *J. Mar. Res.*, **54**, 763–792.

Wheatcroft, R.A., Sommerfield, C.K., Drake, D.E., Borgeld, J.C. and Nittrouer, C.A. (1997) Rapid and widespread dispersal of flood sediment on the northern California margin. *Geology*, **25**, 163–166.

Wiberg, P.L. (2000) A perfect storm: formation and potential for preservation of storm beds on the continental shelf. *Oceanography*, **13**, 93–99.

Wiberg, P.L., Drake, D.E. and Cacchione, D.A. (1994) Sediment resuspension and bed armoring during high bottom stress events on the northern California inner continental shelf: measurements and predictions. *Cont. Shelf Res.*, **14**, 1191–1219.

Wiberg, P.L., Drake, D.E., Harris, C.K. and Noble, M. (2002) Sediment transport on the Palos Verdes shelf over seasonal to decadal time scales. *Cont. Shelf Res.*, **22**, 987–1004.

Wright, L.D. (1977) Sediment transport and deposition at river mouths: a synthesis. *Geol. Soc. Am. Bull.*, **88**, 857–868.

Wright, L.D., Kim, S.-C. and Friedrichs, C.T. (1999) Across-shelf variations in bed roughness, bed stress and sediment suspension on the northern California continental shelf. *Mar. Geol.*, **154**, 99–115.

Wright, L.D., Friedrichs, C.T., Kim, S.C. and Scully, M.E. (2001) Effects of ambient currents and waves on gravity-driven sediment transport on continental shelves. *Mar. Geol.*, **175**, 25–45.

Wynn, R.B., Weaver, P.P.E., Ercilla, G., Stow, D.A.V. and Masson, D.G. (2000) Sedimentary processes in the Selvage sediment-wave field, NE Atlantic; new insights into the formation of sediment waves by turbidity currents. *Sedimentology*, **47**, 1181–1197.

Xu, J.P., Noble, M.A. and Rosenfeld, L.K. (2004) In-situ measurements of velocity structure within turbidity currents. *Geophys. Res. Lett.*, **31**, L09311, doi:10.1029/2004GLO19718.

Yun, J.W., Orange, D.L. and Field, M.E. (1999) Subsurface gas offshore of northern California and its link to submarine geomorphology. *Mar. Geol.*, **154**, 357–368.

Zhang, Y., Swift, D.J.P., Fan, S., Niederoda, A.W. and Reed, C.W. (1999) Two-dimensional numerical modeling of storm deposition on the northern California shelf. *Mar. Geol.*, **154**, 155–167.

Sediment delivery to the seabed on continental margins

PAUL S. HILL*, JASON M. FOX*, JOHN S. CROCKETT†, KRISTIAN J. CURRAN*,
CARL T. FRIEDRICHS‡, W. ROCKWELL GEYER§, TIMOTHY G. MILLIGAN¶,
ANDREA S. OGSTON†, PERE PUIG**, MALCOLM E. SCULLY‡,
PETER A. TRAYKOVSKI§ and ROBERT A. WHEATCROFT††

*Department of Oceanography, Dalhousie University, Halifax, Nova Scotia B3H 4J1, Canada (Email: paul.hill@dal.ca)
†School of Oceanography, University of Washington, Seattle, WA 98195, USA
‡Virginia Institute of Marine Science, College of William and Mary, Gloucester Point, VA 23062-1346, USA
§Department of Applied Physics and Engineering, Woods Hole Oceanographic Institution, Woods Hole, MA 02543, USA
¶Habitat Ecology Division, Fisheries and Oceans Canada, Bedford Institute of Oceanography, Dartmouth, Nova Scotia B2Y 4A2, Canada
**Department of Marine Geology and Physical Oceanography, Institut de Ciencies del Mar (CSIC), Barcelona E-08003, Spain
††College of Oceanic and Atmospheric Sciences, Oregon State University, Corvallis, OR 97331-5503, USA

ABSTRACT

On river-influenced continental margins, terrigenous muds tend to accumulate in the middle of the continental shelf. The common occurrence of mid-shelf mud belts has been attributed to three basic across-margin transport mechanisms. Muds either diffuse to the mid-shelf under the influence of storms, or they are advected there by oceanographic currents, or they arrive at the mid-shelf in dense suspensions that flow across the margin under the influence of gravity. Until recently, observations generally favoured the hypothesis that ocean currents are responsible for advecting dilute suspensions of mud to the mid-shelf. Transport by dense gravity flows was widely rejected, based primarily on the arguments that the bathymetric gradients of continental shelves are too small to sustain gravity flows, and that sediment concentrations cannot grow large enough to cause suspensions to flow down gradient. Observations conducted on the Eel River continental shelf off northern California, however, demonstrate that cross-margin transport by dense suspensions can be an important mechanism for the emplacement of muds on the mid-shelf. Dense suspensions form near the seabed when sediment in the wave boundary layer cannot deposit because of stress exerted on the bottom by waves, and when sediment does not diffuse out of the wave boundary layer because of relatively weak current-induced turbulence. In the future, the importance of these flows on other margins needs to be assessed.

Keywords Flocculation, particle settling velocity, bottom boundary layer, sediment transport, plumes, fluid mud, gravity flows, mid-shelf mud deposit, nepheloid layer.

INTRODUCTION

During the 1960s research on the sedimentology of continental shelves underwent dramatic transformation. The complexity of spatial patterns of sediment composition and size made it clear that purely descriptive studies and simple conceptual models (Fig. 1) were inadequate because they failed to probe systematically or treat adequately the mechanisms and rates of **sediment transport**. Without a comprehensive knowledge of sediment transport, formation of the veneer of sediments on continental shelves was impossible to explain mechanistically. This lack of understanding was a fundamental concern to sedimentologists and sedimentary geologists, because sedimentary rocks formed on shelves and other nearshore areas represent a major portion of the stratigraphic record.

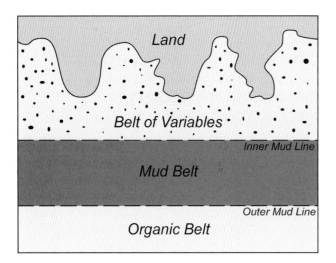

Fig. 1 An early conceptual model of sediment distribution on continental shelves. An inshore region called the 'Belt of Variables' is characterized by variable but generally coarse (> 63 μm) sediment sizes. This region gives way at the inner mud line to a mid-shelf 'Mud Belt' with mean sediment < 63 μm because of decreasing energy offshore. At the outer mud line, terrigenous mud gives way to pelagic biogenic deposits in the 'Organic Belt'. This outer transition is caused by the depletion of suspended terrigenous sediment due to its deposition shoreward of the transition. (Redrawn from Marr, 1929.)

An inability to explain the present was leaving geologists ill-equipped to unlock the secrets of the past stored in the stratigraphic record.

Two new general methodologies supplanted traditional descriptive sedimentology on continental shelves during the 1960s. First, models were developed that cast continental-shelf sediment transport in quantifiable, mechanistically based terms. Second, instrumentation was developed that made it possible to monitor sediment concentrations, waves and currents near the seabed over long time periods, thus enabling systematic characterization of the mechanisms, pathways and rates of sediment movement on continental shelves. An emerging philosophy among marine geologists was that progress in understanding the stratigraphic record depended on building an understanding of the formation of bedding at the scale of individual events such as storms, floods, debris flows and turbidity currents. Knowledge gained through event-scale studies would be applied to the sweeping time-scales of the rock record by judicious use

of emerging models of shelf sediment transport. This philosophy was summarized in the preface to Swift *et al.*'s 1972 monograph on shelf sediment transport which stated 'Geological oceanographers and marine geologists will hopefully never lose their unique sense of the vastness of geological time, which gives them a special insight into their studies, but they stand to gain much from the increased sensitivity to short-term processes which when integrated through geological time and preserved, yield the stratigraphic record.'

The decades following the 1960s witnessed dramatic advances in measurements and models of shelf sediment transport (e.g. Grant & Madsen, 1986) and continental-margin stratigraphy (e.g. Mitchum *et al.*, 1977). These efforts, in large part, however, evolved separately, and the fundamentally different time-scales considered by process sedimentologists and stratigraphers posed considerable challenges to building an integrated understanding of strata formation, from the event scales considered by sedimentologists, to the million-year time-scales considered by stratigraphers.

With the goal of meeting these challenges, the US Office of Naval Research developed and funded the programme entitled *Strata Formation on Continental Margins* (STRATAFORM). STRATAFORM brought together sedimentologists, stratigraphers and modellers with the explicit goal of using investigations of short-term (< 100 yr) sedimentary processes to place better constraint on longer time-scale (10^4–10^6 yr) stratigraphic interpretations (Nittrouer, 1999). The overall approach encompassed detailed event-scale observations of sediment delivery and deposition, investigations of longer-term sediment accumulation, seismic imaging of strata, and extensive coring of recent and ancient (Ma) deposits (Nittrouer, 1999). Vital to the integration of these various efforts into a coherent framework were modelling studies designed to bridge the gap between the time-scales of sedimentary processes and sequence stratigraphy.

The Eel River margin on the coast of northern California (Fig. 2) was one of two study sites in STRATAFORM and was the exclusive site for studying short-term sedimentary processes, which are the focus of this paper. The margin is tectonically active and prone to seismically triggered mass wasting (Lee *et al.*, this volume, pp. 213–274). Intense winter storms batter the coast, generat-

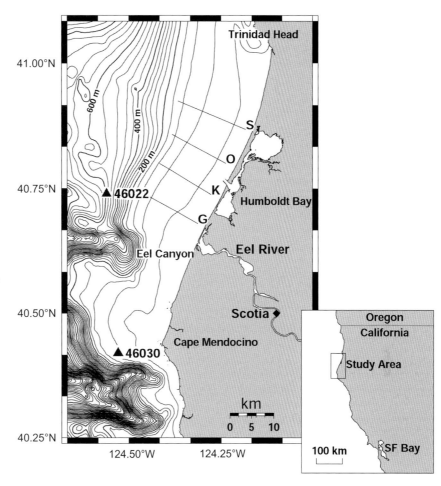

Fig. 2 Location map for the Eel River margin. Lines G, K, O and S indicate positions of a subset of cross-shelf transects that extend from the Eel Canyon (just south of the river mouth) to Trinidad Head (at the northern limit of the Eel margin). Triangles marked with numbers identify the positions of NOAA's National Data Bouy Center (NDBC) oceanographic buoys. The diamond labelled 'Scotia' marks the position of the Scotia River monitoring station. Contour interval is 40 m.

ing large waves at sea and episodic flooding on land. The active processes on the margin enhance the possibility of observing significant sediment-transporting events.

In 1995 a series of storms resulted in prolonged and intense rainfall over the entire Eel River basin. The ensuing flood was one of the largest recorded in the 85 yr of hydrographic monitoring on the river, and it delivered an estimated 25×10^6 t of fine-grained (< 63 μm) sediment to the coastal ocean (Wheatcroft *et al.*, 1997). A month after the flood, extensive coring revealed a distinct layer of flood-derived mud on the shelf. The oblong deposit was up to 8.5 cm thick, 30 km long in an along-shelf direction, 8 km wide across-shelf, and centred on the 70-m isobath north of the river mouth (Wheatcroft *et al.*, 1997). Thus, the STRATAFORM programme was initiated by the formation of a distinct event bed that could be probed and whose fate could be tracked.

The goal of this paper is to evaluate mechanisms that deliver sediment to continental margins by focusing on the Eel dispersal system, which received substantial input during the 1995 flood, during an ensuing larger flood in 1997, and during a series of smaller floods in 1998. An essential aspect of this synthesis is to place results from the Eel margin firmly into context with the large body of work that preceded them.

The paper begins with a review that is guided by the question of how well the fate of Eel River flood sediment could have been predicted given the state of knowledge in the early 1990s. Next, the observations are presented, with particular attention being paid to where these results support or refute reigning continental-shelf sediment-transport paradigms. Finally, the paper summarizes current understanding of sediment delivery to the seabed and provides new insight into which processes deserve greater attention in the future.

REVIEW OF PREVIOUS WORK

Early conceptual models

Interest in the physical environment of continental shelves flourished early for economic and strategic reasons (Emery, 1969). Ninety per cent of the world's marine food resources and nearly 20% of the world's petroleum and natural gas were being extracted from continental shelves. Shelves also promised to fill rapidly growing demand for sand and gravel and to provide a rich source of minerals. Strategically, shelves were key to the operation of submarines because the complex acoustic environment made it easy to conceal underwater objects. This upsurge of interest in continental shelves motivated several seminal papers that laid the conceptual foundations for the next three decades of research on continental-shelf sedimentology (Curray, 1965; Moore, 1969; Swift, 1970; McCave, 1972).

The most fundamental challenge for researchers of the time was developing sound physical models to explain the distribution of various grain sizes on the continental shelf. Geologists for some time had realized that simple equilibrium models (e.g. Fig. 1) failed to explain the offshore progression of grain sizes commonly observed on the Pacific Coast of North America. On the west coast, sands typically blanket inner shelves, muds occupy the middle shelves, and sand covers the outer shelves (Shepard, 1932; Emery, 1952). Emery (1952) proposed that inner-shelf sands and mid-shelf muds were currently being supplied from the continent and that outer-shelf sands were **relict** in the sense that they were not connected to modern supply and dispersal systems. More specifically, relict sands on the outer shelf were deposited when sea level was lower during the last ice age. The rapidity of sea-level rise inhibited adjustment of underlying sediment texture to rising waters.

The notion of **modern** and **relict** sediments took hold (Curray, 1965). Attention turned to explaining why modern sands were retained nearshore, why muds bypassed the inner shelf to form a mid-shelf Holocene mud blanket, and why relict sands on the outer shelf had not been covered by muds as well. Curray (1965) developed a simple model for sedimentation of river-derived sediments that divided the total sediment load into two parts. The

sand, or **bedload**, is carried close to the seabed and parallel to shore, where it deposits in a linear wedge. He proposed that, in general, wave action is too weak to transport significant quantities of sand in water depths greater than 10 m. The mud, or **suspended load**, is carried continuously or intermittently in suspension farther seaward but also parallel to shore, where it deposits in a mid-shelf mud blanket. The mud blankets typically lie in water depths deeper than 10 m. Curray (1965) suggested that when mid-shelf mud deposits are significantly deeper than 10 m, relict sands separate the modern sands and muds.

Curray's admittedly simple model left some key gaps that others proceeded to fill. In particular, Swift (1970) explicitly addressed the mechanisms by which mud bypassed the inner shelf and emphasized the importance of storm sediment transport. Swift (1970) viewed the shoreline as a sediment source, and because of the non-linear increase in sediment-transport rate with stress on the seabed, he identified storms as the key agent for moving sediment seaward. This focus on storms produced the realization that sand movement in water depths greater than 10 m is achieved easily. Swift (1970) drew on the work of Dunbar & Rodgers (1957) to hypothesize that sediment moves offshore by **diffusion**. In essence, these workers felt that currents and waves associated with storms are not organized enough to produce a strong directionality in transport. Instead, sediment moves short distances during storms, first in one direction, then in another. With the shoreline acting as a sediment source and the shelf break acting as a sediment sink, cross-shelf gradients of sediment concentration form during storms. These gradients produce a diffusive flux of sediment, especially fine sediment, across the shelf. The appearance of muds on the mid-shelf arises from preferential deposition of the coarse fraction during the intermittent transport events (Fig. 3a).

McCave (1972) focused more closely on the mechanisms by which fine sediment deposits form. He defined the issue as one of supply versus removal. Muds accumulate where supply overwhelms the ability of waves and currents to resuspend and remove them. Mid-shelf mud belts form because suspended fine sediment supplied from the coast has concentrations high enough over the mid-shelf to allow depositional flux to exceed

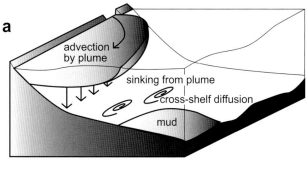

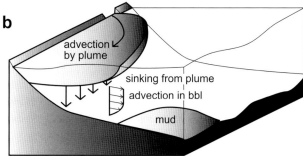

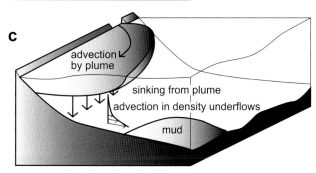

Fig. 3 Three conceptual models for formation of mid-shelf mud belts. All three assume that the removal of mud from advective buoyant plumes occurs rapidly. They differ in proposed mechanisms for seaward transport of muds that have sunk from the plume: (a) wave-generated diffusion (Swift, 1970); (b) advection in oceanographic currents (McCave, 1972) – bbl, bottom boundary layer; (c) seaward transport occurs in wave-supported, gravity-driven underflows (Moore, 1969).

erosional flux, which can be relatively small in mid-shelf water depths. The small erosional flux is mostly due to diminishing wave stress with increasing water depth.

McCave (1972) favoured advective rather than diffusive transport of fine sediment across the inner shelf. He surmised that if sediment is diffusing away from the coast to the shelf break, as proposed by Swift (1970), then most of the sediment escaping coastal seas should be accumulating on the continental slope and rise and in the abyss. Sediment budgets, however, show that most sediment escaping the shelves ends up in the great fans and cones of sediment at the bases of submarine canyons and other major supply points. He concluded that **advection** transports mud from major supply points to the canyons, and then off the shelf. Semi-permanent currents caused by wind, density and inertia were identified as the dominant means of advective transport of muds (Fig. 3b).

McCave's (1972) case was compelling for advection to dominate diffusion as the mechanism for moving fine sediments across shelf. His proposed mechanisms, however, did little to explain why sediment deposition was focused at the base of canyons. Instead, his mechanisms produce broadly distributed loss of sediment from the shelf at points downstream of major supply. Several years before McCave's work, Moore (1969) confronted the same issue of sediment focusing addressed by McCave. Working in California's Borderland Basins, Moore noticed on seismic-reflection profiles that sediments tended to dip away from submarine canyons, gullies and channels rather than away from centres of coastal drainage. He observed that sills within a basin commonly separate thick deposits near a canyon mouth from thin deposits farther away, that some nearshore basins nearly devoid of sediment are bordered seaward by basins with thick deposits, and that typically the only basins with thick sedimentary fill are integrated into a distributary system of canyons, valleys and channels. These patterns of sediment thickness led Moore (1969) to reject the concept of broadly distributed sediment loss from continental shelves. Moore offered an alternative model for how fine sediments migrate across continental shelves into submarine canyons.

Moore (1969) considered the fate of riverine sediments from their point of entry into the coastal ocean to their point of exit from the shelf at the heads of submarine canyons. At river mouths, sand and mud embark on different transport pathways. Sand sinks from buoyant riverine plumes rapidly, and is entrained by coast-parallel longshore transport in the surf zone and on the inner shelf. Where this transport system intersects canyons, sand is introduced directly into canyon heads. Mud at the river mouth remains temporarily in suspension as lower-density river water flows over basin waters.

The bulk of this mud settles rapidly from buoyant plumes, often advecting only several kilometres from the river mouth. Subsequently, or contemporaneously, during periods of energetic wave activity, mud is resuspended, and turbid layers develop over the seafloor. Under the influence of coastal currents and downslope gravity, these wave-supported layers then move across the seafloor as wide, relatively thin sheets. Muds accumulate where they escape wave stresses large enough to resuspend them, either in canyon heads or in water deep enough to inhibit large, wave-induced bed stresses (Fig. 3c).

In the late 1960s, then, several competing conceptual models of continental-shelf sediment transport emerged to guide subsequent decades of research. All essentially agreed that sand deposits rapidly at river mouths and moves alongshore in the surf zone and inner shelf. All envisioned mud residing temporarily within buoyant riverine discharge plumes. The models diverged in their proposed rates and mechanisms of transport once sediments reach the seafloor. Swift (1970) proposed that fine sediment diffuses seaward in a series of storm-generated events. McCave (1972) proposed that advection in inertially, buoyantly or atmospherically driven currents moves sediment seaward. Moore (1969) argued wave-driven erosion produces near-bottom suspensions dense enough to flow across shelf under the influence of gravity (Fig. 3).

Two key questions emerged from these competing conceptual models. Where on the shelf does fine sediment separate from buoyant discharge plumes via sedimentation, and how does fine sediment move across shelf to modern mid-shelf mud deposits?

Sediment loss from discharge plumes

Observations of suspended-sediment concentration collected near the mouths of rivers around the globe provide clear support for the hypothesis that mud and sand both sink rapidly from discharge plumes. As summarized by Drake (1976), studies around the Mississippi (Wright & Coleman, 1974), the Po (Nelson, 1970), and the Santa Barbara and Santa Clara Rivers (Drake, 1972) all showed that fine silt and clay disappear from surface waters and appear in **bottom nepheloid layers** within kilometres of river mouths. Later studies produced similar results in, for example, the dispersal systems of the Zaire, Columbia and Ebro rivers (Nittrouer & Sternberg, 1981; Eisma & Kalf, 1984; Palanques & Drake, 1990).

The rapid removal of fine sediment from discharge plumes on continental shelves requires some form of particle repackaging into larger aggregate particles, because fine silts and clays simply sink too slowly to account for observed loss rates. To demonstrate this, consider the arguments of Drake (1976) regarding the 1969 flood deposit near the mouths of the Santa Barbara and Santa Clara Rivers. Just after a large flood, more than 80% of the discharged sediment could be accounted for in water depths of less than 50 m, at distances < 20 km from the river mouths (Drake *et al.*, 1972). Given that shelf currents typically fall in the range of $10-20$ cm s^{-1} ($\sim 10-20$ km day^{-1}), these observations suggested that particles must have been sinking at speeds of approximately $25-50$ m day^{-1}. These speeds translate to tenths of 1 mm s^{-1}, which are typical of medium silts but exceed settling velocities of clay particles by several orders of magnitude. Similar results have been found in other environments, including tropical rivers and fjords (Eisma & Kalf, 1984; Syvitski *et al.*, 1985).

The hypothesis that particle repackaging causes rapid loss of fine sediment from river plumes was widely proposed and generally accepted. Mechanisms and rates of particle repackaging became a topic of research, and two mechanisms were proposed (e.g. Stumm & Morgan, 1981; McCave, 1984). The increasing ionic strength of water caused by the addition of salt compresses the **ion clouds** that surround charged particles, like fine-sediment grains, in water. In freshwater, ion clouds are thick, so when particles approach one another, their clouds cause repulsion at relatively large separation distances. In seawater, the ion clouds are compressed to such an extent that particles can approach one another quite closely before their ion clouds repel. At small separation distances the powerful yet distance-limited **van der Waals' force** of attraction can overwhelm the repulsive force between ion clouds, causing particles to cohere in a process called **electrochemical coagulation** (e.g. Stumm & Morgan, 1981). Pioneering experiments demonstrated that coagulation occurs at low salinities (Whitehouse *et al.*, 1960; Krone, 1962), a fact that was used to explain the trapping of sediment in estuaries (e.g. Postma, 1967; Kranck, 1973, 1981; Edzwald *et al.*, 1974). Similar processes were invoked to explain rapid disappearance of sediment from river plumes on the continental shelf (McCave, 1972; Drake, 1976; Boldrin *et al.*, 1988).

Biogenic aggregation refers to the agglomeration by organisms of mineral matter into faecal pellets (Drake, 1976; McCave, 1984). In some environments it probably plays an important role in speeding removal of sediment from plumes (e.g. Schubel *et al.*, 1978). The remarkable consistency of sediment removal rates in a variety of settings, however, suggests that biogenic aggregation alone cannot explain rapid particle sinking.

Demonstration of coagulation in the laboratory paired with its hypothesized role in nearshore sedimentation of muds sparked efforts to measure the size and settling velocity of particle aggregates. Rather quickly the tendency of invasive sampling methods to disrupt fragile aggregates was documented (Gibbs, 1982a,b; Gibbs & Konwar, 1983), leading to the development of non-invasive methods for measuring aggregate properties *in situ*. Photography proved most effective (Syvitski & Murray, 1981; Eisma *et al.*, 1983, 1991, 1996; Kranck, 1984; Johnson & Wangersky, 1985; Syvitski *et al.*, 1991; Kranck & Milligan, 1992). Other methods also emerged, such as gentle capture paired with microscopy (Kranck *et al.*, 1992; Droppo & Ongley, 1994) and instruments that link the angular distribution of scattered laser light to particle-size distribution (Bale & Morris, 1987; Agrawal & Pottsmith, 1994).

These studies yielded apparently conflicting results regarding the importance of coagulation as a particle repackaging mechanism. According to the coagulation hypothesis, particles in freshwater are dispersed. Upon entering the sea, river waters mix with salty ocean water. A small rise in salinity to a few parts per thousand induces enough compression of ion clouds to allow aggregates to form. Maximal aggregate size then is set either by sedimentation or by disaggregation resulting from turbulence (Kranck, 1973). In estuaries, however, the expected increase in aggregate size at the interface between fresh and salt water failed to materialize. Instead, aggregate sizes showed no dependence on salinity, with typical diameters of several hundred micrometres in river and seawater alike (Eisma, 1986; Eisma *et al.*, 1991; Kranck *et al.*, 1992). In fjords, however, observations of aggregates did indicate

rapid formation when freshwater suspensions met the sea (Hoskin & Burrell, 1972; Syvitski & Murray, 1981; Cowan & Powell, 1990; Cowan, 1993).

The contrasting results from fjords and estuaries can be reconciled by considering the role of organic matter in aggregation. In a process called **flocculation**, organic molecules can bridge the gap between two particles by bonding to both surfaces. Aggregates produced in this manner are known as **flocs**. The efficacy of organic matter as a bonding agent depends on the composition, configuration and concentration of organic matter, all of which vary with environment and salinity (Eisma *et al.*, 1991; O'Melia & Tiller, 1993). Rivers discharging into temperate estuaries are likely to contain higher concentrations of large organic molecules than glacial meltwaters flowing into fjords, so flocculation predominates in estuaries, whereas coagulation controls particle packaging in the headwaters of fjords.

Scant observations exist of aggregate size in plumes extending from river mouths to the continental shelf, so it remains unclear how much aggregation modifies the *in situ* size distribution of plume sediments. Berhane *et al.* (1997) observed no dependence of aggregate size on salinity in the Amazon plume, but it was admitted that a lack of low-salinity observations may have masked evolution of aggregate size near the river mouth. Prior to the mid-1990s, then, repackaging of sediment was viewed widely as critical to producing rapid removal of fine sediment from surface plumes on the continental shelf. A dominant mechanism for repackaging could not be identified, however. The contribution of coagulation, biogenic aggregation and flocculation mediated by organic matter varied among environments (e.g. Syvitski & Murray, 1981; Berhane *et al.*, 1997).

For determining sediment fluxes, aggregate size is important insofar as it affects sediment settling velocity. Early laboratory work demonstrated clearly the significant enhancement to mud settling velocity caused by aggregation (Krone, 1962; Kranck, 1980), so attention turned to characterizing settling velocity *in situ*. Two main approaches were developed. **Owen tubes** and related devices (Burt, 1986; Dyer *et al.*, 1996) monitor concentration through time in a tube. The tube is lowered to a desired depth in a horizontal position with its ends open. Upon retrieval, the tube closes and flips into a vertical position, thus presumably capturing without severe disturbance a sample of suspension. Bulk clearance rate of the suspension is used to calculate a representative settling velocity. No direct observations of particle sinking are made. The other approach is to observe directly and *in situ* the descent of particles in an enclosure that prevents horizontal advection of particles through the viewing volume. Vertical displacements over set time intervals are used to calculate particle settling velocities (Fennessy *et al.*, 1994; ten Brinke, 1994; Syvitiski *et al.*, 1995; Dyer *et al.*, 1996; Hill *et al.*, 1998).

These two different approaches yielded distinctly different results. In Owen tubes clearance rate increases with increasing concentration (Burt, 1986; Dyer *et al.*, 1996). The explanation given for this result was that aggregation is faster at higher concentrations. Faster aggregation arguably begets larger aggregates, producing the observed increases in clearance rates. Direct observations failed to support this explanation. Across a range of environments and sediment concentrations, aggregate settling velocities are typically in the range of 1 mm s^{-1} (ten Brinke, 1994; Hill *et al.*, 1998).

The observed increase in clearance rate with sediment concentration in Owen tubes has been linked conceptually and through direct and indirect observations to aggregation within the tubes (Milligan, 1995; Dearnaley, 1996; Milligan & Hill, 1998). These workers proposed that removal of sediment from a settling tube proceeds in several steps. First, *in situ* aggregates are disrupted to an unknown degree during sampling. Second, large aggregates form in the quiescent environment of the tube at a rate dependent on concentration. Last, aggregates sink out of suspension at approximately 1 mm s^{-1}. Concentration dependence of removal rate arises due to the concentration dependence of re-aggregation rate, not because aggregates become larger and sink faster at higher concentrations.

Turbulence probably influences aggregate size and settling velocity (Milligan & Hill, 1998), yet observations leave its role unclear. Theory suggests that aggregate size varies with an inverse power of the **turbulent-kinetic-energy dissipation rate** (e.g. Hunt, 1986). Limited experimentation with natural aggregates, however, showed that dependence of size on turbulent-kinetic-energy dissipation rate is either not significant or weaker than predicted (Alldredge *et al.*, 1990).

Time can also influence aggregate size and, by implication, aggregate settling velocity. If sediment grains are dispersed as they enter the sea, then a finite amount of time is required for aggregates to grow to an equilibrium size. If sediment concentration is high, then less time is required for aggregates to form (e.g. McCave, 1984; Hill, 1992). It is difficult to specify an actual time required for aggregation due to uncertainties regarding particle contact, adhesion and break-up rate (Hill, 1992, 1996; Hill & Nowell, 1995).

Prior to the mid-1990s, then, the variables controlling aggregate size and settling velocity were not clear. The most robust result of *in situ* studies was that settling velocities of 1 mm s^{-1} are typical of many marine environments. Therefore, settling velocities of this magnitude could be expected in the Eel River plume, as long as turbulence or lack of time did not prevent aggregates from attaining sizes large enough to sink at this rate.

Advective transport in river plumes

Plume direction, speed, thickness and width are the hydrographic parameters that, along with sediment settling velocity, determine where sediment discharged by a river will reach the seafloor. Research into the dynamics of plumes blossomed in the 1970s, with investigations framed increasingly in quantitative terms. Two subdisciplines were at the forefront of plume research at the time: marine sedimentology and physical oceanography. These disciplines focused their investigations somewhat differently, with the sedimentologists naturally more interested in processes close to river mouths where the bulk of fluvial sediment deposits, and the physical oceanographers more concerned with transport and mixing of river waters that occur both near to and far from river mouths. The work of these groups was complementary and, taken as a whole, provides both solid theoretical and observational frameworks on which to build a conceptual model of plume hydrography on the Eel shelf.

An issue recognized early as important to the direction followed by plumes is **plume buoyancy** (Bates, 1953; Wright, 1977). When the density of inflowing, sediment-laden water is much less than the basin water, the plume rides over the seawater and spreads under the influence of gravity. These plumes are called **hypopycnal**. When inflowing suspensions have approximately the same density as basin water, the plume behaves much like a turbulent inertial jet. These plumes are called **homopycnal**. Under these two scenarios the steering of the plume is dominated either by Earth's rotation or by oceanographic and atmospheric forcing such as winds, currents and tides (Bates, 1953; Scrutton & Moore, 1953; Wright & Coleman, 1974; Wright, 1977; Eisma & Kalf, 1984; Garvine, 1987; Palanques & Drake, 1990; Geyer *et al.*, 1996). In contrast, when river waters are so laden with sediment that the inflowing plume exceeds the density of basin water, a gravity current forms and flows along the seafloor in the direction of maximal gradient (Bates, 1953; Mulder & Syvitski, 1995; Parsons *et al.*, this volume, pp. 275–337). The course of such **hyperpycnal** plumes also is affected by Earth's rotation and oceanographic forcing by tides and currents.

For many years, hyperpycnal plumes were not thought to be possible in marine settings, because the sediment concentrations required to make river water denser than seawater were too high to ever be realized under natural conditions (Bates, 1953; Drake, 1976). A systematic analysis of 150 rivers by Mulder & Syvitski (1995), however, indicated that some rivers do indeed carry sediment concentrations ≥ 40 kg m^{-3} required to overcome typical seawater densities. The conditions for such high concentrations are most common in small- and medium-sized mountainous drainage basins. Mulder & Syvitski (1995) suggested that during major floods, the Eel may reach high enough density to form hyperpycnal underflows. If so, then shelf topography would be important in determining the dispersal pathway of plume sediment. Unfortunately, the uncommon and unpredictable nature of hyperpycnal plumes makes them difficult to observe directly.

Proceeding under the assumption that the Eel plume is less dense than the receiving waters on the shelf leads to the prediction that the plume is steered up the coast to the right as it leaves the mouth. This prediction is relatively safe because both Earth's rotation and oceanographic processes during floods of the Eel force the plume northward along the coast. In the northern hemisphere, currents veer to the right under the influence of the **Coriolis force**, which is towards the north for the westward-discharging Eel (e.g. Garvine, 1987). The Eel discharges 90% of its sediment during and immediately following winter storms (Brown & Ritter, 1971). The cyclonic circulation of the storms produces strong winds blowing from the south during peak discharge. The attendant wind stress on the ocean's surface, combined with the Coriolis force, pushes water to the right, or shoreward in the case of the Eel margin. In response, the sea surface develops a seaward slope that in turn produces a **barotropic flow**. This flow is deflected to the right, again by the effect of the Earth's rotation. In short, winds blowing from the south during storms force a northward flow along the coast (Smith & Hopkins, 1972). Finally, reworking of rivermouth sands by waves associated with winter storms produces a northward littoral drift that has formed an oblique entry of the Eel into the Pacific. This mouth geometry also favours northward transport (Wright, 1977; Garvine, 1987).

The speed of the plume is not as easy to predict as the direction. The deceleration of a plume upon entering the sea depends on the inertia of the outflow, the density contrast between the river and basin waters, and the degree to which plume interaction with the seabed extracts momentum from the flow (Wright, 1977). Inertia dominates plume behaviour when outflow velocity is large, and the density contrast between river and basin waters is small. Buoyancy dominates plume behaviour when outflow velocity is small and the density contrast is large. **Inertia-dominated plumes** decelerate due to turbulent mixing with ambient fluid along the plume's edges and base, but **buoyancy-dominated plumes** decelerate due to spreading and thinning of plume waters as they flow over basin waters. These mechanisms of deceleration differ fundamentally, so it is essential to identify which one dominates in a particular plume.

The **densimetric Froude number** (*Fr*) characterizes the importance of inertia relative to buoyancy. It is dimensionless and defined by the equation

$$Fr = \frac{u}{(g'h_{\mathrm{p}})^{1/2}} \tag{1}$$

where u (m s^{-1}) is the mean outflow speed, h_{p} (m) is plume thickness and g' (m s^{-2}) is modified gravity, which is defined by

$$g' = \frac{\Delta\rho}{\rho} g \tag{2}$$

In Eq. 2, $\Delta\rho$ (kg m^{-3}) represents the density contrast between plume and basin water, ρ (kg m^{-3}) is the density of basin water, and g (m s^{-2}) is gravitational acceleration. If the Froude number is much greater than unity, then inertial forces dominate plume dynamics. If it is much less than unity, then buoyancy dominates plume dynamics (e.g. Wright, 1977).

Before the mid-1990s, the variables required to calculate Fr had not been measured explicitly on the Eel River margin, but data that made it possible to estimate them were available. Turning first to outflow speed, it is approximately equal to river discharge, Q (m^3 s^{-1}), divided by channel depth h_c (m) and channel width W_c (m). During typical, annual floods, Eel discharge is ~5000 m^3 s^{-1} (Brown & Ritter, 1971). Channel width is approximately 1000 m and channel depth is approximately 5 m. The outflow speed during floods, therefore, is ~1 m s^{-1}. The density contrast between the plume and basin water, based on observations elsewhere (e.g. Wright & Coleman, 1974), is probably ~10 kg m^{-3}, and density of basin water is ~1025 kg m^{-3}. With these inputs, $(g'h_p)^{1/2}$ is approximately equal to 0.7 m s^{-1}. The outflow Froude number is therefore larger than unity, so plume dynamics at the mouth are dominated by inertia.

Inertia-dominated plumes do not ride up over basin water to the extent that buoyancy-dominated plumes do, so they can be slowed by frictional interaction with the seabed. Wright (1977) noted that small bottom gradients and depths less than or equal to channel depth seaward of the mouth produce conditions for which bottom friction plays a key role in plume deceleration and spreading. Bottom gradient on the Eel shelf is relatively steep (0.007 m m^{-1} or 0.4°; Leithold, 1989), and the mouth

region has not formed a significant subaqueous delta because of vigorous wave reworking of river-mouth deposits. Bed friction probably does not alter plume dynamics markedly.

Based on this information, it is possible to surmise that the Eel plume during floods is inertia-dominated and not affected strongly by bottom friction. It discharges into the Pacific where oceanographic conditions force a general northward transport. In addition, mouth geometry tends to direct the river outflow northward and along the coast. Interestingly, these general conditions resemble those assumed by Garvine (1987) in a numerical model of plume dynamics. The results of that model can be of use in elucidating plume structure and geometry in the vicinity of the Eel.

Garvine's (1987) model produces a plume with a distinct anticyclonic turning region near the mouth (Fig. 4). The dimensions of this gyre are set by the internal **Rossby radius of deformation**, defined as

$$Li = \frac{u}{f} \tag{3}$$

where Li is the Rossby radius (m), u is outflow speed (m s^{-1}) and f is the Coriolis frequency (s^{-1}). Assuming that outflow speed is approximately equal to 1 m s^{-1} and that the Coriolis frequency is 10^{-4} s^{-1} yields a Rossby radius of approximately 10 km. Garvine's model predicts that downstream of this bulge there is a sharp transition to cyclonic turning into a coastal current. This turning is forced by locally high pressure gradients created by water being forced against the coast. The coastal current that forms is in approximate geostrophic balance for the cross-shelf component.

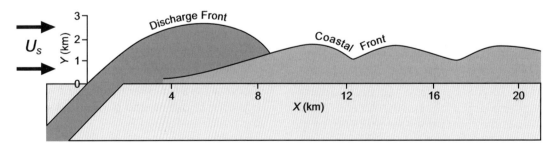

Fig. 4 Schematic of a numerical model for evolution of a buoyant discharge plume. The variable U_s represents the velocity of a poleward-directed ambient current. (Redrawn from Garvine, 1987.)

If the Eel plume behaves like Garvine's model plume, then the fine sediment delivered by the Eel should deposit primarily under the anticyclonic turning region at the mouth. To demonstrate this, consider a simple calculation of the **residence time** of water in the gyre by approximating its volume as a half cylinder with radius Li (10^4 m) and depth h_c (5 m). This volume equals $\sim 10^9$ m^3. Assuming an inflow to this volume equal to the Eel discharge during annual floods (5000 m^3 s^{-1}), the residence time of a water parcel is $\sim 2 \times 10^5$ s, which is just over 2 days. Given typical bulk settling rates of 25–50 m day^{-1} observed in a variety of environments (Drake, 1976; Eisma & Kalf, 1984; Syvitski *et al.*, 1985) and individual aggregate settling velocities of 100 m day^{-1} (ten Brinke, 1994; Hill *et al.*, 1998), fine sediment has ample time to sink out of the plume and reach the seabed before being carried beyond the anticyclonic gyre at the mouth. Sediment therefore should reach the seabed within approximately 10 km of the river mouth and several kilometres from shore.

This simple prediction does not address explicitly the existence of mudstreams extending hundreds to thousands of kilometres downstream of some river mouths (McCave, 1972). Observations of rapid sinking and laterally extensive mudstreams can be reconciled by considering the role of resuspension. Near the coast, turbulence and downwelling can destroy water-column stratification and exert considerable stress on the seabed. In combination, these effects can prevent the deposition of plume sediment and lead to its retention in the plume and associated coastal current (Smith & Hopkins, 1972). The Eel margin typically experiences large waves and downwelling during floods, so a significant amount of fine sediment may be forced northward in nearshore regions. This nearshore flux is difficult to constrain because it depends on the fraction of the plume width under which resuspension occurs, and on northward flow speeds.

Despite uncertainty over how much sediment moves north on the Eel margin in a shore-attached mudstream, the alongshore position of the Eel mud deposit on the shelf indicates that a substantial fraction of Eel mud separates from the plume and its associated coastal current within the $\sim$10-km distance suggested by the previous calculations. Decadal accumulation rates based on vertical profiles of ^{210}Pb in the seabed show that maximal accumulation rates occur 15 km north of the river

mouth (Leithold, 1989). In fact, Leithold (1989) used the distribution of accumulation rates to suggest that the plume flows directly over the mid-shelf mud deposit and loses sediment due to deposition directly to the seabed. The region of maximum accumulation measured by Leithold (1989) is centred 15–20 km offshore, yet the anticyclonic bulge at the mouth should extend to less than 10 km offshore and the associated coastal current should be even thinner (Garvine, 1987; Fig. 4). Therefore, sediment sinking from the plume must move across shelf either by diffusion during storms (Swift, 1970), by advection in coastal currents (McCave, 1972), or by advection in wave-supported, gravity-driven undercurrents (Moore, 1969).

Bottom-boundary-layer transport of flood sediment

Driven by competing hypotheses and rapid technological advances, understanding of benthic-boundary-layer sediment transport advanced dramatically during the 1970s and 1980s. The earliest deployments of current meters in continental-shelf bottom boundary layers documented quite clearly the dominant role of storms in sediment transport (Smith & Hopkins, 1972; Sternberg & McManus, 1972; Sternberg & Larsen, 1976). These measurements supported Swift's (1970) hypothesis that storms dominated transport, but they failed to support the hypothesis that storm-driven transport was diffusive and produced no net along-shelf transport. Instead, correlations were observed between storm resuspension and the direction and strength of near-bottom currents.

On the Washington shelf, storms cause significant across-shelf and along-shelf transport of fine sediment. This advection produces northward dispersal due to the prevalence of northward-flowing near-bed currents during storms (Smith & Hopkins, 1972; Sternberg & McManus, 1972; Sternberg & Larsen, 1976). Similar results were obtained in Norton Sound, Alaska, where wave-induced bottom currents associated with local storms were seen as critical to the northward dispersal of the fine sediment emanating from the Yukon River (Drake *et al.*, 1980). On the Russian River shelf just south of the Eel margin, sediment transport throughout a year is dominated by a few storms that generate strong northward currents with a substantial seaward component (Drake & Cacchione, 1985;

Sherwood *et al.*, 1994). On the Ebro margin in Spain, oceanographic currents push storm-resuspended sediment southward and seaward (Cacchione *et al.*, 1990). These studies and others favour the hypothesis that advection in bottom currents dominates the transport of fine sediment once it sinks from surface plumes. The generally similar forcing on the Russian River margin, the Columbia River margin and the Eel margin suggested that advective transport in the bottom boundary layer would occur primarily during winter storms and would on average be directed northward and seaward. This prediction is consistent with the position of maximum accumulation on the Eel shelf just north and seaward of where sediment would be likely to sink from the plume.

The clear documentation of advective transport by storm- and wave-generated near-bed flows diverted attention away from wave-supported, gravity-driven underflows as a plausible mechanism for across-shelf transport of muds. Furthermore, such underflows were deemed unlikely due to the extraordinarily large sediment concentrations required to overcome typical ocean stratification (Drake *et al.*, 1972). Nonetheless, observations accumulated slowly suggesting that density underflows remained a viable transport mechanism in shelf settings.

Density underflows were first recognized in the form of **turbidity currents** flowing down slopes that were steep compared with the gradients found in nearshore and continental-shelf settings (see Walker, 1973, for review). These steeper slopes allowed turbidity currents to flow rapidly enough to erode sediment from the seabed, thereby maintaining or enhancing their motive force (Bagnold, 1962). The maintenance of a dense suspension capable of flowing downslope seemed unlikely on low gradients until **fluid muds** were observed in estuaries such as the Gironde and Severn (Migniot, 1968; Kirby & Parker, 1977). These dense suspensions were the product of sediment trapping by estuarine flow that produced locally high fluxes of sediment to the seabed. The high fluxes overwhelmed local removal rates and produced concentrations of sediment great enough to hinder particle settling. These highly concentrated layers of mud are mobile and can move under the influence of gravity or currents (Migniot, 1968; Kirby & Parker, 1977).

Fluid muds were considered a unique byproduct of the circulation within estuaries until research in the Amazon and Huanghe rivers demonstrated that they can form at density fronts on the continental shelf. Flow convergence at fronts leads to sediment trapping akin to that observed in estuaries. Sediment trapping produces high concentrations and hindered settling, and it can lead to downslope advection under the influence of gravity (Wright *et al.*, 1988; Kineke *et al.*, 1996). These observations suggested that strong density fronts are a key factor in the formation of gravity-driven flows, and, in a sense, they refuted implicitly Moore's (1969) hypothesis that waves alone can produce concentrations high enough to generate fluid muds.

Seymour (1986) addressed explicitly the possibility that concentrations of sediment great enough to flow downslope under the influence of gravity can form under waves. Taking a theoretical approach, he concluded that velocities, sediment size and supply, and bottom gradients on a typical inner continental shelf are more than adequate to produce wave-supported, gravity-driven underflows. He went on to explain some anomalous observations of other studies in the context of his proposed mechanism.

Sedimentary geologists also struggled to define the mechanisms underlying the formation of **tempestites**, which are sedimentary layers deposited during storms. These storm layers are common in the geological record of past continental-shelf sedimentation. They are curious in that they often show evidence of strong, seaward-directed, near-bottom flow. This evidence led to ongoing support for the hypothesis that wave-supported, dense, near-bed suspensions flowed downslope under the influence of gravity, much as Moore (1969) envisioned (e.g. Hamblin & Walker, 1979; Myrow & Southard, 1996).

By the mid-1990s, therefore, Moore's (1969) hypothesis that sediment moves across shelf under the influence of gravity was not widely recognized in the oceanographic and marine-geology communities. It could not be rejected, however, based on available theory and data. Furthermore, the geological record of storm sedimentation was difficult to explain without it.

Summary of past research

Based on past research, a coherent conceptual model for sedimentation on the Eel River shelf can be constructed. The river plume enters the coastal ocean dominated by inertia. An anticyclonic bulge with radius of 10 km forms at the mouth. This bulge transforms into a northward flowing coastal current with a width of order 10 km or less. Much of the fine sediment in the plume, under the influence of aggregation, sinks from the plume at rates of 25–100 m day^{-1}. This sediment leaves the plume primarily beneath the anticyclonic bulge. Upon leaving the plume, near-bed currents advect sediment northward and seaward in dilute suspensions. This near-bed advection explains qualitatively the location of maximum sediment accumulation on the shelf 15 km north of the river mouth and 15–20 km from shore. The remainder of this paper describes how recent observations of sediment delivery during floods on the Eel margin support or refute elements of this simple conceptual model.

SEDIMENT DELIVERY TO THE EEL MARGIN

Site description

The Eel shelf extends from Cape Mendocino in the south to Trinidad Head in the north (Fig. 2). The shelf is relatively narrow and steep. The shelf break occurs in water depths of 150 m, and it is located approximately 20 km from shore, indicating a slope of slightly greater than 0.4°. In addition to this relatively steep bathymetric gradient, two other physiographical features may play important roles in processing Eel River sediment on the shelf. The Eel Canyon incises the shelf just south of the river mouth. Its proximity to the mouth makes it a potentially important sink for sediment discharged by the Eel River. Humboldt Bay is a long, broad bay with a narrow inlet 15 km north of the river mouth. It, too, may affect sediment dynamics because of significant tidal exchange between the bay and the shelf (Geyer *et al.*, 2000).

The Coast Range rises to elevations > 2000 m over distances of 80 km in the Eel watershed. This steep topography leads to large erosion rates in the Eel basin. Large erosion rates also are favoured by the erodibility of the underlying Mesozoic Franciscan Complex, a **mélange** of intensely deformed sedimentary, low-grade metamorphic and igneous rocks. Much of the Franciscan is so highly sheared that it cannot maintain a slope of greater than

10–15°, commonly failing by shallow landslides following periods of heavy rainfall (Brown & Ritter, 1971; Nolan *et al.*, 1995). The topography also forces intense orographic precipitation as moist ocean air flows in from the west.

After trending inland perpendicular to the shore for 10 km, the main stem of the Eel River turns roughly shore-parallel, draining the heart of the Coast Range south of the river mouth. This interesting morphology arises in part due to uplift in the vicinity of the Mendocino Triple Junction to the south. The result is that the entire watershed often receives intense precipitation contemporaneously during storms, producing rapid and large increases in streamflow.

The regional-scale climate produces essentially two seasons. In the summer a broad area of high pressure is located over the ocean, with its centre well to the west of the California coast. From April to November, clockwise circulation around the high causes winds to blow from the north, and precipitation is minimal. During winter, the Aleutian Low develops in the north Pacific and pushes the high-pressure centre to the east. This shift exposes northern California to intense low-pressure systems moving onshore from the Pacific (Nunn, 1999). These lows have counter-clockwise circulation, and their approach is heralded by strong winds blowing from the south. After the passage of the lows, winds often shift to blow from the north. This stormy, wet period typically extends from November through March.

The average annual precipitation in the Eel basin is 1.26 m. The drainage area of the river is ~8000 km^2, so the mean annual discharge of the river is ~10 km^3 of water (Morehead & Syvitski, 1999). This figure translates to a mean annual discharge of approximately 300 m^3 s^{-1}. The episodicity of precipitation, however, leads to peak discharges well in excess of this value. A typical large annual flood can last about a week and produce peak discharges of 5000 m^3 s^{-1}. Larger, rare flood events produce peak discharges in excess of 8000 m^3 s^{-1} (Morehead & Syvitski, 1999; Sommerfield *et al.*, this volume, pp. 157–212).

Observational programme

The field efforts for the sediment-transport-and-accumulation component of STRATAFORM extended primarily over four flood seasons between 1994 and 1998. The observations can be grouped broadly into seabed observations, plume observations and bottom-boundary-layer observations.

Seabed observations

Seabed sampling was carried out primarily with a 20 × 30 cm box corer (e.g. Wheatcroft & Borgeld, 2000). In 1997–98 a hydraulically damped piston corer was used to collect cores in inner shelf sandy sediments. Coring took place on nine cruises: February, May and September 1995; March and July 1996; January and May 1997; and March and July 1998. In general, during a cruise, 40–70 stations were sampled, extending along-shelf from just south of the river mouth to just south of Trinidad Head 50 km to the north. Stations extended across-shelf to the upper slope (Wheatcroft & Borgeld, 2000).

Sediments within cores were characterized with a variety of techniques. To assess and quantify sediment layering within the seabed, sediment slabs were X-rayed onboard, generally within 30 min of collection, thus limiting the effects of subsequent compaction or bioturbation on internal bedding. Vertical distribution of grain size within cores was characterized in several ways. Sediment was wet-sieved, then size distribution was measured with a Coulter Multisizer (Drake, 1999). In other analyses, discrete organic matter was separated from the sediment. The remaining inorganic sediment was disaggregated, and the size distribution was measured with a Multisizer (e.g. Milligan & Kranck, 1991). Grain size also was characterized with sieve-and-pipette analysis (e.g. Folk, 1977). Resistivity as a function of depth in core was measured as a proxy for sediment porosity (Wheatcroft & Borgeld, 2000). Organic geochemical characterization of the shelf sediments was also undertaken (Leithold & Hope, 1999). Carbon-to-nitrogen ratios and the isotopic ratios of ^{13}C to ^{12}C were used to identify sediment containing terrestrial organic matter introduced by the river onto the continental shelf.

Accumulation rates in sediment cores were measured over a range of time-scales by using a suite of radioisotopes (Sommerfield et al., this volume, pp. 157–212). Most relevant to the short-term, event-scale focus of this paper is ^{7}Be (e.g. Sommerfield

et al., 1999). This isotope is formed by cosmic-ray spallation of nitrogen and oxygen in the Earth's atmosphere, where it adsorbs onto aerosols and then can reach the Earth's surface by wet or dry deposition. In the vicinity of turbid rivers, virtually all ^{7}Be remains adsorbed to particle surfaces. It is, therefore, an excellent tracer of particles recently supplied to the ocean. Its utility in constraining short-term deposition rates derives from the facts that ^{7}Be is concentrated in the surface of sub-aerially exposed soils and that ^{7}Be has a half-life of only 53.3 days. The appearance of ^{7}Be in the seafloor therefore indicates that those sediments have resided in the coastal ocean < 8 months. The Eel River dominates sediment discharge onto the shelf and most of that discharge occurs during 4 months in late autumn and winter, so sediments with measurable ^{7}Be can be linked unambiguously to discharge events during the preceding year (Sommerfield et al., 1999).

Plume observations

Plume observations can be divided into two categories. Between 1996 and 2000, rapid-response helicopter surveys were conducted in association with floods of the Eel River. In 1996–97 and 1997–98 these helicopter observations were paired with time series collected from moorings placed on the G and K transects (see Fig. 2; Geyer et al., 2000).

A helicopter-based sampling programme was developed for STRATAFORM because of the typically extreme sea conditions that accompany floods. By monitoring the discharge of the Eel River via the internet, it was possible to deploy equipment and scientific personnel to sample the plume within 24 h of a discharge threshold on the Eel River. With support and assistance of US Coast Guard Group Humboldt Bay, a profiling instrument package was lowered from a search-and-rescue helicopter through the water column on a grid of 12 stations that extended from the river mouth northward 30 km along the shelf. Sampling generally was shoreward of the 40-m isobath, because the sediment plume did not extend farther seaward than this. The instrument package comprised a CTD (conductivity, temperature, depth device), a camera for observing in situ aggregate size, two pressure-actuated Niskin bottles designed to collect sediment suspensions at 2 m and 10 m below the sea surface, and an optical backscatter sensor (OBS) to monitor the vertical distribution of sediment concentration (Hill *et al.*, 2000; Curran *et al.*, 2002a).

In the first deployment, moorings were located in 30 m and 60 m of water on the G line just north of the river mouth. The path followed by the plume was generally shoreward of these positions so, in the following year, moorings were located on the K line, which is 10 km up-coast from the river mouth. Moorings in 20 m, 40 m and 60 m of water on the K line carried temperature, salinity and OBS sensors at 0.5-m and 4.5-m water depths on the moorings. Current meters were placed at 2 m and 6 m below the surface (Geyer *et al.*, 2000).

Bottom-boundary-layer observations

Bottom tripods and quadrapods, hereafter referred to generically as tripods, were used to monitor waves, currents and suspended-sediment concentration in the bottom boundary layer. Throughout the programme, a tripod was maintained on the S line at 60-m water depth (S60) (Ogston & Sternberg, 1999; Ogston *et al.*, 2000). The configuration of the deployment arrays changed from year to year. Tripods were deployed across the shelf on the S line, extending from 55 m to 70 m, to investigate the role of cross-shelf flux convergence in determining the cross-shelf position of the mud deposit (Cacchione *et al.*, 1999; Ogston & Sternberg, 1999; Wright *et al.*, 1999). Tripods were placed at G65, K63 and S60 to investigate along-shelf flux convergence in determining the along-shelf position of the mud deposit (Cacchione *et al.*, 1999; Ogston & Sternberg, 1999; Wright *et al.*, 1999). An array of tripods was emplaced on the K line with instruments located at 20-, 40- and 60-m water depths to gain more information about the cross-shelf sediment flux in the bottom boundary layer at the along-shelf position where loss of sediment from the plume was maximal (Traykovski *et al.*, 2000; Wheatcroft & Borgeld, 2000).

In general, tripods were equipped with vertical arrays of OBS sensors and electromagnetic current meters. These arrays characterized flow velocity and sediment concentration from heights of 10–30 cm above bottom (cmab) to 1–2 m above bottom (mab) (e.g. Ogston *et al.*, 2000). The tripods also

generally had upward-looking acoustic Doppler current profilers (ADCP) to characterize flow above the tripods. Important additions to this general suite of sensors were acoustic backscatter sensors (ABS) mounted on two tripods along the K line. These downward-looking sensors were deployed to measure profiles of acoustic-backscatter intensity between the seabed and 1.28 mab. The data from these sensors can be used as a proxy for suspended-sediment concentration. These sensors provide observations below the lowermost OBSs (Traykovski *et al.*, 2000).

Results

Environmental conditions during study period

Large floods with peak discharges in excess of 10,000 $m^3\,s^{-1}$ occurred on the Eel River in January 1995 and January 1997 (Figs 5–8). In terms of peak discharge observed since the 1930s, the 1997 and 1995 floods rank second and third behind the remarkable event of 1964, which produced a peak discharge of 21,000 $m^3\,s^{-1}$ (Wheatcroft & Borgeld, 2000). A moderate flood with a peak discharge of ~5000 $m^3\,s^{-1}$ occurred in March 1995, and a series of moderate floods marked the La Niña winter of 1997–98 (Geyer *et al.*, 2000; Wheatcroft & Borgeld, 2000). The winter of 1995–96 was relatively dry.

During the peak discharge months, winds typically blow out of the south, with brief periods out of the north. The same pattern holds true during flood events. As intense low-pressure systems move onshore, winds blow strongly from the south, with typical wind speeds of 20 $m\,s^{-1}$. After the fronts move onshore, winds during flood events decrease and switch to blow from the north.

Large waves are typical of the Eel shelf during winter. Mean wave height is 2.4 m, and 1% of the time wave heights exceed 5.5 m. Waves as high as 12 m have been observed (Ogston & Sternberg, 1999). The largest waves tend to occur in winter. During flood events, the margin typically is exposed to large waves due to the association of high winds and precipitation with low-pressure systems (Cacchione *et al.*, 1999; Ogston & Sternberg, 1999; Geyer *et al.*, 2000; Traykovski *et al.*, 2000). Not all periods of large waves, however, occur during flood events.

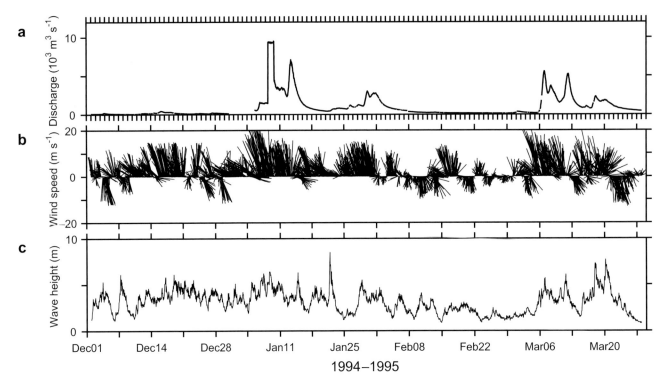

Fig. 5 Environmental conditions on the Eel margin during winter 1994–95. (a) Hourly mean discharge at the Scotia station on the Eel River. (b) Wind velocity vectors, with positive indicating winds blowing toward the north. Winds were measured at NOAA's National Data Buoy Center (NDBC) buoy 46022. (c) Wave heights measured at the same buoy. Note the large flood in January 1995, and the moderate flood in March 1995. The greatest peak in January is truncated in the graph, because the monitoring equipment failed at the highest turbidity levels.

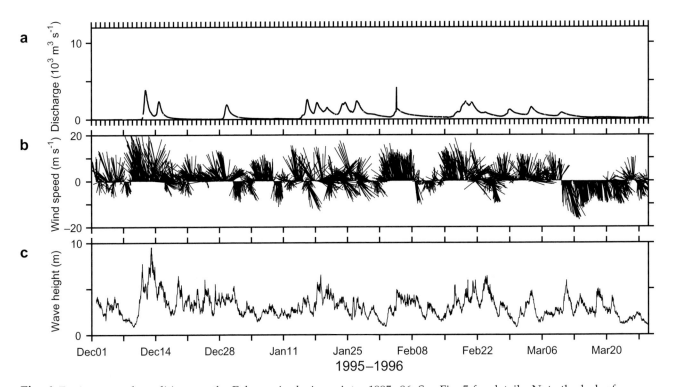

Fig. 6 Environmental conditions on the Eel margin during winter 1995–96. See Fig. 5 for details. Note the lack of significant discharge events during this flood season.

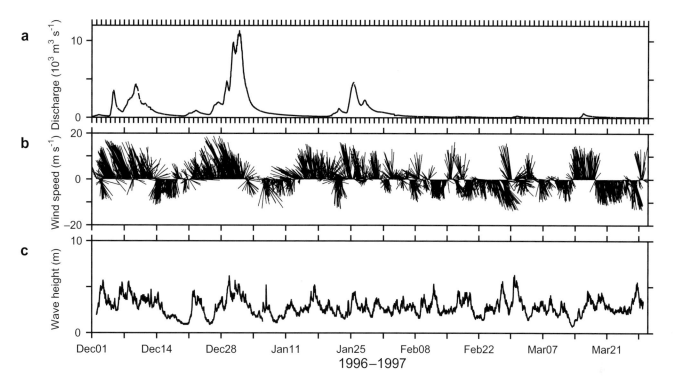

Fig. 7 Environmental conditions on the Eel margin in winter 1996–97. See Fig. 5 for details. Data in the middle and bottom panels from December 1996 were measured by NOAA's National Data Buoy Center (NDBC) buoy 46030. These data were used when buoy 46022 was not functioning. Note the large discharge event in January 1997.

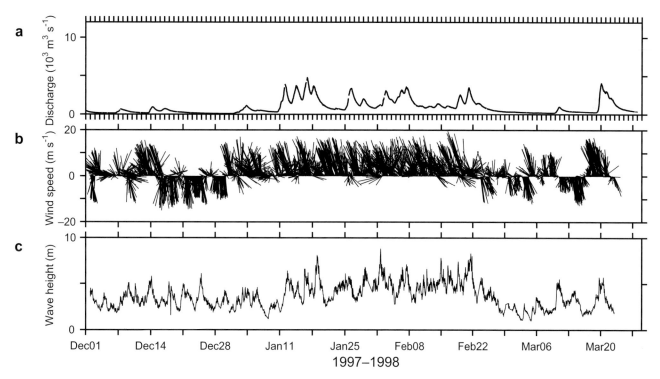

Fig. 8 Environmental conditions on the Eel margin during the winter of 1997–98. See Fig. 5 for details. Note the series of moderate discharge events starting in mid-January.

Description of the flood deposits

Floods of the Eel produce mud deposits that differ physically and chemically from the ambient shelf sediments (Fig. 9). In X-radiographs, sharp wavy contacts separate flood layers from underlying sediment. The flood layers tend to be relatively X-ray transparent and rich in physical structure, including laminations and cross-bedding (see also Nittrouer *et al.*, this volume, pp. 1–48; Wheatcroft *et al.*, this volume, pp. 101–155; Sommerfield *et al.*, this volume, pp. 157–212; Wheatcroft & Borgeld, 2000). In 1995, the flood layer possessed two to six alternating X-ray transparent and X-ray opaque layers.

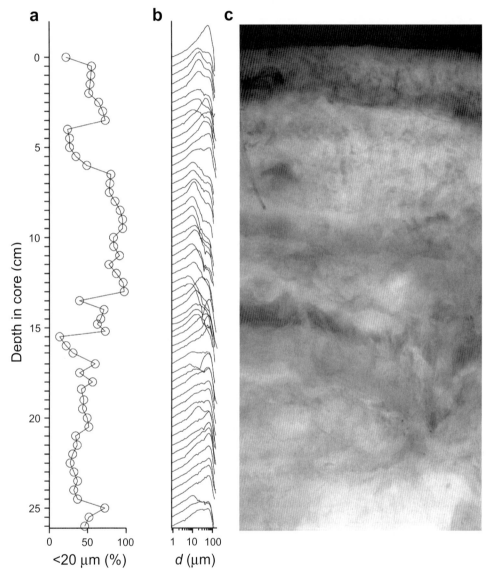

Fig. 9 Vertical distribution of grain size in an X-rayed slab from a box core collected at site S60 in 1997 (also see Nittrouer *et al.*, this volume, pp. 1–48; Wheatcroft *et al.*, this volume, pp. 101–155; Sommerfield *et al.*, this volume, pp. 157–212). (a) Disaggregated inorganic grain-size distributions plotted as a percentage volume < 20 µm equivalent spherical diameter. (b) Fully disaggregated inorganic grain-size distributions are plotted. The data are relative volume versus equivalent spherical diameter plotted on logarithmic axes. Individual sample plots are displaced by amounts proportional to depths in core, which are shown in centimetres along the vertical axis. (c) X-ray negative of the slab. Bright areas correlate generally with coarser sediment. The 1997 flood layer appears at the top of the core (top ~3 cm), and the January 1995 flood layer sits between 7 and 13 cm depth in core.

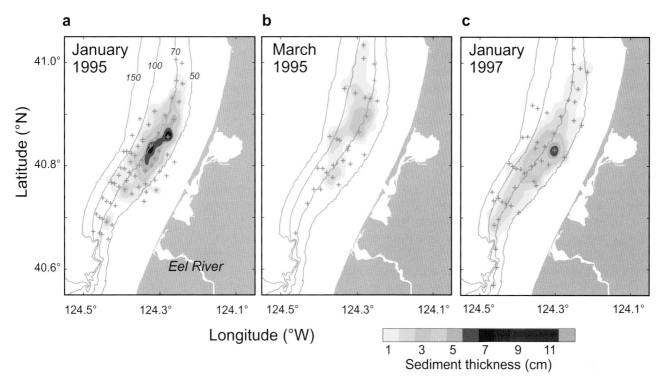

Fig. 10 Isopach maps from three flood deposits: (a) January 1995; (b) March 1995; (c) January 1997. Station locations are indicated. (Redrawn from Wheatcroft & Borgeld, 2000.)

Laminations appeared at some sites in the flood layer produced in January 1997, but at other sites the layer was massive and X-ray transparent. The cores with massively bedded flood layers were located nearer to the river mouth than the cores with laminated flood layers. The spatial differences in bedding within the 1997 flood layer presumably reflect differences in near-bed depositional dynamics (Wheatcroft & Borgeld, 2000).

The differences in X-ray density are tied closely to changes in grain size (Fig. 9). Sediment delivered to the mid-shelf region by floods tends to be finer than the ambient shelf sediment. The flood layers have > 90% of their mass in particles smaller than 20 μm, while ambient sediment contains < 50% mass in the < 20-μm fraction (Drake, 1999). In the January 1995 and January 1997 flood layers, [7]Be was detected uniformly throughout the layers in water depths greater than 50 m. Below the layers, no [7]Be was detected (Sommerfield *et al.*, 1999). The presence of [7]Be in the layers attests to their rapid emplacement and terrestrial source. Flood layers tended to have higher carbon to nitrogen ratios and more negative $\delta^{13}C$ values than ambient sediment,

again indicating a terrestrial source (Leithold & Hope, 1999).

The various distinct physical and chemical signatures of the flood layers made it possible to identify them and map their spatial extent (Fig. 10). Interestingly, the large and moderate floods of 1995 generated distinct flood layers, as did the large flood of 1997, but the series of moderate floods in 1998 produced none (Wheatcroft & Borgeld, 2000). The floods of 1998 poured large volumes of water and sediment into the coastal ocean over the course of the flood season, but none of the events was particularly large. Furthermore, 1998 experienced a generally more energetic wave climate than other years (Fig. 8), suggesting that sediment dispersal and reworking by waves made flood layers indistinguishable from surrounding shelf sediments.

The areal distributions of the flood layers formed during the January and March 1995 and January 1997 floods were ellipsoidal (Wheatcroft & Borgeld, 2000; Fig. 10). Their major axes extended along shelf and were approximately 35–50 km long. Minor axes were 10 km wide and oriented across shelf. Deposits thinned away from central loci of maximal thickness.

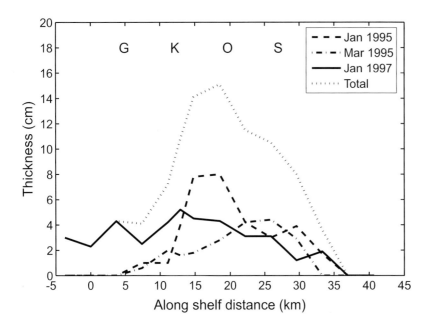

Fig. 11 Flood-layer thickness versus along-shelf distance from the river mouth for three flood deposits. The letters at the top of the panel denote positions of cross-shelf transects (Fig. 2). (Redrawn from Wheatcroft & Borgeld, 2000.)

The centres of mass of the various flood deposits were centred on the 70-m isobath, 15 km north-east of the river mouth. Recognizable flood layers were found in water depths as shallow as 50 m and as deep as 110 m (Fig. 10). In terms of maximal thickness, the layer associated with the January 1995 flood was thickest (8 cm), followed by the January 1997 and March 1995 layers (~5 cm) (Fig. 11). This thickness ranking differs from a ranking based on integrated flood-sediment discharge, for which the 1997 flood ranks first and the January 1995 event ranks second. This reversed ranking emphasizes the importance of both sediment delivery and sediment dispersal in determining the thickness of mid-shelf flood deposits (Wheatcroft & Borgeld, 2000).

The extensive coring of flood layers combined with their distinctiveness made it possible to estimate the total mass of sediment in each layer. Wheatcroft & Borgeld (2000) applied moving-average least-squares and inverse-distance algorithms to calculate the thickness and the volume of the flood layers. Then, using an average porosity based on resistivity measurements and an assumed quartz density for the sediments, they estimated the total mass of each layer. The January 1995 layer holds 6.2×10^9 kg of sediment, the March 1995 layer holds 2.5×10^9 kg and the 1997 layer contains 6.7×10^9 kg of sediment. The larger mass of the 1997 layer is consistent with its ranking as a larger discharge event.

The mid-shelf flood layers account for a relatively small fraction of the total sediment amount delivered to the coastal ocean by the river during floods. The Eel has a long record of measurements of suspended-sediment concentration as a function of water discharge (Brown & Ritter, 1971; Fig. 12). These data have been used to link concentration to discharge mathematically, either through empirical relationships (Wheatcroft *et al.*, 1997) or through process-based mechanistic models that address the stochastic behaviour of hydrological systems (Syvitski *et al.*, 2000; Morehead *et al.*, 2003). The empirical approach indicates that sediment discharge on the Eel, which is the product of water discharge and suspended-sediment concentration, varies approximately with the square of water discharge, thus highlighting the importance of floods to sediment accumulation on the Eel margin.

Wheatcroft & Borgeld (2000) used an empirical relationship between discharge and suspended-sediment concentration and the record of discharge for each flood event (Figs 5, 7 & 12) to estimate the total mass of sediment discharged during the course of the event. Estimates for the mass of sediment delivered to the ocean during the 17-day January 1995 event range from ~22 to ~29 × 10^9 kg. Estimates for the March 1995 flood range from ~10 to ~15 × 10^9 kg, and for the 1997 event from ~29 to ~45 × 10^9 kg. These predictions are imprecise, in large part because natural variability produces

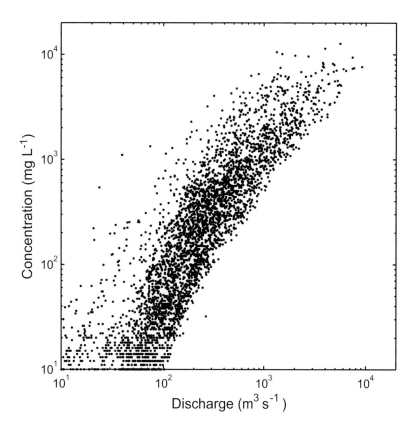

Fig. 12 Suspended-sediment concentration versus river discharge for the Eel River.

a wide range of possible concentrations for a given discharge, and because data during large discharges are scant.

The percentages of the total flood sediment contained within the flood layers average about 25% (Wheatcroft & Borgeld, 2000). The ranges for each flood are 22–31% for January 1995, 17–24% for March 1995 and 15–30% for January 1997. These percentages indicate that the Eel margin is dispersive, retaining only a fraction of the total sediment in mid-shelf, muddy flood deposits. The remainder of the flood sediment must be either stored somewhere on the inner shelf, transported off-shelf, or carried along-shelf beyond the study area.

Plume hydrography

During floods, the Eel plume was observed to flow northward, as expected (Geyer *et al.*, 2000; Fig. 13). Near-surface speeds averaged 0.5 m s^{-1} during periods of elevated discharge. During low discharge, along-shelf flow to the north was weaker, averaging only 0.1 m s^{-1}. Plume speeds as high as 1.3 m s^{-1} were associated with the January 1997

flood. In 1998, maximum plume speeds of up to 0.8 m s^{-1} were observed. Plume speeds often fell markedly during the waning stages of flood events. For example, on 3 January 1997, at the end of the flood, plume speeds fell to 0.2 m s^{-1}.

Flood plumes typically did not extend beyond the 40-m isobath in a seaward direction. When the speed of the plume was large, salinity at the 40-m isobath on the K line was similar to seawater. When plumes slowed at the end of some high-discharge events, salinity decreased at K40 to below 20. The spread of low-salinity plume water to the 40-m isobath was not accompanied by an increase in sediment concentration, however. These observations of plume velocity and extent led Geyer *et al.* (2000) to propose a division of plumes into 'fast and narrow', 'slow and wide' or 'rough' (Fig. 14).

Wind forcing played a dominant role in determining the velocity and cross-shelf extent of Eel flood plumes. Strong winds from the south accompanied precipitation in the Eel basin, and these winds exerted northward-directed stress on the sea surface, thus contributing to the large, northward plume velocities during floods. When winds

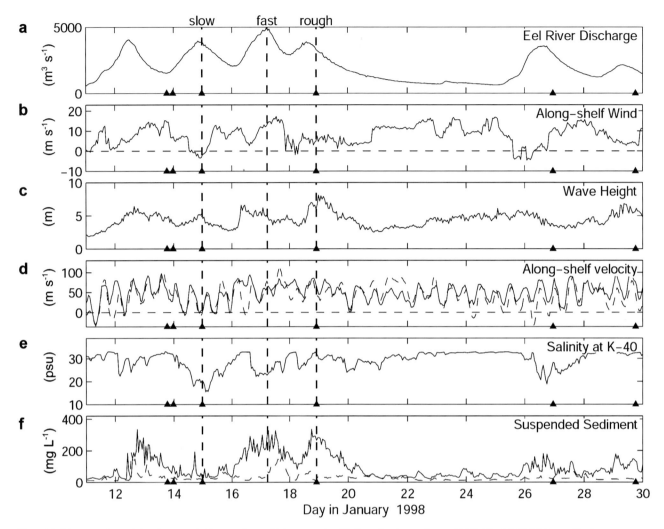

Fig. 13 Time series of forcing variables and conditions at the K line during winter 1997–98. (a) Eel River discharge. (b & c) Along-shelf wind speed (+ northward) and wave height at buoy 46022. (d) Plume velocity along-shelf (+ northward) 2 m below the surface at K20 (solid line) and K40 (dashed line). (e) Salinity at K40 0.5 m below the surface. (f) Estimated suspended-sediment concentration 0.5 m below the surface at K20 (solid line) and at K40 (dashed line). The lines labelled 'slow', 'fast' and 'rough' mark the times represented by plume cross-sections plotted in Fig. 14. Times of helicopter surveys are marked by triangles. (Redrawn from Geyer *et al.*, 2000.)

blew from the north at the close of some events, the sign of the wind stress on the sea surface changed, thereby slowing but not halting the northward flow of the plume. Based on observations of winds at NOAA buoy 46022 and plume velocities at the moorings on the K line in 1998, Geyer *et al.* (2000) proposed that along-shelf currents had a response of 1 m s⁻¹ Pa⁻¹ of along-shelf wind stress. During winter 1998 the average northward wind stress when Eel discharge exceeded 800 m³ s⁻¹ was 0.15 Pa, yielding an average of 0.15 m s⁻¹ of wind-induced, along-shelf flow at the sea surface. During periods

of low discharge, wind direction was variable and wind speed was less than during periods of high discharge. As a result, average wind stress was not significantly different from zero, so winds during these times exerted no net effect on along-shelf transport.

The cross-shelf extent of the plume also was affected strongly by the wind (Geyer *et al.*, 2000; Fig. 14). Winds from the south produced a landward flow of surface waters due to Ekman transport. The landward flow caused a build up of surface waters against the coast, which deepened

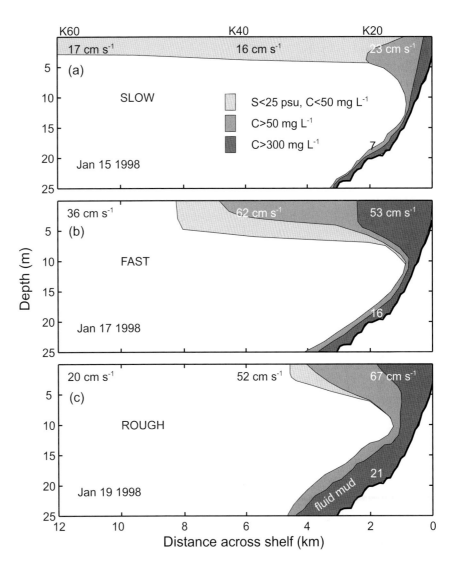

Fig. 14 Cross-sections of suspended-sediment and salinity structure for the plume at the K line during the three times indicated in Fig. 13. The light shading shows where the low-salinity plume has less than 50 mg L^{-1} suspended-sediment concentration. Darker shading corresponds to higher concentrations of suspended sediment. Numbers represent along-shelf plume speeds (cm s^{-1}) 2 m below the surface or speed in the bottom boundary layer. (a) A plume that moves slowly and spreads offshore under the influence of northerly winds. (b) Plume structure when downwelling-favourable winds blowing from south to north pushed surface waters along-margin and constrained the plume to shallower water. (c) Rough plumes occur when large waves resuspend sediment lost from the plume, forming dense near-bed fluid-mud suspensions. Note that fluid-mud layers occupy only the lower ~10 cm of the dark-shaded region in panel c. (Redrawn from Geyer *et al.*, 2000.)

the pycnocline there. The build up produced geostrophic flow that drove surface water along margin to the north. The aggregate effect of these processes was to keep the surface plume confined to relatively shallow water, inshore of the 40-m isobath, during periods when the wind blew strongly from the south. Fast plumes thus were narrow as well. Winds blowing from the north generated seaward flow in surface waters and shoaling of the pycnocline at the coast. Under these conditions, the plume thinned as it spread seaward. Weak wind forcing from the north associated with the trailing edges of low-pressure systems thereby produced slow, wide plumes.

Tides introduced significant variability into plume velocities. When diurnal and semi-diurnal tides were in phase, variations in along-shelf velocity of up to 0.5 m s^{-1} were observed (Geyer *et al.*, 2000; Fig. 13). This variability equalled the along-shelf average plume speed in 1998. Tidal flow in and out of Humboldt Bay probably also caused variability in along-shelf flow. On an ebb tide, flow out of Humboldt Bay introduced an offshore-directed source of momentum into the coastal ocean that was equivalent to the momentum associated with the Eel River when it discharges 10^4 m^3 s^{-1} (Geyer *et al.*, 2000). This large momentum source potentially could have distorted and slowed the plume, as well as provided energy for plume mixing and sediment resuspension. Unfortunately, the measurement programme was not designed in a way that allowed any systematic quantification for the effect of Humboldt Bay tidal exchange on plume structure and velocity.

Observations of plume density and velocity are consistent with the hypothesis that the Eel's dynamics are governed by inertia at the river mouth (Geyer *et al.*, 2000). Using mouth geometry, discharge and salinity during floods, Geyer *et al.* (2000) calculated that Froude numbers at the mouth exceeded unity. Using measured velocities at the K line and estimated mouth velocities to constrain the size of the anticyclonic bulge at the mouth, they calculated that the inertial radius for the Eel in 1998 was ~10 km. Inertia could have pushed the plume seaward to approximately the 50-m isobath, and it could have influenced plume dynamics to the K line. Mouth geometry, however, affected the extent of inertial influence. The river mouth directed flow northward into the coastal ocean at an angle of 20° from the coast. Such an angle of entry probably reduced the seaward extent of the inertial bulge and increased the along-shelf distance over which inertial effects influenced the plume (Garvine, 1987; Geyer *et al.*, 2000).

Theoretically, a buoyant river plume evolves along-shelf into a coastal current with a cross-shelf width determined by a geostrophic force balance. Buoyant water near the coast spreads seaward over more saline, denser basin water. As it spreads, it is deflected to the right in the northern hemisphere by the effect of Earth's rotation. In the limit of a current flowing along a vertical wall and therefore not experiencing any frictional drag from the seabed, the width, W_p (m), of a coastal current in semi-geostrophic balance is defined by (Lentz & Helfrich, 2002)

$$W_p = \frac{(g'h_p)^{1/2}}{f} \tag{4}$$

where, as before (Eqs 2 & 3), g' and f are modified gravity and the Coriolis frequency, respectively, and h_p is plume thickness (m). Assuming that the plume density during floods was 1012 kg m^{-3} and that the density of basin water was 1023 kg m^{-3} (Geyer *et al.*, 2000), modified gravity equalled 0.11. Assuming that a plume thickness of 5 m was typical during floods (Geyer *et al.*, 2000) yields a plume width of ~7 km. Given the regional bathymetric gradient of 0.4°, this width suggests that the plume remained inshore of the 50-m isobath

in 1998 due to the effects of Earth's rotation on the northward-flowing coastal current.

The final factor that can affect plume width and speed is bottom friction. Lentz & Helfrich (2002) proposed a complete expression for plume width that includes the effect of bottom friction:

$$W_p = \frac{(g'h_p)^{1/2}}{f}\left[1 + \frac{(2Qg'f)^{1/4}}{\alpha g'/f}\right] \tag{5}$$

Assuming a discharge of 5000 m^3 s^{-1} and a bottom gradient α of 0.007 m m^{-1} yields a value for the second term inside the brackets of ~0.1. This result indicates that during floods the plume was perhaps only 10% wider than it would have been in the absence of bottom friction. Given the idealizing assumptions that underlie Eqs 4 and 5, this difference is not significant.

In summary, a variety of forces acted to confine the Eel plume landward of the 40-m isobath during floods. The flood deposits were found seaward of the 40-m isobath, so subplume transport in the bottom boundary layer must have transported flood sediment seaward. Understanding of the cross-shelf position of the flood deposits therefore requires an understanding of boundary-layer transport; understanding of the along-shelf position of the flood deposits requires an understanding of the rate at which sediment sank out of the northward-advecting surface plumes during floods.

Sediment loss from the plume

Helicopter-based observations of sediment in the Eel plume indicate that sediment removal occurred more rapidly than by single-grain sinking alone, yet not as rapidly as has been observed in other, less energetic systems (Hill *et al.*, 2000; Curran *et al.*, 2002a). In general, 40–75% of the sediment delivered to the ocean by the river during floods sank from the plume between the river mouth and the K line 10 km to the north (Geyer *et al.*, 2000; Hill *et al.*, 2000). Therefore, sediment sank rapidly enough from the plume to account for the position and mass of flood deposits.

Helicopter- and mooring-based observations revealed that sediment sank from surface waters inshore of the 40-m isobath (Geyer *et al.*, 2000; Hill *et al.*, 2000). When winds blew strongly from the south, the brackish waters of the plume did not extend beyond the 40-m isobath at the surface, so naturally sediment from the river did not extend beyond this either. When winds shifted to northerly at the end of some storms, the salinity signature of river water did appear as far offshore as the 40-m isobath, yet sediment was not associated with the lower salinity water (Geyer *et al.*, 2000; Figs 13 & 14). Sediment did not make it far seaward during northerly winds because the plume slowed and thinned under such forcing. Slower plumes took longer to reach the K line, and thinner plumes reduced the residence time of particles within them. These factors combined to cause removal of sediment from the plume landward of the 40-m isobath, even when upwelling-favourable winds allowed the plume to spread farther offshore (Geyer *et al.*, 2000).

Sediment clearance rates from the Eel plume exceeded single-grain clearance rates, yet they were not as large as observed in other environments. Clearance rates can be parameterized with an **effective settling velocity**. The effective settling velocity is the term w_e (m s^{-1}) in the equation

$$C_s(x) = C_s(0)\exp\left(-\frac{w_e}{h_p u}x\right) \qquad (6)$$

which is a representation of the one-dimensional spatial decay of sediment concentration with distance from the river mouth. The term $C_s(x)$ (kg m^{-3}) is sediment mass concentration at along-shelf position x (m), and $C_s(0)$ is sediment mass concentration at the river mouth. The terms h_p (m) and u (m s^{-1}), as before, represent plume thickness and velocity, respectively (Hill *et al.*, 2000).

A regression of the logarithm of $C_s(x)/C_s(0)$ on the logarithm of x from various helicopter surveys in 1998 provided estimates of $w_e/h_p u$. By inserting values for h_p and u into this estimate, effective settling velocities were calculated. For five flood events in 1998, values for w_e ranged from 0.06 to 0.1 mm s^{-1} (Hill *et al.*, 2000). These values exceed the value of 0.04 mm s^{-1} that would have resulted if particles sank as single grains, but they are well below the ~1 mm s^{-1} settling velocity of individual aggregates and the 0.3–0.6 mm s^{-1} effective settling velocities observed elsewhere (Drake, 1972; Syvitski *et al.*, 1985).

The observed clearance rates probably exceeded single-grain clearance rates due to aggregation. Aggregates were observed in the plume on all surveys. Median aggregate equivalent circular diameter, averaged over all surveys, was 232 µm and varied from 125 µm to as high as 405 µm (Curran *et al.*, 2002a; Fig. 15). Large aggregate size was described with the parameter d_{25}, which is the equivalent circular diameter at the lower boundary of the upper quartile of particle areas in an image (Curran *et al.*, 2002a). The mean value of d_{25} was 280 µm among all surveys. Although aggregate settling velocities were not measured directly, Sternberg *et al.* (1999) did measure them in the bottom boundary layer on the Eel shelf (Fig. 16). Using the relationship from that study for aggregates in the plume yields a settling velocity of 1 mm s^{-1} for the median aggregate diameter, which is a value consistent with other observations of aggregate settling velocity in a wide range of environments (Hill *et al.*, 1998, 2000; Curran *et al.*, 2002a). Interestingly, aggregate size and inferred aggregate settling velocity showed no systematic variation with discharge, concentration, winds, waves, turbulent-kinetic-energy dissipation rate, or time from river mouth (Hill *et al.*, 2000; Curran *et al.*, 2002a; Fig. 15).

The observation that values for effective settling velocity fell between single-grain values and values for individual aggregates arguably indicated that sediment in the plumes was partially aggregated. Hill *et al.* (2000) calculated that observed clearance rates could be reproduced by packaging 75% of the sediment mass at the river mouth in aggregates, and assigning the remaining 25% to single grains. This calculation assumed that dilution of the plume with seawater at the river mouth effectively stopped any subsequent aggregation down-current of the river mouth, by reducing the sediment concentration to the point where interparticle collisions became too rare to affect particle packaging significantly. This extent of particle packaging in suspension is consistent with other studies (Syvitski *et al.*, 1995; Dyer & Manning, 1999).

Curran *et al.* (2002a) examined the above hypothesis regarding particle packaging in two ways. First, they examined along-shelf evolution of the **aggregate fraction**, defined as the proportion of the total suspended mass that is packaged within

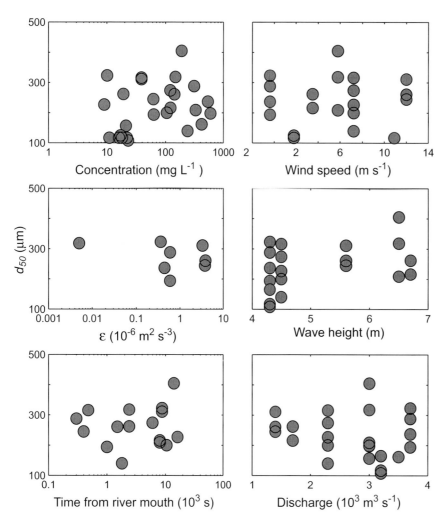

Fig. 15 Aggregate size in the Eel plume versus various environmental variables. Aggregate size is represented by d_{50}, which is the equivalent circular diameter of the median aggregate area in an image. Variable ε is turbulent-kinetic-energy dissipation rate at the depth of an image. Systematic variation is not observed. (Redrawn from Curran *et al.*, 2002a.)

aggregates. They calculated this fraction by estimating the mass concentration of aggregates in plume photographs, and dividing it by the total suspended-sediment mass in suspension. To estimate aggregate mass concentration from photographs, they used aggregate settling velocity versus size data from the Eel shelf (Sternberg *et al.*, 1999) and Stokes Law to estimate aggregate mass as a function of diameter (Fig. 16). This relationship was applied to each aggregate in an image to generate an estimate of total aggregate mass concentration.

If the plume was partially aggregated and too dilute to allow any subsequent aggregation downcurrent of the river mouth, then the aggregate fraction would have decreased along-shelf because aggregates sink faster than the single grains found in the Eel plume (Curran *et al.*, 2002a). Along-shelf

evolution of the aggregate fraction, however, was not observed, thus refuting the Hill *et al.* (2000) hypothesis (Fig. 17). Like aggregate size, aggregate fraction showed no dependence on sediment concentration, wave height, river discharge, winds, turbulent-kinetic-energy dissipation rate, or time from the river mouth. The variability of aggregate properties across a wide range of environmental conditions indicated that some other factor determined aggregate size and the aggregate fraction in the Eel plume.

The second method used by Curran *et al.* (2002a) to examine the Hill *et al.* (2000) hypothesis regarding packaging was analysis of the differential sedimentation of individual grain sizes. In a fully aggregated suspension, all particle sizes should be removed from the plume at the same rate.

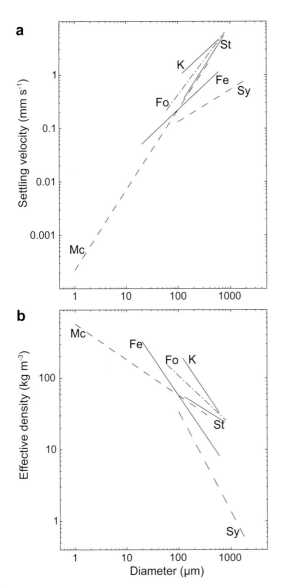

Fig. 16 Relationships between aggregate properties and aggregate size from six different studies: Mc (McCave, 1975); K (Kranck *et al.*, 1992); Fe (Fennessy *et al.*, 1994); Sy (Syvitski *et al.*, 1995); St (Sternberg *et al.*, 1999); Fo (Fox *et al.*, 2004). (a) Settling velocity versus size. (b) Estimated effective density versus size. (Redrawn from Fox, 2003.)

In a disaggregated suspension, effective settling velocity would scale with the square of particle diameter. For partially aggregated suspensions, size-dependent effective settling velocities would vary in a way that is predictable and bracketed by the end-member values for the disaggregated and fully aggregated cases. Curran *et al.* (2002a) found that size-dependent effective settling velocities

did not conform to a relationship derived under the assumptions that the aggregate fraction at the river mouth was 0.75 and no further aggregation took place beyond the river mouth. Instead, the size-dependent values of w_e indicated that aggregate fractions in the plume were higher.

Curran *et al.* (2002a) reconciled the observations of high and non-evolving aggregate fractions with low removal rates by invoking sediment resuspension in the surf zone. Helicopter observations demonstrated that sediment did not enter the plume by resuspension in water depths of 20 m or greater, but the same could not be said of shallower waters. During surveys, wave breaking occurred out to water depths as great as 15 m. During floods, the surf zone was grey with suspended mud. Breaking waves offer substantial energy for resuspension. Therefore, Curran *et al.* (2002a) argued that sediment that sank to the seabed in waters shallower than 15 m was re-entrained into the plume. This resuspension reduced effective settling velocities. Furthermore, horizontal diffusion of sediment-laden surface waters supplied the plume seaward of the 15-m isobath with aggregated sediment that had been resuspended in the surf zone.

In summary, sediment separation from the plume in the cross-shelf direction occurred between approximately 15 m and 40 m water depths. Inshore of the 15-m isobath, wave breaking re-entrained sediment into the plume. Sediment did not sink from the plume seaward of the 40-m isobath primarily because inertial dynamics, Ekman transport and geostrophy all conspired to limit the spread of the plume to no more than 10 km from shore.

The along-shelf position of sediment loss agrees well with the along-shelf location of the flood deposits. The **e-folding distance** for sediment loss is the distance over which sediment concentration in the plume falls to $1/e$ of its initial value (Hill *et al.*, 2000). It is calculated for the Eel plume with the equation

$$x_e = \frac{h_p u}{w_e} \tag{7}$$

Assuming, as before, that plume thickness was approximately 5 m, plume velocity was 0.5 m s^{-1} and that effective settling velocity was 10^{-4} m s^{-1}, the e-folding distance of sediment in the plume

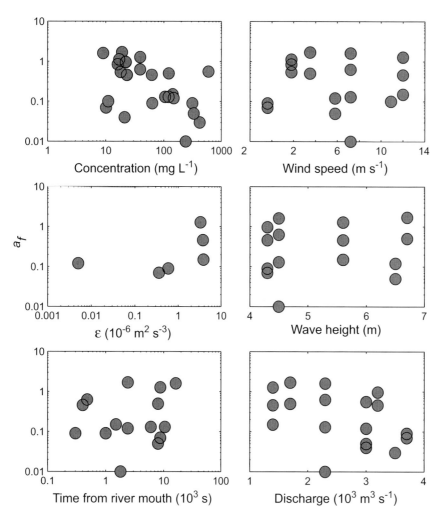

Fig. 17 Aggregate fraction (a_f) in the Eel plume versus various environmental variables. Variable ε is turbulent-kinetic-energy dissipation rate at the depth of an image. Systematic variation is not observed. Fractions exceed unity in some cases, indicating that the drag law used to calculated aggregate mass overestimates drag for permeable aggregates. (Redrawn from Curran *et al.*, 2002a.)

was approximately 25 km. This distance exceeds somewhat the 15-km along-shelf position of the centres of mass of the flood layers, but this is to be expected because the e-folding distance describes the distance required for almost two-thirds of the sediment to fall from the plume. In general, this calculation indicates that sediment did not move much farther along-shelf after leaving the surface plume.

This relatively crude calculation for the along-shelf loss of sediment agrees with calculations for along-shelf distribution of sediment loss from the plume that were based on moored velocity measurements at K20 (Geyer *et al.*, 2000; Fig. 18). These authors modelled advection of sediment away from the river mouth for 10 days (10–20 January 1998) during a period of elevated discharge (Fig. 18). Along-shelf velocity as a function of depth below

the sea surface was based on ADCP measurements of water-column velocity profiles. The advective flux was assumed to be distributed evenly between the shore and the 40-m isobath. In one set of calculations, sediment was assigned a settling velocity of 1 mm s^{-1}, typical of aggregates. In another set, sediment was assigned a settling velocity of 0.1 mm s^{-1}, approximately equal to the observed effective settling velocity. In the model runs with settling velocity equal to 1 mm s^{-1}, sediment arrived at the seabed close to the river mouth. The position of maximum deposition was < 5 km from the mouth (Fig. 18). In the calculations with settling velocity equal to 0.1 mm s^{-1}, sediment arrived at the seabed < 80 km from the river mouth, with maximum flux at ~40 km (Fig. 18). The flood deposits, interestingly, extended from ~5 km to 40 km from the river mouth (Figs 10 & 11). Plume velocity, plume

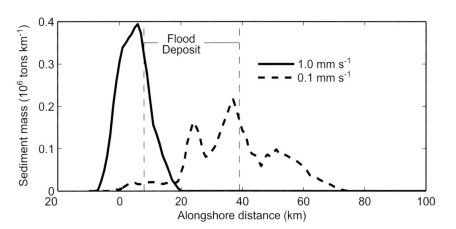

Fig. 18 Model estimates for along-shelf distribution of sediment leaving the Eel plume between 10 January and 20 January 1998. Calculations were carried out with two settling velocities: 1 mm s^{-1} (solid line) and 0.1 mm s^{-1} (dashed line). The peaks in the 0.1 mm s^{-1} curve record pulses in river discharge. (Redrawn from Geyer *et al.*, 2000.)

thickness and aggregation-influenced settling velocities therefore combined to deliver sediment to the seabed at an along-shelf position similar to the flood deposits. The subsequent bottom-boundary-layer processes that transported sediment from the nearshore region beneath the plume to the mid-shelf, where the deposits were found, probably did not produce substantial along-shelf advection of sediment.

Several processes that potentially affect plume transport of sediment were not well documented in the field. As already discussed, resuspension of plume-derived sediment in the surf zone probably retarded the removal rate of sediment from the plume. Data to support or refute this hypothesis directly were not available, however. The surf zone on the Eel margin during winter storms is wide and extremely energetic, making measurements of suspended-sediment concentration and vertical distribution nearly impossible. Tidal exchange with Humboldt Bay was not well documented either. As noted previously, enormous quantities of water flow in and out of Humboldt Bay with the tide. Given the position of the flood deposits directly off-shore of the Bay mouth, it is possible that deceleration and deflection of the plume by waters flowing out of the Bay affected the fate of plume sediment during the floods. Unfortunately, measurements were not available to explore the effect of the Bay in any systematic fashion. Finally, laboratory experiments indicate that **convection** can speed the removal of sediment from hypopycnal plumes. In short, sediment concentration can build at the interface between the sediment-laden, relatively fresh fluid on the surface and the salty basin waters below. The build-up occurs because the density

gradient slows sediment sinking. If concentration grows large enough, then convection ensues, leaking sediment into the lower layer (Parsons *et al.*, 2001). This intriguing mechanism may speed the loss of fine sediment from river plumes, but observations did not resolve it in the field.

Transport in the bottom boundary layer

Boundary-layer measurements of velocity and sediment concentration collected with vertical arrays of current meters and optical backscatter sensors supported the conceptual model of transport proposed many years ago by McCave (1972). This view states that sediment is maintained in the bottom boundary layers by waves and that the sediment is advected by near-bottom currents.

Observations from the tripod at S60 clearly demonstrated the importance of waves (Ogston *et al.*, 2000; Fig. 19). Whenever concentration 30 cmab was elevated, wave orbital velocities were also large. A correlation between sediment concentration and current speed was not evident (Fig. 19). Waves, therefore, were the primary supplier of suspended sediment to the bottom boundary layer.

As on other margins, the importance of waves for supplying sediment to the bottom boundary layer gave storms overriding importance in determining sediment transport rates and pathways. The importance of storms was amplified by the common co-occurrence of elevated discharge and large waves (Figs 5–8). For example, in 1995–96, virtually all transport occurred during a 20-day period in winter. Almost three-quarters of the net along-shelf transport took place during just three storms (Ogston & Sternberg, 1999).

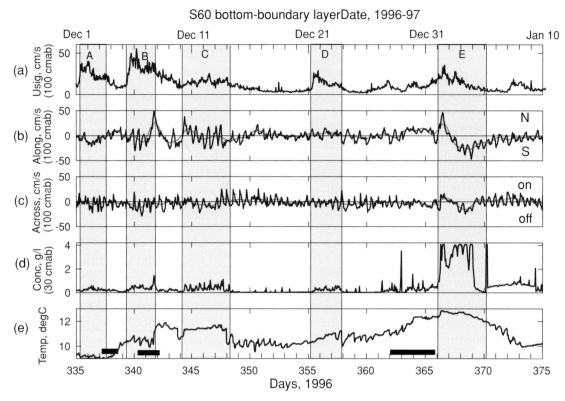

Fig. 19 Time-series, bottom-boundary-layer data from a tripod at location S60. The hourly averaged data shown are: (a) significant wave orbital velocity; (b) mean along-shelf currents, with the blue line depicting mean current (positive to the north and along bathymetry) and the cyan line showing the low-frequency component; (c) mean across-shelf currents, with the blue line representing mean current (positive landward across bathymetry) and the cyan line representing the low-frequency component; (d) suspended-sediment concentration; (e) temperature. All instruments were located 100 cmab except for the suspended-sediment sensor (d), which was 30 cmab. Pale yellow bands (A, B, C, D, E) bracket periods of elevated suspended-sediment concentration. Black bars indicate periods of downwelling-favourable winds. (Redrawn from Ogston *et al.*, 2000.)

Waves alone generate little net transport because they are oscillatory; bottom currents fill that role. Bottom currents on the Eel followed a regular pattern during storms. At the beginning of storms, currents near the seabed in the vicinity of the flood deposit flowed north. After passage of the lows, along-shelf currents switched direction and flowed to the south (Cacchione *et al.*, 1999; Wright *et al.*, 1999; Fig. 19). This pattern of current reversal inhibited along-shelf dispersal during storms, helping to explain why the along-shelf positions of sediment loss from the plume and the flood layers were similar.

Flow convergence in the bottom boundary layer helps to explain the cross-shelf location of the flood deposits. In winter 1995–96, mean cross-shelf current velocity at S60 was seaward at 2.5 cm s^{-1}, but at S70 it was landward at 0.5 cm s^{-1} (Wright *et al.*, 1999). This near-bed convergence of flow presumably produced a net cross-shelf influx of sediment to the boundary layer above the deposit, which in turn increased deposition rates there. Similarly, Ogston & Sternberg (1999) demonstrated that along-shelf velocity and suspended-sediment concentration in 1996–97 were correlated at low frequency, and they argued that large-scale oceanographic forcing contributes to accumulation of sediment in the flood deposits.

In summary, early tripod observations supported the generally accepted view of boundary-layer transport on wave-dominated margins. The comfortable fit between these observations and the accepted paradigm of shelf sediment transport was shaken profoundly by tripod observations made in the winter of 1997–98.

In winter 1997–98, the tripods at K60 and K20 collected data indicating that thin layers with high sediment concentrations occasionally appeared just above the seabed (Fig. 20). Furthermore, these layers apparently accounted for significant sea-

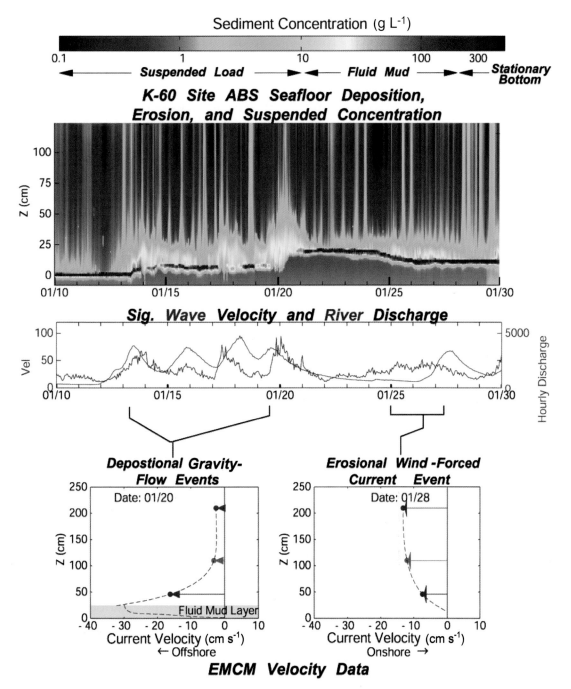

Fig. 20 Data from a tripod at the K60 site that show an acoustic-backscatter record of deposition associated with gravity-flow events that occur during periods of high energy with sediment input from the Eel River. Velocity profiles clearly show the difference between gravity-flow events and events forced by mean currents. (From Traykovski *et al.*, 2000.)

ward transport as well as observed rapid changes in seabed elevation (Traykovski *et al.*, 2000). Close scrutiny of these occasional events at the K line in 1997–98, and at K63 and S60 in 1996–97 (Ogston *et al.*, 2000), led to the surprising conclusion that the near-bed layers were actually wave-supported

fluid muds that flowed across-shelf under the influence of gravity.

The first observation that failed to conform to the conventional shelf sediment-transport paradigm was a rapid drop in the intensity of acoustic backscatter from an upward-looking ADCP deployed

50 cmab at K20 (Traykovksi *et al.*, 2000). After the intensity drop, the instrument continued to collect data that looked reasonable. It appeared simply as if outgoing and returning acoustic pulses suffered significantly greater attenuation after the drop in acoustic intensity than before. This curious result led Traykovski *et al.* (2000) to consider the possibility that the instrument had been buried by fine sediment. Assuming relatively high porosities consistent with recently deposited muds, Traykovski *et al.* (2000) estimated that burial of the tripod by 2 m of mud was consistent with the observed decrease in acoustic backscatter intensity.

The burial hypothesis received support from acoustic-backscatter observations at K60 (Traykovski *et al.*, 2000; Fig. 20). Acoustic-backscatter sensors transmit sound pulses toward the seabed, and the intensity of return as a function of time is used to construct vertical profiles for sediment concentration down to the sediment–water interface. During periods of high wave energy, layers appeared that were ~10–15 cm thick. Concentrations within these layers were so high that normally strong acoustic returns from the seabed were obscured, indicating sediment concentrations in excess of 10 kg m^{-3}. These values are large enough to qualify the layers as fluid muds, that is, they were dense enough to flow downslope under the influence of gravity. Above these layers, estimated sediment concentrations decreased abruptly with height to values ~0.1 kg m^{-3}. The steep concentration gradients at the top of the layers suggested that turbulence was suppressed by suspended-sediment stratification (Trowbridge & Kineke, 1994; Traykovski *et al.*, 2000).

Large changes in bed elevation accompanied the periods of elevated near-bed concentration (Traykovski *et al.*, 2000). During two events at K60 in 1998, the bed level increased by a total of 19 cm (Fig. 20). These depositional events occurred during periods of elevated wave orbital velocities, which was inconsistent with conventional transport models that predict net erosion when wave energy increases and net deposition when wave energy decays. Changes in bed height therefore also implicated alternative transport mechanisms in deposition of sediment on the Eel shelf.

The fingerprints of fluid muds were found at S60 and K63 as well (Ogston *et al.*, 2000). Following the January 1997 flood, an OBS located 30 cmab at S60 experienced sediment concentrations that

exceeded the maximum for which the instrument was designed (Fig. 19). The OBSs at 10 cmab and 23 cmab on the tripod at K63 also went off scale following the January 1997 flood. Ogston *et al.* (2000) estimated that near-bed sediment concentrations may have reached > 100 kg m^{-3} at K63, and were ~10 kg m^{-3} at S60. At K63, the elevated sediment concentrations led to a rapid 10–15 cm increase in bed elevation. Fluid muds, therefore, apparently formed near the seabed at several different times during the study period, and these times were associated with large waves and elevated river discharge.

Perhaps the most compelling evidence for fundamentally different transport dynamics during periods of elevated near-bed sediment concentration on the Eel shelf came from near-bed velocity profiles. When near-bed, high concentration layers were absent, velocity increased with distance above the seabed, as predicted by conventional wave-current boundary-layer theory (Fig. 20). When near-bed, high-concentration layers were present, however, velocity very close to the seabed was directed seaward, while higher in the water column velocities were smaller and at times even directed onshore (Traykovski *et al.*, 2000; Fig. 20). These inverted velocity profiles could not be explained with conventional boundary-layer theory. They could be explained by invoking gravity-driven seaward flow of sediment-laden layers (Ogston *et al.*, 2000; Traykovski *et al.*, 2000).

The final pieces of evidence implicating fluid muds in transport of sediment on the Eel shelf were Richardson numbers with near critical values. The **Richardson number**, *Ri*, is a dimensionless number equal to the ratio of stabilizing force of density stratification to the mixing force induced by velocity shear. In a sediment-stratified boundary layer, the Richardson number is defined by the equation

$$Ri = g' \frac{\partial C / \partial z}{(\partial u / \partial z)^2} \tag{8}$$

in which g' is modified gravity (m s^{-2}), C is sediment volume concentration (m^3 m^{-3}) and u is fluid velocity (m s^{-1}). When *Ri* is small, shear is sufficient to overcome the stabilizing effects of suspended-sediment stratification, so sediment diffuses upward under the influence of turbulent eddies. As *Ri* grows, eventually shear cannot overcome suspended-sediment stratification, and upward diffusion of suspended sediment is hindered because of suppressed turbulence. The critical value of the Richardson number at which the transition occurs is 0.25 (e.g. Trowbridge & Kineke, 1994; Friedrichs *et al.*, 2000; Wright *et al.*, 2001).

Friedrichs *et al.* (2000) estimated *Ri* by using simultaneous measurements of velocity and suspended-sediment concentration. Calculations were made for a range of heights above bottom at S60 from January to March 1996 and at G65 from November 1996 to January 1997. They observed that whenever waves and suspended-sediment concentrations were large, the Richardson number equalled approximately 0.25. When waves were small, *Ri* was less than 0.25 (Friedrichs *et al.*, 2000). Based on these results, the authors argued that large waves provided enough sediment to the near-bed region via resuspension that Richardson numbers reached critical values. As the flow stratified, sediment became trapped near the bed in the wave boundary layer, unable to diffuse upward due to the suppression of turbulence. Limited upward diffusion allowed sediment concentrations to increase enough to induce downslope transport as gravity flows.

The bottom-boundary-layer observational programme provided clear evidence that dense fluid-mud suspensions can form and flow downslope on an open continental shelf, even in the absence of convergent flow at strong density fronts. This evidence supported directly the mechanism for across-shelf sediment transport hypothesized decades earlier (Moore, 1969). The surprising findings led to two lines of further investigation. First, the quantitative importance of gravity-driven near-bed transport was explored. Next, the mechanisms of formation, flow and deposition of near-bed dense suspensions were investigated, with an eye toward answering the question of why, for so long, this mode of transport had been discounted as not viable on open continental shelves.

In winter 1996–97 cross-shelf flux of sediment was dominated by the near-bed high concentration events following the January 1997 flood (Ogston *et al.*, 2000; Fig. 21). At S60, sediment flux 30 cmab exceeded the average annual flux by two orders of magnitude. Net sediment transport during the 3-day flood equalled 75% of the total annual transport from the previous year. These calculations underestimated the total flux because they did not include the near-bed region where concentrations would have been highest but were not measured. Nonetheless, they clearly documented the dominance of high-concentration events in cross-shelf transport on the Eel margin (Ogston *et al.*, 2000).

Wright *et al.* (2001) indicated that the sediment concentration within a dense layer could be described adequately based on the critical Richardson number criterion (Parsons *et al.*, this volume, pp. 275–337). Using linear wave theory, the thickness of the layer and concentration within the layer were predicted as functions of water depth and wave properties. The gravitational downslope velocity was estimated by balancing frictional drag at the seabed against the gravitational force on the excess density of the high-concentration suspension. This balance was able to model the observed downslope velocities of 10–30 cm s^{-1} at K60 in January 1998 (Fig. 20).

To predict the cross-shelf flux for a 2-week period of elevated discharge in January 1998, sediment was input to the K line based on an analysis of plume sedimentation (Geyer *et al.*, 2000). Sediment lost from the plume was placed in the inner-shelf wave boundary layer. It was then allowed to flow downslope with velocities predicted by the frictional/gravitational balance. For 10–24 January 1998 the model transported 8×10^4 kg m^{-1} of sediment past K60 (Traykovski *et al.*, 2000). Transport of this magnitude accounted for a large fraction of the sediment delivered to the shelf by the river and exceeded transport by the more conventional pathway of advection by near-bed currents. These results argue strongly that Moore's mechanism of cross-shelf transport of sediment by density underflows is not only viable but is also the dominant pathway for seaward transport on the Eel shelf. The dominance of gravity-driven near-bed transport on the Eel shelf raises the question of why Moore's hypothesis remained dormant for so long. This question can be reduced to two pertinent questions that have long been considered by sedimentary geologists in attempting to explain tempestites in the geological record (e.g. Myrow & Southard, 1996).

1 How can sediment concentrations grow large enough to induce downslope, gravity-driven transport

P.S. Hill et al.

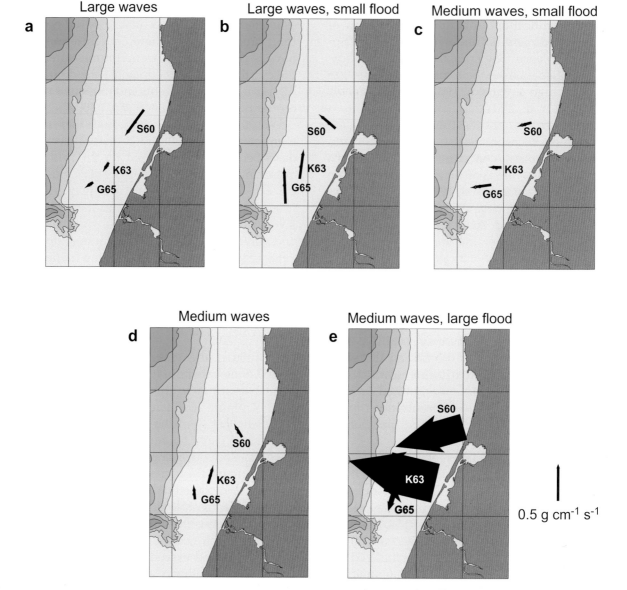

Fig. 21 Vectors of time-averaged, vertically integrated sediment flux in the lower 120 cm of the water column during five transport events (a–e) in 1996–97 (shaded areas in Fig. 19). Vectors are scaled in length and width by the magnitude of the flux. (Redrawn from Ogston *et al.*, 2000.)

in the highly dispersive environment of a storm-wracked open continental shelf?

2 How can gravity-driven flows persist on the low gradients typical of continental shelves?

Modelling of wave-supported fluid muds

To produce large near-bed sediment concentrations on open continental shelves, there must be a large supply of easily resuspended sediment. In the absence of strong density fronts, there also must be wave energy to resuspend that sediment. Under these conditions, especially in the absence of strong currents that cause sediment to diffuse out of the wave boundary layer, sediment concentration within the wave boundary layer can grow large enough to induce suspended-sediment stratification (Friedrichs *et al.*, 2000). Stratification retards upward turbulent diffusion of sediment, but it also reduces the transmission of turbulent, wave-induced stress to the seabed. As a result, suspended sediment starts to deposit. Deposition reduces stratification, increases stress and leads to resuspension. This negative feedback is responsible for keeping Richardson numbers near critical when waves are large and sediment is available (Friedrichs *et al.*, 2000). These conditions apparently occurred frequently on the Eel shelf due to the co-occurrence of precipitation, high discharge and large waves. Plume-derived sediment loaded the wave boundary layer either by direct sedimentation or by storm resuspension after temporary deposition on the seabed.

The earlier failure of continental shelf sedimentologists to recognize fluid muds in open shelf settings is attributable to the fact that the layers are thin, generally residing below the lower-most sediment sensors on tripods. Without data from downward-looking ABSs, the gravity-driven flows arguably would have been missed on the Eel shelf. Perhaps such flows exist elsewhere and simply have not been observed due to limits of instrumentation. Counter to this hypothesis is the argument that the Eel shelf is at present particularly well suited to producing gravity-driven flows, because of simultaneous production of large sediment discharge by a river and large waves by storms (Wheatcroft & Borgeld, 2000). High-resolution boundary-layer observations on other river-influenced shelves will help to resolve this issue.

Another issue that caused Moore's (1969) hypothesis to languish was concern over how gravity currents were maintained on continental shelves, which have small bathymetric gradients. A source of stress at the seabed is required to maintain the suspended sediment in a gravity-driven flow. A suspension flowing downslope under the influence of gravity generates stress on the seabed, and if the flow is fast enough, stress imparted to the seabed is large enough to maintain sediment in suspension (Bagnold, 1962). The small gradients typical of continental shelves cannot support this so-called **autosuspension** because bathymetric gradient affects the magnitude of downslope-directed gravitational force on a dense suspension.

Wright *et al.* (2001) demonstrated this argument quantitatively by making use of the critical Richardson number concept. They reformulated the Richardson number into a bulk parameter (Ri_B)

defined by the equation (Trowbridge & Kineke, 1994)

$$Ri_B = \frac{(B/\ell^2)}{(U_{max}/\ell)^2} \tag{9}$$

In Eq. 9, B is the vertically integrated buoyancy anomaly (m^2 s^{-2}), and U_{max} is the maximum near-bed velocity (m s^{-1}). The buoyancy anomaly, B, is defined by

$$B = \int_0^\ell g'C\,dz \tag{10}$$

and ℓ is the height of the top of the dense layer. The term U_{max} is defined by

$$U_{max} = (U_w^2 + U_g^2 + V_c^2)^{1/2} \tag{11}$$

where U_w is the near-bed velocity associated with waves, U_g is the near-bed velocity associated with the gravity current itself and V_c is the near-bed along-shelf current velocity. Under the condition of $U_g \gg U_w$, V_c, Eq. 9 reduces to

$$Ri_B = \frac{B}{U_g^2} \tag{12}$$

Invoking a balance between frictional drag at the seabed and the gravitational force on the dense suspension, U_g can be written

$$U_g = \sqrt{B \sin\theta / C_D} \tag{13}$$

where θ is the angle between the seabed and a horizontal line, and C_D is a dimensionless drag coefficient with a typical value ~0.003. Substituting Eq. 13 into Eq. 9 arrives at an expression for the bulk Richardson number of

$$Ri_B = \frac{C_D}{\sin\theta} \tag{14}$$

When ample sediment is available for resuspension, the bulk Richardson number takes a value of 0.25, and Eq. 14 can be solved for critical angle. The value is 0.7°. On seabeds with smaller bathymetric gradients, gravity currents flow too slowly to maintain sediment in suspension, deposition occurs, and the current stops. Larger gradients allow gravity flows to maintain sediment in suspension and flow indefinitely (Wright et al., 2001).

The gradient of the Eel shelf apparently precludes gravity-driven transport, if the role of waves and currents in resuspending sediment is neglected. Wright et al. (2001) pointed out that when $U_{\max}$ is enhanced by waves or currents, gravity-driven currents can be maintained on lower gradients. So, the ideal conditions for gravity-driven sediment transport on open shelves are strong waves and/or currents with a large supply of easily resuspended sediment.

The concepts of critical Richardson number and wave-supported near-bed dense suspensions can be extended to a consideration of depositional dynamics. Maximal cross-shelf flux according to this conceptual model can be expressed as (Wright et al., 2001; Scully et al., 2002)

$$Q_{\mathrm{gmax}} = \frac{\alpha \rho_s U_{\max}^3}{4 C_D g'} \tag{15}$$

where Q_{gmax} is the maximal gravity-induced cross-shelf flux (kg m^{-1} s^{-1}), α is the bathymetric gradient, and other variables are as defined previously. The flux of sediment to the seafloor is defined by the cross-shelf gradient in Q_{gmax}:

$$-\frac{\partial Q_{\mathrm{gmax}}}{\partial x} = -\frac{\rho_s}{16 C_D g'} \frac{\partial \alpha U_{\max}^3}{\partial x} \tag{16}$$

Under the simplifying assumption that $U_{\max}$ is set entirely by monochromatic waves impinging on the shelf, Eq. 16 can be solved:

$$J_g = \frac{\rho_s}{16 C_D g'} \frac{\alpha^2 U_{\max}^3}{h} \left[\frac{3kh}{\tanh kh} - \frac{h}{\alpha^2} \frac{\partial \alpha}{\partial x} \frac{(2\beta^2 + 1)}{(1 - \beta^2)} \right] \tag{17}$$

In Eq. 17, J_g is sediment flux to the seabed from the gravity current, k is wave number (m^{-1}), h is water depth and β equals $\alpha/16 C_D$ (Scully et al., 2002).

Equation 17 indicates that the sediment flux to the seafloor under wave-supported gravity-driven flows depends strongly on wave energy. Furthermore, it demonstrates the importance of water depth and bathymetric gradient. The first term inside the brackets originates from the seaward

decay in wave orbital velocity near the seabed. It always favours deposition. It also indicates that, at some water depth, the flux becomes insignificant. The second term inside the brackets characterizes the effect of bathymetric gradient. It favours deposition when the gradient decreases in the seaward direction, and erosion when the gradient increases seaward. In other words, concave slopes are depositional and convex slopes are erosional (Scully et al., 2002).

Scully et al. (2002) evaluated the predictive capabilities of the above model by comparison with data from the tripods. First, they compared predicted and observed cross-shelf velocities at S60 for January 1997 and at K60 for January 1998. The predictions assumed that all cross-shelf transport was by wave-supported, gravity-driven flows. During periods of low discharge, predictions failed to match observations. During periods of elevated discharge from the river, however, predicted and observed velocities were well correlated. These results suggest that the wave-supported, gravity-driven model of cross-shelf flow captures the underlying physics, but only when enough sediment is available to stratify flow within the wave boundary layer. During low discharge periods, conventional boundary-layer transport by the combined effects of waves and currents dominates.

Scully et al. (2002) also compared predicted deposition with mid-shelf cores collected after the floods in January and March 1995 and January 1997, as well as with the observed timing and magnitude of changes in bed elevation observed with the ABS at K60 in 1998. The model was successful in this test as well, and highlighted an important behaviour of wave-supported, gravity-driven flows: rapid deposition coincides with short periods of highest wave energy associated with storms, because deposition depends on the cube of $U_{\max}$. As a result, even though the January 1997 flood had the highest estimated sediment discharge, the January 1995 flood had the thickest mid-shelf deposit because it had the greatest associated wave energy. This counter-intuitive result was further highlighted in 1998, when, in spite of large wave orbital velocities, no erosion was observed or predicted at K60, and significant deposition occurred.

Scully et al. (2003) created a numerical model of wave-supported, gravity-driven flows on the Eel shelf in order to compare observed and predicted spatial distribution of flood deposits for the various years of observation. The conversion to a numerical model made it possible to include realistic shelf geometry and along-shelf variation in sediment supply in the analysis. The numerical model produced thicknesses and along- and across-

shelf distributions of flood sediment that agree in general with observed deposit geometries (Fig. 22). Maximal mid-shelf deposition was predicted between 10 km and 30 km north of the river mouth in water depths of ~50–70 m. The predicted deposits extended along-shore for ~40 km and seaward to

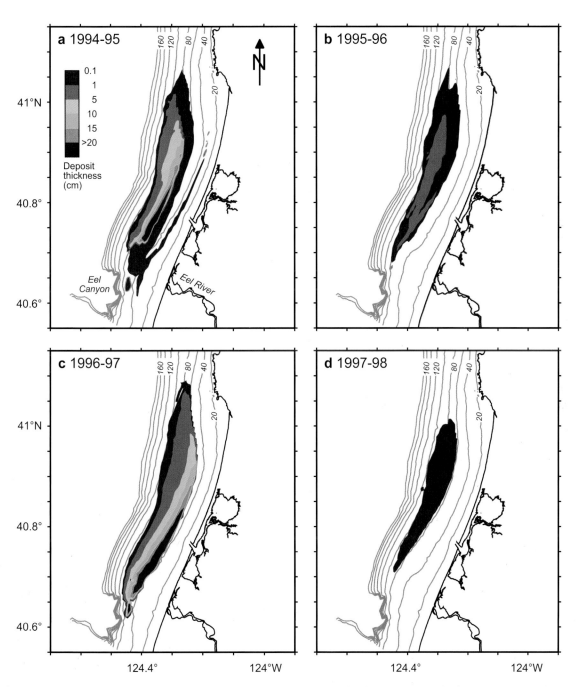

Fig. 22 Predicted wave-supported, gravity-driven deposit thicknesses for four winters. Thicknesses are based on an assumed porosity of 0.75. (Redrawn from Scully *et al.*, 2003.)

the 100-m isobath. Following the floods of 1995 and 1997, predicted mid-shelf deposition agreed favourably with observations, accounting for roughly 29% and 39% of the sediment discharge, respectively. The model predicted significantly less mid-shelf deposition during 1995–96 and 1997–98 when only 16% and 3% of the sediment load, respectively, was predicted to deposit on the mid-shelf. The agreement between the model's predictions and observations is encouraging, and suggests that the physics of wave-supported, gravity-driven flows underlie observed patterns of flood sediment deposition on the Eel shelf.

The results of Scully *et al.* (2003) suggest that the along- and across-shelf position of the flood deposits was controlled both by the across-shelf morphology and by sediment delivery rate to the wave boundary layer. Across-shelf position was determined by water depth, wave energy and bathymetry. The seaward limit of deposition was determined by both the seaward decline in wave energy, which limited rate of supply from shoreward portions of the shelf, and the increasing convexity of the shelf profile, which caused bypassing of available sediment. Similarly, the convex morphology near the river mouth limits gravity-driven mid-shelf deposition despite the high inshore sediment delivery. Greatest mid-shelf deposition was consistently predicted and observed to occur north of the river mouth because the mid-shelf is concave in this region.

Moore's (1969) hypothesis that wave-supported, gravity-driven flows moved sediment across the continental shelf languished for so long primarily because of a lack of understanding of suspended-sediment stratification. Stratification limits turbulent diffusion out of the wave boundary layer, thereby allowing sediment concentration near the bed to build to extraordinary levels whenever sediment is abundant and waves are large enough to resuspend it. Under these conditions, suspensions move downslope, even when the bathymetric gradient is small. Moore's (1969) ideas were bred of compelling geophysical observations, but his proposed mechanisms of transport did not fit the understanding of boundary-layer hydrodynamics during his day. Only when direct physical observations failed to conform to accepted boundary-layer theory did it become clear that other physical processes were at work.

Fate of missing sediment

Seabed observations indicate that only ~25% of the mud delivered by the river to the shelf accumulates in the mid-shelf flood deposit. So, while the plume and shelf boundary-layer observations on the Eel margin paint a novel and compelling picture of how sediment is delivered to the flood deposit, they leave unresolved the fate of the majority of the mud emanating from the river. A variety of studies conducted as part of the overall programme shed some light on the fate of the missing sediment.

Inner-shelf storage

Inner-shelf storage may account for some of the missing sediment, but it is difficult to quantify because of the energetic nature of the environment in shallow water depths. Breaking waves make near-bed observations difficult. Extensive sediment reworking by waves complicates the interpretation of cores. Nonetheless, studies of sediment grain size in the Eel plume, and limited coring, both suggest that inner-shelf storage does trap some flood sediment.

The disaggregated inorganic grain-size distributions gathered in the plume at the river mouth and at the 20-m isobath on the G line indicate that coarse silt sank rapidly from the plume in the vicinity of the river mouth (Fig. 23). At the mouth, approximately 15–20% of the suspended load was coarser than 30 μm, yet by the G line only about 1% of the grains in suspension were so large. This coarse silt does not appear in the flood deposit. The surficial-sediment grain-size distribution at G70 following the January 1997 flood was similar to the plume grain-size distribution at the 20-m isobath. It contained few grains coarser than 30 μm. These observations reveal rapid dumping of coarse silt and sand near the river mouth. This material was presumably redistributed by waves in shallow water. It did not follow the same transport pathway as the finer plume sediment, which moved across shelf to the flood deposit.

Cores demonstrated that muds as well as sands can accumulate on the inner shelf. In a core from I45, a mud-rich layer was detected 45 cm below the surface (Fig. 24; Crockett & Nittrouer, 2004). Based on an analysis of ^{137}Cs and ^{210}Pb, the mud layer probably formed during the massive 1964

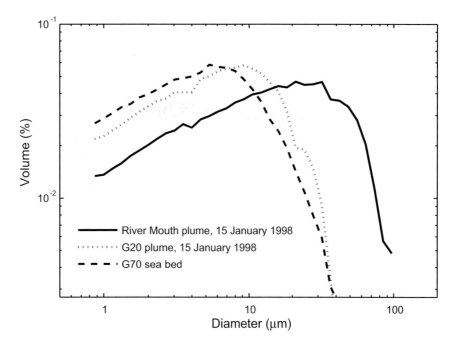

Fig. 23 Disaggregated inorganic grain-size distributions from the Eel margin. Percentage volume concentration versus diameter is shown for the plume at the river mouth (solid line) and at G20 (dotted line) on 15 January 1998. The dashed line shows the size distribution in the January 1997 flood deposit at G70. Coarse silt and sand are lost from the plume between the river mouth and G20.

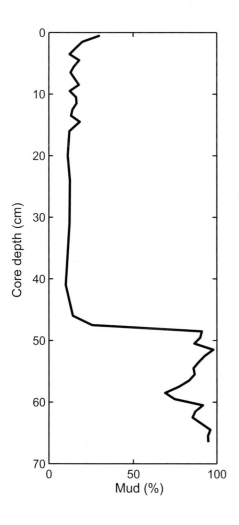

flood (Crockett & Nittrouer, 2004). The results of Scully *et al.* (2002) suggested that even during the relatively energetic January 1997 flood, more sediment was supplied to the inner shelf than could be removed by gravity-driven transport. This notion that mud can accumulate in the energetic nearshore environment of the inner Eel shelf was supported by observations of measurable [7]Be in grab samples collected from 30–50-m water depths in 1997 (Sommerfield *et al.*, 1999). Crockett & Nittrouer (2004) estimated that up to 10% of the mud delivered to the shelf by the Eel may end up in the predominantly sandy deposits of the inner shelf.

The accumulation of mud in energetic nearshore environments may appear paradoxical. However, whenever supply of mud overwhelms the ability of waves, currents, or gravity to remove it, mud can deposit (McCave, 1972). Periods of large supply, i.e. floods, clearly are required on the energetic Eel shelf. Low wave energy during high discharge, although rare, would further enhance the potential for mud deposition on the inner shelf.

Fig. 24 (*left*) Mud (silt and clay) percentage versus depth in core at I45. The abrupt increase in mud percentage at ~45 cm core depth is interpreted as a deposit from the 1964 flood on the Eel River. (Redrawn from Crockett & Nittrouer, 2004.)

Along-shelf bypassing

A significant fraction of sediment may have bypassed the Eel shelf in the along-shelf direction. Two factors favour along-shelf bypassing. First, resuspension in the surf zone reintroduces sediment into the northward flowing plume, thus retarding sedimentation losses from the plume (Curran *et al.*, 2002a). Direct evidence for this mechanism is lacking, but it does help to explain the relatively low values for effective settling velocity in the plume. Visual observations from the helicopter on one plume survey revealed a sediment-laden plume streaming past Trinidad Head at the northern limit of the Eel shelf.

The second factor that favours along-shelf bypassing is sediment stranding within the plume. Sediment aggregation rate in a suspension is a second-order function of sediment concentration. As concentration falls due to sedimentation and dilution, aggregation rate slows markedly. Eventually, aggregation time-scales grow so long that aggregation no longer affects sinking dynamics (Curran *et al.*, 2002b). The remaining fine sediment, with its low, single-particle settling velocities, stays in the plume for long periods before sinking out. The **stranded** sediment can be transported great distances (e.g. Geyer *et al.*, 2000). This process explains the common appearance on satellite images of large sediment plumes emanating from river mouths that account for only a small fraction of the sediment delivered to the coastal ocean by the river (Mertes & Warrick, 2001). The stranded sediment possesses a distinct optical signature visible to satellite sensors, yet these plumes may carry only 1–2% of the total sediment load.

Measurements from the STRATAFORM programme do not place tight constraint on the quantity of sediment exported north of the Eel shelf by the plume. Geyer *et al.* (2000) used observations at the K line to estimate that 30–40% of the Eel's load was transported more than 10 km beyond the river mouth in the along-shelf direction, but Mertes & Warrick (2001) indicated that far-field plumes contain a much smaller fraction of a river's initial sediment load. More observations are required to resolve the fate of mud that does not sink from plumes in the immediate vicinity of river mouths.

Transport from shelf to slope

Flood-derived sediment may have been transported beyond the shelf to the slope. Observations of water-column light attenuation and deposition rates of sediment in traps and on the seafloor suggest that sedimentation on the open slope during the study period was comparable in magnitude with inner-shelf storage, and may in fact have rivalled accumulation in the flood deposit (Alexander & Simoneau, 1999; McPhee-Shaw *et al.*, 2004).

A series of water-column transmissometer profiles was collected over the Eel shelf and slope in 1996 and 1998–99 (McPhee-Shaw *et al.*, 2004). These surveys documented the common occurrence of **intermediate nepheloid layers** (INLs), which are layers of fluid with higher sediment concentration than surrounding water, and which extend from the margin into the basin interior. Some of the observed layers originated from mid-shelf depths similar to the depth of the flood deposits. Others detached from the continental slope in water depths > 150 m, i.e. beyond the shelf break. The shelf INLs had sediment concentrations as much as 0.012 kg m^{-3} above background, and they extended as far as 25 km into the basin from their detachment points. The slope INLs carried less sediment and only extended basinward 3–8 km. The shelf INLs were prevalent during periods when waves were large and winds were favourable for downwelling. These conditions were associated with winter and spring. Slope INLs showed no marked seasonal variability, although they were most pronounced in August. These differences suggest different mechanisms of formation of INLs on the shelf and slope.

The detachment of shelf INLs over the mid-shelf mud deposit and the correlation of INLs with large waves and downwelling point to wave resuspension of recently deposited flood sediments as a likely source for shelf INLs (McPhee-Shaw *et al.*, 2004). Observations from the tripod at K60 in 1998 revealed the removal of flood sediment by wave resuspension, thereby supporting the hypothesis of wave resuspension as a source for INLs (Traykovski *et al.*, 2000). By estimating typical concentrations and spatial scales of shelf INLs, McPhee-Shaw *et al.* (2004) proposed that approximately 10% of the total annual load of the Eel could be transported from the shelf to the slope by shelf INLs.

The lack of seasonal variability in slope INLs adjacent to the highly seasonal Eel shelf suggests that slope INLs are decoupled from shelf processes (McPhee-Shaw *et al.*, 2004). Their formation is more likely to be tied to resuspension by internal tides. The stratification of the water column at the shelf edge of the Eel margin is such that over a range of depths on the upper slope, energy associated with M2 internal tides propagates upslope parallel to the seabed (Cacchione *et al.*, 2002; McPhee-Shaw *et al.*, 2004). The propagation leads to intensification of near-bed flow, probably strong enough to cause sediment resuspension. In general, slope INLs appeared where bathymetric gradients favoured this focusing of internal-tide energy (McPhee-Shaw *et al.*, 2004). Also in these regions, seabed grain sizes are relatively coarse (Alexander & Simoneau, 1999), suggesting active winnowing of surficial sediments.

Shelf-to-slope transport of fluvial sediment helps to explain sediment-trap data on the Eel slope. From September 1995 to January 1997, a mooring carrying three sediment traps was deployed in 450 m of water on the Y line, 50 km north of the river mouth (Walsh & Nittrouer, 1999). The traps were located at depths of 60 m, 220 m and 435 m. Flux to the traps was predominately lithogenic sediment, accounting for 53%, 70% and 83% of the flux to the top, middle and bottom traps, respectively. Sediment flux was greater than on other open slopes, and it was episodic. For example, six of 33 sampling intervals accounted for more than 50% of the flux to the middle trap. Walsh & Nittrouer (1999) argued that the episodic flux implicated INLs as an important transport pathway for Eel River sediment crossing the continental shelf. The timing and magnitude for pulses of sediment flux on the slope varied with sediment resuspension by waves, river discharge and shelf circulation (Walsh & Nittrouer, 1999). These results are consistent with those from the water-column surveys of light attenuation (McPhee-Shaw *et al.*, 2004).

The appearance of [7]Be in slope sediments lends credence to the hypothesis that river-derived sediment can be transported rapidly across the shelf to the upper slope of the Eel margin. Following the January 1997 flood, and at no other time during the 394-day observation period, the middle sediment trap at Y450 contained measurable [7]Be (Walsh & Nittrouer, 1999). Similarly, following the January

1995 flood and again after the January 1997 event, [7]Be was detected in cores collected on the upper slope (Sommerfield *et al.*, 1999). Detection of [7]Be in upper slope sediments reveals rapid bypassing of the continental shelf by a significant amount of Eel sediment (Sommerfield *et al.*, 1999), because the half-life of [7]Be is only 53.3 days and its source in coastal waters is primarily terrestrially derived sediments. Furthermore, the modelling results of Scully *et al.* (2003) suggested that significant gravity-driven transport of flood-derived sediment can reach the upper slope following floods that have large associated wave energy, and this transport may account for a significant fraction of the total fine sediment input.

Other radionuclides can be used to constrain whether short- and long-term loss rates of sediment from the shelf to the slope are similar. The amount of sediment bypassing the shelf to the slope over decadal time-scales was estimated by Alexander & Simoneau (1999), who used ^{210}Pb and ^{137}Cs to constrain sediment accumulation rates in 60 cores collected from a range of depths on the upper slope. Accumulation rates were relatively high for the open slope, ranging from 0.2 to 1.3 g cm^{-2} yr^{-1}. These high accumulation rates indicate that over longer time-scales, redistribution of fluvial sediment is widespread. When integrated across the slope and through time, these accumulation rates can account for as much as 20% of the sediment discharge of the river (Alexander & Simoneau, 1999). These various slope studies on the Eel margin, despite the fact that they address sediment loss from shelf to slope on different time-scales, agree that 10–20% of the discharge reaches the slope. Together, deposits on the inner shelf, the mid-shelf, and the slope may account for approximately half of the sediment discharged by the Eel.

Transport to the Eel Canyon

Coring studies conducted in the Eel Canyon implicate down-canyon losses as a major sink for sediment delivered to the Eel shelf (Mullenbach & Nittrouer, 2000; Mullenbach *et al.*, 2004). Cores were collected from the canyon in January and March 1998. Profiles of [7]Be and ^{210}Pb were generated from each core. The winter of 1998 was characterized by large waves and multiple moderate discharge events starting in mid-January. January cores, which were collected prior to the onset of flooding on the river, exhibited only slightly elevated ^{7}Be in the top 1 cm and no detectable ^{7}Be below. In the top 2 cm of the January cores, ^{210}Pb was depleted slightly. March cores, by contrast, contained significantly elevated ^{7}Be inventories down to 8-cm depth in core. Profiles of ^{210}Pb activity were non-steady state, suggesting rapid episodic accumulation of sediment. Finally, cores collected in March contained more clay in surface sediments than did the January cores. These differences suggest that thick layers of flood sediment can form annually in the Eel Canyon. This hypothesis is supported by the modelling of Scully *et al.* (2003) that predicted a significant gravity-driven flux into the canyon following floods of the Eel River. Over 2.0×10^8 kg of sediment was predicted to enter Eel Canyon during the January and February floods of 1998, potentially explaining the presence of ^{7}Be observed in the cores collected in March 1998.

The inventory of ^{7}Be decreased downward through the top 8 cm of the March cores. This distribution differed from cores taken in the flood deposit on the shelf, in which ^{7}Be was constant to the bottom of distinct flood layers but then decreased abruptly. This pattern of ^{7}Be in canyon cores suggests that the tops of cores taken in March comprised mixtures of older shelf sediment and recently discharged muds that were emplaced throughout the winter. The recently discharged sediment, with high ^{7}Be, became more abundant through the course of the winter. Emplacement was likely to have been episodic and associated with periods characterized by large waves (Mullenbach & Nittrouer, 2000).

Deeper in the cores collected from the Eel Canyon head were two to three layers characterized by fine sediment and depleted ^{210}Pb. These properties are consistent with rapid emplacement of fluvially derived sediment. The layers therefore probably record periods of enhanced sedimentation associated with past floods (Mullenbach & Nittrouer, 2000). Although the unsteady ^{210}Pb profiles made it difficult to constrain accumulation rates, minimum rates of 0.4 cm yr^{-1} were required to explain the presence of excess ^{210}Pb at the bottom of the cores. Using this value to reconstruct the accumulation rate of the fine-grained layers led to the conclusion that a large fraction of the sediment delivered by the Eel River is transported from shelf to slope through the canyon, potentially closing the sediment budget. This intriguing result is poorly constrained due to a lack of core coverage and time control within the cores, yet it clearly implicates down-canyon transport as a major sediment sink.

Near-bed observations at the head of the Eel Canyon suggest that Moore's (1969) proposed mechanism for transport of sediment into canyons is sound. From January to April 2000 a tripod was deployed in 120 m of water in the northern thalweg of the Eel Canyon (Puig *et al.*, 2003, 2004). The tripod carried two current meters located at 30 cmab and 100 cmab, a pressure sensor, a sonic altimeter, two OBSs at the same heights as the current meters and a seabed-imaging video camera. Unfortunately, the OBSs failed, making estimation of suspended-sediment concentration difficult. To fill this void, camera opacity was used as a proxy for suspended-sediment concentration.

Tripod data showed clearly that elevated camera opacity, and by implication suspended-sediment concentration, was correlated with large waves and not with Eel River discharge (Fig. 25; Puig *et al.*, 2003, 2004). When opacity was so high as to render images black for several hours, velocity was larger near the bed, as observed with wave-supported, gravity-driven flows on the shelf (Fig. 25). Waves were implicated in the maintenance of the dense suspension by fluctuations in down-canyon current at the same frequencies as fluctuations in pressure. Puig *et al.* (2004) argued that these observations demonstrated that waves liquefied fine sediment at the canyon head and that the sediment flowed down-canyon as wave-supported, gravity-driven underflows (Lee *et al.*, this volume, pp. 213–274; Parsons *et al.*, this volume, pp. 275–337).

The crude sediment budget presented here relies on a small number of observations and uses tools with intrinsically different time-scales. It must be regarded, therefore, as highly speculative. Nonetheless, the estimates of accumulation rate in the various regions produce a relatively consistent, closed sediment budget. Inner shelf storage accounts for 10% of the annual discharge of fine sediment by the river. The mid-shelf flood deposit contains 20–25%, and the slope stores 10–20%. An unknown quantity of sediment exits the Eel margin to the north, carried by the buoyant coastal current. This loss term probably amounts to a few per cent of the Eel discharge. If along-shelf export

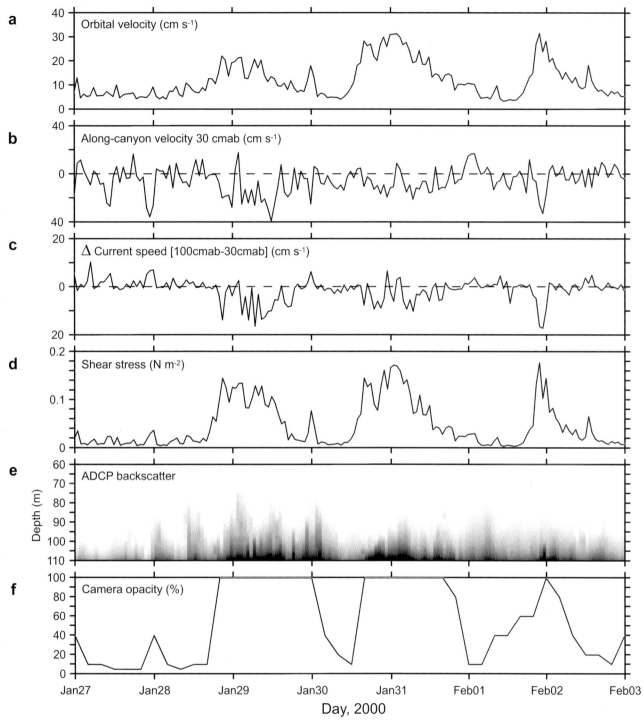

Fig. 25 Tripod measurements from the head of the Eel Canyon in 2000. (a) Wave orbital velocity. (b) Along-canyon velocity at 30 cmab, with positive values corresponding to up-canyon flow. (c) Difference between along-canyon current velocities at 100 and 30 cmab. (d) Boundary shear stress. (e) Intensity of acoustic backscatter measured by an ADCP as a function of height above bottom. Darker greys indicate higher sediment concentration. (f) Opacity on a seabed-imaging video camera. High opacity indicates dark images caused by high suspended-sediment concentration. Three periods of elevated wave orbital velocities (a) and shear stresses (d) were accompanied by increased acoustic backscatter near the seabed (e) and periods of large camera opacity (f), reflecting increased suspended-sediment concentration. During these periods, near-bed flow was directed down the canyon (b), and flow was faster at 30 cmab than it was at 100 cmab (c). These results indicate that sediment resuspension from waves resulted in formation of gravity-driven flows. (Redrawn from Puig *et al.*, 2004.)

beyond the margin is assigned a value of 5% of the discharge, 45–55% of the Eel's load must be accounted for. This quantity is similar to the estimated storage of sediment in the Eel Canyon.

SUMMARY AND CONCLUSION

Summary of STRATAFORM sediment delivery

Eel margin studies that focused on delivery of flood sediment to the seabed produced a picture of shelf sediment processes remarkably consistent with the speculative conceptual model proposed by Moore (1969) in the early days of process-based investigation of continental-margin sediment transport. So, it is fitting to summarize this chapter with Moore's own words:

'Silt- and clay-sized particles introduced by streams and rivers are initially distributed by floating surface layers of fresh turbid water; this distribution is not widespread, and the lutum mostly settles to the bottom relatively near the river mouth. Subsequently or contemporaneously, during occurrence of long-period swell, the lutum is resuspended by the flattened orbital motion of the swell as it impinges on the bottom. Where this motion is long enough and where a sufficient supply of loose silts and clays exists, a turbid layer will develop over the seafloor. Turbid layers move across the seafloor as wide, relatively thin sheets under the influence of coastal current and downslope gravity flow superimposed on the to-and-fro movement of the swell. These two component forces should result in a net movement of most of the river-supplied lutum, diagonally across the shelf, nearly parallel to the coast, but with a small offshore component.'

In the next paragraph, Moore proposed that if suspensions are dense enough, then they can develop into autosuspending, channelized turbidity currents when they encounter a canyon. Otherwise,

'If the turbid layer is not of sufficient thickness, density, and duration to form a channelized, low-density turbidity current, it should after flowing into the canyon gully, or depression, lose its wave-generated orbital component of velocity and perhaps also its coastal-current component. Thus, it will flow downslope, depositing lutum until it dissipates. This process may cause relatively rapid accumulation of fine sediment on canyon and valley walls and on the floors of the canyons as well. Where no canyon interception occurs and the turbid layer flows over the shelf edge, it will dissipate by deposition on the mainland basin slope. Because deposition is over a broader front on the open slope and is believed to accumulate slowly relatively to canyon walls, it forms a largely stable deposit not easily induced to fail by slumping.'

Although envisioned many years ago, confirmation of this conceptual model required the observational and modelling efforts described in this chapter.

Questions for future research

Delivery of sediment to the seabed is complex, and many new questions have arisen from studies on the Eel margin. Perhaps the most important question is: 'Do wave-supported, gravity-driven flows dominate cross-shelf sediment transport in other systems?' The conditions required to form such flows are ample fine sediment, combined with large waves. Sediment must be abundant enough to induce suspended-sediment stratification. Exactly how much is required depends on the velocity shear in the water column. The Eel shelf supports this type of flow because the river produces a high yield of sediment and discharges episodically, and because waves typically accompany floods.

Aspects of the Eel arguably are unique. The sediment yield of the river is high because of the underlying, easily eroded geological substrate. Added to this property of the basin are land-use practices that favour high sediment yields. In northern California and elsewhere in western North America, excess sediment production has been linked to forestry operations (e.g. Best *et al.*, 1995). Paired basin analysis within the Redwood Creek watershed just north of the Eel Basin showed that sediment yield from logged tributaries was up to an order of magnitude greater than from nearby forested tributaries (Nolan & Janda, 1995). Extensive logging in the Eel watershed probably contributes to its high sediment load and distinguishes it from less impacted systems and from systems in the recent geological past. During the latter half of the 20th century, precipitation in the Eel Basin increased, also increasing the sediment yield (Sommerfield *et al.*, this volume, pp. 157–212).

The close timing of peak sediment discharge and large waves also distinguishes the Eel system from other systems (Wheatcroft, 2000). The trend of the watershed parallel to the coast is a manifestation of the complex plate geometry in the region, and it is a factor in the occurrence of basin-wide intense precipitation during storms. The watershed geometry also produces short lag times between precipitation and discharge into the ocean, so sediment arrives in the coastal ocean when waves are large.

Together, these features of the Eel system and other similar systems lead to the formation of open-shelf, wave-supported, gravity-driven flows. Extreme conditions, however, may not be necessary. Moore (1969) formed his conceptual model of these flows based on observations in California's Borderland Basins to the south, where sediment yields and discharges are lower. Close timing between sediment discharge and waves may not be necessary. If sediment does not compact too rapidly, then it may reside on the seafloor for days to weeks before being entrained into wave-supported, gravity-driven flows during storms. The canyon observations of Puig *et al.* (2003, 2004) support the hypothesis that the seabed can provide enough sediment to a wave boundary-layer to fuel this type of flow. Perhaps open-shelf, wave-supported, gravity-driven flows form near many mountainous, coastal rivers that drain collisional margins. Observations that can characterize thin, near-bed flows on other shelves will reveal the generality of the fluid-mud processes at work on the Eel shelf.

A second question is: 'How is fine sediment from rivers delivered to canyon heads during sea-level highstands?' Modelling studies suggest that hyperpycnal inflows of sediment-laden river waters can deliver sediment rapidly and efficiently to canyon heads (Parsons *et al.*, this volume, pp. 275–337; Syvitski *et al.*, this volume, pp. 459–529), as do observations from the Sepik River in Papua New Guinea (Kineke *et al.*, 2000). Limited observations, however, indicate that oceanographic processes deliver sediment to the Eel Canyon head, where waves subsequently generate wave-supported, gravity-driven flows via liquefaction of recently deposited sediment (Mullenbach & Nittrouer, 2000;

Puig *et al.*, 2003, 2004). Future studies in complex canyon systems will help clarify which pathways dominate under various forcings.

A third question is: 'How does fine sediment become trapped in inner-shelf sands?' Historically, on wave-dominated coasts, muds were thought to bypass the inner shelf completely. Yet cores from the Eel shelf suggest that layers of mud can deposit and persist under energetic forcing. The mechanisms of mud burial within sandy deposits are unclear, but perhaps relate to suspended-sediment stratification and suppression of turbulence. Resolving these mechanisms as well as rates of fine-sediment accumulation on the inner shelf poses considerable challenges. Equipment placed on the inner shelf is at high risk, and establishing age control in coarse, frequently resuspended, inner-shelf sands is difficult. Nonetheless, improved understanding of this transition region is vital to building integrated models of stratigraphy on continental margins.

A final question is: 'What is the importance to overall sediment budgets of far-field suspended-sediment transport in buoyant coastal currents?' Satellites offer compelling visual documentation of this transport pathway. The plumes may be visually spectacular, but they may not be important quantitatively due to relatively low sediment concentrations within them. If they are supplied constantly with sediment, however, either by river discharge or resuspension of bottom sediments, coastal currents hold the potential, over time, to redistribute considerable quantities of sediment along the coast. As with the other questions, more observations are needed to improve understanding of this sediment-transport pathway.

In closing, the pattern of discovery regarding sediment delivery to the seabed on continental shelves is classic. Early inquiry generated competing hypotheses that motivated subsequent investigations. These investigations strongly favoured one hypothesis over the others, namely that advection by near-bed currents was the dominant cross-shelf transport mechanism for muds. The data that supported this hypothesis, however, were inadequate for examining the alternative hypothesis for cross-shelf transport proposed by Moore (Moore, 1969), because they did not characterize the region of the flow within the wave boundary layer. The diverse and sophisticated array of sensors deployed during STRATAFORM was able to characterize this near-bed region, and in so doing exposed the dominance of wave-supported, gravity-driven flow on this open, wave-impacted shelf. These findings challenge the reigning paradigm of continental-shelf sediment transport, and undoubtedly will fuel research for years to come.

ACKNOWLEDGEMENTS

Joe Kravitz took active interest in the work described here, from inception of the research to final publication of this volume. We recognize his contribution with simple and sincere thanks. We are all grateful to the Office of Naval Research for years of support for our research. The editors and reviewers shouldered a heavy load in producing this volume. In particular, this paper benefited from comments and criticisms from editors Chuck Nittrouer, Pat Wiberg, James Syvitski and reviewers Steve Goodbred, Lonnie Leithold, Don Wright, Harry Roberts and Nick McCave.

NOMENCLATURE

Symbol	Definition	Dimensions
a_f	aggregate fraction in suspension	
B	integrated buoyancy anomaly	$L^2 T^{-2}$
C_D	drag coefficient	
C_s	suspended-sediment mass concentration	$M L^{-3}$
d_{25}	upper quartile diameter of aggregates	L
d_{50}	median diameter of aggregates	L
Fr	Froude number	
f	Coriolis frequency	T^{-1}
g	gravitational acceleration	$L T^{-2}$
g'	modified gravity	$L T^{-2}$
h	water depth	L
h_c	river channel depth	L
h_p	plume thickness	L
Li	internal Rossby radius of deformation	L
ℓ	thickness of dense layer	L
Q	river discharge	$L^3 T^{-1}$
Ri	Richardson number	
Ri_B	bulk Richardson number	
U_{max}	maximum nearbed velocity	$L T^{-1}$
U_w	nearbed wave velocity	$L T^{-1}$
U_g	nearbed gravity-current velocity	$L T^{-1}$
u	plume speed	$L T^{-1}$
V_c	nearbed alongshore current velocity	$L T^{-1}$
W_c	river channel width	L
W_p	width of coastal current	L
w_e	effective settling velocity	$L T^{-1}$
x_e	e-folding distance	L
α	bathymetric gradient	
ρ	density of basin water	$M L^{-3}$
ρ_s	sediment density	$M L^{-3}$
$\Delta\rho$	density contrast between plume and basin water	$M L^{-3}$
θ	angle of bathymetric gradient	degrees

REFERENCES

Agrawal, Y.C. and Pottsmith, H.C. (1994) Laser diffraction particle sizing in STRESS. *Cont. Shelf Res.*, **14**, 1101–1121.

Alexander, C.R. and Simoneau, A.M. (1999) Spatial variability in sedimentary processes on the Eel continental slope. *Mar. Geol.*, **154**, 243–254.

Alldredge, A.L., Granata, T.C., Goschalk, C.G. and Dickey, T.D. (1990) The physical strength of marine snow and its implications for particle disaggregation in the ocean. *Limnol. Ocean.*, **35**, 1415–1428.

Bagnold, R.A. (1962) Auto-suspension of transported sediment; turbidity currents. *Proc. Roy. Soc., London, Ser. A*, **265**, 315–319.

Bale, A. and Morris, A.W. (1987) *In situ* measurement of particle size in estuarine waters. *Estuarine Coast. Shelf Sci.*, **24**, 253–263.

Bates, C.C. (1953) Rational theory of delta formation. *Bull. Am. Assoc. Petrol. Geol.*, **37**, 2119–2162.

Berhane, I., Sternberg, R.W., Kineke, G.C., Milligan, T.G. and Kranck, K. (1997) The variability of suspended aggregates on the Amazon continental shelf. *Cont. Shelf Res.*, **17**, 267–285.

Best, D.W., Kelsey, H.M., Hagans, D.K. and Alpert, M. (1995) Role of fluvial hillslope erosion and road construction in the sediment budget of Garrett Creek, Humboldt County, California. In: *Geomorphic Processes and Aquatic Habitat in the Redwood Creek Basin, Northwestern California* (Eds K.M. Nolan, H.M.

Kelsey and D.C. Marron), pp. M1–M11. *U.S. Geol. Surv. Prof. Pap.*, 1454.

Boldrin, A., Bortoluzzi, G., Frascari, F., Guerizoni, S. and Rabitti, S. (1988) Recent deposits and suspended sediments off the Po Della Pila (Po River, main mouth), Italy. *Mar. Geol.*, **79**, 159–170.

Brown, W.M. and Ritter, J.R. (1971) Sediment transport and turbidity in the Eel River basin, California. *U.S. Geol. Surv. Wat. Supply Pap.*, **1986**, 70 pp.

Burt, T.N. (1986) Field settling velocities of estuary muds. In: *Estuarine Cohesive Sediment Dynamics* (Ed. A.J. Mehta), pp. 126–150. Springer-Verlag, New York.

Cacchione, D.A., Drake, D.E., Losada, M.A. and Medina, R. (1990) Bottom-boundary-layer measurements on the continental shelf off the Ebro River, Spain. *Mar. Geol.*, **95**, 179–192.

Cacchione, D.A., Wiberg, P.L., Lynch, J., Irish, J. and Traykovski, P. (1999) Estimates of suspended-sediment flux and bedform activity on the inner portion of the Eel continental shelf. *Mar. Geol.*, **154**, 83–97.

Cacchione, D.A., Pratson, L.F. and Ogston, A.S. (2002) The shaping of continental slopes by internal tides. *Science*, **296**, 724–727.

Cowan, E.A. (1993) Characteristics of suspended particulate matter and sedimentation of organic carbon in Glacier Bay fjords. In: *Proceedings of the Third Glacier Bay Science Symposium, 1993* (Ed. D.R. Engstrom), pp. 24–28. National Park Service, Anchorage, Alaska.

Cowan, E.A. and Powell, R.D. (1990) Suspended sediment transport and deposition of cyclically interlaminated sediment in a temperate glacial fjord, Alaska, U.S.A. In: *Glacimarine Environments: Processes and Sediments* (Eds J.A. Dowdeswell and J.D. Scourse), pp. 75–89. Special Publication 53, Geological Society Publishing House, Bath.

Crockett, J.S. and Nittrouer, C.A. (2004) The sandy inner shelf as a repository for muddy sediment: an example from Northern California. *Cont. Shelf Res.*, **24**, 55–73.

Curran, K.J., Hill, P.S. and Milligan, T.G. (2002a) Fine-grained suspended sediment dynamics in the Eel River flood plume. *Cont. Shelf Res.*, **22**, 2537–2550.

Curran, K.J., Hill, P.S. and Milligan, T.G. (2002b) The role of particle aggregation in size-dependent deposition of drill mud. *Cont. Shelf Res.*, **22**, 405–416.

Curray, J.R. (1965) Late Quaternary history, continental shelves of the United States. In: *The Quaternary of the United States* (Ed. H.E. Wright Jr. and D.G. Frey), pp. 723–735. Princeton University Press, Princeton, NJ.

Dearnaley, M.P. (1996) Direct measurements of settling velocities in the Owen Tube: a comparison with gravimetric analysis. *J. Sea Res.*, **36**, 41–47.

Drake, D.E. (1972) Distribution and transport of suspended matter, Santa Barbara Channel, California. *Deep-sea Res.*, 18, 763–769.

Drake, D.E. (1976) Suspended sediment transport and mud deposition on continental shelves. In: *Marine Sediment Transport and Environmental Management* (Eds D.J. Stanley and D.J.P. Swift), pp. 127–158. John Wiley & Sons, New York.

Drake, D.E. (1999) Temporal and spatial variability of the sediment grain-size distribution on the Eel shelf: the flood layer of 1995. *Mar. Geol.*, **154**, 169–182.

Drake, D.E. and Cacchione, D.A. (1985) Seasonal variation in sediment transport on the Russian River shelf, California. *Cont. Shelf Res.*, **4**, 495–514.

Drake, D.E., Kopack, R.L. and Fischer, P.J. (1972) Sediment transport on Santa Barbara-Oxnard shelf, Santa Barbara Channel, California. In: *Shelf Sediment Transport: Process and Pattern* (Eds D.J.P. Swift, D.B. Duane and O.H. Pilkey), pp. 307–332. Dowden, Hutchinson and Ross, Stroudsburg, Pennsylvania.

Drake, D.E., Cacchione, D.A., Meunch, R.D. and Nelson, C.H. (1980) Sediment transport in Norton Sound, Alaska. *Mar. Geol.*, **36**, 97–126.

Droppo, I.G. and Ongley, E.D. (1994) Flocculation of suspended sediment in rivers of southeastern Canada. *Water Res.*, **28**, 1799–1809.

Dunbar, C.O. and Rodgers, J. (1957) *Principles of Stratigraphy*. John Wiley & Sons, New York, 356 pp.

Dyer, K.R. and Manning, A.J. (1999) Observation of the size, settling velocity and effective density of flocs and their fractal dimensions. *J. Sea Res.*, **41**, 87–95.

Dyer, K.R., Cornelisse, J., Dearnaley, M.P., *et al.* (1996) A comparison of *in situ* techniques for estuarine floc settling velocity measurements. *J. Sea Res.*, **36**, 15–29.

Edzwald, J.K., Upchurch, J.B. and O'Melia, C.R. (1974) Coagulation in estuaries. *Environ. Sci. Technol.*, **8**, 58–63.

Eisma, D. (1986) Flocculation and de-flocculation of suspended matter is estuaries. *Neth. J. Sea Res.*, **20**, 183–199.

Eisma, D. and Kalf, J. (1984) Dispersal of Zaire River suspended matter in the estuary and the Angola Basin. *Neth. J. Sea Res.*, **17**, 385–411.

Eisma, D., Boon, J., Groenewegen, R., Ittekkot, V., Kalf, J. and Mook, W.G. (1983) Observations on macroaggregates, particle size and organic composition of suspended matter in the Ems estuary. *Mitt. Geol. Palaontol. Inst. Univ. Hamburg*, **55**, 295–314.

Eisma, D., Bernard, P., Cadee, G.C., *et al.* (1991) Suspended-matter particle size in some West-European estuaries; Part I: particle size distribution. *Neth. J. Sea Res.*, **28**, 193–214.

Eisma, D., Bale, A.J., Dearnaley, M.P., Fennessy, M.J., van Leussen, W., Maldiney, M.-A., Pfeiffer, A. and Wells, J.T. (1996) Intercomparison of *in situ* suspended matter (floc) size measurements. *J. Sea Res.*, **36**, 3–14.

Emery, K.O. (1952) Continental shelf sediments off southern California. *Geol. Soc. Am. Bull.*, **63**, 1105–1108.

Emery, K.O. (1969) The continental shelves. *Sci. Am.*, **221**, 106–122.

Fennessy, M.J., Dyer, K.R. and Huntley, D.A. (1994) INSSEV: an instrument to measure the size and settling velocity of flocs *in situ*. *Mar. Geol.*, **117**, 107–117.

Folk, R.L. (1977) *Petrology of Sedimentary Rocks*. Hemphill Publishing Company, Austin, Texas, 186 pp.

Fox, J.M. (2003) *Flocculation on the Po River delta*. MSc thesis, Dalhousie University, Halifax, Nova Scotia, Canada, 74 pp.

Fox, J.M., Hill, P.S., Milligan, T.G., Ogston, A.S. and Boldrin, A. (2004) Floc fraction in the waters of the Po River prodelta. *Cont. Shelf Res.*, **24**, 1699–1715.

Friedrichs, C.T., Wright, L.D., Hepworth, D.A. and Kim, S.C. (2000) Bottom-boundary-layer processes associated with fine sediment accumulation in coastal seas and bays. *Cont. Shelf Res.*, **20**, 807–841.

Garvine, R.W. (1987) Estuary plumes and fronts in shelf waters: a layer model. *J. Phys. Ocean.*, **17**, 1877–1896.

Geyer, W.R., Beardsley, R.C., Candela, J., *et al.* (1996) Physical oceanography of the Amazon Shelf. *Cont. Shelf Res.*, **16**, 575–616.

Geyer, W.R., Hill, P.S., Milligan, T.G. and Traykovski, P. (2000) The structure of the Eel River plume during floods. *Cont. Shelf Res.*, **20**, 2067–2093.

Gibbs, R.J. (1982a) Floc breakage during HIAC light-blocking size analysis. *Environ. Sci. Technol.*, **16**, 298–299.

Gibbs, R.J. (1982b) Floc stability during Coulter counter sizer analysis. *J. Sediment. Petrol.*, **52**, 657–660.

Gibbs, R.J. and Konwar, L. (1983) Disruption of mineral flocs using Niskin bottles. *Environ. Sci. Technol.*, **17**, 374–375.

Grant, W.D. and Madsen, O.S. (1986) The continental-shelf bottom boundary layer. *Ann. Rev. Fluid Mech.*, **18**, 265–305.

Hamblin, A.P. and Walker, R.G. (1979) Storm-dominated shallow marine deposits: the Fernie-Kootenay (Jurassic) transition, southern Rocky Mountains. *Can. J. Earth Sci.*, **16**, 1673–1690.

Hill, P.S. (1992) Reconciling aggregation theory with observed vertical fluxes following phytoplankton blooms. *J. Geophys. Res.*, **97**(C2), 2295–2308.

Hill, P.S. (1996) Sectional and discrete representations of floc breakage in agitated suspensions, *Deep-sea Res.*, I, **43**(5), 679–702.

Hill, P.S. and Nowell, A.R.M. (1995) Comparison of two models of aggregation in continental-shelf bottom boundary layers. *J. Geophys. Res.*, **100**(C11), 22749–22763.

Hill, P.S., Syvitski, J.P., Cowan, E.A. and Powell, R.D. (1998) *In situ* observations of floc settling velocities in Glacier Bay, Alaska. *Mar. Geol.*, **145**, 85–94.

Hill, P.S., Milligan, T.G. and Geyer, W.R. (2000) Controls on effective settling velocity in the Eel River flood plume. *Cont. Shelf Res.*, **20**, 2095–2111.

Hoskin, C.M. and Burrell, D.C. (1972) Sediment transport and accumulation in a fjord basin, Glacier Bay, Alaska. *J. Geol.*, **80**, 539–551.

Hunt, J.R. 1986. Particle aggregate breakup by fluid shear. In: *Estuarine Cohesive Sediment Dynamics* (Ed. A.J. Mehta), pp. 85–109. Springer-Verlag, New York.

Johnson, B.D. and Wangersky, P.J. (1985) A recording backward scattering meter and camera system for examination of the distribution and morphology of macroaggregates. *Deep-sea Res.*, **32**, 1143–1150.

Kineke, G.C., Sternberg, R.W., Trowbridge, J.H. and Geyer, W.R. (1996) Fluid-mud processes on the Amazon continental shelf. *Cont. Shelf Res.*, **16**, 667–696.

Kineke, G.C., Woolfe, K.J., Kuehl, S.A., *et al.* (2000) Sediment export from the Sepik River, Papua New Guinea: evidence for a divergent sediment plume. *Cont. Shelf Res.*, **20**, 2239–2266.

Kirby, R.R. and Parker, W.R. (1977) The physical characteristics and environmental significance of fine-sediment suspensions in estuaries. *Symposium on Estuaries, Geophysics and the Environment*, National Academy of Science, Washington, pp. 110–120.

Kranck, K. (1973) Flocculation of suspended sediment in the sea. *Nature*, **246**, 348–350.

Kranck, K. (1980) Experiments on the significance of flocculation in the settling of fine-grained sediment in still water. *Can. J. Earth Sci.*, **17**, 1517–1526.

Kranck, K. (1981) Particulate matter grain-size characteristics and flocculation in a partially mixed estuary. *Sedimentology*, **28**, 107–114.

Kranck, K. (1984) The role of flocculation in the filtering of particulate matter in estuaries. In: *The Estuary as a Filter* (Ed. V. Kennedy), pp. 159–175. Academic Press.

Kranck, K. and Milligan, T.G. (1992) Characteristics of suspended particles at an 11-hour anchor station in San Francisco Bay, California. *J. Geophys. Res.*, **97**, 11373–11382.

Kranck, K., Petticrew, E., Milligan, T.G. and Droppo, I.G. (1992) *In situ* particle size distributions resulting from flocculation of suspended sediment. In: *Nearshore and Estuarine Cohesive Sediment Tranport* (Ed. A.J. Mehta), pp. 60–74. Coastal and Estuarine Studies Series, Vol. 42, American Geophysical Union, Washington, DC.

Krone, R.B. (1962) *Flume Studies of the Transport of Sediment in Estuarial Shoaling Processes*. Hydraulic Engineering Laboratory and Sanitary Engineering Research Laboratory, University of California, Berkeley.

Leithold, E.L. (1989) Depositional processes on an ancient and modern muddy shelf, northern California. *Sedimentology*, **36**, 179–202.

Leithold, E.L. and Hope, R.S. (1999) Deposition and modification of a flood layer on the northern California shelf: lessons from and about the fate of terrestrial particulate organic carbon. *Mar. Geol.*, **154**, 183–195.

Lentz, S.J. and Helfrich, K.R. (2002) Buoyant gravity currents along a sloping bottom in a rotating fluid. *J. Fluid Mech.*, **464**, 251–278.

Marr, J.E. (1929) *Deposition of the Sedimentary Rocks*. Cambridge University Press, 245 pp.

McCave, I.N. (1972) Transport and escape of fine-grained sediment from shelf areas. In: *Shelf Sediment Transport: Process and Pattern* (Eds D.J.P. Swift, D.B. Duane and O.H. Pilkey), pp. 225–248. Dowden, Hutchinson and Ross, Stroudsburg, PA.

McCave, I.N. (1975) Vertical flux of particles in the ocean. *Deep-sea Res.*, **22**, 491–502.

McCave, I.N. (1984) Size spectra and aggregation of suspended particles in the deep ocean. *Deep-sea Res.*, **31**, 329–352.

McPhee-Shaw, E.E., Sternberg, R.W., Mullenbach, B. and Ogston, A.S. (2004) Observations of intermediate nepheloid layers on the northern California continental margin. *Cont. Shelf Res.*, **24**, 693–720.

Mertes, L.A.K. and Warrick, J.A. (2001) Measuring flood output from 110 coastal watersheds in California with field measurements and SeaWiFS. *Geology*, **29**, 659–662.

Migniot, C. (1968) Etude des proprietes physiques de differents sediments tres fins et de leur comportement sous des action hydrodynamiques. *La Houille Blanche*, **23**, 591–620.

Milligan, T.G. (1995) An examination of the settling behaviour of flocculated suspensions. *Neth. J. Sea Res.*, **33**, 163–171.

Milligan, T.G. and Hill, P.S. (1998) A laboratory assessment of the relative importance of turbulence, particle composition and concentration in limiting maximal floc size. *J. Sea Res.*, 39, 227–241.

Milligan, T.G. and Kranck, K. (1991) Electro-resistance particle size analyzers. In: Principles, Methods and Applications of Particle Size Analysis (Ed. J.P.M. Syvitski), pp. 109–118. Cambridge University Press, New York.

Mitchum, R.M. Jr., Vail, P.R. and Thompson III, S. (1977) Seismic stratigraphy and global changes in sea level, Part 2: the depositional sequence as the basic unit for stratigraphic analysis. In: *Seismic Stratigraphy – Applications to Hydrocarbon Exploration* (Ed. C.E. Payton), pp. 53–62. Memoirs 26, American Association of Petroleum Geologists, Tulsa, OK.

Moore, D.G. (1969) *Reflection Profiling Studies of the California Continental Borderland: Structure and Quaternary Turbidite Basins*. Geol. Soc. Am. Spec. Pap., **107**, 142 pp.

Morehead, M.D. and Syvitski, J.P. (1999) River-plume sedimentation modeling for sequence stratigraphy: application to the Eel margin, northern California. *Mar. Geol.*, **154**, 29–41.

Morehead, M.D., Syvitski, J.P.M., Hutton, E.W.H. and Peckham, S.D. (2003) Modeling the inter-annual and intra-annual variability in the flux of sediment in ungauged river basins. *Global Planet. Change,* **39**(1/2): 95–110.

Mulder, T. and Syvitski, J.P.M. (1995) Turbidity currents generated at river mouths during exceptional discharges to the world oceans. *J. Geol.*, **103**, 285–299.

Mullenbach, B.L. and Nittrouer, C.A. (2000) Rapid deposition of fluvial sediment in the Eel Canyon, northern California. *Cont. Shelf Res.*, **20**, 2191–2212.

Mullenbach, B.L., Nittrouer, C.A., Puig, P. and Orange, D. (2004) Sediment deposition in a modern submarine canyon: Eel Canyon, northern California. *Mar. Geol.*, **211**, 101–119.

Myrow, P.M. and Southard, J.B. (1996) Tempestite deposition. *J. Sediment. Res.*, **66**(5), 875–887.

Nelson, B.W. (1970) Hydrography, sediment dispersal and recent historical development of the Po River delta. In: *Deltaic Sedimentation – Modern and Ancient* (Eds J.P. Morgan and R.H. Shaver), pp. 152–184. Special Publication 15, Society of Economic Paleontologists and Mineralogists, Tulsa, OK.

Nittrouer, C.A. (1999) STRATAFORM: overview of its design and synthesis of its results. *Mar. Geol.*, **154**, 3–12.

Nittrouer, C.A. and Sternberg, R.W. (1981) The formation of sedimentary strata in an allochthonous shelf environment: the Washington continental shelf. *Mar. Geol.*, **42**, 201–232.

Nolan, K.M. and Janda, R.J. (1995) Impacts of logging on stream-sediment discharge in the Redwood Creek Basin, northwestern California. In: *Geomorphic Processes and Aquatic Habitat in the Redwood Creek Basin, Northwestern California* (Eds K.M. Nolan, H.M. Kelsey and D.C. Marron), pp. L1–L8. *U.S. Geol. Surv. Prof. Pap.*, **1454**.

Nolan, K.M., Kelsey, H.M. and Marron, D.C. (1995) Summary of research in the Redwood Creek basin, 1973–1983. In: *Geomorphic Processes and Aquatic Habitat in the Redwood Creek Basin, Northwestern California* (Eds K.M. Nolan, H.M. Kelsey and D.C. Marron), pp. A1–A6. *U.S. Geol. Surv. Prof. Pap.*, **1454**.

Nunn, P.D. (1999) *Environmental Change in the Pacific Basin.* John Wiley & Sons, Chichester, 357 pp.

Ogston, A.S. and Sternberg, R.W. (1999) Sediment-transport events on the northern California continental shelf. *Mar. Geol.*, **154**, 69–82.

Ogston, A.S., Cacchione, D.A., Sternberg, R.W. and Kineke, G.C. (2000) Observations of storm and river flood-driven sediment transport on the northern California continental shelf. *Cont. Shelf Res.*, **20**, 2141–2162.

O'Melia, C.R. and Tiller, C.L. (1993) Physiochemical aggregation and deposition in aquatic environments. In: *Environmental Particles*, Vol. 2 (Eds J. Buffle and H.P. van Leeuwen), pp. 353–386. Lewis Publishers, Boca Raton, FL.

Palanques, A. and Drake, D.E. (1990) Distribution and dispersal of suspended particulate matter on the Ebro continental shelf, northwestern Mediterranean Sea. *Mar. Geol.*, **95**, 193–206.

Parsons, J.D., Bush, J.W.M. and Syvitski, J.P.M. (2001) Hyperpycnal plumes with small sediment concentrations. *Sedimentology*, **48**, 465–478.

Postma, H. (1967) Sediment Transport and Sedimentation in the Estuarine Environment. In: Estuaries (Ed. G.H. Lauff), pp. 158–179. A.A.A.S. pub 83.

Puig, P., Ogston, A.S., Mullenbach, B.L., Nittrouer, C.A. and Sternberg, R.W. (2003) Shelf-to-canyon sediment-transport processes on the Eel continental margin (northern California). *Mar. Geol.*, **193**, 129–149.

Puig, P., Ogston, A.S., Mullenbach, B.L., Nittrouer, C.A., Parsons, J.D. and Sternberg, R.W. (2004) Storm-induced sediment-gravity flows at the head of the Eel submarine canyon. *J. Geophys. Res.*, **109**, C03019, doi:10.1029/2003JC001918.

Schubel, J.R., Wilson, R.E. and Okubo, A. (1978) Vertical transport of suspended sediment in upper Chesapeake Bay. In: *Estuarine Transport Processes* (Ed. B. Kjerfve), pp. 161–175. University of South Carolina Press, Columbia, SC.

Scrutton, P.C. and Moore, D.G. (1953) Distribution of surface turbidity off Mississippi delta. *Bull. Am. Assoc. Petrol. Geol.*, **37**, 1067–1074.

Scully, M.E., Friedrichs, C.T. and Wright, L.D. (2002) Application of an analytical model of critically stratified gravity-driven sediment transport and deposition to observations from the Eel River continental shelf, northern California. *Cont. Shelf Res.*, **22**, 1951–1974.

Scully, M.E., Friedrichs, C.T. and Wright, L.D. (2003) Numerical modeling of gravity-driven sediment transport and deposition on an energetic continental shelf: Eel River, northern California. *J. Geophys. Res.*, **108**, 17-1–17-14.

Seymour, R.J. (1986) Nearshore auto-suspending turbidity flows. *Ocean Eng.*, **13**, 435–447.

Shepard, F.P. (1932) Sediments of the continental shelves. *Geol. Soc. Am. Bull.*, **43**, 1017–1040.

Sherwood, C.R., Butman, B., Cacchione, D.A., *et al.* (1994) Sediment-transport events on the northern California continental shelf during the 1990–91 STRESS experiment. *Cont. Shelf Res.*, **14**, 1063–1099.

Smith, J.D. and Hopkins, T.S. (1972) Sediment transport on the continental shelf off of Washington and Oregon in light of recent current measurements. In: *Shelf Sediment Transport: Process and Pattern* (Eds D.J.P. Swift, D.B. Duane and O.H. Pilkey), pp. 143–180. Dowden, Hutchinson and Ross, Stroudsburg, PA.

Sommerfield, C.K., Nittrouer, C.A. and Alexander, C.R. (1999) ^{7}Be as a tracer of flood sedimentation on the northern California continental margin. *Cont. Shelf Res.*, **19**, 335–361.

Sternberg, R.W. and Larsen, L.H. (1976) Frequency of sediment movement on the Washington continental shelf: a note. *Mar. Geol.*, **21**, 37–47.

Sternberg, R.W. and McManus, D.A. (1972) Implications of sediment dispersal from long-term, bottom-current measurements on the continental shelf of Washington. In: *Shelf Sediment Transport: Process and Pattern* (Eds D.J.P. Swift, D.B. Duane and O.H. Pilkey), pp. 181–194. Dowden, Hutchinson and Ross, Stroudsburg, Pennsylvania.

Sternberg, R.W., Berhane, I. and Ogston, A.S. (1999) Measurement of size and settling velocity of suspended aggregates on the northern California continental shelf. *Mar. Geol.*, **154**, 43–53.

Stumm, W. and Morgan, J.J. (1981) *Aquatic Chemistry*, 2nd edn. Wiley-Interscience, New York, 780 pp.

Swift, D.J.P. (1970) Quaternary shelves and the return to grade. *Mar. Geol.*, **8**, 5–30.

Swift, D.J.P., Duane, D.B. and Pilkey, O.H. (1972) *Shelf Sediment Transport: Process and Pattern.* Dowden, Hutchinson and Ross, Stroudsburg, PA, 656 pp.

Syvitski, J.P.M. and Murray, J.W. (1981) Particle interaction in fjord suspended sediment. *Mar. Geol.*, **39**, 215–242.

Syvitski, J.P.M., Asprey, K.W., Clattenburg, D.A. and Hodge, G.D. (1985) The prodelta environment of a fjord: suspended particle dynamics. *Sedimentology*, **32**, 83–107.

Syvitski, J.P.M., Asprey, K.W. and Heffler, D.E. (1991) The Floc Camera: a three-dimensional imaging system of suspended particulate matter. In: *Microstructure of Fine-grained Sediments* (Eds R.H. Bennett, W.R. Bryant and M.H. Hulbert), pp. 281–289. Springer-Verlag, New York.

Syvitski, J.P.M., Asprey, K.W. and LeBlanc, K.W.G. (1995) *In situ* characteristics of particles settling within a deep-water estuary. *Deep-sea Res.*, II, **42**, 223–256.

Syvitski, J.P.M., Morehead, M.D., Bahr, D. and Mulder, T. (2000) Estimating fluvial sediment transport: the rating parameters. *Water Resour. Res.*, **366**, 2747–2760.

ten Brinke, W.B.M. (1994) Settling velocities of mud aggregates in the Oosterchelde tidal basin (the Netherlands), determined by a submersible video system. *Estuar. Coast. Shelf Sci.*, **39**, 549–564.

Traykovski, P., Geyer, W.R., Irish, J.D. and Lynch, J.F. (2000) The role of wave-induced density-driven fluid mud flows for cross-shelf transport on the Eel River continental shelf. *Cont. Shelf Res.*, **20**, 2113–2140.

Trowbridge, J.H. and Kineke, G.C. (1994) Structure and dynamics of fluid muds on the Amazon continental shelf. *J. Geophys. Res.*, **99**, 865–874.

Walker, R. (1973) Mopping up the turbidite mess. In: *Evolving Concepts in Sedimentology* (Ed. R.N. Ginsburg), pp. 1–37. Studies in Geology, No. 21, Johns Hopkins University, Baltimore.

Walsh, J.P. and Nittrouer, C.A. (1999) Observations of sediment flux to the Eel continental slope, northern California. *Mar. Geol.*, **154**, 55–68.

Wheatcroft, R.A. (2000) Oceanic flood sedimentation: a new perspective. *Cont. Shelf Res.*, **20**, 2059–2066.

Wheatcroft, R.A. and Borgeld, J.C. (2000) Oceanic flood deposits on the northern California shelf: large-scale distribution and small-scale physical properties. *Cont. Shelf Res.*, **20**, 2163–2190.

Wheatcroft, R.A., Sommerfield, C.K., Drake, D.E., Borgeld, J.C. and Nittrouer, C.A. (1997) Rapid and widespread dispersal of flood sediment on the northern California margin. *Geology*, **25**, 163–166.

Whitehouse, U.G., Jeffrey, L.M. and Debrecht, J.D. (1960) Differential settling tendencies of clay minerals in saline waters. *Proceedings of the 7th National Conference on Clays and Clay Minerals*, pp. 1–79.

Wright, L.D. (1977) Sediment transport and deposition at river mouths: a synthesis. *Geol. Soc. Am. Bull.*, **88**, 857–868.

Wright, L.D. and Coleman, J.M. (1974) Mississippi River mouth processes: effluent dynamics and morphologic development. *J. Geol.*, **82**, 751–778.

Wright, L.D., Wiseman, W.J., Bornhold, B.D., *et al.* (1988) Marine dispersal and deposition of Yellow River silts by gravity-driven underflows. *Nature*, **332**, 629–632.

Wright, L.D., Kim, S.-C. and Friedrichs, C.T. (1999) Across-shelf variations in bed roughness, bed stress and sediment suspension on the northern California continental shelf. *Mar. Geol.*, **154**, 99–115.

Wright, L.D., Friedrichs, C.T., Kim, S.C. and Scully, M.E. (2001) Effects of ambient currents and waves on gravity-driven sediment transport on continental shelves. *Mar. Geol.*, **175**, 25–45.

Post-depositional alteration and preservation of sedimentary strata

ROBERT A. WHEATCROFT*, PATRICIA L. WIBERG†, CLARK R. ALEXANDER‡, SAMUEL J. BENTLEY§, DAVID E. DRAKE¶, COURTNEY K. HARRIS** and ANDREA S. OGSTON††

*College of Oceanic and Atmospheric Sciences, Oregon State University, Corvallis, OR 97331, USA (Email: raw@coas.oregonstate.edu)
†Department of Environmental Sciences, University of Virginia, Charlottesville VA 22904, USA
‡Skidaway Institute of Oceanography, Savannah, GA 31411, USA
§Earth Sciences Department, Memorial University of Newfoundland, St John's NL A1B 3X5, Canada
¶Drake Marine Consulting, Ben Lomond, CA 95005, USA
**Virginia Institute of Marine Science, Gloucester Point, VA 23062, USA
††School of Oceanography, University of Washington, Seattle, WA 98195, USA

ABSTRACT

All sediment that is delivered to the seabed is subject to post-depositional alteration before it becomes part of the preserved stratigraphic record. Physical and biological processes occurring in the upper few decimetres of the seabed alter newly deposited sediment, thereby creating the fine-scale sedimentary record. The geographical region of focus is the Eel margin of northern California, where a combination of high precipitation, frequent winter storms and intense benthic biological processes exert fundamental controls on the fine-scale stratigraphy. Major findings are that:

1 although the wave conditions on the Eel shelf are highly energetic, only a few millimetres to one or two centimetres of bed material are typically put in suspension during storms;
2 despite a high initial porosity (i.e. low strength), flood beds quickly become resistant to erosion;
3 shelf macrofauna show considerable resilience to deposition of flood sediment;
4 storm deposits have a low preservation potential on open continental margins;
5 episodic sedimentation is key to preservation of strata and signals produced by flood and storm events.

Future insight into post-depositional alteration will benefit from focused observations and theoretical modelling that explicitly couple physical and biological processes on multiple time-scales.

Keywords Consolidation, porosity, waves, currents, bedforms, bioturbation, macrofauna, storm deposits, flood deposits, sedimentary structures.

INTRODUCTION AND SCOPE

All sediment that is delivered to the seabed, whether as discrete particles, as multigrain aggregates or as *en-masse* layers, is subject to post-depositional alteration before it becomes part of the preserved stratigraphic record. Sources of alteration are diverse in origin, involving physical, chemical and biological processes. Also known as **diagenesis**, post-depositional alteration has been discussed in many monographs (e.g. Bathurst, 1971; Rieke & Chilingarian, 1974; McCave, 1976; Berner, 1980; Boudreau, 1997; Boudreau & Jørgensen, 2001) that highlight its importance and underscore its exceptional breadth. Any attempt to discuss concisely post-depositional alteration processes must therefore be limited in scope, and this contribution is no exception. This review is restricted to a consideration of 'earliest' diagenesis; that is, time-scales of change extend from hours to years and vertical length scales below the sediment–water interface range from millimetres to 1 or 2 dm (note that Berner (1980) published an influential book, *Early Diagenesis*, which encompassed length scales of a few hundred metres, and therefore time-scales of 10^3 to 10^6 yr that are well in excess of those covered herein). In addition, because the primary focus of this paper is sedimentary strata, for the most part, chemical alteration processes such as organic carbon degradation or calcite dissolution (see e.g. Berner, 1980) are not considered. That is not to say that chemical processes are unimportant; rather, compared with physical and (macro) biological processes, they are likely to have a less direct impact on the preserved stratigraphic record, especially that which may be sensed seismo-acoustically (e.g. Stoll, 1989).

Post-depositional alteration and hence preservation are considered to be a 'local' issue. That is, the controls on various processes, whether it is gravity as in the case of consolidation, or benthic biomass as in the case of bioturbation intensity, depend only on the magnitude of that parameter at the site in question. Certainly, spatial gradients occur in forcing and static variables, but there is little long-range interaction; that is, the seabed roughness at one site has no impact on the roughness of a site 5 km along shelf. This local perspective means that it is possible to take a one-dimensional view, in which significant changes occur only in the vertical dimension (where the z-axis is positive into the bed).

The subject matter includes physical and biological processes that lead to the alteration and preservation of sedimentary strata. What exactly is being altered and what is being preserved? As detailed herein, post-depositional alteration acts in two ways. First, various primary seabed properties, such as grain size, bottom roughness and porosity, can be modified by **physical and biological processes** (e.g. consolidation, storm reworking, deposit feeding). These bed properties are often dynamically important (they affect bottom shear stress and erosion thresholds), and they can influence subsequent sediment resuspension or deposition. Thus, there are strong linkages among various alteration processes necessitating a multifaceted approach. Second, post-depositional alteration may replace primary, typically physiogenic signals (bedding, physical sedimentary structures) with new sedimentary fabrics or structures that are usually,

but not exclusively, biogenic. In terms of preservation, the concern of this paper is mainly with high-frequency sedimentary signals related to specific events, such as a flood or storm. Thus, the subjects addressed are, for example, whether a grain-size signature associated with a storm-event bed has a chance of preservation (i.e. recognition) in the sedimentary record. Issues associated with preservation also have an impact on facies indicators because it is ultimately the combination of primary and secondary sedimentary signals or fabrics that constitute the sedimentary record.

Like many of the other papers in this volume that discuss processes (e.g. Hill *et al.*, pp. 49–99; Sommerfield *et al.*, pp. 157–212), the focus is on alteration and preservation processes on the Eel River margin of northern California (Fig. 1). In particular, the concern is mainly with the deeper portions of the sandy inner shelf and with the muddy middle–outer shelf, which collectively extend from ~50 m to ~130 m. Some attention is devoted to the non-canyon upper slope as well. The reason for this focus is threefold. First, it is on the middle to outer shelf that arguably the most diverse set of alteration processes occur. There, both physical and biological processes are involved in post-depositional alteration, whereas at other locations on the margin some important processes are excluded. For example, in deep water (i.e. > 130 m) wave reworking is much less frequent and biological processes dominate, whereas in shallow portions of the shelf the bed is almost always mobile, there is little net deposition, and storm-driven bedform reworking dominates. Second, the data set is most complete on the muddy portion of the Eel shelf. There, a comprehensive coring campaign involving more than 15 cruises over a 5-yr period provided an unprecedented view of the shallow seabed. In addition, bottom boundary-layer tripods were deployed at several 50-, 60- and 70-m sites during the winter storm season, and at a 60-m site for a 5-year period, thereby providing important time-series information on physical forcing. Although the geographical and bathymetric restrictions may seem limiting, both the types and rates of alteration processes present on the muddy Eel shelf have considerable general significance.

The third reason for devoting most attention to the middle to outer shelf is that a unique oppor-

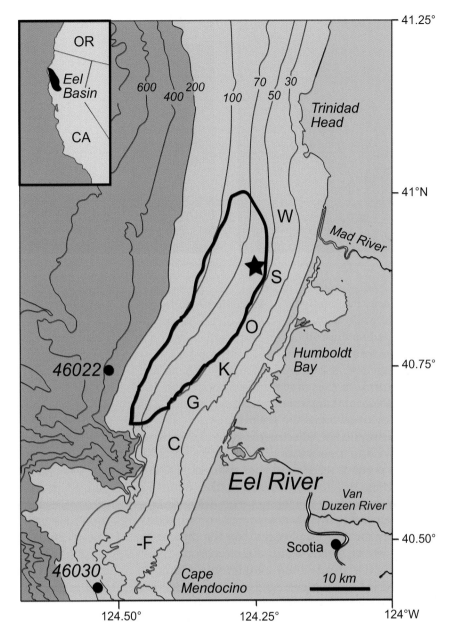

Fig. 1 Map showing the bathymetry (in metres), the location of major rivers and key features on the Eel River margin. The solid circles mark the location of a US Geological Survey gauging station at Scotia, and NOAA's National Data Buoy Center (NDBC) meteorological buoys discussed herein (46030 and 46022). The star marks the location of the long-term monitoring tripod at station S60 (e.g. Ogston *et al.*, 2004) and the heavy black line encloses the approximate location of the 1995 and 1997 flood deposits (Wheatcroft & Borgeld, 2000). Station locations discussed in the text are referenced by a letter designation corresponding to shore-normal transects and nominal water depths (i.e. station K60 is at ~60 m on the K-line).

tunity occurred in that location. As described in detail elsewhere (Wheatcroft *et al.*, 1996; Wheatcroft *et al.*, 1997; Ogston *et al.*, 2000; Traykovski *et al.*, 2000; Wheatcroft & Borgeld, 2000; Hill *et al.*, this volume, pp. 49–99), historically significant flooding by the Eel River created several flood deposits on the adjacent shelf (Fig. 2). These flood deposits had diagnostic sedimentological and geochemical properties (Wheatcroft *et al.*, 1996; Drake, 1999; Leithold & Hope, 1999; Sommerfield *et al.*, 1999; Wheatcroft & Borgeld, 2000) that allowed precise

monitoring of various changes over time. The importance of capturing event layers at an early stage of their evolution cannot be overstated. In most studies of post-depositional alteration, an investigator is faced with the unfortunate situation of making observations and measurements at some undetermined point in a layer's evolution, thereby introducing considerable uncertainty to many aspects of the problem (e.g. initial conditions, absolute rates). In the STRATAFORM study several events were 'caught' while they were happening,

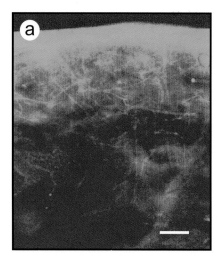 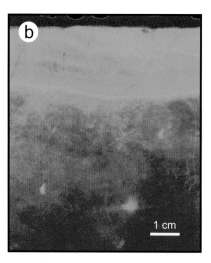

Fig. 2 False-colour representations of scanned X-radiographs showing (a) the steady-state seabed characterized by abundant, multiscale biogenic sedimentary structures and a general lack of bedding, and (b) a 3-cm-thick flood deposit formed in January 1997 at station C70. In each X-radiograph, the colours red, green and blue represent water, X-ray transparent sediment and relatively X-ray opaque sediment, respectively. Scale bars = 1 cm.

thereby avoiding many of these problems. Although a certain amount of luck was involved in recording the event deposits from the start, data interpretation was helped immeasurably by taking an event-response sampling approach (Wheatcroft, 2000). Future studies of margin sedimentation will benefit from making event-response sampling an explicit component of the science plan.

The approach to discussing post-depositional alteration and preservation is to begin with the simplest possible set of conditions and build in complexity (Fig. 3). Thus **consolidation**, the sequential decrease in porosity that occurs with depth or time in marine sediments, is considered first. Initially, the treatment is focused mainly on the theoretical framework of steady-state, abiotic consolidation (Fig. 3a). The steady-state assumption is quickly relaxed (Fig. 3b), however, and the observed short-term changes in the porosity (or wet bulk density) of the Eel flood deposits described. Unsteadiness is a particular focus of the subsequent section on **physical alteration processes**. In particular, circumstances leading to erosion and the response of the seabed to physical forcing (e.g. winnowing) are discussed in detail. In addition, modes and rates of deposition are described. **Biological alteration processes** are discussed next, with subsections on alteration of key dynamical bed properties and sediment bioturbation (Fig. 3c). The section on **strata preservation** integrates the physical and biological processes described earlier. Last, the Eel margin research is summarized *vis-à-vis* post-depositional processes. For simplicity, whenever possible issues

are discussed separately, however, it is important to note that feedbacks and linkages are the rule, and that identifying, quantifying and modelling these interactions remains a formidable future challenge in marine sedimentary geology.

CONSOLIDATION

Consolidation is discussed first because it is ubiquitous; all sediment deposited on the seafloor is subject to post-depositional consolidation. Other processes such as physical reworking or bioturbation may be absent from a particular site, but sediment consolidation will always occur. The essence of consolidation (herein viewed as equivalent to **compaction**) is that an overburden (e.g. overlying sediment) causes excess pore pressures that force pore water to move upward and the volume fraction of pore space within the seabed to decrease over time. For steady-state sediment accumulation, depth below the seabed is directly proportional to time. Longer consolidation times and larger overburdens with increasing depth lead to a vertical porosity gradient in the upper 10 cm of the seabed. Most often, this gradient can be described as an exponential function that asymptotically approaches some uniform value (Fig. 4). If grain density remains uniform with depth, then decreases in porosity result in corresponding increases in **bulk density**.

Understanding controls on the **porosity** (φ) of marine sediment and how it changes vertically and

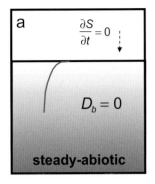

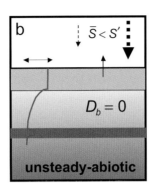

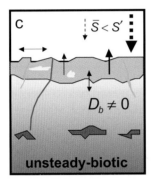

Fig. 3 A schematic representation of the dominant processes and products discussed in the text. (a) In the simplest case, physical processes are steady and no biological effects are included (i.e. sediment bioturbation, represented by D_b, is zero). The fluctuating component of sediment deposition (S') is much smaller than the mean sediment accumulation ($\bar{S}$). Waves and currents result in bottom stresses below threshold values for erosion, hence resuspension is lacking, and the vertical profile of porosity (blue line) is a decreasing exponential curve typical of a normally consolidated bed. The seafloor is flat and, except for porosity, vertical and horizontal properties of the seabed are uniform. (b) In the unsteady, abiotic case, the fluctuating component of sediment deposition is large relative to the mean sediment accumulation. As a result, event layers are formed and these perturb the vertical porosity profile, and cause vertical variations in seabed properties (e.g. grain size, organic carbon). Episodic wave or current events are now important and these result in resuspension and/or winnowing of bottom sediment. Biological processes are still absent, so small-scale horizontal variability in seabed properties and hence erodibility are lacking. The seafloor may be flat or remoulded into bedforms in sandy settings. (c) In the most complex and realistic situation, the system is unsteady (as described above) and biologically modulated. Activities of macrobenthos cause: the redistribution and blurring of sedimentary signals; the destruction of bedforms and the creation of complex biogenic microtopography; and significant small-scale horizontal and vertical variability in seabed properties.

through time (i.e. consolidation) is important for a number of reasons. First, critical shear stresses and erosion rates of fine-grained sediment are a strong function of sediment porosity (Dade *et al.*, 1992), whereby higher-porosity, less-consolidated sediment is more readily eroded. Second, models that couple shelf sediment-transport with bed evolution (Harris & Wiberg, 2002) must take into account the **solid fraction** ($\varphi_s = 1 - \varphi$) of sediment in the near-surface portion of the seabed. Therefore, having information on the initial porosity and how it changes over time is important for building better sediment-transport and bed-evolution models. Third, proper interpretation of geochemical measurements (e.g. [210]Pb; Mulsow *et al.*, 1998) depends on knowledge of vertical porosity gradients. Fourth, the interaction of high-frequency sound with the seabed depends in part on the sediment bulk density (Hamilton, 1980; Stoll, 1989). Last, bulk-density changes of discrete layers must be accounted for, if accurate mass-balance calculations (Drake, 1999) or time-series interpretation of radiographs (Bentley & Nittrouer, 2003) are to be made.

Owing to the importance of sediment consolidation to many fields of study (e.g. civil engineering, soil science, acoustics), there has been much prior

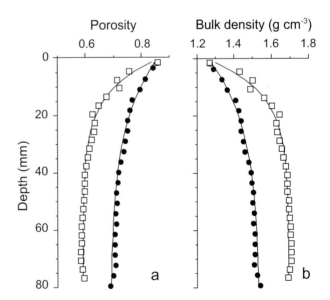

Fig. 4 Representative steady-state porosity (φ) and wet-bulk-density profiles (ρ_b) from a 70-m ($\square$) and a 150-m ($\bullet$) station on the Eel margin. Solid lines are empirical fits of the form $\varphi = (\varphi_0 - \varphi_\infty) \exp(-az) + \varphi_\infty$ (see text) with grain density (ρ_s) assumed to be 2.65 g cm^{-3}.

research in this area. Most consolidation studies, however, have either been theoretical (Gibson *et al.*, 1967; Toorman, 1996; Boudreau & Bennett, 1999) or laboratory based (Skempton, 1970; Been & Sills, 1981), and therefore many factors potentially important in marine environments, such as wave pumping and bioturbation, have been neglected. For the most part, this situation is not redressed herein, although, as shown below, the measurements from the Eel shelf provide field data on consolidation of flood deposits and insight into complicating issues, such as bioturbation (see also Lee *et al.*, this volume, pp. 213–274).

Theoretical framework

Within a seabed deposit, overlying sediment and/or an external load exerts a stress on the underlying sediment. A portion of the downward force per unit area (total stress) due to the weight of the overlying material is supported by the solid framework of the bed (i.e. grains are in contact). The remainder is supported by pore-fluid pressure. In a fully consolidated deposit, the pore-fluid pressure is **hydrostatic** (i.e. pressure imposed by overlying water). In deposits that have not fully consolidated, **excess pore pressures** exist (pressures in excess of hydrostatic) that drive an upward pore-water flow. This flow reduces the excess pore pressure while increasing the load supported by the solid framework. The transfer of load leads to a slow collapse of the framework, resulting in a decrease in porosity (increase in bulk density) with time and also with depth below the surface (because load increases with depth). Consolidation proceeds until pore pressures reach hydrostatic values and the buoyant weight of the overlying load is completely supported by the solid framework. The rate of consolidation and the steady-state porosity profile depend on the nature of the load and the permeability and porosity of the bed.

A mathematical representation of consolidation can be derived by combining equations for conservation of solid and fluid mass in the bed with Darcy's law for flow through porous media, and partitioning of the total stress (σ) into pore pressure (p) and the **effective stress** (σ_e, the stress carried by the solid framework). Here, only the case of self-weight consolidation is considered, in which the load is the weight per unit area of the sediment

layer overlying a specified depth within a sediment deposit. For this case, the total stress is given by

$$\sigma = \int_{-z}^{0} [\rho_s(1 - \varphi) + \rho\varphi]g\,dz = \int_{-z}^{0} \rho_b g\,dz \tag{1}$$

where ρ_s is sediment density, ρ is fluid density, ρ_b is bulk density, g is acceleration of gravity and z is depth below the sediment surface (taken to be at $z = 0$).

As sediment is deposited on the seabed, the total stress at any depth within the bed increases in response to the (buoyant) weight of the newly added material, $\Delta\sigma = (\rho_s - \rho)g\Delta z(1 - \varphi) = \rho_b g\Delta z - \rho g\Delta z$, where Δz is the thickness of the new layer and $\rho g\Delta z$ is the hydrostatic fluid pressure (p_h) at the base of the new layer. When a new layer is deposited, the additional stress is initially supported by an increase in pore-water pressure above hydrostatic values. The rate of pore-water flow due to the excess pore pressure, $p_x = p - p_h$, depends on the hydraulic conductivity, K (or **permeability** $k = K\rho g/\mu$, where μ is fluid viscosity; see Middleton & Wilcock, 1994), of the sediment deposit and the excess-pore-pressure gradient. Pore-water velocity, v, is given by Darcy's law

$$\varphi v = -\frac{K}{\rho g}\frac{\partial p_x}{\partial z} \tag{2}$$

The velocity in the consolidation problem is really the relative velocity of the pore fluid (v_w) and the solid grains (v_s); i.e. $v = v_w - v_s$. Substituting $p_x = \Delta\sigma - \sigma_e = (\sigma - p_h - \sigma_e)$ into Darcy's law and using the definition of $\Delta\sigma$ from above, yields (Toorman, 1996)

$$\varphi = (v_w - v_s) = -\frac{K}{\rho g}\left(\frac{\partial\Delta\sigma}{\partial z} - \frac{\partial\sigma_e}{\partial z}\right)$$

$$= -\frac{K}{\rho g}\left[(\rho_s - \rho)g(1 - \varphi) - \frac{\partial\sigma_e}{\partial z}\right] \tag{3}$$

To relate flow to change in porosity, the conservation of mass for the solid and the fluid fractions of the bed must be used. For the solid fraction $\varphi_s = 1 - \varphi$, the conservation of mass equation is

$$\frac{\partial\varphi_s}{\partial t} + \frac{\partial v_s\varphi_s}{\partial z} = 0 = \frac{\partial(1 - \varphi)}{\partial t} + \frac{\partial v_s(1 - \varphi)}{\partial z} \tag{4}$$

For the fluid fraction,

$$\frac{\partial \varphi}{\partial t} + \frac{\partial v_w \varphi}{\partial z} = 0 \qquad (5)$$

Combining these equations gives

$$\frac{\partial}{\partial z} [v_w \varphi + v_s (1 - \varphi)] = 0 \qquad (6)$$

Following Toorman (1996), $U = v_w \varphi + v_s (1 - \varphi)$, and $\varphi(v_w - v_s)$ is rewritten in Darcy's law as $U - v_s$. Furthermore, for saturated sediment, U, the average velocity of the fluid and solid components of the bed, is zero (Toorman, 1996). This leaves

$$
\begin{aligned}
v_s &= \frac{K}{\rho g}\left((\rho_s - \rho)g(1 - \varphi) - \frac{\partial \sigma_e}{\partial z}\right) \\
&= \frac{K(\rho_s - \rho)}{\rho}\varphi_s - \frac{K}{\rho g}\frac{\partial \sigma_e}{\partial z}
\end{aligned} \qquad (7)
$$

Substituting this expression for v_s into the conservation of mass equation for the solid fraction,

$$\frac{\partial \varphi_s}{\partial t} = -\frac{\partial v_s \varphi_s}{\partial z} = \frac{\partial}{\partial z}\left(\frac{-K\varphi_s^2(\rho_s - \rho)}{\rho} + \frac{K\varphi_s}{\rho g}\frac{\partial \sigma_e}{\partial z}\right) \qquad (8)$$

The differential equation for φ_s (Eq. 8) shows that the rate of change for porosity ($\varphi = 1 - \varphi_s$) depends on hydraulic conductivity, porosity and effective stress, and their variation with depth. For high permeability sediment, such as sand, the upward flow of pore water is fast enough that excess pore pressures do not build up, and the load due to newly deposited sediment is transferred rapidly to the solid matrix. Thus, porosity profiles in sands display extremely sharp gradients to the theoretical minimum porosity of approximately 0.35 (Fraser, 1935; Wheatcroft, 2002). In contrast, for low-permeability sediment such as mud, relatively large excess pore pressures can develop if a layer of new sediment is added rapidly. In this case, the time required for consolidation of the new layer can be days to weeks, as shown in a number of laboratory experiments (Been & Sills, 1981; Berlamont *et al.*, 1992). Bioturbation will also affect the rate of change of porosity in fine-grained sediments (Rhoads, 1970, 1974; Mulsow *et al.*, 1998), but the proper way to

quantify its effects in diagenetic models is a continuing topic of discussion (Meysman *et al.*, 2003).

Hydraulic conductivity and effective stress are functions of the solid fraction (φ_s) or porosity (φ). This functional dependence must be made explicit before Eq. 8 can be solved for φ_s. Observations by Berlamont *et al.* (1992) indicate that the relationship between K and φ_s can be approximated as $\log(K) = \alpha \rho_b + \beta$, where α and β are empirically determined and $\rho_b = \rho_s \varphi_s + \rho(1 - \varphi_s)$. They suggest a similar form for σ_e, $\log(\sigma_e/\rho_b) = \alpha' \rho_b + \beta'$ (Berlamont *et al.*, 1992). Alternative formulations are possible (Boudreau & Bennett, 1999, described below). In any case, Eq. 8 is non-linear and also involves a moving upper boundary (the sediment surface as it moves downward during consolidation). The equation cannot be solved analytically or with simple numerical techniques, although a number of numerical approaches for solving Eq. 8 (and related equations expressed in terms of void ratio rather than φ_s) have been proposed (Kynch, 1952; Toorman, 1999; Papanicolaou & Diplas, 1999).

Under equilibrium conditions, $v_w = v_s = 0$, and Eq. 3 reduces to $\partial \sigma_e/\partial z = \varphi_s(\rho_s - \rho)g$. Boudreau & Bennett (1999) combined this with an empirical relationship between σ_e and φ_s to derive an expression describing the steady-state profile of φ_s in a consolidated sediment deposit. Using observations from the California margin off Cape Mendocino, Boudreau & Bennett (1999) proposed a simple constitutive relationship between φ_s and σ_e

$$d\varphi_s = A \exp(-b\sigma_e)d\sigma_e \qquad (9)$$

or, after integration

$$\varphi_s = \frac{A}{b}(1 - e^{-b\sigma_e}) + \varphi_{s0} \qquad (10)$$

where φ_{s0} is the initial solid fraction. Boudreau & Bennett (1999) noted that the ratio $A/b = \varphi_{s\infty} - \varphi_{s0}$ at steady state, where $\varphi_{s\infty}$ is the asymptotic value of the solid fraction at large stress (a large distance below the surface). Combining the steady-state equation for φ_s with the above constitutive equation leads to

$$\varphi_s = \frac{\varphi_{s0} - \varphi_{\infty s}}{\varphi_{s0} + (\varphi_{s\infty} - \varphi_{s0})\exp(-\gamma z)} \qquad (11)$$

where $\gamma \equiv b\varphi_{s\infty}(\rho_s - \rho)g$, a depth attenuation coefficient (Boudreau & Bennett, 1999). By fitting this expression for φ_s to a porosity profile, it is possible to estimate the parameters A and b, thereby completely specifying the relationship between the solid fraction (φ_s) and the stress carried by the sediment matrix (σ_e). This formulation is similar, though not identical, to another commonly used form of the equilibrium solid-concentration or porosity profile, $\varphi_s = (\varphi_{s0} - \varphi_{s\infty})\exp(-az) + \varphi_{s\infty}$ (see Boudreau & Bennett (1999) for a discussion of the two profiles and their differences).

Observations

The primary source of data regarding spatial and temporal variability in porosity (or wet bulk density) on the Eel River continental shelf is shipboard microresistivity profiles. Described in detail in several publications (Manheim & Waterman, 1974; Andrews & Bennett, 1981; Bennett *et al.*, 1990; Martin *et al.*, 1991; Wheatcroft & Borgeld, 2000; Wheatcroft, 2002), shipboard **microresistivity** profiling consisted of inserting a small (~5-mm diameter) probe vertically into sediment subcores and logging the resistance at 0.25-mm to 3-mm depth intervals. Empirical calibrations between the **formation factor** (the ratio of bottom-water resistivity to sediment resistivity) and gravimetrically measured water-content samples permit the calculation of high-resolution porosity profiles (Archie, 1942). Resistivity profiles of the upper 10 cm of the seabed were measured in hundreds of box cores collected on the Eel shelf, thereby permitting an examination of spatial variability in porosity and of temporal changes associated with the consolidation of the 1995 and 1997 flood deposits.

Average porosity profiles collected in September 1995 outside of the 1995 flood deposits (Wheatcroft & Borgeld, 2000) are shown for various depths in Fig. 5. The resulting profiles illustrate two important points. The profiles are for the most part consistent with the theoretical, steady-state shape (Fig. 4) with sharp gradients in the upper 3 cm of the seabed merging into a uniform or slowly decreasing zone below. There is also a clear trend with water depth, whereby shallow stations (i.e. 60–70 m) have an asymptotic porosity of ~0.6, and deeper stations (i.e. 110–130 m) are closer to 0.7 in the subsurface. This pattern probably reflects

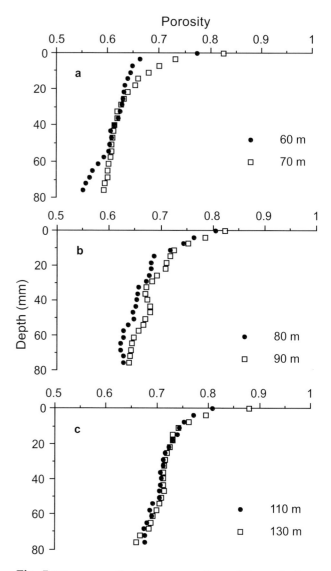

Fig. 5 Mean porosity in the upper 8 cm of the seabed measured in September 1995 at sites outside of the 1995 flood deposits (see Wheatcroft & Borgeld, 2000). Each profile is an average of three to seven stations.

the seaward decrease in mean grain size (Borgeld, 1985), which has an important impact on porosity (Bennett *et al.*, 1990).

In contrast to the steady-state equilibrium condition, porosity profiles in fresh flood deposits on the Eel shelf are characterized by high surficial porosity (Wheatcroft *et al.*, 1996; Wheatcroft & Borgeld, 2000). Near-surface porosity values for the January 1995 and January 1997 flood deposits are in the range of 0.8 to 0.9 compared with a porosity of 0.6–0.7 in ambient (or underlying) sediment (Fig. 6). In addition, the shape of the near-surface

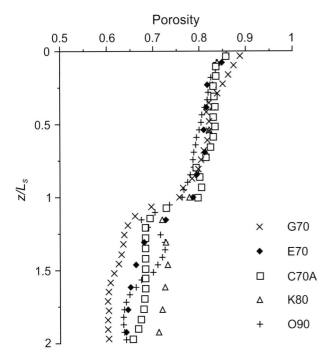

Fig. 6 Porosity as a function of depth (z) normalized by flood-deposit thickness (L_s) measured about 2 weeks after the January 1997 flood. Data points where $z/L_s < 1$ are within the flood deposit, whereas data points where $z/L_s > 1$ are below the flood layer (Modified from Wheatcroft & Borgeld, 2000.)

porosity profile is markedly different in the fresh flood deposits, which consist of a several-centimetre-thick (dependent on the flood-deposit thickness) zone of quasi-uniform porosity, compared with the characteristic convex-upward profile of marine sediments (e.g. Fig. 5, in which there is an abrupt porosity decrease in the upper 1–3 cm). These important contrasts, that is, a high and gradually decreasing porosity (flood sediment) compared with a lower and abruptly decreasing porosity (ambient sediment), have potentially important implications for sediment erodibility. A key unknown is therefore the rate at which the flood bed consolidates.

The resistivity data provide some important insight into temporal change in seabed porosity. Direct seabed information, including profiles of resistivity, was obtained approximately every 120 days for the period winter 1995 to summer 1998. Shown in Fig. 7 is a time series of porosity measured at station C70, which includes replicate profiles numerous times before and after deposition of a 3-cm-thick flood deposit. Prior to deposition of the flood sediment, the porosity profile is at steady state with an asymptotic value of ~0.6 and little spatial variability. In contrast, the fresh ~3-cm-thick flood deposit (January 1997) has a nearly uniform porosity of 0.83, with a sharp decrease to pre-flood values. Four

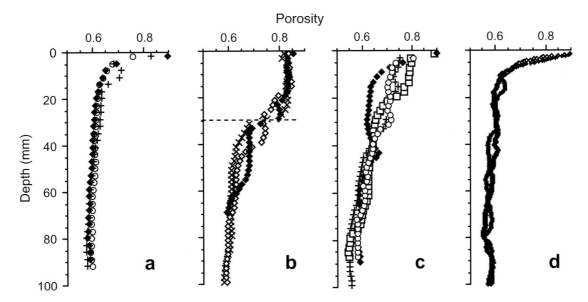

Fig. 7 Replicate porosity profiles measured at station C70: (a) 6 months before the 1997 flood, (b) 2 weeks after, (c) 4 months after (□, ◆, ○) and 10 months after (+), and (d) 2.5 yr after deposition of the January 1997 flood deposit. The dashed horizontal line in (b) denotes the estimated thickness of the flood deposit from X-radiographs (Fig. 2b; see Wheatcroft & Borgeld, 2000).

months later, the porosity within the flood sediment shows considerable small-scale variability that is likely to be a reflection of intense bioturbation at this site. The upper portions for two of the profiles are starting to develop the sharp gradient characteristic of the steady-state profile, whereas the third profile is more uniform with depth. There is also a suggestion that porosity in the 4-to-6-cm depth range has increased over time, consistent with interphase bioturbation (Boudreau, 1986; Mulsow et al., 1998) that acts to smooth the gradient (i.e. the high-porosity sediment in the flood bed is mixed downward). Several years later (August 1999) the porosity profile has returned to the pre-flood condition. Nearly identical sequences occur at many other stations (e.g. Fig. 8).

Consolidation also can be examined by calculating so-called sedimentation-compression curves (Skempton, 1970; Burland, 1990), which plot **void ratio** (ratio of void volume to solid volume) as a

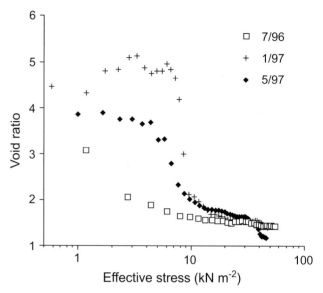

Fig. 9 Sedimentation-compression curves (Skempton, 1970; Burland, 1990) for porosity profiles collected at station C70 six months before, 2 weeks after, and 4 months after deposition of a 3-cm-thick flood bed in January 1997. The effective stress calculations assume a depth-independent grain density of 2.65 g cm^{-3}. A normally consolidated sediment column is typified by a roughly straight line on a plot of void ratio versus $\log_{10}$ effective stress (7/96), whereas an underconsolidated seabed has abnormally high void ratios for a given effective stress (1/97 and 5/97).

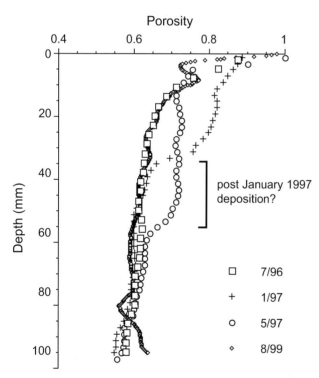

Fig. 8 Porosity profiles measured at station G70 six months before the 1997 flood, and 2 weeks, 4 months and 2.5 yr after deposition of a 3-cm-thick deposit. The apparent 2-cm increase in the flood-bed thickness between January and May 1997 could be due to subsequent deposition or may simply reflect small-scale spatial variability.

function of the logarithm for effective stress, σ_e. In **normally consolidated** sediment, the resultant data form an approximately straight line, with void ratio decreasing as effective stress increases (Skempton, 1970). Departures of points below the straight line (i.e. low void ratio) are indicative of **overconsolidated** sediment (due, for example, to erosion removing sediment), whereas **underconsolidated** sediment has higher void ratios for a given effective stress. The latter is the case for the Eel shelf sediments immediately after deposition of flood beds (Fig. 9). These data suggest that consolidation of the flood beds proceeds *en-masse*. That is, rather than the flood bed consolidating from the base upwards, it appears that the void ratio decreases by approximately the same amount throughout the bed (e.g. Fig. 9).

Collectively, these results suggest the following:

1 it takes several months to years for a porosity profile perturbed by flood deposition to return to the equilibrium shape;

2 despite the slow return to an equilibrium shape for the entire profile, the upper 5–10 mm of the seabed approaches pre-flood conditions more rapidly (< 4 months);

3 bioturbation increases the small-scale variability and, through interphase mixing, disrupts the equilibrium profile (Figs 7 & 8).

The second point is especially important as it suggests that the strength of the sediment at the seawater interface, where it is subject to fluid forcing (see below), is relatively steady. Higher frequency sampling is necessary to make a closer link between theoretical (e.g. Eq. 8) and observed evolution of seabed porosity. At present, the scientific community's ability to measure high-frequency seabed change lags considerably our ability to measure water-column processes associated with sediment transport (e.g. turbidity, velocity). Newly developed instrumentation, including a time-lapse camera that measures surficial grain size (Rubin, 2004; Rubin *et al.*, in press), and an autonomous resistivity profiler designed to measure seabed porosity at 30-min intervals for several months (R.A. Wheatcroft, unpublished manuscript), promises to redress this imbalance.

PHYSICAL ALTERATION

Sediment erosion, transport and deposition contribute to post-depositional alteration of sediment beds in two important ways. First, they can translate existing strata toward or away from the seabed surface via net erosion or deposition, thereby increasing or decreasing their susceptibility to physical and biological reworking (Fig. 3b). Second, they can modify the grain size and fabric of the bed through processes such as winnowing, ripple migration and deposition of graded beds. Here, these post-depositional alterations produced by sediment transport are referred to as **physical reworking** of the seabed. Biological alteration will be discussed in a later section.

Understanding the impact of physical reworking on marine sediment deposits requires knowledge about the environmental conditions causing sediment transport, particularly waves and currents, and the bed properties that determine the response of the seabed to environmental forcing, such as sediment size, density, porosity and erodibility. Environmental forcing is episodic over most of the shelf; only the shallowest parts of the inner shelf and the nearshore region persistently experience bed stresses exceeding the threshold of motion. The magnitude and frequency of transport events are critical controls on the intensity of physical reworking. In addition, processes such as winnowing, consolidation, transport and bioturbation produce changes in bed properties through time, which in turn affect the response of the seabed to subsequent fluid forcing.

Fluid forcing

In open-shelf environments, waves are usually the dominant control on the frequency and duration of sediment resuspension and transport (Komar *et al.*, 1972; Drake & Cacchione, 1985; Ogston *et al.*, 2004). Current-generated bed stresses (tidal or wind driven) typically are not large enough to mobilize sediment in the absence of wave-generated bed stresses. Currents, however, dictate the direction and rate of transport during resuspension events on the continental shelf (Sherwood *et al.*, 1994). Knowledge of both is required to characterize physical reworking and sediment redistribution on the shelf.

Waves

Wave-generated near-bed flows and bed shear stresses are functions of wave height, H, period, T, and water depth, h. Wave heights increase episodically in response to storm winds. Storms are typically seasonal, so that most sediment transport and bed reworking occurs during the stormy, usually winter, months (Fig. 10a; Drake & Cacchione, 1985). In addition, longer time-scale climatic variations such as the El Niño–Southern Oscillation (ENSO) or Pacific Decadal Oscillation (PDO) can modulate wave heights (Seymour, 1998; Allan & Komar, 2002). For example, the winters of 1997–98, an El Niño period, and 1998–99, a La Niña period, were marked by a particularly large number of extreme storms along the Pacific Northwest (Oregon and Washington) margin (Allan & Komar, 2002). This is also evident on the Eel shelf, where more than 30 days during each of these winters had daily significant wave heights greater than two standard

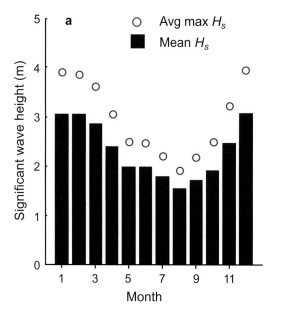
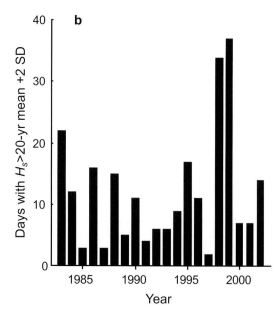

Fig. 10 Temporal variations in wave characteristics at NDBC Buoy 46022 on the Eel shelf for the period 1982–2002. (a) Seasonal variation in mean monthly significant wave height (H_s) and average monthly maximum significant wave height. The most energetic waves occur during the period from November through April. (b) Interannual variation in the number of days with large waves, defined here as having daily average significant wave heights in excess of two standard deviations (SD) above the mean of the full 20-year record (mean = 2.4 m and SD = 1.1 m).

deviations above the 20-yr mean (1982–2002; NDBC Buoy 46022; Fig. 10b). [**Significant wave height** (H_s) is the average of the highest one-third of waves recorded during a sampling interval.]

In regions such as the USA West Coast, with narrow shelves exposed to large ocean storms, wave height and period remain relatively constant across much of the continental shelf. Wave-generated shear stresses on the seabed, however, vary as a function of water depth. Surface waves produce an oscillatory flow at the seabed, if water depth is less than about half the wavelength, L, of the wave. Long-period swell (waves with periods greater than about 10 s for the northern California shelf) will generate oscillatory (orbital) motion at the seabed in water depths as great as 100–200 m. **Near-bed wave orbital velocities** (U_b) at the seabed are usually characterized by the maximum velocity reached during each passing wave,

$$U_b = \frac{H}{T \sinh(2\pi h/L)} \tag{12}$$

For a 4-m wave with a period of 12 s, near-bed wave orbital velocity would be 0.2 m s^{-1} at a depth of 80 m and increase to 0.6 m s^{-1} in depths of 40 m.

When spectral wave data are available (e.g. from an NDBC wave buoy), near-bed significant orbital velocity and average period can be calculated using a modified form of Eq. 12 (Madsen, 1994; Harris & Wiberg, 2001). Good agreement has been found between measured and calculated U_b for several sites along the California shelf (Sherwood *et al.*, 1994; Wiberg, 2000; Wiberg *et al.*, 2002). As water depths become shallower, especially on broad shelves, wave-shoaling effects also must be taken into account in calculations of near-bed wave orbital velocities (Ardhuin *et al.*, 2003).

Wave-generated bed shear stresses (τ_{bw}) can be related to U_b through a wave friction factor, $\tau_{bw} = 0.5\rho f_w U_b^2$. A number of empirical relationships for the friction factor (f_w) are available, such as $f_w = 0.04(a/k_s)^{-1/4}$ for values of $a/k_s > 50$ (Fredsøe & Deigaard, 1992), where k_s represents the physical roughness length of the bed and a is the orbital excursion amplitude near the bed ($a = U_b T/(2\pi)$). The dependence of U_b on water depth means that τ_{bw}, like U_b, increases with decreasing depth across the shelf for a given set of wave conditions.

Time series of near-bed orbital velocity or wave-generated bed shear stress can be used to identify wave-driven sediment transport events at a site given representative erosion thresholds. Observations at inner to mid-shelf depths on the Russian, Eel and Palos Verdes shelves, California (Wiberg *et al.*, 1994, 2002; Wiberg, 2000), indicate that U_b > 0.14 m s^{-1}, corresponding to a bed shear stress of about 0.12 N m^{-2}, is required for a resuspension event (defined by near-bed suspended-sediment concentrations about 10 times background values;

e.g. Fig. 11). The resulting sets of transport events can be used to estimate the frequency and duration of physical reworking across the shelf (Wiberg *et al.*, 2002).

The Eel shelf has the highest energy waves recorded along the northern California margin between San Francisco and the California–Oregon border (Fig. 12a), based on a return-period analysis

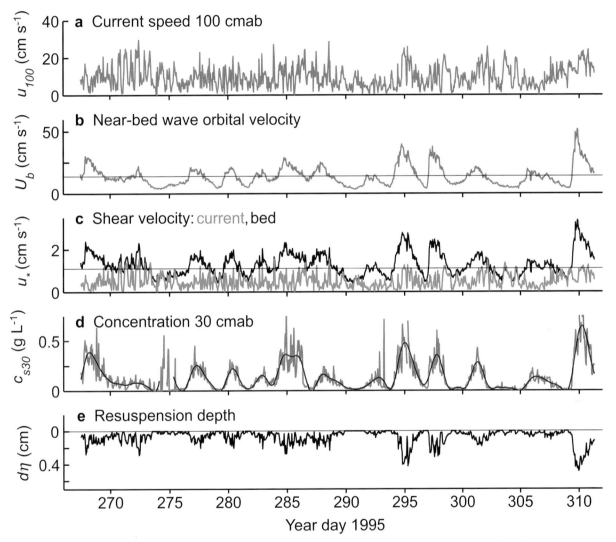

Fig. 11 Time series of: (a) hourly currents 100 cmab (cm above bottom); (b) near-bed wave orbital velocity; (c) calculated current and bed (combined wave-current skin friction) shear velocities; (d) suspended-sediment concentration 30 cmab (grey line), with a low-pass, 35-hr filter (red line); (e) calculated resuspension depth (the depth of the layer that would be formed if all sediment in suspension were deposited on the bed with the average porosity of the seabed) during the autumn–winter of 1995 at station S60 (e.g. Ogston *et al.*, 2004). The horizontal lines at an orbital velocity of 0.14 m s^{-1} in (b) and a shear velocity of 1.1 cm s^{-1} in (c) represent an estimate of the threshold for resuspension. Suspended-sediment concentrations are well correlated with orbital velocity; the correlation with currents is much weaker, consistent with threshold shear velocity being exceeded almost exclusively owing to the contribution of waves to the bed stress; current shear velocity alone seldom exceeds threshold conditions. Calculations were performed using a one-dimensional shelf sediment transport model (Wiberg *et al.*, 1994; Harris & Wiberg, 2001).

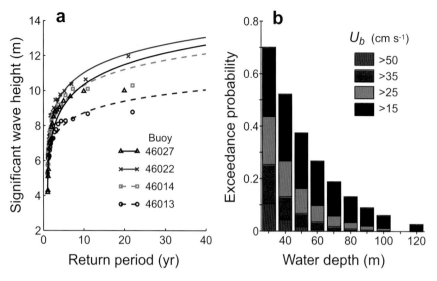

Fig. 12 Spatial variations in wave characteristics at NDBC Buoy 46022 on the Eel shelf for 1982–2002. (a) Return periods for annual maximum significant wave heights at four locations between San Francisco and the California–Oregon border (in north to south order downward in the legend). The highest wave conditions occur on the Eel shelf. (b) Across-shelf variations in exceedence probabilities for values of near-bed orbital velocity (U_b) ranging from 0.15 to 0.50 m s^{-1}.

of annual maximum wave heights performed using a Gumbel distribution (Gumbel, 1958; Borgman & Resio, 1982). Peak storm wave heights on the Eel shelf exceeded 10 m six times during 1995–2000, even though 10-m waves have a 5-yr return period based on the 20-yr record (Fig. 12a). A 20-yr time series (1983–2002) of near-bed orbital velocity and wave-generated bed shear stress for specified water depths across the shelf was calculated from measured wave spectra recorded on the Eel margin (NDBC Buoy 46022). Probabilities of near-bed wave orbital velocities exceeding a range of specified values show that the relative intensity of near-bed wave conditions decreases seaward across the shelf (Fig. 12b; Wiberg, 2000). Characteristics of wave-driven transport events, obtained by applying a resuspension threshold of $U_b > 0.14$ m s^{-1} to the orbital velocity time series, are summarized in Fig. 13. The number and duration of events drop off in water deeper than 50 m, owing to the decrease in near-bed wave orbital velocity (U_b) with increasing water depth for a given set of wave conditions. The increase in U_b (Fig. 12b) in shallow water (< 50 m) results in fewer (Fig. 13a), but longer (Fig. 13b), events. The total number of days of transport (the product of the number of events and event duration) increases monotonically with decreasing depth (not shown).

Currents

Current-generated shear stresses interact with wave-generated shear stresses to enhance sedi-

ment transport on the continental shelf (Smith, 1977; Grant & Madsen, 1979; Cacchione & Drake, 1990). Currents also act as the advection mechanism for sediment resuspended under combined wave–current flows (Drake & Cacchione, 1985). Spatial and temporal variations in currents lead to convergence and divergence in sediment flux, thereby contributing to patterns of net deposition and erosion. Temporal variations in currents occur on tidal and higher-frequency time-scales, in addition to subtidal and interannual time-scales. The subtidal currents are affected by large-scale atmospheric forcing as well as regional pressure gradients, large-scale coastline morphology and input of freshwater.

The relative contributions of tidal, wind-driven and other components of currents to sediment transport and physical reworking of the seabed are site dependent. For example, on the Eel shelf, measurements of current 15 m above bottom (mab) at a depth of 90 m on the Eel shelf during 1988–1989 (Largier *et al.*, 1993) indicated mean current speeds of 0.15 m s^{-1}, with tidal and higher frequency currents contributing 89% and 59% of the across-shelf and along-shelf velocity variance, respectively. Similarly, harmonic analysis of near-bed current and pressure time series at a depth of ~50 m on the Eel shelf showed that semidiurnal tidal motions are most significant, with along-shelf semidiurnal tidal energy almost twice as large as the across-shelf component (Cacchione *et al.*, 1999). Although tidal currents comprise a large portion of the velocity variance on the Eel shelf, the net sediment flux associated with them may be relatively

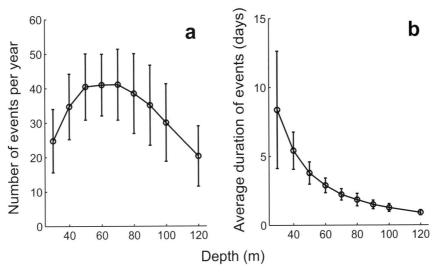

Fig. 13 Resuspension events across the Eel shelf. (a) Average number of resuspension events per year, and (b) average duration of events, calculated using 20 yr of spectral wave data from NDBC Buoy 46022. Circles and bars indicate mean values and standard deviation, respectively. Events are defined as periods with peak orbital velocities exceeding 0.14 m s^{-1}and orbital velocities exceeding 0.1 m s^{-1} for at least three hours. The number and duration of events drop off in water deeper than 50 m owing to the decrease in near-bed wave orbital velocity (U_b) with increasing water depth for a given set of wave conditions. The increase in U_b (Fig. 12b) in shallow water (< 50 m) results in fewer (a), but longer (b), events. The total number of days of transport (the product of the number of events and event duration) increases monotonically with decreasing depth (not shown).

small owing to their directional reversal over periods that are short compared with the duration of typical resuspension events (i.e. there may be a high instantaneous flux, but little net flux over the tidal cycle; Ogston *et al.*, 2004). However, increases in suspended-sediment concentrations in the bottom boundary layer associated with tidally induced turbulent mixing can significantly increase the volume of sediment in suspension and therefore the net flux due to subtidal currents.

During the summer, the northern California continental shelf experiences persistent southward winds, while winter winds have a near-zero or slightly northward mean velocity and fluctuate in magnitude and direction as storm systems pass over the region (Dever & Lentz, 1994). Resulting summer along-shelf mean currents are strongest and southward near the surface, decreasing to near-zero mean values near the bottom (Dever & Lentz, 1994). During winter, mean currents tend to be more uniform through the water column, yet their direction and magnitude vary along and across the shelf. Mean across-shelf flows are small and seaward on average, owing to dominantly downwelling conditions in the winter, although signi-

ficant spatial variability in subtidal shelf currents has been observed. For example, the presence of Cape Mendocino just to the south of the Eel shelf results in reduced upwelling during periods with southward winds (Largier *et al.*, 1993). Additionally, topographic features can induce mesoscale circulation features (eddies and meanders) due to flow disruption, deflection and possibly flow separation (Pullen & Allen, 2000). Shelf circulation also can be impacted by freshwater input to the coastal ocean, through the formation of fronts and the addition of momentum to the shelf circulation (Kourafalou *et al.*, 1996; Pullen & Allen, 2000).

Interannual variations in the frequency structure of currents on the continental shelf may be a reflection of climate-scale fluctuations (e.g. ENSO and PDO), causing patterns of storm tracks to shift spatially and in magnitude. This level of variability can be of great importance to the ultimate fate of suspended particles by changing the locations of sediment flux convergences/divergences. For example, on the Eel shelf, net sediment flux along the shelf appears to be controlled by the frequency structure in the very low frequency bands (weeks to months). This structure changes from year to year, hypothetically due to changes in the eddy structure formed over the shelf and slope as the result of pressure gradients and topographic interferences (Ogston *et al.*, 2004).

Bed properties

The response of the seabed to fluid forcing depends on a number of bed properties, including: grain size, sorting, density, porosity, surface roughness, consolidation and cohesion. On the inner portion of the continental shelf, landward of the **sand–mud transition** (located at a depth of ~55 m on the Eel shelf), the seabed is a non-cohesive silty sand, with wave-generated ripples and possibly larger-scale bedforms at the surface (Cacchione *et al.*, 1999; Hanes *et al.*, 2001). Seaward of the sand–mud transition, the fine-grained bed sediment behaves cohesively and surface roughness is dominantly biogenic (Wheatcroft, 1994). Grain-size-related differences in consolidation, critical shear stress and settling rates across the shelf cause differences in sediment erodibility and availability that affect the volume of sediment in transport and the depth of physical reworking at the bed surface.

Critical shear stress and sediment erodibility

The fluid shear stress at the seabed that must be exceeded for significant transport to occur is often termed the **critical shear stress (τ_{cr})**. Values of critical shear stress depend on the size and density distribution of the sediment, porosity and, for fine-grained beds, **cohesion** (including organic binding, consolidation, aggregate structure). In the absence of cohesion, critical shear stresses decrease with grain diameter as indicated by the Shields curve (see, e.g. Miller *et al.*, 1977); for fine to very-fine sand, $\tau_{cr} \approx 0.1$–0.2 N m^{-2}. The Shields curve was developed for well sorted, non-cohesive sediment. On poorly sorted, non-cohesive beds, the finer fractions are sheltered by the larger grains on the bed, making them harder to move than if the bed was one of uniform grain size. Similarly, the coarser fractions are more exposed than they would be on a uniform bed, making them relatively easier to move. As a result, the range of critical shear stresses necessary to initiate transport of all grain sizes present in the bed is considerably smaller than

applying the Shields curve to each size would indicate (Wiberg & Smith, 1987; Wilcock, 1993).

For sediments typical of muddy regions of the continental shelf, the effect of grain size on critical shear stress becomes less important than the effects of cohesion and consolidation. Critical shear stress for consolidated beds (see above) is generally a function of porosity (or bulk density; e.g. Torfs & Mitchener, 1996). With increasing consolidation, porosity decreases (solid concentration increases) and critical shear stress increases (Fig. 14). Thus, critical shear stress for consolidated beds increases with depth in the seabed and with time since deposition, until a steady-state consolidation profile (Fig. 4) is reached. As noted above, predicting porosity for a consolidated bed is challenging even under the most ideal circumstances. Similarly, there is no general, predictive relationship between porosity and critical shear stress. Additional factors that affect critical shear stress for fine-grained beds, such as organic binding and aggregated structure, are even harder to quantify. As a result, determining critical shear stresses for consolidated, cohesive beds requires site-specific measurements using

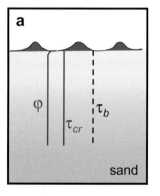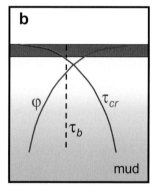

Fig. 14 Vertical variations in porosity, φ (blue line), and critical shear stress, τ_{cr} (red line), in the surface layer of: (a) a well sorted, non-cohesive sand; (b) a normally consolidated, cohesive mud. Porosity and critical shear stress are relatively uniform for a well-sorted sandy bed (a), although a thin layer of higher porosity has been observed at the surface of sand beds (Wheatcroft, 2002). The surface active layer for sand beds (brown) is related to the size and migration rate of bedforms. In mud beds (b), φ decreases and τ_{cr} increases with depth owing to effects of consolidation. For a given applied bed shear stress (τ_b), the layer above the intersection of τ_b and τ_{cr} (brown layer in b) defines the thickness of the sediment layer available for resuspension.

some sort of *in situ* flume (Maa *et al.*, 1993), erosion chamber (Gust & Müller, 1997), laboratory measurements (McNeil *et al.*, 1996; Johansen *et al.*, 1997), or near-bed measurements of suspended-sediment concentration through resuspension events (Sternberg & Larsen, 1975; Chang *et al.*, 2001; Wiberg *et al.*, 2002; Fig. 11).

Sediment **erosion or entrainment rates** (E) can be expressed in the general form,

$$E = m(\tau_{\rm b} - \tau_{\rm cr})^n \tag{13}$$

(Sanford & Maa, 2001), where m and n are empirical coefficients. For well sorted, non-cohesive sands, $\tau_{\rm cr}$ is a function of grain size only and n is typically taken to be 1. If the flow is steady, this suggests that entrainment rates will be constant provided the bed shear stress exceeds the critical value ('Type II' erosion, Sanford & Maa, 2001). More commonly, there is a range of sizes present in the bed. In this case, the entrainment rate of each size also depends on its abundance at the bed surface. Fine sediment is more readily suspended when mobilized than coarse sediment, so the former tends to become depleted at the bed surface. As the mass or volume fraction of fine sediment approaches zero in the surface-active layer (see below), E also goes to zero. This **winnowing** of fine sediment out of the bed surface layer leaves a coarsened, **armoured** layer at the bed surface that must be mobilized before additional fine sediment in the bed can become available for suspension (Kachel & Smith, 1986; Wiberg *et al.*, 1994; Reed *et al.*, 1999). Therefore, for non-cohesive beds, the grain-size distribution and active-layer thickness of the bed surface control sediment availability for resuspension and transport, and thus the volume of sediment in suspension.

For cohesive mud, the same entrainment-rate expression (Eq. 13) can be applied, but with $\tau_{\rm cr}$ now a function of depth and time rather than grain size (Sanford & Maa, 2001; Fig. 14b). Typical measured critical shear stress profiles for marine and estuarine fine-grained beds show a relatively abrupt increase in critical shear stress with depth (Johansen *et al.*, 1997; Sanford & Maa, 2001). The effect of this increase is to limit the volume or mass of sediment in the bed that is available to be suspended for any applied wave-current shear stress ('Type I' erosion; Fig. 14b). Under most flow conditions, only the upper few millimetres or less of a fine-grained, consolidated bed has a critical shear stress less than the applied stress.

Bedforms and controls on sediment availability

Small-scale, wave-generated ripples (length ~10–100 cm, height ~1–10 cm) are nearly ubiquitous on sandy beds in marine environments. These ripples, which are symmetrical and sharp-crested, are active when the shear stress acting on the bed (**skin friction**) exceeds the critical shear stress and sediment is moving as bedload ($w_{\rm s}/u_* > 1$ where u_* is shear velocity, defined as $\sqrt{\tau_{\rm b}/\rho}$, and $w_{\rm s}$ is settling velocity). At higher values of shear stress, sand is suspended and ripples become washed out. However, as shear stress drops, ripples quickly reform and persist after bed shear stresses once again drop below the threshold of motion (Li & Amos, 1999). The height and spacing of these ripples are related to wave height and period, water depth and grain size (Clifton, 1976; Miller & Komar, 1980; Wiberg & Harris, 1994). Larger-scale ripples or other larger bedforms have also been observed at some inner-shelf sites (Cacchione *et al.*, 1987, 1999; Traykovski *et al.*, 1999; Hanes *et al.*, 2001).

When ripples (or larger bedforms) are active, the sediment comprising them is reworked within a layer extending down to the base of the ripple troughs (Fig. 14a). This **surface-active layer**, typically in the order of a few centimetres in thickness, controls the availability of sediment for resuspension in sandy beds (Kachel & Smith, 1986; Wiberg *et al.*, 1994; Reed *et al.*, 1999). Larger active bedforms have the potential to rework the bed to greater depths as they migrate, but the time required for complete reworking of the surface layer (the time required for a bedform to migrate one complete wavelength) depends on the bedform migration rate, which can be slow for large bedforms and bedforms in oscillatory flows with little net transport.

Muddy beds are generally characterized by biogenic surface roughness, although if the sand or coarse-silt content is sufficient, small ripples can at times be present (Wheatcroft, 1994). For example, both two- and three-dimensional ripples were observed at the 60-m, long-term monitoring site on the Eel shelf following transport events (Fig. 15a & b). Mud-sized primary particles are typically suspended as soon as they are mobilized, so they

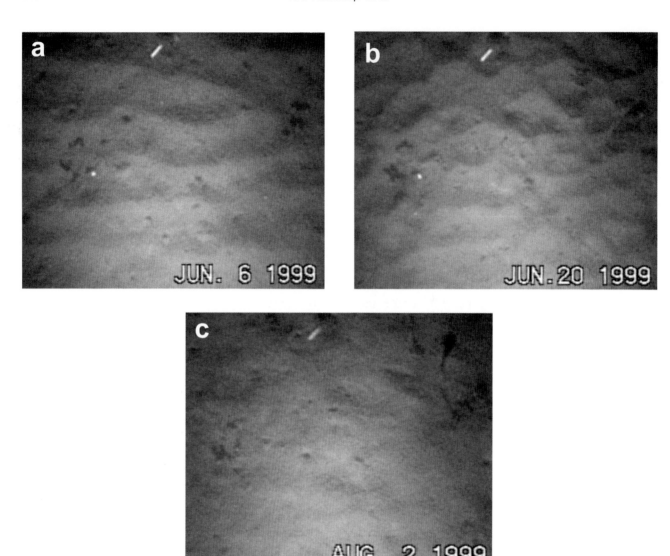

Fig. 15 Representative images from video at the long-term monitoring site, S60, showing (a) two-dimensional and (b) three-dimensional ripples in June 1999 and (c) biogenic microtopography in August 1999. A key for understanding bed roughness is the rate at which the bed morphology changes from the rippled to the biogenic microtopographic state.

never move as bedload. Breakage-resistant mud aggregates such as faecal pellets may move as bedload, but they are usually not abundant or resilient enough under high-flow conditions to dominate the dynamics of the bed surface (Taghon *et al.*, 1984). As a result, roughness on muddy beds does not migrate in the same way that ripples do on sandy beds. Instead, availability of muddy sediment for suspension is controlled by the thickness of the layer of sediment near the bed surface in which the critical shear stress is less than the imposed bed shear stress (Fig. 14b).

Seabed response to physical forcing

When waves and currents produce shear stresses at the bed surface in excess of the critical shear stress, sediment is put in motion. Sands will be transported as bedload (contributing to bedform migration) or, under more intense flows, as suspended load; fine sediment (clay and fine–medium silt) is generally transported in suspension once it is mobilized. For relatively uniform flow conditions with no significant input of new sediment (e.g. no flood events), the volume of sediment in the water column depends only on the bed properties and local stresses in the water column. Under these conditions the amount of sediment deposited after the transport event is roughly equal to the amount that was resuspended (eroded) from the bed during the event, with no net change in bed elevation. This type of transport event leaves a relatively thin, graded layer of reworked sediment at the bed surface (**storm bed**), that may be armoured at its base and include ripple cross-bedding if the bed is sandy (Reineck & Singh, 1972; Nittrouer & Sternberg, 1981; Myrow & Southard, 1996). When flow and bed conditions are non-uniform, spatial variations in flux can lead to net erosion and deposition, and to net changes in the grain-size distribution of the bed. Bed alterations due to sediment transport under uniform and non-uniform conditions are considered in this section.

Resuspension and deposition under uniform conditions

The volume of sediment in suspension depends on the bottom stress, turbulent mixing profile and bed sediment characteristics (entrainment threshold, settling rate and porosity). Under conditions in which the flow and bed sediment properties are uniform along the transport pathway, **resuspension depth** can be defined as the volume of sediment in suspension at the peak of transport divided by the concentration of sediment in the bed ($\varphi_s = 1 - \varphi$). Although waves, currents, and bed conditions are rarely uniform spatially, observations and modelling of near-bed flow and suspended-sediment concentration on the shelf indicate that suspended-sediment concentrations are largely a function of local conditions (Harris, 1999). Therefore resuspension depth may be used as a measure of the depth of physical reworking during storms. This depth will be augmented by the depth of mixing associated with bedform migration in sandy beds and by any net deposition under non-uniform conditions.

Measurements and models of shelf sediment transport allow us to quantify the volume of sediment in suspension, and therefore resuspension depth, during specific wave and current conditions at a site with known bed properties. For example, the 60-m, long-term tripod on the Eel shelf (Site S60; Ogston *et al.*, 2004) provides measurements of hourly average currents and suspended-sediment concentration 0.3 m and 1.0 m above the bed (mab)

and hourly near-bed wave orbital velocity (e.g. Fig. 11). A one-dimensional shelf sediment-transport model (Grant & Madsen, 1979; Wiberg *et al.*, 1994; Styles & Glenn, 2000; Li & Amos, 2001) can be used with the measured current, wave and sediment conditions to calculate suspended-sediment profiles in the bottom boundary layer. These profiles can be compared with measured suspended-sediment concentrations and can be used to compute suspended-sediment volumes and fluxes (Lyne *et al.*, 1990; Wiberg *et al.*, 1994; Cacchione *et al.*, 1999; Wright *et al.*, 1999; Wiberg *et al.*, 2002, Traykovski *et al.*, 2007).

Resuspension depths calculated for the transport events at Site S60 during the autumn of 1995 (Fig. 11e) are a few millimetres or less. Wave conditions during that autumn were fairly modest. Storms with larger waves result in greater resuspension depths. Additionally, for given wave conditions, resuspension depths are greater in shallower water where bed stresses are higher. Estimated resuspension depth during the largest recorded wave and current conditions on the Eel shelf (December 1995; see Cacchione *et al.*, 1999) was less than 2 cm at a 50-m-deep site, despite the high wave and current speeds that characterized this event (Wiberg, 2000). Under the most extreme combination of recorded (but not simultaneous) hourly waves ($U_b = 1.35$ m s^{-1}) at 50 m and currents (0.60 m s^{-1} at 1 mab) on the Eel shelf, the calculated resuspension depth was 12 cm (Wiberg, 2000). The probability of these occurring simultaneously, however, is much lower than the probability of either occurring individually, particularly given the poor correlation between waves and currents observed at several sites along the California shelf (Harris & Wiberg, 1997; Wiberg *et al.*, 2002). Taken together, available wave and current data and suspended-sediment calculations for the Eel shelf suggest that storms capable of producing a 5-cm-thick storm bed in water depths of 50–60 m have recurrence intervals greater than 100 yr, while those able to produce a 10-cm-thick storm bed have recurrence intervals in excess of 1000 yr (Wiberg, 2000). The NDBC wave data indicate that wave conditions on the Eel shelf are more energetic than at other locations on the northern California margin (Fig. 12a). Therefore, wave recurrence intervals for storm beds of comparable thickness in similar water depths at other sites along the northern California shelf would be longer.

Net changes in bed elevation under non-uniform conditions

Under non-uniform flow and/or bed conditions, suspended-sediment flux will vary spatially. Spatial gradients in flux (flux divergences) are related to net changes in bed elevation through the erosion equation

$$\frac{\partial \eta}{\partial t} = -\frac{1}{\varphi_s}\left(\frac{\partial Q_s}{\partial x} + \frac{\partial Q_s}{\partial y} + \frac{\partial V_s}{\partial t}\right) \qquad (14)$$

(Smith, 1977), where η is bed elevation, Q_s is volume flux of suspended sediment and V_s is the volume of sediment in suspension. Owing to complex three-dimensional flow fields, variations in bed shear stress and sediment properties with water depth, and the need to evaluate small changes in large signals, it is difficult to reliably observe gradients in the magnitude of sediment flux and related deposition, although flux convergence (Ogston *et al.*, 1998; Wright *et al.*, 1999) and preferential deposition (Dyer & Huntley, 1999) due to persistent circulation features have been documented in shelf environments.

On the continental shelf, where wave action provides the primary resuspending force, gradients in wave-generated bed shear stresses can cause across-shelf variations in flux. Everything else being equal (sediment texture, current velocity), across-shelf gradients in wave-generated bed shear stresses that result from changes in water depth mean that flows on the inner shelf are capable of suspending more sediment than flow on the outer shelf. Sediment flux should therefore decrease as water depth increases. While the volume of sediment in suspension at any one time usually amounts to only millimetres of bed material per unit area, across-shelf gradients in sediment flux are capable of removing or adding in the order of 1 cm of bed material during a single storm event (Fig. 16; Zhang *et al.*, 1999; Harris & Wiberg, 2002). Seaward transport events are characterized by inner-shelf erosion and mid-outer-shelf deposition (Fig. 16). Although shoreward transport events can reverse this pattern, sediment carried into shallower, more energetic depths is likely to be resuspended during subsequent events and transported seaward.

Wave shear stresses and sediment grain sizes vary over shorter distances across the narrow, steep shelves of the USA Pacific coast compared with the wider mid-Atlantic shelf. As a result, wave-driven sediment flux gradients are larger and more effective for creating net erosion and deposition on narrow shelves (Harris & Wiberg, 2002). On wide shelves, and shelves with larger-scale circulation features, including eddies and fronts, it is likely that spatially varying flow will be a dominant cause of sediment flux gradients. Cookman & Flemings (2001), using an across-shelf transport model with wind-driven and pressure-gradient-driven currents, found that cross-shelf gradients in current velocity on the mid-Atlantic shelf dominated the erosional–depositional regime of the inner shelf. Even on the steep, wave-dominated Eel River shelf, Harris (1999) found that gradients in current velocity (Wright *et al.*, 1999) dictated the locus of deposition, and were capable of increasing deposition thicknesses during a transport event by up to a factor of three.

Large-scale shelf circulation features, if persistent during winter storm periods, may influence the location of longer-term depocentres. On the Eel margin, along-shelf flow convergence was observed on average between two sites ~8 km apart along the 60-m isobath throughout the winter of 1996–97 (Ogston *et al.*, 1998; Hill *et al.*, this volume, pp. 49–99). A similar flux convergence was obtained in calculations using a shelf circulation model for that same deployment period (Pullen & Allen, 2000). These two sites bracket the depocentre of the mid-shelf mud deposit on the Eel shelf (Sommerfield & Nittrouer, 1999). The distances between along-shelf observation sites are large, so deposition or erosion may not simply be the result of the observed gradients in flux. However, the pattern is consistent with the long-term accumulation on the shelf (Sommerfield *et al.*, this volume, pp. 157–212).

Modifications to seabed grain size and fabric

Sediment transport can alter grain size vertically within the bed and horizontally across the shelf. As bed shear stresses initially increase during a resuspension event, fine sediment is resuspended from the surface layer of the bed. This leaves the coarsest sediment as a lag or armouring layer on the bed. When flow conditions wane, coarse material in suspension will settle out first, owing to its higher settling velocity and the fact that it is carried close to the seafloor. The resulting redeposited layer will fine upward (e.g. Fig. 16e;

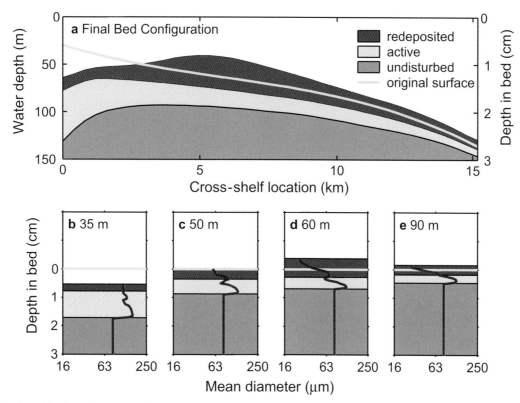

Fig. 16 Calculated bed configuration for a narrow, steep shelf acted on by a 10-day realistic time series of winter waves and currents from the Eel shelf calculated using a two-dimensional (cross-shelf), time-dependent shelf sediment-transport model (Harris & Wiberg, 2001, 2002). Initially, the seabed across the shelf is a uniform poorly sorted silty sand. Sediment that was resuspended (dark grey), reworked but not resuspended (light grey) and unaffected by reworking (medium grey) is indicated. (a) Cross-shelf variations in erosion and deposition (right axis); the solid yellow line indicates the original sediment–water interface and hence water depth (left axis). (b–e) Vertical profiles of mean grain size (red line) at depths of 35 m, 50 m, 60 m and 90 m at the end of the 10 days.

Nittrouer & Sternberg, 1981; Kachel & Smith, 1986; Leithold, 1989). These **graded beds** can form in the absence of any net change in bed elevation or sediment size, but are usually limited to layers of just a few millimetres in thickness, as described above. Biogenic mixing tends to erase these graded beds on time-scales of days to weeks (Harris & Wiberg, 1997; Bentley & Nittrouer, 2003) unless they are buried by a significant depositional event.

Horizontal gradients in sediment flux will alter the size distribution of the bed if the flux divergence for each size is not proportional to the abundance of that size class in the bed (Cui & Parker, 1998). This particle-size-dependent flux divergence can lead to winnowing and armouring when the flux of the fine sediment into a site is smaller than the flux out of the site. Differences in availability and transport of coarser and finer fractions in the seabed,

and across-shelf variations in wave-generated bed shear stresses, will tend to produce spatial segregation of sediment sizes, even if the seabed is initially spatially homogeneous (Fig. 16). These tendencies are accentuated by differences in sources and delivery of sands (shoreline erosion and river bedload) and finer sediment (flood plumes and possibly fluid-mud flows that can carry sediment to mid-shelf depths; Hill *et al.*, this volume, pp. 49–99).

Modifications to bed grain size can occur during a single storm event, but the changes appear to be restricted to a relatively thin (a few centimetres or less), surficial layer of reworked sediment (e.g. Fig. 16; Zhang *et al.*, 1999; Harris & Wiberg, 2002). Over time-scales of a few years, however, biogenic mixing and successive resuspension events can imprint cross-shelf variations in size to depths of tens of centimetres. As a result, there is typically less fine sediment available for resuspension on the sandy inner shelf than a flow can carry (i.e. that region is supply-limited). Farther offshore, in muddy mid-shelf depths, cohesion and consolidation limit the availability of fine sediment for resuspension. Thus, transport of fine sediment over most of the shelf is typically supply limited. Among other considerations, the effect of this limitation is to moderate sediment redistribution (including net erosion and deposition) during transport events on the shelf.

Evolution of grain size in a flood bed

The impact of post-depositional physical reworking on the preservation of event beds is best illustrated with an example. Event-response sampling of flood layers formed in January 1995 and January 1997 on the Eel shelf was followed by several years of frequent re-sampling at a few standard shelf sites to monitor subsequent changes. The initial characteristics of the layers are discussed in detail in Hill *et al.* (this volume, pp. 49–99). Here, the temporal and spatial changes that occurred to the grain-size distributions of the flood beds during the months and years following deposition are considered.

More than 85 km^2 of shelf north of the Eel River was covered by high-porosity, fine-grained sediment with mean diameters ranging from just 2 µm to 7 µm following the 1995 and 1997 floods (Drake, 1999; Wheatcroft & Borgeld, 2000). The mean weight percentage and standard deviation (SD) of the

dominant size class (< 20 µm) in 11 samples from within the flood deposit were 94% and 1.8% in February 1995 (Drake, 1999). By May 1995, a similar set of regional samples documented a general coarsening of the sediment surface (uppermost 0.5 cm) and a large increase in SD to 19.1% (Table 1; Drake, 1999). The regional picture in September 1995 revealed a further decrease in the mean percentage of < 20-µm grains at the seafloor, while the SD remained uniformly high. By July 1996, after a winter of low river discharge and minimal new sediment contributions, the mean surface grain size was approximately the same as in September 1995 and the SD continued to be high at 17.7% (Table 1; Drake, 1999).

Frequent box-core sampling along the O-line, a sampling transect that crossed the thickest part of the 1995 flood deposit, and at O70 (70-m water depth) in particular, permitted an examination of temporal changes in the grain size, porosity and sediment mass of the 1995 flood layer. The observed changes included: a general coarsening of the layer; development of a gradually thickening bioturbated layer at the seafloor; and consolidation of the sediment (Drake, 1999). Porosity and thickness were determined for each flood-bed observation, enabling absolute changes in the total mass of the flood layer and in the mass of various grain-size fractions to be computed. Grain-size distributions were determined using two methods, one (wet sieving) that preserved resistant aggregates such as faecal pellets and one that completely disaggregated the sediment. Comparison of these

Table 1 Surface-sediment characteristics of the Eel shelf

	Month	Number	Mean % of < 20-µm class	SD
All stations within the January and March 1995 flood layers	February 1995	11	93.4	1.8
	May 1995	11	68.0	19.1
	September 1995	11	56.2	16.3
	July 1996	6	52.5	17.7
Box cores at station O70 only	May 1995	5	68.6	11.6
	July 1996	8	47.1	10.7
Subcores in one O70 box core	July 1996	9	52.0	2.8

distributions allowed monitoring of the faecal-pellet production within a bed that had very few faecal pellets when it was fresh. Although flocculation was an important factor in the transport and accumulation of flood sediment in both 1995 and 1997 (Hill *et al.*, this volume, pp. 49–99), the wet sieving technique used for grain-size analysis did not preserve the flocs.

Change in mass of four grain-size fractions within the 1995 flood layer at site O70 during 1995 and 1996 are shown in Fig. 17 (Drake, 1999). The winter of 1995–96 was one of very low rainfall and river flow, so it can be assumed that observed changes in mass at O70 were caused largely by reworking of the 1995 flood material. The results suggest that the flood layer added more mass throughout the observation period (Fig. 17). Simple calculations indicate that the only way to coarsen the bed and also maintain its total mass is to add new coarse material (silt and very fine sand) more rapidly than finer fractions were winnowed away. It is interesting to note that the finest fraction (< 20 μm) of the 1995 flood layer did not lose

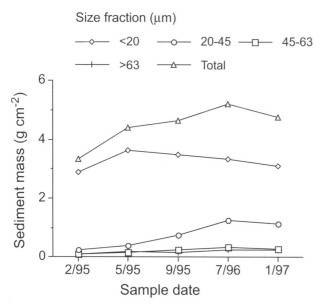

Fig. 17 Mass changes of various sediment grain-size fractions (in μm) in the January 1995 flood layer at station O70 on the Eel shelf. These results, as well as time-series X-radiographs, demonstrate that storms did not erode the 1995 flood layer, and that the change in texture (coarsening) was not caused by winnowing of fines, but rather was mostly caused by the advection and deposition of medium to coarse silt (presumably from the inner shelf).

mass in 1995 and 1996, but instead held relatively steady at about 3.5 g, suggesting that the finest fraction, which was by far the most plentiful in the flood layer, was replenished as rapidly as it was winnowed from O70 and redistributed by waves and currents.

Despite the occurrence of several strong storms in March, April and December 1995, the fine-grained and very high-porosity (> 80%) flood sediment at O70 was not aggressively eroded (Fig. 17). Given the high porosity and almost soupy consistency of the fresh flood sediment on the central shelf, this resistance to erosion was surprising. This suggests that early consolidation of the flood sediment increased critical shear stresses within the bed sufficiently to resist large wave-generated currents at depths of 60 m and 70 m. The January 1997 flood layer (Hill *et al.*, this volume, pp. 49–99) subsequently buried the 1995 flood layer, thereby greatly slowing subsequent post-depositional changes in the grain size of the older layer.

To place the mass inventory data at O70 into a wider context, changes in the mass of several grain-size fractions were tracked from early 1995 through August 1999 at O60, O70 and O90 (Fig. 18). The results document the strong flood contributions In 1995 and 1997, but they also reveal a significant accumulation of sediment at all three sites during the 'non-flood' winter seasons of 1995–96, 1997–98 and 1998–99. The non-flood-year contributions amounted to about 2.7 g cm^{-2} at O60, 5.3 g cm^{-2} at O70 and 6.0 g cm^{-2} at O90, with a significant part of those values coming in 1998 and 1999 at the deepest site. The mass separations show that, whereas the flood-year contributions were most important at the shoreward end (60 m) of the O-line, the non-flood-year sedimentation seemed to favour the deeper part of the O-line. This might be explained in part by temporal variation in wave energy; waves were typically more energetic in the period 1998–2000, and that was especially true in winter 1998–99 (Fig. 10b; Ogston *et al.*, 2004).

It is concluded that flood sediment deposited in relatively shallow water (< 60 m) is transferred seaward during subsequent months and years, consistent with the processes described above. Interestingly, none of the sites on the O-line (the shallowest being 50 m) ever became net erosional, even though the textures were changing significantly and material was being redistributed.

124   R.A. Wheatcroft et al.

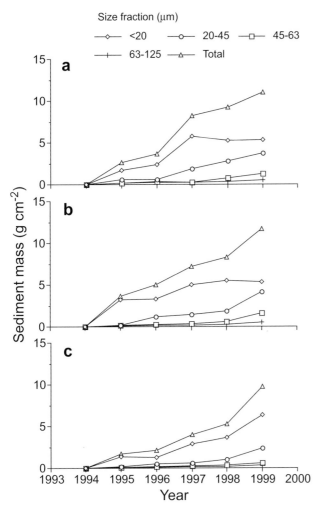

Fig. 18 Mass changes in grain size (in μm) at sites (a) O60, (b) O70 and (c) O90 on the Eel shelf during the period 1994–1999. All sites show an increase in mass, largely comprising sediment in the < 20-μm and 20–45-μm size fractions. The total mass increases at O60 and O70 were relatively linear, although increases in the finest fraction were more important initially. Increases in the 20–45-μm fraction were more important toward the end of the period, which was marked by low flooding but large waves (see Fig. 10). Increases in mass at site O90 occurred mostly toward the end of the period when storms were more energetic.

Deposition

Although deposition may not commonly be thought of as a form of physical reworking, deposition of new sediment moves all subjacent sediment downward at a rate that has a bearing on preservation of signals, and also affects the pace of depth-dependent altera-

tion (e.g. pelletization, consolidation). Whether deposition is steady or episodic, it has an impact on preservation. More complete discussions of deposition modes and rates can be found elsewhere in this volume (Hill *et al.*, pp. 49–99; Sommerfield *et al.*, pp. 157–212; Syvitski *et al.*, pp. 459–529).

Episodic deposition increases event-bed thickness, thereby increasing its potential for preservation. Flood deposition is treated elsewhere in Hill *et al.* (this volume, pp. 49–99), but observational evidence from Traykovski *et al.* (2000) indicates that during times of flooding on the Eel shelf, deposition on the middle shelf occurs during the waning phases of energetic wave intervals. The location and thickness of these depositional packets depend on cross-shelf location, wave energy and a fresh supply of flood sediment. By monitoring grain-size changes in the seabed, Drake (1999) showed that the dominant source of sediment to the middle shelf between flood events seems to be inner-shelf fine sands and coarse silts that are probably transported across the shelf during times of intense storm waves. Long-term accumulation patterns (Sommerfield *et al.*, this volume, pp. 157–212) represent the time-integration of many discrete events, each of which results from an individual storm, tidal cycle or flood event.

Attempts to observe discrete depositional events on continental shelves are hampered by the difficulty in obtaining undisturbed samples of the sediment bed surface. Two-dimensional, cross-shelf sediment-transport models able to account directly for flux divergence and thereby resolve event-time-scale depositional processes while preserving realistic cross-shelf behaviour (Zhang *et al.*, 1999; Harris & Wiberg, 2001, 2002; Fan *et al.*, 2004) indicate that cross-shelf transport convergence during wave-current resuspension events will generally not produce enough net deposition (< 2 cm) to substantially improve preservation potential. However, across-shelf gradients in sediment flux associated with wave-supported gravity flows appear capable of generating thicker deposits (Traykovski *et al.*, 2000, 2006; Scully *et al.*, 2003; Hill *et al.*, this volume, pp. 49–99). Three-dimensional models of shelf sediment transport (Harris *et al.*, 2002, 2005) provide better representations of net deposition associated with convergent flow, although at present these cannot resolve both event-time-scales and longer-term accumulation patterns.

BIOLOGICAL ALTERATION

On most continental margins the activities of sediment-associated animals (i.e. benthic invertebrates, demersal fishes and mammals) cannot be neglected (Fig. 3c). Through their day-to-day activities, benthic organisms can play three important roles in the post-depositional alteration and preservation of strata. First, they can alter dynamically important properties of bottom sediments, for example porosity and grain size, which in turn can influence physical processes such as consolidation and erosion. Second, **bioturbation**, the stirring of bulk sediment due to organisms (Richter, 1952), has a first-order impact on the destruction of strata, and hence the preserved stratigraphic record. In addition, bioturbation may redistribute material (e.g. sand-sized particles) within the seabed, thereby reorganizing the bulk properties of the bed. Third, animals produce sedimentary structures such as burrows and feeding traces that replace the primary physical sedimentary structures and bedding with a bioturbate texture or 'ichnofabric' (Frey & Pemberton, 1990). Discrete traces as well as the overall ichnofabric may have ethological and palaeoenvironmental significance (Schäfer, 1972; Frey, 1975), and are therefore studied in detail by palaeontologists.

Alteration of key dynamical bed properties

Some of the earliest studies of animal–sediment relations recognized clearly the important modifying effect that benthic organisms could exert on marine sediments (Moore, 1931; Dapples, 1942). Much ensuing research, summarized by Rhoads (1970, 1974), Rhoads & Boyer (1982) and others (Jumars & Nowell, 1984; Aller *et al.*, 2001; Paterson, 2001), has documented the manifold ways that seabed properties are altered biogenically. In a broad sense, but with some overlap, organism effects can be broken into two categories: biophysical and biochemical. Biophysical alteration involves typically changes to the small-scale packing of particles (e.g. pelletization) or the location of particles within the seabed (i.e. grain or macroscale roughness; Jumars & Nowell, 1984). Biochemical alteration involves mainly the production of exopolymeric substances that bind sediments together (Decho, 1990; Paterson, 2001), thereby altering their critical erosion and

deposition thresholds (Jumars & Nowell, 1984; Dade *et al.*, 1990).

Potential for benthic microalgal production

Although both metazoans and microbes (e.g. bacteria and benthic diatoms) produce **exopolymeric substances** (EPS), the latter, especially benthic microalgae (or **microphytobenthos**), have been shown to be particularly relevant to sediment dynamics (Paterson, 2001). For example, many studies (Miller *et al.*, 1996; Black *et al.*, 2002; Lund-Hansen *et al.*, 2002) have shown that EPS secretion by microphytobenthos can exert the dominant control on the erodibility of intertidal and shallow-subtidal sediments. A relevant question is then: Are light levels high enough to support microphytobenthos on the Eel shelf? Cahoon (1999) compiled data on the lowest light intensities needed to support detectable gross primary production by benthic microalgae, and found that growth at very low light intensities ($\sim 1 \mu E\ m^{-2}\ s^{-1}$ or 0.1% of surface incident radiation) is possible. Using Secchi depth data (a measure of light attenuation) from $\sim 18,000$ neritic stations and the hypsometry of continental margins, he concluded that the requisite light intensities would be found at roughly a third of the stations examined, which translated to depths of $\leq 65\ m$ (Cahoon, 1999). Of course benthic light intensity will depend on local conditions, which vary in time due to such factors as seasonal insolation, river discharges and phytoplankton production; thus direct measurements are needed.

As part of the long-term seabed monitoring effort (Ogston *et al.*, 2004), the downward irradiance of **photosynthetically active radiation** (PAR; Kirk, 1994) was measured at the S60 tripod using a cosine collector mounted 2 m off the bed. Although fouling became a problem in the latter stages of each deployment, analysis of the first 4 weeks of sensor output allowed a first-order look at the benthic light field at 60 m. Results indicate that during spring and summer, the downward irradiance was often high enough to support benthic microalgae growth, with daytime values in the $1–5\ \mu E\ m^{-2}\ s^{-1}$ range (Fig. 19). In contrast, during the winter the combination of low solar elevation, frequent cloud cover and higher terrestrial runoff resulted in very low irradiance at the 60-m-deep seafloor. Collectively, these preliminary results suggest that sufficient light is present

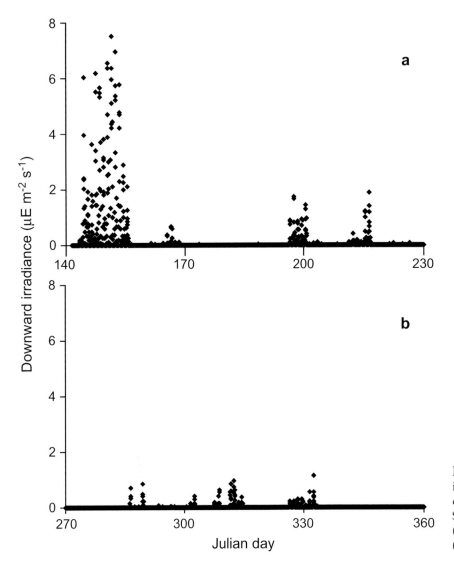

Fig. 19 Time-series of downward irradiance PAR (µE m^{-2} s^{-1}) at 58-m depth on the Eel shelf (sensor on the S60 tripod; Ogston *et al.*, 2004) during (a) the summer solstice period and (b) the autumn equinox.

to support microphytobenthic production at depths ≤ 60 m during at least part of the year. Whether benthic algae were actually present, and whether they had a measurable impact on sediment erodibility remains unclear, but future shelf sediment-transport studies should not ignore this potentially important factor. In addition, given that the lower 2 m of the water column typically has high concentrations of suspended sediment (Ogston *et al.*, 2000; Traykovski *et al.*, 2000), which contributes substantially to light attenuation, future measurements of irradiance need to be made closer to the bottom.

Pelletization

Pelletization, the repackaging of primary particles into faecal pellets by animals, is a common phenomenon in muddy sediments. Many benthic invertebrates, especially deposit and suspension feeders, produce pellets, and these are often deposited at or near the sediment–water interface. In addition, under some conditions zooplankton pellets can be found in extremely high abundances in the seabed (Moore, 1931). Benthic faecal pellets are typically 200 µm to > 1 mm in length, and thus may be orders of magnitude larger than primary grains. Pellets are often encased in organic coatings, so they can persist for significant periods of time and can have an important impact on sediment transport (Rhoads & Boyer, 1982; Taghon *et al.*, 1984; Drake *et al.*, 2002).

Faecal pellets change the response of the muddy seabed to applied stresses in complex and uncertain ways (Nowell *et al.*, 1981). Individual pellets are larger and less dense than primary particles, and have settling velocities that can be several orders of magnitude higher than primary grains (Taghon *et al.*, 1984; Komar & Taghon, 1985). An increase in diameter also leads to greater exposure to the flow, hence pellets are often observed to roll along the bed during otherwise subcritical flow (Nowell *et al.*, 1981). Further complicating the net effect of pellets is that they almost always are associated with mucous that causes adhesion between pellets and the bed, thereby decreasing their erodibility (Nowell *et al.*, 1981; Jumars & Nowell, 1984). The net effect of multiple pellets is similarly complex. On the one hand, high densities of pellets may armour the bed, thereby limiting resuspension flux (Wiberg *et al.*, 1994; Wheatcroft & Butman, 1997). On the other hand, pelletization may result in a shift from a cohesive bed to one that behaves in a cohesionless manner, which is more easily eroded and produces bedforms (Andersen, 2001; Drake *et al.*, 2002).

In order to document the degree of pelletization in surficial sediments and determine whether it varied systematically in time (Drake, 1999), bottom sediment samples collected during several cruises over a 6-year period were analysed to determine the weight percentage of 'aggregates' using a gentle wet-sieving method (Fuller & Butman, 1988; Wheatcroft & Butman, 1997; Drake *et al.*, 2002). Most of these samples were from the region impacted by the January 1995 and January 1997 flood layers (Wheatcroft & Borgeld, 2000), but a number of pre-1995 sediment samples were also examined. None of the samples from the shelf contained more than 6% aggregated particles in the size classes > 45 μm, and there was essentially no aggregate fraction in the finer size classes. Mean aggregate content for all samples was a relatively low 2.6%. Although the average pellet content appears to have increased slightly from February 1995 to September 1995 following the emplacement of the January 1995 flood layer, the increase was small and not statistically significant in light of the large spread in the data. Compared with similar analyses of shelf sediments collected off central (Wheatcroft & Butman, 1997) and southern (Drake *et al.*, 2002) California, the Eel shelf sediments are considerably less pelletized (Fig. 20).

Pellet content on the Eel shelf is thus unlikely to be large enough to materially change the physical properties of the bed. The low content of durable aggregates on the Eel shelf is significant because it suggests that the sediment was not repackaged into larger entities that might behave as non-cohesive coarse silt or fine sand grains. It is possible that this lack of biogenic pelletization may have been a factor in the considerable resistance to erosion that the 1995 flood sediment showed after it was deposited (Drake, 1999). A probable reason for the low abundance of pellets in the Eel shelf sediments is that the benthic macrofaunal community has few surface-deposit feeders (see below), which are thought to play a disproportionate role in pelletization.

A cautionary note regarding measurements of *in situ* aggregated grain size is warranted: namely, the wet-sieving technique is not without error (Wheatcroft & Butman, 1997; Drake, 1999). Many of the fragile pellets and probably all flocs are destroyed during the analysis. Thus, the existence and properties of the benthic fluff-layer (Stolzenbach *et al.*, 1992; Jago & Jones, 1998; Thomsen & Gust, 2000) are unlikely to be resolved using current methods. This shortcoming represents a substantive limitation in the ability to predict and model the erodibility of fine-grained sediment.

Biogenic roughness

Although it is often portrayed as a flat surface, the seafloor has significant small-scale relief. In fact, the seabed surface, especially in sands, is often moulded by wave- and current-generated stresses into distinct bedforms (e.g. ripples, dunes) that have an important impact on diverse aspects of sediment transport (Huettel *et al.*, 1996) and boundary-layer fluid mechanics (Grant & Madsen, 1986). Biological processes such as burrowing and foraging also result in seafloor **roughness** that may have horizontal and vertical length scales comparable to physical bedforms (Briggs, 1989; Wheatcroft, 1994), but respond to different forcings and change at different rates (Nichols *et al.*, 1989; Yager *et al.*, 1993; Wheatcroft, 1994). In fact, it may be safely stated that for all but the most energetic environments (or anoxic sites) biogenic roughness is the 'normal' bed state. Despite this importance, there are few data on **biogenic microtopography**, and theoretical models linking biogenic roughness to fluid and sediment transport remain elusive (see Grant & Madsen, 1986).

Qualitative information on biogenic bottom roughness on the Eel shelf was obtained in two

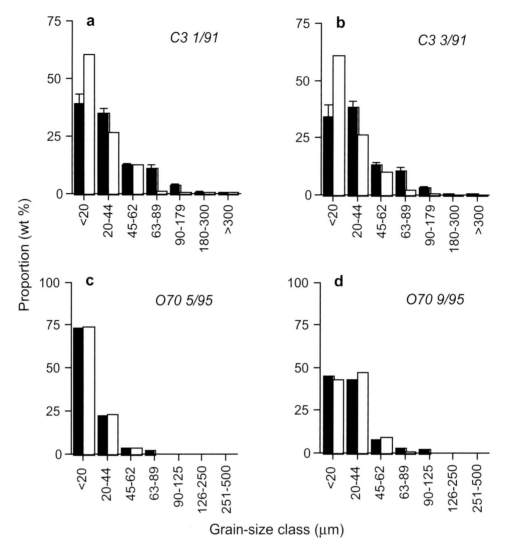

Fig. 20 Weight percentage of disaggregated sediment (open bars) and aggregated sediment (solid bars) as a function of grain-size class. (a & b) For a 90-m site on the Russian River shelf. (From Wheatcroft & Butman, 1997.) (c & d) For a 70-m site on the Eel shelf. (From Drake, 1999.) There is little evidence for aggregates on the Eel shelf (i.e. the disaggregated and aggregated mass fractions are approximately the same for all size classes), whereas the much higher mass fraction in the < 20-μm disaggregated size fraction on the Russian River shelf implies significant pelletization.

ways. Cutter & Diaz (2000) used a sediment profile camera (Rhoads & Cande, 1971) to collect images of seafloor structure at a limited number of stations along the S-line (~30 m to 80 m) in December 1995. The main conclusion from this study was that root-mean-square (RMS) roughness decreased seaward. The RMS values ranged from 5.5 ± 3 mm (mean ± SD) at the inner-shelf stations (~30 m), to 3.7 ± 1.5 mm at ~50-m sites and 3.2 ± 0.6 mm in seaward stations (60–83 m) (Cutter & Diaz, 2000); however, because of the large variability and small sample size, the bathymetric trend was not statistically

significant. They also confirmed that ripples were present in the sandy sediments (< 55 m) and biogenic microtopography dominated in the more seaward muddy sediments. Although useful in re-affirming a priori expectations regarding broad-scale patterns of seafloor roughness, this study had limited utility due to the lack of any temporal coverage, which is known to be important (Wheatcroft, 1994; Cacchione et al., 1999).

To address temporal change, a time-lapse video camera was deployed on the long-term monitoring tripod at station S60 (Ogston et al., 2004). Although quantitative analyses of video data are challenging due to the two-dimensional picture obtained and limited visibility during times of interest (e.g. storm events), some qualitative observations are possible. The main finding is that seabed roughness at the S60 site is due to both physical and biological processes. The former results in mainly two- and three-dimensional ripples (Fig. 15a & b), whereas the latter are a diverse set of feeding, locomotion and dwelling structures created by benthic invertebrates and demersal fish (Fig. 15c). A key issue that is motivated by these observations is to determine how rapidly physical bedforms, formed during a sediment transport event, are destroyed by biological processes. This problem is a difficult one, as bedform destruction is probably due to a combination of steady, but potentially patchy, biodiffusive mixing (see next section) and transient events (e.g. fish foraging). Nevertheless, the next generation of sediment-transport models and data analyses will need to incorporate explicitly temporal changes in bottom roughness during non-transport as well as transport conditions.

Bioturbation

The second major way that biology impacts sedimentary strata alteration and preservation is through bioturbation. Bioturbation is caused by virtually all animals that come into contact with bottom sediment, from millimetre-sized meiofauna to large demersal fish and grey whales. These animals are engaged in a diverse array of activities such as foraging, locomotion and dwelling construction (Schäfer, 1972; Frey, 1975; Thayer, 1983; Bromley 1990). Only in areas of low bottom-water oxygen content (e.g. oxygen minimum zones, restricted basins), near-continuous physical sediment transport (e.g. beaches, tidal channels) or extremely rapid sediment accumulation (e.g. delta fronts of major rivers, proximal glacial environments) are bioturbating animals excluded. Everywhere else in the ocean, particles are displaced many times before they are moved out of the surface mixing layer (see below). It is this particle displacement that destroys physically produced sedimentary structures or layers, and mixes transient signals to a state where they are no longer recognizable. Also, as noted above, bioturbation can preferentially redistribute particles within the seabed, thereby destroy-

ing or creating (Rhoads & Stanley, 1965) graded beds. In considering sedimentary strata alteration and preservation, it is therefore important to obtain some knowledge of the intensity of biologically driven particle displacement, its depth of operation within the seabed and its dominant mode(s).

Excess ^{234}Th biodiffusivities

The primary approach to quantifying bioturbation intensity – and the one used on the Eel margin (Bentley & Nittrouer, 2003; Wheatcroft & Drake, 2003; Wheatcroft, 2006) – was the measurement and modelling of vertical profiles of the short-lived (24.1-day half-life) radionuclide, ^{234}Th. This naturally occurring radionuclide, which reflects processes occurring during roughly the previous 75 days, has proven to be particularly useful in bioturbation studies on continental margins (Aller & Cochran, 1976; Aller *et al.*, 1980; Martin & Sayles, 1987; Wheatcroft & Martin, 1996; Green *et al.*, 2002). Excess ^{234}Th activities in the upper 15 cm of the seabed were determined by γ-ray spectroscopy of dried sediment using standard techniques (Cutshall *et al.*, 1983; Buesseler *et al.*, 1992; Gilmore & Hemingway, 1995).

In the general case (neglecting non-local exchange), the temporal derivative of excess ^{234}Th activity is determined by diffusive bioturbation, sediment accumulation and radioactive decay (Berner, 1980; Boudreau, 1997):

$$\frac{\partial A}{\partial t} = D_b \left(\frac{\partial^2 A}{\partial z^2} \right) - S \left(\frac{\partial A}{\partial z} \right) - \lambda A \qquad (15)$$

where A is excess activity in disintegrations per minute per mass (dpm g^{-1}), z is depth into the sediment (cm), D_b is the **vertical biodiffusion coefficient** (cm^2 yr^{-1}), S (cm yr^{-1}) is the sediment accumulation rate and λ is the ^{234}Th decay constant (10.5 yr^{-1}). To apply Eq. 15 in studies of bioturbation two simplifications are necessary: (i) the time derivative is set to zero and (ii) the sediment accumulation term is neglected. In the case of the Eel margin, where episodic sediment transport and delivery (Hill *et al.*, this volume, pp. 49–99) are important, both of these simplifications must be made carefully.

To permit the steady-state assumption, periods of time immediately following the creation of flood deposits have been neglected. Thus, ^{234}Th

profiles measured in the winters of 1995 and 1997 were not used to compute biodiffusivities, because they were clearly impacted by non-steady sediment accumulation during the previous 75 days. During other parts of the year, and during the winter of 1996, there was little evidence for the input of significant quantities of sediment to the seabed at the relevant study sites. To rule out sediment accumulation, it is necessary for $S^2 \ll 4 D_b \lambda$ (see Sommerfield *et al.*, this volume, pp. 157–212). If this inequality is met, then the sediment accumulation rate is unlikely to affect the ^{234}Th profile (Aller, 1982). In the case of the Eel margin, maximal sediment accumulation rates are ~1 cm yr^{-1} (Sommerfield & Nittrouer, 1999), thus biodiffusivities would need to be < 0.1 cm^2 yr^{-1} to have an impact on the ^{234}Th profile. This value is considerably lower than what was observed, so omission of the sediment accumulation term in Eq. 15 is justified.

The above two simplifications, along with neglecting vertical gradients in porosity, lead to a balance between biodiffusive mixing and radioactive decay

$$D_b \frac{\partial^2 A}{\partial z^2} = \lambda A \tag{16}$$

Note that including porosity in the calculations results in an ~10% change in biodiffusivities, which is unimportant given the large amount of variability. For the boundary conditions $A = A_0$ at $z = 0$ and $A \to 0$ as $z \to \infty$ the solution to Eq. 16 is

$$A_z = A_0 \exp\left[-\left(\sqrt{\frac{\lambda}{D_b}}\right)z\right] \tag{17a}$$

or

$$\ln A_z = \ln A_0 - \left(\sqrt{\frac{\lambda}{D_b}}\right)z \tag{17b}$$

where A_z is the excess ^{234}Th activity at a depth z and all other terms are as before. A linear regression to a plot of $\ln A_z$ (Y) versus z (X) yields a slope that is equal to the square root of λ divided by D_b and a Y intercept that is $\ln A_0$. Knowing λ, D_b can be estimated.

The main objectives of the bioturbation studies on the Eel continental shelf were to document temporal variability in mixing intensity at a small number of sites, and relate any detected changes to variability in important forcing terms (e.g. macrofaunal abundance, organic-carbon flux, bottom-water temperature). Two broadly similar but independent studies were conducted. Bentley & Nittrouer (2003) studied three stations on an across-isobath transect (S50, S60 and S70) between September 1995 and July 1997, with the majority of sampling conducted over the winter of November 1995 to March 1996. The second study focused on four stations along the 70-m isobath (C70, I70, L70 and O70) that were occupied at approximately 4-month intervals from May 1995 to August 1999 (Wheatcroft & Drake, 2003; Wheatcroft, 2006). In combination, the two studies measured excess ^{234}Th profiles on > 120 separate cores, thereby providing a comprehensive look at temporal variation in bioturbation intensity. In addition, both studies augmented the excess ^{234}Th measurements by quantifying, to varying degrees, the abundance, species composition and vertical distribution of the benthic macrofauna.

Broadly speaking, the shelf bioturbation studies yielded similar results. Namely, excess ^{234}Th profiles were characterized by surface activities of 5–15 dpm g^{-1}, log-linear decreases with depth and penetration of excess ^{234}Th to depths of 4–8 cm (Fig. 21). Resultant mixing intensities on the Eel shelf are moderately high (Figs 21 & 22; Table 2), ranging from 3 to > 100 cm^2 yr^{-1}, with an average of 27 ± 24 cm^2 yr^{-1} at the 70-m sites and 21 ± 10 cm^2 yr^{-1} at the shallower sites (S50 and S60). An important aspect of the shelf data is that there was appreciable within-site variability at all sites and times. For example, Wheatcroft (2006) found that biodiffusivities varied by factors of two to five for a particular site and sampling time and by over an order of magnitude for a particular station over all sampling periods (Table 2). The high level of small-scale spatial variability meant that it was not possible to detect significant between-site variability (either along- or across-shelf). In addition, lumping all of the shelf data onto a single year (Fig. 23) indicates little evidence for seasonal variation (i.e. the regression explains $< 10\%$ of the variance). Note that both the magnitude and variability of the measured biodiffusivities are similar to those observed in other well-sampled continental-shelf settings (e.g. Wheatcroft & Martin, 1996; Gerino *et al.*, 1998).

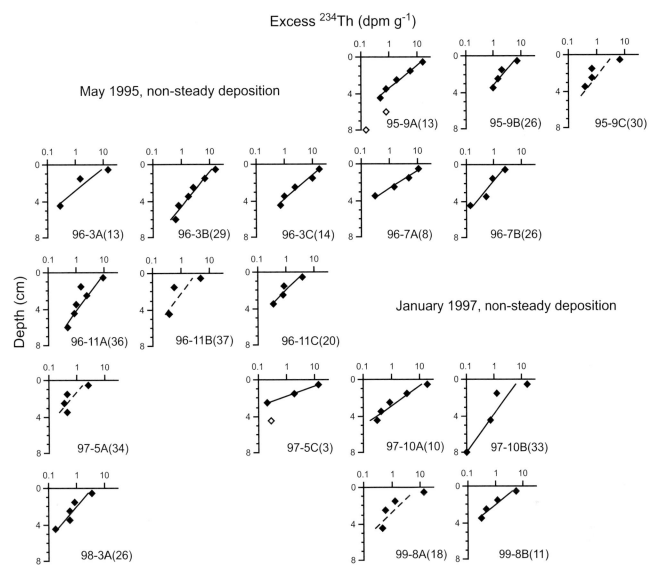

Fig. 21 Representative excess ^{234}Th profiles collected at station O70. Codings indicate the year and month of collection, the replicate letter designation and (in parentheses) the resultant biodiffusivity (in cm^2 yr^{-1}) based on Eq. 17. Solid diamonds and open diamonds denote data points used and not used, respectively, in the fits. Fits with an $r^2 < 0.85$ have dashed lines. Profiles obtained during cruises in May 1995 and January 1997 were impacted by non-steady deposition during the preceding weeks to months. (Modified from Wheatcroft, 2006.)

To help interpret a broad-scale study of sediment accumulation patterns on the Eel continental slope (Alexander & Simoneau, 1999), samples from roughly 40 box cores were collected during several cruises between September 1995 and August 1999 at water depths from 150 m to 1000 m. These were analysed for ^{234}Th in a similar fashion to the shelf cores described above. As with the shelf results, biodiffusive mixing of the seabed on the upper slope (i.e. < 500 m) is relatively vigorous (Fig. 24), with a mean ($\pm$ SD) D_b of 31 ± 27 cm^2 yr^{-1} and a range from < 1 to 100 cm^2 yr^{-1}. From cores ($N = 8$) collected between 550 m and 1000 m, there is a factor of three lower mixing intensity ($D_b = 9.1 \pm 8$ cm^2 yr^{-1}) (Fig. 24b), but the difference is not statistically significant.

Nevertheless, there is broad evidence for a decrease in sediment mixing intensities in deeper

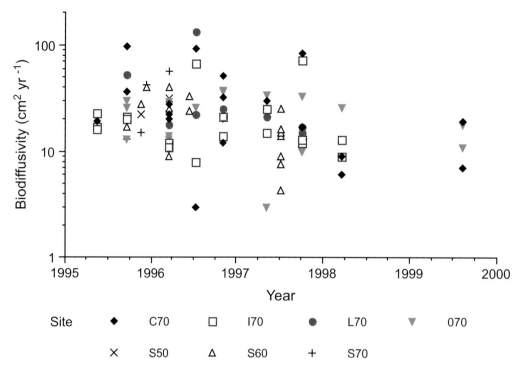

Fig. 22 Estimated biodiffusivities (cm^2 yr^{-1}) from excess ^{234}Th profiles measured at 50-, 60- and 70-m sites on the Eel shelf over a 5-year period (from Bentley & Nittrouer, 2003; Wheatcroft, 2006). The labelled tick marks represent 1 January of the year indicated.

Table 2 Summary of bioturbation intensities measured on the Eel shelf (from Bentley & Nittrouer, 2003; Wheatcroft, 2006): CV, coefficient of variation (i.e. standard deviation divided by mean, multiplied by 100)

Station	Mean (cm^2 yr^{-1})	Standard deviation	CV (%)	Number
S50	26	6.4	24	2
S60	20	11	55	12
S70	28	21	77	4
O70	21	11	50	18
L70	32	36	111	11
I70	22	19	85	17
C70	32	30	92	18
All 70-m stations	27	24	90	67
All shelf stations	25	22	89	82

water, which is consistent with observations made on other continental slopes (Carpenter *et al.*, 1982; DeMaster *et al.*, 1994; Green *et al.*, 2002). In particular, the Eel data are similar to ^{234}Th-derived biodiffusivities computed from samples collected on the upper slope off North Carolina (Fig. 23a; Green *et al.*, 2002), and a global compilation (Fig. 24a; Middelburg *et al.*, 1997). The latter is somewhat surprising because the global compilation was based on biodiffusivities calculated using the much longer lived (half-life of 22.3 yr) radioisotope, ^{210}Pb (Middelburg *et al.*, 1997). That radioisotope

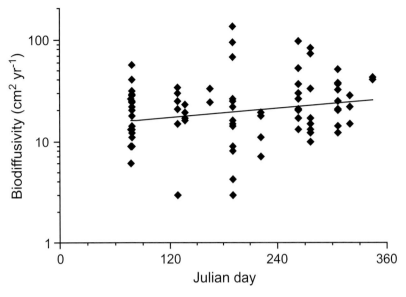

Fig. 23 Biodiffusivity (cm² yr⁻¹) as a function of Julian day for the 5 yr of data depicted in Fig. 22. The regression line is a statistically insignificant fit that explains < 5% of the variance. (Modified from Wheatcroft, 2006.)

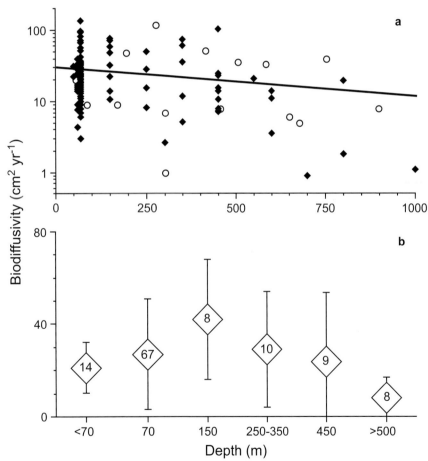

Fig. 24 Biodiffusivity (cm² yr⁻¹) as a function of depth on the Eel margin. (a) Including all data points collected from the Eel margin (solid diamonds), along with data from the North Carolina slope (open circles; Green *et al.*, 2002), and a regression curve (solid black line) based on a global compilation of ²¹⁰Pb-derived biodiffusivities (Middelburg *et al.*, 1997). (b) Mean and standard deviation of Eel margin data binned according to depth. Sample size of each depth interval is listed in the diamond. (From Bentley & Nittrouer, 2003; Wheatcroft, 2006; C.R. Alexander, unpublished data.)

has been shown to yield generally lower biodiffusivities, due to a phenomenon known as 'age-dependent mixing' (Smith *et al.*, 1993). In addition, based on a limited number of profiles (N = 11), no

evidence was found for a significant relationship between water depth and biodiffusivities based on ²³⁴Th (Middelburg *et al.*, 1997). In both instances, the large amount of variability that appears to be inherent in estimates of the biodiffusivity may require significant replication to identify meaningful large-scale patterns.

Macrofauna

Ultimately, the abundance, biomass and species composition of the macrofauna at a particular location must have an impact on the bioturbation intensity at that site (Guinasso & Schink, 1975; Aller, 1982; DeMaster *et al.*, 1994; Wheatcroft & Martin, 1996). Therefore, in association with the excess [234]Th measurements discussed above, samples were collected for enumeration of the **macrofauna** (herein defined as animals retained on a 300-μm sieve). The primary question addressed using these data was whether the flood deposits caused widespread mortality of the benthic community, and hence depressed sediment-mixing intensity (Wheatcroft, 2006). To address this question, triplicate box-cores were collected at two 70-m sites (C70 and L70) roughly every 4 months over a 3-yr period. Subsamples (10 cm by 10 cm) from the box cores were sieved through 1-mm and 300-μm sieves and the retained macrofauna sorted and identified to the lowest possible taxonomic level.

The January 1997 flood provided the best opportunity to examine the effect of flood layer deposition on benthic macrofaunal abundance, because, unlike the January 1995 flood, pre- and post-event sampling was conducted. The results indicate that, with the exception of the 1-mm fraction at L70, there was a significant decrease (Mann-Whitney Test, $\alpha = 0.05$; Conover, 1980) in the mean number of individuals in each size fraction and the overall abundance following deposition of the January 1997 flood deposit (Fig. 25). Individuals in the 300-μm fraction at station L70, where roughly 7 cm of sediment was deposited (Wheatcroft & Borgeld, 2000), suffered the greatest mortality, decreasing by roughly half. At station C70, where only 3 cm of sediment was deposited (Wheatcroft & Borgeld, 2000), abundance in the 300-μm fraction decreased by a third. Taken alone, these data suggest that flood deposition can lead to significant mortality of the shelf macrofauna, and that the level of mortality (at least for the 300-μm fraction) is positively related to layer thickness (Wheatcroft, 2006).

Examination of the macrofaunal data over the 3-yr study (Fig. 26), however, suggests that the

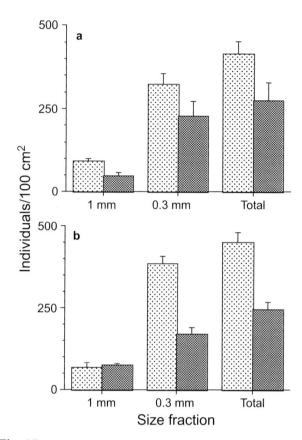

Fig. 25 Histogram of average (± standard deviation) macrofaunal (1- and 0.3-mm fractions) density at two 70-m sites 6 weeks before (light stippling) and 2 weeks after (dark stippling) deposition of the January 1997 flood deposit. (a) Station C70 where 3 cm of sediment was deposited. (b) Station L70 where 7 cm of sediment was deposited. (Modified from Wheatcroft, 2006.)

observed decrease in abundance between November 1996 and January 1997 was part of an annual cycle. In common with other temperate, soft-bottom environments (Buchanan & Moore, 1986), macrobenthos abundance on the Eel River shelf is lowest in the late winter to early spring, increases through the summer and early autumn period as recruitment occurs, and then decreases over the winter. There is little difference between the decreases in abundance observed during the winter of 1996–97, when there was a major flood, and the winter of 1995–96, when there was not (Fig. 26). Therefore, in terms of total abundance, deposition of 3–7 cm of sediment appears to have had little impact on the shelf macrofauna of at least two sites (Wheatcroft, 2006).

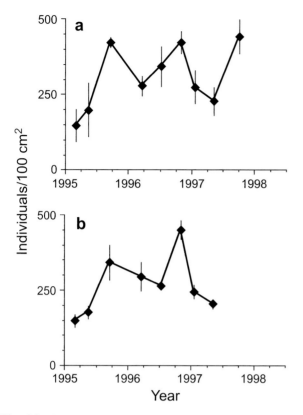

Fig. 26 Three-year time series of average (± standard deviation) total (1- and 0.3-mm fractions) macrofaunal density at stations (a) C70 and (b) L70. (Modified from Wheatcroft, 2006.). The labelled tick marks represent 1 January of the year indicated.

For the most part, a similar conclusion was reached when the data were viewed at different taxonomic levels (e.g. species or family) or functional groupings (e.g. surface versus subsurface deposit feeders). For example, the subsurface deposit-feeding polychaete worms *Cossura pygodactylata* and *Levinsenia gracilis* display either seasonally fluctuating abundance patterns (*C. pygodactylata*) or relatively constant abundance (*L. gracilis*) that is insensitive to flood deposition. Similarly, the abundance of the tube-dwelling, surface deposit-feeding polychaete, *Magelona longicornis*, is also relatively constant, although at station C70, where 3 cm of sediment was deposited, there was a temporary decrease following the January 1997 flood. Only for the spionids (a family of tube-dwelling, 'interfacial' feeding polychaetes; Dauer *et al.*, 1981) is there a suggestion of a flood impact on abundance, although in this case it is also difficult to rule out simple seasonal variability (Wheatcroft, 2006).

In light of previous research on the impact of episodic sedimentation (natural or artificial) on macrobenthos, the apparent resilience of the Eel River shelf macrofauna to flood deposition is a bit surprising. Both laboratory (Kranz, 1974; Mauer *et al.*, 1986) and field (Norkko *et al.*, 2002) sediment-burial experiments have found that significant (in some cases 100%) mortality can occur following as little as 3 cm of deposition. There are several possible explanations for the difference between the results from the Eel and the controlled burial experiments. The rate of sediment delivery in the burial experiments was significantly more rapid than the deposition of sediment following the Eel River floods. The former were essentially instantaneous deposition events, whereas the Eel flood deposits probably formed over periods of hours to days (Bentley & Nittrouer, 2003; Wheatcroft & Drake, 2003). The slower rate of burial may permit the Eel macrofauna to more easily adjust their living positions. In addition, many of the burial experiments observed that mortality was greatest if the 'capping' sediment was finer grained than the ambient sediment (Mauer *et al.*, 1986; Norkko *et al.*, 2002). In the case of the Eel shelf, the difference between the flood deposit and the ambient sediment is small compared with the burial experiments, which in some cases were quite extreme (e.g. clay overlying sand, Norkko *et al.*, 2002). Thus, the Eel macrofauna probably did not encounter problems with burrowing mechanics, as was inferred in some of the manipulative experiments. Third, as described next, the species and functional group composition of the Eel River shelf macrofauna may be pre-adapted to contend with frequent burial events.

The Eel River shelf macrofauna is strongly dominated by polychaete worms. Roughly 80% of the combined 300-µm and 1-mm fractions are polychaetes, with a 7% to 9% (each) contribution from arthropods (mainly amphipods) and molluscs (mainly bivalves), < 1% echinoderms and 2–3% in a miscellaneous category comprising mainly nemerteans, sipunculans and cnidarians. Polychaetes (phylum Annelida) are even more dominant relative to arthropods when only the 1-mm fraction is considered (Table 3). This pattern is further illustrated by examination of the ten most abundant species, all of which are polychaetes. No significant changes in species composition occurred in response to the flood events. In terms of

Table 3 Percentage of major phyla sampled in the 1-mm fraction at sites along the western USA continental shelf (Sources: WA, Washington, Lie (1969); OR, Oregon, Richardson *et al.* (1977); CODE, Russian River shelf, C.A. Zimmer (unpublished data); LA, Palos Verdes shelf, Wheatcroft & Martin (1996))

| | Site | | | | |
Phylum	WA	OR	Eel	CODE	LA
Annelida	48.2	55.1	86.1	61.8	65.6
Mollusca	27.0	38.0	7.8	17.4	0
Arthropoda	18.1	3.2	5.7	2.2	13.0
Echinodermata	6.6	3.6	0.4	18.5	21.6

functional groupings (Fauchald & Jumars, 1979), the polychaete fauna is always dominated by subsurface deposit feeders (Wheatcroft, 2006).

The bioturbation and macrofauna data defined two important characteristics of the Eel shelf seabed. First, the abundance of macrofauna on the Eel River shelf exhibited little extraordinary change in response to the January 1997 flood, which deposited up to 7 cm of sediment at one of the study sites. This resilience to sediment deposition suggests there was unlikely to be a decrease in bioturbation intensity due to the floods, a result confirmed by the ^{234}Th data (Fig. 22). Second, the Eel River shelf macrofauna is strongly dominated by subsurface deposit-feeding polychaetes that are likely to mix sediment to appreciable depths. Both of these attributes have important implications for the preservation of sedimentary strata.

Biogenic sedimentary structures

The study of biogenic sedimentary structures, either modern or ancient, is known as **ichnology** (Frey, 1975; Bromley, 1990 and references therein). Much prior work, beginning in the early part of the 20th century in Germany and continuing to the present, has demonstrated that animal traces possess considerable information. In particular, both **ethological** (i.e. behavioural) knowledge about the trace maker (Bromley, 1990) and diagnostic palaeoenvironmental information (Seilacher, 1967) can be

gleaned from a study of discrete biogenic structures and the overall ichnofabric (Frey & Pemberton, 1990). Ethological considerations were beyond the scope of the Eel margin research, whereas environmental indicators were available from more direct sources, such as *in situ* instrumentation. However, a brief investigation of biogenic sedimentary structures was conducted (Bentley & Nittrouer, 2003), and it shed important light on post-depositional alteration processes.

Bentley & Nittrouer (2003) recognized three classes of biogenic sedimentary structures in X-radiographs and thin-sections (Kuehl *et al.*, 1988) taken of the January 1995 flood deposit that can be broadly grouped by their depth of occurrence within the seabed. In the upper 3–5 cm of the seabed, millimetre-scale unlined burrows appear within weeks following a depositional event, a result that is consistent with the resilience of the Eel shelf macrofauna to flood sedimentation (Wheatcroft, 2006). These early burrows are commonly filled with sediment coarser than that of the burrow walls (Fig. 27c). In X-radiographs, this style of burrowing produces the mottling characteristic of the upper few centimetres of most cores (Fig. 27a & b) and the ambient pre-flood-deposit sediment (e.g. Fig. 2b).

The next class of biogenic structures includes lined burrows, vertical to inclined, which penetrate the upper ~15 cm of the seabed, and are up to 1 cm in diameter. Burrow geometries include isolated shafts, as well as U- and Y-shaped forms (Fig. 27a & b). These structures are most widespread in cores collected during 1996 and later, particularly at S60 and deeper stations. The low attenuation of X-rays evident in X-radiographs of these structures suggests that many of these burrows in the upper seabed are at least partially open (Fig. 27a). The final class of biogenic structure includes large, filled, unlined burrows up to 10 cm across, penetrating the upper 15–20 cm of the seabed (Fig. 27a & b). Burrow geometry includes vertical shafts, as well as horizontal or inclined forms. The fill in vertical shafts tends to be either homogeneous or **retrusive** (concave up) backfill, similar to that produced by burrowing actinian anemones (Frey, 1970). Horizontal and inclined burrows typically exhibit either flat laminations (probably the result of passive fill) or retrusive *spreiten*, similar to the ichnogenus *Teichichnus* (Fig. 27).

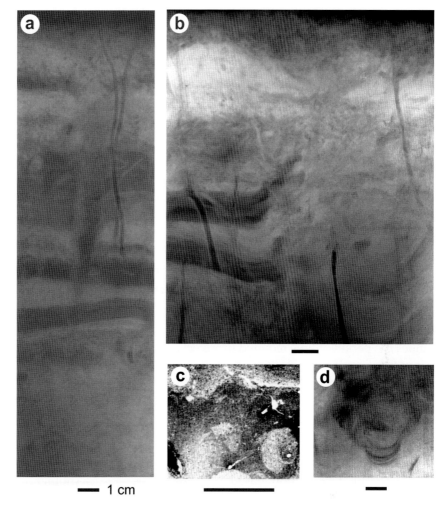

Fig. 27 Representative examples of biogenic sedimentary structures on the Eel shelf. (a) X-radiograph negative (S60, July 1996) showing a lined U-shaped burrow (on right) and a large, unlined burrow (centre) both cross-cutting most of the 1995 flood deposits. (b) X-radiograph negative (S60, July 1996) showing a very large, unlined burrow with meniscate backfill (right) that cross-cuts and has destroyed much of the 1995 flood deposits. Several lined and apparently open burrows are also present. (c) Thin-section photomicrograph showing small burrows with coarse fill. (d) X-radiograph negative of an incipient *Teichichnus* trace at S70, 25–30 cm below the sediment–water interface. Scale bars below each panel = 1 cm. (Modified from Bentley & Nittrouer, 2003.)

PRESERVATION

The general processes related to preservation of strata have been recognized for many years. Early research (Moore & Scruton, 1957; Reineck, 1967; Howard & Reineck, 1972; Howard & Frey, 1975; Howard, 1978; Berger *et al.*, 1979), although qualitative in nature, demonstrated clearly the relative importance of biological and physical processes in shaping the stratigraphic record. More recent research, based primarily on radiochemical measurements, has identified the key terms in the preservation problem to be (Fig. 28): sediment accumulation rate (S), biological mixing intensity (e.g. D_b), surface mixing layer thickness (L_m), and signal or layer thickness (L_s) (Nittrouer & Sternberg, 1981; Wheatcroft, 1990; Bentley & Nittrouer, 2003; Wheatcroft & Drake, 2003). While there is consensus regarding the importance of various terms,

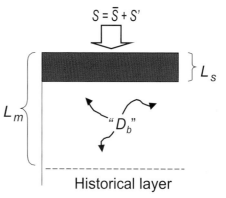

Fig. 28 Conceptual diagram with key terms in the preservation of sedimentary strata and signals. The mean and fluctuating sediment accumulation rates advect a signal or layer of thickness L_s through the surface mixing layer (L_m) and into the historical layer. Mixing within L_m is most often biogenic (i.e. $L_p \ll L_b$) and is therefore represented by a biodiffusivity (D_b). (After Wheatcroft, 1990.)

a testable quantitative theory of strata/signal preservation remains elusive.

In a seminal contribution, Nittrouer & Sternberg (1981), building on ideas expressed by Guinasso & Schink (1975), provided scaling arguments that suggested a non-dimensional combination, G (where $G = D_b/L_mS$), of the above terms could be used to identify dominance by mixing (large G) versus burial (small G). Although the three terms that comprise G are relevant to signal preservation, Wheatcroft (1990) subsequently showed that inclusion of L_m in the denominator was problematic. This is because when L_m is large, an event layer is subject to mixing for a *longer* period of time hence it should have a lower preservation potential. Under the G formulation, however, a large L_m results in a small G, which favours burial or preservation. Wheatcroft (1990) went on to suggest that focus should be placed on time-scales. In this formulation, **transit time** ($T_t = (L_m - L_s/2)/S$) determines the period of time required to advect an event layer (or signal) through the surface mixing layer, whereas **dissipation time** (T_d) represents the time necessary to destroy a signal or event bed. In cases where $T_t < T_d$, some portion of the event bed is preserved intact in the historical record.

Although conceptually straightforward and internally consistent, there are two challenges to applying the formulation based on time-scales. First, there is no theory that allows prediction of dissipation time, although recent efforts appear to be promising (Boudreau *et al.*, 2001; Bentley & Sheremet, 2003). Second, empirical knowledge of dissipation time is lacking, due in part to an absence of well characterized event beds (i.e. those with 'time zero' information). Moreover, dissipation time has been shown to vary as a function of the signal in question (Wheatcroft & Drake, 2003). In the following, each of the governing terms (Fig. 28) in the preservation problem is examined briefly in the light of measurements on the Eel shelf, and new data pertaining to dissipation and transit time are summarized.

Controlling parameters

Bioturbation intensity

Particles are displaced in the marine environment due to a variety of animal activities (e.g. deposit feeding, burrowing, tube building). It is this particle displacement that destroys physically produced sedimentary structures or layers, and mixes transient signals to a state where they are no longer recognizable. Clearly, the intensity of bioturbation must have an impact on the preservation of strata. The challenge is to encapsulate all of the myriad activities that contribute to bioturbation into a small number of measurable terms that ultimately determine dissipation time.

Bentley & Nittrouer (2003) used a volume-based approach to quantify a **biogenic turnover time** (T_b), which represents the time required to displace all the sediment in a unit volume. [Note that this term is not equivalent to dissipation time (Wheatcroft, 1990), as the latter depends in part on the signal of interest (see below).] The T_b was estimated at station S60 by examining X-radiographs and quantifying the volume (assuming cylindrical burrows) of discrete biogenic structures emplaced since the flood deposit was formed. As noted by Bentley & Nittrouer (2003), T_b represents a minimum estimate of bioturbation intensity because it was assumed that all burrows were formed in unburrowed sediment (i.e. there is no cross-cutting) and only discrete structures were measured. In general, this approach yields estimates of T_b that are quite long (10 to 10^2 yr).

Another approach to quantifying bioturbation is to estimate various transport terms from profiles of natural and artificial radionuclides. Most often bioturbation is parameterized as a quasi-diffusive process that can be represented by a biodiffusivity, D_b (Goldberg & Koide, 1962; Guinasso & Schink, 1975). Although conceptualizing bioturbation as a diffusive process is a gross oversimplification that has many shortcomings (Boudreau, 1986; Wheatcroft *et al.*, 1990), it is likely that dissipation time for a given layer or signal will scale inversely with the biodiffusivity; thus the magnitude of D_b provides a *relative* measure of dissipation time. In addition, time-series measurements of biodiffusivity permit determination of whether bioturbation was depressed temporarily due to deposition of an event bed.

Sediment biodiffusivities on the Eel River margin are typically in the range of 5 to > 100 cm^2 yr^{-1} (Bentley & Nittrouer, 2003; Wheatcroft, 2006), in agreement with compilations from continental-shelf (Wheatcroft & Drake, 2003) and slope (Boudreau, 1994; Middleburg *et al.*, 1997) environments. Moreover, for the Eel shelf stations, there is not strong evidence of systematic temporal or spatial variability. For slope stations, there is a hint of a decrease in bioturbation intensity at depths > 500 m (Fig. 23), however, the decrease is small and statistically insignificant. Therefore, as a first approximation we can assume that biological mixing intensity on the Eel shelf is relatively high (~25 cm^2 yr^{-1}), steady and uniform (Figs 21 & 22; Table 2).

Owing to the well documented fact that macrofaunal abundance is greatest at the sediment–water interface and decreases with depth in the sediment, bioturbation intensity is also likely to display within-sediment depth dependency (Guinasso & Schink, 1975; Jumars, 1978), whereby sediment is more intensively displaced near the sediment–water interface. There are substantial difficulties, however, in actually identifying and quantifying that depth dependency based on radionuclide profiles (Boudreau, 1986). These difficulties are due mainly to the rapid decay rate of biologically useful radionuclides, such as ^{234}Th, and the relatively gradual decrease in mixing intensity with depth. Although the radionuclide analyses conducted on the Eel margin do not directly address this issue, indirect evidence of depth-dependent biodiffusion is provided by the variable dissipation times of event beds.

Surface mixing-layer thicknesses

Adjacent to the sediment–water interface, physical and biological sediment displacement occurs; the depths to which these processes act determine the extent of the physical and biological **surface mixing layers** (L_p and L_b, respectively). Note that the term 'mixing' is used rather than 'mixed', because the latter term implies that all properties are uniform within the layer, and that is not the case. Below these layers there is no translocation of particles, and properties are maintained in their relative positions (the so-called **historical layer**, Berger *et al.*, 1979). An accurate knowledge of the vertical extent of these mixing layers and whether they vary systematically in space or time is important in determining transit time.

Physical mixing of the seabed surface occurs in association with bedload and suspended-load transport. The depths of physical reworking might reach several centimetres in the most extreme

conditions, but are generally in the order of millimetres (Wiberg, 2000). Thus, the thickness of the biological surface mixing layer is likely to be more important for strata preservation.

Wheatcroft & Drake (2003) recently discussed definitional inconsistencies associated with the biological surface mixing layer (L_b). Briefly, three independent meanings of L_b were shown to exist in the literature. First, as originally defined, L_b was a region over which D_b was constant (Goldberg & Koide, 1962). Second, it was a region where properties, in particular radionuclides, were mixed uniformly. Third, L_b has been equated with the depth of penetration of a particular radionuclide (Thomson *et al.*, 2000; Smith & Rabouille, 2002). Although, the latter two definitions permit estimation of L_b from radionuclide profiles, they clearly have different meanings. These problems suggest that recent attempts (Boudreau, 1994, 1998) to assign a worldwide average (~10 cm) to the surface mixing layer thickness are questionable. In addition, some of the concepts that are embodied in radiochemical studies (e.g. uniform D_b) are simply not appropriate for the problem at hand, namely the alteration and preservation of sedimentary signals.

Sediment displacement, irrespective of its mode (i.e. diffusive or non-local), destroys bedding, so the vertical distribution of sediment-dwelling macrofauna is critical for the estimation of L_b. On the Eel margin, a surface-mixing-layer thickness of 10 cm is an underestimate, because there is direct and indirect evidence of macrofauna at depths considerably greater than 10 cm. Numerous macrofaunal assays at several sites on the shelf (Bentley & Nittrouer, 2003; Wheatcroft, 2006) demonstrate that ~5% of the total number of individuals are found below 10 cm (Fig. 29). In addition, there is some evidence that following deposition of the flood deposits, the depth distribution of the macrofauna shifts *downwards*, perhaps in response to buried organic matter (Wheatcroft, 2006). Numerous burrows and feeding structures extending from the seabed surface to depths well below 10 cm provide further indirect evidence of deep mixing (e.g. Fig. 27; Bentley & Nittrouer, 2003).

Event-layer thickness

The thickness of an event layer is critically important in determining its preservation potential (Larson

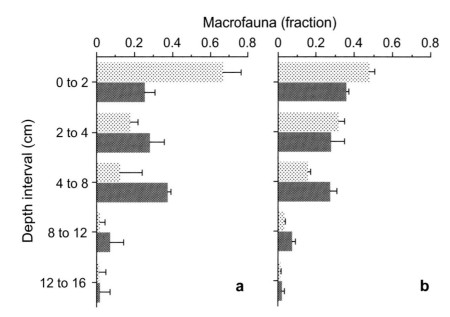

Fig. 29 Vertical distribution of macrofauna at stations (a) L70 and (b) C70 before (light stippling) and after (dark stippling) the January 1997 flood. (Modified from Wheatcroft, 2006.)

& Rhoads, 1983; Wheatcroft, 1990) because it influences transit *and* dissipation time. Transit time is impacted by event-layer thickness because the deposition of a layer represents a short-term increase in sediment accumulation rate. Dissipation time is influenced by event-layer thickness because thin layers present less sediment volume to displace, as well as extend less far into the surface mixing layer. Therefore, there is less of a thin bed to destroy (T_b is less; Bentley & Nittrouer, 2003) and it spends more time in a region of high sediment displacement (note that this assumes there is a decrease in sediment mixing intensity with depth into the seabed).

Time-lapse X-radiographs collected on the Eel margin underscore the importance of event-layer thickness, and hence position within the surface mixing layer, in controlling preservation and alteration (Fig. 30). These X-radiographs track the fate of two flood deposits formed by the Eel River in January 1995 and 1997 (Wheatcroft & Drake, 2003). In the upper sequence (Fig. 30a–c) a 3-cm-thick bed formed in January 1997 is destroyed over a 15-month time period (i.e. $T_d < 1.25$ yr), whereas in the lower sequence (Fig. 30d–f) a 5-cm-thick bed that was buried to 14 cm by the March 1995 flood (Wheatcroft & Borgeld, 2000) soon after formation (January 1995) persists for > 55 months. Although such differences in dissipation time could be due to variable sediment bioturbation intensity at the

two sites, independent estimates of biodiffusivities (Bentley & Nittrouer, 2003; Wheatcroft, 2006) suggest there is little along-shelf variation in mixing intensity on the Eel shelf (e.g. Fig. 22).

Sediment accumulation rate

Accumulation of sediment on the seafloor moves all material downward so that, by definition, the sediment–water interface is always at zero depth. Together with the surface-mixing-layer thickness, accumulation rate determines transit time. The sediment accumulation rate on continental margins is a function of delivery and redistribution processes (Nittrouer & Sternberg, 1981; Hill *et al.*, this volume, pp. 49–99; Sommerfield *et al.*, this volume, pp. 157–212). These processes operate on multiple time and space scales, and they impart considerable complexity to the resultant accumulation-rate patterns. Nevertheless, there is a growing body of data indicating that upper-slope and shelf sediment accumulation rates based on ^{210}Pb fall mostly in the range of 0.1–1 cm yr^{-1} (Nittrouer *et al.*, 1979, 1985; Carpenter *et al.*, 1982, 1985; Lesueur *et al.*, 2001; Wheatcroft & Sommerfield, 2005). Greater rates are typically associated with large rivers, such as the Amazon, Ganges-Brahmaputra and Changjiang, where sediment accumulation rates may be an order of magnitude higher (Nittrouer *et al.*, 1985; Lesueur *et al.*, 2001). Steady-state sedi-

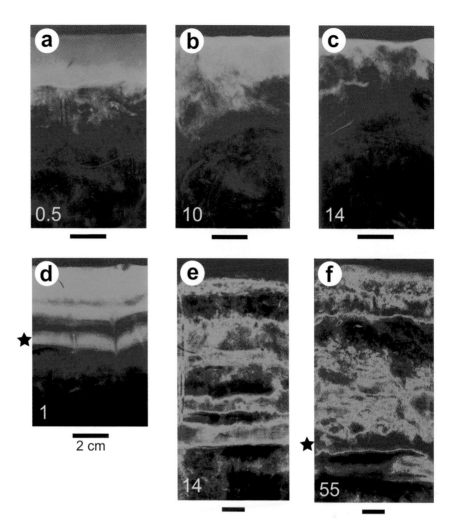

Fig. 30 False-colour representations of scanned X-radiographs. (a–c) Illustrate the dissipation of a 3-cm thick bed formed at station C70 in January 1997 over a 14-month period. (d–f) Illustrate the partial preservation of a 5-cm-thick bed (denoted by a star at its base) at station S60 that was advected to a depth of ~14 cm within months of formation. The number in the lower left of each panel denotes the time in months since formation of the layers and the scale bars under each panel = 2 cm. (Adapted from Wheatcroft & Drake, 2003.)

ment accumulation rates measured on the Eel margin show similar ranges, with an area-averaged mean of 0.4 cm yr^{-1} and upper and lower ranges of ~1.3 and 0.1 cm yr^{-1}, respectively (Alexander & Simoneau, 1999; Sommerfield & Nittrouer, 1999; Wheatcroft & Sommerfield, 2005; Sommerfield *et al.*, this volume, pp. 157–212).

Dissipation times

The 1995 and 1997 Eel shelf flood deposits were captured soon after their formation and tracked forward in time, thereby permitting estimates of dissipation time. In general, X-radiography data from the Eel shelf indicate that layer dissipation times are in the range of 2–30 months (Bentley & Nittrouer, 2003; Wheatcroft & Drake, 2003). For example, the January 1997 flood deposit, which ranged up to 6 cm in thickness (Wheatcroft &

Borgeld, 2000), was no longer recognizable in X-radiographs *anywhere* on the shelf 2.5 yr later. As previously noted (Wheatcroft, 1990), however, dissipation time is a function of the signal in question, whereby signals that depend on diagnostic fabrics (e.g. layers seen in X-radiographs) are more quickly destroyed than those that are defined by some geochemical (Leithold & Hope, 1999; Sommerfield & Nittrouer, 1999) or physical attributes. Therefore, it is likely that the grain-size signature of the flood deposits should persist for longer periods of time.

The dissipation time of the grain-size signal of the January 1995 and January 1997 flood layers was determined by comparing the peak weight-percentage of the < 20-µm-size fraction to the mean weight-percentage of that size fraction at a site (determined from piston- or box-core samples; Drake, 1999; Wheatcroft & Drake, 2003). When the peak value was less than two standard deviations above (for flood layers) or below (for storm layers) the mean of the < 20-μm fraction at the site, it was no longer distinguishable from ambient sediment and thus was classified as dissipated. As long as a flood layer remained within the surface mixing layer, there was an easily measured temporal change in the peak concentration of the < 20-μm fraction as the signal progressed toward destruction. In general, the rate of that change was inversely related to the initial thickness of the flood layer. Layers that were 1–3 cm thick and remained in the surface mixing layer tended to be completely destroyed or were very close to being destroyed in a few years. The grain-size signatures of thicker layers, such as the January 1995 flood layer at station O70 (8 cm) or the January 1997 layer at station O60 (5 cm), were still recognizable at the time of the last sampling (October, 2001).

By plotting peak concentrations in an event layer as a function of time, it was possible to determine the trend of the layer toward destruction. To do this, a linear regression was fit to the data and the trend line extrapolated to the point representing two standard deviations above the long-term mean. In this way, grain-size-signal dissipation time was determined for the January 1995 and 1997 flood layers at nine separate sites where the flood-layer thickness ranged from 1.5 cm to 7.5 cm. The results were plotted as a function of initial layer thickness (Fig. 31). The strong positive correlation ($r^2 = 0.95$) between layer thickness and signal dissipation time for the two flood beds at several widely distributed mid-shelf sites suggests that the mechanisms that are responsible for destroying the beds operate with similar intensities on a regional scale. This result is in agreement with the extensive bioturbation data that show little spatial variability in sediment mixing intensity on the Eel shelf. In addition, the fact that dissipation time is linearly related to layer thickness, and not a square root function of time, is intriguing and implies that the particle transport mechanism governing grain dispersion is not strictly diffusive (see Kirwan & Kump, 1987; Boudreau, 1989).

As expected, dissipation time for the grain-size signature of an event layer is appreciably longer than the period of time the layer is recognizable in X-radiographs. For example, the grain-size signature of a 3-cm-thick flood layer persists for roughly

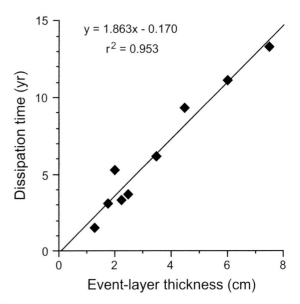

Fig. 31 Grain-size-signal dissipation time (yr) as a function of initial bed thickness (cm) for flood layers on the Eel shelf. A linear regression to the data yields a highly significant regression coefficient. (Modified from Wheatcroft & Drake, 2003.)

5 yr (Fig. 31), whereas the layer itself is not recognizable after ~1.3 yr (Fig. 30d–f). Despite the longer persistence of the grain-size signal, it is still short relative to the steady-state transit time (10–100 yr).

The key role of episodic sedimentation

The large number of studies on the Eel margin yielded a comprehensive database for the investigation of event-layer alteration and preservation (Table 4). Although few other margins have been studied in such detail, the data available indicate that the magnitude and range of important variables (i.e. biodiffusivity, mixing-layer thickness, event-layer thickness and sediment accumulation rate) observed on the Eel margin are broadly representative of many other shelf–upper-slope settings.

The parameter estimates listed in Table 4 suggest that there is a low probability for transient sedimentary signals or event layers formed on the Eel shelf to be preserved intact in the stratigraphic record. That is because the transit time of, for example, a 4-cm-thick event bed is 9–65 yr (accumulation rate range and L_b from Table 4). Yet

Table 4 Parameter estimates for the Eel River shelf

Parameter (units)	Average	Range	Sources
Biodiffusivity ($cm^2\,yr^{-1}$)	25	4–135	Wheatcroft, 2006; Bentley & Nittrouer, 2003
Mixing-layer thickness (cm)		5–15	Wheatcroft, 2006; Bentley & Nittrouer, 2003
Event-bed thickness (cm)		< 1–8	Wheatcroft & Borgeld, 2000
Sediment accumulation rate ($cm\,yr^{-1}$)	0.4	0.2–1.4	Sommerfield & Nittrouer, 1999

dissipation times are considerably less (months to 1–2 yr) (Bentley & Nittrouer, 2003; Wheatcroft & Drake, 2003). Strata and signals are clearly preserved in the Eel River shelf sediments (Leithold, 1989; Sommerfield & Nittrouer, 1999; Sommerfield *et al.*, 2002; Bentley & Nittrouer, 2003; Wheatcroft & Drake, 2003), requiring consideration of an additional preservation mechanism, i.e. **episodic sedimentation events**. Short-lived events, such as the January 1995 and 1997 oceanic floods, can deposit several centimetres of sediment over hours to days (Ogston *et al.*, 2000; Traykovski *et al.*, 2000; Hill *et al.*, this volume, pp. 49–99), thereby rapidly advecting layers and signals through the surface mixing layer. Transit times can decrease by more than one order of magnitude due to episodic sedimentation events, a key factor in determining preservation potential.

Further appreciation of the role of episodic sedimentation in event-layer preservation can be obtained by examining stratigraphic sequences formed on the Eel shelf during the past 50 yr. During that period, the Eel River has had major floods roughly once a decade (1955, 1964, 1974, 1986, 1995 and 1997). Of these floods, the 1964 and 1955 events had the highest discharge ($2.1 \times 10^4\,m^3\,s^{-1}$ and $1.5 \times 10^4\,m^3\,s^{-1}$, respectively), whereas the others were roughly equivalent ($\sim 1 \times 10^4\,m^3\,s^{-1}$). If all things were equal, then one might expect that the largest events would have left the most obvious record, whereas evidence for the others would be of lesser extent and approximately equal in magnitude. Wheatcroft & Drake (2003) found that intact layers, typically 1-to-5-cm-thick beds with relatively sharp lower contacts and biologically disturbed upper contacts, were distributed fairly evenly over the upper 40 cm of the sediment column (Fig. 32a). However, when converted to age, the layers cluster

into essentially two groups that correspond to the 1964 and 1974 floods (Fig. 32b).

Given its extreme magnitude, one might expect that the 1964 flood layer would remain recognizable in the sediment column. However, it is

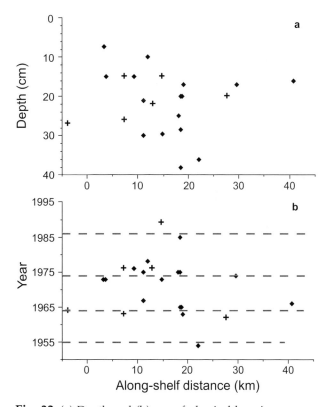

Fig. 32 (a) Depth and (b) age of physical layering observed in X-radiographs taken along the 70-m isobath on the Eel River shelf in September 1995. The along-shelf position of the river mouth is referenced to zero on the abscissa. Solid diamonds denote obvious layers (1–4 cm thick), whereas crosses represent vague contacts. The dashed red lines in (b) denote the years of major (> 8000 $m^3\,s^{-1}$) Eel River floods between 1950 and 1990. (Modified from Wheatcroft & Drake, 2003.)

surprising that the 1974 flood layer is readily apparent, especially considering that there is little record of the 1955 and 1986 floods, which were of equal or greater magnitude. It was postulated that variable transit times following the floods caused differences in preservation (Wheatcroft & Drake, 2003). Thus, substantial river discharges soon after (months to year) deposition of a flood bed may bury signals deep enough in the surface mixing layer that they are partially preserved. The best example of that phenomenon was burial of the January 1995 flood deposit by the 1997 event (Wheatcroft & Borgeld, 2000).

Insight obtained from the X-radiographs and the river-discharge history was also used to predict the fate of the flood-layer grain-size signal over the past 50 yr. In particular, because it was a larger flood, the 1974 flood deposit is expected to be more readily recognizable than the 1986 flood bed, and at two sites for which deep grain-size samples exist, the data match expectations. Stations O60 and O70, located roughly 20 km north of the Eel River mouth near the centre of mass for the 1995 and 1997 flood deposits (Wheatcroft & Borgeld, 2000), reveal a detailed record of grain-size fluctuations over the past millennium (Sommerfield et al., 2002). Focusing on the upper 60 cm of the sediment column at O60, there are four regions of fine-grained sediment with < 20-μm peak values appreciably greater than two standard deviations (Fig. 32). On the basis of geochronological data from ^{210}Pb and ^{137}Cs (Sommerfield & Nittrouer, 1999; Sommerfield et al., 2002), as well as information on the thickness of 1995 and 1997 flood deposits (Wheatcroft & Borgeld, 2000), the approximate age of each fine-grained zone can be estimated. Thus, the upper two layers correspond to the 1997 and 1995 flood deposits, whereas the lower two layers (at ~26 and ~35 cm depth) correspond to the 1974 and 1964 flood deposits, respectively (Fig. 33). The region between 15 cm and 23 cm is roughly where the 1986 flood bed should reside, but instead there is a zone of relatively coarse sediment. The probable reason for this pattern is that the transit time of the 1986 flood bed was greater than its dissipation time, hence its grain-size signature was destroyed completely. At even longer time-scales (e.g. 1500 yr), piston cores suggest that the clustering of events may have an impact on the preserved stratigraphic record (Sommerfield et al., 2002).

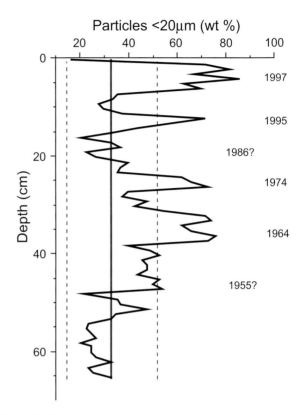

Fig. 33 Weight percentage of particles < 20 μm as a function of depth in the upper 60 cm of a piston core collected at station O60. The solid and dashed vertical lines are the average (± two standard deviation) weight percentage of particles < 20 μm determined over the full depth of the piston core. High percentages of fine sediment correspond to historically significant river floods, with some exceptions (1986, possibly 1955). (Modified from Wheatcroft & Drake, 2003.)

SUMMARY

Before discussing key findings regarding post-depositional alteration and preservation of strata on the Eel margin it is worth revisiting the critical importance of capturing geological and oceanographic events as they happen. In the absence of the Eel River's winter 1995 floods, and especially the January 1997 event, the focus of the seabed *and* bottom-boundary layer studies (Hill *et al.*, this volume, pp. 49–99) would have been quite different. Lacking well-documented events for which the initial large- and small-scale properties of the resultant products were measured, studies probably would have focused on establishing spatial patterns on the margin, and conducting time-series sampling at somewhat arbitrary locations. Such studies have significant merit and were done to some extent. Seabed studies typically lack a clear time horizon, so researchers are forced to consider processes only in a relative sense or, more commonly, assume steady-state conditions. If there is an overarching lesson to be learned from the Eel shelf seabed efforts, it is that temporal variation in forcing and sedimentary products can be substantial. Therefore, there are clearly periods when the steady-state assumption is violated. Determining the frequency of these periods and the 'recovery' to steady conditions is a key unknown. An equally important lesson is that event-response sampling can yield tremendous payoffs, and the research community should determine ways to foster adaptive sampling.

The seabed studies summarized in this paper have shed light on several important aspects of post-depositional alteration and preservation of strata. In particular, the following key findings were made.

1 Although the wave conditions on the Eel shelf are the most energetic on the northern California shelf, only a few millimetres to 1 or 2 cm of bed material is typically put in suspension during storms.
2 Despite having a high initial porosity (i.e. low strength) the Eel flood beds quickly became resistant to erosion.
3 The shelf macrofauna showed considerable resilience to deposition of flood sediment.
4 Storm deposits have a low preservation potential on open continental margins.
5 Episodic sedimentation is key to the preservation of strata and the signals produced by flood and storm events.

Large winter waves that characterize the Eel shelf produce seabed stresses exceeding the threshold for resuspension of fine-grained sediment 10% of the time at a depth of 90 m (Fig. 12b); because most of these waves occur during the months of November–April, the threshold of resuspension at 90 m is exceeded about 20% of the time during that season. Less frequent resuspension extends beyond the shelf break, whereas in shallower water, resuspension events are more intense and last longer (Fig. 13b). At a depth of 60 m, there are an average of 40 resuspension events each year

(Fig. 13a). Given this highly energetic wave environment and the fine-grained texture of the seabed, one might expect that a relatively thick layer of the seabed could be eroded (and then perhaps redeposited) during a large storm event. Instead, measured near-bed suspended-sediment concentrations, detailed analysis of the texture and fabric of the uppermost seabed, and sediment-transport modelling all indicate that storm resuspension and reworking affects only the uppermost few millimetres to few centimetres of the seabed.

The limitation on the volume of fine sediment in suspension is largely attributed to the development of high critical shear stresses within the bed due to consolidation, particle aggregation and biological binding of sediment; water-column processes such as flocculation and near-bed stratification are also important. When consolidation-related, depth-varying critical shear stresses are included in calculations of resuspension by waves and currents on the shelf, the results, which agree well with measured near-bed suspended-sediment concentrations, indicate resuspension depths on the order of millimetres during all but the most extreme storms. If the effect of consolidation on erodibility is not accounted for, calculated resuspension volumes are much larger than near-bed water-column measurements and near-surface bed sediment analyses indicate. Although the importance of consolidation and cohesion for fine-grained beds has been recognized for some time, many short- and long-term shelf sediment transport models have not accounted for it directly.

Initial porosities of the deposits created by the 1995 and 1997 floods of the Eel River were 80% or higher throughout the flood deposit (Fig. 6). After a period of several years of consolidation, porosities in the flood bed decreased to pre-flood values (Fig. 7) that presumably represent the equilibrium consolidation profile. The intervening years included the very stormy winter of 1998–99. Calculations of resuspension during these storms indicate that only the upper 1–2 cm of the bed or less were ever in suspension. This is best explained by consolidation processes increasing erosion resistance (critical shear stresses) within the flood layer as a result of the decrease in porosity. This increase in erosion resistance appears to have occurred quickly – the largest wave-current-generated bed shear stresses during the STRATAFORM experiment occurred during the winter of 1995–96, within a year of the emplacement of the 1995 flood deposit. Of the sediment that was resuspended, an inventory of mass in the bed (Drake, 1999) indicates that essentially all of it was returned to the seabed during subsequent periods of deposition.

Two-dimensional sediment-transport models suggest that there is a general redistribution of fine sediment seaward across the shelf, but no evidence of significant net erosion is present in the mass inventories determined at 60 m, 70 m and 90 m depths during the period from 1994 to 1999 (Fig. 18). This suggests that net erosion is inhibited by increases in erosion resistance associated with consolidation. Long-term (> 1 yr) calculations of sediment redistribution on the shelf will have to account for the effect of consolidation on limiting net erosion to yield reasonable rates of change in bed elevation and bed texture.

Net erosion will also be inhibited when seaward fluxes of sediment from inner to mid-shelf depths exceed sediment fluxes from mid- to outer shelf depths, as appears to be the case on the Eel shelf (Drake, 1999). Observations of the textural composition of the seabed at mid-shelf depths show a distinct increase in the mass of medium–coarse silt fractions (20–63 μm) during a period when the mass of the finer fractions (< 20 μm) remained constant (Fig. 17). These increases are most apparent between 1997 and 1999, a period characterized by unusually large storms. This grain-size specific flux convergence increased the percentage of medium–coarse silt at the surface of the seabed, and may have further inhibited erosion of the finer silt and clay by armouring the underlying sediment or by otherwise increasing erosion resistance in some fashion not yet understood.

The a priori expectation was that deposition of a several-centimetre-thick (up to 8 cm) flood layer on the mid-shelf would constitute a major disturbance to the resident benthic macrofauna, particularly since a number of benthic ecological studies (Brenchley, 1981; Thayer, 1983) have shown that localized, low-intensity biogenic reworking of sediment can structure macrofaunal communities. The much more severe and widespread burial caused by flood deposition should therefore have resulted in significant mortality (Norkko et al., 2002), and, potentially, a short-term depression in bioturbation intensity. The time-series data indicate

(Fig. 26), however, that the winter-time decreases in macrofaunal abundance were simply part of a seasonal cycle common to many temperate shelf settings (Buchanan & Moore, 1986). It was not possible to distinguish between years when major flood deposition occurred on the shelf (1997) and those when there was no deposition (1996). In retrospect, this outcome is not surprising given that the most abundant macrobenthos on the Eel shelf are almost always subsurface, deposit-feeding polychaetes that live several centimetres in the bed. The resilience of the macrofauna to flood deposition meant that there was no short-term depression in bioturbation intensities (Bentley & Nittrouer, 2003; Wheatcroft, 2006), a point that is important for strata preservation.

Storm sedimentation and resultant storm deposits or 'tempestites' have become a hallmark of continental-margin stratigraphy, with whole texts and many papers devoted to their description, recognition and formation (Ager, 1981; Dott, 1983; Aigner, 1985, Myrow & Southard, 1996). Given this prevailing view of margin sedimentation, it might be expected that on a high-energy margin such as the Eel shelf, where the wind-driven wave climate is extreme (Wiberg, 2000; Fig. 12a), the observed sedimentary record would be dominated by storm deposits. In fact, except in the shallow-water sand facies (Crockett & Nittrouer, 2004), there is little evidence for storm deposition in the Eel shelf sediment column. Storms certainly play a role in the post-depositional alteration and redistribution (Sommerfield et al., this volume, pp. 157–212) of sediment on the Eel margin, but they do not leave a preserved record on the muddy shelf except in association with wave-supported gravity flows that occur during floods, when the availability of fine-grained sediment is much higher than usual (Traykovski et al., 2000). The probable reason for the lack of storm deposits on the open shelf is that they are never thick enough to escape total destruction by bioturbation. Both stochastic (Zhang et al., 1997) and deterministic (Harris & Wiberg, 2002) models with realistic forcing have difficulty producing horizontal flux convergences that are large enough to create storm beds much thicker than a few centimetres. The lack of storm deposits on the Eel muddy shelf contrasts markedly with the situation in many ancient sequences, and forces consideration of whether the genesis of preservable storm deposits depends on mechanisms that focus sediment deposition, thereby leading to thick beds, or whether ancient examples represent non-actualistic events (see Myrow & Southard (1996) for a more general discussion of this issue).

Finally, studies on the Eel shelf (Wiberg, 2000; Sommerfield *et al.*, 2002; Bentley & Nittrouer, 2003; Wheatcroft & Drake, 2003) indicate that, in the absence of episodic sedimentation events, sedimentary layers and grain-size signals of event beds on continental margins will be destroyed before burial. The basis of this statement is that an analysis of all terms relevant to preservation (Fig. 28) shows that transit time through the mixing layer is long relative to layer or signal dissipation time (Wheatcroft, 1990). Therefore, layers and signals should not be preserved. There is direct evidence, however, for preserved flood beds on the Eel margin, including the January 1995 flood deposit (Sommerfield *et al.*, 2002; Bentley & Nittrouer, 2003; Wheatcroft & Drake, 2003). The explanation for this seeming paradox is that episodic sedimentation events can instantaneously advect layers and signals through the surface mixing layer. Analyses of the 100-yr-long instrumental record of river discharge and core X-radiographs suggest that the **sequential timing** of sedimentation events plays a critical role in determining what is preserved in shelf sediments (Wheatcroft & Drake, 2003). Furthermore, there is strong evidence that episodicity in sedimentation can be clustered on all time-scales up to at least 10^3 yr (Sommerfield *et al.*, 2002). Determining the generality of these results must await future intensive studies of continental-shelf sediment dynamics.

ACKNOWLEDGEMENTS

We thank our many STRATAFORM colleagues, from the programme manager at ONR, Joe Kravitz, and the chief scientist, Chuck Nittrouer, who conceived and directed the programme (i.e. herded the cats), to the many research vessel crew members, technicians, graduate and undergraduate students who helped at sea and in the laboratory – the research detailed herein would not have happened without your involvement. An earlier version of the manuscript was materially improved by comments from Bob Aller, Jamie Austin, Jeff Borgeld, John Jaeger, Tim Milligan and Chuck Nittrouer. We apologize for the length of our discourse and for those cases where we did not take your advice. Last, we are grateful for the patience, fortitude and leadership exhibited by our flexible leader, Chuck.

NOMENCLATURE

Symbol	Definition	Dimensions
a	orbital excursion amplitude	L
A	excess activity	$T^{-1}M^{-1}$
A_0	activity at the sediment–water interface	$T^{-1}M^{-1}$
A_z	activity at a depth z	
D_b	vertical biodiffusion coefficient	$L^2 T^{-1}$
Δz	sediment layer thickness	L
e	void ratio	
E	entrainment rate	$M L^{-2} T^{-1}$
f_w	friction factor	
g	acceleration of gravity	$L T^{-2}$
G	Guinasso number $(D_b/L_m S)$	
H	wave height	L
h	water depth	L
H_s	significant wave height	L
K	hydraulic conductivity	$L T^{-1}$
k	permeability	L^2
k_s	physical roughness length scale	L
L	wave length	L
L_b	biological surface-mixing-layer thickness	L
L_m	surface-mixing-layer thickness	L
L_p	physical surface-mixing-layer thickness	L
L_s	layer or signal thickness	L
p	pore pressure	$M L^{-1} T^{-2}$
p_x	excess pore pressure	$M L^{-1} T^{-2}$
p_h	hydrostatic pressure	$M L^{-1} T^{-2}$
Q_s	volume flux of sediment in suspension	$L^2 T^{-1}$
S	sediment accumulation rate	$L T^{-1}$
S'	fluctuating component of sediment deposition	$L T^{-1}$

T	wave period	T
T_b	biogenic turnover time	T
T_d	dissipation time	T
T_t	transit time	T
u_*	shear velocity	L T^{-1}
U_b	wave orbital velocity	L T^{-1}
v_s	grain velocity	L T^{-1}
V_s	volume of sediment in suspension per unit area	L
v_w	pore-fluid velocity	L T^{-1}
w_s	particle settling velocity	L T^{-1}
z	sediment depth	L
γ	depth attenuation coefficient	
η	bed elevation	L
λ	decay constant	T^{-1}
μ	fluid viscosity	M L^{-1} T^{-1}
ρ	fluid density	M L^{-3}
ρ_b	bulk density	M L^{-3}
ρ_s	sediment density	M L^{-3}
σ	total stress	M L^{-1} T^{-2}
σ_e	effective stress	M L^{-1} T^{-2}
τ_b	bed shear stress	M L^{-1} T^{-2}
τ_{bw}	wave-generated bed shear stress	M L^{-1} T^{-2}
τ_c	critical shear stress	M L^{-1} T^{-2}
φ	porosity	
φ_s	solid fraction $(1 - \varphi)$	
$\varphi_{s\infty}$	asymptotic solid fraction	
φ_{s0}	initial solid fraction	

REFERENCES

Ager, D.V. (1981) *The Nature of the Stratigraphic Record*, 2nd edn. Wiley, New York, 122 pp.

Aigner, T. (1985) *Storm Depositional Systems: Dynamic Stratigraphy in Modern and Ancient Shallow-Marine Sequences*. Springer-Verlag, Berlin, 174 pp.

Alexander, C.R. and Simoneau, A.M. (1999) Spatial variability in sedimentary processes on the Eel continental slope. *Mar. Geol.*, **154**, 243–254.

Allan, J. and Komar, P.D. (2002) Extreme storms on the Pacific Northwest coast during the 1997–98 El Niño and 1998–99 La Niña. *J. Coast. Res.*, **18**, 175–193.

Aller, J.Y., Woodin, S.A. and Aller, R.C. (Eds) (2001) *Organism–Sediment Interactions*, University of South Carolina Press, Columbia, SC, 403 pp.

Aller, R.C. (1982) The effects of macrobenthos on chemical properties of marine sediment and over-lying water. In: *Animal–Sediment Relations* (Eds P.L. McCall and M.J.S. Tevesz). Plenum Press, New York, pp. 53–102.

Aller, R.C. and Cochran, J.K. (1976) ^{234}Th/^{238}U disequilibrium in near-shore sediment, particle reworking and diagenetic time scales. *Earth Planet. Sci. Lett.*, **29**, 37–50.

Aller, R.C., Benninger, L.K. and Cochran, J.K. (1980) Tracking particle-associated processes in nearshore environments by the use of ^{234}Th/^{238}U disequilibrium. *Earth Planet. Sci. Lett.*, **47**, 161–175.

Andersen, T.J. (2001) Seasonal variation in erodibility of two temperate, microtidal mudflats. *Estuarine Coast. Shelf Sci.*, **53**, 1–12.

Andrews, D. and Bennett, A. (1981) Measurements of diffusivity near the sediment–water interface with a fine-scale resistivity probe. *Geochim. Cosmochim. Acta*, **45**, 2169–2175.

Archie, G.E. (1942) The electrical resistivity log as an aid in determining some reservoir characteristics. *Trans. Am. Inst. Mining Metall. Eng.*, **146**, 54–62.

Ardhuin, F., O'Reilly, W.C., Herbers, T.H.C. and Jessen, P.F. (2003) Swell transformation across the continental shelf. Part I: Attenuation and directional broadening. *J. Phys. Ocean.*, 33, 1921–1939.

Bathurst, R.G.C. (1971) *Carbonate Sediments and their Diagenesis*. Elsevier, Amsterdam, 620 pp.

Been, K. and Sills, G.C. (1981) Self-weight consolidation of soft soils: an experimental and theoretical study. *Géotechnique*, **31**, 519–535.

Bennett, R.H., Li, H., Lambert, D.N., Fischer, K.M., *et al.* (1990) *In situ* porosity and permeability of selected carbonate sediment: Great Bahama Bank Part 1: measurements. *Mar. Geotech.*, **9**, 1–28.

Bentley, S.J. and Nittrouer, C.A. (2003) Emplacement, modification and preservation of event strata on a flood-dominated continental shelf, Eel shelf, northern California. *Cont. Shelf Res.*, **23**, 1465–1493.

Bentley, S.J. and Sheremet, A. (2003) A new model for the emplacement, bioturbation and preservation of sedimentary strata. *Geology*, **31**, 725–728.

Berger, W.H., Ekdale, A.A. and Bryant, P.B. (1979) Selective preservation of burrows in deep-sea carbonates. *Mar. Geol.*, **32**, 205–230.

Berlamont, J., Van den Bosch, L. and Toorman, E. (1992) Effective stresses and permeability in consolidating mud. In: *Proceedings of the 23rd International Conference on Coastal Engineering*, Venice, pp. 2962–2975.

Berner, R.A. (1980) *Early Diagenesis: a Theoretical Approach.* Princeton University Press, Princeton, NJ, 241 pp.

Black, K.S., Tolhurst, T.J., Paterson, D.M. and Hagerthey, S.E. (2002) Working with natural cohesive sediments. *J. Hydraul. Eng.*, **128**, 1–7.

Borgeld, J.C. (1985) *Holocene stratigraphy and sedimentation on the northern California continental shelf.* Unpublished PhD thesis, University of Washington, Seattle, WA, 177 pp.

Borgman, L.E. and Resio, D.T. (1982) Extremal statistics in wave climatology. In: *Topics in Ocean Physics* (Eds A. Osborne and P.M. Rizzoli), pp. 439–471. International School of Physics 'Enrico Fermi', Course 80, North-Holland Publishing Company, Amsterdam.

Boudreau, B.P. (1986) Mathematics of tracer mixing in sediments, I. Spatially-dependent, diffusive mixing. *Am. J. Sci.*, **286**, 161–198.

Boudreau, B.P. (1989) The diffusion and telegraph equations in diagenetic modeling. *Geochim. Cosmochim. Acta*, **53**, 1857–1866.

Boudreau, B.P. (1994) Is burial velocity a master parameter for bioturbation? *Geochim. Cosmochim. Acta*, **58**, 1243–1249.

Boudreau, B.P. (1997) *Diagenetic Models and their Implementation.* Springer-Verlag, Berlin, 414 pp.

Boudreau, B.P. (1998) Mean mixed depth of sediments: The wherefore and the why. *Limnol. Ocean.*, **43**, 524–526.

Boudreau, B.P. and Bennett, R.H. (1999) New rheological and porosity equations for steady-state compaction. *Am. J. Sci.*, **299**, 517–528.

Boudreau, B.P. and Jørgensen, B.B. (Eds) (2001) *The Benthic Boundary Layer.* Oxford University Press, Oxford, 404 pp.

Boudreau, B.P., Choi, J. and François-Carcaillet, F. (2001) Diffusion in a lattice-automaton model of bioturbation by small deposit feeders. *J. Mar. Res.*, **59**, 749–768.

Briggs, K.B. (1989) Microtopographical roughness of shallow-water continental shelves. *IEEE J. Ocean. Eng.*, **14**, 360–367.

Brenchley, G.A. (1981) Disturbance and community structure: an experimental study of bioturbation in marine soft-bottom environments. *J. Mar. Res.*, **39**, 767–790.

Bromley, R.G. (1990) *Trace Fossils: Biology and Taphonomy.* Unwin Hyman, London, 280 pp.

Buchanan, J.B. and Moore, J.J. (1986) Long-term studies at a benthic station off the coast of Northumberland. *Hydrobiologia*, **142**, 121–127.

Buessler, K.O., Cochran, J.K., Bacon, M.P., *et al.* (1992) Determination of thorium isotopes in seawater by nondestructive and radiochemical procedures. *Deep-sea Res.*, **39**, 1103–1114.

Burland, J.B. (1990) On the compressibility and shear strength of natural clays. *Géotechnique*, **40**, 329–378.

Cacchione, D.A. and Drake, D.E. (1990) Sediment transport: an overview with applications to the northern California continental shelf. In: *The Sea*, Vol. 9, *Ocean Engineering Science* (Eds B. LeMehaute and D. Hanes), pp. 729–773. Wiley Interscience, New York.

Cacchione, D.A., Field, M.E., Drake, D.E. and Tate, G.B. (1987) Crescentic dunes on the inner continental shelf off northern California. *Geology*, **15**, 1134–1137.

Cacchione, D.A., Wiberg, P.L., Lynch, J.F., Irish, J.D. and Traykovski, P. (1999) Estimates of suspended-sediment flux and bedform activity on the inner portion of the Eel continental shelf. *Mar. Geol.*, **154**, 83–98.

Cahoon, L.B. (1999) The role of benthic microalgae in neritic systems. *Ocean. Mar. Biol. Ann. Rev.*, **37**, 47–86.

Carpenter, R., Peterson, M.L. and Bennett, J.T. (1982) ^{210}Pb-derived sediment accumulation and mixing rates for the Washington continental slope. *Mar. Geol.*, **48**, 135–164.

Carpenter, R., Peterson, M.L. and Bennett, J.T. (1985) ^{210}Pb-derived sediment accumulation and mixing rates for the greater Puget Sound region. *Mar. Geol.*, **64**, 291–312.

Chang, G.C., Dickey, T.D. and Williams, A.J. (2001) Sediment resuspension over a continental shelf during hurricanes Edouard and Hortense. *J. Geophys. Res.*, **106**, 9517–9531.

Clifton, H.E. (1976) Wave-formed sedimentary structures: a conceptual model. In: *Beach and Nearshore Sedimentation* (Eds R.A. Davis, Jr. and R.L. Ethington), pp. 126–148. Special Publication 24, Society of Economic Paleontologists and Mineralogists, Tulsa, OK.

Conover, W.J. (1980) *Practical Nonparametric Statistics.* John Wiley & Sons, New York, 493 pp.

Cookman, J.L. and Flemings, P.B. (2001) STORMSED1.0: hydrodynamics and sediment transport in a 2-D, steady-state, wind- and wave-driven coastal circulation model. *Comput. Geosci.*, **27**, 647–674.

Crockett, J.S. and Nittrouer, C.A. (2004) The sandy inner shelf as a repository for muddy sediment: an example from northern California. *Cont. Shelf Res.*, **24**, 55–73.

Cui, Y. and Parker, G. (1998) The arrested gravel front: stable gravel–sand transitions in rivers. Part: general numenrical solutions. *J. Hydraul. Res.*, **36**, 159–182.

Cutshall, N.H., Larsen, I.L. and Olsen, C.R. (1983) Direct analysis of Pb–210 in sediment samples: self-absorption corrections. *Nucl. Instrum. Meth.*, **206**, 309–312.

Cutter, G.R. and Diaz, R.J. (2000) Biological alteration of physically structured flood deposits on the Eel margin, northern California. *Cont. Shelf Res.*, **20**, 235–253.

Dade, W.B., Davis, J.D., Nichols, P.D., *et al.* (1990) Effects of bacterial exopolymer adhesion on the entrainment of sand. *Geomicrobiol. J.*, **8**, 1–16.

Dade, W.B., Nowell, A.R.M. and Jumars, P.A. (1992) Predicting the erosion resistance of muds. *Mar. Geol.*, **105**, 285–297.

Dapples, E.C. (1942) The effect of macro-organisms upon near shore marine sediments. *J. Sediment. Petrol.*, **12**, 118–126.

Dauer, D.M., Maybury, C.A. and Ewing, R.M. (1981) Feeding behavior and general ecology of several spionid polychaetes from the Chesapeake Bay. *J. Experiment. Mar. Biol. Ecol.*, **54**, 21–38.

Decho, A.W. (1990) Microbial exopolymer secretions in ocean environments: their role(s) in food webs and marine processes. *Ocean. Mar. Biol. Ann. Rev.*, **28**, 73–153.

DeMaster, D.J., Pope, R.H., Levin, L.A. and Blair, N.E. (1994) Biological mixing intensity and rates of organic carbon accumulation in North Carolina slope sediments. *Deep-sea Res.*, II, **41**, 735–753.

Dever, E.P. and Lentz, S.J. (1994) Heat and salt balances over the northern California shelf in winter and spring. *J. Geophys. Res.*, **99**, 16001–16017.

Dott, R.H. (1983) Episodic sedimentation – How normal is average? How rare is rare? Does it matter? *J. Sediment. Petrol.*, **53**, 5–23.

Drake, D.E. (1999) Temporal and spatial variability of the sediment grain-size distribution on the Eel shelf: the flood layer of 1995. *Mar. Geol.*, **154**, 169–182.

Drake, D.E. and Cacchione, D.A. (1985) Seasonal variation in sediment transport on the Russian River shelf, California. *Cont. Shelf Res.*, **4**, 495–514.

Drake, D.E., Eganhouse, R. and McArthur, W. (2002) Physical and chemical effects of grain aggregates on the Palos Verdes margin, southern California. *Cont. Shelf Res.*, **22**, 967–986.

Dyer, K.R. and Huntley, D.A. (1999) The origin, classification and modeling of sand banks and ridges. *Cont. Shelf Res.*, **19**, 1285–1330.

Fan, S.J., Swift, D.J.P., Traykovski P., *et al.* (2004) River flooding, storm resuspension and event stratigraphy on the northern California shelf: observations compared with simulations. *Mar. Geol.*, **210**, 17–41.

Fauchald, K. and Jumars, P.A. (1979) The diet of worms: a study of polychaete feeding guilds. *Ocean. Mar. Biol. Ann. Rev.*, **16**, 193–284.

Fraser, H.J. (1935) Experimental study of the porosity and permeability of clastic sediments. *J. Geol.*, 43, 910–1010.

Fredsøe, J. and Deigaard, R. (1992) *Mechanics of Coastal Sediment Transport*. Advanced Series in Ocean Engineering, Vol. 3, World Scientific, Singapore, 369 pp.

Frey, R.W. (1970) The lebensspuren of some common marine invertebrates near Beaufort, North Carolina II. Anemone burrows. *J. Paleontol.*, **44**, 308–311.

Frey, R.W. (Ed) (1975) *The Study of Trace Fossils*. Springer-Verlag, New York, 562 pp.

Frey, R.W. and Pemberton, S.G. (1990) Bioturbate texture or ichnofabric? *Palaios*, **5**, 385–386.

Fuller, C.M. and Butman, C.A. (1988) A simple technique for fine-scale, vertical sectioning of fresh sediment cores. *J. Sediment. Petrol.*, **58**, 763–768.

Gerino, M., Aller, R.C., Lee, C., *et al.* (1998) Comparison of different tracers and methods used to quantify bioturbation during a spring bloom: 234-thorium, luminophores and chlorophyll *a*. *Estuarine Coast. Shelf Sci.*, **46**, 531–547.

Gibson, R.E., England, G.L. and Hussey, M.J.L. (1967) The theory of one-dimensional consolidation of saturated clays. I. Finite non-linear consolidation of thin homogeneous layers. *Géotechnique*, **17**, 261–273.

Gilmore, G. and Hemingway, J.D. (1995) *Practical Gamma-ray Spectrometry*. John Wiley & Sons, New York, 314 pp.

Goldberg, E.D. and Koide, M. (1962) Geochronological studies of deep-sea sediments by the ionium/thorium method. *Geochim. Cosmochim. Acta*, **26**, 417–450.

Grant, W.D. and Madsen, O.S. (1979) Combined wave and current interaction with a rough bottom. *J. Geophys. Res.*, **84**, 1797–1808.

Grant, W.D. and Madsen, O.S. (1986) The continental-shelf bottom boundary layer. *Ann. Rev. Fluid Mech.*, **18**, 265–305.

Green, M.A., Aller, R.C., Cochran, J.K., Lee, C. and Aller, J.Y. (2002) Bioturbation in shelf/slope sediments off Cape Hatteras, North Carolina: the use of ^{234}Th, Chl-*a* and Br$^-$ to evaluate rates of particle and solute transport. *Deep-sea Res.*, II, **49**, 4627–4644.

Guinasso, N.L. and Schink, D.R. (1975) Quantitative estimates of biological mixing rates in abyssal sediments. *J. Geophys. Res.*, **80**, 3032–3034.

Gumbel, E.J. (1958) *Statistics of Extremes*. Columbia University Press, New York, 375 pp.

Gust, G. and Müller, V. (1997) Interfacial hydrodynamics and entrainment functions of currently used erosion devices. In: *Cohesive Sediments* (Eds N. Burt, R. Parker and J. Watts), pp. 149–174, John Wiley & Sons, New York.

Hamilton, E.L. (1980) Geoacoustic modeling of the sea floor. *J. Acoust. Soc. Am.*, **108**, 1313–1340.

Hanes, D.M., Alymov, V., Chang, Y.S. and Jette, C. (2001) Wave-formed sand ripples at Duck, North Carolina. *J. Geophys. Res.*, **106**, 22575–22592.

Harris, C.K. (1999) *The importance of advection and flux divergence in the transport and redistribution of continental shelf sediment*. Unpublished PhD thesis, University of Virginia, Charlottesville, VA, 155 pp.

Harris, C.K. and Wiberg, P.L. (1997) Approaches to quantifying long-term continental shelf sediment transport with an example from the northern California STRESS mid-shelf site. *Cont. Shelf Res.*, **17**, 1389–1418.

Harris, C.K. and Wiberg, P.L. (2001) A two-dimensional, time-dependent model of suspended sediment transport and bed reworking for continental shelves. *Comput. Geosci.*, **27**, 675–690.

Harris, C.K. and Wiberg, P.L. (2002) Across-shelf sediment transport, interactions between suspended sediment and bed sediment. *J. Geophys. Res.*, **107**(C1), 10.1029/2000JC000634.

Harris, C.K., Butman, B. and Traykovski, P. (2002) Winter-time circulation and sediment transport in the Hudson Shelf Valley. *Cont. Shelf Res.*, **23**, 801–820.

Harris, C.K., Traykovski, P.A. and Geyer, W.R. (2005) Flood dispersal and deposition by near-bed gravitational sediment flows and oceanographic transport: A numerical modeling study of the Eel River shelf, northern California. *J. Geophys. Res.*, **110**, C09025.

Howard, J.D. (1978) Sedimentology and trace fossils. In: *Trace Fossil Concepts* (Ed P.B. Basan), pp. 11–42. Short Course 5, Society of Economic Paleontologists and Mineralogists, Tulsa, OK.

Howard, J.D. and Frey, R.W. (1975) Estuaries of the Georgia Coast, U.S.A.: Sedimentology and biology. II. Regional animal-sediment characteristics of Georgia estuaries. *Senckenbergiana Marit.*, **7**, 33–103.

Howard, J.D. and Reineck, H.-E. (1972) Georgia coastal region, Sapelo Island, U.S.A.: Sedimentology and biology. IV. Physical and biogenic sedimentary structures of the nearshore shelf. *Senckenbergiana Marit.*, **4**, 81–123.

Huettel, M., Ziebis, W. and Forster, S. (1996) Flow-induced uptake of particulate matter in permeable sediments. *Limnol. Ocean.*, **41**, 309–322.

Jago, C.F. and Jones, S.E. (1998) Observation and modeling of the dynamics of benthic fluff resuspended from a sandy bed in the southern North Sea. *Cont. Shelf Res.*, **18**, 1255–1283.

Johansen, C., Larsen, T. and Petersen, O. (1997) Experiments on erosion of mud from the Danish Wadden Sea. In: *Cohesive Sediments* (Eds N. Burt, R. Parker and J. Watts), pp. 305–314. John Wiley & Sons, New York.

Jumars, P.A. (1978) Spatial autocorrelation with RUM (Remote Underwater Manipulator), vertical and horizontal structure of a bathyal benthic community. *Deep-sea Res.*, **25**, 589–604.

Jumars, P.A. and Nowell, A.R.M. (1984) Effects of benthos on sediment transport, difficulties with functional grouping. *Cont. Shelf Res.*, **3**, 115–130.

Kachel, N.B. and Smith, J.D. (1986) Geological impact of sediment transporting events on the Washington continental shelf. In: *Shelf Sands and Sandstones* (Eds R.J. Knight and J.R. McLean), pp. 145–162. Memoir II, Canadian Society of Petroleum Geology, Calgary.

Kirk, J.T.O. (1994) *Light and Photosynthesis in Aquatic Ecosystems*, 2nd edn. Cambridge University Press, Cambridge, 509 pp.

Kirwan, A.D. and Kump, L.R. (1987) Models of geochemical systems from mixture theory: diffusion. *Geochim. Cosmochim. Acta*, **51**, 1219–1226.

Komar, P.D. and Taghon, G.L. (1985) Analyses of the settling velocities of fecal pellets from the subtidal polychaete *Amphicteis scaphobranchiata*. *J. Mar. Res.*, **43**, 605–614.

Komar, P.D., Neudeck, R.H. and Kulm, L.D. (1972) Observations and significance of deep-water oscillatory ripple marks on the Oregon continental shelf. In: *Shelf Sediment Transport: Process and Pattern* (Eds D.J.P. Swift, D.B. Duane and O.H. Pilkey), pp. 601–624. Dowden, Hutchinson and Ross, Stroudsburg, PA.

Kourafalou V.H., Oey L.Y., Wang J.D. and Lee, T.N. (1996) The fate of river discharge on the continental shelf. 1. Modeling the river plume and the inner shelf coastal current. *J. Geophys. Res.*, **101**, 3415–3434.

Kranz, P.M. (1974) The anastrophic burial of bivalves and its paleoecological significance. *J. Geol.*, **82**, 237–265.

Kuehl, S.A., Nittrouer, C.A. and DeMaster, D.J. (1988) Microfabric study of fine-grained sediments, observations from Amazon subaqueous delta. *J. Sediment. Petrol.*, **58**, 12–23.

Kynch, G.J. (1952) A theory of sedimentation. *Trans. Faraday Soc.*, **48**, 166–176.

Largier, J.L., Magnell, B.A. and Winant, C.D. (1993) Subtidal circulation over the northern California shelf. *J. Geophys. Res.*, **98**, 18147–18179.

Larson, D.W. and Rhoads, D.C. (1983) The evolution of infaunal communities and sedimentary fabrics. In: *Biotic Interactions in Recent and Fossil Benthic Communities* (Eds M.J.S. Tevesz and P.L. McCall), pp. 627–648. Plenum Press, New York.

Leithold, E.L. (1989) Depositional processes on an ancient and modern muddy shelf, northern California. *Sedimentology*, **36**, 179–202.

Leithold, E.L. and Hope, R.S. (1999) Deposition and modification of a flood layer on the northern California shelf: Lessons from and about the fate of terrestrial particulate organic carbon. *Mar. Geol.*, **154**, 183–195.

Lesueur, P., Jouanneau, J.-M., Boust, D., Tastet, J.-P. and Weber, O. (2001) Sedimentation rates and fluxes in the continental shelf mud fields in the Bay of Biscay (France). *Cont. Shelf Res.*, **21**, 1383–1401.

Li, M.Z. and Amos, C.L. (1999) Field observations of bedforms and sediment transport thresholds of fine sand under combined waves and currents. *Mar. Geol.*, **158**, 147–160.

Li, M.Z. and Amos, C.L. (2001) SEDTRANS96: the upgraded and better calibrated sediment-transport model for continental-shelves. *Comput. Geosci.*, **27**, 619–645.

Lie, U. (1969) Standing crop of benthic infauna off the coast of Washington. *J. Fish. Res. Board Can.*, **26**, 55–62.

Lund-Hansen, L.C., Laima, M., Mouritsen, K., Lam, N.N. and Hai, D.N. (2002) Effects of benthic diatoms, fluff layer and sediment conditions on critical shear stress in a non-tidal coastal environment. *J. Mar. Biol. Assoc.*, **82**, 929–936.

Lyne, V.D., Butman, B. and Grant, W.D. (1990) Sediment movement along the U.S. east coast continental shelf– II. Modeling suspended sediment concentrations and transport rates during storms. *Cont. Shelf Res.*, **10**, 429–460.

Maa. J.P.-Y., Wright, L.D., Lee, C.-H. and Shannon, T.W. (1993) VIMS Sea Carousel: a field instrument for studying sediment transport. *Mar. Geol.*, **115**, 271–287.

Madsen, O.S. (1994) Spectral wave-current bottom boundary layer flows. In: *Coastal Engineering 1994*. Proceedings, 24th International Conference, Coastal Engineering Research Council/ASCE, Kobe, Japan, pp. 384–397.

Manheim, F.T. and Waterman, L.S. (1974) Diffusimetry (diffusion constant estimation) on sediment cores by resistivity probe. In: *Initial Reports of the Deep Sea Drilling Project*, Vol. XXII, pp. 663–670. College Station, TX.

Martin, W.R. and Sayles, F.L. (1987) Seasonal cycles of particle and solute transport processes in nearshore sediments: ^{222}Rn/^{226}Ra and ^{234}Th/^{238}U disequilibrium at a site in Buzzards Bay, MA. *Geochim. Cosmochim. Acta*, **51**, 927–943.

Martin, W.R., Bender, M., Leinen, M. and Orchardo, J. (1991). Benthic organic carbon degradation and biogenic silica dissolution in the central equatorial Pacific. *Deep-sea Res.*, **38**, 1481–1516.

Mauer, D., Keck, R.T., Tinsman, J.C., Leathem, W.A., Wethe, C., Lord, C. and Church T.M. (1986) Vertical migration and mortality of marine benthos in dredged material: a synthesis. *Int. Rev. Ges. Hydrobiol.*, **71**, 49–63.

Meysman, F.J.R., Boudreau, B.P. and Middelburg, J.J. (2003) Relations between local, nonlocal, discrete and continuous models of bioturbation. *J. Mar. Res.*, **61**, 391–410.

McCave, I.N. (Ed) (1976) *The Benthic Boundary Layer.* Plenum Press, New York, 323 pp.

McNeil, J., Taylor, C. and Lick, W. (1996) Measurements of erosion of undisturbed bottom sediments with depth. *J. Hydraul. Eng.*, **122**, 316–324.

Middelburg, J.J., Soetaert, K. and Herman, P.M.J. (1997) Empirical relationships for use in global diagenetic models. *Deep-sea Res.*, I, **44**, 327–344.

Middleton, G.V. and Wilcock, P.R. (1994) *Mechanics in the Earth and Environmental Sciences.* Cambridge University Press, Cambridge, 459 pp.

Miller, D.C., Geider, R.J. and MacIntrye, H.L. (1996) Microphytobenthos: the ecological role of the 'Secret Garden' of unvegetated, shallow-water marine habitats. II. Role in sediment stability and shallow-water food webs. *Estuaries*, **19**, 202–212.

Miller, M.C. and Komar, P.D. (1980) A field investigation of the relationship between oscillation ripple spacing and the near-bottom wave orbital motions. *J. Sediment. Petrol.*, **50**, 183–191.

Miller, M.C., McCave, I.N. and Komar, P.D. (1977) Threshold of motion under unidirectional currents. *Sedimentology*, **24**, 507–527.

Moore, D.G. and Scruton, P.C. (1957) Minor internal structures of some recent unconsolidated sediments. *Bull. Am. Assoc. Petrol. Geol.*, **41**, 2723–2751.

Moore, H.B. (1931) The muds of the Clyde Sea area. III. Chemical and physical conditions; rate and nature of sedimentation; and fauna. Journal of the Marine Biological Society UK, **17**, 325–358.

Mulsow, S., Boudreau, B.P. and Smith, J.N. (1998) Bioturbation and porosity gradients. *Limnol. Ocean.*, **43**, 1–9.

Myrow, P.M. amd Southard, J.B. (1996) Tempestite deposition. *J. Sediment. Res. Sect. A*, **66**, 875–887.

Nichols, F.H., Cacchione, D.A., Drake, D.E. and Thompson, J.K. (1989) Emergence of burrowing urchins from California continental shelf sediments – a response to alongshore current reversals? *Estuarine Coast. Shelf Sci.*, **29**, 171–182.

Nittrouer, C.A. and Sternberg, R.W. (1981) The formation of sedimentary strata in an allochthonous shelf environment: application to the Washington continental shelf. *Mar. Geol.*, **42**, 201–232.

Nittrouer, C.A., Sternberg, R.W., Carpenter, R. and Bennett, J.T. (1979) The use of ^{210}Pb geochronology as a sedimentological tool: application to the Washington continental shelf. *Mar. Geol.*, **31**, 297–316.

Nittrouer, C.A., DeMaster, D.J., Kuehl, S.A., McKee, B.A. and Thorbjarnarson, K.W. (1985) Some questions and answers about accumulation of fine-grained sediment in continental margin environments. *Geo-Mar. Lett.*, **4**, 211–213.

Norkko, A., Thrush, S.F., Hewitt, J.E., *et al.* (2002) Smothering of estuarine sandflats by terrigenous clay, the role of wind-wave disturbance and bioturbation in site-dependent macrofaunal recovery. *Mar. Ecol. Progr. Ser.*, **234**, 23–41.

Nowell, A.R.M., Jumars, P.A. and Eckman, J.E. (1981) Effects of biological activity on the entrainment of marine sediments. *Mar. Geol.*, **42**, 133–153.

Ogston, A.S., Cacchione, D.A. and Sternberg, R.W. (1998) River flood sediment flux, flux convergence and shelf circulation on the northern California shelf. *Eos Transactions, American Geophysical Union, 1998 Fall Meeting*, San Francisco.

Ogston, A.S., Cacchione, D.A., Sternberg, R.W. and Kineke, G.C. (2000) Observations of storm and river flood-driven sediment transport on the northern California continental shelf. *Cont. Shelf Res.*, **20**, 2141–2162.

Ogston, A.S., Guerra, J.V. and Sternberg, R.W. (2004) Interannual variability of nearbed sediment flux on the Eel River shelf, northern California. *Cont. Shelf Res.*, **24**, 117–136.

Papanicolaou, A.N. and Diplas, P. (1999) Numerical solution of a non-linear model for self-weight solids settlement. *Appl. Math. Model.*, **23**, 345–362.

Paterson, D.M. 2001. The fine structure and properties of the sediment surface. In: *The Benthic Boundary Layer* (Eds B.P. Boudreau and B.B. Jørgensen), pp. 127–143. Oxford University Press, Oxford.

Pullen, J.D. and Allen, J.S. (2000) Modeling studies of the coastal circulation off northern California: Shelf response to a major Eel River flood event. *Cont. Shelf Res.*, **20**, 2213–2238.

Reed, C.W., Niederoda, A.W. and Swift, D.J.P. (1999) Modeling sediment entrainment and transport processes limited by bed armoring. *Mar. Geol.*, **154**, 143–154.

Reineck, H.-E. (1967) Parameter von schichtung und bioturbation. *Geol. Rundsch.*, **56**, 420–438.

Reineck, H.-E. and Singh, I.B. (1972) Genesis of laminated sand and graded rhythmites in storm-sand layers of shelf mud. *Sedimentology*, **18**, 123–128.

Rhoads, D.C. (1970) Mass properties, stability and ecology of marine muds related to burrowing activity. In: *Trace Fossils* (Eds T.P. Crimes and J.C. Harper), pp. 391–406. Seel House Press, Liverpool.

Rhoads, D.C. (1974) Organism-sediment relations on the muddy seafloor. *Ocean. Mar. Biol. Ann. Rev.*, **12**, 263–300.

Rhoads, D.C. and Boyer, L.F. (1982) The effects of marine benthos on physical properties of sediments, a successional perspective. In: *Animal–Sediment Relations* (Eds P.L. McCall and M.J.S. Tevesz), pp. 3–52. Plenum Press, New York.

Rhoads, D.C. and Cande, S. (1971) Sediment profile camera for *in-situ* study of organism–sediment relations. *Limnol. Ocean.*, **16**, 110–114.

Rhoads, D.C. and Stanley, D.J. (1965) Biogenic graded bedding. *J. Sediment. Petrol.*, **35**, 956–963.

Richardson, M.D., Carey, A.G. and Colgate, W.A. (1977) The effects of dredged material disposal on benthic assemblages. In: *Aquatic Disposal Field Investigations Columbia River Disposal Site, Oregon*. Technical Report D-77-30, U.S. Army Corps of Engineers, 208 pp.

Richter, R. (1952) Fluidal-Textur in Sediment-Gesteinen und über Sedifluktion überhaupt. *Notizbl. Hess. Landesamtes Bodenforsch. Wiesbaden*, **6**, 67–81.

Rieke, H.H. and Chilingarian, G.V. (1974) *Compaction of Argillaceous Sediments*. Elsevier, Amsterdam, 424 pp.

Rubin, D.M. (2004) A simple autocorrelation algorithm for determining grain size from digital images of sediment. *J. Sediment. Res.*, **74**, 160–165.

Rubin, D.M., Chezar, H., Topping, D.J., Melis, T.S. and Harney, J.N. (in press) Two new approaches for measuring spatial and temporal changes in bed-sediment grain size: observations using a new underwater bed-sediment microscope and calculations from suspended sediment. *Sediment. Geol.*

Sanford, L.P. and Maa, J.P.-Y. (2001) A unified erosion formulation for fine sediments. *Mar. Geol.*, **179**, 9–23.

Schäfer, W. (1972) *Ecology and Palaeoecology of Marine Environments*. University Chicago Press, Chicago, 568 pp.

Scully, M.E., Friedrichs, C.T. and Wright, L.D. (2003) Numerical modeling of gravity-driven sediment transport and deposition on an energetic continental shelf: Eel River, northern California. *J. Geophys. Res.*, **108**, 17.1–17.14.

Seilacher, A. (1967) Bathymetry of trace fossils. *Mar. Geol.*, **5**, 413–428.

Seymour, R.J. (1998) Effects of El Niño on the west coast wave climate. *Shore Beach*, **66**, 3–6.

Sherwood, C.R., Butman, B., Cacchione, D.A., *et al.* (1994) Sediment transport events on the northern California shelf during the 1990–1991 STRESS experiment. *Cont. Shelf Res.*, **14**, 1063–1099.

Skempton, A.W. (1970) The consolidation of clays by gravitational compaction. *Q.J. Geol. Soc. London*, **125**, 373–411.

Smith, C.R. and Rabouille, C. (2002) What controls the mixed-layer depth in deep-sea sediments? The importance of POC flux. *Limnol. Ocean.*, **47**, 418–426.

Smith, C.R., Pope, R.H., DeMaster, D.J. and Maggard, L. (1993) Age-dependent mixing in deep-sea sediments. *Geochim. Cosmochim. Acta*, **57**, 1473–1488.

Smith, J.D. (1977) Modeling of sediment transport on continental shelves. In: *The Sea*, Vol. 6 (Eds E.D. Goldberg, I.N. McCave, J.J. O'Brien and J.H. Steele), pp. 539–577. Wiley-Interscience, New York.

Sommerfield, C.K. and Nittrouer, C.A. (1999) Modern accumulation rates and a sediment budget for the Eel shelf, a flood-dominated depositional environment. *Mar. Geol.*, **154**, 227–241.

Sommerfield, C.K., Nittrouer, C.A. and Alexander, C.R. (1999) [7]Be as a tracer of flood sedimentation on the northern California continental margin. *Cont. Shelf Res.,* **19**, 335–361.

Sommerfield, C.K., Drake, D.E. and Wheatcroft, R.A. (2002) Shelf record of climatic changes in flood magnitude and frequency, north-coastal California. *Geology,* **30**, 395–398.

Sternberg, R.W. and Larsen, L.H. (1975) Threshold of sediment movement by open ocean waves: observations. *Deep-sea Res.,* 22, 299–309.

Stoll, R.D. (1989) *Sediment Acoustics.* Springer-Verlag, New York, 153 pp.

Stolzenbach, K.D., Newman, K.A. and Wong, C.S. (1992) Aggregation of fine particles at the sediment water interface. *J. Geophys. Res.,* **97**, 17889–17898.

Styles R. and Glenn, S.M. (2000) Modeling stratified wave and current bottom boundary layers on the continental shelf. *J. Geophys. Res.,* **105**, 24119–24139.

Taghon, G.L., Nowell, A.R.M. and Jumars, P.A. (1984) Transport and breakdown of fecal pellets, biological and sedimentological consequences. *Limnol. Ocean.,* **29**, 64–72.

Thayer, C.W. (1983) Sediment-mediated biological disturbance and the evolution of marine benthos. In: *Biotic Interactions in Recent and Fossil Communities* (Eds M.J.S. Tevesz and P.L. McCall), pp. 479–625. Plenum Press, New York,

Thomsen, L. and Gust, G. (2000) Sediment erosion thresholds and characteristics of resuspended aggregates on the western European continental margin. *Deep-sea Res.,* I, 47, 1881–1897.

Thomson, J., Brown, L., Nixon, S., Cook, G.T. and Mackenzie, A.B. (2000) Bioturbation and Holocene sediment accumulation fluxes in the north-east Atlantic Ocean (Benthic Boundary Layer experiment sites). *Mar. Geol.,* **169**, 21–39.

Toorman, E.A. (1996) Sedimentation and self-weight consolidation: general unifying theory. *Géotechnique,* **46**, 103–113.

Toorman, E.A. (1999) Sedimentation and self-weight consolidation, constitutive equations and numerical modeling. *Géotechnique,* **49**, 709–726.

Torfs, H. and Mitchener, H. (1996) Erosion of mud/sand mixtures. *Coast. Eng.,* **29**, 1–25.

Traykovski, P., Hay, A.E., Irish, J.D. and Lynch, J.F. (1999) Geometry, migration, and evolution of wave orbital ripples at LEO-15. *J. Geophys. Res.,* **104**(C1), 1505–1524.

Traykovski, P., Geyer, W.R., Irish, J.D. and Lynch, J.F. (2000) The role of wave-induced density-driven fluid mud flows for cross-shelf transport on the Eel River continental shelf. *Cont. Shelf Res.,* **20**, 2113–2140.

Traykovski, P., Wiberg, P.L. and Geyer, W.R. (2007) Observations and modeling of wave-supported sediment gravity flows on the Po prodelta and comparison to prior observations from the Eel shelf. *Cont. Shelf Res.* doi: 10.106/j.csr.2005.07.008

Wheatcroft, R.A. (1990) Preservation potential of sedimentary event layers. *Geology,* **18**, 843–845.

Wheatcroft, R.A. (1994) Temporal variation in bed configuration and one-dimensional bottom roughness at the mid-shelf STRESS site. *Cont. Shelf Res.,* **14**, 1167–1190.

Wheatcroft, R.A. (2000) Oceanic flood sedimentation, a new perspective. *Cont. Shelf Res.,* **20**, 2059–2066.

Wheatcroft, R.A. (2002) *In situ* measurements of near-surface porosity in shallow-marine sands. *IEEE J. Ocean. Eng.,* **27**, 561–570.

Wheatcroft, R.A. (2006) Time-series measurements of macrobenthos and sediment bioturbation intensity on a flood-dominated shelf. *Progr. Ocean.,* **71**, 88–122.

Wheatcroft, R.A. and Borgeld, J.C. (2000) Oceanic flood layers on the northern California margin: Large-scale distribution and small-scale physical properties. *Cont. Shelf Res.,* **20**, 2163–2190.

Wheatcroft, R.A. and Butman, C.A. (1997) Spatial and temporal variability in aggregated grain-size distributions, with implications for sediment dynamics. *Cont. Shelf Res.,* **17**, 367–390.

Wheatcroft, R.A. and Drake, D.E. (2003) Post-depositional alteration and preservation of sedimentary event layers on continental margins, I. The role of episodic sedimentation. *Mar. Geol.,* **199**, 123–137.

Wheatcroft, R.A. and Martin, W.R. (1996) Spatial variation in short-term (^{234}Th) sediment bioturbation intensity along an organic-carbon gradient. *J. Mar. Res.,* **54**, 763–792.

Wheatcroft, R.A. and Sommerfield, C.K. (2005) River sediment flux and shelf sediment accumulation rates on the Pacific Northwest margin. *Cont. Shelf Res.,* **25**, 311–332.

Wheatcroft, R.A., Jumars, P.A., Smith, C.R. and Nowell, A.R.M. (1990) A mechanistic view of the particulate biodiffusion coefficient: Step lengths, rest periods and transport directions. *J. Mar. Res.,* **48**, 177–207.

Wheatcroft, R.A. Borgeld, J.C., Born, R.S., *et al.* (1996) The anatomy of an oceanic flood deposit. *Oceanography,* **9**, 158–162.

Wheatcroft, R.A., Sommerfield, C.K., Drake, D.E., Borgeld, J.C. and Nittrouer, C.A. (1997) Rapid and widespread dispersal of flood sediment on the northern California continental margin. *Geology,* **25**, 163–166.

Wiberg, P.L. (2000) A perfect storm: Formation and potential for preservation of storm beds on the continental shelf. *Oceanography,* **13**, 93–99.

Wiberg, P.L. and Harris, C.K. (1994) Ripple geometry in wave-dominated environments. *J. Geophys. Res.*, **99**, 775–789.

Wiberg, P.L. and Smith, J.D. (1987) Calculations of the critical shear stress for motion of uniform and heterogeneous sediments. *Water Resour. Res.*, **23**, 1471–1480.

Wiberg, P.L., Drake, D.E. and Cacchione, D.A. (1994) Sediment resuspension and bed armoring during high bottom stress events on the northern California inner continental shelf: measurements and predictions. *Cont. Shelf Res.*, **14**, 1191–1219.

Wiberg, P.L., Drake, D.E., Harris, C.K. and Noble, M. (2002) Sediment transport on the Palos Verdes shelf over seasonal to decadal time scales. *Cont. Shelf Res.*, **22**, 987–1004.

Wilcock, P.R. (1993) Critical shear stress of natural sediments. *J. Hydraul. Eng.*, **119**, 491–505.

Wright, L.D., Kim, S.-C. and Friedrichs, C.T. (1999) Across-shelf variations in bed roughness, bed stress and sediment suspension on the northern California shelf. *Mar. Geol.*, **154**, 99–116.

Yager, P.L., Nowell, A.R.M. and Jumars, P.A. (1993) Enhanced deposition to pits: a local food source for benthos. *J. Mar. Res.*, **51**, 209–236.

Zhang, Y., Swift, D.J.P., Fan, S., Niederoda, A.W. and Reed, C.W. (1999) Two-dimensional numerical modeling of storm deposition on the northern California shelf. *Mar. Geol.*, **154**, 155–167.

Oceanic dispersal and accumulation of river sediment

CHRISTOPHER K. SOMMERFIELD*, ANDREA S. OGSTON†, BETH L. MULLENBACH‡,
DAVID E. DRAKE§, CLARK R. ALEXANDER¶, CHARLES A. NITTROUER†, JEFFRY C.
BORGELD**, ROBERT A. WHEATCROFT†† and ELANA L. LEITHOLD‡‡

*College of Marine Studies, University of Delaware, Lewes, DE 19958, USA (Email: cs@udel.edu)
†School of Oceanography, University of Washington, Box 357940, Seattle, WA 98195, USA
‡Department of Oceanography, Texas A&M University, TAMU-3146, College Station, TX 77843, USA
§Drake Marine Consulting, Ben Lomond, CA 95005, USA
¶Skidaway Institute of Oceanography, 10 Ocean Circle, Savannah, GA 31411, USA
**Department of Oceanography, Humboldt State University, Arcata, CA 95521, USA
††College of Oceanic and Atmospheric Sciences, Oregon State University, Corvallis, OR 97331, USA
‡‡Department of Marine, Earth and Atmospheric Sciences, North Carolina State University, Raleigh, NC 27695, USA

ABSTRACT

The sedimentology and stratigraphy of river-fed continental margins reflect a diverse range of tectonic, climatic and hydrodynamic conditions that moderate the supply, transport and accumulation of terrigenous sediment in the coastal ocean. This paper describes a study of the modern and late Holocene northern California shelf and slope. The aim is to elucidate fundamental processes of land-to-ocean sediment dispersal and accumulation relevant to active margins worldwide. Annually, northern California rivers deliver a total of $30-40 \times 10^6$ tons of suspended sediment to the coast, nearly 70% of which bypasses the shelf in association with oceanic storms and major floods. Off-shelf export is highly time dependent, and is maximal when periods of peak river discharge and across-shelf flow coincide. Sediment not exported is sequestered on the shelf through dynamic trapping mechanisms in the bottom-boundary layer, as well as by static conditions related to tectonically produced topography. Significantly, interactions between shelf bathymetry and near-bottom flows influence patterns and rates of strata formation on a wide range of temporal and spatial scales. The shelf displays a rich sedimentary record that archives signatures related to Holocene transgression, climatic variations in continental runoff, and land-use change during historical times.

Keywords Fluvial sediment, sediment dispersal systems, sediment accumulation rates, radioisotopes, tectonic sedimentation, continental-shelf bypassing.

INTRODUCTION

The ability to predict the flux and fate of terrigenous sediment delivered to coastal waters has enormous societal and economic ramifications to nations worldwide. Sediments and sedimentary processes are central to chronic issues such as shoreline erosion, filling of navigable channels and dispersal of particle-borne contaminants – significant problems in terms of environmental impact and mitigative outlay (National Research Council, 1997). As the effects of coastal urbanization press ever seaward, there is a growing need to forecast the transport of sedimentary matter from observational data and through numerical modelling. It is apparent that many sediment-related problems are local consequences of regional hydrological and oceanic regimes, and that a **systems** perspective is needed to fully characterize source-to-sink movement of terrigenous sediment. A process-oriented understanding of sediments is also essential to decipher the sedimentary record, an invaluable history of natural hazards, including sea-level rise, extreme storms and other phenomena, that affect life at the coast. Among coastal and marine environments there is a particular need to understand river-fed

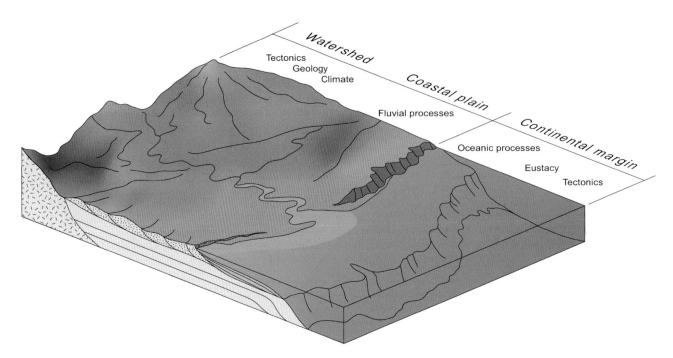

Fig. 1 Idealized sediment dispersal system for an active margin showing subregions referred to in the text, and major processes of sediment production, transport, and accumulation.

sedimentary systems, given their far-reaching effects on material flux, biochemical cycling and environmental quality.

The purpose of this paper is to elevate our general knowledge of river sediment dispersal and accumulation in the coastal ocean by presenting a case study of the northern California margin, a focus area of the STRATAFORM programme (Nittrouer, 1999). This interdisciplinary effort involved a wide range of research studies with a common goal of illuminating fundamental mechanisms of strata formation and preservation on time-scales ranging from depositional events to many millennia. The distinctive quality of STRATAFORM was the breadth and depth of observation and modelling of sediment transport, deposition and accumulation. The northern California margin provided the ideal study area because its geological and coastal oceanographic properties are characteristic for **active margins**, both in terms of tectonic activity and terrigenous sediment flux. The mountainous river drainages are of the type responsible for a large fraction of the global delivery of terrigenous sediment to the ocean (Milliman & Syvitski, 1992; Syvitski *et al.*, 2005), thus the relevance of this case study extends far beyond its geographical boundaries.

This paper focuses on sedimentary processes and products of the latest Holocene and modern times, emphasizing mechanisms of terrigenous sediment transport and trapping on the continental shelf and slope. The scales of investigation considered are necessarily bounded by the range and resolution of instrument and sedimentary records, but they are germane to long-term stratigraphy and modelling discussed in this volume by Mountain *et al.* (pp. 381–458) and Syvitski *et al.* (pp. 459–529). To organize a wide range of observations, a **margin sediment dispersal system** is used in this paper to denote a succession of hydraulically contiguous sedimentary environments with upstream and downstream ends delimited by the watershed and continental slope, respectively (Fig. 1). Naturally, these boundaries are artificial, but the northern California margin is well suited to systemization, as described in this paper.

Margin sediment dispersal systems: a Holocene perspective

The sedimentology and stratigraphy of modern continental margins (shelves and slopes) reflect a diverse range of tectonic, climatic and oceanographic conditions that moderate the supply and accumulation of terrigenous sediment. Most recently, humans have become an additional geomorphological agent in sediment dispersal systems. Under natural conditions, climate and tectonics control the production and delivery of fluvial sediment to the first order, whereas depositional base-level (sea-level) controls the locus of accumulation within downstream fluvial, coastal and margin environments (Posamentier & Allen, 1993; Schumm, 1993; also see Mountain *et al.*, this volume, pp. 381–458). In general, rivers discharge at the shelf edge during sea-level lowstands, favouring slope- and rise-centred deposition in the form of slope drapes and submarine fans (Piper & Normark, 2001). Conversely, decreased land-to-ocean hydraulic gradients during times of rising sea level cause sediments to become entrapped landward of the shelf edge, perhaps to the point of backfilling river channels and estuaries (Posamentier & Vail, 1988; Blum & Törnqvist, 2000). Marine flooding of the coastal plain and its incised valleys forms **accomodation space**, the amount of space available for sediment accumulation between the sea surface and seafloor. The relative rates of sea-level rise, subsidence and sediment aggradation dictate the amount of accommodation space available and, in the long term, influence river-mouth evolution toward an estuarine or deltaic end-member environment. As a general rule, **estuaries** are present on modern coasts where rates of fluvial sediment supply and accumulation have not kept up with creation of accommodation space by relative sea-level rise, whereas **deltas** form where the available space has been filled (Boyd *et al.*, 1992).

The origin of modern sediment dispersal systems can be traced to decelerated eustatic sea-level rise during the late Holocene (Fig. 2). Stratigraphic reconstructions indicate that the rate of eustatic sea-level rise decreased from ~5–6 mm yr^{-1} to the present rates of 1–3 mm yr^{-1} between about 7.5 ka and 6 ka (Fleming *et al.*, 1998). During this time, accommodation space created by marine flooding of coastal plains was rapidly filled, and transition to transgressive-highstand conditions fostered delta formation on a global scale (Stanley & Warne, 1994). As rates of sediment accumulation and sea-level rise converged during the late Holocene, **dynamic equilibrium** was established in many dispersal systems. In this condition, net accretion

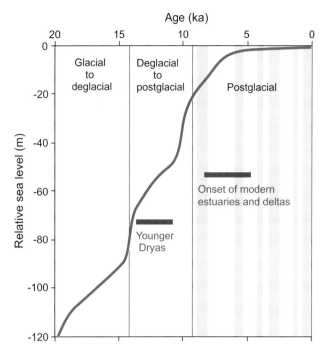

Fig. 2 Global sea-level curve from the compilation of Fleming *et al.* (1998). Climatic periods are from Knox (1995). Intervals of rapid climate change identified by Mayewski *et al.* (2004) are noted by vertical bars.

of the seabed is limited by tidal erosion depth or storm wave base, ultimately as a consequence of sea-level height. Only in the most sediment-rich or protected settings will accretion extend to the intertidal or supratidal zone, perhaps in the form of tidal flats and vegetated marsh (Nichols, 1989). One effect of increased sediment trapping landward of the shelf edge was a substantial reduction in the mass of river sediment delivered directly to the continental slope. Indeed, modern sediment accumulation rates on slopes worldwide are generally lower compared with the Last Glacial Maximum and subsequent transgression, although the timing of this transition appears to have varied widely among margin systems (Caddah *et al.*, 1998; Carter *et al.*, 2002; Sommerfield & Lee, 2004).

With equilibration of many estuarine and deltaic environments during the late Holocene, river sediment was increasingly bypassed seaward where it accumulated atop older coastal strata. Non-deltaic shelf mud accumulations occur in various forms ranging from isolated patches to continuous belts, and in cross-section form aggradational wedges and progradational **clinoforms**. These sediment units are fed by advective mud streams generated by large deltaic rivers, or through diffusive transport of suspensions from smaller rivers to distal shelf sites (McCave, 1972; Nittrouer & Wright, 1994; McKee et al., 2004). Whereas the presence of shelf mud reflects a particular balance between suspended-sediment concentration, particle settling velocity and limiting fluid shear stress at the bed (see Hill et al., this volume, pp. 49–99), the loci and orientation of depocentres are related to boundary-layer conditions that control the time-averaged fields of these variables. In general, muddy strata will form when and where the depositional flux exceeds the erosional flux generated by stress-induced resuspension. For this reason, the geometry of shelf mud bodies conveys information on quasi-stationary gradients in sediment accumulation rate, and patterns of aggradation and progradation (Pirmez et al., 1998; Driscoll & Karner, 1999).

Holocene climates and land-surface conditions played an important role in moderating the flux of terrigenous sediment to coasts. The period 14–10 ka was transitional between glacial and postglacial conditions, and unstable climates brought about dramatic changes in continental hydrology that moderated sediment production and delivery (Meybeck & Vörösmarthy, 2005). For example, lake records document extreme changes in precipitation runoff during this period (Benson et al., 1996), and the failure of ice dams in temperate regions produced catastrophic discharges to the coastal ocean (Donnelly et al., 2005). A climatically elevated land-to-ocean terrigenous flux was particularly pronounced for South Asian systems impacted by the south-west monsoon (Goodbred & Kuehl, 2000). Fluvial stratigraphic records from North America and Europe reveal that rapid warming following the **Younger Dryas** (~11–10 ka) led to increased vegetative cover and a decrease in the magnitude and frequency of extreme floods (Knox, 1995). On the North American west coast, this transition led to reduced sediment delivery from hillslopes to river channels (Reneau et al., 1990).

Climates after 10 ka were considerably more stable, yet not completely devoid of fluctuations relevant to continental runoff and erosion. Proxy records of Holocene climate indicate global shifts between warm–wet and cool–dry conditions apparently forced by solar variability and changes in insolation (Mayewski et al., 2004). Temperate lake records suggest that the frequency of large rainstorms and extreme runoff events increased during the Holocene (Noren et al., 2002). This observation is consistent with North American fluvial records of late Holocene 'flood epochs', a consequence of change from dominantly zonal to meridional jet-stream flow, coincident with cool, wet climates (Knox, 2000). These records suggest that rivers were particularly flood-prone during 5000–3600 yr BP and 1100–900 yr BP, with an increasing trend from 500 yr BP to the present day (Ely et al., 1993). Interestingly, it has been postulated that large increases in flood magnitude and frequency can result from modest changes in climate, for example, variations in mean annual temperature of merely 1–2°C and precipitation of ≤ 10–20% (Knox, 1993). Recurrent climatic phenomena such as **El Niño–Southern Oscillation** (ENSO) and the **Pacific Decadal Oscillation** (PDO) are now known to significantly increase runoff and terrigenous sediment flux over mean conditions (Inman & Jenkins, 1999; Farnsworth & Milliman, 2003; Warrick & Milliman, 2003); there is marine stratigraphic evidence to suggest that El Niño has been a persistent climatic feature of the late Quaternary (Bull et al., 2000).

Since the inception of agriculture, and particularly after the early 19th century, humans have altered fluvial sediment loads worldwide. Land clearing for farming, grazing and timber harvesting is regarded as the most significant cause for elevated sediment loads due to its destabilizing effects on soils (Meade, 1990). Globally, the pre-human suspended-sediment load of rivers worldwide has been estimated at 14 Gt yr^{-1} whereas the present load is 16.3 Gt yr^{-1}, an increase of ~10% over 'natural' conditions (Syvitski et al., 2005). On a single-basin scale, however, loads of some rivers have increased by up to an order of magnitude due to land-cover change (Meybeck & Vörösmarthy, 2005). At the same time, as much as 25–30% of sediment transported by rivers worldwide is trapped by reservoirs and dams, never making it to the coast (Vörösmarty et al., 2003). Consequently, a net 12.6 Gt yr^{-1} of terrigenous sediment is delivered to the world ocean, a global *reduction* of 1.4 Gt over pre-human levels (Syvitski et al., 2005).

The downstream sedimentary effects of river-flow regulation and deforestation vary widely, but in extreme cases the outcome of these practices is unmistakable at the coast. On account of sediment starvation, river damming and diversion have led to accelerated shoreline retreat (Carriquiry & Sánchez, 1998), delta abandonment (Stanley & Warne, 1998) and loss of estuarine marshland (White *et al.*, 2002). In the face of continued sea-level rise, coastal erosion by wholesale reduction in sediment supply is a global concern. Conversely, deforestation in drainage basins can generate a surplus of sediment at the coast with consequences that vary with a system's ability to disperse excess loads. At one end of the spectrum, land-cover change in small catchments has been shown to result in a nominal increase in sediment accumulation rate within downstream coastal seas (Goff, 1997). At the other end, widespread land clearing and cultivation has forced the expansion of deltas at some river mouths, as exemplified by the Huanghe and Changjiang rivers of China (Milliman *et al.*, 1987; Saito *et al.*, 2001).

Although it is generally regarded that climate plays a subordinate role to humans in moderating sediment loads in small- to medium-sized rivers, separating climatic and anthropogenic agents of change presents challenges when interpreting sedimentary records of fluvial–marine sedimentation (Dearing & Jones, 2003). Climate and land clearing can work in the same direction to moderate sediment transport and storage such that downstream variations in accumulation may be difficult to interpret (Lang *et al.*, 2003). Moreover, river systems have an inherent capacity to buffer sedimentary signals through channel or floodplain storage; whether an upstream disturbance is perceptible at the coast depends on factors including perturbation scale, sediment routing and basin storage capacity (Walling, 1999). The Gironde-shelf mud fields described by Lesueur *et al.* (2002) provide an interesting example of how natural and anthropogenic change in a watershed can converge toward a similar sedimentary product at the coast. These deposits originated by highstand estuarine bypassing after 2 ka, but expanded with anthropogenic increases in oceanic sediment delivery during the past ~500 yr. Challenges notwithstanding, our ability to decipher coastal–marine geological records continues to improve with knowledge gained through observation and modelling of sedimentary processes.

Research background

The STRATAFORM programme was rooted in at least three decades of prior oceanographic and geological research on continental margins worldwide, providing a scientific basis to address advanced questions regarding the dispersal and accumulation of river sediment on an open shelf and slope. Early conceptual models of shelf sedimentation are reviewed in Hill *et al.* (this volume, pp. 49–99), and topical review articles of note include: sediment transport on shelves (Cacchione & Drake, 1990; Nittrouer & Wright, 1994; Ogston *et al.*, 2005); sediment dispersal by coastal flows (Geyer *et al.*, 2004); seabed processes associated with major rivers (McKee *et al.*, 2004); geometry of shelf sediment bodies (Orton & Reading, 1993; Pirmez *et al.*, 1998; Paola, 2000); and transgressive sedimentology and stratigraphy (Cattaneo & Steel, 2003). Case studies of modern mud-dominated shelves are too numerous to mention, though this topic is ably covered in Johnson & Baldwin (1996).

Among the topics addressed on the northern California margin, the effect of episodic sediment delivery on strata formation was a particular focus. In his seminal review of this topic, Dott (1983) concluded that 'episodic deviations from average or steady-state conditions deserve more emphasis in interpreting the sedimentary record.' It is well recognized that the effects of rare events (e.g. storms, river floods), rather than typical events, are most apt to be preserved in the long term; the geological record is replete with sedimentary event deposits. On the other hand, Dott's (1983) assertion poses obvious observational challenges in modern sedimentary environments. A related aspect of episodicity is stratigraphic completeness, which, for shallow-marine sections, is greatly reduced on account of **hiatuses**, periods of non-deposition or erosion. Quantifying missing time (and space) in stratigraphic records is complicated by the fact that hiatuses, when recognizable, are difficult to date accurately. Sadler (1981) was first to estimate levels of stratigraphic completeness for a wide range of sedimentary environments using empirical relations of sediment accumulation rate versus averaging time span. Trends attributable to wave base variation were documented on the short term, and to crustal subsidence on time-scales greater than a millennium. However, a mechanistic understanding of stratigraphic completeness has been elusive, despite the ever increasing use of sedimentary records to document environmental change. To this end, research on the northern California margin attempted to relate short-term transport phenomena to strata formation and preservation on a range of time and space scales, from river-flood events to sea-level cycles.

The influence of episodic terrigenous flux on the sediment budget of a margin dispersal system provided an additional focus. Sediment budgets are commonly developed to identify how and where river sediments are partitioned among hydraulically linked environments (Nittrouer & Wright, 1994). Milliman & Syvitski (1992) hypothesized that much of the sediment delivered to tectonically active coasts bypasses narrow shelves in association with extreme discharges that depart from mean conditions. This notion is consistent with geological observations on mud-dominated, active margins (Goodbred & Kuehl, 1999), yet the actual mechanisms and time-scales of across-shelf transport are rather poorly known. The narrow-shelf concept was critically examined in the northern California

study area by developing sediment budgets averaged over flood-event and centennial time-scales. These budgets offer a holistic view of sediment dispersal during wide-ranging conditions of sediment loading and shelf energy.

QUANTIFYING SEDIMENTARY PROCESSES

Suspended-sediment transport

Time-series instrumentation observations of water properties and sediment transport provide a means to quantify water-column sediment flux, which is needed to understand how particles are deposited, eroded and eventually buried in the seabed (Fig. 3). Measurements of water velocity from electromagnetic current meters, acoustic doppler velocimeters and acoustic doppler profilers provide information on the magnitude and direction of mean, tidal and wave-induced currents (Dickey et al., 1998), all of which bear upon sediment transport. Water properties, such as temperature and salinity, are measured using CTD (conductivity,

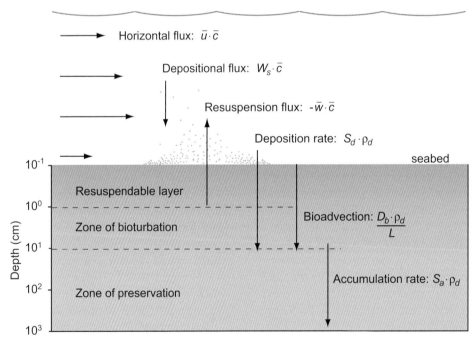

Fig. 3 Continuum of sediment flux in dispersal systems. Symbols are current velocity (u, w), suspended-sediment concentration (c), particle-setting velocity (W_s), biological mixing coefficient (D_b), mixing length scale (L), deposition rate (S_d), accumulation rate (S_a) and dry-bulk density (ρ_d). Overbar denotes a time-averaged property. See text for description of terms and methods of determination.

temperature, depth) sensors either in a fixed position or in a profiling mode to determine temporal and spatial changes in the density structure of the water column. Variations in suspended-sediment concentration can be documented through light transmission, optical backscatter and acoustic backscatter (Sternberg & Nowell, 1999), and these indirect measures are generally calibrated against filtered water samples.

Modern oceanographic instruments have the ability to make measurements at high temporal resolutions (1–25 Hz), providing a very detailed time series of sediment fluxes and water properties. Remotely deployed systems have finite battery power, which restricts the total measurement period. Of course there is a trade-off between operation time and deployment length; a longer deployment time requires shorter sampling bursts. With the recent advances of computer technology, only instruments that sample at very high frequencies are limited by memory of their data logger. Measurements from *in situ* sensors also are limited in terms of spatial resolution, as detailed temporal information is recorded at only one geographical location. Ideally, multiple instrument packages are simultaneously deployed over an area of interest and at multiple water depths to obtain detailed synoptic data. In reality, however, deployment locations must be chosen prudently owing to the expense of instrumentation, as well as the potential for damage by fishing and shipping activities.

Simultaneous measurements of currents and suspended-sediment concentrations allow the calculation of horizontal sediment flux (Fig. 3), as well as shear stress within the bottom boundary layer. Vertical sediment flux (deposition and resuspension) can be estimated in environments with low wave energy using sediment traps that collect sinking particles, and this material can be compositionally analysed to estimate component-specific fluxes (e.g. organic versus inorganic). Some sediment traps sample temporal variations in vertical flux during deployments using multiple sample collectors. Although sediment traps have been used extensively, trap-derived fluxes must be interpreted with caution due to sample biasing (U.S. Global Ocean Flux Study, 1989). Vertical sediment flux also can be estimated from profiles of suspended-sediment

concentration using eddy diffusivity models such as the Rouse Equation (Orton & Kineke, 2001).

Current and sediment sensor packages can be deployed on benthic frames or taut-line moorings. As an example, the tripod in Fig. 4a is a basic design for long-term monitoring of the bottom boundary layer. Benthic instrumentation packages typically carry a pressure sensor, current meters and suspended-sediment sensors positioned at multiple heights above the bed. Data from vertical arrays of sensors are used to compute horizontal and vertical sediment fluxes, as well as current shear stress from the logarithmic velocity profile (Cacchione & Drake, 1990). Other types of sensors are available to compute shear stress from the turbulent properties of boundary-layer flow (Williams *et al.*, 1997). More sophisticated tools such as the sector-scanning sonar can be employed to monitor seabed roughness, and a variety of camera and laser systems have been used to examine particle size and settling velocity *in situ* (Sternberg & Nowell, 1999; Thorne & Haynes, 2002). As part of the new generation of high-frequency acoustic instrumentation, the Acoustic Backscatter Sensor (ABS) is an important tool that combines measurements of suspended-sediment profiles and precision measurement of seabed erosion and deposition (Traykovski *et al.*, 2000). The instrument transmits pulses of high-frequency sound through as many as four transducers of different frequency, and measures the sound backscattered by suspended particulate material in vertical bins that range from millimetres to several centimetres. The main advantage of the ABS over traditional methods is its ability to track the movement of sediment between bottom waters and the seabed.

Taut-line moorings are designed to quantify sediment transport throughout the water column on continental margins. A typical mooring configuration is shown in Fig. 4b. Electromagnetic current meters, transmissometers, CTDs, temperature sensors and sediment traps are strategically positioned to capture variations in sediment flux (horizontal and vertical) and water properties. Surface moorings are deployed to track sediment flux associated with river discharge and storms, whereas subsurface moorings are used to study deeper hemipelagic transport and deposition (Biscaye & Anderson, 1994).

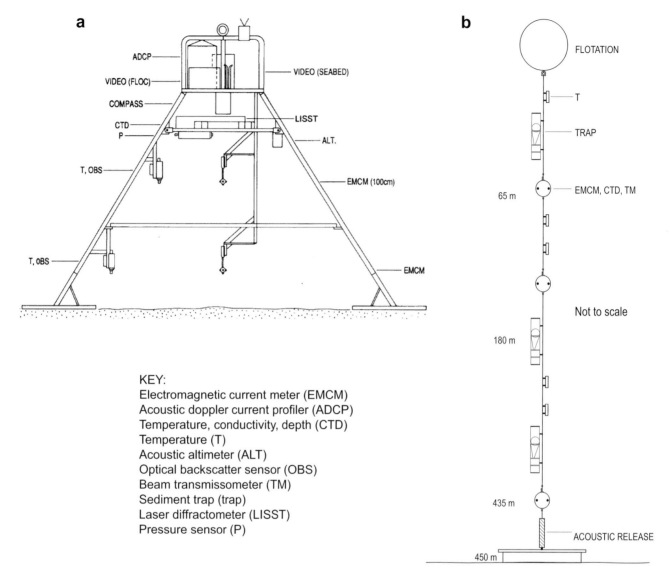

Fig. 4 Typical instrumentation systems. (a) A benthic tripod for studies of the shelf bottom-boundary layer. (From Sternberg & Nowell, 1999.) (b) A taut-line mooring with subsurface flotation for continental-slope studies.

Sediment deposition and accumulation

Sediment accumulation (burial) is the final transport pathway for suspended particles in the coastal ocean. Sediment accumulation integrates conditions of sediment supply and dispersal over a given time span, so spatial gradients in accumulation rates convey information on preferential transport pathways and depocentres. Rates of sediment accumulation can be determined from sedimentary records or high-resolution seismic profiles, but age control provided by radioisotopic or biostratigraphic dating is necessary. Modern methods in sediment coring are reviewed by Hebbeln (2003) and are not discussed here.

Sediment accumulation rates are most frequently defined in terms of (i) a spatial reference frame and (ii) time span of averaging. **Sediment accumulation rate** is simply the linear velocity of seafloor aggradation expressed in length per unit time (e.g. cm yr^{-1}), and is equivalent to a linear sedimentation rate. Linear rates are widely used in basin analysis because they are most intuitive and can be computed with limited data. On the other hand, failure to take

Table 1 Radioisotope tracers of sediment accumulation and biological mixing

Radioisotope	Half-life	Range	Source	Input
^{234}Th	24.1 days	Days–4 months	Oceanic	Continuous
^{7}Be	53.3 days	Days–8 months	Cosmogenic	Continuous
^{210}Pb	22.3 yr	Years–100 yr	Atmospheric	Continuous
^{137}Cs	30.1 yr	Post-1954	Anthropogenic	Pulse
^{14}C	5730 yr	400–40,000 yr	Cosmogenic	Continuous
^{239,240}Pu	2.4×10^4, 6563 yr	Post-1954	Anthropogenic	Pulse

into account downsection variations in porosity can lead to erroneous accumulation rates (Behrens, 1980). The preferred form in studies of sediment flux and mass balance is **mass accumulation rate**, the mass of sediment buried per unit area per unit time (e.g. g cm^{-2} yr^{-1}). Mass accumulation rate can be computed as the product of linear accumulation rate and mean bulk density of the sediment column, or from a profile of radioisotope activity plotted versus accumulated mass (g cm^{-2}) when detailed porosity data are available. Upon establishing the bulk mass accumulation rate for a section, burial fluxes for specific sediment constituents (e.g. biogenic silica) can be computed from concentration profiles.

It is useful to further distinguish accumulation rates based on the averaging period, given that river sediments undergo enumerable depositional and erosional episodes before becoming permanently emplaced on margins (Wright & Nittrouer, 1995). It has become customary to reserve **deposition rate** for describing a short-term or seasonal average and to use **accumulation rate** when reporting data averaged over 100 yr and longer (McKee *et al.*, 1984). This distinction is essential, because sediment accumulation rates are a strong negative function of measurement time span, as a result of hiatuses in sedimentation (Sadler, 1981; Sommerfield, 2006). For example, the deposition rate of a storm-produced bed is not representative for the underlying sediment column, which has a much lower rate of net accumulation. In modern sedimentary environments this time-span dependence is demonstrated by the observation that ^{137}Cs- and ^{210}Pb-based accumulation rates (decadal-centennial average) are almost always higher than ^{14}C rates (centennial–millennial average) for the same sedimentary section.

In part, the difference arises because ^{14}C averages over a much wider range of sedimentary conditions, with time and space lost through non-deposition and erosion. The time-variant nature of sediment accumulation rates will be discussed further in the context of stratigraphic completeness later in this paper.

Natural and anthropogenic radioisotopes are the most widely applied method of sediment geochronology in studies of modern-to-Holocene estuarine and marine systems, because their sources and half-lives are well known (Turekian & Cochran, 1978). The most commonly employed radioisotopes in continental-margin sedimentology are ^{234}Th, ^{7}Be, ^{137}Cs, ^{239,240}Pu, ^{210}Pb and ^{14}C (Table 1). With the exception of ^{14}C, application of these tracers is limited to fine-grained sediments (silt and clay), which scavenge particle-reactive radioisotopes and deliver them to the seabed through deposition. Once permanently removed, radioisotope concentration (activity) decays through time as a function of the **half-life**. The decay gradient of the activity–depth profile in sediments provides the information needed to compute a sediment accumulation rate, or bioadvection rate, whatever the case may be. Implicit in radioisotope geochronology is a knowledge of the sources of the radiotracer in question, as well as the sedimentary processes that determine its distribution in the sediment column.

Thorium-234 is produced continually in the water column by the decay of its ^{238}U grandparent and is removed to the seabed by scavenging and deposition at levels in excess of those produced *in situ* by ^{238}U decay in the sediment column. Thorium-234 is limited to brackish and saline sedimentary environments as a particle tracer, because ^{238}U activity varies directly with salinity. Removal of ^{234}Th to the seabed generally increases with suspended-sediment concentration, and its theoretical inventory in bottom sediments can be well constrained knowing the salinity and depth of the water column. In the most rapidly accumulating shelf environments, ^{234}Th has been used as a tracer of short-term deposition (McKee *et al.*, 1983; Smoak *et al.*, 1996). However, because the rate of particle bioadvection exceeds that of sediment accumulation on all but the most rapidly accumulating estuaries and shelves, ^{234}Th is more typically employed as a tracer of biological mixing intensity (Wheatcroft & Martin, 1996). In this case, sediment profiles of excess ^{234}Th activity are used to compute the biological mixing coefficient, D_b (Wheatcroft *et al.*, this volume, pp. 101–155).

Cosmogenic ^{7}Be is supplied to Earth's surface largely via precipitation and is most commonly used as a tracer of short-term deposition in shallow turbid estuaries, where particle removal is rapid (Olsen *et al.*, 1986). Transfer of atmospheric ^{7}Be to coastal waters occurs through direct deposition, dissolved-phase river runoff, and erosion/transport of activity-laden surficial soils. ^{7}Be can be an effective tracer of short-term deposition on margins when a large influx of terrigenous particles has rapidly transferred activity from the water column to the seabed (Sommerfield *et al.*, 1999; Allison *et al.*, 2000). Beryllium-7 is for the most part transient in fine-grained shelf and slope sediments, so its presence is a useful indicator of new sediment deposited on a time-scale within several half-lives of measurement.

The anthropogenic nuclides ^{137}Cs and ^{239,240}Pu were first introduced to the environment through nuclear weapons testing and reactor releases around 1954, after which fallout peaked in 1963–1964, and thereafter dropped to insignificant levels by about 1980. This pulse source function provides 'absolute' time markers for the sediment column, provided the atmospheric source function is faithfully preserved. Under conditions of steady-state sediment accumulation, sediment depths corresponding to the basal and peak activity are concordant with 1954 and 1964, respectively. Anthropogenic radioisotopes have been used to estimate sediment accumulation rates and biological mixing intensity in a wide range of estuarine and marine environments, although the application depends on the relative importance of accumulation and

bioadvection in particle burial (Anderson *et al.*, 1988).

A member of the ^{238}U decay series, ^{210}Pb is derived from decay of ^{222}Rn in the atmosphere and is delivered to Earth via precipitation. The primary sources of ^{210}Pb in shelf sediments in order of importance include: advected oceanic waters; direct atmospheric flux to shelf waters; and river input (Carpenter *et al.*, 1981). The most widely employed ^{210}Pb model specifies that the activity of sediment arriving at the seabed is constant through time (i.e. constant initial concentration). In this way the slope of the activity–depth profile reflects a balance between steady-state sediment accumulation and radioactive decay, in the absence of biological mixing. Unless the deep biological mixing can be completely ruled out, which is not the case for many margin environments, ^{210}Pb-derived accumulation rates are generally taken as maximum estimates. Failure to take into account bioadvection in ^{210}Pb burial can result in two- to threefold overestimates of the true sediment accumulation rate (Benninger *et al.*, 1979; Carpenter *et al.*, 1982; DeMaster *et al.*, 1985). Commonly, ^{137}Cs or ^{239,240}Pu measurements are made in conjunction with ^{210}Pb to validate rate estimates by evaluating the predicted versus observed penetration below the zone of bioturbation (Nittrouer *et al.*, 1984), or numerically using a time-dependent solution of the advection-diffusion equation (Sherwood *et al.*, 2002).

Carbon-14 geochronology involves dating an organic (^{14}C-C$_{org}$) or inorganic (^{14}C-CaCO$_3$) carbonaceous substrate in the sediment column for establishing absolute ages or long-term sediment accumulation rates. The ^{14}C-CaCO$_3$ approach is based on the notion that carbonate-secreting organisms living near the sediment–water interface preserve the ^{14}C composition of bottom waters upon death and burial. After correcting for transfer time of atmospheric ^{14}C to oceanic waters (Stuvier *et al.*, 1998), the measured ^{14}C-CaCO$_3$ age approximates the true stratigraphic age. The chief assumption in this method is that the dated material has not been reworked and redeposited, in which case the measured age post-dates the true stratigraphic age. In ^{14}C-C$_{org}$ geochronology, the fundamental assumption is that the initial ^{14}C age of sedimentary organic matter arriving at the seafloor is constant through time. Both methods involve computing mean sediment accumulation rates based on the ^{14}C

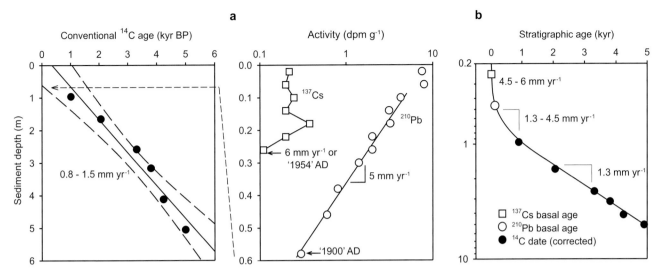

Fig. 5 (a) Example of composite geochronology (^{137}Cs, ^{210}Pb and ^{14}C) for a depositional site on the Eel shelf (E-95). (b) The same data plotted in composite form as a continuous age-model curve. Note how the accumulation rate appears to accelerate up-section, a consequence of time-variant sediment accumulation and stratigraphic incompleteness. (From Sommerfield, 2006.)

age–depth profile in a sediment core (Thomas *et al.*, 2002; Sommerfield & Lee, 2003).

The most restrictive aspect in dating modern-to-Holocene strata is the gap in coverage provided by ^{210}Pb and ^{14}C methods, which, unfortunately, coincides with the late 19th century. Whereas ^{210}Pb has a resolution of several years and a maximum range of ~100 yr, ^{14}C dating has a minimum resolution of ~200–400 yr (the cumulative analytical and calibration errors). Composite geochronology is an approach in which ^{137}Cs or ^{210}Pb and ^{14}C data are abutted to develop a continuous-age model for a sediment column (Fig. 5; Colman & Bratton, 2003). The major attraction of this technique is the potential to tie ^{210}Pb and ^{14}C chronologies: an advantage when investigating modern-to-Holocene sediment accumulation history. However, because all forms of sediment geochronology have inherent limitations, caution must be exercised when manipulating accumulation rates based on different dating methods.

THE NORTHERN CALIFORNIA MARGIN

Previous work and insight

Experiments undertaken on the shelf off the Russian River in northern California from the late 1970s to

the early 1990s shed new light on the role of short, energetic winter storms on shelf sedimentology, laying a foundation for the STRATAFORM programme. These earlier programmes included Coastal Ocean Dynamics Experiments (CODE; Beardsley & Lentz, 1987) and Sediment Transport Events on Shelves and Slopes (STRESS; Trowbridge & Nowell, 1994). Sensors deployed at 45–90 m water depths during CODE provided time series that were statistically robust enough to test new combined wave–current sediment transport models (Smith, 1977; Grant & Madsen, 1979). The data demonstrated that the models could adequately predict key dynamical features of the **bottom boundary layer**, and provided estimates of bed shear stress that were in good agreement with expected values at the threshold of sediment motion (Grant & Madsen, 1986). The models also appeared to give reasonable predictions of suspended-sediment concentrations during storm events, but the field observations of particle concentrations and characteristics were not yet sufficient to properly test these new formulations. The STRESS experiments added critical new observations on seafloor microtopography, bottom-boundary-layer flow, sediment dynamics and bed armouring during storms. In addition, observations revealed that much of the suspended load occurs in relatively large, rapidly settling flocs and faecal pellets. By the early 1990s, the capability was

Fig. 6 Landsat Multispectral Scanner image (bands 1, 2 and 3) for a segment of the northern California margin, showing a turbid buoyant plume emanating from the mouth of the Eel River (28 March 1975). In this example, the plume is deflected southward under the influence of northerly winds.

available to successfully model bottom-boundary-layer flows and sediment transport over shelves characterized by a variety of bedforms and sediment types (see reviews of Cacchione & Drake, 1990; Nittrouer & Wright, 1994).

Early work on the northern California margin employed a range of research methods to track the dispersal of river sediment in the coastal ocean. In the 1970s, the Landsat Multi-Spectral Scanner provided striking images of turbid surface plumes emanating from the mouths of California coastal rivers (Fig. 6). From time series of these images, it was concluded that buoyant river plumes are a chief mechanism of sediment dispersal (Pirie & Stellar, 1977; Griggs & Hein, 1979), a notion that was later dispelled by oceanographic measurements on the Eel shelf (Geyer *et al.*, 2000; Hill *et al.*, this volume, pp. 49–99). Satellite remote sensing continues to provide valuable information on regional sediment dispersal (Mertes & Warrick, 2001), although it is important to note that imaged surface plumes do not represent the mass efflux of coastal rivers. The margin-wide pattern of dispersal was first illuminated by Karlin (1980), who examined the provenance of clay minerals in continental slope sediments off Oregon and Washington, and identified a northward transport pathway from the northern California coast. This finding provided the first circumstantial evidence that the shelf environment was highly dispersive, and that a significant fraction of river sediment was bypassed to the continental slope.

The pioneering sedimentological studies of Kulm *et al.* (1975) and Borgeld (1985) showed that bed sediments on the shelves of northern California and Oregon are hydraulically sorted by wave-generated currents to form prominent mid-shelf mud belts. These continuous mud deposits extend from the transition with inner-shelf sands at 50–60 m water depths to the outer shelf. Importantly, they observed storm- and flood-produced layers that could be linked to specific hydrological and oceanic transport phenomena of geological significance. Indeed, ancient analogues of these types of deposits are preserved in late Quaternary mudrocks of the margin (Leithold, 1989). The first seismic-reflection profiles of the northern California shelf revealed thick (10–50 m) accumulations of latest Quaternary strata onlapping tectonically produced topography present at the Last Glacial Maximum (Field *et al.*, 1980).

Seismic stratigraphy indicated that transgressive sedimentation almost completely filled topographic lows to create a morphologically smooth shelf seafloor, a testament to the enormous sediment supply by rivers.

To summarize, early research on the northern California margin showed that large quantities of modern river sediment bypass the continental shelf in association with oceanic storms and river floods, but also that the material trapped accumulates at rates high enough to modify tectonically produced topography created on time-scales of many millennia. With this knowledge in hand, research studies on the Eel margin could pose advanced questions concerning the magnitude and variability of sediment transport and accumulation.

SEDIMENT PRODUCTION AND COASTAL DELIVERY

Tectonics and sediment yield

Tectonics, watershed geology and climate are first-order controls on the production and downstream transport of fluvial sediment to the coastal ocean (Milliman & Syvitski, 1992; Hovius, 1998). On the northern California margin, these factors converge in the extreme to produce an active setting both in terms of tectonics and sediment accumulation. The margin possesses a number of characteristics typical for systems of the greater Pacific Rim, such as: high rates of coastal uplift and erosion; small, mountainous rivers with minor estuaries or deltas; and massive amounts of suspended sediment delivered to the coast during intense rainstorms (Hill *et al.*, this volume, pp. 49–99). The structural grain has been influenced by compression related to convergence of the Gorda and North American plates, and by deformation related to northward migration of the Mendocino Triple Junction (Fig. 7). The triple junction represents the transition from strike-slip deformation associated with the San Andreas Fault system in the south, to subduction-related deformation associated with convergence of the Gorda and North American Plate in the north (Clarke, 1992; Mountain *et al.*, this volume, pp. 381–458).

Tectonics control sediment production on terrestrial and marine portions of the Eel margin by:

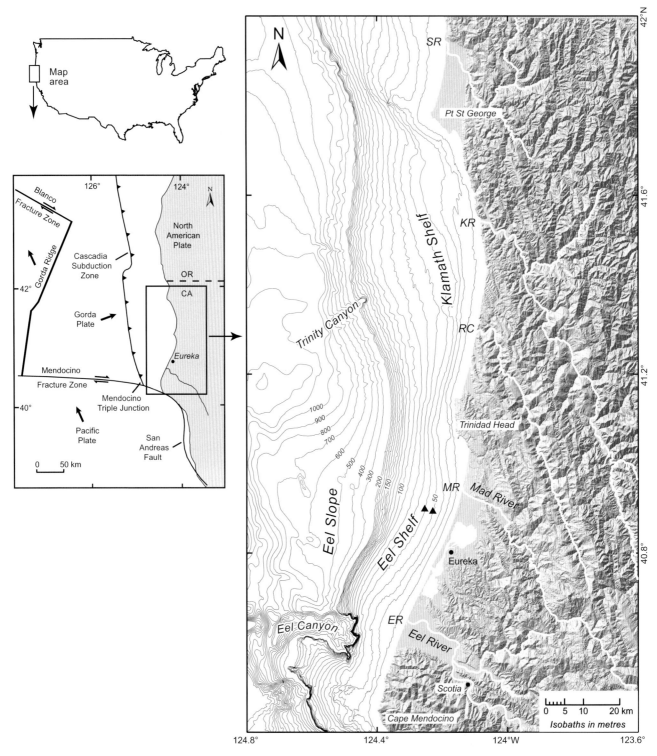

Fig. 7 Location maps for the northern California margin showing geographical features referred to in the text. Rivers noted are Eel River (ER), Mad River (MR), Klamath River (KR), Redwood Creek (RC) and the Smith River (SR). Triangles on Eel shelf mark locations of benthic instrumentation sites S50 and S60. Bathymetric contour interval is 10 m from 20 m to 200 m and 100 m from 200 to ≥ 1000 m. (Plate boundary map is adapted from Clarke, 1992.)

1 orographically enhancing precipitation and runoff from higher mountains;
2 controlling river drainage pattern and channel form;
3 modulating rates of channel incision, sediment storage and production (Merritts & Vincent, 1989; Burger *et al.*, 2001; Snyder *et al.*, 2003).

In the vicinity of the Mendocino Triple Junction, long-term Quaternary uplift rates based on marine terraces and subsided coastal deposits alternate from subsidence to uplift along the coast (Kelsey *et al.*, 1994). Short-term rates based on geodetic data suggest a broad region of uplift along the greater northern California–Oregon coast at rates of ≤ 3 mm yr^{-1} (Mitchell *et al.*, 1994). Coastal uplift and subsidence vary along-coast at rates of 0.24 to 2.5 mm yr^{-1} and -0.24 to -3.3 mm yr^{-1}, respectively (Orange, 1999). The disparity between the long- and short-term rates of tectonic movement is believed to result from tectonic subsidence and rebound in the years following major seismic events (Kelsey *et al.*, 1994; Mitchell *et al.*, 1994).

Partly as consequence of rapid uplift, northern California watersheds exhibit high rates of erosion and river sediment loads. Five rivers between Cape Mendocino and Crescent City (Fig. 7) collectively discharge an estimated 30×10^6 tons of suspended sediment to the coast annually (Table 2). The Eel

and Mad Rivers are particularly high yielding, due to the poorly consolidated and thus easily erodable bedrock geology of the California Coast Ranges (Cenozoic marine sediments), compared with the more stable igneous and metamorphic terrains of the Klamath River Basin (Janda & Nolan, 1979). Earth-flow landslides are a primary mechanism of hillslope-to-channel sediment delivery, and this form of erosion yields a large proportion of the suspended sediment transported in flood waters (Janda & Nolan, 1979). Most of the sediment is discharged between the months of December and March during intense rainfall events generated by Pacific storms. The Eel has a large **basin yield** (i.e. sediment load divided by basin area) which, at 1900 tons km^{-2} yr^{-1}, is roughly 30 times that of the Mississippi River basin. Approximately 25% of the total load is sand, whereas the remainder is nearly equal parts of silt and clay (Brown & Ritter, 1971).

Hydroclimatology and recorded streamflow

On seasonal to decadal time-scales, variation in coastal sediment delivery is controlled by **hydroclimatology**, the influence of climate upon the waters of the land. Establishing a hydroclimatic history for a watershed is useful for understanding

Table 2 Properties of major coastal rivers in northern California

River	Drainage area (km²)	Flow record period	Mean annual flow (m³ s⁻¹)	Mean peak flow (m³ s⁻¹)	Sediment record period	Sediment load (10⁶ t yr⁻¹)	Yield (t km⁻² yr⁻¹)
Smith at Crescent City*	1577	1931–2000	107	1400	ND	0.5	320
Klamath at Orleans (+ Trinity at Hoopa)	29,453	1912–2000	390	3700	1957–1980	10.1	340
Redwood Creek at Orick	720	1953–2000	32	280	1971–1980	1.6	2200
Mad near Arcata	1256	1953–2000	40	420	1958–1974	2.5	1990
Eel at Scotia (+ Van Duzen at Bridgeville)	8638	1911–2000	227	3500	1958–1980	16.5	1900

ND, not determined.
*Smith River sediment load is estimated (see Griggs & Hein, 1979).

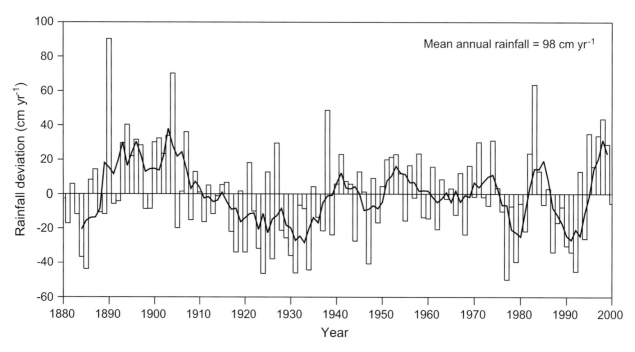

Fig. 8 Rainfall record for Eureka, California (1880–2000), showing the annual deviation from the long-term mean (98 cm yr⁻¹). Bold line is a 5-yr running average. The trend toward wetter and more variable conditions since 1935 mirrors similar patterns observed throughout western North America (see text). (Data are from the National Weather Service.)

sediment flux variations in dispersal systems, provided suitable instrument or proxy records of precipitation or streamflow exist. As is generally the case, hydrological records for the Eel River watershed do not extend back beyond the middle 1800s, and earlier accounts of weather patterns are limited to the most extreme events (Engstrom, 1996). On the basis of a 100-yr-long precipitation record for Eureka, California, the period 1885–1915 was wetter than average, 1915–1935 was dryer than average, and variable rainfall prevailed between 1935 and 2000 (Fig. 8). This precipitation trend is manifest in streamflow throughout northern California (Cayan & Peterson, 1989) and mirrors streamflow trends for the greater USA (McCabe & Wolcock, 2002). According to reconstructions based on correlations of rainfall and tree-ring records for western North America, the last half of the 19th century and first half of the 20th century were respectively among the wettest and driest periods of the last millennium (Earle, 1993; Meko et al., 2001). Interestingly, several reports have suggested that the frequency of extreme rainfall events and floods has increased during the 20th century in North America (Kunkel et al., 1999) and

worldwide (Milly et al., 2002), a trend with major implications to coastal sedimentation.

A significant climatic phenomenon relevant to streamflow and sediment discharge in some dispersal systems is ENSO, which has two recurrent phases, **El Niño** and **La Niña** (Cayan et al., 1999). During a strong El Niño, triggered by a weakening of easterly winds in the equatorial Pacific and the arrival of warm ocean water off western South America, winter storms off North America tend to track farther south, with heavy and persistent rainfall in southern California. In contrast, El Niño usually brings relatively dry conditions to regions northward (Oregon and Washington). During La Niña, major winter storms track through these regions, producing cool, dry weather in southern California, and wet, stormy weather in the Pacific North-west. In association with La Niña, atmospheric blocking of westerly storm tracks has produced some of the most intense flood-producing rainstorms on record in western North America (Hirschboeck, 1987; Ely et al., 1994), including the 1955 and 1964 megafloods described later. The northern California coast happens to be located midway between the regions where ENSO imparts characteristic rainfall and stream-flow signals (Schonher & Nicholson, 1989; Cayan *et al.*, 1999). For instance, the wettest (1983) and driest (1977) years on record for the Eel River watershed were both El Niño years. Accordingly, a characteristic ENSO signal is not apparent in sediment discharge records for coastal rivers of northern California and Oregon (Wheatcroft & Sommerfield, 2005), unlike southern California (Inman & Jenkins, 1999).

Floods, land use and sediment delivery

Flood magnitude and frequency is by far the most important control on the flux of terrigenous sedi-

ment to the northern California shelf. Suspended-sediment discharge scales exponentially with water discharge (Syvitski *et al.*, 2000), so interannual variation in suspended-sediment load correlates more closely with river peak flow than the annual mean flow (Fig. 9). Major floods ($\geq$ 10-yr recurrence interval) have a disproportionate influence on suspended-sediment transport as these events activate stream-bank landsides, an important sediment source in northern California rivers (Kelsey, 1980). Flood-generated landslides charge channels with sediment, and increase suspended-sediment concentration per unit water discharge for years or even decades thereafter. This lingering effect

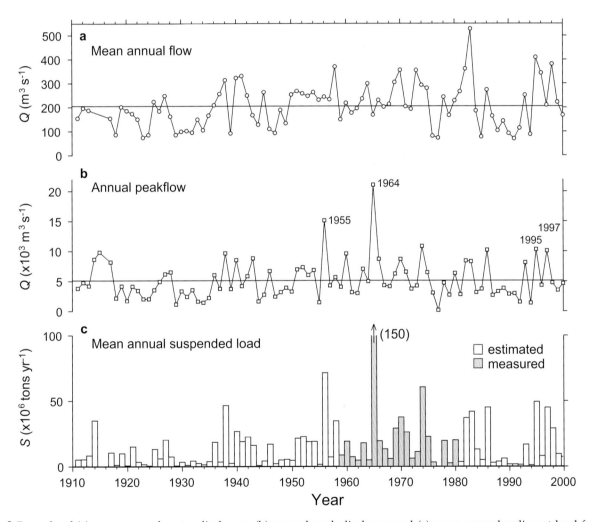

Fig. 9 Records of (a) mean annual water discharge, (b) annual peak discharge and (c) mean annual sediment load for Eel River at Scotia, 1911–2000. Note the close correspondence of peak discharge and annual sediment load. Major floods mentioned in the text are noted. Sediment loads were computed from daily suspended-sediment-concentration data for 1959–1980 (measured), and a sediment rating curve relating concentration and daily water discharge (estimated). (Data are from US Geological Survey.)

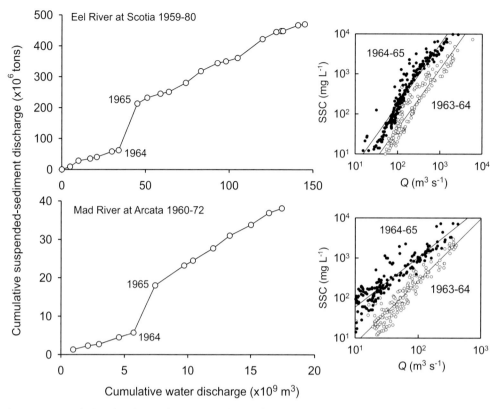

Fig. 10 Cumulative suspended-sediment and water discharge for the Eel and Mad Rivers (US Geological Survey data). Sediment discharge is the product of daily mean water discharge (Q) and daily mean suspended-sediment concentration (SSC). The marked increase in cumulative discharge triggered by the 1964 flood was caused by widespread landsliding into the channel, with increased suspended-sediment concentration (per unit Q) during subsequent years. (From Sommerfield *et al.*, 2002.)

occurred in the wake of catastrophic floods in 1955 and 1964 (Fig. 10), which, by many accounts, were the most significant hydrological events recorded in western North America (Hofmann & Rantz, 1963; Waananen *et al.*, 1971; Sloan *et al.*, 2001). There is geological evidence to suggest that major floods have been an important geomorphological agent in north coastal California during the latest Holocene. For example, Helley & LaMarche (1973) identified five palaeoflood events (approximately 1600, 1750, 1862, 1881, 1890) that they ranked in magnitude equal to or greater than the 1964 flood. The 1955 and 1964 megafloods occurred during a climatic period that was wetter than average and more flood-prone (Ely *et al.*, 1993, 1994), but also followed several decades of widespread deforestation in the watershed. Changing land cover is presumed to have intensified the geomorphological impacts of these events, as discussed below.

Ranching, farming and logging have been important industries in north coastal California since the mid-1800s, but the impacts of these changes were tempered by low population density. However, starting in the 1940s, and continuing through the 1960s, significant changes in land-cover took place as many of the watersheds were heavily logged (Best, 1995). Studies of the Eel River watershed suggest that the excess sediment yields (above the predicted natural yield) of some tributary basins increased markedly in the 1950s, ~20% to > 60% of which has been attributed to timber harvesting practices (U.S. Environmental Protection Agency, 1999a,b). The average excess yield for the South Fork Eel and the Van Duzen River was roughly 33%, and the average for nine other coastal basins was 48%. Redwood Creek basin, the most heavily logged and intensively studied, was found to have the largest (> 60%) land-use sediment component (Nolan &

Janda, 1995). In sum, roughly half of post-1950 river sediment loads in north coastal California rivers may be human-produced. Logging practices are further implicated in the initiation or reactivation of landslides, which in many high-relief basins are the single most important source of sediment (Montgomery *et al.*, 2000). There is no doubt that most of the landslides are generated by extreme rainstorms, but there is evidence that land-cover change associated with logging tends to amplify the geomorphological effects of storms (Kelsey, 1980; Marron *et al.*, 1995). Separating climatic and human factors in sediment delivery to the coast is made difficult by lack of sediment-discharge records before 1950, although the shelf sedimentary record provides several clues regarding the timing and scale of change. These dual influences in sediment delivery are considered in the context of shelf sedimentation later in this paper.

Coastal ocean circulation

The regional physical oceanography of the northern California shelf and slope is conducive to widespread dispersal of fine-grained river sediment. The major oceanic circulation north of Cape Mendocino consists of the southward-flowing California Current, and the northward-flowing Davidson Current (Fig. 11). The perennial California Current flows between the shelf break and ~1000 km seaward of the coast at mean speeds of about 10 cm s^{-1} (Hickey, 1979, 1998). The Davidson Current flows in autumn and winter at speeds of 5–10 cm s^{-1} over the continental shelf (Strub *et al.*, 1987). The annual mean density field is characterized by isopycnals that tilt upward toward the coast, indicating upwelling and a corresponding southward mean flow (Bray & Greengrove, 1993). Consistent with this overall picture, measurements by a benthic tripod and mooring on the Eel shelf and slope during 1995–2000 showed annual mean currents to be small, southward and seaward, and temporal variations in the current exhibited fluctuations in the tidal, wind and very-low-frequency bands (Ogston *et al.*, 2004). Winter downwelling in the bottom boundary layer caused persistence of seaward flow with important implications for sediment transport across-shelf (Ogston *et al.*, 2004).

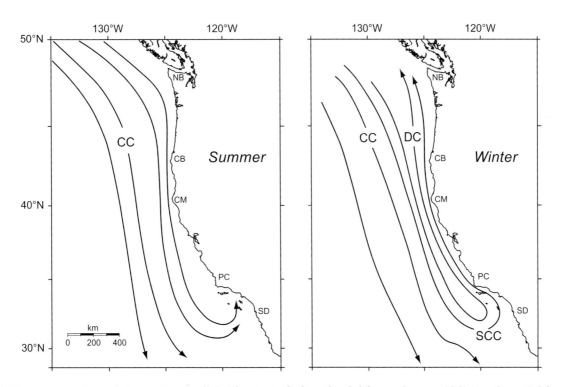

Fig. 11 Mean seasonal circulation patterns off California, including the California Current (CC), Southern California Current (SCC) and Davidson Current (DC). Shoreline references include Neah Bay (NB), Cape Blanco (CB), Cape Mendocino (CM), Point Conception (PC) and San Diego (SD). (Adapted from Hickey, 1998.)

On a regional scale, the shelf flow field is influenced by topographic effects, namely, flow deflection and separation at Cape Mendocino and Trinidad Head (Pullen & Allen, 2000).

In terms of suspended-sediment flux, it cannot be overstated that the continental shelf bottom boundary layer is the most critical zone of the water column. The bottom boundary layer forms by consequence of frictional drag and turbulent diffusion, and is composed of a mean-current boundary layer and thinner, transient wave boundary layer (Grant & Madsen, 1986). On the northern California shelf, the mean-current bottom boundary layer extends ~5–15 m above the bottom (Lentz & Trowbridge, 1991), decreasing in thickness on the innermost shelf where it merges with the surface, wind-mixed boundary layer. The oscillatory wave boundary layer is much thinner at ~10 cm (Cacchione & Drake, 1990). The significance of the bottom boundary layer in sediment dispersal is that it provides a site for across-isobath transport in conjunction with along-isobath mean flows. Cross-shelf transport within the bottom boundary layer is facilitated by frictional Ekman turning, typically ~10–20° to the left of the mean flow (in the Northern Hemisphere), and the amount of turning increases as the bottom boundary layer thickness decreases (Lentz & Trowbridge, 1991). Consequently, northward, along-isobath mean flow is most favourable for transporting suspended matter across-shelf (and perhaps off-shelf).

It is important to point out that the magnitude and direction of currents over the shelf and slope are highly variable seasonally, and that storm-synoptic patterns generally deviate from annual means (see review of Hickey (1998) and references therein). The majority of sediment discharge by the coastal rivers takes place during storm-generated rainstorm events in winter, so the wind-driven shelf flow during these times has a disproportionate impact on the initial fate of suspended matter. For example, mean annual winds over the Eel shelf are northerly, which force southward surface currents, yet southerly winds are statistically most prevalent during times of river peak flow and floods (Fig. 12a). During these times, wind-driven northward flow of shelf surface water prevails on the shelf and slope of the northern California margin (Fig. 12b), transporting river sediment far northward from points of discharge. Consequently, the coincidence of southerly storm winds and elevated river discharge tends to produce large sediment fluxes to the north (at least initially) during flood events (Hill *et al.*, this volume, pp. 49–99). It is not uncommon for the wind direction to switch to northerly as storm cells pass over land, forcing southward shelf flow and sediment transport (Cacchione *et al.*, 1999). In addition to wind forcing, the presence (or absence) of buoyant river plumes during storms contributes to along-shelf variability in current magnitude (Pullen & Allen, 2000).

Storms contribute to highly energetic swell and sea waves on the northern California coastline, typically in the months of November through to March, and wave-generated orbital flow plays an important role in reworking the shelf seabed (significantly more so than tidal currents) in ≤ 50–60 m water depths (Wheatcroft *et al.*, this volume, pp. 101–155). On the Eel shelf, mean annual significant wave height is ~2.4 m with a dominant period of 10.9 s, but annual peak wave heights are significantly larger at 6–8 m (Wiberg, 2000). Under the latter conditions, wave-orbital currents in water depth up to ~110 m exceed the threshold for sediment resuspension (i.e. 10–15 cm s^{-1}; Wiberg *et al.*, 2000). Tides on the northern California shelf are characterized as mixed semi-diurnal with a mean tidal and maximum diurnal range of 1.4 m and 2.0 m, respectively (Emery & Aubrey, 1986). Tidal currents can cause variations in the along-shelf velocity of up to 50 cm s^{-1}, and the principal axis of flow is generally oriented along shelf, with the minor axis being very small near the coast, increasing offshore (Lentz & Trowbridge, 1991; Geyer *et al.*, 2000). On the middle shelf, along-shelf tidal variance is nearly twice that of the across-shelf component (Sherwood *et al.*, 1994; Cacchione *et al.*, 1999).

SEDIMENT TRANSPORT AND ACCUMULATION

In the coastal ocean, both dynamic and static factors influence the dispersal and sequestration of fluvial sediment. A useful concept that will be used to organize the following discussion is **trapping efficiency**, the quantity of sediment accumulated in a basin divided by the fluvial influx. **Dynamic trapping** is related to the properties of the water column and suspended particles, for example,

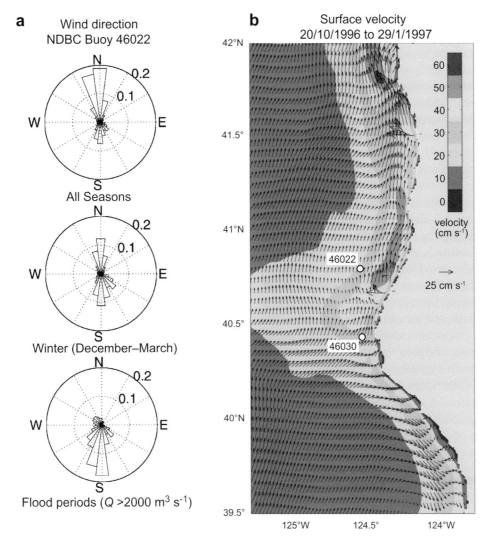

Fig. 12 (a) Wind-rose diagrams based on data from National Data Buoy Center Buoy 46022 for 1981–1990 (0.1 = 10% frequency of occurrence). Wind roses denote direction from which wind blew. Wind speed and direction averaged annually are variable but predominately from the north. Winds averaged over the winter season are more variable, whereas strong winds from the south are most frequent during times of river floods (i.e. wet storms). (b) Southerly storm winds force strong northward surface currents over the northern California shelf and slope as shown in this numerical simulation from Pullen & Allen (2000). In this example, northward flow is enhanced by buoyant river outflows. (Reprinted with permission from Elsevier.)

density stratification, hydrodynamics and particle flocculation, and generally takes place on short time-scales (minutes to days). **Static trapping** is caused by topographic and morphological properties of the receiving basin, and functions on a much wider range of temporal scales (days to millennia). Coastline orientation, capes, canyons and banks are perhaps the most evident topographic features that influence coastal flows (Trowbridge *et al.*, 1998) and sediment suspensions by extension.

The long-term stratigraphic consequences of static trapping are most conspicuous at larger geological scales, for example, sediment ponding within margin structural troughs (Bohannon *et al.*, 2004). Tectonics can be considered a 'dynamic' process on margins in which rates of crustal movement equal or exceed that of sediment accumulation, but the initial deposition of sediment is only indirectly influenced by tectonic morphology. Dynamic and static trapping mechanisms are interrelated at some level, but it is instructive to consider them separately in order to highlight specific process–product relationships.

Mechanisms of sediment transport

A variety of mechanisms that transport and resuspend sediment on the northern California margin have been documented directly by observations. In general, shelf sediment is resuspended from the bed due to energetic wave processes and is transported by the currents in the water column, and those currents have a distinct frequency structure that may or may not be correlated with the resuspension of sediment (Ogston *et al.*, 2004). During winter storm events, close association between strong winds, energetic wave events, seaward mean flow, along with a large supply of fine-grained river sediment all contribute to significant sediment fluxes across-shelf. Along-shelf dispersal of sediment is more variable in time and space as winds reverse with passing storm systems, and as storm waves and currents interact with prevailing circulation and local morphology. In addition, the quantities of sediment transported across and along the shelf vary depending on the separate impacts from fluvial supply and local erosion. Below, two end-member storm types are described to illustrate the variable influences of river discharge and wave-driven resuspension on sediment transport:

1 dry storms, characterized by large waves with minimal continental runoff;
2 wet storms, large waves coincident with peak river flow and sediment discharge.

In terms of suspended sediment, dry and wet storms are low- and high-concentration regimes, respectively, that create characteristic sedimentary event deposits (Fan *et al.*, 2004).

The sedimentary response of a dry-storm event was observed by benthic instrumentation deployed on the Eel shelf (60-m water depth) in September 1995. In response to surface-gravity waves, wave-orbital velocity felt at the seabed started to rise at day 283 in the record, and caused increased suspended-sediment concentration (Fig. 13a & b). The sediment concentration at both 30 and 100 cm above bottom (cmab) was highly correlated with the wave-orbital velocity, and this signal was modu-

lated by the relatively strong tidal currents. As the wave-orbital velocity returned to low levels (below the threshold for bed erosion), the suspended-sediment concentration also returned to a low background level (a smaller storm event followed on day 286.5). The sediment flux in response to the first storm shows the pattern of a typical North Pacific storm. As the storm approaches the coastline, strong southerly winds force northward flows, as can be seen in days 283.4–284.4 of the low-pass filtered sediment-flux record (Fig. 13d), and downwelling. As the storm passes, winds weaken and switch direction, such that southward geostrophic flows return and force a southward sediment flux, as shown in the low-pass filtered record for days 284.4–285.7 (Fig. 13d). The *net* flux of sediment in response to this event is to the south and seaward. The exact location of flow and sediment transport depends on whether the storm centre passes directly over the study area, and not all storm events exhibit these characteristics (Ogston *et al.*, 2004). As detailed below, dry storms rework the seabed on the inner shelf and produce sand-rich event beds near the inner-middle shelf transition.

Suspended-sediment flux due to wave resuspension and transport by currents can be enhanced by advected sediment input from fluvial sources during wet storms. As an example, in December 1995 a large wet storm passed across the Eel margin and was recorded by benthic instrumentation at a 50-m site (Fig. 14). During this event, significant wave height reached 9.5 m at a dominant period of 17 s, and Eel River discharge reached 4000 m^3 s^{-1} (i.e. 2–3 yr recurrence interval for peak flow). Wave-orbital velocity of 60 cm s^{-1} and mean current speeds > 40 cm s^{-1} occurred, and resulting bed stresses yielded suspended-sediment concentrations approaching 0.70 g L^{-1} at 100 cmab (Fig. 14a–c). The highest concentrations occurred after the peak in bottom stress, suggesting an advective component. Indeed, correlation was strong between the concentration record and Eel River discharge (Fig. 14f), which suggests that river sediment is advected quickly to the middle shelf, dominating the turbidity signal at times. In response to the local currents, the sediment flux was initially directed to the north and seaward (days 345.4–346.8), then turned to the south and landward with the passage of the storm (days 346.8–348.5 in Fig. 14d & e). Advection of sediment from the inner to middle shelf can occur not only

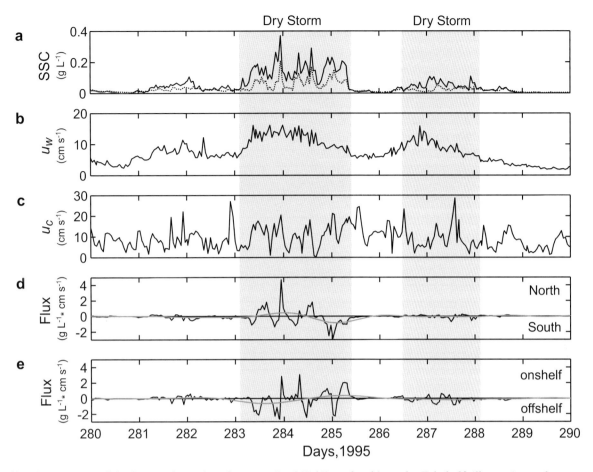

Fig. 13 Observations of the bottom boundary layer at site S60 (60-m depth) on the Eel shelf, illustrating a dry storm. See Fig. 7 for instrument location. River discharge during this period was low (~6 m³ s⁻¹). Hourly averaged data consist of: (a) suspended-sediment concentration at 30 cm (solid line) and 100 cm (dotted line) above bottom; (b) mean wave-orbital velocity; (c) current speed; (d) along- and (e) across-shelf sediment-flux components, hourly average (black line) and 38-h low-pass filtered (grey line). (Adapted from Ogston *et al.*, 2000.)

due to suspended-sediment transport, but also as gravity-driven **fluid-mud** flows (Traykovski *et al.*, 2000; Wright *et al.*, 2001; Hill *et al.*, this volume, pp. 49–99; Parsons *et al.*, this volume, pp. 275–337). Fluid muds are relatively dense, near-bed suspensions of fine-grained sediment and salt water, with concentrations that are in excess of 10 g L⁻¹ (i.e. sediment bulk density of 1.01 g cm⁻³). Regardless of the actual transport mechanism, the wet storms produce characteristic beds of clayey silt on the Eel shelf and in Eel canyon.

Nepheloid layers are low-concentration lithogenic and biogenic particle suspensions, typically a few to tens of milligrams per litre, found over continental shelves and slopes (McCave, 1986). Elevated concentrations observed at various depths

in the water column form surface (SNL), intermediate (INL) and bottom (BNL) nepheloid layers. The SNLs are prevalent in areas affected by river sediment plumes and primary production (Hill *et al.*, this volume, pp. 49–99), whereas BNLs are widespread over both continental shelves and slopes; they can be composed of material from multiple sources such as SNL fallout and *in situ* resuspended material. The INLs are important in the dispersal of sediment from shallow shelf environments to deeper open-slope and canyon environments (Walsh & Nittrouer, 1999; McPhee-Shaw *et al.*, 2004). In particular, INLs that emanate from the shelf break and extend seaward along isopycnal surfaces are important shelf-to-slope conduits; INLs are most common over continental

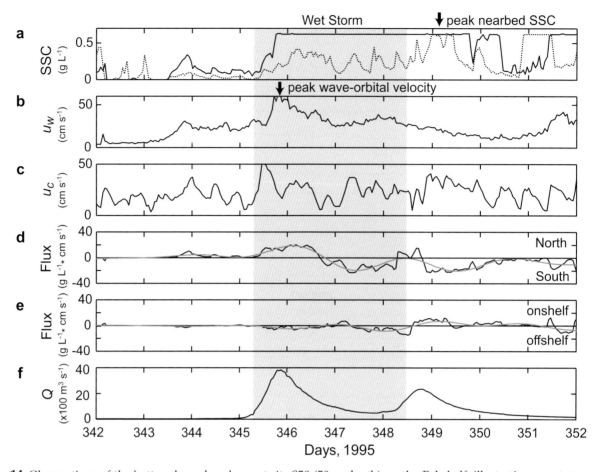

Fig. 14 Observations of the bottom boundary layer at site S50 (50-m depth) on the Eel shelf, illustrating a wet storm. Hourly averaged data consist of: (a) suspended-sediment concentration at 30 cmab (solid line) and 100 cmab (dotted line); (b) mean wave-orbital velocity; (c) current speed; (d) along-shelf and (e) across-shelf sediment-flux components, hourly averaged (black line) and 38-h low-pass filtered (grey line); (f) Eel River discharge. Note that the suspended-sediment concentration (SSC) sensor became saturated at a sediment concentration of ~0.7 g L^{-1}. (Adapted from Cacchione *et al.*, 1999.)

slopes where material is supplied by horizontal flux from shelf BNLs.

Intermediate nepheloid layers have been consistently observed on the Eel margin in open-slope and canyon environments, both during winter periods with elevated fluvial sediment discharge and energetic wave conditions, and summer periods with no fluvial sediment discharge and quiescent conditions (Fig. 15). The INL sedimentation observed on the Eel slope is characteristic for the present sea-level conditions in which the fluvial sediment source is relatively distant from the shelf edge. However, the steady-state depositional flux of INL sediment (~4.5 g m^{-2} day^{-1} at 400–700 m depths) on the Eel slope (Walsh & Nittrouer, 1999) is typically higher than on sediment-starved slopes of wider continental margins (Biscaye & Anderson, 1994). The morphology of the Eel shelf and slope (narrow, steep) and limited delta-plain storage of fluvial sediment favour shelf bypassing and export to the slope and deeper waters. Hemipelagic sedimentation is particularly enhanced within the Eel Canyon, as demonstrated by consistently higher suspended-sediment concentrations of INLs, compared with the adjacent open-slope areas (Fig. 15). Based on water-column and seabed data, high-concentration INLs (up to 15 mg L^{-1}) may be important in the rapid dispersal of fluvially derived sediment to the canyon seabed (McPhee-Shaw *et al.*, 2004).

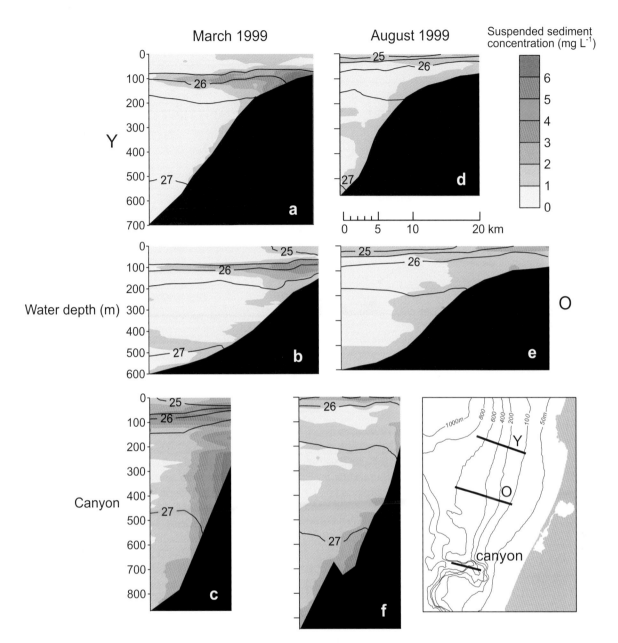

Fig. 15 Hydrographic sections of suspended-sediment concentration (coloured) and density (σ_t) on the Eel margin. Shown are data for March (a–c) and August (d–f) of 1999. Sediment concentration is generally higher in the canyon compared with the open slope during both seasons. Shelfbreak intermediate nepheloid layers (INLs) are strongest during winter due to resuspension of shelf sediment. In contrast, slope INLs are strongest in summer as a consequence of increased density stratification and bed resuspension by shoaling internal waves.

Internal waves and tides are energetic, ubiquitous and persistent oceanic phenomena (Pratson *et al.*, this volume, pp. 339–380). The interactions of these waves with bottom topography generate strong bottom flows that inhibit deposition and resuspend bottom sediment, particularly near the shelf-slope break (Cacchione *et al.*, 2002). Internal-

wave dynamics, with waves breaking at critical angles on the slope and subsequent ejection of BNL particles along isopycnal surfaces (McPhee-Shaw & Kunze, 2002), have been shown to produce INLs, providing a mechanism for the transfer of sediment from the continental shelf to the deep ocean. Internal-wave energy has been proposed

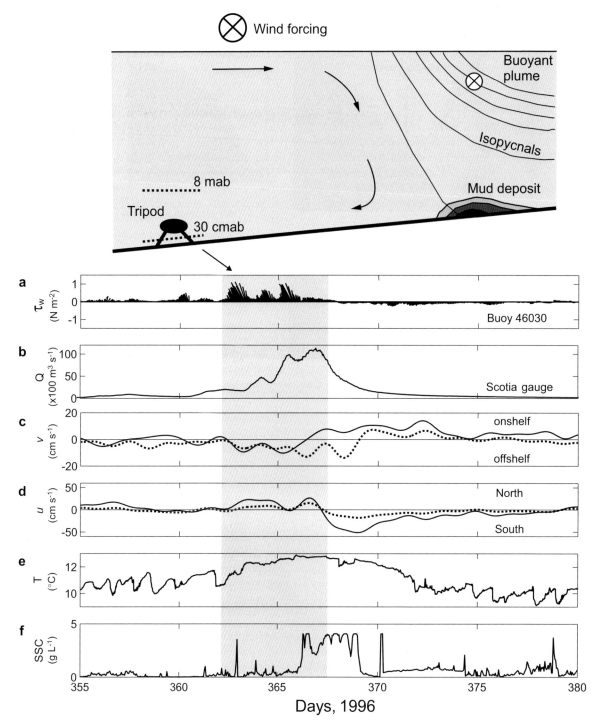

Fig. 16 Schematic of sediment trapping by a buoyant freshwater plume and downwelling-favourable winds. Benthic observations at site S60 (see Fig. 13) north of the Eel River mouth during the 1997 flood are consistent with this interpretation: (a) wind vectors (toward the north-west, see Fig. 12 for buoy location); (b) Eel River discharge; (c) low-frequency across-shelf currents (solid line, 8 mab; dashed line, 30 cmab); (d) low-frequency along-shelf currents (solid line, 8 mab; dashed line, 30 cmab); (e) temperature at 10 cmab; (f) suspended-sediment concentration at 30 cmab.

as a mechanism controlling the slope of the continental slope by limiting deposition, and perhaps eroding the slope when the critical angle for internal waves is approached (Pratson *et al.*, this volume, pp. 339–380). Upper-slope INLs (200–500 m) on the Eel margin are relatively turbid during the summer (2–3 mg L^{-1}; Fig. 15), and are maintained by internal-wave resuspension over the slope. Observations of locally coarse grain size and low sediment accumulation rates on the upper Eel slope also support internal-wave resuspension as a mechanism of redistributing fine-grained sediment to deeper water (Alexander & Simoneau, 1999).

Dynamic trapping mechanisms

A good example of dynamic trapping is the estuarine turbidity maximum, in which suspended matter derived from distal sources is concentrated by particle flocculation, convergent circulation and stratification effects (Dyer, 1995). Similarly, sediment trapping within the continental-shelf bottom boundary layer congregates large quantities of particles that form cohesive strata upon deposition (Wheatcroft *et al.*, this volume, pp. 101–155). These particles may settle into the wave boundary layer from a surface plume, a converging flow or the overlying mean-current boundary layer. High wave shear stresses and, potentially, hindered settling are effective mechanisms for trapping sediment within the wave boundary layer. On shelves where wave action is the primary agent of resuspension, across-shelf flux convergence may be related to the gradient in wave energy felt on the seabed. For instance, results from an across-shelf array of tripods on the Eel shelf indicate that mud accumulation is favoured by across-shelf sediment-flux convergence between the 60-m and 70-m isobaths (Wright *et al.*, 1999). If these results are representative for the shelf in general, they imply that the cross-shelf gradient in resuspension by waves may dictate the depth of fine-sediment deposition as suggested by Harris & Wiberg (2002).

In the eastern North Pacific, oceanic fronts develop due to upwelling and downwelling along the coastline, and in association with buoyant river plumes on the shelf. A conceptual model for trapping of suspended sediment behind a coast-parallel front associated with downwelling (Fig. 16) has

been proposed for the Eel shelf by Ogston *et al.* (2000). Rapid input of sediment into the bottom boundary layer beneath the surface plume coupled with a weak density front associated with the Eel River discharge plume contribute to this process (Pullen & Allen, 2000). Dynamic trapping and deposition of suspended sediment on the inner shelf would be likely to occur until the weak front dissipated or the density of the thin bottom layer increased sufficiently to induce gravity-driven transport across-shelf.

Large-scale circulation phenomena can also provide along-shelf sediment flux convergence. If fronts and eddies are persistent features during winter-storm or resuspension periods, they can influence sediment accumulation patterns and rates. On the Eel shelf, an along-shelf flow convergence between two sites at 60-m water depth and approximately 8 km apart (bracketing the mid-shelf mud depocentre) was observed on average throughout the winter of 1996–1997 (Ogston & Sternberg, 1999; Wright *et al.*, 1999), and was similarly modelled by Pullen & Allen (2000) for that same deployment period. The pattern of convergence and divergence on the Eel shelf may also be caused by eddy-like circulation on the shelf, which has been observed in satellite images (Washburn *et al.*, 1993; Walsh & Nittrouer, 1999). Pullen & Allen (2000, 2001) attributed this eddy to flow separation at Cape Mendocino, enhanced by river input from the Eel River. In a model of shelf flow (averaged over 100 days of winter 1996–1997), they found the anticyclonic eddy centred on the Eel shelf to be persistent – the residual circulation retains the eddy feature created during individual storms (Fig. 17).

Static trapping mechanisms

Static trapping of river mud on the northern California margin is associated with both subtle and pronounced topographic elements that bear upon the trajectories of particle-laden flows. These morphologies can range in scale from nearshore scour depressions to canyon re-entrants (Cacchione & Drake, 1990). Although larger-scale morphologies are ultimately a consequence of tectonic movements on geological time-scales, on dynamical scales they influence sediment transport pathways and the loci of depocentres. Indeed, the same tectonic

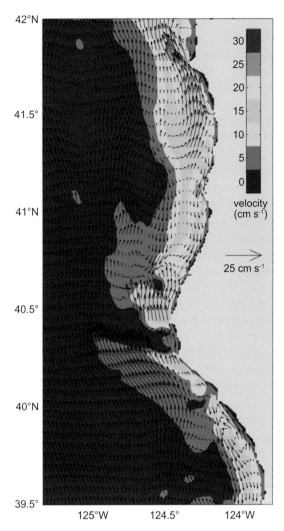

Fig. 17 Numerical simulation of depth-averaged current velocity during wet-storm conditions between 20 October 1996 and 29 January 1997 on the northern California shelf. Convergent flow is predicted over the Eel mid-shelf with an eddy-like feature present. Note that the convergence is spatially coherent with the 1995 and 1997 flood depocentre (see Fig. 21). (After Pullen & Allen, 2000.)

regime that controls sediment production on land creates bathymetric 'traps' for sediments in the coastal ocean.

The modern Eel shelf is segmented by NW–SE trending anticlines and synclines that have had a major influence on shelf accumulation patterns since the Last Glacial Maximum (Mountain *et al.*, this volume, pp. 381–458). At lowstand, with sea level ~120 m below present, the Eel shelf was bifurcated by an anticline that created a drainage divide between the modern Eel and Mad rivers (Fig. 18). Transgressive sedimentation after ~15 ka was greatly impacted by the antecedent topography, with thick successions emplaced in structural lows, and condensed intervals forming atop structural highs. As mentioned previously, subsidence and uplift rates along the shelf are in the order of millimetres per year (averaged over several millennia), approximating long-term sediment accumulation rates (1–6 mm yr^{-1}) determined by ^{14}C geochronology. The thickest section (50–60 m) is centred between 50-m and 100-m water depths ~25 km north of the Eel River mouth, and overlies the axes of anticlinal and synclinal folds. Not surprisingly, ^{14}C-based sediment accumulation rates are high (6 mm yr^{-1}) where the Holocene sediment package is thickest (Fig. 18), and low (< 1 mm yr^{-1}) at the shelf edge where relict, late Pleistocene strata are exposed. This parallel between structure and sediment thicknesses suggests that antecedent topography exerted a major influence on sediment dispersal and accumulation patterns during the Holocene (Orange, 1999; Burger *et al.*, 2002; Spinelli & Field, 2003).

Eel Canyon is both a depocentre and conduit for sediment derived from the Eel River (Mullenbach & Nittrouer, 2000; Puig *et al.*, 2003, 2004). The canyon incises the shelf to within ~10 km of the Eel River mouth (Fig. 19), with three major re-entrants believed to have been produced by multiple episodes of incision during the past ~500 kyr (Burger *et al.*, 2001). For a normal winter season in 2000, several down-canyon gravity flows were recorded during a period of peak riverflow and wave activity – evidence of storm-resuspension events on the shelf and/or wave-induced liquefaction at the canyon head (Puig *et al.*, 2004; Hill *et al.*, this volume, pp. 49–99). Subsequent coring revealed new sedimentary deposits within the upper thalwegs (Fig. 19). Larger-scale transport events, such as slope failures, appear to be active within the canyon on decadal time-scales (Mullenbach & Nittrouer, 2006). Although sediment movement through submarine canyons is generally considered most important during sea-level lowstands, these findings mirror observations on other active margins showing that canyon incisions indeed convey modern river sediment to the deep sea (Kudrass *et al.*, 1998; Kineke *et al.*, 2000; Johnson *et al.*, 2001).

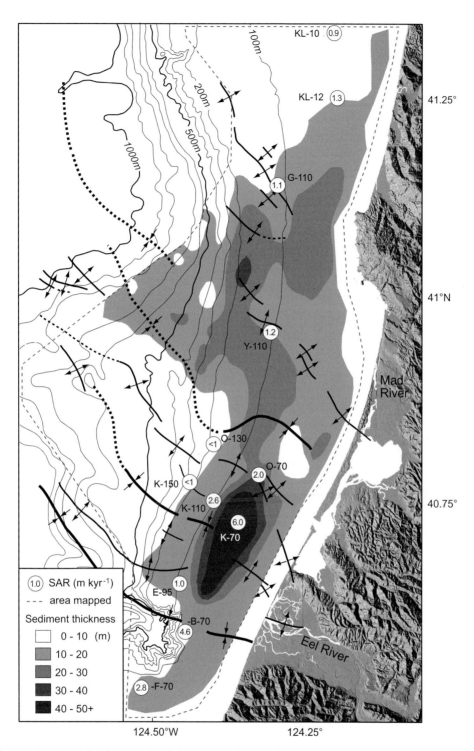

Fig. 18 Isopach map for the Eel–Klamath shelf showing sediment thickness between the 18 ka lowstand surface of erosion and the modern seafloor (from Spinelli & Field, 2003). Anticline (diverging arrows) and syncline (converging arrows) locations as mapped by Clarke (1992) are shown. Also plotted are locations of ^{14}C-dated piston cores with sediment accumulation rates from Sommerfield & Wheatcroft (in press). The highest accumulation rates correspond to the region of thickest Holocene sediment, which overlies a syncline.

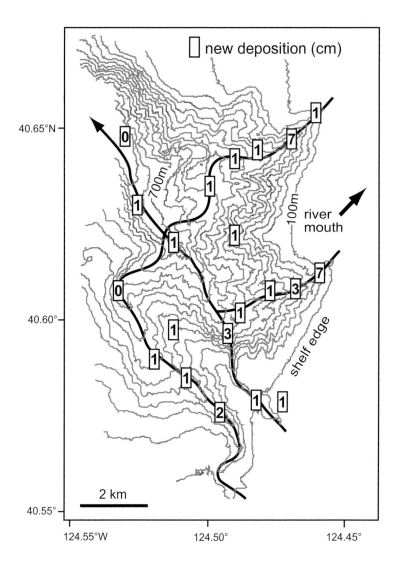

Fig. 19 Map of Eel Canyon head showing the thickness of sediment layers deposited during a 3-month period in 2000. Sediment is preferentially trapped within thalwegs of the two river-proximal re-entrants. (Adapted from Mullenbach *et al.*, 2004.)

SEDIMENTATION PATTERNS AND RATES

Comparative sedimentology and stratigraphy convey a wealth of information on time-averaged sediment supply and dispersal conditions on continental margins. Depth-associated sediment types, the underpinning of facies models, allow process–product relationships to be gleaned from sediment cores and outcrops (Stow & Piper, 1983; Swift & Thorne, 1991). On the other hand, establishing links between sediment transport-trapping and strata formation requires knowing the actual rates of processes. Although not always feasible, the most effective approach to this end is coordinated oceanographic and geological observations, as described below.

Sedimentary event deposition

Major wet storms hit the northern California shelf in early January 1995, and again in January 1997, providing a rare opportunity to characterize the land-to-ocean sediment flux and sedimentology of resulting event beds. Respectively, the 1995 and 1997 floods of the Eel River delivered an estimated 24×10^6 and 29×10^6 tons of suspended sediment to the coast, some of which was trapped on the shelf to form a widespread deposit (Wheatcroft *et al.*, 1997; Wheatcroft & Borgeld, 2000). Small-scale properties of flood-produced layers that helped distinguish new deposition from ambient sediment included:

1 high activities of ⁷Be derived from storm rainfall and continental runoff;

2 anomalously low ²¹⁰Pb activity due to limited particulate scavenging during rapid deposition;

3 elevated (> 30% weight) clay content, an indicator of rapidly deposited river sediment or fluid mud (Fig. 20a).

Flood-produced beds were massive to interlaminated with silt, evidence of multiple depositional pulses, and generally had a sharp basal contact with subjacent strata. Similar signatures were observed in Eel Canyon during studies of seasonal deposition in 1998–2000 (Fig. 20b). The footprint of flood deposition in 1995 and 1997 was broadly similar: ~50-km along-shelf extent with an elliptical centre

of mass situated between the 50-m and 100-m isobaths (Fig. 21; Hill *et al.*, this volume, pp. 49–99). The flood depocentre (10–12 cm layer thickness) was situated ~25 km north-west of the river mouth, consistent with the northward wind-driven shelf flow characteristic of wet storms.

The quantity of fluvial sediment trapped on the shelf in association with the 1995 and 1997 floods was estimated by mapping the spatial extent of the deposit through rapid-response coring, and measurements of porosity to convert sediment volume to mass (Wheatcroft *et al.*, 1997; Wheatcroft & Borgeld, 2000). Roughly 20–30% of the 1995-flood suspended load (as measured at the Scotia gauge) was trapped on the shelf, and the presence of ⁷Be in upper slope sediments revealed that much

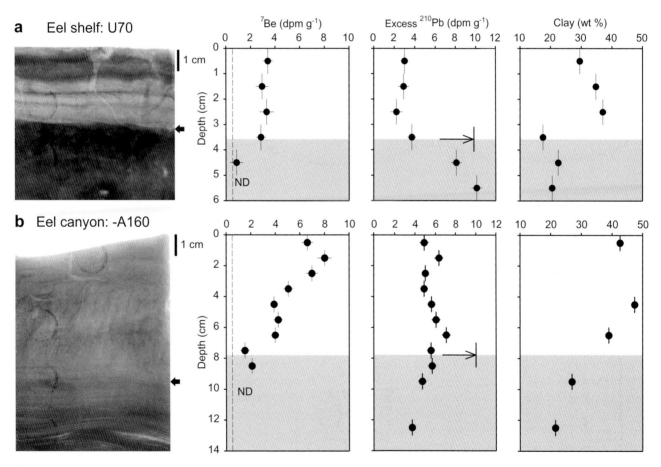

Fig. 20 (a) Example of flood-produced layer on the Eel shelf observed several weeks after the February 1995 flood (arrow marks base). (b) Similar signatures were observed in the upper Eel Canyon after a period of wet and dry storms in 1998. Evidence of rapid land-to-ocean sediment dispersal includes presence of ⁷Be in the flood layer, low excess ²¹⁰Pb activity (arrows denote typical surface activity) and high clay content. As shown, internal silt laminae and some evidence of bioturbation were characteristic sedimentary structures. See Fig. 21 for core locations. (Redrawn from Sommerfield *et al.* (1999) and Mullenbach & Nittrouer (2000).)

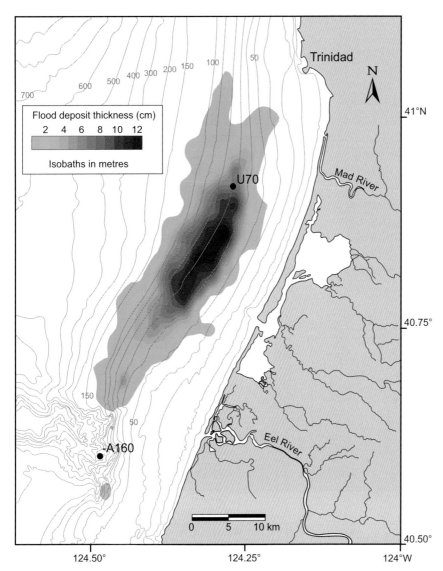

Fig. 21 Composite mud deposit formed by major floods of the Eel River from February 1995 to January 1997. The deposit was mapped by box coring, and layers were delineated using criteria described in Fig. 20. Note that the region of thickest flood deposition corresponds to the zone of maximum ^{210}Pb accumulation rate (see Fig. 25).

of the remaining material was broadcast seaward (Sommerfield *et al.*, 1999). By comparison, approximately 15–30% of the 1997 flood load was retained on the shelf. Although Eel Canyon was not sampled after the 1995 and 1997 events, observations of flood-derived sediment in the canyon during subsequent years make it clear that some of the balance was trapped in the canyon (Mullenbach & Nittrouer, 2000; Mullenbach *et al.*, 2004). In sum, flood-response studies on the Eel margin revealed that the greatest proportion ($\geq 70\%$) of river sediment bypassed the shelf within days of major flood events. These observations shed light on the

workings of the longer-term sediment budget for the dispersal system.

Among the distinguishing properties of the shelf flood deposit was its organic-carbon signature, which, remarkably, could be linked to specific sediment-source areas within the Eel River watershed (Leithold & Hope, 1999; Leithold & Blair, 2001; Blair *et al.*, 2003). Along with mineral particles, rivers transport terrestrial organic matter that is geochemically distinct from that generated by marine productivity, allowing for its use as a tracer of sediment source. The ratio of the two stable carbon isotopes, ^{13}C and ^{12}C, is a useful tool for tracing the dispersal of terrestrial organic matter in margin environments when the δ^{13}C signatures of local carbon sources are known (Hedges & Parker, 1976; Showers & Angle, 1986; Keil *et al.*, 1994; Prahl *et al.*, 1994). End-member δ^{13}C values for modern carbon of terrigenous and marine origins are typically −26‰ and −20‰, respectively, sufficiently distinct to quantify their fractional concentrations in bulk sediments by isotopic mass balance.

On the Eel shelf, the distribution of δ^{13}C values for surface sediments following both 1995 and 1997 floods closely matched the outline of mapped flood deposits (Fig. 22), with the most negative values at the centres of the flood deposits − presumably reflecting concentrations of C_3 terrestrial plant debris (associated with vascular plants). The more positive δ^{13}C values observed toward the fringes

of the deposit must be interpreted with caution, however, because the organic carbon associated with fine sedimentary particles delivered by the Eel River is derived not only from terrestrial C_3 plants, but also from the erosion of uplifted sedimentary rocks of the Mesozoic–Tertiary Franciscan Complex (Leithold & Blair, 2001; Blair *et al.*, 2003), which contain organic carbon of mixed terrestrial and marine origin (δ^{13}C = −24.3‰). On the Eel shelf, therefore, there are three important end-members to account for in interpretation of δ^{13}C values − modern terrestrial (plant-derived) carbon, ancient sedimentary carbon and marine carbon. Within the flood deposits sampled in 1995 and 1997, δ^{13}C values more negative than about −24‰ are consistent with an entirely terrigenous source of organic matter.

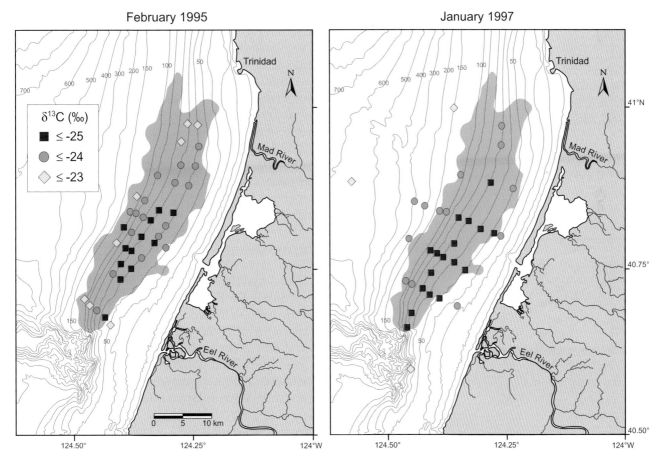

Fig. 22 Values of δ^{13}C for the clay-size fraction of surface sediments (0–1 cm depth) on the Eel margin after the 1995 and 1997 floods. The footprint of the composite deposit is shown. Terrestrial organic carbon predominates in the flood depocentre, whereas marine organic carbon increases toward the edges of the deposit and beyond.

Adjacent to the shelf flood deposits, $\delta^{13}C$ values between −24‰ and −23‰ provide evidence for the addition of modern marine organic carbon to bottom sediments that have resided for greater lengths of time prior to burial. For example, by isotopic mass balance, Blair *et al.* (2003) calculated that the organic-carbon fraction associated with clay-sized (< 4 µm) surface sediments at 150-m and 550-m depths on the Eel slope in January 1997 (Fig. 22) was approximately 20% and 50% marine in origin, respectively, suggesting progressive addition of marine carbon with distance from the coast. Mixing of organic carbon end-members on the Eel shelf is also time-dependent, as revealed by

downcore $\delta^{13}C$ records (Fig. 23). Whereas rapid deposition of fine-grained fluvial sediment during major floods limits the co-burial of marine carbon, slower deposition during more moderate discharge events provides enough time for marine and terrigenous carbon sources to become admixed. Accordingly, the clay-sized fraction of flood layers has more negative $\delta^{13}C$ values than does the same fraction of more extensively reworked sediment beds, providing a carbon-isotopic marker of flood deposition.

Centennial to millennial accumulation

The cumulative effect of sediment dispersal on the Eel margin is evident by the **progressive sorting** of surface sediments along the dispersal system. Overall, mean grain size decreases with distance along- and across-margin from the Eel River mouth, illustrating preferential accumulation of coarser material proximal to source (Fig. 24). On the shelf, mean grain size decreases across-isobath from 125 µm (fine sand) at 30 m, to 8–16 µm (medium–fine silt) at 100 m. North-westward fining is observed along isobath and continues seaward of the shelf break, as upper-slope sediments are considerably finer north of the anticline (4–16 µm) than to the south (4–63 µm) for the same depth range. Across-slope, bed sediments are fine at the shelf edge, coarsen between 250 and 350 m water depth and then progressively fine again across a broad mid-slope plateau, where grain size is finest within the dispersal system (2–4 µm). As mentioned previously, the coarsening could result from the winnowing effects of shoaling internal waves.

Clues that flood sedimentation is a recurrent process on the Eel shelf include: the spatial coherence of the 1995–1997 flood depocentre; the region of maximum ^{210}Pb-derived (100-yr) sediment accumulation rate (compare Figs 21 & 25); and the presence of well-preserved palaeoflood layers in the deeper sediment column (Sommerfield & Nittrouer, 1999). Centennial accumulation rates increase across-shelf from 0.2–0.4 g cm^{-2} yr^{-1} at the sand–mud transition, to a maximum of 0.6–1.7 g cm^{-2} yr^{-1} on the middle shelf, and decrease to ≤ 0.2 g cm^{-2} yr^{-1} at the shelf edge (relict late Pleistocene strata are exposed in places). Flood-produced beds comprise much of the most rapidly accumulating shelf strata, thus the short-term depositional episodes have a direct

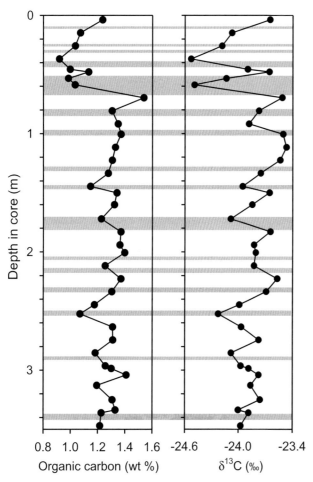

Fig. 23 Weight percentage organic carbon and $\delta^{13}C$ values for the clay-sized fraction in a piston core from Eel shelf site K-70 (see Fig. 18 for location). Grey bars denote clay-rich flood layers in the core. Note correspondence of low $\delta^{13}C$ values (larger fraction of terrestrial organic carbon) and flood-produced layers. (Adapted from Leithold *et al.*, 2005.)

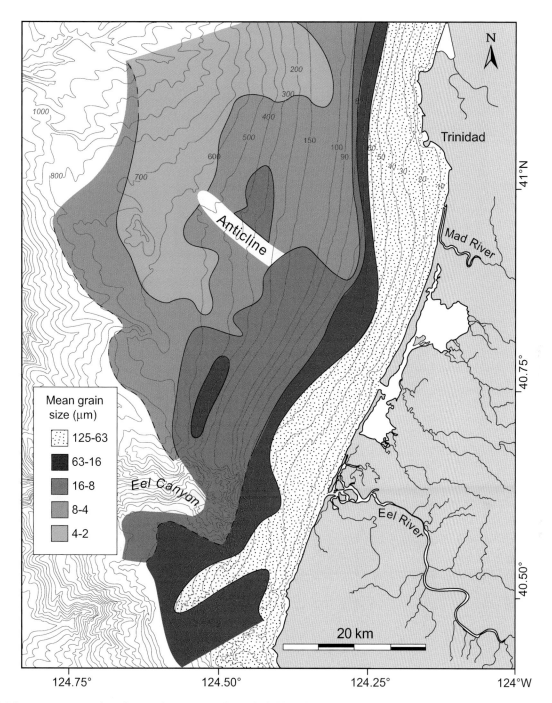

Fig. 24 Mean grain size of surface sediments on the Eel shelf and slope. Sediments fine across-shelf and also along the dispersal to the north-west. (Adapted from Borgeld (1985) and Alexander & Simoneau (1999).)

bearing upon longer-term accumulation rates. The similarity of flood deposition and 100-yr accumulation rates contrasts with other shelf systems, in which accumulation patterns develop from seasonal or longer sediment-transport conditions (Nittrouer & Wright, 1994).

Modern sediment accumulation on the open Eel slope is also coupled with oceanic storms through bypass of resuspended shelf deposits or flood-produced sediment gravity flows. Although there is radioisotopic (Sommerfield *et al.*, 1999) and sediment-trap evidence (Walsh & Nittrouer, 1999)

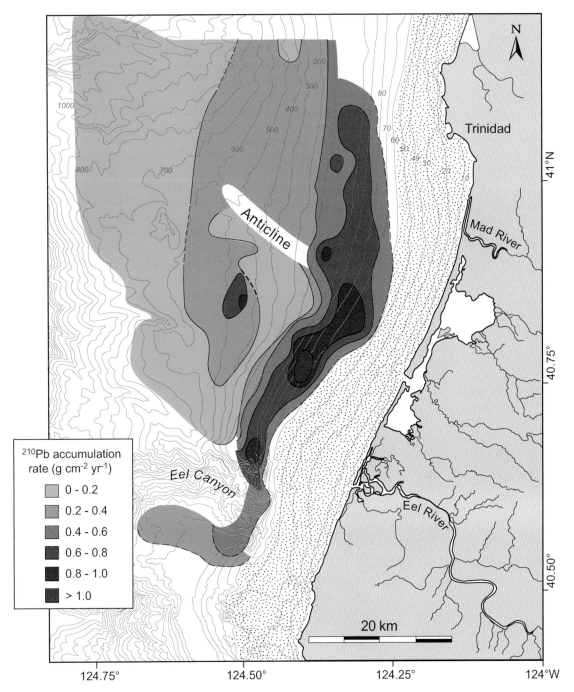

Fig. 25 Lead-210-derived sediment accumulation rates from the mid-shelf to the slope of the Eel margin. Sands (stipple pattern) on the inner shelf preclude ^{210}Pb dating. Maximum rates occur within a mid-shelf depocentre north-west of the Eel River mouth, which is nearly coincident with the region of thickest Holocene sedimentary cover (see Fig. 18.)

of flood-related slope deposition, the flux is generally too slow to form characteristic beds (Alexander & Simoneau, 1999). Slope sediment accumulation rates based on ^{210}Pb geochronology are greatest near the shelf-slope break and in depths greater than 450 m (0.2–0.6 g cm^{-2} yr^{-1}), decreasing to a minimum in the 250–350 m depth range (0.02–0.2 g cm^{-2} yr^{-1}) possibly due to internal waves. North and south of a slope bifurcating anticline, accumulation rates differ by a factor of two, with higher rates to the north (Fig. 25). A region of anomalously high accumulation rates is present in the floor of the Humboldt Slide, reaching values as high ($1.7\ \mathrm{g\ cm^{-2}}$ $\mathrm{yr^{-1}}$) as observed on the shelf. This region probably receives large quantities of sediment from episodic gravity flows (Lee *et al.*, 2002).

The sedimentology and bed-scale stratigraphy of shelf deposits is related to supply and transport conditions specific to dry and wet storms. Upon reaching the coast, the total suspended load of the Eel and Mad rivers fractionates as a function of particle settling velocity, turbulence and time-variant shelf flow (Hill *et al.*, this volume, pp. 49–99). Coarse silt and fine sand discharged by the Eel initially settle on the inner shelf near the mouth, but are subsequently remobilized and transported northward and southward as bedload or in suspension, depending on the trajectory and magnitude of prevailing along-shelf currents (Ritter, 1971). Storm waves and currents redistribute this material seaward by a combination of wave-orbital, Ekman-

veering and downwelling flows, forming beds enriched with fine sand and coarse silt on the middle shelf. Upon bioturbation by macrofauna, these beds become admixed with finer muds that accumulate below fair-weather wave base (Drake, 1999; Bentley & Nittrouer, 2003; Wheatcroft *et al.*, this volume, pp. 101–155). This form of sedimentation, characteristic for dry storms and minor wet ones, comprises the greater proportion of the late Holocene shelf succession (Fig. 26a).

High suspended-sediment concentration and the presence of a buoyant plume are the chief factors that distinguish typical deposition from that related to major wet storms. During these events, fine silts and clays reach the mid-shelf region directly through fallout from buoyant river plumes, and, more importantly, indirectly by way of the inner shelf (Fig. 26b). Observations show that strong south-westerly winds characterizing the early stages of Pacific storms confine river plumes to the inner shelf, where much of the suspended-sediment

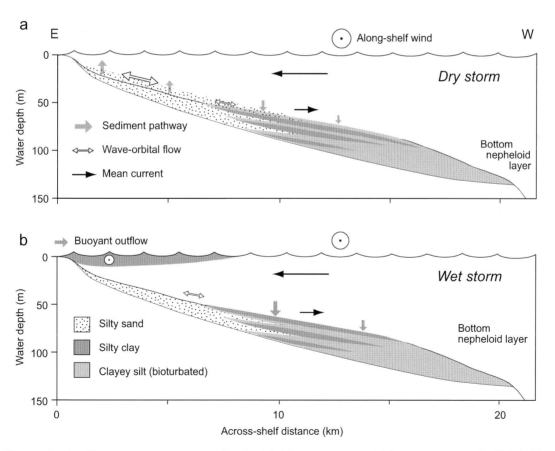

Fig. 26 Schematic of sedimentary processes associated with (a) dry storms and (b) wet storms on the Eel shelf. See text for interpretation. (Adapted from Fan *et al.*, 2004.)

inventory settles to form temporary fluid-mud beds (Geyer *et al.*, 2000; Hill *et al.*, 2000, this volume, pp. 49–99). Although some of this mud becomes permanently trapped through burial by coarse silts and fine sands (Crockett & Nittrouer, 2004), the greater proportion moves seaward within the bottom boundary layer as a high-concentration suspension. These suspensions are maintained by wave-generated turbulence, and, when sufficiently concentrated, flow across shelf under the influence of gravity (Traykovski *et al.*, 2000). In spite of wave-generated turbulence, clay-rich flood deposits formed in this manner are nearly devoid of inner-shelf sands and silts (Drake, 1999). Accordingly, the unique sedimentology of wet-storm deposits allows their recognition in the deeper sedimentary record.

Storm-specific variations in fluvial-sediment discharge and the wave field may explain the abrupt transition from sandy silt to clayey silt at 50–60 m water depths on the Eel shelf. This idea is supported by sediment transport models showing that the amount of sediment available for deposition is as important as waves in setting the locus and thickness of event beds (Harris & Wiberg, 2002; Fan *et al.*, 2004). Landward of 50–60 m water depths, event beds are in general poorly preserved owing to physical mixing by wave-orbital flows. Seaward, storm waves are less effective in bed reworking, both because of a weaker wave-orbital flow (due to deeper water depth), and because of cohesive shear strength imparted by the muddy strata. Consequently, physical reworking of middle–outer shelf strata is low compared with the more energetic inner shelf. Instead, there is greater potential for post-depositional bioturbative destruction of bed contacts and interlaminae. In general, layer preservation potential increases with the magnitude (layer thickness) and frequency of depositional events, i.e. with net sediment accumulation rate (see Wheatcroft *et al.*, this volume, pp. 101–155). In sum, the cumulative product of sedimentation is a distinct across-margin gradient in sediment texture and bedding: homogenous, fine sands and coarse silts on the inner shelf; interbedded sands and muds at the inner–middle shelf transition; and bioturbated muds on the middle–outer shelf and beyond (Fig. 26).

It is tempting to attribute the shape of the Holocene sediment depocentre (Fig. 18) to oceanic dispersal processes in light of its concentric shape and general similarity to accumulation patterns generated at flood-event and 100-yr time-scales. On the other hand, post-glacial shelf accumulations were initially bound by tectonically produced topography, and it was not until the late transgression or highstand stage that sediment accumulation blanketed the shelf more-or-less uniformly (Spinelli & Field, 2003). The antecedent morphology is still seen in the modern bathymetry: topographic highs seaward of the Eel and Mad river mouths, and a low centred near the along-shelf inflection of isobaths, appear to coincide with underlying anticlinal and synclinal structures (Goff *et al.*, 1999). The bathymetric low is coincident with the area of highest 100-yr sediment accumulation and thickest 1995–1997 flood deposits as though exerting an influence on the locus of deposition. This association begs the question: to what extent is modern sediment accumulation controlled by dynamical processes versus antecedent morphology?

Observations of sediment-gravity flows on the Eel shelf (Traykovski *et al.*, 2000) made clear that subtle topographic variations influence the trajectory and depositional footprint of flood-generated suspensions; an unexpected link between dynamic and static trapping mechanisms. Unlike turbidity currents that flow downslope by way of internally generated turbulence, sediment gravity flows on shelves require an external turbulence source such as energetic waves (Hill *et al.*, this volume, pp. 49–99; Parsons *et al.*, this volume, pp. 275–337). Using a transport model that considers the wave-energy field, gravity flow and bathymetry, Scully *et al.* (2003) modelled deposition following the 1995 and 1997 floods. The modelled pattern captured many attributes of the observed deposit, but, more importantly, the results confirmed that the locus of deposition is at least partly governed by shelf bathymetry. Given that the morphology of the Eel shelf is controlled by interactions between tectonics and sedimentation on the long term, these findings can explain the broad spatial coherence of sedimentation patterns produced on flood-event, centennial and millennial time-scales.

Latest Holocene sedimentary record

The sedimentary record of the northern California shelf documents how the dispersal system evolved during late Holocene to modern times in the face of sea-level rise, climatic variations in terrigenous sediment export and, most recently, human activities in the watershed. Cores from the Eel (site E-95) and Klamath (site KL-0) segments of the shelf document ~4 kyr of accumulation at the foot of the Eel's subaqueous delta and north-west of Klamath River mouth, respectively (Fig. 27). These deposits are similarly composed of sandy, clayey silt with poor preservation of fine-scale bedding. The core sediments accumulated at long-term rates of 1.1 mm yr^{-1} (E-95) and 0.60 mm yr^{-1} (KL-0). Both records show a net up-section fining due to decreasing sand and increasing silt content after 3500–4500 yr BP, with smaller-scale variations superimposed. The upward fining is consistent with stratigraphic models that call for increased bypass-

ing of fines as sediment accommodation space is filled in the coastal plain (Posamentier & Vail, 1988; Lesueur *et al.*, 2002). Estuaries initially formed through marine flooding of incised valleys on the northern California coast were small and thus filled rapidly, becoming estuarine deltas by late Holocene times (Dingler & Clifton, 1994). The lithological records for sites E-95 and KL-0 indicate that burial of river silt increased at the expense of fine-grained sand, which comprises a smaller fraction of the fluvial suspended load (Brown & Ritter, 1971) and is indirectly supplied to the middle shelf after storage on the inner shelf.

Superimposed on this background trend are sedimentological changes suggestive of accelerated terrigenous efflux starting around the time the drainage basins were settled by Europeans in the

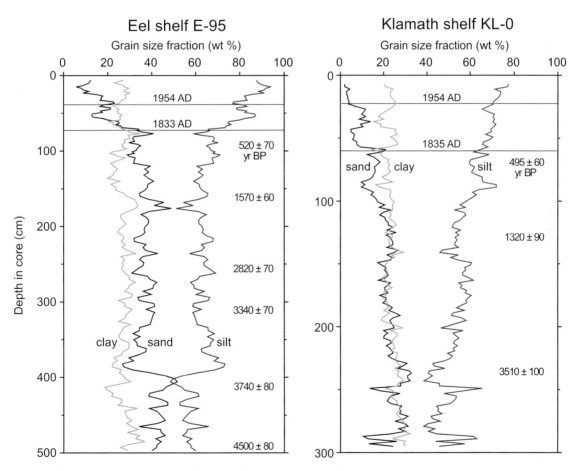

Fig. 27 Late Holocene sedimentary records for mid-shelf mud deposits off the Eel and Klamath Rivers (site E-95: 40°42.25′N, 124°28.39′W, 103 m depth; site KL-0, 41 23.48′N, 12°13.67′W, 70 m depth). The fining-upward trend is attributed to increased river-mouth bypassing of silt since 3500–4500 yr BP, and elevated river-sediment loads since the middle 19th century.

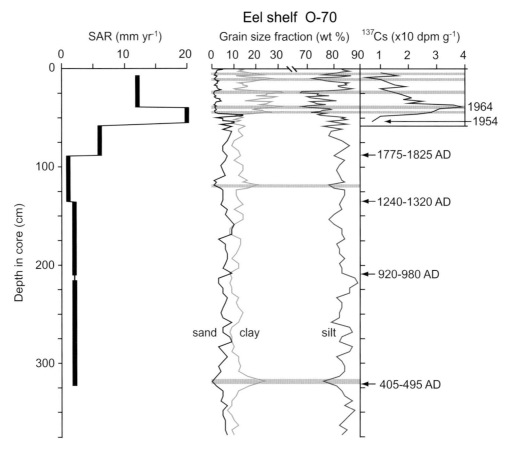

Fig. 28 Sedimentary record for Eel shelf site O-70 showing a change in sedimentation conditions after ~1950. See Fig. 18 for core location. Increased frequency of clay-rich beds (shaded), decrease in sand content and accelerated sediment accumulation rate (SAR) indicate that river floods have had a greater impact on shelf sedimentation in recent times due to climatic and anthropogenic factors. (From Sommerfield *et al.*, 2002.)

early 19th century. The concentrations of sand and silt at sites E-95 and KL-0 changed abruptly after about 1830, with sand decreasing and silt increasing by ~20 wt% (Fig. 27). Although the absolute concentration of clay did not change much, interestingly, its covariation with silt content switched from inverse prior to ~1830, to direct thereafter. The close association of fluvial silt and clay after ~1830 can be explained by an increase in the concentration of river-generated suspensions and accelerated deposition of mud aggregates. Highly concentrated suspensions promote flocculation and rapid deposition, so the grain-size signature of these suspensions is more-or-less preserved upon burial (Hill *et al.*, this volume, pp. 49–99). Accordingly, the increase in the depositional flux of mud (transported directly from the river) led to a decrease in the concentration of fine sand (supplied indirectly from the inner

shelf) as recorded by the grain-size profiles. In this manner, the post-1830 shelf record documents an increase in river-sediment loading, most probably due to land-cover change and accelerated erosion in the watershed (Sommerfield & Wheatcroft, in press). As noted for other mountainous drainage basins, forest clearing can elevate basin sediment-yields and overland runoff (Montgomery *et al.*, 2000), ultimately increasing suspended-sediment discharge per unit riverflow.

At the same time, the shelf record provides evidence for more recent hydroclimatic increases in sediment supply after ~1950 that are consistent with the instrumental record of sediment discharge (Fig. 10). By comparing sediment loads for the Eel River (1911–2000) and flood-layer stratigraphy, Sommerfield *et al.* (2002) found that major floods after ~1950 had a more pronounced impact on sedi-

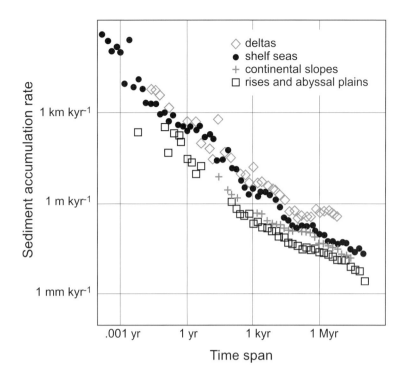

Fig. 29 Mean sediment accumulation rates for terrigenous sediments plotted as a function of measurement time span. The inverse relationship between rate and time span of averaging is a consequence of stratigraphic incompleteness. (Redrawn from Sadler, 1999.)

ment export and accumulation than earlier during the 20th century. Supporting evidence includes a step-increase in the concentration of clay (and concomitant decrease in sand and silt), and accelerated sediment accumulation rate (Fig. 28). Most of the clay is packaged within flood-produced layers, which are best-preserved on the middle shelf in 60–80 m water depths. They attributed this change to several interrelated factors including: an increase in regional flood magnitude and frequency since 1950; elevated sediment concentration in the river immediately following the 1955 and 1964 floods; and increased river-sediment loads in the wake of intense, mid-century timber harvesting. The respective increase and decrease in clay and silt burial after 1950 is not nearly as pronounced at site E-95, but the stratigraphic fidelity of that site is comparatively low due to its lower sediment accumulation rate. For the same period, Leithold *et al.* (2005) documented changes in the organic-carbon composition for a nearby site (K-70), and attributed them to accelerated erosion following land-use change in the watershed. It is likely that both climatic and anthropogenic factors were responsible for the observed changes, but separating their specific influences is complicated by the fact that the catastrophic 1955 and 1964 floods occurred during

a period of widespread landscape disturbance by clear-cutting and construction of logging roads, which may have intensified the geomorphological impacts of these and subsequent floods (Kelsey, 1980; Harden, 1995). As a complicating factor, river-discharge signals are imperfectly preserved on much of the Eel shelf, as a consequence of depositional and post-depositional processes that reduce the fidelity of the sedimentary record (Wheatcroft & Drake, 2003).

Accumulation rates and stratigraphic completeness

Much of the preceding discussion concerning short-term sediment deposition rates versus longer-term accumulation rates can be recast in terms of **stratigraphic completeness**: a measure of gaps in a sediment column. Hiatuses pervade strata at scales ranging from grain boundaries to regional unconformities, and arise from periods of non-deposition or erosion. The corollary is that sediment accumulation rates are typically time-variant, decreasing with increasing period of averaging due to accrual of hiatuses (Fig. 29). This fundamental truism of stratigraphy has been shown for a range of sedimentary environments (Sadler, 1981, 1999; Sadler & Strauss, 1990), although it is commonly overlooked by researchers interpreting modern sediment accumulation rates from dated cores. The time-span dependence of sediment accumulation rates should not be confused with temporal variability, which arises from changes in sediment supply and depositional flux in an environment. Rather, it is a consequence of averaging accumulation rates over an incomplete time series of particle burial (Crowley, 1984).

Following Sadler (1981), the stratigraphic completeness of a sediment column can be approximated by the ratio of a long-term rate of sediment accumulation (S_a) of a sediment column to the rate for a smaller time span (S_d) as follows:

$$\frac{S_a}{S_d} = \left(\frac{t_d}{t_a}\right)^{-m} \tag{1}$$

where t_a is the whole-section time span, t_d is the time span at a specified level of resolution, and m is the slope of regression for S_a versus t_a, varying between -1 and 0. This ratio provides a gross measure of stratigraphic completeness at the time span of the short-term rate, but it cannot distinguish non-depositional from erosional hiatuses. In a log–log plot of accumulation rate (y axis) versus time span of averaging (x axis), the slope of the power function increases with increasing incompleteness, in other words, accumulation rates correlate inversely with the averaging period. A sediment column produced by steady-state deposition would have a slope of zero, as would a column with equally distributed hiatuses. Only under these circumstances are accumulation rates completely independent of time span.

The concept of stratigraphic completeness is useful for illustrating various mechanisms of strata formation and preservation in sediment dispersal systems (Sommerfield, 2006). Sadler-type plots based on nested geochronology ([7]Be, ^{210}Pb and ^{14}C) clearly show that the accumulation rate on the Eel shelf is a strong negative function of time span (Fig. 30a). Note that this trend is not a manifestation of the aforementioned historical increases in shelf accumulation rate, which are small (e.g. factor of two to three) compared with the range exhibited by the greater sediment column. Instead, accumulation rates vary with time span due to stratigraphic incompleteness. The slopes of the regression lines in Fig. 30a are consistent with deltaic and terrigen-

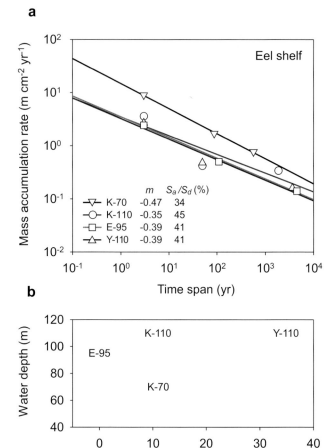

Fig. 30 (a) Sadler-type plots for four depositional sites on the Eel shelf showing levels of stratigraphic completeness (S_a/S_d) at the 100-yr level of resolution (t_d in Eq. 1, see text). Trendlines are based on [7]Be, ^{137}Cs, ^{210}Pb and ^{14}C geochronologies at each site. (b) Site water depth and distance from Eel River mouth. (From Sommerfield, 2006.)

ous shelf environments as determined by Sadler (1981), although there is within-environment variability due to localized sedimentary processes. For example, the sites most proximal (E-95) and distal (Y-110) to the river mouth exhibit an identical level of completeness at 41% (Fig. 30b), revealing that sediment dispersal overprints source-proximity trends. Similarly, the shallowest site (K-70) is comparatively incomplete at 34%, despite having the highest short-term deposition rate. At this site, some combination of depositional episodicity and post-depositional erosion and sediment redistribution is responsible for reducing stratigraphic completeness relative to the other sites.

The convergence of regression trendlines in Fig. 30a to a relatively narrow range of rates suggests that stratigraphic completeness on the shelf is controlled by an overarching mechanism. At the intra-annual time-scale, mass accumulation rates vary by nearly an order of magnitude (9–50 g cm^{-2} yr^{-1}), but this variation decreases to a factor of two at 10^4 yr (0.1–0.2 g cm^{-2} yr^{-1}). When converted to linear values, the long-term rates (0.1–0.6 mm yr^{-1}) equal or exceed that of relative sea-level rise reported for the northern California coast (Emery & Aubrey, 1986). Therefore, it is likely that the rate of relative sea-level rise controls the net accretion rate (by way of wave base), keeping in mind that accommodation space on the Eel shelf is created by both vertical tectonics and eustasy (Burger *et al.*, 2002; Spinelli & Field, 2003). Regardless, the narrow range of long-term accumulation rates implies that wave-related sediment bypass or erosion is effective across the greater width of the shelf.

SEDIMENT BUDGETS OF DISPERSAL SYSTEMS

Development of sediment budgets

A sediment budget is a quantitative statement of the relations between sediment production, transport, storage and permanent burial for a sediment dispersal system. In addition to fluvial sediment, coastal erosion, seafloor scour and biological production provide source materials that may be quantitatively significant on shelves. As defined earlier, sediment-trapping efficiency provides a measure of material sequestration (or alternatively dispersion) within a dispersal system, and insight about how it is partitioned among fluvial, shelf and slope subenvironments. Naturally, the usefulness of a sediment budget rests on the accuracy of the source and sink terms, the quantification of which almost always requires some degree of spatial and temporal averaging. The Eel dispersal system is especially amenable to sediment budgeting due to its quantifiable sediment loads and negligible onshore sediment storage.

Special attention was paid to the estimated river sediment loads, because river discharge is by far the most important local source of fine-grained sediment on the northern California margin. In general, because continuous records of river discharge (Q) are more common than suspended-sediment concentration (C) data, sediment rating curves are used to predict sediment discharge (S) on the basis of at-station correlations between Q and C (Syvitski *et al.*, 2000). Although this approach is widely applied due to ease of use, there are at least two well-known limitations that are worth noting:

1 lack of sediment concentration data for extreme values of Q;
2 dependence on gauge location.

First, sediment-rating relations for some rivers are non-linear across the full range of Q, thus extrapolating empirical curves based on typical peak flows to extreme flows can lead to erroneous estimates of S. Fortunately, a large quantity of sediment concentration measurements that span nearly the full range of measured Q are available for the Eel River at Scotia (available through the US Geological Survey). Second, location is an issue when the seaward-most gauge (typically located at the head of tides) is not representative for the river mouth, owing to downstream sediment production or storage. This can be an intractable problem when developing sediment budgets for large rivers that possess extensive delta plains (Nittrouer *et al.*, 1995; Goodbred & Kuehl, 1999). In contrast, the Eel River gauge at Scotia is merely 25 km upstream of the coast (Fig. 7), and the intervening delta plain traps insignificant quantities of fine-grained sediment. Further, measurements of suspended-sediment concentration at the mouth by Geyer *et al.* (2000) during flood events in 1997 and 1998 revealed concentrations that were comparable to values predicted by the sediment rating relation for the Eel at Scotia.

Worldwide, many river basins are undergauged or ungauged altogether, and historically this has greatly limited our ability to predict terrigenous sediment fluxes to the ocean. Sediment discharge can now be estimated from empirical relationships between climate and sediment production for gauged basins of similar geological and morphological properties (Syvitski *et al.*, 1998, 2005; Morehead *et al.*, 2003). In recent years there have been major advances in theoretical modelling of sediment flux in river basins (reviewed by Paola, 2000), and these models hold further promise for estimating sediment flux in ungauged basins. A particularly exciting application of these empirical and theoretical approaches is hindcasting fluvial palaeoflux for the geological past (Syvitski *et al.*, this volume, pp. 459–529).

For the Eel margin sediment budget, suspended-sediment rating curves were computed using the flow-duration, rating-curve technique (Crawford, 1991; Hill *et al.*, this volume, pp. 49–99; Syvitski *et al.*, this volume, pp. 459–529). This standard method is based on an empirical relation between daily suspended-sediment load (S, in tons per day) and mean daily river discharge (Q, in $m^3 s^{-1}$) usually expressed as a power function $S = aQ^b$ where a and b are the power-law intercept and slope, respectively (Nash, 1994). Values of S are computed as the product of daily cross-sectionally averaged suspended-sediment concentration and Q at the same station.

A bias-corrected, logarithmic transformation was used to linearize the power function

$$\ln S = \beta_0 + \beta_1 \ln Q + \varepsilon \qquad (2)$$

where β_0 and β_1 are the slope and intercept of the rating curve, and ε is the residual error. In its unbiased form, the relationship between S and Q is given by

$$S = \exp(\beta_0)Q^{\beta_1}bc \qquad (3)$$

where bc is the correction for transformation bias, required to avoid underestimating the sediment load (Ferguson, 1986). This method was used to predict S on the basis of Q for years when concentration data are not available for direct calculation of daily loads (~80% of the time), and is considered most accurate when the sampling interval is shorter than the period over which S varies significantly. The major sources of error in estimates of S are: (i) the standard error of its regression against Q, and (ii) temporal variation in C independent of Q. The magnitude of error varies considerably among river systems due to variable data quality and quantity, and can range from within a factor of two to several orders of magnitude (Syvitski *et al.*, 2003).

The regional limits of a dispersal system must be delineated before the sink term in a sediment budget can be determined. This is accomplished by mapping the lateral distribution of a system-related property such as clay or heavy minerals, organic

and inorganic pollutants, radioisotopes and gradients in sediment accumulation rate. In general, there are two different ways to quantify the sink term in a sediment budget. The first, the **stratigraphic method**, is carried out by seismic profiling and/or seabed sampling to map the distribution of sediment thickness in a basin (Grützner & Meinert, 1999). The sedimentary units in question must have laterally traceable bounding surfaces of known age, for example, a lowstand surface of erosion and the modern seafloor. The basin sink term (M_H) is calculated as

$$M_H = \sum_1^n \frac{H \cdot A \cdot \rho_d}{T} \qquad (4)$$

where H is the stratal thickness, A is the mapped or interpolated extent of H, ρ_d is the dry bulk density (measured or estimated) and T is the age difference between the reference surfaces. For a number (n) of characteristic subregions, the stratigraphic method quantifies the sediment sink term in units of mass per time, or mass alone if T is unknown.

The **burial flux method** quantifies the basin sediment sink based on measurements of linear sediment accumulation rate or mass accumulation rates at multiple sites within a dispersal system. This approach was pioneered by Nittrouer (1978), who employed ^{210}Pb geochronology to quantify modern sediment accumulation rates on the Washington shelf. Importantly, this method allows the sediment sink term to be determined over the period for which river discharge data may be available, allowing direct comparison to the source term. Sediment budgets based on burial flux address the sink term by spatially interpolating accumulation rates measured at high density as follows

$$M_f = \sum_1^n MAR \cdot A \qquad (5)$$

where MAR is the at-site mass accumulation rate (mass/area/time) for a particular region (n) of area A. The primary source of error in M_f lies with mass accumulation rates based on radioisotope geochronology. If it is known or assumed that these rates are accurate, then the intrinsic error in M_f is simply the propagated error associated with the radioisotope method (i.e. radioisotope counting errors, error of age model). The major source of error in sediment geochronology, in turn, is the assumption of steady state, and that bioadvection (if present) has not elevated the true accumulation rate. If these factors cannot be completely ruled out, then the sink term must be bracketed by confidence limits that span the possible range of accumulation rates.

A hypothetical sediment mass-balance for a river-shelf system with local erosional and biological sources of sediment is described by

$$\bar{S}_R + \bar{E} + \bar{P} = \bar{M}_{f(shelf)} \qquad (6)$$

where S_R is fluvial sediment input, E is the erosional sediment supply, P is the supply of biological particles and the overbar denotes a time-averaged value. In general, the specific source and sink terms are based on the data available and some a priori knowledge of the sedimentary system. For example, in urban-impacted coastal waters it may be necessary to add wastewater particle loading as a source term. Fluvial, erosional and biological sediment fluxes are quantified by different methods, each with operational time spans and assumptions. As such, the cumulative source can be difficult to establish, and it is common for small or poorly constrained terms to be excluded altogether. Commonly it is the degree of mass *imbalance* that provides information on the significance of particular source and sink terms, and the mechanisms that control sediment trapping efficiency (Nittrouer & Wright, 1994).

Eel margin sediment budget

The modern sediment budget for the Eel margin was developed from several independent studies of the shelf (Sommerfield & Nittrouer, 1999; Crockett & Nittrouer, 2003), open slope (Alexander & Simoneau, 1999) and canyon environments (Mullenbach & Nittrouer, 2006). River gauge records of mean daily sediment discharge provided estimates of fluvial influx, and similar ^{210}Pb or ^{137}Cs methods were used to quantify sediment burial averaged over the past 100 yr. Margin subenvironments were delimited and partitioned according to water depth (Fig. 31), whereas the lateral extent of the receiving basin was delimited by the Eel Canyon to the south and the area seaward of Trinidad Head to the north, where the influence of the Klamath River on shelf and slope sedimentation increases.

In addition to sediment input from the Eel and Mad rivers (Table 2), the shelf and slope sequester material from coastal bluff erosion and primary production; however, these contributions are negligible (< 1%) compared with the terrigenous supply and therefore were not included in the overall budget. The simplified mass balance for the Eel margin is then

$$\bar{S}_{Eel} + \bar{S}_{Mad} = \bar{M}_{f(shelf)} + \bar{M}_{f(slope)} + \bar{M}_{f(canyon)} \qquad (7)$$

where S is the river-specific source term, M_f is the environment-specific sink term and the overbar denotes a time average (~100 yr). The error associated with the source terms is ±15% based on the standard error of least-square regressions of S versus Q, whereas each sink term has an intrinsic error of 10–15% (i.e. the cumulative error of radioisotope measurements and spatial interpolation of accumulation rates).

Of the total 19×10^6 tons yr^{-1} of suspended sediment supplied to the coast, on average, ~10% is trapped on the sandy inner shelf, 20% is trapped on the middle–outer shelf, and another 20% can be accounted for on the open slope (between 150 m and 600 m). Last, at least 12% is sequestered in the upper Eel Canyon. Accordingly, the unaccounted ~38% must be dispersed along-margin to northward and southward, and farther seaward, especially to deeper portions of Eel Canyon. Recall that ~15–30% of the flood-generated load in 1995 and 1997 was trapped on the shelf, which is comparable to the 100-yr shelf trapping efficiency of ~20–30%. The similarity of these results suggests that most of the fluvial sediment load escapes the shelf during wet-storms, rather than through subsequent redistribution of temporary shelf deposition. This latter condition was observed by Drake *et al.* (1972) following the 1969 flood of the Santa Clara River (southern California), the only similarly documented case study available for comparison. The key insight is that the decadal–centennial sediment mass balance of the dispersal system is set at the time-scale of oceanic storms and river floods, because of the disproportionate influence of these conditions on sediment delivery and across-shelf transport.

It is interesting to note that the trapping efficiency of the Eel shelf at 20–30% is considerably lower than that determined for neighbouring systems

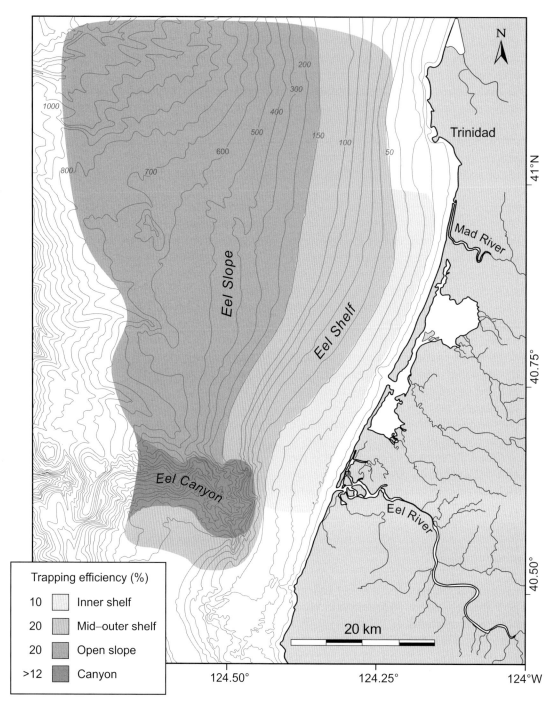

Fig. 31 Sediment trapping efficiency for subenvironments of the Eel shelf and slope. Percentages are the fraction of fine-grained Eel (16.5×10^6 tons yr^{-1}) and Mad (2.5×10^6 tons yr^{-1}) river sediment retained on a 100-yr time-scale.

including the Washington shelf (60%; Nittrouer, 1978), the Russian River shelf (> 50%; Demirpolat, 1991) and the Santa Cruz shelf (> 75%; Eittreim *et al.*, 2002). There are at least three general factors that may account for this difference: (i) timing of

sediment delivery and pulse amplitude; (ii) wave climate; and (iii) shelf geometry (Wheatcroft & Sommerfield, 2005). The seasonality of sediment delivery with respect to the shelf energy and flow field is important, because small, mountainous watersheds such as the Eel exhibit hydrographs that are dominated by rainstorm runoff events in winter, placing sediment on the shelf during the period of peak wave energy. In the face of large storm waves and turbulence, deposition on the shelf is reduced by resuspension and breakup of flocs (decreasing settling velocity), and elevated currents (due to buoyancy effects or strong winds) disperse suspended matter over appreciable distances. During these times, deposition on the middle shelf occurs largely because the bottom boundary layer is overwhelmed by high concentrations of suspended sediment. Conversely, snowmelt runoff during spring and early summer in the Columbia River basin carries sediment to the shelf during the low-energy season. In this case, the times of peak river flow and maximum oceanic dispersal are out of phase by nearly 6 months, favouring sediment deposition on the shelf followed by consolidation and preservation (Nittrouer & Sternberg, 1981).

Perhaps the overarching control on the shelf trapping efficiency is geometry, including width, gradient and presence of indenting canyons. Across-shelf current speed is roughly a quarter of along-shelf speed (Sherwood *et al.*, 1994; Ogston *et al.*, 2004), so minor differences in shelf width along a margin should influence shelf-to-slope sediment exchange. Submarine canyons, particularly those that incise to the littoral zone, have potential to intercept bedload and suspended load, and to provide a terminal sink for sediment that would otherwise accumulate on the outer shelf. It is worth noting that among shelves of the northern USA Pacific coast, the Eel at ~12–17 km width is considerably narrower than the Russian River shelf to the south (~30 km) and Columbia River shelf to the north (~40 km). Narrow width typically provides a steeper shelf surface, and this can facilitate gravity flows reaching the continental slope when hyperpycnal or fluid-mud conditions occur (Friedrichs & Wright, 2004). Indeed, considering the dispersive nature of the coastal ocean during winter storms, shelf geometry is likely to be a chief factor in the transfer of terrigenous sediment beyond the shelf break.

CONCLUSIONS

Some fundamental mechanisms of sediment dispersal and accumulation on continental margins have been reviewed through a study of the northern California shelf and slope. The broad sketch of this system is that tectonic, hydrological and coastal ocean processes come together in the extreme to yield a large land-to-ocean sediment flux, most of which escapes to the continental slope in association with Pacific storms. Building on the notion that mountainous rivers supply a large fraction of the terrigenous flux to the coast (Milliman & Syvitski, 1992), the present case study reveals that delivery to the deeper ocean can be highly time-dependent, occurring when the times of maximum sediment discharge and across-shelf transport coincide. On the northern California margin, this arises during south-westerly winter winds characterized by energetic waves and intense runoff, an ideal combination for sweeping terrigenous sediment to the open slope and canyons.

In spite of export processes, sediment is sequestered on the shelf by means of dynamic trapping mechanisms related to particle aggregation and hydrodynamic processes in the bottom boundary layer. The intensity of trapping is such that characteristic event deposits form within days of major floods, a process that leaves a conspicuous stratigraphic imprint. On longer time-scales, static trapping creates thick sedimentary successions, smoothing tectonically produced topography. One of the most unexpected discoveries in STRATA-FORM was that dynamic and static trapping are coupled at the scale of depositional events. Indeed, subtle bathymetric variations, ultimately a consequence of tectonic subsidence, modulate transport and deposition of storm-generated sediment gravity flows. The product is a distinct shelf depocentre that is stable on centennial-to-millennial time-scales. In this manner, the Eel shelf study confirms that the geological/tectonic framework plays an important role in sedimentary processes, even on dynamical time-scales.

The completeness of the shelf sedimentary record varies on small spatial scales (km) in association with coastal ocean processes on the short term, and relative sea level on longer time-scales. Analysis of stratigraphic completeness reveals that the sediment column is merely 34–45% complete (on a 100-yr time span) as the result of hiatuses related to episodic deposition and erosion. The Eel shelf exhibits levels of stratigraphic incompleteness consistent with deltas and terrigenous shelves in general, although there is significant within-system variability attributable to preferential sediment-transport pathways and deposition. Ultimately, however, storm wave base appears to be the over-arching control on the long-term rates of sediment accumulation.

Late Holocene sedimentation on the northern California shelf has responded to rising relative sea level, climatic variations in river runoff and, most recently, anthropogenic increases in terrigenous sediment delivery. The shelf sedimentary record displays an upward-fining trend since at least ~4 ka that evokes an increase in silt bypassing of the delta plain, presumably as a consequence of channel and floodplain filling during the late Holocene. Superimposed on this trend is sedimentary evidence of human disturbance in the watershed starting in the early 19th century, when Europeans began to modify the landscape. Additionally, the shelf record shows a step-increase in the concentration and accumulation rate of flood-derived clay after about 1950. The mosaic of evidence suggests that this increase is due to some combination of catastrophic floods in 1955 and 1964, and anthropogenic sediment production in the watershed. Separating these natural and anthropogenic influences from the viewpoint of the shelf record has proven difficult, because, in the present case, both operate to increase terrigenous sediment export over the same time period.

The STRATAFORM programme, like its predecessors, generated many new insights about oceanic sediment dispersal and accumulation, but of course this has led to additional, more complex, questions. Among topics needing special attention in the future are:

1 mechanisms of shelf-to-slope sediment transfer during sea-level highstands;
2 effects of geological/tectonic framework on shelf sediment transport, deposition and erosion;
3 elucidating climatic and anthropogenic influences on terrigenous sedimentation in coastal and marine environments;
4 three-dimensional morphodynamic modelling of river-mouth-shelf sediment bodies.

Progress in the areas above will require much observational and theoretical work.

ACKNOWLEDGEMENTS

We are indebted to the large number of scientists, technicians and students who collectively made the STRATAFORM programme a success. We thank the Office of Naval Research for sponsorship, and offer special thanks to Dr Joseph Kravitz for his unwavering direction of the programme. Reviews of an earlier version of the manuscript by Mead Allison, Sam Bentley, Steve Kuehl, Joe Kravitz, and Peter Traykovski are gratefully acknowledged. The first author was supported by ONR grants N00014-99-1-0047 and N00014-01-1-0689 during the early stage of manuscript preparation.

NOMENCLATURE

Symbol	Definition	Dimensions
a	power-law intercept	
A	area of subregion	L^2
b	power-law slope	
bc	correction for transformation bias	
E	erosional sediment supply	$M\,T^{-1}$
H	stratal thickness	L
m	regression slope	
n	accumulation rate subregion	
MAR	mass accumulation rate	$M\,L^{-2}\,T^{-1}$
M_H, M_f	basin sink term	$M\,T^{-1}$
P	supply of biological particles	$M\,T^{-1}$
Q	river discharge (flow)	$L^3\,T^{-1}$
S	daily suspended-sediment load	$M\,T^{-1}$
S_a	long-term accumulation rate	$L\,T^{-1}$
S_d	short-term accumulation rate	$L\,T^{-1}$
S_R	fluvial sediment input	$M\,T^{-1}$
t_a	long time span	T
t_d	short time span	T
T	age difference between two surfaces	T
β_0	slope of rating curve	
β_1	intercept of rating curve	$M\,L^{-3}$
ε	residual error	
ρ_d	dry bulk density	$M\,L^{-3}$

REFERENCES

Alexander, C.R. and Simoneau, A.M. (1999) Spatial variability in sedimentary processes on the Eel continental slope. *Mar. Geol.*, **154**, 243–254.

Allison, M.A., Kineke, G.C., Gordon, E.S. and Goni, M.A. (2000) Development and reworking of a seasonal flood deposit on the inner continental shelf off the Atchafalaya River. *Cont. Shelf Res.*, **20**, 2267–2294.

Anderson, R.F., Bopp, R.F., Buessler, K.O. and Biscaye, P.E. (1988) Mixing of particles and organic constituents in sediments from the continental shelf and slope off Cape Cod: SEEP I results. *Cont. Shelf Res.*, **8**, 925–946.

Beardsley, R.C. and Lentz, S.J. (1987) The coastal ocean dynamics experiment collection: an introduction. *J. Geophys. Res.*, **92**, 1455–1463.

Behrens, E.W. (1980) On sedimentation rates and porosity. *Mar. Geol.*, 35, M11–M16.

Benninger, L.K., Aller, R.C., Cochran, J.K. and Turekian, K.K. (1979) Effects of biological sediment mixing on the Pb-210 chronology and trace-metal distribution in a Long Island Sound sediment core. *Earth Planet. Sci. Lett.*, **43**, 241–259.

Benson, L.V., Burdett, J.W., Kashgarian, M., *et al.* (1996) Climatic and hydrologic oscillations in the Owens Lake Basin and adjacent Sierra Nevada, California. *Science*, **274**, 746–749.

Bentley, S.J. and Nittrouer, C.A. (2003) Emplacement, modification and preservation of event strata on a flood-dominated continental shelf: Eel shelf, Northern California. *Cont. Shelf Res.*, **23**, 1465–1493.

Best, D.W. (1995) History of timber harvest in the Redwood Creek Basin, northwestern California. *U.S. Geol. Surv. Prof. Pap.*, **1454**, C1–C7.

Biscaye, P.E. and Anderson, R.F. (1994) Fluxes of particulate matter on the slope of the southern Middle Atlantic Bight: SEEP-II. *Deep-sea Res. II*, **41**, 459–509.

Blair, N.E., Leithold, E.L., Ford, S.T., *et al.* (2003) The persistence of memory: the fate of ancient sedimentary organic carbon in a modern sedimentary system. *Geochim. Cosmochim. Acta*, **67**, 63–73.

Blum, M.D. and Törnqvist, T.E. (2000) Fluvial responses to climate and sea-level change: a review and look forward. *Sedimentology*, **47**, 2–48.

Bohannon, R.G., Gardner, J.V. and Sliter, R.W. (2004) Holocene to Pliocene tectonic evolution of the region offshore of the Los Angeles urban corridor, southern California. *Tectonics*, **23**, TC1016, 1–34.

Borgeld, J.C. (1985) *Holocene Sedimentation and Stratigraphy on the Northern California Continental Shelf.* University of Washington, Seattle, 178 pp.

Boyd, R., Dalrymple, R. and Zaitlin, B.A. (1992) Classification of clastic coastal depositional environments. *Sediment. Geol.*, **80**, 139–150.

Bray, N.A. and Greengrove, C.L. (1993) Circulation over the shelf and slope off northern California. *J. Geophys. Res.*, **98**, 18119–18145.

Brown, W.M. and Ritter, J.R. (1971) Sediment transport and turbidity in the Eel River basin, California. *U.S. Geol. Surv. Wat. Supply Pap.*, **1986**, p. 70.

Bull, D., Kemp, A.E.S. and Weedon, G.P. (2000) A 160-k.y.-old record of El Niño–Southern Oscillation in marine production and coastal runoff from Santa Barbara Basin, California, USA. *Geology*, **28**, 1007–1010.

Burger, R.L., Fulthorpe, C.S. and Austin, J.A., Jr. (2001) Late Pleistocene channel incisions in the southern Eel River Basin, northern California: Implications for tectonic vs. eustatic influences on shelf sedimentation patterns. *Mar. Geol.*, **177**, 317–330.

Burger, R.L., Fulthorpe, C.S., Austin Jr., J.A. and Gulick, S.P.S. (2002) Lower Pleistocene to present structural deformation and sequence stratigraphy of the continental shelf, offshore Eel River Basin, northern California. *Mar. Geol.*, **185**, 249–281.

Burger, R.L., Fulthorpe, C.S. and Austin Jr., J.A. (2003) Effects of triple junction migration and glacioeustatic cyclicity on evolution of upper slope morphologies, offshore Eel River Basin, northern California. *Mar. Geol.*, **99**, 307–336.

Cacchione, D.A. and Drake, D.E. (1990) Shelf sediment transport. In: *The Sea*, Vol. 9, *Ocean Engineering Science* (Eds B. LeMehaute and D.M. Hanes), pp. 729–774. John Wiley & Sons, New York.

Cacchione, D.A., Irish, J., Traykovski, P., Wiberg, P.L. and Lynch, J. (1999) Estimates of suspended-sediment flux and bedform activity on the inner portion of the Eel continental shelf. *Mar. Geol.*, **154**, 83–97.

Cacchione, D.A., Pratson, L.F. and Ogston, A.S. (2002) The shaping of continental slopes by internal tides. *Science*, **296**, 724–727.

Caddah, L.F.G., Kowsmann, R.O. and Viana, A.R. (1998) Slope sedimentary facies associated with Pleistocene and Holocene sea-level changes, Campos Basin, southeast Brazilian Margin. *Sediment. Geol.*, **115**, 159–174.

Carpenter, R., Bennett, J.T. and Peterson, M.L. (1981) Pb-210 activities in and fluxes to sediments of the Washington continental slope and shelf. *Geochim. Cosmochim. Acta*, **45**, 1155–1172.

Carpenter, R., Peterson, M.L. and Bennett, J.T. (1982) Pb-210 derived sediment accumulation and mixing rates for the Washington continental slope. *Mar. Geol.*, **48**, 135–164.

Carriquiry, J.D. and Sánchez, A. (1998) Sedimentation in the Colorado River Delta and upper Gulf of California after nearly a century of discharge loss. *Mar. Geol.,* **158**, 125–145.

Carter, L., Manighetti, B., Elliot, M., Trustrum, N. and Gomez, B. (2002) Source, sea level and circulation effects on the sediment flux to the deep ocean over the past 15 ka off eastern New Zealand. *Global Planet. Change,* **33**, 339–355.

Cattaneo, A. and Steel, R.J. (2003) Transgressive deposits: a review of their variability. *Earth-Sci. Rev.,* **62**, 187–228.

Cayan, D.R. and Peterson, D.H. (1989) The influence of North Pacific circulation on streamflow in the west. In: *Aspects of Climate Variability in the Pacific and Western Americas* (Ed. D.H. Peterson), pp. 375–397. Monograph 55, American Geophysical Union, Washington, DC.

Cayan, D.R., Redmond, K.T. and Riddle, L.G. (1999) ENSO and hydrologic extremes in the western United States. *J. Climate,* **12**, 2881–2893.

Clarke, S.H. (1992) Geology of the Eel River Basin and adjacent region: implications for late Cenozoic tectonics of the southern Cascadia Subduction Zone and Mendocino Triple Junction. *Bull. Am. Assoc. Petrol. Geol.,* **76**, 199–224.

Colman, S.M. and Bratton, J.F. (2003) Antropogenically induced changes in sediment and biogenic silica fluxes in Chesapeake Bay. *Geology,* **31**, 71–74.

Crawford, C.G. (1991) Estimation of suspended-sediment rating curves and mean suspended-sediment loads. *J. Hydrol.,* **129**, 331–348.

Crockett, J.S. and Nittrouer, C.A. (2004) The sandy inner shelf as a repository for muddy sediment: an example from northern California. *Cont. Shelf Res.,* **24**, 55–73.

Crowley, K.D. (1984) Filtering of depositional events and the completeness of sedimentary sequences. *J. Sediment. Petrol.,* **54**, 127–136.

Dearing, J.A. and Jones, R.T. (2003) Coupling temporal and spatial dimensions of global sediment flux through lake and marine sediment records. *Global Planet. Change,* **39**, 147–168.

DeMaster, D.J., McKee, B.A., Nittrouer, C.A., Qian, J.C. and Chen, G.D. (1985) Rates of sediment accumulation and particle reworking based on radiochemical measurements from continental shelf deposits in the East China Sea. *Cont. Shelf Res.,* **4**, 143–158.

Demirpolat, S. (1991) Surface and near-surface sediments from the continental shelf off the Russian River, northern California. *Mar. Geol.,* **99**, 163–171.

Dickey, T., Plueddemann, A.J. and Weller, R.A. (1998) Current and water property measurement in the coastal ocean. In: *The Sea,* Vol. 10, *The Global Coastal Ocean: Processes and Methods* (Eds K.H. Brink and A.R. Robinson), pp. 367–398. John Wiley & Sons, New York.

Dingler, J.R. and Clifton, H.E. (1994) Barrier systems of California, Oregon and Washington. In: *Geology of Holocene Barrier Island Systems* (Ed. R.A. Davis), pp. 140–165. Springer-Verlag, New York.

Donnelly, J.P., Driscoll, N.W., Uchupi, E., *et al.* (2005) Catastrophic meltwater discharge down the Hudson Valley: a potential trigger for the Intra-Allerod cold period. *Geology,* **33**, 89–92.

Dott, R.H. (1983) Episodic sedimentation – How normal is average? How rare is rare? Does it matter? *J. Sediment. Petrol.,* **53**, 5–23.

Drake, D.E. (1999) Temporal and spatial variability of the sediment grain-size distribution on the Eel shelf: the flood layer of 1995. *Mar. Geol.,* **154**, 169–182.

Drake, D.E., Kolpack, R.L. and Fisher, P.J. (1972) Sediment-transport on the Santa Barbara-Oxnard shelf, Santa Barbara Channel, California. In: *Shelf Sediment Transport: Process and Pattern* (Eds D.J.P. Swift, D.B. Duane and O.H. Pilkey), pp. 307–331. Dowden, Hutchinson and Ross, Stroudsburg, PA.

Driscoll, N.W. and Karner, G.D. (1999) Three-dimensional quantitative modeling of clinoform development. *Mar. Geol.,* **154**, 383–398.

Dyer, K.R. (1995) Sediment transport processes in estuaries. In: *Geomorphology and Sedimentology of Estuaries* (Ed. G.M.E. Perillo), pp. 423–449. Elsevier, Amsterdam.

Earle, C.J. (1993) Asynchronous droughts in California streamflow as reconstructed from tree rings. *Quat. Res.,* **39**, 290–299.

Eittreim, S.L., Xu, J.P., Noble, M. and Edwards, B.D. (2002) Towards a sediment budget for the Santa Cruz shelf. *Mar. Geol.,* **181**, 235–248.

Ely, L.L., Enzel, Y., Baker, V.R. and Cayan, D.R. (1993) A 5000-year record of extreme floods and climate change in the southwestern United States. *Science,* **262**, 410–412.

Ely, L.L., Enzel, Y. and Cayan, D.R. (1994) Anomalous North Pacific atmospheric circulation and large winter floods in the southwestern United States. *J. Climate,* **7**, 977–987.

Emery, K.O. and Aubrey, D.G. (1986) Relative sea level change from tide gauge records of western North America. *J. Geophys. Res.,* **91**, 13941–13953.

Engstrom, W.N. (1996) The California storm of January 1862. *Quat. Res.,* **46**, 141–148.

Fan, S., Swift, D.J.P., Traykovski, P., *et al.* (2004) River flooding, storm resuspension and event stratigraphy on the northern California shelf: observations compared with simulations. *Mar. Geol.,* **210**, 17–41.

Farnsworth, K.L. and Milliman, J.D. (2003) Effects of climatic and anthropogenic change on small mountainous rivers: the Salinas River example. *Global Planet. Change*, **39**, 53–64.

Ferguson, R.I. (1986) River loads underestimated by rating curves. *Water Resour. Res.*, **22**, 74–76.

Field, M.E., Clarke, S.H. and White, M.E. (1980) Geology and geologic hazards of the offshore Eel River Basin, northern California Continental margin. *U.S. Geol. Surv. Open-file Rep.*, **80–1080**, 80 pp.

Fleming, K., Johnston, P., Zwartz, D., *et al.* (1998) Refining the eustatic sea-level curve since the Last Glacial Maximum using far- and intermediate-field sites. *Earth Planet. Sci. Lett.*, **163**, 327–342.

Friedrichs, C.T. and Wright, L.D. (2004) Gravity-driven sediment transport on the continental shelf: implications for equilibrium profiles near river mouths. *Coast. Eng.*, **51**, 795–811.

Geyer, W.R., Traykovski, P., Hill, P. and Milligan, T. (2000) The structure of the Eel river plume during floods. *Cont. Shelf Res.*, **20**, 2067–2093.

Geyer, W.R., Hill, P.S. and Kineke, G.C. (2004) The transport, transformation and dispersal of sediment by buoyant coastal flows. *Cont. Shelf Res.*, **24**, 927–949.

Goff, J.R. (1997) A chronology of natural and anthropogenic influences on coastal sedimentation, New Zealand. *Mar. Geol.*, **138**, 105–117.

Goff, J.A., Hughes-Clarke, J.E., Orange, D.L. and Mayer, L.A. (1999) Detailed investigation of continental shelf morphology using a high-resolution swath-sonar survey: the Eel margin, northern California. *Mar. Geol.*, **154**, 255–269.

Goodbred, S.L., Jr. and Kuehl, S.A. (1999) Holocene and modern sediment budgets for the Ganges-Brahmaputra river system: evidence for highstand dispersal to flood-plain, shelf and deep-sea depocenters. *Geology*, **27**, 559–562.

Goodbred, S., Jr. and Kuehl, S.A. (2000) Enormous Ganges-Brahmaputra sediment discharge during strengthened early Holocene monsoon. *Geology*, **28**, 1083–1086.

Grant, W.D. and Madsen, O.S. (1979) Combined wave and current interaction with a rough bottom. *J. Geophys. Res.*, **84**, 1797–1808.

Grant, W.D. and Madsen, O.S. (1986) The continental shelf bottom boundary layer. *Ann. Rev. Fluid Mech.*, **18**, 265–305.

Griggs, G.B. and Hein, J.R. (1979) Sources, dispersal and clay mineral composition of fine-grained sediment off central and northern California. *J. Geol.*, **88**, 541–566.

Grützner, J. and Mienert, J. (1999) Lateral changes of mass accumulation rates derived from seismic reflection profiles: an example from the Western Atlantic. *GeoRes. Forum*, **5**, 87–108.

Harden, D.R. (1995) A comparison of flood-producing storms and their impacts in northeastern California. *U.S. Geol. Surv. Prof. Pap.*, **1454**, D1–D9.

Harris, C.K. and Wiberg, P. (2002) Across-shelf sediment transport: Interactions between suspended sediment and bed sediment. *J. Geophys. Res., C: Oceans*, **107**, 8-1–8-12.

Hebbeln, D. (2003) State of the art and future prospects of scientific coring and drilling of marine sediments. In: *Ocean Margin Systems* (Eds G. Wefer, D. Billet, D. Hebbeln, *et al.*), pp. 57–66. Springer-Verlag, Berlin.

Hedges, J.I. and Parker, P.L. (1976) Land-derived organic matter in surface sediments from the Gulf of Mexico. *Geochim. Cosmochim. Acta*, **40**, 1019.

Helley, E.J. and LaMarche, V.C.J. (1973) Historic flood information for northern California streams from geological and botanical evidence. *U.S. Geol. Surv. Prof. Pap.*, **485-E**, E1–E16.

Hickey, B.M. (1979) The California Current system – hypotheses and facts. *Progr. Ocean.*, **8**, 191–279.

Hickey, B.M. (1998) Coastal oceanography of western North America from the tip of Baja California to Vancouver Island. In: *The Sea*, Vol. 11, *The Global Coastal Ocean: Regional Studies and Syntheses* (Eds A.R. Robinson and K.H. Brink), pp. 345–393. John Wiley & Sons, New York.

Hill, P.S., Milligan, T.G. and Geyer, W.R. (2000) Controls on effective settling velocity of suspended sediment in the Eel River flood plume. *Cont. Shelf Res.*, **20**, 2095–2111.

Hirschboeck, K.K. (1987) Catastrophic flooding and atmospheric circulation anomalies. In: *Catastrophic Flooding* (Eds L. Mayer and D. Nash), pp. 23–56. Allen and Unwin, New York.

Hofmann, W. and Rantz, S.E. (1963) Floods of December 1955–January 1956 in the far Western States. *U.S. Geol. Surv. Wat. Supply Pap.*, **1650-A**, 156 pp.

Hovius, N. (1998) Controls on sediment supply by large rivers. In: *Relative Role of Eustacy, Climate and Tectonism in Continental Rocks* (Eds K.W. Shanley and P.J. McCabe), pp. 3–16. Special Publication 59, Society of Economic Paleontologists and Mineralogists, Tulsa, OK.

Inman, D.L. and Jenkins, S.A. (1999) Climate change and the episodicity of sediment flux of small California rivers. *J. Geol.*, **107**, 251–270.

Janda, R.J. and Nolan, K.M. (1979) Stream sediment discharge in northwestern California. In: *Guidebook to a Field Trip to Observe Natural and Management-related Erosion in Franciscan Terrane of Northern California*, Vol. IV, pp. 1–27. Cordilleran Section, Geological Society of America, Boulder, CO.

Johnson, H.D. and Baldwin, C.T. (1996) Shallow silici-clastic seas. In: *Sedimentary Environments and Facies*, 3rd edn (Ed. H.G. Reading), pp. 229–282. Blackwell Science, Oxford.

Johnson, K.S., Paull, C.K., Barry, J.P. and Chavez, F.P. (2001) A decadal record of underflows from a coastal river into the deep sea. *Geology*, **29**, 1019–1022.

Karlin, R. (1980) Sediment sources and clay mineral distributions off the Oregon coast. *J. Sediment. Petrol.*, **50**, 543–559.

Keil, R.G., Hedges, J.I., Tsamakis, E., Fuh, C.B. and Giddings, J.C. (1994) Mineralogical and textural controls on the organic composition of coastal marine sediments: hydrodynamic separation using SPLITT-fractionation. *Geochim. Cosmochim. Acta*, **58**, 879–893.

Kelsey, H.M. (1980) A sediment budget and an analysis of geomorphic processes in the Van Duzen River basin, north coastal California, 1941–1975. *Geol. Soc. Am. Bull.*, **91**, 1190–1216.

Kelsey, H.M., Engebretson, D.C., Mitchell, C.E. and Ticknor, R.L. (1994) Topographic form of the coast ranges of the Cascadia margin in relation to coastal uplift rates and plate subduction. *J. Geophys. Res.*, **99**, 12245–12255.

Kineke, G.C., Woolfe, K.J., Kuehl, S.A., *et al.* (2000) Sediment export from the Sepik River, Papua New Guinea: evidence for a divergent sediment plume. *Cont. Shelf Res.*, **20**, 2239–2266.

Knox, J.C. (1993) Large increase in flood magnitude in response to modest changes in climate. *Nature*, **361**, 430–432.

Knox, J.C. (1995) Fluvial systems since 20,000 years BP. In: *Global Continental Paleohydrology* (Eds K.J. Gregory, L. Starkel and V.R. Baker), pp. 87–108. John Wiley & Sons, Chichester.

Knox, J.C. (2000) Sensitivity of modern and Holocene floods to climate change. *Quat. Sci. Rev.*, **19**, 439–457.

Kudrass, H.R., Michels, K.H., Wiedicke, M. and Suckow, A. (1998) Cyclones and tides as feeders of a submarine canyon off Bangladesh. *Geology*, **26**, 715–718.

Kulm, L.D., Roush, R.C., Harlett, J.C., *et al.* (1975) Oregon continental shelf sedimentation: Interrelationships of facies distribution and sedimentary processes. *J. Geol.*, **83**, 145–175.

Kunkel, K.E., Andsager, K. and Easterling, D.R. (1999) Long-term trends in extreme precipitation events over the conterminous United States and Canada. *J. Climate*, **12**, 2515–2527.

Lang, A., Bork, H.R., Mäckel, R., *et al.* (2003) Changes in sediment flux and storage within a fluvial system: some examples from the Rhine catchment. *Hydrological Processes*, **17**, 3321–3334.

Lee, H.J., Syvitski, J.P.M., Parker, G., *et al.* (2002) Distinguishing sediment waves from slope failure deposits: Field examples, including the 'Humboldt slide' and modelling results. *Mar. Geol.*, **192**, 79–104.

Leithold, E.L. (1989) Depositional processes on an ancient and modern muddy shelf, northern California. *Sedimentology*, **36**, 179–202.

Leithold, E.L. and Blair, N.E. (2001) Watershed control on the carbon loading of marine sedimentary particles. *Geochim. Cosmochim. Acta*, **65**, 2231–2240.

Leithold, E.L. and Hope, R.S. (1999) Deposition and modification of a flood layer on the northern California shelf: Lessons from and about the fate of terrestrial particulate organic carbon. *Mar. Geol.*, **154**, 183–195.

Leithold, E.L., Perkey, D.W., Blair, N.E. and Creamer, T.N. (2005) Sedimentation and carbon burial on the northern California continental shelf: the signatures of land-use change. *Cont. Shelf Res.*, **25**, 349–371.

Lentz, S.J. and Trowbridge, J.H. (1991) The bottom boundary layer over the northern California shelf. *J. Phys. Ocean.*, **21**, 1186–1201.

Lesueur, P., Tastet, J.P. and Weber, O. (2002) Origin and morphosedimentary evolution of fine-grained modern continental shelf deposits: the Gironde mud fields (Bay of Biscay, France). *Sedimentology*, **49**, 1299–1320.

Marron, D.C., Nolan, K.M. and Janda, R.J. (1995) Surface erosion by overland flow in the Redwood Creek Basin, northwestern California – effects of logging and rock type. *U.S. Geol. Surv. Prof. Pap.*, **1454**, H1–H6.

Mayewski, P.A., Rohling, E.E., Stager, J.C., *et al.* (2004) Holocene climate variability. *Quat. Res.*, **62**, 243–255.

McCabe, P.J. and Wolcock, D.M. (2002) A step increase in streamflow in the conterminous United States. *Geophys. Res. Lett.*, **29**, doi:10.1029/2002GL015999.

McCave, I.N. (1972) Transport and escape of fine-grained sediment from shelf areas. In: *Shelf Sediment Transport: Process and Pattern* (Eds D.J.P. Swift, D.B. Duane and O.H. Pilkey), pp. 225–248. Dowden, Hutchinson & Ross, Stroudsburg, PA.

McCave, I.N. (1986) Local and global aspects of the bottom nepheloid layers in the world ocean. *Neth. J. Sea Res.*, **20**, 167–181.

McKee, B.A., Nittrouer, C.A. and DeMaster, D.J. (1983) Concepts of sediment deposition and accumulation applied to the continental shelf near the mouth of the Yangtze River (China). *Geology*, **11**, 631–633.

McKee, B.A., Aller, R.C., Allison, M.A., Bianchi, T.S. and Kineke, G.C. (2004) Transport and transformation of dissolved and particulate materials on continental margins influenced by major rivers: Benthic boundary layer and seabed processes. *Cont. Shelf Res.*, **24**, 899–926.

McPhee-Shaw, E.E. and Kunze, E. (2002) Boundary layer intrusions from a sloping bottom: a mechanism for generating intermediate nepheloid layers. *J. Geophys. Res.*, **107**, 10.1029/2001JC000801.

McPhee-Shaw, E.E., Sternberg, R.W., Mullenbach, B. and Ogston, A.S. (2004) Observations of intermediate nepheloid layers on the northern California continental margin. *Cont. Shelf Res.*, **24**, 693–720.

Meade, R.H., Yuzyk, T.R. and Day, T.J. (1990) Movement and storage of sediment in rivers of the United States and Canada. In: *Geology of North America: Surface Water Hydrology* (Eds M.G. Wolman and H.C. Riggs), pp. 255–280. Geological Society of America, Boulder, CO.

Meko, D.M., Therrell, M.D., Baisan, C.H. and Hughes, M.K. (2001) Sacramento river flow reconstructed to A.D. 869 from tree rings. *J. Am. Wat. Resour. Assoc.*, **37**, 1029–1038.

Merritts, D. and Vincent, K.R. (1989) Geomorphic response of coastal streams to low, intermediate and high rates of uplift, Mendocino junction region, northern California. *Geol. Soc. Am. Bull.*, **101**, 1373–1388.

Mertes, L.A.K. and Warrick, J.A. (2001) Measuring flood output from 110 coastal watersheds in California with field measurements and SeaWiFS. *Geology*, **29**, 659–662.

Meybeck, M. and Vörösmarty, C. (2005) Fluvial filtering of land-to-ocean fluxes: from natural Holocene variations to Anthropocene. *C. R. Geosci.*, **337**, 107–123.

Milliman, J.D. and Syvitski, J.P.M. (1992) Geomorphic/tectonic control of sediment discharge to the ocean: the importance of small mountainous rivers. *J. Geol.*, **100**, 525–544.

Milliman, J.D., Qin, Y.-S., Ren Mei, E. and Saito, Y. (1987) Man's influence on the erosion and transport of sediment by Asian Rivers: the Yellow River (Huanghe) example. *J. Geol.*, **95**, 751–762.

Milly, P.C.D., Wetherald, R.T., Dunne, K.A. and Delworth, T.L. (2002) Increasing risk of great floods in a changing climate. *Nature*, **415**, 514–517.

Mitchell, C.E., Vincent, P., Weldon, R.J.I. and Richards, M.A. (1994) Present-day vertical deformation of the Cascadia margin, Pacific northwest United States. *J. Geophys. Res.*, **99**, 12,257–12,277.

Montgomery, D.R., Schmidt, D.R., Greenberg, H.M. and Deitrich, W.E. (2000) Forest clearing and regional landsliding. *Geology*, **28**, 311–314.

Morehead, M.D., Syvitski, J.P., Hutton, E.W.H. and Peckham, S.D. (2003) Modeling the temporal variability in the flux of sediment from ungauged river basins. *Global Planet. Change*, **39**, 95–110.

Mullenbach, B.L. and Nittrouer, C.A. (2000) Rapid deposition of fluvial sediment in the Eel Canyon, northern California. *Cont. Shelf Res.*, **20**, 2191–2212.

Mullenbach, B.L. and Nittrouer, C.A. (2006) Decadal record of sediment export to the deep sea via the Eel Canyon. *Cont. Shelf Res.*, **26**, 2157–2177.

Mullenbach, B.L., Nittrouer, C.A., Puig, P. and Orange, D.L. (2004) Sediment deposition in a modern submarine canyon: Eel Canyon, northern California. *Mar. Geol.*, **211**, 101–119.

Nash, D.B. (1994) Effective sediment-transporting discharge from magnitude-frequency analysis. *J. Geol.*, **102**, 79–95.

National Research Council (1997) *Contaminated Sediments in Ports and Waterways: Cleanup Strategies and Technologies*. National Academy Press, Washington, DC, 295 pp.

Nichols, M.M. (1989) Sediment accumulation rates and relative sea-level rise in lagoons. *Mar. Geol.*, **88**, 201–219.

Nittrouer, C.A. (1978) *The process of detrital sediment accumulation in a continental shelf environment: example from the Washington shelf.* Thesis, Univeristy of Washington, Seattle, WA, 243 pp.

Nittrouer, C.A. (1999) STRATAFORM: overview of its design and synthesis of its results. *Mar. Geol.*, **154**, 3–12.

Nittrouer, C.A. and Kravitz, J.H. (1995) Integrated continental margin research to benefit ocean and earth sciences. *Eos (Trans. Am. Geophys. Union)*, **76**, 121–126.

Nittrouer, C.A. and Sternberg, R.W. (1981) The formation of sedimentary strata in an allochthonous shelf environment: the Washington continental shelf. *Mar. Geol.*, **42**, 201–232.

Nittrouer, C.A. and Wright, L.D. (1994) Transport of particles across continental shelves. *Rev. Geophys.*, **32**, 85–113.

Nittrouer, C.A., DeMaster, D.J., McKee, B.A., Cutshall, N.H. and Larsen, I.L. (1984) The effect of sediment mixing on Pb-210 accumulation rates for the Washington continental shelf (Columbia River, USA). *Mar. Geol.*, **54**, 201–221.

Nittrouer, C.A., Kuehl, S.A., Sternberg, R.W., Figueiredo, A.G. and Faria, L.E.C. (1995) An introduction to the geological significance of sediment transport and accumulation on the Amazon continental shelf. *Mar. Geol.*, **125**, 177–192.

Nolan, K.M. and Janda, R.J. (1995) Impacts of logging on stream-sediment discharge in the Redwood Creek basin, northwestern California. *U.S. Geol. Surv. Prof. Pap.*, **1454**, L1–L8.

Noren, A.J., Bierman, P.R., Steig, E.J., Lini, A. and Southon, J. (2002) Millennial-scale storminess variability in the northeastern United States during the Holocene epoch. *Nature*, **419**, 821–824.

Ogston, A.S. and Sternberg, R.W. (1999) Sediment-transport events on the northern California continental shelf. *Mar. Geol.*, **154**, 69–82.

Ogston, A.S., Cacchione, D.A., Sternberg, R.W. and Kineke, G.C. (2000) Observations of storm and river flood-driven sediment transport on the northern California continental shelf. *Cont. Shelf Res.*, **20**, 2141–2162.

Ogston, A.S., Guerra, J.V. and Sternberg, R.W. (2004) Interannual variability of nearbed sediment flux on the Eel River shelf, northern California. *Cont. Shelf Res.*, **24**, 117–136.

Ogston, A.S., Sternberg, R.W. and Nittrouer, C.A. (2005) New advances in fine-sediment transport. In: *The Sea*, Vol. 10, *The Global Coastal Ocean: Multi-scale Interdisciplinary Processes* (Eds K.H. Brink and A.R. Robinson), pp. 101–128. John Wiley & Sons, New York.

Olsen, C.R., Nichols, M.M., Larsen, I.L., Lowry, P.D. and Cutshall, N.H. (1986) Geochemistry and deposition of Be–7 in river-estuarine and coastal waters. *J. Geophys. Res.*, **91**, 896–908.

Orange, D.L. (1999) Tectonics, sedimentation and erosion in northern California: submarine geomorphology and sediment preservation potential as a result of three competing processes. *Mar. Geol.*, **154**, 369–382.

Orton, G.J. and Reading, H.G. (1993) Variability of deltaic processes in terms of sediment supply, with particular emphasis in grain size. *Sedimentology*, **40**, 475–512.

Orton, P.M. and Kineke, G.C. (2001) Comparing calculated and observed vertical suspended-sediment distributions from a Hudson River Estuary turbidity maximum. *Estuarine Coast. Shelf Sci.*, **52**, 401–410.

Paola, C. (2000) Quantitative models of sedimentary basin infilling. *Sedimentology*, **47**, 121–178.

Piper, D.J.W. and Normark, W.R. (2001) Sandy fans – from Amazon to Hueneme and beyond. *Bull. Am. Assoc. Petrol. Geol.*, **85**, 1407–1438.

Pirie, D.M. and Stellar, D.D. (1977) *California Coastal Processes Study: Landsat II, Final Report*. NTIS no. E77–10158, 164 pp.

Pirmez, C., Pratson, L.F. and Steckler, M.S. (1998) Clinoform development by advection-diffusion of suspended sediment: Modeling and comparison to natural systems. *J. Geophys. Res., B: Solid Earth*, **103**, 24,141–24,157.

Posamentier, H.W. and Allen, G.P. (1993) Variability of the sequence stratigraphic model: effects of local basin factors. *Sediment. Geol.*, **86**, 91–109.

Posamentier, H.W. and Vail, P.R. (1988) Eustatic controls on clastic deposition II-sequence and systems tract models. In: *Sea-level Changes; an Integrated Approach* (Eds C.K. Wilgus, B.S. Hastings, C.A. Ross, *et al.*), pp. 125–154. Special Publication 42, Society of Economic Paleontologists and Mineralogists, Tulsa, OK.

Prahl, F.G., Ertel, J.R., Goni, M.A., Sparrow, M.A. and Eversmeyer, B. (1994) Terrestrial organic carbon contributions to sediments on the Washington margin. *Geochim. Cosmochim. Acta*, **58**, 3035–3048.

Puig, P., Ogston, A.S., Mullenbach, B.L., Nittrouer, C.A. and Sternberg, R.W. (2003) Shelf-to-canyon sediment-transport processes on the Eel continental margin (northern California). *Mar. Geol.*, **193**, 129–149.

Puig, P., Ogston, A.S., Mullenbach, B.L., Nittrouer, C.A., Parsons, J.D. and Sternberg, R.W. (2004) Storm-induced sediment gravity flows at the head of the Eel submarine canyon. *J. Geophys. Res.*, **109**, doi:10.1029/2003JC001918.

Pullen, J.D. and Allen, J.S. (2000) Modeling studies of the coastal circulation off Northern California: Shelf response to a major Eel River flood event. *Cont. Shelf Res.*, **20**, 2213–2238.

Pullen, J. and Allen, J.S. (2001) Modeling studies of the coastal circulation off Northern California: Statistics and patterns of wintertime flow. *J. Geophys. Res., C: Oceans*, **106**, 26959–26984.

Reneau, S.L., Dietrich, W.E., Donahue, D.J., Jull, A.J. and Rubin, M. (1990) Late Quaternary history of colluvial deposition and erosion in hollows, central California Coast Ranges. *Geol. Soc. Am. Bull.*, **102**, 969–982.

Ritter, J.R. (1971) Sand transport by the Eel River and its effect on nearby beaches. *U.S. Geol. Surv. Wat. Supply Pap.*, **2009-A**, 38 pp.

Sadler, P.M. (1981) Sediment accumulation rates and the completeness of stratigraphic sections. *J. Geol.*, **89**, 569–584.

Sadler, P.M. (1999) The influence of hiatuses on sediment accumulation rates. *GeoRes. Forum*, **5**, 15–40.

Sadler, P.M. and Strauss, D.J. (1990) Estimation of completeness of stratigraphical sections using empirical data and theoretical models. *J. Geol. Soc. London*, **147**, 471–485.

Saito, Y., Yang, Z. and Hori, K. (2001) The Huanghe (Yellow River) and Changjiang (Yangtze River) deltas: a review on their characteristics, evolution and sediment discharge during the Holocene. *Geomorphology*, **41**, 219–231.

Schonher, T. and Nicholson, S.E. (1989) The relationship between California rainfall and ENSO events. *J. Climate*, **2**, 1258–1269.

Schumm, S.A. (1993) River response to baselevel change: implications for sequence stratigraphy. *J. Geol.*, **101**, 279–294.

Scully, M.E., Friedrichs, C.T. and Wright, L.D. (2003) Numerical modeling of gravity-driven sediment transport and deposition on an energetic continental shelf: Eel River, northern California. *J. Geophys. Res., C: Oceans*, **108**, 17-1–17-14.

Sherwood, C.R., Butman, B., Cacchione, D.A., *et al.* (1994) Sediment-transport events on the northern California continental shelf during the 1990–1991 stress experiment. *Cont. Shelf Res.*, **14**, 1063–1100.

Sherwood, C.R., Drake, D.E., Wiberg, P.L. and Wheatcroft, R.A. (2002) Prediction of the fate of p,p-DDE in sediment on the Palos Verdes shelf, California. *Cont. Shelf Res.*, **22**, 1025–1058.

Showers, W.J. and Angle, D.G. (1986) Stable isotopic characterization of organic carbon accumulation on the Amazon continental shelf. *Cont. Shelf Res.*, **6**, 227–244.

Sloan, J., Miller, J.R. and Lancaster, N. (2001) Response and recovery of the Eel River, California and its tributaries to floods in 1955, 1964 and 1997. *Geomorphology*, **36**, 129–154.

Smith, J.D. (1977) Modeling of sediment transport on shelves. In: *The Sea*, Vol. 6, *Marine Modelling* (Eds E.D. Goldberg, I.N. McCave, J.J. O'Brien and J.H. Steele), pp. 539–577. Wiley-Interscience, New York.

Smoak, J.M., DeMaster, D.J., Kuehl, S.A., Pope, R.H. and McKee, B.A. (1996) The behavior of particle-reactive tracers in a high turbidity environment: ^{234}Th and ^{210}Pb on the Amazon continental shelf. *Geochim. Cosmochim. Acta*, **60**, 2123–2137.

Snyder, N.P., Whipple, K.X., Tucker, G.E. and Merritts, D. (2003) Channel response to tectonic forcing: field analysis of stream morphology and hydrology in the Mendocino triple junction region. *Geomorphology*, **53**, 97–127.

Sommerfield, C.K. (2006) On sediment accumulation rates and stratigraphic completeness: Lessons from Holocene ocean margins. *Cont. Shelf Res.*, **26**, 2225–2240.

Sommerfield, C.K. and Lee, H.J. (2003) Magnitude and variability of Holocene sediment accumulation in Santa Monica Bay, California. *Marine Environmental Research*, **56**, 151–176.

Sommerfield, C.K. and Lee, H.J. (2004) Across-shelf sediment transport since the Last Glacial Maximum, southern California margin. *Geology*, **32**, 345–348.

Sommerfield, C.K. and Nittrouer, C.A. (1999) Modern accumulation rates and a sediment budget for the Eel shelf: a flood-dominated depositional environment. *Mar. Geol.*, **154**, 227–241.

Sommerfield, C.K. and Wheatcroft, R.A. (In press) Late Holocene sediment accumulation on the northern California continental shelf: Oceanic, fluvial and anthropogenic influences. *Geol. Soc. Am. Bull.*

Sommerfield, C.K., Nittrouer, C.A. and Alexander, C.R. (1999) Be–7 as a tracer of flood sedimentation on the northern California continental margin. *Cont. Shelf Res.*, **19**, 335–361.

Sommerfield, C.K., Drake, D.E. and Wheatcroft, R.A. (2002) Shelf record of climatic changes in flood magnitude and frequency, north-coastal California. *Geology*, **30**, 395–398.

Spinelli, G.A. and Field, M.E. (2003) Controls of tectonics and sediment source locations on along-strike variations in transgressive deposits on the northern California margin. *Mar. Geol.*, **197**, 35–47.

Stanley, D.J. and Warne, A.G. (1994) World-wide initiation of Holocene marine deltas by deceleration of sea-level rise. *Science*, **265**, 228–231.

Stanley, D.J. and Warne, A.G. (1998) Nile delta in its destruction phase. *J. Coast. Res.*, **14**, 794–825.

Sternberg, R.W. and Nowell, A.R.M. (1999) Continental shelf sedimentology: scales of investigation define future research opportunities. *J. Sea Res.*, **41**, 55–71.

Stow, D.A.V. and Piper, D.J.W. (1984) Deep-water fine-grained sediments: facies models. In: *Fine-grained Sediments: Deep-water Processes and Facies* (Eds D.A.V. Stow and D.J.W. Piper), pp. 611–646. Special Publication 15, Geological Society of London. Blackwell Scientific Publications, Oxford.

Strub, P.T., Allen, J.S., Huyer, A. and Smith, R.L. (1987) Seasonal cycles of currents, temperatures, winds and sea level over the Pacific continental shelf: 35N to 48N. *J. Geophys. Res.*, **92**, 1507–1526.

Stuiver, M., Reimer, P.J., Bard, E., *et al.* (1998) INTCAL98 radiocarbon age calibration, 24,000-0 cal BP. *Radiocarbon*, **40**, 1041–1083.

Swift, D.J.P. and Thorne, J.A. (1991) Sedimentation on continental margins, I: a general model for continental shelf sedimentation. In: *Shelf Sand and Sandstone Bodies: Geometry, Facies and Sequence Stratigraphy* (Eds D.J.P. Swift, G.F. Oertel, R.W. Tillman and J.A. Thorne), pp. 3–31. Special Publication 14, International Association of Sediemntologists. Blackwell Scientific Publications, Oxford.

Syvitski, J.P.M., Morehead, M.D. and Nicholson, M. (1998) HYDROTREND: a climate-driven hydrologic-transport model for predicting discharge and sediment load to lakes or oceans. *Comput. Geosci.*, **24**, 51–68.

Syvitski, J.P.M., Morehead, M.D., Bahr, D.B. and Mulder, T. (2000) Estimating fluvial sediment transport: the rating parameters. *Water Resour. Res.*, **36**, 2747–2760.

Syvitski, J.P.M., Peckham, S.D., Hilberman, R. and Mulder, T. (2003) Predicting the terrestrial flux of sediment to the global ocean: a planetary perspective. *Sediment. Geol.*, **162**, 5–24.

Syvitski, J.P.M., Vörösmarty, C.J., Kettner, A.J. and Green, P. (2005) Impact of humans on the flux of terrestrial sediment to the global coastal ocean. *Science*, **308**, 376–380.

Thomas, C.J., Blair, N.E., Alperin, M.J., *et al.* (2002) Organic carbon deposition on the North Carolina continental slope off Cape Hatteras (USA). *Deep-sea Res. II*, **49**, 4687–4709.

Thorne, P.D. and Hanes, D.M. (2002) A review of acoustic measurements of small-scale sediment processes. *Cont. Shelf Res.*, **22**, 603–632.

Traykovski, P., Geyer, W.R., Irish, J.D. and Lynch, J.F. (2000) The role of wave-induced density-driven fluid mud flows for cross-shelf transport on the Eel River continental shelf. *Cont. Shelf Res.*, **20**, 2113–2140.

Trowbridge, J.H. and Nowell, A.R.M. (1994) An introduction to the Sediment Transport on Shelves and Slopes (STRESS) program. *Cont. Shelf Res.*, **14**, 1957–1061.

Trowbridge, J.H., Chapman, D.C. and Candela, J. (1998) Topographic effects, straits and the bottom bounday layer. In: *The Sea*, Vol. 10, *The Global Coastal Ocean: Multiscale Interdisciplinary Processes* (Eds K.H. Brink and A.R. Robinson), pp. 63–88. John Wiley & Sons, New York.

Turekian, K.K. and Cochran, J.K. (1978) Determination of marine chronologies using natural radionuclides. In: *Chemical Oceanography*, Vol. 7 (Eds J.P. Riley and R. Chester), pp. 313–360. Academic Press, London.

U.S. Environmental Protection Agency (1999a) *South Fork Eel River Total Maximum Daily Loads, Sediment and Temperature*. EPA Region IX Water Division, San Francisco, CA, pp. 62.

U.S. Environmental Protection Agency (1999b) *Van Duzen and Yager Creek Total Maximum Daily Loads for Sediment*. EPA Region IX Water Division, San Francisco, California, pp. 57.

U.S. Global Ocean Flux Study (1989) *Sediment Trap Technology and Sampling*. U.S. JGOFS Planning Office, Woods Hole Oceanographic Institution, Woods Hole, 94 pp.

Vörösmarty, C.J., Meybeck, M., Fekete, B., *et al.* (2003) Anthropogenic sediment retention: Major global impact from registered river impoundments. *Global Planet. Change*, **39**, 169–190.

Waananen, A.O., Harris, D.D. and Williams, R.C. (1971) Floods of December 1964 and January 1965 in the Far Western States. *U.S. Geol. Surv. Wat. Supply Pap.*, **1866-A**, 265 pp.

Walling, D.E. (1999) Linking land use, erosion and sediment yields in river basins. *Hydrobiologia*, **410**, 223–240.

Walsh, J.P. and Nittrouer, C.A. (1999) Observations of sediment flux to the Eel continental slope, northern California. *Mar. Geol.*, **154**, 55–68.

Warrick, J.A. and Milliman, J.D. (2003) Hyperpycnal sediment discharge from semiarid southern California rivers: Implications for coastal sediment budgets. *Geology*, **31**, 781–784.

Washburn, L., Swenson, M.S., Largier, J.L., Kosro, P.M. and Ramp, S.R. (1993) Cross-shelf sediment transport by an anticyclonic eddy off northern California. *Science*, **261**, 1560–1564.

Wheatcroft, R.A. and Borgeld, J.C. (2000) Oceanic flood deposits on the northern California shelf: Large-scale distribution and small-scale physical properties. *Cont. Shelf Res.*, **20**, 2163–2190.

Wheatcroft, R.A. and Drake, D.E. (2003) Post-depositional alteration and preservation of sedimentary event layers on continental margins, I. Role of episodic sedimentation. *Mar. Geol.*, **199**, 123–137.

Wheatcroft, R.A. and Martin, W.R. (1996) Spatial variation in short-term (^{234}Th) sediment bioturbation intensity along an organic-carbon gradient. *J. Mar. Res.*, **54**, 763–792.

Wheatcroft, R.A. and Sommerfield, C.K. (2005) River sediment flux and shelf sediment accumulation rates on the Pacific Northwest margin. *Cont. Shelf Res.*, **25**, 311–332.

Wheatcroft, R.A., Sommerfield, C.K., Drake, D.E., Borgeld, J.C. and Nittrouer, C.A. (1997) Rapid and widespread dispersal of flood sediment on the northern California margin. *Geology*, **25**, 163–166.

White, W.A., Morton, R.A. and Holmes, C.W. (2002) A comparison of factors controlling sedimentation rates and wetland loss in fluvial-deltaic systems, Texas Gulf coast. *Geomorphology*, **44**, 47–66.

Wiberg, P.L. (2000) A perfect storm: formation and potential for preservation of storm beds on the continental shelf. *Oceanography*, **13**, 93–99.

Williams, A.J., III, Schaffner, L.C. and Maa, J.P.-Y. (1997) SuperBASS tripod for benthic turbulence measurement. *Oceans '97, MTS/IEEE Conference Proceedings*, pp. 524–527.

Wright, L.D. and Nittrouer, C.A. (1995) Dispersal of river sediments in coastal seas: six contrasting cases. *Estuaries*, **18**, 494–508.

Wright, L.D., Kim, S.C. and Friedrichs, C.T. (1999) Across-shelf variations in bed roughness, bed stress and sediment suspension on the northern California shelf. *Mar. Geol.*, **154**, 99–115.

Wright, L.D., Friedrichs, C.T., Kim, S.C. and Scully, M.E. (2001) Effects of ambient currents and waves on gravity-driven sediment transport on continental shelves. *Mar. Geol.*, **175**, 25–45.

Submarine mass movements on continental margins

HOMA J. LEE*, JACQUES LOCAT†, PRISCILLA DESGAGNÉS†, JEFFREY D. PARSONS‡,
BRIAN G. McADOO§, DANIEL L. ORANGE¶, PERE PUIG**, FLORENCE L. WONG*,
PETER DARTNELL* and ERIC BOULANGER†

*United States Geological Survey, 345 Middlefield Road, Menlo Park, CA 94025, USA (Email: hjlee@usgs.gov)
†Département de Géologie et de Génie Géologique, Laval University, Québec G1K 7P4, Canada
‡School of Oceanography, University of Washington, Seattle, WA 98195, USA
§Department of Geology and Geography, Vassar College, Poughkeepsie, NY 12604, USA
¶AOA Geophysics, Inc., Moss Landing, CA 95039, USA
**Institut de Ciències del Mar, Barcelona 08003, Spain

ABSTRACT

Submarine landslides can be important mechanisms for transporting sediment down sloping seabeds. They occur when stresses acting downslope exceed the available strength of the seabed sediments. Landslides occur preferentially in particular environments, including fjords, active river deltas, submarine canyons, volcanic islands and, to a lesser extent, the open continental slope. Evaluating the relative stability of different seabeds requires an understanding of driving stresses and sediment strength. Stresses can be caused by gravity, earthquakes and storm waves. Resisting strength can be reduced by pore water and gas pressures, groundwater seepage, rapid sediment deposition, cyclic loading and human activity. Once slopes have become unstable or have failed, strength may continue to decrease, leading to sediment debris flows and possibly turbidity currents. Recent submarine landslide research has: shown that landslides and sediment waves may generate similar deposits, which require careful interpretation; expanded our knowledge of how strength develops in marine sediment; improved techniques for predicting sediment rheology; and developed methodologies for mapping and predicting the medium- to large-scale regional occurrence of submarine landslides.

Keywords Landslides, earthquakes, rheology, shear strength, Humboldt Slide, sediment waves.

INTRODUCTION

This paper considers **submarine mass movements**, which result when marine sediment or rock is loaded by the environment until it fails. These events are also referred to as **landslides**, although not all involve actual sliding of one mass of material over another. Mass movements can result either from an increase in the environmental loads, a decrease in the strength of the sediment/rock, or a combination of load increase and strength decrease. The resulting types of gravity-driven sediment transport events differ from such processes as fluid-mud migration in that the moving sediment has previously been formed into consolidated material possessing strength and stiffness. Submarine landslides can subsequently transform into debris flows or even turbidity currents (Hampton, 1972), but such transformations do not always occur. Likewise, turbidity currents and sediment flows can occur by means other than an initial landslide (e.g. hyperpycnal flows from rivers, Mulder & Syvitski, 1995). Parsons *et al.* (this volume, pp. 275–337) review flows including fluid mud, debris flows and turbidity currents, and this paper focuses on submarine landslides: how and when these events produce subsequent flows. This paper also focuses on how shear strength develops in marine sediment, because knowledge of shear strength is essential in predicting slope stability.

HISTORIC DEVELOPMENT OF UNDERSTANDING

Knowledge about submarine landslides extends back at least to the 1929 Grand Banks incident (Heezen & Ewing, 1952), when a magnitude 7.2 earthquake set off a sequence of events that led to an orderly progression of submarine cable breaks south of Newfoundland. Following the event, the scientific community began a discussion that has continued to the present concerning exactly what caused the cable breaks and how the resulting processes related to the initial earthquake loading. Terzaghi (1956) proposed a mechanism that is now known to be almost certainly false: that the cable breaks resulted from an advancing front of liquefaction within the seafloor sediment. That is, the sediment did not move but rather a liquefaction wave passed through the sediment. Heezen & Ewing (1952) presented the more convincing argument that the earthquake 'jarred the continental slope and shelf, setting landslides and slumps in motion'. These mass movements 'raced downward, and by the incorporation of water, the moving sediment was transformed into turbidity currents'. The currents started in different submarine canyons; as the flows down each canyon joined, the currents grew larger. Ultimately the combined current covered the floor of a 300-km-wide bight. The currents easily snapped the cables and proceeded toward the south for at least 600 km.

The model suggested by Heezen & Ewing (1952) has generally withstood the test of time, but the details of the event are not yet fully understood. At one point scientists thought that the initial failure might have been one or two very large landslides. Heezen & Drake (1964) found what appeared to be a 500-m-thick displaced block on the upper part of the Laurentian Fan, and Emery *et al.* (1970) found a similar scale feature 40 km to the west. Later investigations (Piper & Normark, 1982) showed that these features were not displaced blocks at all, but were parts of eroded terrain that only appeared to be blocks when viewed along individual tracklines. Recent studies involving high-resolution sidescan sonar and sub-bottom profiles show there were probably many small failures in the source region having a variety of scales and morphologies. These include rotational, retrogressive slumps that passed downslope to form debris flows that carried blocks and formed channels. Piper *et al.* (1999) inferred that the debris flows, in turn, transformed into turbidity currents (Fig. 1). A problem that has still not been fully resolved is the difference in sediment types between that of the

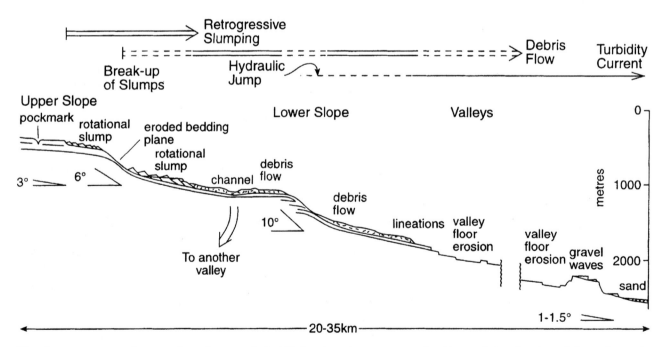

Fig. 1 Interpretative drawing from Piper *et al.* (1999) showing the sequence of sediment facies developed downslope of the Grand Banks margin.

source region, where the sediment is dominantly fine grained, and that of the Sohm Abyssal Plain, where the massive turbidite deposited in 1929 is predominantly sand (Piper & Aksu, 1987). The continuing discussion of the Grand Banks event, which occurred 75 yr ago, is an indication of the difficulty in understanding events that occur out of sight and in an environment that has similarities to the subaerial world but displays many differences.

Another burst of interest in submarine landslides occurred in 1969 when Hurricane Camille struck the coast of Louisiana. Wave heights of 21–23 m were recorded near the South Pass portion of the Mississippi Delta, and three offshore drilling platforms were badly damaged or destroyed. One of the platforms was found, half-buried in mud and displaced 30 m downslope. Investigations after the hurricane showed that bottom relief had changed by as much as 12 m. The mode of damage to the platforms (piles bent rather than pulled out) led investigators to believe that sediment failure was a primary cause for the platform damage (Sterling & Strohbeck, 1973; Bea *et al.*, 1983). A period of active research followed, leading to models of storm-wave-induced slope failure. There also was renewed interest in the deltaic sediment strength, which resulted in a new appreciation for the role of rapid sediment accumulation in generating low strength and excess pore-water pressures, the importance of weak layers, and the development of instrumentation to measure excess pore pressures (Garrison, 1977).

In the late 1980s (Moore *et al.*, 1989; Holcombe & Searle, 1991), following improvements in seafloor mapping techniques (e.g. long-range sidescan sonar system GLORIA), the marine community became aware that the largest landslides on Earth occur under water. These giant landslides, having run-out distances over 200 km and volumes exceeding 5000 km^3, occur most commonly off volcanic islands, and are particularly well displayed off Hawaii and the Canary Islands. However, giant landslides also occur on continental margins and include the massive Storegga landslide complex off Norway (Bryn *et al.*, 2003). The mechanics are not well understood of how these landslides are triggered; however, strength development in active volcanoes and gas hydrate dissociation on continental margins appear to play a part (Kayen & Lee, 1991).

The role of submarine landsides in producing **tsunamis** has recently received attention following the 1998 magnitude 7.0 earthquake that occurred off the northern coast of Papua New Guinea. Two thousand people died from tsunami waves that washed ashore after the earthquake. This event sparked considerable discussion and debate (Tappin *et al.*, 1999; Geist, 2000), but the majority of scientists presently believe that the large sea waves were not generated directly by the earthquake motions but rather by a large underwater landslide initiated by seismic shaking. Marine surveys have identified an amphitheatre-shaped region and complex topography, which have been interpreted, respectively, as the erosion scar and a series of large, displaced blocks having a thickness of up to 700 m (Tappin *et al.*, 2003).

Landslide-induced tsunamis also were observed during the 1964 Alaska earthquake in Resurrection Bay and Port Valdez and many other locations off the southern coast of Alaska (Coulter & Migliaccio, 1966; Lemke, 1967). Coastal and submarine slope failures caused loss of shoreline and coastal facilities. Tsunamis generated by the landslides repeatedly inundated the coastal areas, causing many fatalities and considerable property damage to the communities. In addition to earthquake-induced landslides, fjords are susceptible to slope failures, particularly those caused by low tides. The coastline and seafloor of Kitimat Arm, British Columbia, failed in 1975 just after a low tide. The deposits remaining after the failure were mapped by Prior *et al.* (1982a) and these images showed a wide range of recognizable features, including outrunner blocks that probably raced ahead of the rest of the landslide debris through a process of hydroplaning. Similar failures occurred in Skagway Alaska in 1994 (Rabinovich *et al.*, 1999) and Howe Sound, British Columbia in 1955.

CLASSIFICATION

The discussion above provides a background for the scale, importance and continuing debate related to the field of submarine landslides. However, a classification system is needed to consider these events in greater detail. This paper follows the terminology recommended by Varnes (1958) with some modification (Fig. 2). Other classification

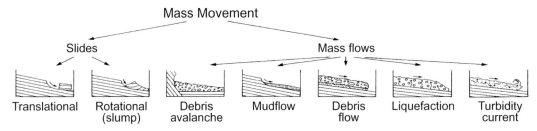

Fig. 2 General landslide classification. See text for further details. (Modified from Varnes, 1958.)

schemes have been provided by Prior (1984), Norem *et al.* (1990) and Mulder & Cochonat (1996). **Slope failure** occurs when the downslope driving forces acting on the material composing the seafloor are greater than the forces acting to resist major deformations. Following slope failure, the failed mass moves downslope under the influence of gravity and possibly other forces. Thus, **mass movement** is defined as the movement of the failed material driven directly by gravity or other body forces, rather than tractive stresses associated with fluid motion. If the moving sediment takes a form that resembles a viscous fluid, the feature is termed a **mass flow (gravity flow)**. Such a failure has considerable internal deformation with innumerable invisible or short-lived internal slip surfaces. **Slides** are movements of essentially rigid, internally undeformed masses along discrete slip planes. In the literature, all forms of mass movement are occasionally referred to as **slumps**. Correctly, slumps are a kind of slide in which blocks of failed material rotate along curved slip surfaces. The other kind of slide involves movement on a planar surface and is referred to as a **translational slide**. In each of these types, movement can be fast or slow. Extremely slow movement is called **creep**.

Submarine slides can become mass flows (gravity flows) as the failed mass progressively disintegrates and continuous downslope movement occurs (Morgenstern, 1967; Hampton, 1972). End-member products of disintegrating slides have been given special names. **Debris flows** are flows in which the sediment is heterogeneous and may include larger clasts supported by a matrix of fine sediment. **Mud flows** involve predominantly muddy sediment. **Turbidity currents** involve the downslope transport of a relatively dilute suspension of sediment grains that are supported by an upward component of fluid turbulence (Parsons *et al.*, this volume, pp. 275–337). Turbidity currents are often generated

by the disintegration and dilution of slides or debris flows, although they also may be generated independently of other mass wasting events. **Liquefaction flows** occur when a loosely packed sediment collapses under environmental conditions such as cyclic loading from earthquakes or waves. The grains temporarily lose most contact with one another, and the particle weight is temporarily transferred to the pore fluid, producing excess porewater pressures. The material may flow downslope under the influence of gravity or spread laterally under the influence of stresses induced by earthquakes or perhaps storm waves.

As discussed above, recent surveys have revealed giant submarine landslides that involve the failure of thousands of cubic-kilometres of rock and sediment (Moore *et al.*, 1989). When these landslides have disintegrated into smaller pieces (which may still be quite large) and have clearly moved very rapidly without a channel, they are referred to as **debris avalanches** (Varnes, 1958).

ENVIRONMENTS

Submarine landslides are not distributed uniformly over the world's oceans, but instead they tend to occur commonly where there are thick bodies of soft sediment, where the slopes are steep, and where the loads exerted by the environment are high. These conditions are met in fjords, deltas, submarine canyons and on the continental slope.

Fjords

Fjords with high sediment accumulation rates are one of the environments most susceptible to failure, both in terms of the proportional areal extent of deposits that can become involved in mass movement and also in terms of the recurrence interval of slope failures at a given location. **Fjords** are glacially eroded steep-walled valleys that have been inundated by the sea and are typically fed by sediment-laden rivers and streams that drain glaciers. These factors lead to environmental conditions that are highly conducive to slope failure. There is typically a delta at the head of the fjord, formed by streams draining the remnant of the glacier that initially eroded the valley, with foreset beds that dip at 5° to 30° between 10 m and 50 m water depth. Below this, the prodelta dips at angles of 0.1° to 5° to the flat basin floor, typically at depths between 100 m and 1000 m (Syvitski *et al.*, 1987).

The glacial streams feeding fjord deltas carry both rock flour and coarse sediment, which form deposits that easily lose strength when shaken by an earthquake. In addition, the sediment may be deposited so rapidly that pore-water pressures cannot dissipate completely, resulting in an under-consolidated state and abnormally low strength. Abundant organic matter brought down with the glacial debris also can decay and produce bubble-phase gas within the seabed that can lead to elevated pore-water pressure and low strength. Some fjord-delta deposits are so near instability that they fail during greater than average low tides, during which the supporting forces of the fjord waters themselves are temporarily removed from the sediment. Many of these steep slopes are composed of weak sediment that is susceptible to cyclic loading, and may fail seasonally or semi-continuously via numerous small-scale slope failures. Slope failures of a seasonal or semi-continuous nature have been reported in many Canadian fjords, including Bute Inlet (Syvitski & Farrow, 1983; Prior *et al.*, 1986a), Knight Inlet (Syvitski & Farrow, 1983) and North Bentinck Arm (Kostaschuk & McCann, 1983). In some situations, fjord-head delta slopes may fail infrequently and produce catastrophic effects, such as occurred in Valdez (Coulter & Migliaccio, 1966), Seward (Lemke, 1967) and Whittier (Kachadoorian, 1965) during the 1964 Alaska earthquake or in Kitimat Arm, Canada (Prior *et al.*, 1982a, 1982b).

In addition to the fjord-delta deposits, the side-wall slopes of fjords also can be unstable. Deposition of suspended sediment on the steep (10° to 90° overhangs) submerged valley sides can frequently lead to small slope failures (Farrow *et al.*, 1983). Even more important are slope failures on side-entry deltas that build out rapidly onto the side-

wall slopes (e.g. Howe Sound, Canada; Terzaghi, 1956; Prior *et al.*, 1981).

Finally, the deep fjord basins, which tend to have slopes of less than 0.1°, commonly receive failed sediment masses and flows from the side walls and fjord-head deltas. If these landslides incorporate enough water during their movement, they can evolve into gravity flows and turbidity currents. Submarine channels can feed these gravity flows and turbidity currents into and across the basins (Syvitski *et al.*, 1987).

Fjords are commonly found in rugged mountainous terrain, and the unstable fjord-head deltas and side-entry deltas are commonly the only flat land available for coastal development. Not only do these developments become vulnerable to natural slope failure, but human activities also can lead to additional slope failures. For example, a river channel stabilization programme at Howe Sound (Terzaghi, 1956) caused rapid delta growth to be localized and probably contributed to ultimate slope failure on the delta.

Active river deltas on the continental shelf

Active river deltas are another likely site for slope instability. Rivers contribute large quantities of sediment to relatively localized areas on the continental margins. Depending upon a variety of environmental factors, including rate of sediment influx, wave and current activity, and the configuration of the continental shelf and coastline, thick deltaic deposits can accumulate fairly rapidly. These sediment wedges can become the locations of sediment instability and landsliding. To create large, deep-seated landslides, a thick deposit containing comparatively low-strength sediment or containing weak zones is needed. Most of the continental shelves were subaerially exposed during the last glacial cycle, so most sediment on the shelves from that time or before has been eroded, desiccated or otherwise diagenetically altered. Accordingly, the strengths of these older deposits are commonly high enough to resist downslope gravitational stresses on the gentle shelves, and all but the very greatest storm- and earthquake-induced stresses. As a result of rapid sedimentation, young deposits tend to have relatively low strength. In addition, decaying organic matter can produce bubble-phase gas that can further reduce strength. These locations may fail under gravitational loading (due to the slope angle alone) or during storms or earthquakes.

The locations of the major sedimentary depocentres provide some information on where undersea landslides might be expected on the continental shelf. Glacially fed rivers debouching into the Gulf of Alaska or adjacent sounds and inlets contribute 450×10^6 tons of sediment per year (Milliman & Meade, 1983). Slope failures have been identified in modern sediment all along the margin including within the Kayak Trough (Molnia et al., 1977) and Alsek prodelta (Schwab & Lee, 1983; Schwab et al., 1987). Also included are major landslides on the Copper River prodelta (Reimnitz, 1972), off Icy Bay and the Malaspina Glacier (Carlson, 1978; Lee & Edwards, 1986) and off Yakutat (Schwab & Lee, 1983). The high incidence of landsliding arises because of the intensity of earthquake and storm-wave loading (Schwab & Lee, 1983), the thickness of modern sediment, and the tendency of glacial rock flour to lose strength when cyclically loaded.

The Mississippi River contributes the most sediment to the sea of any single river within North America (2.1×10^8 tons of sediment per year; Milliman & Meade, 1983). Most of this sediment is deposited in front of the modern bird-foot or Balize delta, a delta-lobe that has been in existence for only 600–800 yr (Fisk et al., 1954). The distributary mouths of the modern delta build seaward at rates varying from 50 to 100 m yr^{-1} depending on the particular distributary. Seaward of these distributaries, the sediment accumulation rates are very high, reaching 1 m yr^{-1} or more (Coleman et al., 1980). The sediment consists mostly of clay-sized particles, rich in organic matter that is rapidly degraded to gas (mainly methane and carbon dioxide). Rapid sedimentation and gas charging lead to high excess pore-water pressures and a state of extreme underconsolidation. Although the gradients of the delta front are very low (ranging from 0.2° to 1.5°), evidence of slope failure is widespread, including submarine gullies (Shepard, 1955) and extensive fields of sediment instability features all along the delta front (Coleman et al., 1980). Another subaqueous delta that displays considerable landslide activity is the Huanghe (Prior et al., 1986b), where the gradient is so gentle that the sediment collapses upon itself (collapse depression) rather than deforming downslope.

Submarine canyon-fan systems

Submarine canyon-fan systems serve as conduits for passing large amounts of sediment from the continental shelf to the deep sea. The presence of extensive, thick sediment fans and abyssal plains off the coasts of many areas testifies to the importance of mass-movement mechanisms associated with these systems, which are capable of bringing sand-size and even coarser particles to locations hundreds of kilometres from shore. Landsliding appears to be an element that allows the formation of massive submarine fans. According to one model (Hampton, 1972), sediment accumulations in canyon heads begin to move as coherent landslide blocks following some triggering event, such as an earthquake or storm. As the blocks move downslope, the resulting jostling and agitation causes disintegration and subsequent incorporation of water. The debris flow that is produced displays increasingly fluid-like behaviour. As the debris flow continues on its path, further dilution by surrounding water occurs, particularly as sediment is eroded from the front of the flow. Ultimately, a dilute turbulent cloud is created that has a density below 1.1 g cm^{-3}. The resulting turbidity current can flow for long distances (up to hundreds of kilometres) at moderate to high velocities (Parsons et al., this volume, pp. 275–337).

Landsliding, particularly within submarine canyons, appears to be an important, if not essential, part of the process of building deep-sea fans, which are among the most extensive sedimentary features of the Earth's surface. However, the circumstances surrounding these slope failures and their subsequent conversion into turbidity currents are poorly understood. Storms cause sediment movement in canyons, perhaps by inducing failure near the canyon heads (Marshall, 1978; Puig et al., 2003) or perhaps by introducing or resuspending enough sediment to form a gravity current directly (Shepard & Marshall, 1973; Reynolds, 1987). Earthquakes also cause landslides in canyon heads and subsequent turbidity-current flow (Malouta et al., 1981; Adams, 1984), but details of this process are lacking. Major earthquakes and other shocks do not always cause canyon-head landslides (Dill, 1964). However, landslides can occur under aseismic conditions (Shepard, 1951). Landsliding in canyon heads and turbidity-current mobilization were probably more common during glacial stages (Nelson, 1976; Barnard, 1978) because of increased sediment supply, proximity of river sediment source to steep canyon slopes, and possibly increased storm-wave loading resulting from lowered sea level.

The open continental slope

A final common environment for undersea landsliding is the intercanyon area of the continental slope. Landslides have been reported all around the globe along continental slopes removed from submarine canyon-fan systems. Included are slopes off southern California (Buffington & Moore, 1962; Haner & Gorsline, 1978; Field & Clarke, 1979; Nardin *et al.*, 1979a,b; Ploessel *et al.*, 1979; Field & Edwards, 1980; Field & Richmond, 1980; Hein & Gorsline, 1981; Thornton, 1986), central and northern California (Field *et al.*, 1980; Richmond and Burdick, 1981), Alaska (Marlow *et al.*, 1970; Hampton & Bouma, 1977; Carlson *et al.*, 1980), Gulf of Mexico (Lehner, 1969; Woodbury, 1977; Booth & Garrison, 1978; Booth, 1979), and Atlantic coast (Embley & Jacobi, 1977; McGregor, 1977; Knebel & Carson, 1979; McGregor *et al.*, 1979; Malahoff *et al.*, 1980; Cashman & Popenoe, 1985; O'Leary, 1986).

The COSTA (Continental Slope Stability) Project (Canals *et al.*, 2004) investigated large submarine landslides on the open continental slope around the European continent and off Africa. There are several very large features, particularly off Norway. The largest is the Storegga Slide, off central Norway, with a runout distance of 810 km and a volume of 2400 to 3200 km³. Others off Norway include the Traenadjupet Slide, off northern Norway, and the Malenbukta Slide off Svalbard (Haflidason *et al.*, 2004). The occurrence of these large landslides off Norway appears to be related to glacial stages. At Storegga, contrasting sedimentation conditions produced clay layers during interglacial stages overlain by glaciomarine layers produced during glacial stages. Owing to rapid sedimentation, some of these layers contain excess pore pressures. The resulting weak layers provide a plane along which failure can occur, although the failures are still probably triggered by earthquakes (Bryn *et al.*, 2003; Canals, 2004). Large failures have occurred at the site of the Storegga Slide at semi-regular intervals over the past 500 kyr, but the most recent large failure occurred 8200 yr ago. Other large failures investigated by COSTA include the BIG 95 Slide off the east coast of Spain in the Mediterranean and the Canary Slide in the east-central Atlantic off Africa (Canals *et al.*, 2004).

Open-continental-slope failures are found near river mouths and far removed from them, as well as in both arid and humid climates. Ages of the slope failures are seldom known (the Storegga Slide is one of the few exceptions), so investigators cannot determine whether they occurred under glacial or interglacial conditions. Many were probably seismically induced because the typical gradients of continental slopes are 5° or less and the seabed should be statically stable (unless very weak layers are present), and because storm-wave loading is seldom a major factor much below the shelf break (Lee & Edwards, 1986). Some failures appear to be related to the presence of relatively weak sediment layers. The occurrence of failures seems to correlate with sediment accumulation rate, bathymetric gradient, seismicity, and presence of bubble-phase gas and gas hydrate, but the relationships are complex (Field, 1981). The continental slope is an area of extensive mass wasting; however, estimating recurrence intervals for slope failures in this environment is often difficult. Predicting the likelihood of failure (Syvitski *et al.*, this volume, pp. 459–529) at any specific location can be accomplished using three-dimensional seismics and deep sediment boreholes, but the process is very expensive. Landslides on the open slope are probably an important sediment-transport mechanism and pose hazards to offshore development.

STATISTICS OF SUBMARINE LANDSLIDES

Booth *et al.* (1993) presented statistical information on a large suite of submarine landslides on the USA Atlantic margin. The Atlantic margin is representative of passive margins, as far as landslide prevalence is concerned, and includes both glaciated and non-glaciated sections. The authors reviewed the characteristics of 179 individually mapped landslides and prepared a map showing the location of each. The statistical information showed that the features in their study most commonly originated on a seafloor gradient of between

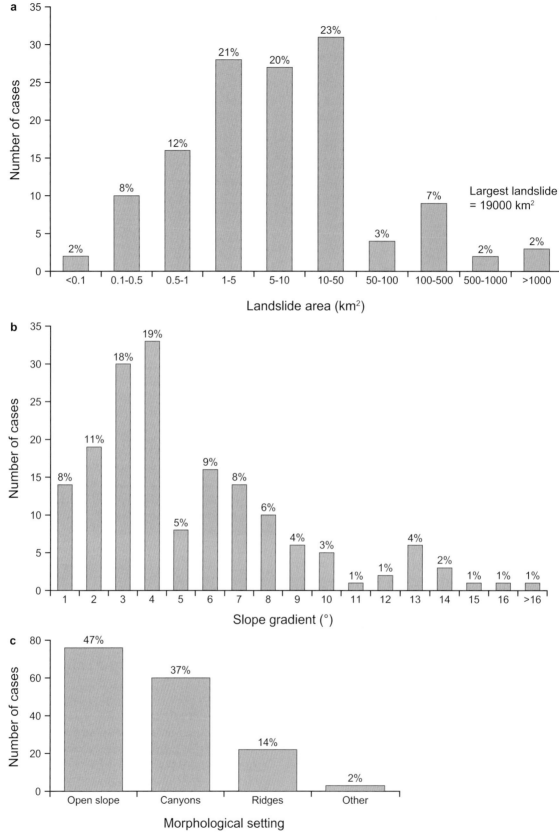

Fig. 3 (a) Distribution of submarine-landslide area on USA Atlantic margin. (b) Distribution of bathymetric gradients at these sites. (c) Distribution of landslide locations for the same sites. (After Booth *et al.*, 1993.)

3° and 4° (Fig. 3b) and had an area of 1–50 km² (Fig. 3a); however, smaller landslides might have been missed in the survey because of a lack of complete data coverage. They found that most landslides occurred on the open slope (Fig. 3c), although almost as many occurred in submarine canyons. Given that canyons cover a smaller percentage of total area than do open-slope environments, frequency of landsliding per unit area is probably higher in canyons than the open slope.

Booth *et al.* (1993) observed that most of the landslides were disintegrative; that is, after initial slope failure, whether translational or rotational, most landslides tend to develop large strains, lose their internal structure, and flow, collapse, or generally break up into debris or rubble. This implies that the sediment tends to weaken considerably once it experiences stresses greater than its original strength can withstand (i.e. after initial failure).

MECHANICS OF SLOPE FAILURE

To understand the mechanics involved in slope failure, it is necessary to consider the following factors: (i) driving stresses, (ii) strength and (iii) potential mobilization into flows.

Driving stress

Generally, a landslide will take place in a given setting if the driving stresses exceed the shearing resistance of the sediment or rock mass. Gravity exerts a downslope driving stress as long as the seafloor is not flat. In fact, even on nearly flat surfaces, where the seafloor gradient is much less than 1°, debris flows continue to move downslope, sometimes for great distances. Bathymetric-map data can be used to give an estimate of this driving stress field, because the gravity-induced shear stresses vary with steepness. This is done by assuming that each point on the seafloor initially can be approximated as part of an infinite surface. In effect, most topographic complexities are ignored. For an infinite surface, the downslope gravitational **shear stress** at any point below the seafloor is then given by

$$\tau_s = \gamma' h \sin \alpha \qquad (1)$$

where τ_s is the downslope shear stress, γ' is the average buoyant (submerged) density of sediment, h is the depth below the seafloor (i.e. thickness of the failed layer) and α is the seafloor gradient. Therefore, if charts of bathymetric gradient and surface-sediment density can be made from multi-beam and core data (Fig. 4), respectively, then the downslope driving stress can be calculated within the context of a geographical information system (GIS) by applying Eq. 1 to the slope and density spatial data (Lee *et al.*, 1999).

Earthquake shaking also produces shear stresses that can cause the seafloor to fail. These stresses are related to accelerations and the dynamic response of the sediment column. Earthquake-induced stresses can add to the ambient gravitational stresses, causing a previously stable slope to deform, fail and potentially transform into a flow. **Earthquake-induced cyclic stresses** also can lead to the development of excess pore-water pressures, and then to a degradation in shear strength. The combination of enhanced stress and degraded strength causes earthquake loading to be a particularly effective mechanism for slope failure. A simple method to account for both earthquake and gravitational loads was presented by Morgenstern (1967), modified by Lee & Edwards (1986), and assumes an infinite surface and a pseudo-static earthquake acceleration.

In continental-shelf depths, large **storm waves** can induce a field of shear stresses on the seafloor. These result from the passage of alternating wave crests and troughs that differentially load the seafloor surface and produce the highest shear stresses halfway between crest and trough (Henkel, 1970). As discussed above, loading from hurricane waves has been known to cause failures in the Mississippi Delta and the subsequent loss of off-shore drilling platforms on the shelf (Bea *et al.*, 1983). A simplified method for predicting the shear stresses resulting from storm waves was presented by Seed & Rahman (1978). Storm-wave-induced shear stresses can combine with the ambient gravitational stresses on slopes and lead to failure (Lee & Edwards, 1986).

Resisting stress (strength)

Submarine mass movements take place either in rock, in sediment or a mixture of both. At one end of the spectrum are hard rocks for which failure

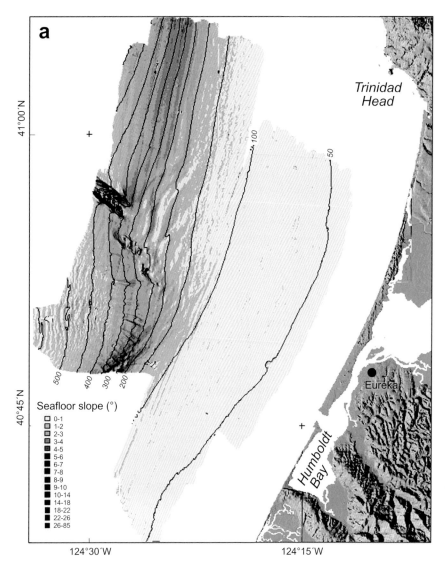

Fig. 4 STRATAFORM study area. (a) Example of regional variability of seabed steepness from multibeam bathymetry. (b) Example of the regional variability of surface-sediment density obtained from cores. Isobaths in metres. (After Lee *et al.*, 1999.)

usually takes place along pre-existing discontinuities (e.g. bedding planes) so the shearing resistance of the intact rock is not mobilized. For soft sediment, the shearing resistance is mobilized throughout the full volume. For soft rocks or hard sediment, the shearing resistance is often mobilized along shear bands or within localized zones of failure.

Material strength may be measured either *in situ*, on shipboard, or in the laboratory. In the laboratory, the shearing resistance of a sediment or rock is obtained by geotechnical tests such as **direct shear**, **simple shear** or **triaxial shear** under confining pressures representative of *in situ* conditions. Simple tests, such as unconfined compression, vane or cone tests, are also useful (see Lee (1985) for details on

measurement techniques and Lambe & Whitman (1969) for further details on shear strength).

The shear strength of sediment represents its ability to resist shear stress. As the shear stress applied to a sediment element steadily increases, the shear strain of the sediment element increases as well. At some point a limiting shear stress is reached and the sediment element strains by a large or unlimited amount. This limiting stress is taken as the **shear strength**. If sediment is considered to be a particular assemblage of grains, the shear strength of that sediment can vary dramatically depending on the way in which shear stresses are applied and the stress history of the sediment. If stresses are applied so rapidly that pore

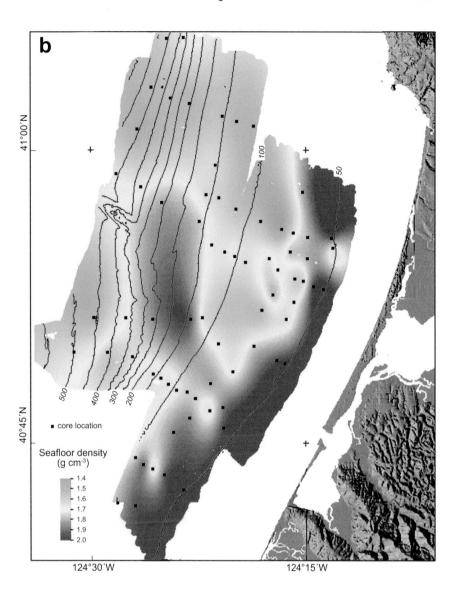

Fig. 4 (*cont'd*)

water cannot leave or enter the sediment framework, conditions are said to be **undrained**. Under these conditions, pressures in the pore water, either positive or negative, commonly build up and have an influence on the strength. On the other hand, if stresses are applied so slowly that no excess pore pressures are developed, loading conditions are said to be **drained**. The rate of loading required to achieve either drained or undrained conditions is highly dependent upon the grain size, sorting and permeability of the sediment. Undrained failure of a sandy sediment commonly occurs during very rapid loading, such as one might encounter during an earthquake. Otherwise, slope failure in sandy sediment usually occurs under drained condi-

tions. In contrast, clayey and silty sediment more commonly fails under undrained conditions. The critical case for slope stability in these sediments is the undrained one, because the undrained shear strength is commonly less than the drained shear strength.

The **stress history** of the sediment also is an important factor influencing shear strength. When sediment particles first come together in a flocculating environment and are deposited on the seafloor, the effective stress acting on them is very low. Here the effective stress, σ', is taken to represent the total stress, σ, minus the pore-water pressure, u. On the seafloor the high **hydrostatic pressure**, corresponding to the overall water depth, contributes to both the total stress and the pore-water pressure. The difference between the total stress and the pore-water pressure is small and includes only the accumulated submerged weight per unit area of the overlying sediment particles (e.g. near the sediment surface the effective stress resulting from sediment overburden is practically zero). As deposition continues and the sediment element becomes buried to greater depths in the sediment column, the **effective overburden stress** increases nearly linearly with sub-bottom depth. Under these conditions, the sediment compacts and dewaters, and the shear strength increases. **Normal consolidation** occurs when sediment accumulation is slow and steady.

For normal consolidation, the shear strength measured under either drained or undrained conditions increases linearly with the increasing effective overburden stress. Such a response is distinctly frictional and for drained loading is represented by

$$\tau_f = \sigma' \tan \phi' \tag{2}$$

where τ_f is the shear stress at failure, σ' is the effective stress and ϕ', which relates the two, is defined as the **friction angle**. For undrained loading, the shear strength is represented by

$$s_u = \sigma_v' S \tag{3}$$

where s_u is the undrained shear strength, σ_v' is the vertical effective overburden stress and S is a sediment constant (often equal to about 0.3 for fine-grained marine sediment; Lee & Edwards, 1986). The vertical effective stress is

$$\sigma_v' = \gamma' z - u \tag{4}$$

where γ' is the average buoyant weight of sediment, z is the depth in the sediment column and u is the excess pore-water pressure (in excess of hydrostatic).

Equations 2 and 3 show a simple application of the concept of **normalized shear-strength behaviour**, that is, shear strength normalized by effective stress can be related to simple expressions that involve a limited number of sediment constants.

When overburden is removed from normally consolidated sediment (by erosion, for example), it becomes **overconsolidated**. The largest stress that was ever reached is termed the maximum past stress, σ_{vm}', and the **overconsolidation ratio** (OCR) is σ_{vm}' divided by the present overburden stress or σ_v'. The normalized strength formulation has been determined through extensive experimental work (Ladd *et al.*, 1977; Lee & Edwards, 1986) to be

$$s_u = \sigma_v' S (\text{OCR})^m \tag{5}$$

where m is a sediment constant that can be determined by experiment but commonly is equal to about 0.8.

A problem with marine sediment is that classic overconsolidation such as that resulting from erosion, which is expressed by Eq. 5, is not the only mechanism by which sediment can become densified and strengthened. Other factors such as bioturbation, repeated seismic loading and cementation also can play a role. Strength that is higher than expected in a sediment that appears to have a normal-consolidation history is termed **pseudo-overconsolidation** (Silva, 1974).

Slope stability analysis

Most simplistically, slopes fail when the driving stresses exceed the resisting stresses. In the marine environment the **factor of safety** of a slope is often calculated by simply dividing the shear strength (e.g. obtained from Eq. 5) by the shear stress exerted on an infinite surface (Eq. 1). Sometimes an earthquake-acceleration term is included to allow for seismic loading (Morgenstern, 1967; Lee & Edwards, 1986). Other, more complex techniques that allow for the geometry of real surfaces are given in Syvitski *et al.* (this volume, pp. 459–529).

PORE-WATER PRESSURE

As can be seen in Eq. 4, the effective stress is strongly impacted by pore-water pressure, which in turn has an impact on the shear strength through Eqs 2–4. Accordingly, much of the literature and research in marine geotechnology and marine slope stability have been directed toward estimating or measuring pore-water pressure.

Sangrey (1977) provided a review of the various mechanisms that can produce excess pore-water pressures in marine sediment. Perhaps the most common cause is rapid sedimentation. When sediment is initially loaded by **overburden**, all of the new load is carried by pressure in the pore water. This occurs because the sediment mineral framework cannot increase its load carrying capacity (**effective stress**) without compressing, and compression cannot occur instantaneously in a fully saturated medium (i.e. water must flow out for the mineral framework to compress). If the sediment has a low permeability or a high compressibility, considerable time may be required for the excess pore-water to drain away. A measure of the sediment's ability to discharge water in a short amount of time is the geotechnical engineering parameter, c_v, termed the **coefficient of consolidation**. Rate of drainage varies directly with c_v, which is the ratio of permeability to compressibility (change in thickness per unit change in vertical effective stress). If the sedimentation rate is high and c_v is low, then potentially large excess pore-water pressures can form in the sediment column.

Gibson (1958) developed a theoretical relation for predicting the level of excess pore pressure that might result from rapid consolidation. High pore-water pressures, approaching **lithostatic** (the pressure exerted by the buoyant weight of the overlying sediment), can exist for the combinations of accumulation rate and c_v found in many active river deltas, including the Mississippi. In fact, the large numbers of landslides on very gentle gradients found in the Mississippi Delta are generally attributed to pore pressures produced by rapid sedimentation (Coleman & Garrison, 1977).

A second source of pore-water pressure results from gas charging. **Bubble-phase gas** can develop in marine sediment through a variety of geochemical and physical processes. These include the decay of organic matter, migration from other locations and dissociation of hydrates during temperature or pressure changes (Kayen & Lee, 1991; Sultan *et al.*, 2003). Whatever the cause, the expanding bubbles pressurize the water around them. Again, if this pressured water cannot flow away fast enough, the pore-water pressure builds up, the effective stress drops and the shear strength decreases (Esrig & Kirby, 1977). For a study in Norton Sound, Alaska, Hampton *et al.* (1982) found that *in situ* penetration resistance (directly related to strength) varied considerably with apparent

gas charging. In areas of gas anomalies, the penetration resistance was as much as 10 times lower than that in sediment only 1 km away that did not show gas anomalies.

Investigators have attempted to determine the influence of gas charging on sediment strength behaviour in the laboratory. Recently Grozic *et al.* (2000) evaluated the ability of gassy sediment to resist dramatic strength loss (liquefaction) using cyclic triaxial tests. Standard sands were artificially charged with gas in a controlled manner. They were then tested using standard techniques to determine the **cyclic stress ratio** (ratio of applied cyclic shear stress to effective stress) necessary to cause liquefaction (termed the **cyclic resistance ratio**). The results of the laboratory experiments showed that the cyclic resistance ratio increases as sediment density and gas content increase. In other words, the higher the gas content, the less susceptible the sediment is to dramatic strength losses during cyclic loading events, including earthquakes and storms. This is likely because the gas compresses and dissolves during cyclic loading, thereby increasing the density and increasing the strength. Such a finding seems to conflict with the general understanding of marine geologists that gas-charged sediment is more susceptible to slope failure (Field & Barber, 1993). Perhaps there are trade-offs; gas-charged sediment is much weaker with respect to static loads, as illustrated by the penetration tests in Norton Sound discussed above. Then, even if the sediment does not lose strength as readily during cyclic loads, the strength is already so low because of its intitial state that failures in gas-charged sediment may be more likely.

A special type of gas charging results from gas-hydrate dissociation. In fact, gas-hydrate dissociation has been suggested as a cause of mega-landslides and has even been considered to be a 'gun' that responds to climate change (Maslin *et al.*, 2004). **Gas hydrates** are solid-solution compounds in which natural gases are caged within a rigid lattice of water ice. These compounds can exist only under conditions of high pressure and low temperature, and they serve as enormous and highly concentrated reservoirs of gaseous hydrocarbons (Kayen & Lee, 1991). When they are intact, gas hydrates form strong layers in the sediment column by virtue of their ice-like structure. However, gas hydrates can disassociate when there is a change in environmental conditions: specifically an increase in temperature or a decrease in pressure. When this occurs, large quantities of free gas can be released into the sediment column. If the pressures produced by this gas release cannot dissipate, perhaps because the coefficient of consolidation, c_v, is too low, pore-fluid (water + gas) pressures will increase dramatically and sediment shear strength will decrease. Such changes could produce massive slope failures. Clearly, global increases in temperature or decreases in pressure (sea-level fall) have the potential for causing seafloor failures on a world-wide scale, thus the notion of a hydrate 'gun'.

A third source of elevated pore-water pressure is **groundwater flow** and artesian pressure conditions. Such pressures can easily result when the water table near the coast is above sea level and water flows out of the seafloor though coastal aquifers (Sangrey, 1977). A somewhat similar situation exists in subduction zones; high porosity sediment is subducted, and the interstitial fluids are forced upward into the accretionary complex. Again, if the fluids cannot be removed fast enough, pore-water pressures increase dramatically, possibly approaching lithostatic (Shi & Wang, 1988). Orange & Breen (1992) and Orange et al. (1997) showed that pore pressures which developed in subduction complexes can contribute to slope failure and the formation of headless canyons.

A fourth source of excess pore pressure is **cyclic loading** of sediment. When cyclically varying shear stresses are applied by either an earthquake or storm waves, sediment grains can become mobile. With increasing numbers of cycles, the grains can gradually move from being in contact with, and supporting, each other to being separated from each other and supported by the pore water. In extreme examples, the grains become totally supported by pore fluid (100% pore pressure response) and the shear strength can decrease to almost zero. This last situation is referred to as **liquefaction**, because the sediment begins to behave as a liquid with almost no shear strength (Seed, 1968). Sands and silts tend to be more susceptible to liquefaction; however, certain classes of fine-grained sediment known as **quick clays** also can lose their strength catastrophically (Bjerrum, 1955).

Strength loss because of cyclic loading can be evaluated either in the laboratory or in the field.

In the laboratory (Fischer et al., 1976; Silver et al., 1976), samples can be conditioned under a state of stress representative of in situ conditions before an earthquake or storm. This is done in either a triaxial cell or simple shear device. Next, cyclically varying shear stresses are applied and both strain and pore-pressure response are monitored. For the sake of consistency and because of laboratory equipment limitations, failure is defined as either a certain strain level (e.g. 15%) or a certain pore-pressure response (e.g. 80%) (Lee & Focht, 1976). The results are then used to define strength loss during cyclic loading, and these loss terms are used to forecast the failure. Lee & Edwards (1986) developed a cyclic-loading strength-reduction factor, A_r, that represents the cyclic stress level (relative to the static shear strength) that causes failure in 10 cycles (representative of a moderate earthquake). This methodology is most suitable for fine-grained sediment because the sediment is less sensitive to coring disturbance.

The potential for liquefaction in sands is generally evaluated with field tests using the methods of Seed & Idriss (1971). The procedure involves an empirical curve on a plot of **cyclic stress ratio** (CSR; cyclic shear stress, τ_c, divided by the overburden stress, σ_v') versus modified blow count from a standard penetration test performed using a geotechnical drilling rig (Fig. 5). The CSR is calculated for an assumed level of earthquake shaking represented by a peak acceleration (Seed & Idriss, 1971).

SEDIMENT MOBILIZATION AND STRENGTH LOSS

Following initial failure, some landslides mobilize into flows whereas others remain as limited deformation slides (Hampton et al., 1996). The mechanisms for mobilization into flows are not well understood but the initial density state of the sediment is likely to be one important factor (Poulos et al., 1985, Lee et al., 1991). If the sediment is initially less dense than an appropriate steady-state condition (**contractive sediment**) the sediment is more likely to flow than one that is denser than the steady-state condition (**dilatant sediment**) (Fig. 6). The **steady state** represents a boundary between contractive and dilatant behaviour and is described by the porosity-effective stress conditions

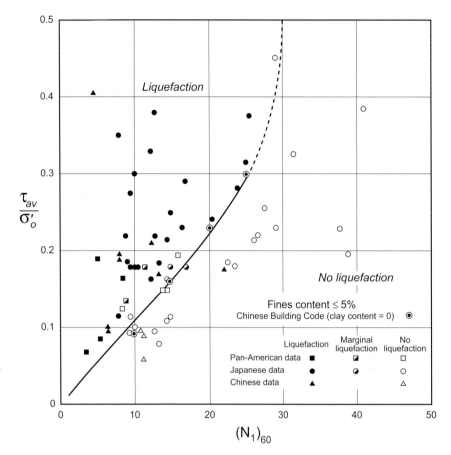

Fig. 5 Common procedure for predicting liquefaction susceptibility: $(N_1)_{60}$ is a measure of the resistance to penetration from a standard penetration test; τ_{av} is the average shear stress anticipated from a design earthquake; σ_o' is the overburden effective stress. The solid line defines the boundary between sediment that will liquefy and sediment that will not. (After Seed & Idriss, 1971.)

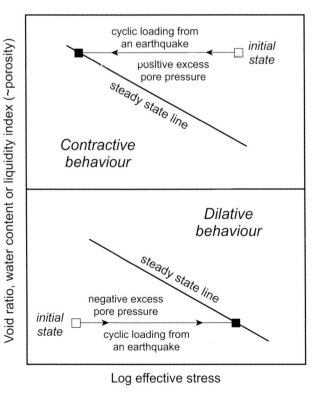

that a sediment will assume when strained by a large amount. This response is another example of pore-pressure generation; contractive sediment generates positive pore pressure during shear and these increased pore pressures reduce the strength. Contractive failure of loose sedimentary deposits can occur at constant porosity. The sediment essentially collapses upon itself and loses much of its strength in the process. Dilatant sediment generates negative pore-water pressures

Fig. 6 (*left*) Line of steady-state deformation. If the initial sediment state lies above the steady-state line, there will be a tendency toward the generation of positive pore pressures during a failure event (e.g. an earthquake). Such pore-pressure generation will lead to a dramatic decrease in effective stress and shear strength and will increase the tendency for sediment flows. An initial sediment state below the steady-state line will produce dilatant behaviour, negative pore-pressure generation, increased effective stress and decreased tendency to flow. (After Lee *et al.*, 1991.)

during shear and becomes stronger in the process. The ability to flow also may be related to the amount of energy transferred to the failing sediment during the failure event (Leroueil *et al.*, 1996; Locat & Lee, 2002).

Strength loss at constant porosity cannot explain the behaviour of some far-reaching debris flows (Locat *et al.*, 1996; Schwab *et al.*, 1996). For example, there is evidence for debris flows reaching the distal lobes of the Mississippi Fan that must have flowed for roughly 500 km on slopes as gentle as 0.06° (Locat *et al.*, 1996). If these deposits represent such flows, then an estimate for the threshold yield strength is 9 Pa. Such a value is three orders of magnitude lower than an estimated remoulded shear strength in the presumed source region (Locat *et al.*, 1996), where the **remoulded shear strength** is the minimum strength value obtained by manually working a sample. Accordingly, the sediment must have taken on additional water during flow, increasing its porosity and decreasing its strength, or the flow must have hydroplaned (Parsons *et al.*, this volume, pp. 275–337; Syvitski *et al.*, this volume, pp. 459–529). The dilution mechanism is consistent with that suggested by Hampton (1972), who described an increase in the sediment water content caused by the jostling and deformation within the sediment during the remoulding phase of the failure event. This greatly reduces the shear strength and provides a fluid-like behaviour to the mixture. During this dilution process, however, the water content cannot become so high that the resulting flow transforms into a turbidity current. In the example of the Mississippi Fan debris-flow deposits, the presence of clasts demonstrates that the sediment flow still retained competence and was not so energetic as to cause the clasts to disintegrate.

TRIGGERS

Submarine landslides are triggered either by an increase in the driving stresses, a decrease in strength, or a combination of the two. The following possible triggers show the interplay of these factors. Note that the relative importance of each of these triggers is not well understood. For example, in some environments one of these triggers will dominate, whereas in others a different trigger will be most significant.

Sediment accumulation

Rapid sediment accumulation contributes to failure in several ways. First, as discussed above, when sediment accumulates rapidly, most of the weight of newly added sediment is carried by pore-water pressures. The shear strength probably increases somewhat because some water will always be squeezed out, even if the coefficient of consolidation (c_v) is low (i.e. relatively low permeability and/or high compressibility). However, the shear stress acting downslope increases more rapidly. As seen in Eq. 1, the shear stress increases with the weight of sediment and is not influenced by pore pressure. The shear stress also may increase because more sediment may be deposited at the head of the sloping surface than at the toe. All three of the following processes push the slope toward failure: retarded strength development, increased development of shear stress because of thickness of the sediment body, and increased development of shear stress because of increases in the slope steepness.

As discussed above, the Mississippi Delta is an ideal example of sediment failure induced by sediment accumulation. Coleman *et al.* (1993), Prior & Coleman (1978) and many other publications document a large variety of sediment failures on the Mississippi prodelta (Fig. 7). Coleman (1988) showed that virtually the entire seafloor surface of the delta front was covered with failure features.

Erosion

Localized erosion by moving water or sediment flows is common in deep-sea channels, submarine canyons and other active sediment-transport systems. When seabed surfaces are undercut, this can decrease the stability by increasing shear stress and in some cases decreasing the shear strength. Monterey Canyon, located off central California, shows many examples of erosion-induced slope failures (Fig. 8). Often these failures dam the canyon so that subsequent turbidity-current flows are diverted, leading to further erosion and second-generation landslides (Greene *et al.*, 2002).

Earthquakes

Earthquakes are called upon as a cause for many unexplained submarine landslide features (Lee &

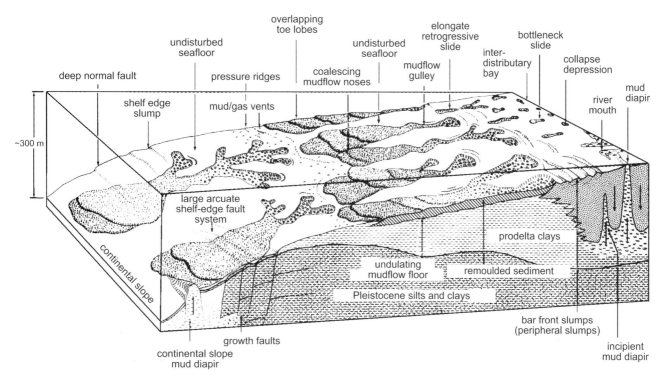

Fig. 7 Schematic block diagram showing the relationship of the various types of submarine sediment instabilities for the Mississippi Delta. (From Coleman *et al.*, 1993.)

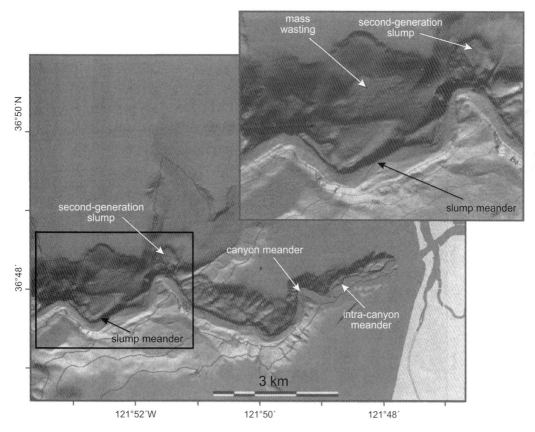

Fig. 8 Multibeam image of the headward part of Monterey Canyon showing canyon and intracanyon meanders, slump-produced meanders, and mass-wasting associated with undercutting along the sides of the canyon axis. Expanded view shows a slump meander and a well-defined second-generation slump resulting from erosion at the apex of a meander. (From Greene *et al.*, 2002.)

Edwards, 1986; Hampton *et al.*, 1996). One reason is that, under water, earthquake-induced shear stresses are quite large relative to shear strength. The seismic shear stress is high because the earthquake must accelerate all of the sediment column including the interstitial water. The shear strength is relatively low because it builds up in proportion to the submerged unit weight of the sediment (Eqs 3 & 4) and may be even lower if there are excess pore pressures (Eq. 4). The ratio of driving stress to resisting strength is high relative to what is usually found on land for the same earthquake. This is because, on land, the water table is seldom at the surface continuously so the strength builds up with the total weight of sediment above the water table. Earthquakes also generate excess pore pressures through cyclic loading as discussed above, which can lower the strength more and possibly induce a state of liquefaction.

Examples of earthquake-induced submarine failures are numerous and include the 1929 Grand Banks event (Piper *et al.*, 1999), multiple failures in Alaskan fjords during the 1964 earthquake (Coulter & Miliaccio, 1966; Fig. 9) and the 1998 Papua New Guinea earthquake and tsunami (Tappin *et al.*, 2003). Earthquake-induced landslides, the resulting turbidity currents, and the turbidites they produce have been used to date major subduction-zone earthquakes in Cascadia (Goldfinger *et al.*, 2003).

Fig. 9 Submarine landslide area at Valdez, Alaska, following the 1964 earthquake. The dashed lines indicate the dock area destroyed by the landslide. (After Coulter & Miliaccio, 1966.)

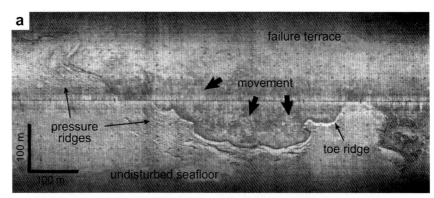

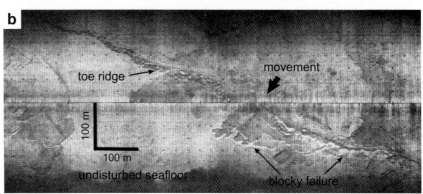

Fig. 10 Sidescan-sonar images of the seafloor off the Klamath River mouth following the 1980 earthquake. (a) The failure terrace has a mottled appearance that is distinctly different from that of the unfailed area. Note the well-defined toe ridge that marks the seaward edge of the landslide and the parallel pressure ridges within the landslide deposit. (b) More complexity occurs near the terminus. The boundary between the landslide zone and the undisturbed seafloor is marked by a series of discrete blocks measuring 10–50 m on each side. (From Field, 1993.)

Another example of earthquake-induced failure is the magnitude 7.0 earthquake that struck the northern California coast in 1980 in an area that had been surveyed previously using sub-bottom profiling equipment. Following the earthquake, local commercial fishermen, who frequently travel the coastal waters, reported the presence of one or more north-west-trending scarps seaward of the Klamath River mouth. The previous surveys had shown the area to be a smooth, featureless depositional environment, and the sudden appearance of scarps suggested a causal relation to the earthquake. Surveys of the area were conducted 2, 7 and 12 months after the earthquake using cameras, sidescan sonar and high-resolution sub-bottom profiling equipment (Field, 1993). The surveys showed a series of features indicative of sediment liquefaction, lateral spreading and flows, all on a gradient of only 0.25°. Pressure ridges and a toe ridge occurred near the seaward boundary of the failed area, which approximated the sand–mud depositional boundary (Fig. 10). The failed area was about 2 km wide and 20 km long. This survey was one of the first to show that continental shelf sands can liquefy readily during earthquakes and

that the northern California margin is susceptible to earthquake-induced failures.

Volcanoes

The existence of giant submarine landslides on the flanks of the Hawaiian Islands has been the subject of debate for at least 50 yr (Normark *et al.*, 1993). Using limited bathymetry data, Moore (1964) interpreted irregular blocky ridges extending downslope from giant amphitheatre-shaped scars on the submarine north flanks of Molokai and Oahu as representing giant landslides. The origin of these deposits was confirmed when complete GLORIA sidescan-sonar data were acquired in the 1980s (Moore *et al.*, 1989). In fact, the GLORIA surveys showed that the Hawaiian Islands were surrounded by many giant submarine landslides (Fig. 11). Further work (Holcombe & Searle, 1991) has shown that the Hawaiian Islands are not alone, and that many, if not most, oceanic volcanoes fail catastrophically during part of their existence.

Some component of oceanic volcanism is clearly a trigger for submarine landslides, but the nature

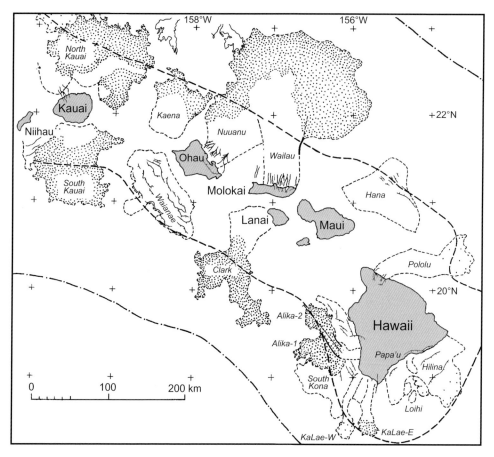

Fig. 11 South-eastern Hawaiian Ridge showing the outline of major landslides identified by name. Dotted pattern indicates hummocky deposits observed on seismic-reflection profiles and GLORIA images; more widely spaced dot pattern indicates subdued relief. Ticked lines indicate major fault scarps. Thin, downslope-directed, irregular lines indicate submarine canyons and their subaerial extensions. Heavy dashed line is the axis of the Hawaiian Trough, and the dashed–dotted line is the crest of the Hawaiian Arch. (Modified from Normark *et al.*, 1993.)

of that component has not been determined. Most of the larger, older landslides seem to have occurred late in the shield-building phase of the host volcano (Moore *et al.*, 1989). At this point, the volcano is still producing significant magma and stands at its highest elevation above sea level. Many factors are present at this point and may combine to produce a trigger. Clearly there are earthquakes and significant gravitational downslope stresses resulting from the great topographic relief of the islands. Such factors are also present on most active continental margins, but giant landslides are infrequent in these locations (McAdoo *et al.*, 2000). Many volcanic islands, including the Hawaiian Islands, are built upon pre-existing pelagic sediment bodies, commonly clay. This could produce a weak basal layer (Dietrich, 1988) that

might contain excess pore-water pressures, even though the island basalts are fairly permeable and were built over millions of years. Magma pressure in the rift zones is too small to trigger landslides wider than a few kilometres (Iverson, 1995). Groundwater forces provide another possible trigger, but are probably not important except under special circumstances (Iverson, 1995). Another intriguing but speculative possibility is that magma fractionates near erupting volcanoes producing a body of olivine cumulates, which has a rheology similar to ice and should flow down from the summit of the volcano, causing a massive submarine landslide (Clague & Denlinger, 1994). The ultimate trigger for the giant submarine landslides cannot be identified at present and probably includes a combination of several of the above factors.

Waves

Storm waves can trigger slope failure, as illustrated by damage to offshore drilling rigs during Hurricane Camille in 1969 (Bea *et al.*, 1983). Storm-wave-induced failure actually involves several elements. As was demonstrated by Henkel (1970), the passage of a wave train subjects the seafloor to alternating water pressure as the crests and troughs pass. This non-uniform pressure field induces the greatest shear stresses between crest and trough. Henkel (1970) considered the situation to be one of a simple moment resulting from alternating zones of positive and negative pressure. Seed & Rahman (1978) improved upon Henkel's approach and developed the following equation for the induced shear stresses

$$\tau_c / \sigma'_v = f_z f_d 2\pi (\gamma_w / \gamma') H/L \qquad (6)$$

where τ_c is the induced peak cyclic shear stress, σ'_v is the vertical effective stress, $f_z = \exp(-2\pi z/L)$, $f_d = 0.5[1/\cosh(2\pi d/L)]$, γ_w is the unit weight of water, γ' is the buoyant unit weight of sediment, H is wave height, z is depth below the seabed, L is wave length and d is water depth.

Equation 6 demonstrates that induced shear stresses vary with the characteristics of the waves, the water depth and the depth below the seafloor. These shear stresses are much like earthquake loads in that they add to pre-existing downslope gravitational stresses and they are cyclic in nature, so that they gradually induce increasing pore-water pressures in the sediment. The sediment can fail after the passage of a wave train, or it can liquefy and flow if the pore pressures reach a high enough value (Van Kessel & Kranenburg, 1998).

Lee & Edwards (1986) showed that there can be a transition in the importance of triggers in environments that are both subjected to large storms and are seismically active. In shallow water, the largest shear stresses may be induced by storm waves, and these would control seabed stability. Seismic loading would be more important in deeper water. For example, in the north-east Gulf of Alaska, storm waves appear to be the dominant trigger in water depths < 80 m, and earthquake loads are more important in greater water depths.

Clukey *et al.* (1985) considered another implication of wave-loading effects. As the storm-wave-induced pore-water pressures build up, the effective stress decreases and the sediment approaches a state of liquefaction. As a result, the current velocity necessary to initiate sediment transport decreases. Accordingly, wave loading, cyclic-shear-stress development and pore-pressure generation lead to slope failure and also to enhanced bottom-current-induced sediment transport.

Gas and gas hydrates

Gas charging of sediment is not so much a trigger as a means by which shear strength may be altered. Gas charging can affect sediment strength either by decreasing it through the development of excess pore pressures, or potentially increasing it by reducing the impact of cyclic loads. In cases where gas charging reduces strength, the actual trigger causing failure is likely to be some other factor such as an earthquake.

Dissociation of gas hydrates can be considered a trigger because it results from environmental changes. Sea-level fall has often been invoked as a means of triggering landslides through destabilization at the base of the gas-hydrate zone, the part of the sediment column closest to gas-hydrate equilibrium. Kayen & Lee (1991) modelled pore-pressure generation on the continental slope of the Beaufort Sea during the last eustatic fall in sea-level. They determined that fluid-diffusion properties dominate the process. Following sea-level fall, pressures develop within the pore space of sediment at the base of the hydrate in response to the liberation of gas. They concluded that excess pressure generated at the base of the gas hydrate zone during Pleistocene falls in sea level was probably sufficient to initiate seafloor landsliding in the unlithified sediment that underlies the continental slope in the Beaufort Sea. This process probably operated at many other locations in the world's oceans at the same time.

Sultan *et al.* (2003) modelled the impact of sea-level rise and warming of the North Atlantic on the stability of the Norwegian continental slope where the giant Storegga Slide occurred about 8200 yr ago (Bryn *et al.*, 2003). Sultan *et al.* (2003) suggested a mechanism by which increases in pressure and temperature associated with the end of the last glacial period could have increased the solubility of methane and induced a dissociation of methane hydrate at the top of the hydrate layer. The resulting excess pore-water pressures could have led to massive slope failure. Such a mechanism for producing the Storegga slide is still being debated. Another mechanism for triggering the slide is rapid sediment accumulation during peak glaciation, followed by earthquake loading due to post-glacial isostatic rebound (Bryn *et al.*, 2003).

Groundwater seepage

Sangrey (1977) speculated, based on experience and proprietary information, that underconsolidation and excess pore pressures resulting from artesian reservoir sources are 'very common offshore and may be the most significant mechanism' for causing slope failure. Orange & Breen (1992) suggested that pore fluids percolating up from subducted sediment could induce slope failure and lead to the development of **headless canyons**, i.e. submarine canyons that are not linked to incised valleys on the shelf. Many others (Saffer & Bekins, 1999) have developed models for the ways in which subducted fluids and the resulting excess pore pressures influence the mechanics of subduction zones.

Groundwater seepage from coastal aquifers also could serve as a trigger for landslides. Based on an examination of morphology, Robb (1984) suggested that spring sapping (i.e. erosion of sediment and rock by underwater springs) may have occurred on the lower continental slope off New Jersey during periods of lowered sea level. In support of this suggestion, it was observed that nearly fresh interstitial water is found beneath the continental shelf ~100 km off the New Jersey coast. Hot fluid seeps also are known to occur (Hampton *et al.*, 2002) on the Palos Verdes continental shelf (southern California) near the head of a very large submarine landslide (Bohannon & Gardner, 2004).

Failures often occur in fjords and other coastal locations during periods of low tides (Prior *et al.*, 1982a,b). These failures occur because of a phenomenon that engineers term **rapid drawdown** (Lambe & Whitman, 1969, p. 477). When water levels fall rapidly, pore pressures within coastal slopes often cannot adjust quickly enough. This results in an elevated water table directly adjacent to the coast and in accelerated seepage of groundwater. This situation can be modelled as **seepage forces**, which effectively add downslope

driving stress, or as excess pore pressures reducing the effective stress and the corresponding sediment shear strength. Regardless of the specific mechanism, the slope becomes less stable and failures can occur. Atigh & Byrne (2004) modelled liquefaction in the Fraser delta resulting from tidal variations, which cause unequal pore-pressure generation.

Diapirism

Any tectonic or diapiric deformation that results in steepened seabed surfaces will lead to a reduction in the factor of safety and increased likelihood of slope failure. This element becomes one of a number of factors that ultimately determine whether or not a slope will fail. The northern Gulf of Mexico is an area in which diapiric deformation is one of the major causes of failure on the continental slope. Martin & Bouma (1982) noted that large diapiric and non-diapiric masses of Jurassic salt and Tertiary shale underlie the northern Gulf of Mexico continental slope and adjacent outer continental shelf. The masses show evidence of being structurally active at present and in the very recent geological past. The vertical growth of these structures causes local steepening of the seafloor and causes many knolls and ridges interspersed by topographic depressions and canyon systems. Large overburden pressures created by sediment accumulation from the late Jurassic to the present have caused the underlying salt sheet to flow and sometimes extrude toward the surface. The movement of the salt sheet, or **halokenesis**, is largely responsible for the surface morphology (Silva *et al.*, 2004).

Human activity

Human-constructed facilities, either along the coastline or on the seafloor, have the potential for causing submarine slope failures. Typically, these facilities increase the downslope stresses. Human influence in causing landslides is hotly debated because fault must be assigned to damages, injuries and even death. The question debated is commonly whether a natural slope failure affected the human development or whether the human development caused the slope to fail.

The role of human activity is clear in the case of the quick-clay failure in Rissa, Norway, during 1978. The landslide was initiated when 700 m^3 of earth fill was placed by the shore of Lake Bottnen to expand the area of a farm. The movement of fill had just been finished when 90 m of shore slid into the lake. The slide then developed retrogressively with each new slide fully liquefying and flowing into the lake. After about 40 min, a very large slide removed an area of about 150 m × 200 m. The sliding took only about 5 min. A house was seen moving down the 'quick clay river' at 30–40 km h^{-1} (Gregersen, 1981).

Two cases that involved coastal failures and tsunamis have been debated as to whether they were human-induced or natural. The first occurred at Nice, France, in 1979, and involved the failure of fill that had been placed near a delta to construct a new airport (Seed *et al.*, 1988). The slide contained ~10^7 m^3 of fill and native material, and occurred over a period of about 4 min. The debris moved down the sloping face of the delta deposit, into a submarine canyon, and onto an abyssal plain, eventually rupturing two sets of cables as far as 120 km offshore. A tsunami struck the coastline with a maximum amplitude of 3 m. Several lives were lost and considerable damage was done to local communities and harbours.

Two hypotheses were advanced to explain the failure at Nice.

1 The construction area failed first, perhaps because of the construction activity. The sliding material moved downslope, undercut the canyon walls and caused continued failure of considerably more natural material.
2 There was a large natural underwater landslide that caused a tsunami. The tsunami caused a 'rapid drawdown' condition that produced a failure in the newly constructed fill (Seed *et al.*, 1988).

Considerable debate has followed both in the scientific literature and in court over the true cause of the disaster.

A similar case occurred in Skagway Alaska, in 1994, when a dock that was under construction slid into a fjord. A particularly low tide was accompanied by a series of tsunami waves estimated to be as high as 11 m. Subsequent surveys showed that a submarine landslide had occurred. Again, the debate focused on whether the dock construction caused the landslide or a large natural landslide caused the dock to fail. Rabinovich *et al.* (1999)

developed a model that showed that the tsunami was caused by the dock failure and not by an external submarine landslide. However, others have argued to the contrary (Mader, 1997).

A final example of a human-induced submarine landslide occurred in 1985 near Duwamish Head in Seattle, Washington (Kraft *et al.*, 1992). Dredging operations were undertaken to extend sewage effluent pipes about 3 km into Puget Sound (Fig. 12). The slide occurred during low tide, involved 400,000 m^3 of sediment, and resulted from liquefaction. The landslide was clearly triggered by the low tide, but the dredging operation was an underlying cause.

CONTRIBUTIONS TO SUBMARINE LANDSLIDE RESEARCH FROM THE STRATAFORM PROJECT

The Eel margin study area of the STRATAFORM programme is an excellent place to study submarine mass-movement processes. The study area includes a sediment-laden river entering the sea, a major submarine canyon (Eel Canyon), and the area is extremely active seismically. The repeated seismic events could induce landslides. Perhaps most importantly for landslide research, previous maps had shown a submarine landslide deposit, the 'Humboldt Slide', situated within the study area (Field *et al.*, 1980; Lee *et al.*, 1981).

The New Jersey study area also showed evidence of mass movements. Locat *et al.* (2003) conducted a geotechnical analysis of a failure on the Hudson Apron slope using results of testing on ODP cores (Austin *et al.*, 1998; Dugan *et al.*, 2002) and long Calypso piston cores. The analysis showed that high excess pore pressures were required to cause the failure and that they probably acted in the context of a layered system with groundwater seepage.

In addition to the possibility of investigating landslide features, the STRATAFORM programme (in particular the Eel margin study area) provided an abundance of closely spaced and dated sediment cores that could be analysed for physical and geotechnical properties. These analyses led to a better understanding about the variability of sediment physical properties (Goff *et al.*, 2002) and how marine sediment acquires shear strength as a function of increasing burial (Locat *et al.*, 2002). The closely spaced cores also provided a basis for

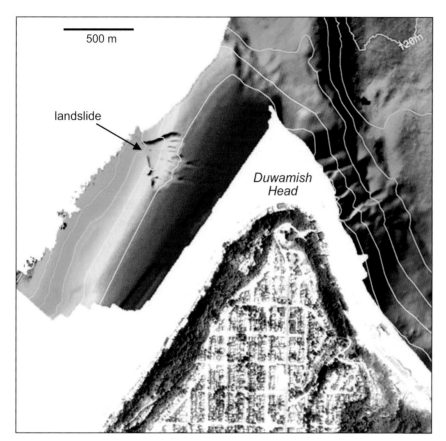

Fig. 12 Multibeam image and aerial photography in the vicinity of Duwamish Head, Seattle, Washington. The landslide feature to the north-west of Duwamish Head occurred in 1985 during dredging operations associated with construction of two sewage effluent pipes (parallel depressions extending east from the slope failure). Two shore-parallel pits are also shown, which were produced independently by removing sediment for beach nourishment. (US Geological Survey and NOAA data.)

regional mapping of sediment properties and the development of GIS-based landslide-susceptibility predictions (Lee *et al.*, 1999, 2000).

The sections below report on application of the findings to submarine landslide research, beginning with a discussion of the 'Humboldt Slide' and the controversy that has developed as to whether or not it actually represents a continental-slope failure. Next considered are failures in Eel Canyon and the factors that caused them. The following sections on gas charging and pore pressures, and the development of shear strength and rheology in marine sediment, discuss the elements that determine shear strength, which is often the most important factor in predicting instability. In fact, determining how shear strength develops in marine sediment as a result of sedimentation, stress history, biological activity and geochemical changes has been one of the most important topics investigated by marine geotechnical engineers over the past 40 yr, and has relevance to seafloor engineering as well as slope stability. The final sections discuss submarine landslide occurrence and prediction over medium to large regional scales.

'Humboldt Slide' controversy

Background

Hummocky terrain extending over at least 90 km^2 was discovered on the Eel margin during offshore hazards studies conducted in the late 1970s (Field *et al.*, 1980). This terrain was interpreted to be a giant submarine landslide and was subsequently named the 'Humboldt Slide'. The occurrence of giant landslides in this area is reasonable, given the many potential triggers: the rate of sediment accumulation is high owing to the proximity of the Eel River; numerous areas of gas charging and pockmarks are present (Field & Barber, 1993; Yun *et al.*, 1999); and the rate of seismicity is one of the highest in the USA (Clarke, 1992). In addition, massive liquefaction failures were recorded in the nearby Klamath Delta during the 1980 earthquake.

A series of papers described the details of the 'Humboldt Slide' and attempted to analyse its cause quantitatively (Lee *et al.*, 1981, 1991; Lee & Edwards, 1986; Field & Barber, 1993). Given the results of strength tests on sediment cores taken in the deposit during 1979, geotechnical analyses showed that a lateral pseudostatic earthquake acceleration of 0.12 g could have produced the failure (Lee & Edwards, 1986). Such accelerations are expected to occur frequently on the Eel margin (Frankel *et al.*, 1996; Lee *et al.*, 1999) as a result of the high level of seismicity known to exist near the Mendocino triple junction. Lee *et al.* (1991) performed an analysis of sediment mobility and determined that the density state (Fig. 6) was too dense to allow an initial landslide to convert into a debris flow. Rather, the sediment would fail during an earthquake, produce limited deformation along shear planes, and then maintain its stability after the earthquake shaking had stopped. This would lead to a limited deformation slide, with blocks that would not evacuate the source region. This appeared to be the morphology observed in the 'Humboldt Slide'.

Multibeam (Goff *et al.*, 1999) and high-resolution sub-bottom profiles (Huntec Deep Tow System (DTS); Gardner *et al.*, 1999) were obtained early in the STRATAFORM Project and provided far superior information on the 'Humboldt Slide' feature than had been available previously. Multibeam data (Fig. 13) show a crenulated surface contained within a bowl-like depression, which has been identified as an amphitheatre. The amphitheatre has a steep eastern face that is highly gullied and is separated from the crenulated surface by a smooth sloping surface. In profile (Fig. 14), the crenulated surface consists of a series of identifiable units containing reflectors that dip shoreward. The zones separating units of shoreward-dipping reflectors (Fig. 15) show that there is rough connection of reflectors between units. The zones both above and below the crenulated surface contain slope-parallel beds. There is no headwall at the boundary separating crenulated beds from slope-parallel beds.

Discussions following the acquisition of these data led to a division among participants in the project that has not yet been settled. Some participants feel that the data confirm the earlier interpretations of the feature as a giant submarine landslide (Gardner *et al.*, 1999). Other participants feel that the data are more suggestive of migrating sediment waves; that is, a depositional feature formed slowly under the influence of bottom water or turbidity currents (Lee *et al.*, 2002). The details of each argument follow.

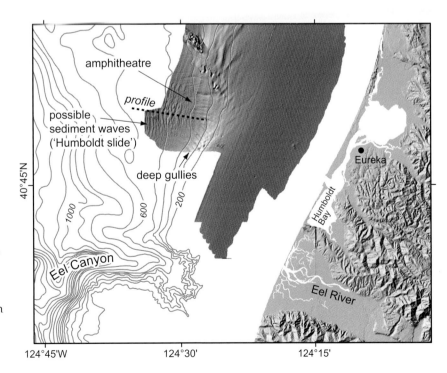

Fig. 13 Overview of Eel River STRATAFORM area (after Field *et al.*, 1999) and the 'Humboldt Slide'. Data include US Geological Survey on-shore digital elevation model, as well as shaded relief bathymetry from Simrad EM-1000 (Goff *et al.*, 1999). Location of track line shown in Fig. 14 is identified.

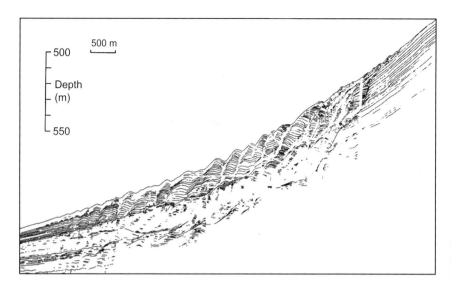

Fig. 14 Simplified line representation of acoustic stratigraphy for the 'Humboldt Slide', obtained using a Huntec deep-tow, seismic-reflection system operating at a narrow peak frequency of about 3.5 kHz. (After Gardner *et al.*, 1999.)

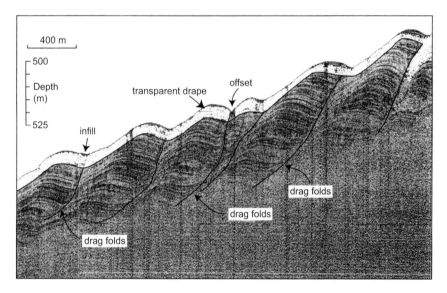

Fig. 15 Portion of a high-resolution (Huntec) sub-bottom profile showing the main body of the 'Humboldt Slide'. Interpretations of folded and back-rotated slide blocks are supportive of a slope-failure origin (from Gardner *et al.*, 1999). Interpreted shear surfaces are traced with black lines. Vertical exaggeration 20×.

The slope-failure interpretation

Gardner *et al.* (1999), interpreting the results of multibeam mapping (Goff *et al.*, 1999), observed that the 'Humboldt Slide' lies within a shallow amphitheatre-shaped depression (Fig. 13). The depression is bounded by the shelf break to the east, by the Little Salmon Fault and an associated plunging anticline on the north, and by a bathymetric high on the south. The east (upslope) side of the depression is located at the 220-m isobath and the downslope limit of crenulation is at the 650-m isobath. The main body of 'Humboldt Slide' is elongate with a length (10 km) and thickness (60 m) that places the feature in the middle of the population of reported slides documented by Woodcock (1979). The overall feature has a thickness/length ratio (expressed as a percentage; following Skempton, 1953) of 0.6%, although the upslope portion of the feature is more elongate with a ratio of 0.2%. These relationships are consistent with those from reported submarine slides (Prior & Coleman, 1979). The eastern boundary of the depression is distinctly steeper (3°–6°) than farther down (1°–2°) in the zone of crenulations. Gullies occur along the steeper eastern side of the depression between water depths of 230 m and 380 m. They have < 20 m of relief and occur within a zone of erosion correlated with high backscatter on the multibeam image.

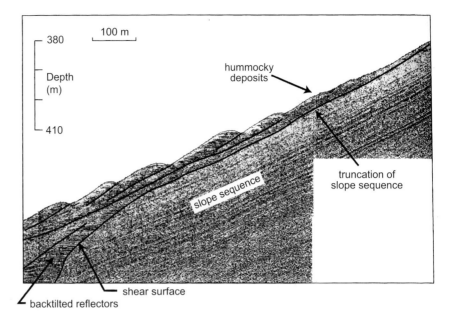

Fig. 16 Part of a high-resolution (Huntec) sub-bottom profile showing a portion of the 'Humboldt Slide'. Interpretations shown are supportive of a slope-failure origin. (From Gardner *et al.*, 1999.)

Below the base of the gullied slope, hummocky deposits of sediment occur with fingers onlapping upslope (Fig. 16). These deposits bury the lower portions of the eroded slope between the 380-m and 430-m isobaths. Farther downslope the seafloor is dominated by distinctive ridges and swales (Fig. 15) that resemble subaerial retrogressive landslides described by Mitchell (1978). The bathymetry reveals an intricate pattern of branching and truncated ridges. The ridge cross-section profiles appear as asymmetric steps with inclined risers of < 8 m relief and back-dipping treads generally 100–300 m wide. In plan view, the ridges appear relatively short and irregular around the lower flanks of the zone where slope erosion occurs (380–430 m). They are longer and more regular toward the centre and downslope portions of the 'slide' zone (Fig. 13). The treads of the ridge crests strike directly across the depression and do not parallel the curvature of the bathymetric contours. The seismic profiles between 560-m and 590-m water depths show that the ridges evolve into rhythmic, undulating forms interpreted as folds, that die out at a water depth of about 650 m (Fig. 17).

Sidescan-sonar images reveal numerous pockmarks on the seafloor throughout the region. The pockmarks on the 'Humboldt Slide' are small depressions with a random and dense distribution. Compared with the adjacent continental slope to the north, the 'Humboldt Slide' zone has a higher concentration of pockmarks (Yun *et al.*, 1999),

possibly indicating a greater tendency toward gas charging and strength reduction.

The erosion and gully zone (Fig. 18) extends between the 230-m and 380-m istobaths where Huntec seismic profiles show truncations of shelf reflectors with an estimated 5–15 m of sediment missing. Geotechnical tests (Lee *et al.*, 1981) from cores collected in the zone show that the sediment is overconsolidated, consistent with the loss of 15 m of sediment by erosion or slope failure. Huntec seismic data show local hummocky deposits (< 5 m thick) that unconformably overlie the slope below the gullies (Fig. 16). A surface defining the upslope limit of crenulations cuts the slope sequence and crops out at the 380-m isobath (Fig. 16). This exposed basal surface separates the sequence of slope-parallel-bedded, seaward-dipping reflectors that lie upslope from crenulated, back-tilted reflectors of what is interpreted to be the main slide mass. The basal surface is locally buried < 5 m near its upslope limit by disrupted reflectors, and dips 3°–4° seaward, parallel to the slope reflectors. Farther downslope, the basal surface dips ~8° and cuts deeply into the slope section (Fig. 16).

According to Gardner *et al.* (1999), the main body of the landslide is composed of a zone of back-tilted and gently folded blocks (Fig. 15). The landward side of each back-tilted block is bounded by a gently warped surface defined by the termination of reflectors. The defining surfaces of each block generally dip seaward ~8° near the seafloor and

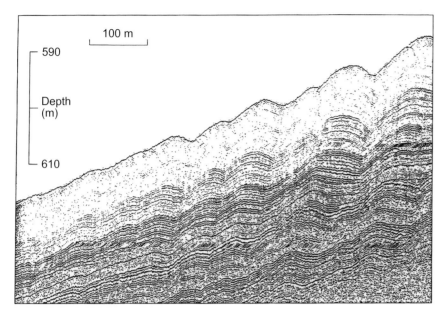

Fig. 17 Portion of a high-resolution (Huntec) sub-bottom profile showing the toe of the 'Humboldt Slide'. According to the slope-failure hypothesis, this figure shows the transition from back-rotated blocks to gently folded to almost undeformed sequences. (From Gardner *et al.*, 1999.)

gently flatten and merge with underlying reflectors having dips < 0.5° at about 65 m below the seafloor. Each tilted block is composed of anticlinally folded reflectors that dip landward 2°–4° and seaward 4°–6°. Landward-dipping reflectors within the blocks have drag folds along the separating surfaces which clearly show growth features (Fig. 15). Gardner *et al.* (1999) interpreted seaward-dipping

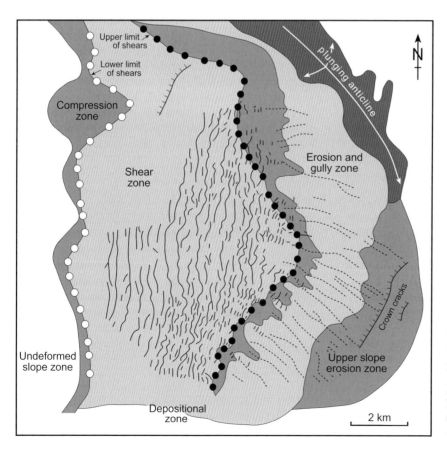

Fig. 18 Map of the surface features of 'Humboldt Slide' derived from multibeam images and Huntec seismic profiles. Assuming a slope-failure origin, shear planes occur between the two lines of circles. Ridge crests are mapped as solid black lines within the shear zone. Gullies are mapped as dashed lines. (From Gardner *et al.*, 1999.)

portions of the folds to be diffraction features from the abrupt edge of rotated blocks. If so, some of the reflectors must be broken, rather than folded, at their anticlinal axes. The diffraction features appear as seaward-dipping reflections, but they also converge and match a model of expected diffraction patterns (Gardner *et al.*, 1999). The main body of the 'slide' has been draped uniformly with a ~10-m-thick surficial unit. The drape unit is almost acoustically transparent, compared with the section it covers. Huntec profiles provide only limited information more than ~65 m below the seabed. Indications of older 'failures' beneath the 'Humboldt Slide' are suggested also by sparker profiles (Field *et al.*, 1980) and recently acquired, high-resolution multichannel seismic-reflection profiles (Fulthorpe *et al.*, 1996) confirm at least four older features with geometries similar to 'Humboldt Slide'.

The downslope transition from the main body of the 'slide' to undisturbed slope sediment occurs over a distance of ~2 km. The downslope portion of the 'slide' is characterized by gentle folds with seafloor relief less than 2 m (Fig. 17). The fold axes generally are 75–150 m apart (Fig. 17) and fold amplitudes tend to increase up-section. The undisturbed sequence beyond the distal toe of the 'slide' can be traced upslope into the increasingly deformed main body of the 'slide'. There is no evidence in the seismic data of a basal surface cropping out in the toe of the 'slide'.

Gardner *et al.* (1999) interpreted the data to show that the 'Humboldt Slide' began with basin subsidence that initiated extension-related shearing of the slope sequence to sub-bottom depths of ~65 m (Fig. 17), followed by rotation and folding of shear-bounded blocks (Fig. 19). Failure began in the middle of the 'slide' and simultaneously progressed upslope (retrogressive) and downslope (progressive). This interpretation is based on the observation that the largest apparent displacements of blocks (both horizontally and vertically) appear in the middle of the feature.

According to the landslide interpretation, the main body of the 'slide' was deformed by a combination of limited downslope translational and shallow rotational movements. The shear-bounded blocks moved short distances downslope over long, shallow, shear surfaces. The shear-surface geometry created more downslope extensional movement than rotational subsidence. The overall displacement was limited, consistent with the relatively intact

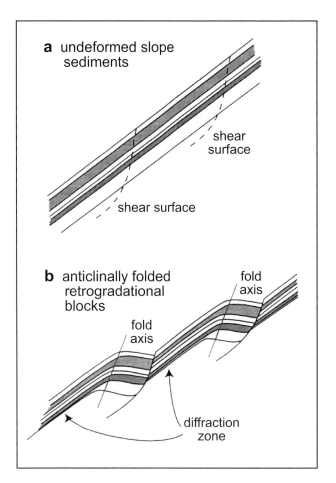

Fig. 19 Conceptual cartoon of deformation (according to the slope-failure interpretation) within the main body of the 'Humboldt Slide'. (From Gardner *et al.*, 1999.)

character of individual shear surfaces. The toe of the slide did not fail and shows only compressional displacement because the slide movements were shear-dominated with relatively small net downslope movement.

The sediment-wave interpretation

Lee *et al.* (2002) compared the morphology of the 'Humboldt Slide' with documented observations of sediment-wave fields in a variety of environments. For example, the sediment wave field off the Selvage Islands, near the west coast of Africa, has recently been described by Wynn *et al.* (2000). As shown in Figs 20 & 21, the sediment-wave field extends over ~15 km and is present between two zones of parallel-stratified sediment consisting of interbedded turbidites and pelagic sediment. The beds within the sediment-wave field appear to

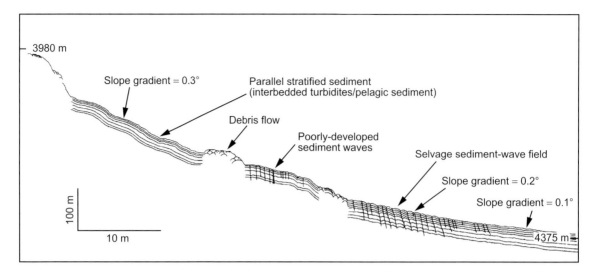

Fig. 20 An interpretative line drawing of high-resolution (TOPAS) acoustic profiles taken across the continental rise north of the Selvage Islands. The profile shows that sediment waves occur only over a narrow range of slope angles (0.1°–0.3°) and that the sediment-wave field has expanded with time (as indicated by sediment-wave deposits near the seabed overlying slope-parallel beds at depth). (After Wynn *et al.*, 2000.)

be continuous with parallel-stratified beds in the zone downslope from the wave field. The upslope transition is less clear because of rock outcrops and debris-flow deposits. However, the trend appears to be one in which individual turbidity currents (corresponding to individual turbidite beds) can contribute to both parallel-stratified bedding and sediment waves during a flow event. Significantly, the general form of the seafloor is concave upward

with the sediment waves existing only within a particular range of slope gradients (0.1°–0.3°). Figure 20 shows that the wave field was less extensive in the past and that sediment waves are expanding both upslope and downslope with time. Based on a study of a number of turbidity-current sediment-wave fields, Wynn *et al.* (2000) concluded that both the wavelength and wave height typically decrease downslope.

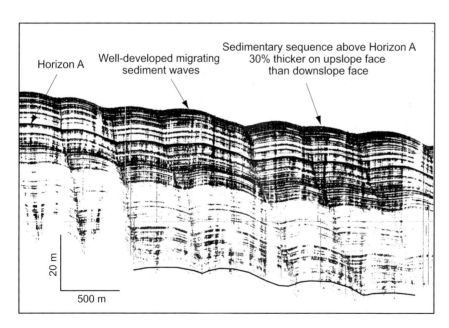

Fig. 21 High resolution (TOPAS) acoustic profile of well-developed sediment waves in the Selvage sediment-wave field. (After Wynn *et al.*, 2000.)

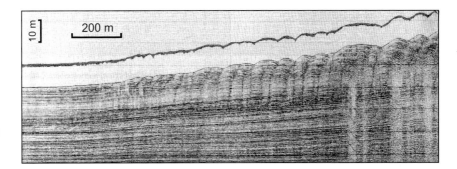

Fig. 22 Features on the Noeick Delta interpreted as sediment waves by Bornhold & Prior (1990).

An enlargement of the central part of the Selvage sediment-wave field (Fig. 21) shows other characteristics. First, based on measurements by Wynn *et al.* (2000), the sedimentary sequence above a defined reflector is 30% thicker on the upslope face of the sediment waves than on the downslope face. This preferential accumulation pattern causes the crests of the sediment waves to migrate upslope. In addition, inflection points of reflections passing from one wave to the next line up in such a way as to mimic bounding surfaces between blocks. Close observation shows that the reflections are continuous across these 'surfaces', which are visual artefacts that provide a false impression of 'blocks' that have moved along shear planes relative to each other. These apparent surfaces can be either concave upward, concave downward or linear, depending upon the details of deposition within the sediment-wave field. The waves in Fig. 21 show all three of these shapes, occasionally within a single bounding 'surface'. The overall appearance of the sediment-wave field is one of regular, rhythmic waves.

Another feature interpreted as a migrating sediment-wave field has been reported for the Noeick River prodelta (Fig. 22, Bornhold & Prior, 1990), in British Columbia, Canada. This field is similar to, but smaller than, the Selvage sediment-wave field. The gradient of the slope is ~0.1°–1.4°, wavelengths are 50–100 m and wave heights are 2–5 m. The wave field has a general concave-upward shape in cross-section and the wavelength decreases with distance from the source. Apparent boundaries between separate waves have a generally convex-upward shape. Sub-bottom reflectors show that the wave field was less extensive in the past and that it is building out from its source. Downslope from the sediment-wave field, reflectors appear similar to or continuous with reflectors in

the waves, indicating that turbidity currents in the fjord can produce either slope-parallel bedding or sediment waves. After the slope steepens beyond a certain point (a slope gradient of about 0.1°), waves begin to form; on more gentle slopes, the turbidity currents produce parallel beds.

Based on these and other examples of documented sediment-wave fields, Lee *et al.* (2002) developed the following list of criteria for recognizing sediment waves.

1 Differential accumulation rates. The upstream flanks accumulate sediment more rapidly than the downstream flanks. This effect causes the sediment waves to migrate upslope.
2 Continuous acoustic reflections through the features. Although spacing between reflections may vary as a result of **1**, the reflections are typically continuous throughout the sediment-wave field.
3 In cross-section, the apparent boundaries between sediment waves may be linear, convex upwards or concave upwards. **Listric faulting**, characteristic of rotational slumps, would produce concave-upward boundaries almost exclusively.
4 In cross-section, the overall sediment-wave field commonly has a concave-upward surface.
5 If there is a trend, the wavelength and wave height of the sediment waves appear to decrease with distance from the source of sediment (Wynn *et al.*, 2000). This may be related to a slowing of the turbidity current as it passes onto progressively more gentle gradients (Normark *et al.*, 1980).
6 Beds can be traced through sediment-wave fields into areas of parallel reflectors. This suggests that the same sequence of turbidites can produce or not produce sediment waves depending upon changes in environmental conditions (e.g. slope gradient). This effect may be related to the range of Froude numbers over which sediment-wave formation can occur (Wynn *et al.*, 2000).

7 Within many sediment-wave fields, the structure of internal reflectors appears similar from one wave to the next, that is, the waves display regularity. For example, beds within the upstream flanks tend to have generally the same dip throughout the sediment-wave field.

8 Migrating sediment-wave fields do not require a headwall scarp or a zone of evacuation, such as is usually found in a landslide.

9 Sediment-wave fields are constructed over a long time period and involve deposits from many turbidity currents (e.g. the Amazon fan, Shipboard Scientific Party, 1995). Accordingly, profiles through the deposits often show rhythmic bedding extending all the way to the sediment surface.

Within the 'Humboldt Slide' there are alternating bathymetric highs and lows that create block-like units (or 'sediment waves') with a wavelength of 400–1000 m, and a wave height of 2–10 m. Individual 'wave' crests can be traced for up to 4.5 km along isobaths (Fig. 13). Within the main body of the feature, internal reflections can be traced across the crests and troughs of each wave (Figs 23, 24a & 25). The downslope flank of one wave meets the upslope flank of the next lower wave within

a zone that can vary from sharp and acoustically incoherent to broad and traceable across the features (Fig. 23). In some cases, these zones become broader up-section, whereas in other cases they become broader down-section. The latter observation provides a critical constraint on the origin of these features. If the block-like units were in fact slide blocks, then there would be a discontinuity (fault) between the units. Growth faults, which would be active at the time the sediment is being deposited, should decrease in slip up-section. Faults rarely decrease in slip down-section. In Fig. 24a, apparent displacement between units decreases down-section, so strain would have to be accommodated down-section some other way, for example, through a low-angle **décollement**. No such décollements are apparent. Given these observations, the suggestion can be made that in many of the block-like units ('waves') sediment beds are continuous, favouring a sediment-wave origin for the feature. Note that the scale of the wavelengths and wave heights is comparable to that observed in recognized sediment-wave fields, and that wave length and wave height decrease with distance seaward (Fig. 14), in accordance with the usual trend observed in sediment-wave fields elsewhere (Wynn *et al.*, 2000).

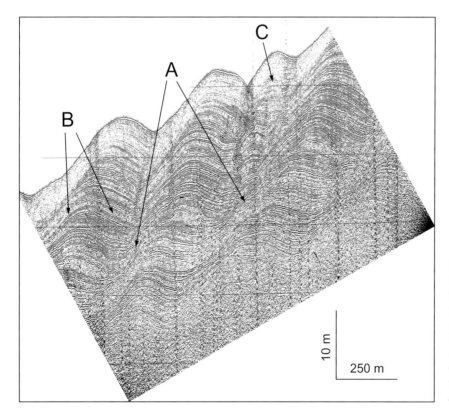

Fig. 23 Huntec sub-bottom image from the main body of the 'Humboldt Slide', showing: A, internal reflections that can be traced from one wave to another; B, layers in the downslope flanks that are thinner than the same layers in the upslope flanks; C, a wave that merges down-section with the upslope wave. (From Lee *et al.*, 2002.)

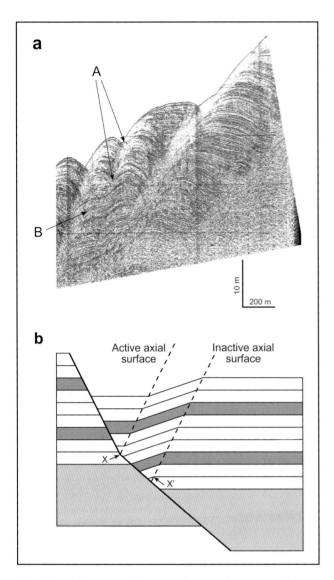

Fig. 24 (a) Huntec sub-bottom image of waves within the main body of the 'Humboldt Slide', showing: A, a zone that separates two waves decreases in offset down-section, counter to most structural faults; B, at depth, internal reflectors can be traced across the two units, whereas up-section at A the continuity cannot be traced through the hyperbola. If a fault is present up-section, the displacement must be transferred parallel to bedding and over the 'folded' strata above B. (From Lee *et al.*, 2002.) (b) Diagram illustrating the deformation that would develop for a block sliding from X to X' along a listric fault. (From Xiao & Suppe, 1992.) The active and inactive axial planes of the block dip seaward, in contrast with the axes of the wave-like features in the 'Humboldt Slide', which dip landward.

In contrast, the slope gradient is steeper for the 'Humboldt Slide' than for deep-water sediment waves, and the overall environment is different (continental slope versus deep-sea fans).

Individual 'waves' off the Eel River have upslope flanks that display thicker beds than the downslope flanks. This results in a landward (upslope) migration of the 'wave' crest. If the landward-migrating packages were in fact slide blocks, a concave-upward basal shear surface (slip plane) would produce a landward-dipping axial plane and an apparent seaward migration of the block's crest (Xiao & Suppe, 1992; Fig. 24b). Such geometry is, in fact, the opposite of what is observed in the 'Humboldt Slide', where the boundaries between block-like units commonly appear to be convex upward or linear (Fig. 24a). If the linear boundaries between blocks were slip planes, then the intersection of these internal slip planes and the basal surface would be concave upward (and the axial surface of the fold would dip landward). The surface that joins the crest of the 'waves' dips seaward, however, contrary to geometry of an axial plane that would form in a fold above a slip plane. The geometry of the internal structure for the 'Humboldt Slide' is not compatible with the kinematics for a series of moving blocks.

Figure 25 shows neither a headwall scarp nor zone of evacuation at the landward margin of the 'Humboldt Slide'. In fact, this figure allows the tracing of individual reflections from the slope-parallel region above the feature to the undulating block-like units observed within the feature. Such an interpretation implies a depositional origin for the feature rather than shear deformation and faulting at the landward margin. Near the downslope edge of the feature, undulating reflections pass into a zone of slope-parallel reflections without interruption.

On the basis of the above arguments, Lee *et al.* (2002) concluded that the 'Humboldt Slide' is not a landslide deposit, but rather is a field of migrating sediment waves. Lee *et al.* (2002) suggested that a plausible explanation for the formation of this field is that turbidity currents form at or near the mouth of the Eel River (perhaps related to hyperpycnal flows from the river) and flow through a series of deep gullies into the bowl-shaped depression (amphitheatre) that contains the 'sediment wave field' (Fig. 13). In the steeper area of the exposed gullies, on the upper part of the continental slope,

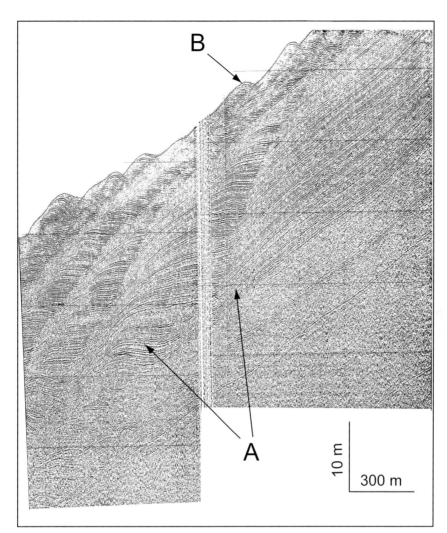

Fig. 25 Huntec sub-bottom image of the upper portion of the 'Humboldt Slide' (Fig. 16). Note that strata A can be traced across the 'shear surface' interpreted by Gardner *et al.* (1999; Figs 16 & 19). The uppermost 'wave' field B represents the most recent period of 'sediment wave' accumulation. (From Lee *et al.*, 2002.)

the sediment is overconsolidated (Lee *et al.*, 1999). This may indicate bypassing or erosion by traversing turbidity currents. In greater water depths than this overconsolidated zone, turbidity currents would accumulate slope-parallel beds over a short (~1 km) distance (Fig. 14). The turbidity currents could then begin to deposit their load as sediment waves (Fig. 25). Sediment-wave accumulation continues for 6 km downslope, with distal sediment waves having lower wave height and wave length than the proximal waves (Fig. 14). To the west of the wave field, the turbidity currents again deposit slope-parallel beds.

Recent studies have provided further information on the formation of the 'Humboldt Slide'. Schwehr & Tauxe (2003) developed a technique that measures the **anisotropy of magnetic susceptibility** (AMS) of sediment as an indicator of deformation.

The technique has been applied to known slump deposits on land and adjacent undeformed sediment, and has shown that the AMS records are clearly different and that deformed sediment presents a clear AMS signal. The same technique has been applied to core samples from the 'Humboldt Slide'. Based on examining cores from the centre and top of the 'Humboldt Slide' structure, Schwehr *et al.* (2003) found no evidence for deformation. The cores are from areas that are clearly free from drape, and thus the authors were sure that they were sampling the structure seen in high-resolution sub-bottom profiles.

A similar controversy in the Adriatic Sea

Crenulated features similar to the 'Humboldt Slide' have also been identified in the Adriatic Sea and described by Correggiari *et al.* (2001), who interpreted the features as examples of failures in a late Holocene highstand prodelta wedge. Lee *et al.* (2002) noted the resemblance to sediment waves and questioned the landslide origin proposed by Correggiari *et al.* (2001). Lee *et al.* (2002) applied the criteria for recognizing sediment waves listed above, and concluded that the features probably are sediment waves. Subsequently, Cattaneo *et al.* (2004) revisited the subject and asserted that there are differences between the crenulated features in the Adriatic and the deep-water sediment waves that Lee *et al.* (2002) reviewed in developing the list of criteria. The differences listed by Cattaneo *et al.* (2004) include:

1 the Adriatic features occur in shallow water (30–70 m water depth);
2 they are several orders of magnitude smaller than the deep-water waves and the sediment accumulation rates are as much as three orders of magnitude higher;
3 they do not develop upslope-dipping limbs;
4 they do not show any consistent trend in downslope variation of undulation parameters;
5 they show a great morphological variability with minor changes in water depth away from the offlap break.

The differences discussed by Cattaneo *et al.* (2004) are important, but they do not disprove a sediment wave origin for the features. Rather, they may imply that turbidity-current sediment waves can occur in a variety of environments, including shallow water, and that the resulting morphologies may be somewhat different. The Adriatic crenulations do not show clear shear planes or a headwall scarp that would confirm a landslide origin. They do, however, show some evidence of deformation at the base of their section and also possible fluid-escape deformations within the crenulated sediment body. Accordingly, the situation in the Adriatic (and possibly at 'Humboldt Slide' as well) may not be one of pure landslide or sediment-wave origin, but rather a hybrid containing elements of each (similar to features described by Faugères *et al.* (2002) in the Bay of Biscay). Another factor affecting the growth of crenulations in the Adriatic may be bottom sediment transport, possibly influenced by internal waves (Puig *et al.*, in press).

The controversy in the Adriatic Sea may be resolved at least partially by careful examination of deep cores that were drilled through the crenulations in the summer of 2004 by the PROMESS Project. Similar cores in the 'Humboldt Slide' could also contribute to a resolution. Finally, a kinematic model of the 'landslide' blocks would be useful to determine if the resulting deposits are geometrically possible.

Liquefaction failures in Eel Canyon

Studies of sediment input to Eel Canyon demonstrate that considerable amounts of sediment are being supplied on annual and century timescales (Mullenbach & Nittrouer, 2000; Mullenbach *et al.*, 2004). During the winter of 1999–2000, an instrumented mooring and a benthic tripod were installed in the northern thalweg of the canyon, and a tripod was installed on the shelf (Puig *et al.*, 2003). The instruments showed that Eel Canyon acts as a preferential conduit of sediment to the deep sea. Sediment fluxes within the canyon were not directly related to the Eel River discharge, but they were linked to the occurrence of major storms that generated down-canyon density-driven flows, carrying large amounts of sediment toward deeper parts of the margin (Puig *et al.*, 2003).

The mechanism for mobilizing flows down-canyon probably involves components of mass movement, where recently deposited sediment fails or liquefies during storms and the failed sediment is easily eroded, entrained into the water column and transported down-canyon as a sediment gravity flow (Puig *et al.*, 2004). Another possibility might be the migration of fluid muds off the shelf and into the canyon, but measurements from one of these events demonstrated that it would take 12 h for fluid mud to move from 60 to 65 m water depth. Such a flow would arrive at the outer shelf many hours after being generated by wave action in shallower locations. Tripod data showed that the sediment gravity flows occurred almost simultaneously with an increase in orbital velocity at the bed of the canyon head. The flows did not coincide with a major flood event, and fluid muds were not observed on the shelf when sediment gravity flows were observed in the canyon head (Puig *et al.*, 2003). The rapid formation of sediment gravity flows, immediately after the increase of wave-orbital velocity, suggests that such flows could not be initiated from wave-current resuspension alone. Entrainment of sediment into suspension requires a period of time (hours) to fill the boundary layer with enough particles to generate fluid-mud and develop a gravity flow (Traykovski *et al.*, 2000).

The more likely mechanism for mobilization is wave-load-induced **liquefaction** (Clukey *et al.*, 1985; Puig *et al.*, 2004). During a given storm, infiltration pressures oscillate with wave pulses, while cyclic-shear-stress-induced excess pore pressures increase progressively. When the bed structure is degraded and the effective stress is lowered, the critical shear stress for sediment erosion decreases significantly and the volume of transportable sediment under large wave stresses can increase considerably. Additional gravity shear stresses imposed by the gradients at the canyon head can help initiate transport of wave-fluidized sediment and gener-

ate sediment gravity flows. The thickest deposits were consistently observed in the upper channel thalwegs (< 500-m water depth) and not deeper (Mullenbach *et al.*, 2004), suggesting that most of the sediment transported down-canyon settles to the seabed at depths just below where wave energy is sufficient to maintain the sediment gravity flow.

Gas charging and pore pressures

Investigations made during the STRATAFORM programme showed that the Eel margin displays plentiful examples of gas-charged sediment. Seismic reflection data (Yun *et al.*, 1999) and ROV dive observations (Orange *et al.*, 2002) show evidence of spatially variable subsurface gas in many areas. Gas distribution is subparallel to isobaths (Fig. 26) and occurs in both near-surface and deep sub-bottom

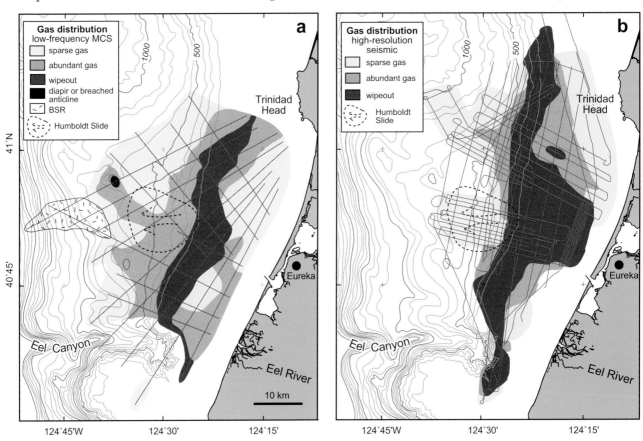

Fig. 26 Gas distribution on the Eel margin. (a) Map based on multichannel-seismic reflection data. The area mapped as BSR is underlain by a bottom-simulating reflector, a possible indicator of gas hydrates. Note that gas-abundance trends are subparallel to isobaths, with a zone of abundant gas in the 'Humboldt Slide'. (b) Gas distribution in upper seabed determined from Huntec high-resolution seismic reflection data. Regions of most abundant gas occur landward of the 400-m isobath. The body of the 'Humboldt Slide' appears nearly free of gas in these data in contrast to (a). (Modified from Yun *et al.*, 1999.)

sediment. An area of acoustic wipeouts and pock-marks is found near the head of the amphitheatre containing the 'Humboldt Slide' and suggests that gas migration is related to the feature, either as a cause or an effect. Field & Barber (1993) suggested that gas migration has played a role in forming the 'Humboldt Slide'.

Yun *et al.* (1999) concluded that gas expulsion through pockmarks is a significant force for redistributing sediment and increasing bed roughness in water depths less than 400 m, where most pockmarks are found. Using an average diameter of a detectable pockmark as 15 m and an excavation depth of 3 m, Yun *et al.* (1999) found that over 6.6×10^5 m^3 of sediment in an area ~2100 km^2 has been excavated and redistributed by gas expulsion. Some pockmarks occur in linear gullies (Fig. 27) suggesting a causative relationship between fluid expulsion in geomorphological lows and gully excavation (Orange *et al.*, 2002). However, an ROV dive that focused on this region showed no evidence for fluid seepage.

In situ pore pressures were measured at five stations on the Eel margin using the Excaliber probe (Christian, 1993, 1998), which reached a maximum penetration depth of 4.6 m. Only small *in situ* pore pressures were measured, with the largest value of 0.5 kPa measured at a sub-bottom depth of 3 m. Such a value represents about 3% of lithostatic pressure and could result in a shear-strength reduction of about that amount. Such a level of pore pressure is probably not a significant factor in affecting the stability of the Eel margin. However, higher excess pore pressures deeper in the seabed or at other sites could exist and could be important.

In general, the results of work on gas charging and pore pressures in the Eel margin are ambiguous. There are reasons (e.g. seismic wipeouts, pockmarks, high accumulation rates) to expect both, but direct proof has been elusive and evidence of slope failure is limited.

Development of shear strength and rheology in marine sediment

The STRATAFORM programme presented an opportunity to develop an improved understanding of how marine sediment acquires shear strength, because the northern California site demonstrated an interesting paradox. The area clearly has many of the triggers needed to cause slope failure, but broad regions of the open slope show no evidence of landslide features (even if the 'Humboldt Slide' is considered to be a failure). Either the driving stresses are lower than expected, or the strength is higher. Accordingly, factors that might lead to higher than expected shear strength were evaluated as a means of explaining the extensive slope regions without landslides.

Physical properties of reconstituted sediment

To understand the development of shear strength, a normally consolidated sediment can be reconstituted in the laboratory and compared with natural samples to determine the effects of various factors on the development of geotechnical properties. Two samples (designated S80 and Y450), corresponding to the maximum range in grain size found on the Eel continental margin, were reconstituted in a large cell that reproduced both the SEDimentation and CONsolidation phases of a sediment (SEDCON test; Locat, 1982; Perret, 1995; Locat *et al.*, 1996). The **liquidity index**, I_L, is an important measure for the **degree of openness** (relative proportion of space filled with fluid,

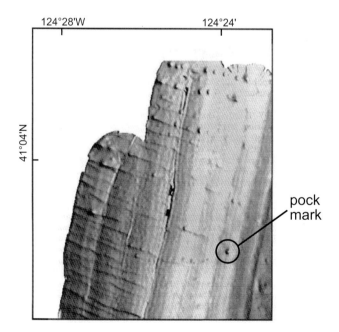

Fig. 27 Shaded relief multibeam bathymetry of the Eel continental slope, showing the occurrence of pockmarks along the axes of gullies. (From Orange *et al.*, 2002.)

analogous to porosity or void ratio but normalized to account for variations in plasticity) in a sedimentary deposit

$$I_{\mathrm{L}} = (w - w_{\mathrm{p}})/I_{\mathrm{p}} \qquad (7)$$

where w is the water content (% dry weight), w_{p} is the plastic limit and I_{p} is the plasticity index (liquid limit minus plastic limit). The liquidity index shows how the water content of the sediment compares with the common geotechnical properties, the **Atterberg limits** (liquid, plastic limits, plasticity index). The **plastic** and **liquid limits** (determined by standard tests) correspond to the water contents at which a remoulded sediment begins to behave as a plastic or liquid. The plasticity index is the difference between the liquid and plastic limits. The liquidity index reflects the sediment's stress history much better than the water content (or porosity or density). That is, sediments with different mineralogies and grain sizes but the same stress histories will have the same liquidity index, although water contents and porosities may differ. Note that a sediment with a water content greater than or equal to the liquid limit does not necessarily behave as a liquid, if it is undisturbed. It will behave as a liquid after remoulding.

In Fig. 28, SEDCON test results for Eel margin sediment (S80 and Y450) are compared with test results for Québec clays from Saguenay Fjord (SF, Perret, 1995). The similarity of all of these SEDCON curves on a broad variety of sediments is due to the normalizing effect obtained from using the liquidity index. For a vertical effective stress (σ'_{v}) greater than 1 kPa, SEDCON curves for most sediment can be described as a power-law function of the following form:

$$I_{\mathrm{L}} = a(\sigma'_{\mathrm{v}})^{-b} \qquad (8)$$

The values of the coefficients a and b are given in Table 1. The range of SEDCON test results provides a good estimate for the degree of openness for normally consolidated sediment. Using the liquidity index allows a broad range of sediment types to be represented with one compression curve. Any other measure of the degree of openness of the sediment (e.g. water content, porosity) would require multiple curves.

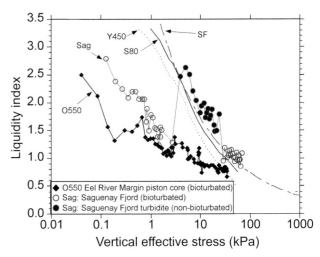

Fig. 28 Relationship between the liquidity index and burial effective stress to show the impact of bioturbation on the consolidation properties of a sediment. Sediments from the Eel River Margin (O550) and Saguenay Fjord (Sag) are compared with respect to their reference curve obtained by SEDCON tests. Curves identified as S80 and Y450 are for two reconstituted samples of the Eel continental margin, whereas the SF curve is for a reconstituted sample from the Saguenay Fjord (Québec). All the SEDCON curves (S80, Y450 and SF) are similar. The Saguenay Fjord sediment contains a 5-m-thick turbidite (black dots), which follows more or less the SEDCON curve. The other portions (open circles), particularly above the turbidite, are at a liquidity index that is clearly below that of the SEDCON curve. The Eel margin sediment (O550) has a trend similar to the bioturbated part of the Saguenay Fjord curve which is well below the SEDCON curve. Accordingly, for these sediments, bioturbation probably reduces the liquidity index (or porosity) at least in the upper part of the sediment column. (After Locat et al., 2002.)

Table 1 Empirical evaluation of coefficients for Eq. 8 developed for three sites (see Fig. 28)

Site	S80	Y450	SF
Coefficient a	3.70	3.25	3.99
Coefficient b	0.38	0.40	0.37

Sediment densification and strengthening from bioturbation

The SEDCON curves (Fig. 28) represent sediments that are normally consolidated and have **sensitivities** (ratio of undisturbed to remoulded shear strength) varying from 5 to 10 (Locat & Lefebvre, 1986). These curves can be used to assess properties of sediments that differ in stress history (e.g. overconsolidation). The profile at O550 illustrates a typical marine sediment that has been subjected to bioturbation, whereas the 'Sag' site (Perret *et al.*, 1995) contains a rapidly deposited layer that was not bioturbated. The liquidity indices for O550 lie well below those obtained using the SEDCON tests, and this is possibly due to sediment deposition rate (time) and bioturbation.

The SEDCON test procedures simulate a sediment deposition rate more rapid than that occurring in nature with the exception of catastrophic events such as debris flows or turbidity currents. These are comparable to, or even faster than, the rate imposed by the SEDCON test, i.e. centimetres of sediment emplaced in minutes to a few hours. This is shown by the 'Sag' profile, which contains a 5-m-thick turbidite (Fig. 28, filled circles). The turbidite is situated between units of bioturbated sediment, with liquidity indices (open circles) that are low relative to the SEDCON curve of Perret (1995; Fig. 28, curve marked SF). The liquidity-index values for the turbidite (effective stress levels of about 3–15 kPa) are very close to those expected from the SEDCON curve. Within the turbidite, consolidation processes are similar to the simple compaction present in the SEDCON test. The more slowly deposited and bioturbated sediment above and below the turbidite is significantly more dense (lower liquidity index).

Under its own weight, a normal sediment deposit rapidly reaches a liquidity index between 2 and 3 near the water–sediment interface. The depth and intensity of bioturbation depend on the ambient fauna, but the net impact of the community can be to aggregate particles through particle repackaging (e.g. faecal-pellet production; Wheatcroft *et al.*, this volume, pp. 101–155). This will increase the bulk density of the sediment while the effective stresses are still low (< 0.5 kPa), so that the liquidity-index effective stress curve is depressed well below the SEDCON curve. At locations where rapid sedimentation occurs, the deposit may not be bioturbated, and would retain a signature characterized by an abnormally high liquidity index and more uniform changes in shear strength (Mucci & Edenborn, 1992; Perret *et al.*, 1995; Maurice *et al.*, 2000).

Other investigators (Bokuniewicz *et al.*, 1975; Richardson *et al.*, 1983; de Deckere, *et al.*, 2001), working mainly with shallow coastal sediments, have not observed densification or strengthening associated with bioturbation. Typically, these investigators found that biological activity has destabilized the seabed surface, resulting in highly remoulded, high-porosity sediment in the upper 10 cm. Bioturbation also increases random variability of physical properties on the scale of a few centimetres. Investigators have also indicated that the response of sediment to bioturbation is related to the composition of the benthic community.

Sediment densification and strengthening from repeated seismic loading

Seismic loading and oversteepening were considered in the early work of Morgenstern (1967), and many procedures for prediction of submarine landslide initiation have since focused on these triggers (Lee *et al.*, 2000). However, recent work on Eel margin sediment (Boulanger *et al.*, 1998; Boulanger, 2000) has shown that repeated, non-failure, seismic events can actually strengthen the sediment column through development of excess pore-water pressures during earthquakes and subsequent drainage, resulting in densification during intervening periods. This effect was observed during a series of cyclic-loading and drainage tests on normally consolidated fine-grained sediment. An example of the test results is given in Fig. 29 for a reconstituted sample that was initially normally consolidated. Here, the sediment begins to exhibit overconsolidation and a significant reduction in void ratio (directly related to porosity and liquidity index) if a period of drainage (~days) is allowed between repeated earthquake simulations. By relating the amount of densification to an equivalent degree of overconsolidation in these samples, the increase of strength that would result from this process could be estimated in response to four significant (simulated) earthquakes. During each earthquake, pore-water pressures increased in the sediment, and then dissipated with time after the earthquake. The sediment slowly densified, and the shear strength increased by about 65%. This seismic strengthening possibly explains, at least in part, the paucity of shallow submarine landslides on the Eel margin, an area that has much seismic activity.

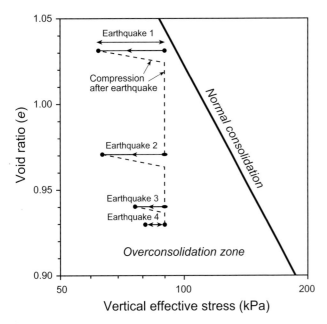

Fig. 29 Effects on void ratio of a few alternating episodes of cyclic loading and drainage (Boulanger *et al.*, 1998; Boulanger, 2000).

Changes in liquidity index, density and strength due to burial

An important component of the STRATAFORM programme was to bridge the gap between early burial (few metres) and deep burial (few hundred metres). In many cases, shear strength data are available only for depths of < 15 m in the sediment column. The SEDCON curves, which simulate the process of increasing stress from sediment accumulation, allow the extrapolation of experimental data to greater depths in a manner similar to standard consolidation tests (Richards, 1976). The SEDCON curves can be expressed in terms of density, water content, or liquidity index versus depth (Locat *et al.*, 2002). Regressions of SEDCON curves S80 and Y450 were used to evaluate the effect of burial at site O550 (Fig. 30).

The liquidity index of a sediment at its natural water content correlates well with the remoulded shear strength, s_{ur}, which can be approximated (Locat & Demers, 1988) by the following relationship

$$s_{ur} = \frac{\delta}{(I_L)^\varepsilon} \qquad (9)$$

where $\delta = 1.167$ kPa and $\varepsilon = 2.44$. Leroueil *et al.* (1983) conducted a similar analysis and obtained values of 1.615 kPa and 2.27 for the empirical terms, δ and ε, respectively. Then, by assuming a value for the sensitivity ($S_t \sim 2$–10 typical of normally consolidated marine sediment; Richards, 1976;

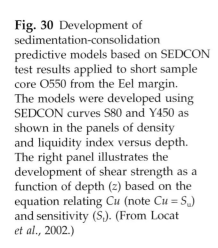

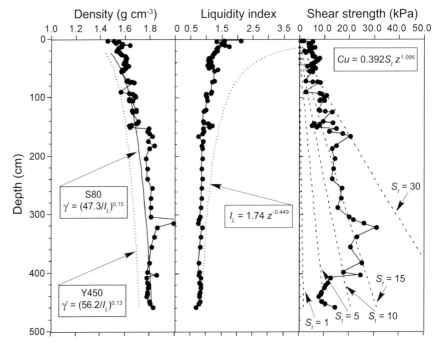

Fig. 30 Development of sedimentation-consolidation predictive models based on SEDCON test results applied to short sample core O550 from the Eel margin. The models were developed using SEDCON curves S80 and Y450 as shown in the panels of density and liquidity index versus depth. The right panel illustrates the development of shear strength as a function of depth (z) based on the equation relating Cu (note $Cu = S_u$) and sensitivity (S_t). (From Locat *et al.*, 2002.)

Locat & Lefebvre, 1986), it is possible to calculate the intact undrained shear strength (s_u) using:

$$s_u = S_t \lambda z^{1.095} \qquad (10)$$

where $\lambda = 0.392$ kPa (a coefficient derived from Eqs 8 & 9) and z is the depth below the seabed. The value of sensitivity introduces a large uncertainty for these calculations, but Eqs 3 & 10 provide two different approaches for estimating shear-strength profiles in normally consolidated sediment.

The smooth lines shown in Fig. 30 are predicted values calculated from SEDCON tests (Eqs 8 & 10). For shear strength, the predicted curves are provided for sensitivity ranging from 1 to 30, an upper limit for sensitivity in marine sediment (Richards, 1976; Locat & Lefebvre, 1986). Note that these predictions are for normal consolidation and ignore strengthening effects.

Considering the previous discussions concerning the effects of bioturbation on sediment density (and shear strength), it is not surprising to see that observed and predicted values are not in very good agreement at shallow burial depths. However, there is some convergence of the results with depth (Fig. 30), indicating that the initial differences due to alternate strengthening processes (e.g. bioturbation, cementation) are minimized, at least for the density and liquidity-index parameters. The strength values are indicative of well-structured sediment, with either a relatively high sensitivity (~15) or a relatively high degree of overconsolidation (Richards, 1976; Locat & Leroueil, 1988).

To check the applicability of these relationships to greater depths, liquidity-index data from core O550 (Eel Margin), Saguenay Fjord turbidite (Sag) and Osaka Bay Kansai clay, Japan (Tanaka & Locat, 1999) were compared (Fig. 31). The Osaka Bay profile represents a sequence of alternating sand and clay layers, and provides a good check because it extends to almost 400 m below the seafloor. An ash layer at a depth of about 250 m has been dated as ~700,000 yr BP, and the sedimentary history is complex due to interactions of basin filling and tectonic movements. Despite the complexity, the overall distribution of Osaka Bay data lies near or slightly above the SEDCON curves, and is roughly in line with the extension of the Saguenay Fjord profile. These results support the validity of Eq. 8.

Sediment rheology

In modelling mobilized sediment flows, the sediment and water mixture can be considered to be a fluid with a yield stress, so that the rheological behaviour of the matrix can be represented by a **yield strength** and **viscosity**. The Bingham model is routinely used to model debris flows (Johnson, 1970). It is defined by

$$\tau = \tau_y + \eta\gamma \qquad (11)$$

where τ_y is the yield strength, η is the viscosity, γ is the shear rate and τ is the corresponding shear stress. According to the model, the flow moves as a plug surrounded by sheared fluid (Fig. 32) and progressively loses thickness. Motion ceases when the thickness of the sediment flow can no longer produce enough shear stress to exceed the yield strength. If the model assumptions are true, the thickness of a debris-flow deposit is a measure of its **rheology**. With an estimation of the viscosity (η) based upon rheometer tests of similar material, the degree of runout can be predicted. The simplicity of the Bingham model has been utilized many times by outcrop geologists seeking an estimation of travel distance for both subaqueous (Hiscott & James, 1985) and subaerial (Whipple & Dunne, 1992) debris flows.

The yield strength and viscosity can be related to the liquidity index (Locat & Demers, 1988; Locat, 1997) as long as the liquidity index is greater than 0 (i.e. for a water content above the plastic limit). Locat (1997) found that the yield strength contributes about 1000 times more than the viscosity to the resistance of the fluid to flow. These relationships have been represented numerically (Locat, 1997) and used by Elverhøi *et al.* (1997) to analyse the behaviour of debris flows along the coast of Norway.

In response to the complexities observed by many researchers (Coussot & Meunier, 1996), yield-strength models have been extended to several different generalized forms. The most common is the Herschel–Bulkley model (Hemphill *et al.*, 1993)

$$\tau = \tau_c + K\gamma^n \qquad (12)$$

where τ_c is the critical shear stress, K is a linear coefficient analogous (although not identical) to viscosity and n is an exponent describing the rate

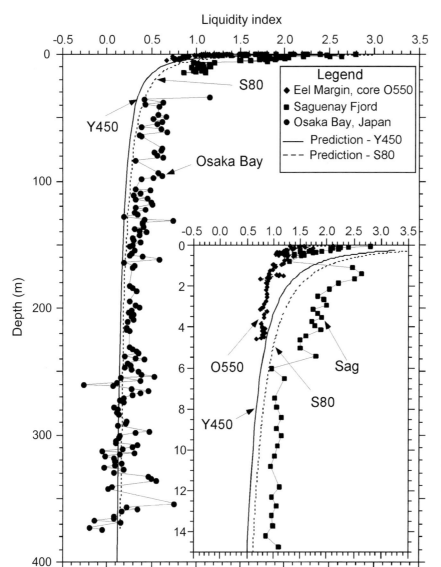

Fig. 31 Application of sedimentation-consolidation predictive models derived from SEDCON tests Y450 and S80, to a deep profile from Osaka Bay, Japan, and to a long core from the Saguenay Fjord Central Basin (Sag). The insert panel expands the first 15 m, where an Eel margin sample (O550) is compared with the Saguenay Fjord sample (Sag) – all with reference to the sedimentation-consolidation models. (From Locat *et al.*, 2002.)

of change of the viscosity with the shear-imposed stress. If $n = 1$, Bingham behaviour is regained. If $K = 1$, the formulation is of a power-law fluid (another common rheological model). In most cases, the viscosity will depend on the shear rate. If the viscosity is decreased by increasing shear, the material is said to behave as a **shear-thinning fluid**; if viscosity is increased, it is **shear thickening**. Another possibility is the bilinear model

$$\tau = \tau_{ya} + \mu_{dh}\gamma - \frac{\tau_{ya}\gamma_0}{\gamma + \gamma_0} \qquad (13)$$

where τ_{ya} is an apparent yield strength, and μ_{dh} and γ_0 are coefficients that regulate behaviour at small

shear rates. Equations 11–13 outline the three most common yield-strength models.

In addition to the above Bingham rheological models, Norem *et al.* (1990) proposed to analyse the mobility of subaqueous mass movements by using a visco-plastic model described by

$$\tau = \tau_c + \sigma(1 - r_u)\tan\phi' + \mu\gamma^n \qquad (14)$$

where σ is the total stress, r_u the pore-pressure ratio ($u/\gamma z$), μ is a viscosity-like term similar to K in Eq. 12, ϕ' the friction angle and $n = 1$ for viscous flow and 2 for inertial or granular flow. This constitutive equation is a hybrid model, similar to that proposed by Suhayda & Prior (1978). The first

a   **b**

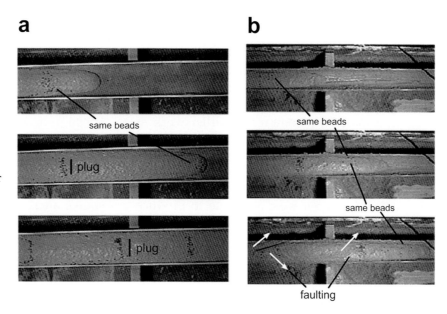

Fig. 32 (a) Debris-flow experiment illustrating unsheared plug and shear bands (near sides of flow). The behaviour of the material is well described by a yield-strength model (shear thinning, Herschel–Bulkley). The sand content in the experiment was 50% by volume. The pipe diameter was 15 cm. (b) Coarse-grain experiment where frictional behaviour was observed. The snout in this flow was driven forward by accumulation of flowing material in the debris-flow body. The sand content in this experiment was 65% by volume. The pipe diameter was 10 cm. All frames were obtained approximately 1 s apart. (From Parsons *et al.*, 2001.)

and third terms are related to the viscous components of the flow, as in Eqs 11–13. The second term is a plasticity term described by the effective stress and the friction angle. This approach can be adjusted to various flow conditions. For example, for a rapid (undrained) granular flow, the third term of Eq. 14 with $n > 1$ is most important. In the case of a mud flow (undrained), terms one and two would dominate with $n = 1$. For flows where the excess pore pressures can dissipate, the second term could dominate and the equation would approach the sliding-consolidation model proposed by Hutchinson (1986). For rock avalanches, the last two terms would be most significant.

Substantial criticism has been levelled against all of these yield-strength models in recent years (Iverson, 1997; Major, 2000). Large-scale experiments have shown that the interstitial pore pressure plays a key role in regulating the fluidity of a sediment mass (Iverson & LaHusen, 1993; Iverson, 1997). Once failure is initiated, pressures within the moving material are increased. Due to the low permeability of most natural materials, these pressures remain elevated and continue to fluidize the material (Major, 2000). The effects of heightened pore pressures are not easily incorporated into a Bingham model. As a result, Iverson (1997) proposed an alternative to the yield-strength-fluid model. The new model attempts to marry the clay-dominated Bingham behaviour with the pore-pressure-modulated granular dynamics observed

in the large-scale experiments, incorporating both yield-strength and frictional behaviour. The model also recovers a yield-strength model for fine-grained systems. The only drawback is that it requires a partitioning of the grain-size distribution into granular material (grains) and matrix (slurry). As a result, experiments are required to characterize the transition of granular behaviour and assess the degree to which sand participates in the formation of a fluid phase.

Previous experiments to assess the rheology of natural materials have been performed in rheometers of various geometries. These devices typically impose shear onto the flow in an artificial manner (i.e. a central spindle turning within a larger cylinder), and have focused on coarser materials relevant to subaerial debris flows (Major & Pierson, 1992; Coussot & Piau, 1995). Major & Pierson (1992) discovered that as the percentage of sand is increased, the fluid becomes increasingly shear thinning. In addition, when shear rates are large ($\gamma > 10 \text{ s}^{-1}$), the material will behave more like a yield-strength fluid. However, using the results of O'Brien & Julien (1988) for shear rates in natural, subaerial flows, Major & Pierson (1992) concluded that debris flows with an excess of 20% sand by volume will behave frictionally.

Parsons *et al.* (2001) sought to examine the rheological transition of a fine-grained slurry to a frictionally dominated mass in a geometry similar to an actual flow. Using profiles of velocity across the surface and the flow rate of sediment in half-pipes of different size, they were able to vary the shear rate on a single sample and derive rheological behaviour (based upon Whipple, 1997). They also altered the grain-size distribution and the clay content to examine effects on rheology of natural materials. Contrary to earlier work in rheometers, the Bingham model predicted the flow rate of material (within experimental error) for all runs with sand contents less than about 50% by volume. Their experiments also showed that clay contents of 2.5% were adequate to produce yield-strength behaviour, while clay contents in excess of 5% produced a Bingham fluid.

Like earlier studies (Major & Pierson, 1992; Coussot & Piau, 1995), Parsons et al. (2001) found that frictional behaviour dominated for shear rates less than $10\,s^{-1}$, whereas strongly sheared flows behaved more like a yield-strength fluid. Unlike earlier studies (Major & Pierson, 1992), shear rates were observed directly within the flows and generally exceeded $10\,s^{-1}$, even for gentle gradients (i.e. $< 10°$), because the shear rate was associated with the shear bands bordering the half-pipes (Fig. 32). The plug did not participate in shearing, and was not used as the length scale in the calculation of the shear rate (unlike O'Brien & Julien, 1988). At the shear rates observed, shear-thinning behaviour was dominant and fine sand participated in the formation of the fluid phase. Within the plug, frictional behaviour most probably dominated, but this region did not regulate travel distances. These results indicate that the boundary between matrix and grain behaviour (Iverson, 1997) is highly complex and flow dependent.

The transition to frictional behaviour consistently began at the snout of the flow, which was found to coarsen and 'dry' quickly (Parsons et al., 2001). Coarse snouts are typical of both subaerial (Whipple & Dunne, 1992) and subaqueous (Hiscott & James, 1985) debris flows. However, the snouts observed by Parsons et al. (2001) did not form because of purely frictional processes. The flow itself caused coarse material to collect there, possibly due to internal circulation in the manner described by Suwa (1988). The bodies of flows remained fluid, so the flow rate caused material to pile up behind the snout and drive the flow forward (Fig. 32). It is uncertain how these mechanics interact with the dynamics of hydroplaning, which is a com-

mon dynamic process associated with subaqueous debris flows and is discussed at length in Parsons et al. (this volume, pp. 275–337).

A key question, which also applies to subaerial mass movement, is how sediment acquires these rheological properties. For example, Locat et al. (1996) indicated that the mobilized yield strength (or remoulded shear strength) back-calculated for Gulf of Mexico debris flows was up to three orders of magnitude lower than the minimum remoulded shear strength that was measured in the potential source area. There must be mechanical processes taking place during the transition from slide to flow that generate a mixture having a very low remoulded shear strength. Understanding this transition, which is accompanied by acceleration of the moving mass, remains one of the major challenges ahead in the study of mass movements.

Submarine landslide geomorphology

Based on multibeam bathymetric data and GLORIA sidescan surveys, McAdoo et al. (2000) identified a total of 83 gravity flows, slides and slumps on the continental slopes of Oregon (Fig. 33a), central California (Fig. 33b), Texas (Fig. 33c) and New Jersey (Fig. 33d). The largest failures occur in the Gulf of Mexico, adjacent to Mississippi Canyon and between salt withdrawal basins (McGregor et al., 1993; Silva et al., 2004). Smaller landslides occur within the basins, and at the base of the Sigsbee Escarpment (Orange et al., 2003; Young et al., 2003). The smaller landslides tend to have higher headscarps than the larger ones and do not mobilize into mass flows as readily, indicating a stronger rheology. The Oregon section has the steepest slopes, but surprisingly few large failures for a seismically active margin, implying that slope angle and seismic activity may not be the most important slope-stability controls. A similar absence of failures in a comparable environment occurs in the Eel margin. The California continental slope is heavily incised, and this makes it difficult to identify landslides. Most of the landslides occur within larger canyons and adjacent to a pockmark field near Point Arena. The majority of the landslides on the New Jersey slope occur on the open slope between two major canyons. The slope in the Gulf of Mexico has the highest percentage (27%) of its surface area covered with failures, followed

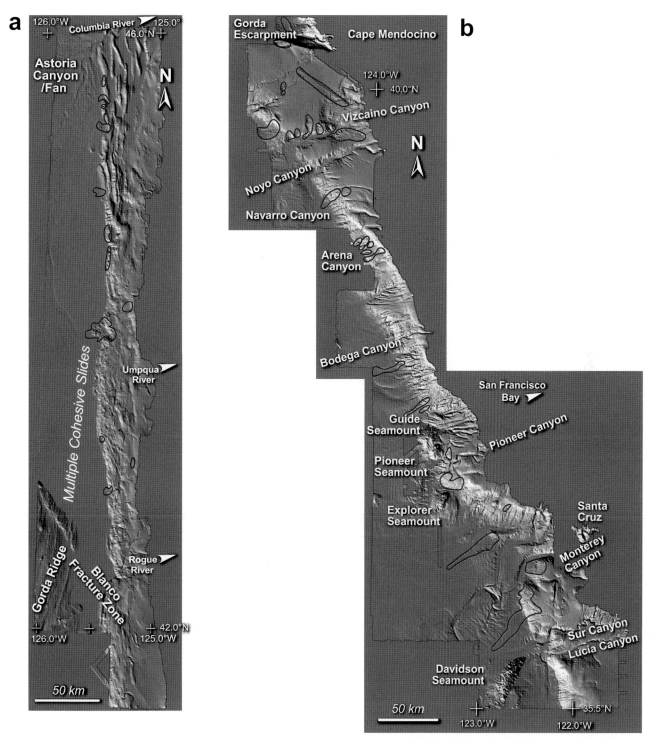

Fig. 33 Gridded NOAA multibeam bathymetric data of the (a) Oregon, (b) California, (c) Gulf of Mexico and (d) New Jersey margins. Landslides are outlined in red. (After McAdoo *et al.*, 2000.)

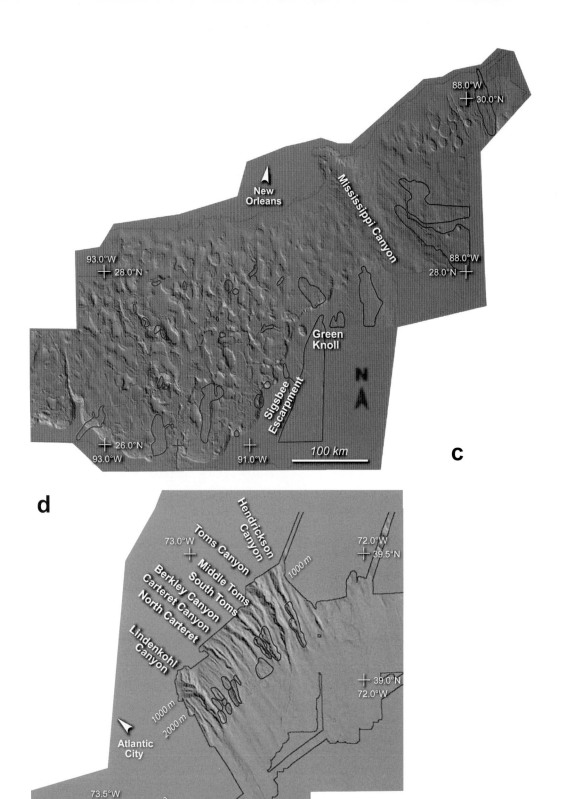

c

88.0°W
30.0°N

New Orleans

Mississippi Canyon

93.0°W
28.0°N

88.0°W
28.0°N

Green Knoll

Sigsbee Escarpment

N

26.0°N
93.0°W

91.0°W

100 km

d

Hendrickson Canyon

Toms Canyon

73.0°W

Middle Toms

South Toms

Berkley Canyon

Carteret Canyon

North Carteret

Lindenkohl Canyon

72.0°W
39.5°N

1000 m

39.0°N
72.0°W

1000 m

2000 m

Atlantic City

73.5°W
38.5°N

2000 m

Wilmington Canyon

N

Baltimore Canyon

73.5°W
38.0°N

73.0°W
38.0°N

25 km

Fig. 33 (*cont'd*)

by New Jersey (9.5%), California (7.1%) and Oregon (3%). Interestingly there is a rough inverse relation between the area covered by landslides and the local seismicity.

McAdoo *et al.* (2000) developed a set of morphometric statistics based on their synthesis. They found that most landslides occur on slopes with gradients < 10°, and that the steepness of the slope adjacent to the failure tends to be inversely proportional to the runout length. In both California and Oregon, slope failures tend to make the local slope steeper, whereas failures in the Gulf of Mexico and New Jersey slopes tend to make the local slope less steep. Landslides with rubble beneath the scar are generally small, deep seated and make the slope steeper. The ratio of headscarp height to runout length can be used as a measure of the failure's dynamic rheology. For submarine landslides, this ratio is orders of magnitude less than it is for subaerial landslides. McAdoo *et al.* (2000) noted that hydroplaning of the failed mass may be responsible for very long runout lengths.

Regional mapping of landslide susceptibility

Multibeam techniques provide detailed maps that describe seafloor topography with a high degree of precision. This information can be used to calculate the bathymetric gradient at any point, and compute gravitationally induced shear stresses throughout a region. This is a strong first step toward predicting the susceptibility of the seafloor to mass movement. In addition, multibeam systems often provide measures of backscatter intensity that contain information about lithology at the water–sediment interface and, in time, will allow evaluation of surface density and grain size. For now, bathymetry and surface character can be used to make predictions about the regional response of the seabed, including shallow-seated seabed stability. These sorts of analyses have been used productively on land (Carrara *et al.*, 1991; Jibson *et al.*, 1998). For the seafloor, several regional schemes to predict landslide susceptibility have been developed; the first of these originated out of the STRATAFORM programme.

GIS mapping

Lee *et al.* (1999, 2000) presented a methodology for applying the infinite-slope method to assess the regional variability of slope-failure potential. A series of layers were used that were operated upon by algorithms within the structure of a geographical information system (GIS). The first two layers were a map of bathymetric gradients from multibeam data (Goff *et al.*, 1999) and a map of surface density derived from analyses of closely spaced sediment cores.

Conducting a regional slope-stability analysis requires estimating an appropriate shear strength, in this case, on the basis of a surface density map. Given that seismic loading may be the critical condition for slope failure, two factors were considered:

1 the short duration of earthquakes will cause failure to occur without any flow of pore water (undrained loading);
2 the cyclic nature of earthquake loading will cause pore-water pressures to increase or decrease and will alter the shear strength.

Both these factors are considered if the strength is evaluated using a **cyclic, undrained triaxial strength test**. In such a test, cylindrical samples are encased in a membrane and consolidated to an initial effective stress, σ'_c, which is equal to the overburden effective stress being simulated. Commonly, consolidation stresses applied in the laboratory are large enough (well beyond the maximum stresses measured) that the sediment sample is forced into the normally consolidated range. Following consolidation, repeated cycles of shear stress are applied in both extension and compression until failure (defined as 15% axial strain) is achieved. For a given sediment, the number of loading cycles required to reach failure varies inversely with the applied cyclic-shear-stress level.

On a semilog diagram, the **cyclic stress ratio** (CSR) is plotted versus the number of cycles to failure. If samples with the same lithology are tested at different levels of CSR, such a plot typically generates a nearly linear relation. Seed & Idriss (1971) reported that a representative number of cycles for a typical strong earthquake is approximately 10. Accordingly, the point at which CSR corresponds to failure in 10 cycles (designated as CSR_{10}) was chosen as a measure of cyclic shear strength in seismically active areas.

A previous geotechnical study of the Eel margin (Lee *et al.*, 1981) included testing of six gravity cores for cyclic shear strength. The goal was to understand the strength properties in the vicinity of the 'Humboldt Slide'. Although 21 cyclic triaxial tests were performed as part of that study, these results do not represent the full variety of sediment lithologies in the study area and cannot be extrapolated to the entire Eel margin.

The previous Eel margin study was part of a much broader series of cyclic triaxial tests conducted at the USGS over a roughly ten-year period. Values of CSR at failure versus the number of cycles to failure were plotted for 144 tests (Fig. 34a). The complete data set forms a broad field with a range of CSR_{10} extending from about 0.25 to 0.60. Data points were grouped according to initial water content with each group extending over a range of about 10% water content (Fig. 34a). Note that water content is defined in the engineering sense as the weight of interstitial water divided by the weight of solids. For each water-content group, a linear regression analysis was performed on the values of CSR versus the log of the number of cycles to failure. The intercept of these regression lines with

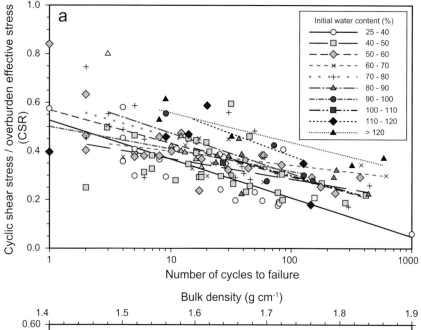

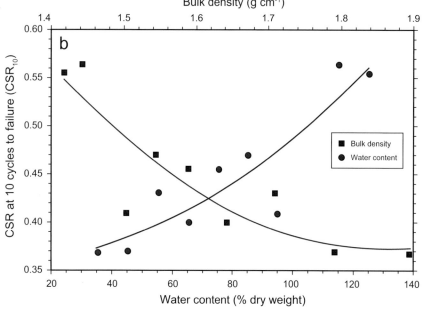

Fig. 34 Laboratory test results used to derive cyclic shear strength from initial sediment-density measurements. (a) Cyclic shear stress normalized by laboratory consolidation stress (CSR) versus number of cycles to failure (15% strain) from 144 cyclic triaxial tests performed on sediment from 10 marine study areas distributed worldwide (see Lee *et al.* (1999) for more information). Data points are identified according to natural water content (*w*) of the sediment tested. (b) The cyclic stress ratio producing failure in 10 cycles versus initial sediment water content and bulk density. Plotted points were obtained from regression fits of data presented in (a).

a value of 10 cycles to failure (CSR_{10}) corresponds to the appropriate midpoint of the water-content range. For this set, CSR_{10} varies consistently with water content and allows a parabolic regression fit of the data (Fig. 34b). For saturated marine sediment, water content and bulk density are directly related to each other at an assumed grain density of 2.7 g cm^{-3}. Accordingly, a parabolic relation between CSR_{10} and bulk density can be obtained (Fig. 34b). This relation provides an algorithm for estimating the cyclic undrained shear strength from a measure of lithology, namely the sediment bulk density. This data synthesis provides a tool for regional mapping of sediment strength.

Acoustic backscatter cannot be used quantitatively to map physical properties at present, but it does allow identification of rock outcrops and over-consolidated sediment (Lee *et al.*, 1999). These areas can often be excluded from consideration as locations for shallow-seated sediment failure.

Figure 35 shows an application of GIS regional mapping techniques to Santa Monica Bay, California (Lee *et al.*, 2000). The GIS layers for slope gradient, density and a measure of anticipated level of seismic shaking (a_p, Frankel *et al.*, 1996) are used in a series of algorithms to calculate k_c, the 'critical acceleration' to cause failure and a stability factor, k_c/a_p, which increases with level of predicted stability. Figure 35 indicates an association between areas that are predicted to be less stable (low k_c/a_p) and the locations of shallow-seated failures. A similar analysis was applied to the Eel margin (Fig. 36; Lee *et al.*, 1999, 2000), where there are few, if any, classic examples of shallow slope failure (Fig. 36b). The 'Humboldt Slide' exists in the south-west part of the area mapped with multibeam, but even if this feature were a slide, it would be so deep seated that the regional mapping approach based on relatively short cores as described above would not be applicable. Elsewhere, the seafloor appears stable except for gullies. These are relatively diffuse and highly pockmarked in the north relative to those farther south, where sharper boundaries and steeper sides are present. The northern gullies are associated with the lowest values of k_c/a_p (0.18–0.26) and those farther south are associated with higher values of k_c/a_p (0.22–0.30). This suggests that the stability of the gully sidewalls is lower for the northern diffuse gullies than it is for the more sharply defined gullies farther south.

Other regional mapping

Mulder *et al.* (1994) presented an infinite-slope formulation of regional-slope stability analysis that is similar to that of Lee *et al.* (1999, 2000), although strength properties were handled differently and earthquake loads were not considered. A more advanced approach was given by Sultan *et al.* (2001), who considered failure planes that are not necessarily parallel to the seafloor, as is the case in infinite-slope stability analysis. Large numbers of arbitrary failure surfaces were evaluated throughout a finely spaced grid and minimum values for factors of safety were selected at each grid point. This method was applied to the same area considered by Mulder *et al.* (1994), and significant quantitative differences were observed in comparisons between the two sets of results. The Sultan *et al.* (2001) method does not consider seismic loads, and undrained-shear-strength properties are applied directly from a limited number of 2-m to 7-m long cores (as opposed to a model, such as the normalized-soil-property approach, Eq. 5).

SUMMARY

This paper provides an introduction to the field of submarine landslides with an emphasis on recent advances due to the STRATAFORM programme.

Overall occurrence and triggers

Much has been learned about submarine landslides over the past 100 yr, driven, in part, by events such as the 1929 Grand Banks earthquake, the 1964 Alaska Earthquake, Hurricane Camille in 1969, the 1979 failure of the Nice airport, the 1980 earthquake in northern California, and the Papua New Guinea tsunami of 1998. The field also has been driven by technological development, including sidescan sonar, GLORIA, multibeam swath mapping, and high-resolution sub-bottom seismic profiling. These studies show that submarine landslides are common in fjords, active river deltas, submarine canyons and the open continental slope. Landslides are triggered by increases in the driving stresses, decreases in the resisting strength, or a combination of the two. Among the important triggers are sediment accumulation, erosion, earthquakes, storm

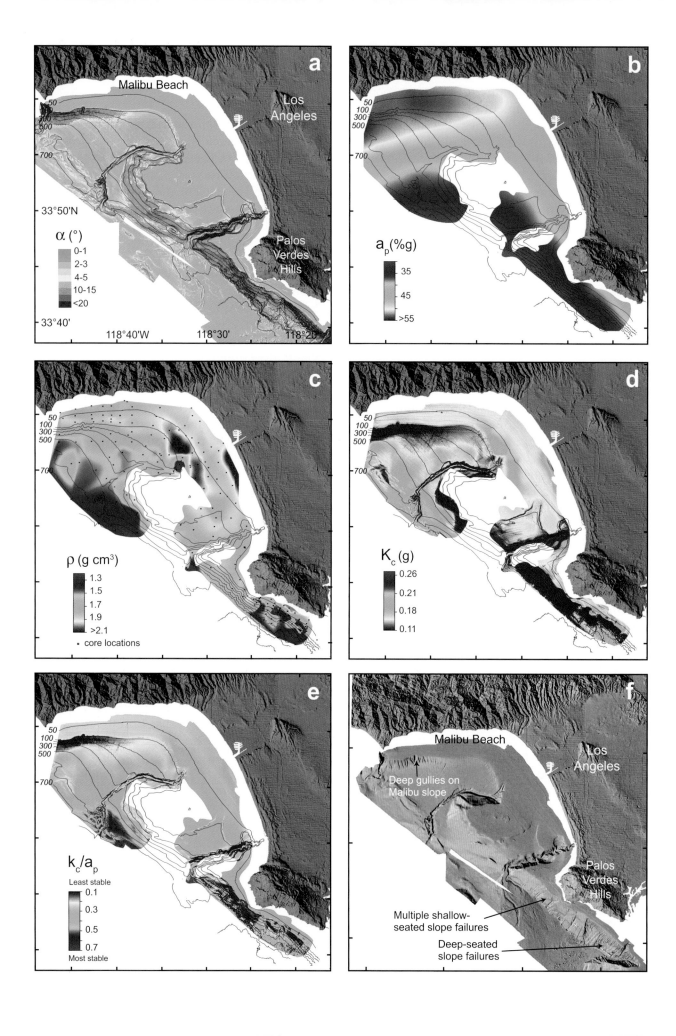

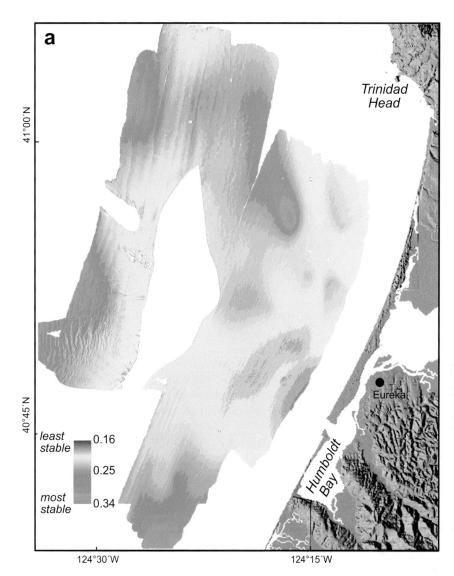

Fig. 36 Sediment failure susceptibility and deformation in the Eel margin study area. (a) Calculated values of k_c/a_p. Lower values represent a greater susceptibility to failure during seismic loading. (b) Shaded bathymetric relief of the Eel margin study area; possible deformation features indicated. (From Lee *et al.*, 1999.)

waves, volcanoes, gas and gas hydrates, ground-water seepage, diapirism and human activity.

Some elements of landslide occurrence are surprising. According to McAdoo *et al.* (2000), the greatest density of landslides occurs in the relatively aseismic Gulf of Mexico, whereas the seismically active Oregon and California margins have a much smaller percentage of their areas covered by landslide deposits. The Atlantic margin, with intermediate seismicity, has an intermediate occurrence of submarine landslides. Despite great seismicity, the Eel margin has few obvious slope-failure features, aside from the controversial 'Humboldt Slide'.

Fig. 35 (*opposite*) Results of a GIS-based analysis of slope-failure susceptibility in Santa Monica Bay, California. (a) Seafloor gradient, α, obtained from interpretations of multibeam bathymetric data. (b) Peak seismic acceleration (%g) with a 10% probability of exceedance in 50 yr, a_p (Frankel *et al.*, 1996). (c) Sediment bulk density, ρ, at 15 cm below the seafloor, interpreted from sediment core logs and contoured using a surface model. (d) Calculated values (g) of the critical horizontal earthquake acceleration, k_c, a measure of the shaking required to cause shallow slides. (e) Calculated values for the ratio of k_c to a_p (lower values represent a greater susceptibility to failure during seismic loading). (f) Shaded bathymetric relief of the Santa Monica Bay study area; possible landslide features noted. Isobaths are in metres. (From Lee *et al.*, 2000.)

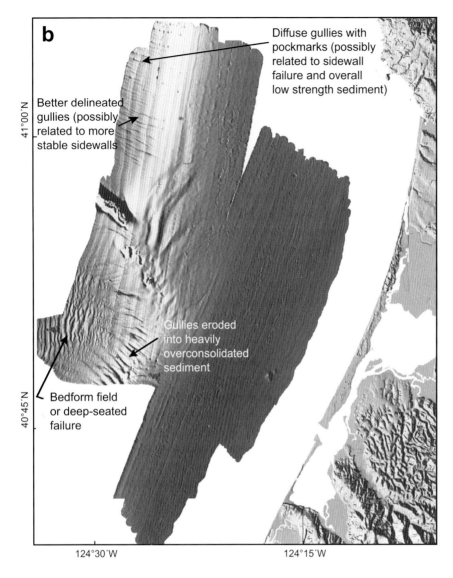

Diffuse gullies with pockmarks (possibly related to sidewall failure and overall low strength sediment)

Better delineated gullies (possibly related to more stable sidewalls

Gullies eroded into heavily overconsolidated sediment

Bedform field or deep-seated failure

Fig. 36 (*cont'd*)

Controversies

Considerable surface morphology and sub-bottom profile information is available on the 'Humboldt Slide', a large area of hummocky relief that some have interpreted as a large landslide. In profile, the feature seems to consist of a series of back-rotated blocks that have moved along discrete planes. However, the regularity of the hummocks, the continuity of beds, and the absence of a head scarp bear a great deal of resemblance to turbidity-current sediment waves observed in other areas around the world (Wynn *et al.*, 2000). Models have been formulated for the 'Humboldt Slide', as

a landslide (Gardner *et al.*, 1999) and as a field of sediment waves (Lee *et al.*, 2002).

Controversy also surrounds the processes that led to the 1929 Grand Banks turbidity current. There were many small to medium-sized failures in fine-grained sediment at the source region. However, the deposit resulting from the turbidity current in the Sohm Abyssal Plain is mainly sand. No clear source of sand has been found. Landslides that involve human structures are also controversial in the cases of failures at both the Nice airport and the Skagway dock. In Papua New Guinea, a large landslide triggered by an earthquake is thought to be the cause for a tsunami, although the earthquake alone might have been sufficient cause (Geist, 2000).

Importance of the liquidity index

The liquidity index (a dimensionless number that relates the sediment water content to the Atterberg limits) is a good representation for the degree of openness in a sediment element, and is preferable to the water content itself or the porosity or void ratio. For normal consolidation, a plot of liquidity index versus depth seems to be almost independent of the sediment type, and one graph can be used to represent many different lithologies. Many important parameters needed for modelling debris flows and slope stability can be obtained from correlations with the liquidity index (Locat & Demers, 1988).

Pore pressures and the development of anomalously weak sediment

Failure occurs when the environmental stresses exceed the strength of the sediment. If the strength of the sediment is low, the tendency toward failure is increased. Low strengths often result from high excess pore pressures developed within the sediment fabric, which can be produced by a number of factors. The degradation of shear strength can be quantitatively determined for cyclic loading due to earthquakes or storm waves (Lee *et al.*, 1999).

Development of anomalously high strength

Many seismically active environments, such as the Eel margin, do not display extensive submarine landsliding. A factor limiting slope failure may be the development of sediment strength. One mechanism for developing great strength (e.g. 50% increase) is repeated cyclic loading followed by pore-pressure dissipation. Certain types of bioturbation of surface sediment also seem to produce anomalously high strengths.

Slope stability analysis and regional assessment of landslide susceptibility

Slope stability analysis generally involves balancing the forces that tend to move sediment masses downslope against those that tend to resist such motion. Many methodologies have been developed (Syvitski *et al.*, this volume, pp. 459–529) and can be used for regional assessment of landslide susceptibility. This application requires maps of the critical input parameters, including bathymetric gradients and geotechnical properties of the sediment. These maps can be operated upon within the context of a geographical information system (GIS) to calculate values of failure susceptibility or the factor of safety (Lee *et al.*, 1999, 2000). Locations of actual failures appeared to be associated with areas calculated to have low values of relative susceptibility. Accordingly, these regional slope-stability assessment maps are recommended as an initial means of identifying the areas most vulnerable to shallow-seated slope failure.

An important contribution

Probably the most important contribution of the STRATAFORM programme to the field of submarine landslide research is the recognition that seafloor features that initially appear to be landslide deposits can in fact be ambiguous. That is, hummocky or crenulated bottom features may be suggestive of sediment failure, but closer examination may introduce questions concerning whether the features may rather be depositional, resulting from turbidity current deposition, or bottom-current modification. Resolving the conflicting interpretations can be a difficult process.

ACKNOWLEDGEMENTS

The authors gratefully acknowledge the support of the Office of Naval Research and, in particular, the long-term encouragement of Dr Joseph Kravitz. The authors also acknowledge support from the US Geological Survey and Laval University. Finally we acknowledge helpful reviews by Charles Nittrouer, Richard Faas, Joseph Kravitz, David Mohrig, Beth Mullenbach, Andrea Ogston and Patricia Wiberg.

NOMENCLATURE

Symbol	Definition	Dimensions
a, b	empirical constants in relation between I_L and σ'_v (Eq. 8)	a, mixed units, depending upon value of b; b dimensionless

a_p	anticipated level of seismic shaking, expressed probabilistically (Frankel *et al.*, 1996)	portion of gravitational acceleration
A_r	cyclic-loading strength-reduction factor	
CSR	cyclic stress ratio (cyclic shear stress/$(\sigma'_v$ or $\sigma'_c))$	
$Cu = s_u$	undrained shear strength	$M\,L^{-1}\,T^{-2}$
c_v	coefficient of consolidation	$L^2\,T^{-1}$
d	water depth	L
f_z	exp $(-2\pi z/L)$ (factor used in predicting wave-induced shear stress)	
f_d	0.5 $(1/\cosh(2\pi d/L))$ (factor used in predicting wave-induced shear stress)	
h, z	depth below the seafloor	L
H	wave height	L
I_L	liquidity index $= (w - w_p)/I_p$	
I_p	plasticity index (liquid limit minus plastic limit)	
K	linear coefficient analogous (though not identical) to viscosity used in the Herschel–Bulkley model	mixed units, depending upon value of n
k_c	critical acceleration (pseudo-static lateral acceleration needed to cause failure)	portion of gravitational acceleration
L	wavelength	L
m	sediment constant used in predicting normalized shear strength (commonly ~0.8)	
n	exponent describing the rate of change of	

	viscosity with imposed stress	
$(N_1)_{60}$	a measure of the resistance to penetration from a standard penetration test	
OCR	overconsolidation ratio (σ'_{vm}/σ'_v)	
r_u	pore-pressure ratio $(u/\gamma z)$	
$s_u = Cu$	undrained shear strength	$M\,L^{-1}\,T^{-2}$
s_{ur}	remoulded shear strength	$M\,L^{-1}\,T^{-2}$
S	ratio of shear strength to vertical effective stress for normal consolidation	
S_t	sensitivity (s_u/s_{ur})	
u	excess pore-water pressure (in excess of hydrostatic pressure)	$M\,L^{-1}\,T^{-2}$
w	water content (% dry weight)	
w_p	plastic limit	
z	depth below seafloor	L
α	seafloor gradient	degrees
γ	shear rate or total unit weight of sediment	T^{-1} or $M\,L^{-2}\,T^{-2}$
γ'	buoyant (submerged) unit weight of sediment	$M\,L^{-2}\,T^{-2}$
γ_0	coefficient that regulates rheological behaviour at small shear rates (bilinear model)	T^{-1}
γ_w	unit weight of water	$M\,L^{-2}\,T^{-2}$
δ	empirical constant in relation between I_L and s_{ur} (Eq. 9)	$M\,L^{-1}\,T^{-2}$
ε	empirical constant in relation between I_L and s_{ur} (Eq. 9)	
η	viscosity	$M\,L^{-1}\,T^{-1}$
λ	empirical constant in relation between s_{ur} and z (Eq. 10)	$M\,L^{-1}\,T^{-2}$

μ	viscosity-like term	mixed units depending on value of n
μ_{dh}	coefficient that regulates rheological behaviour at small shear rates (bilinear model)	$M\,L^{-1}\,T^{-1}$
ρ	total sediment mass density	$M\,L^{-3}$
σ	total stress	$M\,L^{-1}\,T^{-2}$
σ'_c	consolidation stress	$M\,L^{-1}\,T^{-2}$
σ'_v, σ_o	vertical effective overburden stress	$M\,L^{-1}\,T^{-2}$
σ'_{vm}	maximum past stress	$M\,L^{-1}\,T^{-2}$
τ	shear stress	$M\,L^{-1}\,T^{-2}$
τ_{av}	average shear stress anticipated from a design earthquake	$M\,L^{-1}\,T^{-2}$
τ_c	cyclic shear stress	$M\,L^{-1}\,T^{-2}$
τ_c	critical shear stress (Herschel–Bulkley model)	$M\,L^{-1}\,T^{-2}$
τ_f	τ shear stress at failure	$M\,L^{-1}\,T^{-2}$
τ_s	downslope shear stress	$M\,L^{-1}\,T^{-2}$
τ_y	yield strength	$M\,L^{-1}\,T^{-2}$
τ_{ya}	apparent yield strength	$M\,L^{-1}\,T^{-2}$
ϕ'	friction angle	degrees

REFERENCES

Adams, J. (1984) Active deformation of the Pacific northwest continental margin. *Tectonics*, **3**, 449–472.

Atigh, E. and Byrne, P.M. (2004) Liquefaction flow of submarine slopes under partially undrained conditions: an effective stress approach. *Can. Geotech. J.*, **41**, 154–165.

Austin, J.A., Jr., Christie-Blick, N., Malone, M.J., et al. (1998) *Proceedings of the Ocean Drilling Program, Initial Reports*, **174A** (CD-ROM).

Barnard, W. (1978) The Washington continental slope: Quaternary tectonics and sedimentation. *Mar. Geol*, **27**, 79–114.

Bea, R.G., Wright, S.G., Sircar, P. and Niedoroda, A.W. (1983) Wave-induced slides in South Pass Block 70, Mississippi Delta. *J. Geotech. Eng.*, **109**, 619–644.

Bjerrum, L. (1955) Stability of natural slopes in quick clay. *Géotechnique*, **5**, 101–119.

Bohannon, R.G. and Gardner, J.V. (2004) Submarine landslides of San Pedro Escarpment, southwest of Long Beach, California. *Mar. Geol.*, **203**, 261–268.

Bokuniewicz, H.J., Gordon, R. and Rhoads, D.C. (1975) Mechanical properties of the sediment–water interface. *Mar. Geol.*, **18**, 263–278.

Booth, J.S. (1979) Recent history of mass wasting on the upper continental slope, northern Gulf of Mexico as interpreted from the consolidation states of the sediment. In: *Geology of Continental Slopes* (Eds O.H. Pilkey and L.J. Doyle), pp. 153–164. Special Publication 27, Society of Economic Paleontologists and Mineralogists, Tulsa, OK.

Booth, J.S. and Garrison, L.E. (1978) A geologic and geotechnical analysis of the upper continental slope adjacent to the Mississippi Delta. *Proceedings, 10th Offshore Technical Conference*, Houston, TX, pp. 1019–1028.

Booth, J.S., O'Leary, D.W., Popenoe, P. and Danforth, W.W. (1993) U.S. Atlantic continental slope landslides: their distribution, general attributes and implications. In: *Submarine Landslides: Selected Studies in the U.S. EEZ* (Eds W.C. Schwab, H.J. Lee and D.C. Twichell). *U.S. Geol. Surv. Bull.*, **2002**, 14–39.

Bornhold, B.D. and Prior, D.B. (1990) Morphology and sedimentary processes on the subaqueous Noeick River delta, British Columbia, Canada. In: *Coarse-grained Deltas* (Eds A. Colella and D.B. Prior), pp. 169–184. Special Publication 10, International Association of Sedimentologists. Blackwell Scientific Publications, Oxford.

Boulanger, E. (2000) *Comportement cyclique des sédiments de la marge continentale de la rivière Eel: une explication possible pour le peu de glissements sous-marins superficiels dans cette région*. MSc thesis, Department of Geology and Geological Engineering, Laval University.

Boulanger, E., Konrad, J.-M., Locat, J. and Lee, H.J. (1998) Cyclic behavior of Eel River sediments: a possible explanation for the paucity of submarine landslide features. *Eos (Trans. Am. Geophys. Union)*, **79**, 254.

Bryn, P., Solheim, A., Berg, K., et al. (2003) The Storegga slide complex: repeated large scale sliding in response to climatic cyclicity. In: *Submarine Mass Movements and their Consequences* (Eds J. Locat and J. Mienert), pp. 215–222. Kluwer, Dordrecht.

Buffington, E.C. and Moore, D.G. (1962) Geophysical evidence on the origin of gullied submarine slopes, San Clemente, California. *J. Geol.*, **71**, 356–370.

Canals, M., Lastras, G., Urgeles, R., et al. (2004) Slope failure dynamics and impacts from seafloor and shallow sub-seafloor geophysical data: case studies from the COSTA project. *Mar. Geol.*, **213**, 9–72.

Carlson, P.R. (1978) Holocene slump on continental shelf off Malaspina Glacier, Gulf of Alaska. *Bull. Am. Assoc. Petrol. Geol.*, **62**, 2412–2426.

Carlson, P.R., Molnia, B.F. and Wheeler, M.C. (1980) Seafloor geologic hazards in OCS Lease Area 55, eastern Gulf of Alaska. *Proceedings, 12th Offshore Technology Conference*, Houston, TX, Vol. 1, pp. 593–603.

Carrara, A., Cardinali, M., Detti, R., *et al.* (1991) GIS Techniques and statistical models in evaluating landslide hazard. *Earth Surf. Process. Landf.*, **16**, 427–445.

Cashman, K.V. and Popenoe, P. (1985) Slumping and shallow faulting related to the presence of salt on the continental slope and rise off North Carolina. *Mar. Petrol. Geol.*, **2**, 260–272.

Cattaneo, A., Correggiari, A., Marsset, T., *et al.* (2004) Seafloor undulation pattern on the Adriatic shelf and comparison to deep-water sediment waves. *Mar. Geol.*, **213**, 121–148.

Christian, H.A. (1993) *In situ* measurement of consolidation and permeability of soft marine sediments. *Proceedings, 4th Canadian Conference on Marine Geotechnical Engineering*, St John's, Newfoundland, pp. 663–379.

Christian, H.A. (1998) *Excalibur Pore Pressure and Fluid Sampling STRATAFORM Wecoma 9807A Cruise.* Report to Laval University, Vol. 16.

Clague, D.A. and Denlinger, R.P. (1994) Role of olivine cumulates in destabilizing the flanks of Hawaiian volcanoes. *Bull. Volcanol.*, **56**, 425–434.

Clarke, S.H., Jr. (1992) Geology of the Eel River basin and adjacent region: implications for late Cenozoic tectonics of the southern Cascadia subduction zone and Mendocino triple junction. *Bull. Am. Assoc. Petrol. Geol.*, **76**, 199–224.

Clukey, E., Kulhawy, F.H., Liu, P.L.F. and Tate, G.B. (1985) The impact of wave loads and pore-water pressure generation on initiation of sediment transport. *Geo-Mar. Lett.*, **5**, 177–183.

Coleman, J.M. (1988) Dynamic changes and processes in the Mississippi River delta. *Geol. Soc. Am. Bull.*, **100**, 999–1015.

Coleman, J.M. and Garrison, L.E. (1977) Geological aspects of marine slope stability, northwestern Gulf of Mexico. *Mar. Geol.*, **2**, 9–44.

Coleman, J.M., Prior, D.B. and Garrison, L.E. (1980) Subaqueous sediment instabilities in the offshore Mississippi River delta. *U.S. Bur. Land Manag. Open-file Rep.*, **80–01**, 60.

Coleman, J.M., Prior, D.B., Garrison, L.E. and Lee, H.J. (1993) Slope failures in an area of high sedimentation rate: offshore Mississippi River Delta. In: *Submarine Landslides: Selected Studies in the U.S. EEZ* (Eds W.C. Schwab, H.J. Lee and D.C. Twichell). *U.S. Geol. Surv. Bull.*, **2002**, 79–91.

Correggiari, A., Trincardi, F., Langone, L. and Roveri, M. (2001) Styles of failure in late Holocene highstand prodelta wedges on the Adriatic shelf. *J. Sediment. Res.*, **71**, 218–236.

Coulter, H.W. and Migliaccio, R.R. (1966) Effects of the earthquake of March 27, 1964 at Valdez, Alaska. *U.S. Geol. Surv. Prof. Pap.*, **542-C**.

Coussot, P. and Meunier, M. (1996) Recognition, classification and mechanical description of debris flows. *Earth-Sci. Rev.*, **40**, 209–227.

Coussot, P. and Piau, J. (1995) A large-scale field coaxial cylinder rheometer for the study of the rheology of natural coarse suspensions. *J. Rheol.*, **39**, 105–124.

De Deckere, E.M., Tolhurst, T.J. and de Brouwer, J.F.C. (2001) Destabilization of cohesive intertidal sediments by infauna. *Estuar. Coast. Shelf Sci.*, **53**, 665–669.

Dietrich, J.H. (1988) Growth and persistence of Hawaiian rift zones. *J. Geophys. Res.*, **93**, 4258–4270.

Dill, R.F. (1964) Sedimentation and erosion in Scripps submarine canyon head. In: *Papers in Marine Geology* (Ed. R.L. Miller), pp. 232–248. MacMillan, New York.

Dugan, B., Olgaard, D.L., Flemings, P.B. and Gooch, M.J. (2002) Data report: bulk physical properties of sediments from ODP Site 1073. In: *Proceedings of the Ocean Drilling Program, Scientific Results*, Vol. 174A (Eds N. Christie-Blick, J.A. Austin, Jr. and M.J. Malone), pp. 1–62. College Station, TX.

Elverhøi, A., Norem, H., Andersen, E.S., *et al.* (1997) On the origin and flow behavior of submarine slides on deep-sea fans along the Norwegian–Barents Sea continental margin. *Geo-Mar. Lett.*, **17**, 119–125.

Embley, R.W. and Jacobi, R.D. (1977) Distribution and morphology of large submarine sediment slides and slumps on Atlantic continental margins. *Mar. Geotech.*, **2**, 205–228.

Emery, K.O., Uchupi, E., Phillips, J.D., *et al.* (1970) Continental rise of eastern North America. *Bull. Am. Assoc. Petrol. Geol.*, **54**, 44–108.

Esrig, M.I. and Kirby, R.C. (1977) Implications of gas content for predicting the stability of submarine slopes. *Mar. Geotech.*, **2**, 81–100.

Farrow, G.E., Syvitski, J.P.M. and Tunnicliffe, V. (1983) Suspended particulate loading on the macrobenthos in a highly turbid fjord: Knight Inlet, British Columbia. *Can. J. of Fisheries and Aquatic Sci.*, 40(suppl. 1), 273–288.

Faugères, J.-C., Gonthier, E., Mulder, T., *et al.* (2002) Multi-process generated sediment waves on the Landes Plateau (Bay of Biscay, North Atlantic). *Mar. Geol.*, **182**, 279–302.

Field, M.E. (1981) Sediment mass-transport in basins: controls and patterns. In: *Short Course Notes Pacific Section* (Eds R.G. Douglas, E.P. Colburn and D.S.

Gorsline), pp. 61–83. Society of Economic Paleontologists and Mineralogists, Tulsa, OK.

Field, M.E. (1993) Liquefaction of continental shelf sediment: the northern California earthquake of 1980. In: *Submarine Landslides: Selected Studies in the U.S. EEZ* (Eds W.C. Schwab, H.J. Lee and D.C. Twichell). *U.S. Geol. Surv. Bull.*, **2002**, 143–150.

Field, M.E. and Barber, J.H., Jr. (1993) A submarine landslide associated with shallow sea-floor gas and gas hydrates off northern California. In: *Submarine Landslides: Selected Studies in the U.S. EEZ* (Eds W.C. Schwab, H.J. Lee and D.C. Twichell). *U.S. Geol. Surv. Bull.*, **2002**, 151–157.

Field, M.E. and Clarke, S.H. (1979) Small-scale slumps and slides and their significance for basin slope processes, Southern California Borderland. In: *Geology of Continental Slopes* (Eds O.H. Pilkey and L.J. Doyle), pp. 223–230. Special Publication 27, Society of Economic Paleontologists and Mineralogists, Tulsa, OK.

Field, M.E. and Edwards, B.D. (1980) Slopes of the southern California borderland: a regime of mass transport. In: *Proceedings, Quaternary Depositional Environments of the Pacific Coast: Pacific Coast Paleogeography, Symposium No. 4* (Eds M.E. Field, A.H. Bouma, I.P. Colburn, J.J. Douglas and J.C. Ingle), pp. 69–184. Pacific Section, Society of Economic Paleontologists and Mineralogists, Tulsa, OK.

Field, M.E. and Richmond, W.C. (1980) Sedimentary and structural patterns on the northern Santa Rosa–Cortes Ridge, southern California. *Mar. Geol.*, **34**, 79–98.

Field, M.E., Gardner, J.V. and Prior, D.B. (1999) Geometry and significance of stacked gullies on the northern California slope. *Mar. Geol.*, **154**, 271–286.

Field, M.E., Clarke, S.H., Jr. and White, M.E. (1980) Geology and geologic hazards of offshore Eel River Basin, northern California continental margin. *U.S. Geol. Surv. Open-file Rep.*, **80–1080**.

Fischer, J.A., Koutsoftas, D.C. and Lu, T.D. (1976) The behavior of marine soils under cyclic loading. *Proceedings, First International Conference, Behavior of Off-Shore Structures*, Trondheim, Norway, pp. 407–414.

Fisk, H.N., McFarlan, E., Jr., Kold, C.R. and Wilbern, L.J., Jr. (1954) Sedimentary framework of the modern Mississippi Delta. *J. Sediment. Petrol.*, **24**, 76–99.

Frankel, A., Mueller, C., Barnhard, T., *et al.* (1996) National seismic-hazard maps: documentation. *U.S. Geol. Surv. Open-file Rep.*, **96–532**.

Fulthorpe, C.S., Mountain, G.S., Austin, J.A., *et al.* (1996) STRATAFORM high-resolution MCS survey, Eel River basin, northern California margin: shelf/slope stratigraphy and processes (abstr.). *Eos (Trans. Am. Geophys. Union)*, **77**, F330.

Gardner, J.V., Prior, D.B. and Field, M.E. (1999) Humboldt slide – a large shear-dominated retrogressive slope failure. *Mar. Geol.*, **154**, 323–338.

Garrison, L.E. (1977) The SEASWAB Experiment. *Mar. Geotech.*, **2**, 117–122.

Geist, E.L. (2000) Origin of the 17 July 1998 Papua New Guinea tsunami; earthquake or landslide. *Seism. Res. Lett.*, **71**, 344–351.

Gibson, R.E. (1958) The process of consolidation in a clay layer increasing in thickness with time. *Géotechnique*, **8**, 171–182.

Goff, J.A., Orange, D.L., Mayer, L.A. and Hughes Clarke, J.E. (1999) Detailed investigation of continental shelf morphology using a high-resolution swath sonar survey: the Eel margin, northern California, *Mar. Geol.*, **154**, 255–270.

Goff, J.A., Wheatcroft, R.A., Lee, H., *et al.* (2002) Spatial variability of shelf sediments in the STRATAFORM natural laboratory, *Cont. Shelf Res.*, **22**, 1199–1223.

Goldfinger, C., Nelson, C.H., Johnson, J.E. and Shipboard Scientific Party (2003) Holocene earthquake records from the Cascadia subduction zone and the northern San Andreas Fault based on precise dating of offshore turbidites. *Ann. Rev. Earth Planet. Sci.*, **31**, 555–577.

Greene, H.G., Maher, N.M. and Paull, C.K. (2002) Physiography of the Monterey Bay national Marine Sanctuary and implications about continental margin development. *Mar. Geol.*, **181**, 55–82.

Gregersen, O. (1981) The quick clay landslide in Rissa, Norway. *Proceedings, 10th International Conference on Soil Mechanics and Foundation Engineering*, Vol. 3, pp. 421–426.

Grozic, J.L.H., Robertson, P.K. and Morgenstern, N.R. (2000) Cyclic liquefaction of loose gassy sand. *Can. Geotech. J.*, **37**, 843–856.

Haflidason, H., Sejrup, H.P., Nygard, A., *et al.* (2004) The Storegga slide: architecture, geometry and slide development. *Mar. Geol.*, **213**, 201–234.

Hampton, M. (1972) The role of subaqueous debris flow in generating turbidity currents. *J. Sediment. Petrol.*, **42**, 775–993.

Hampton, M.A. and Bouma, A.H. (1977) Slope instability near the shelf break, western Gulf of Alaska. *Mar. Geotech.*, **2**, 309–331.

Hampton, M.A., Lee, H.J. and Beard, R.M. (1982) Geological interpretation of cone penetrometer tests in Norton Sound, Alaska. *Geo-Mar. Lett.*, **2**, 223–231.

Hampton, M.A., Lee, H.J. and Locat, J. (1996) Submarine landslides. *Rev. Geophys.*, **34**, 33–59.

Hampton, M.A., Karl, H.A. and Murray, C.J. (2002) Acoustic profiles and images of the Palos Verdes margin; implications concerning deposition from the White's Point outfall. *Cont. Shelf Res.*, **22**, 841–858.

Haner, B.E. and Gorsline, D.S. (1978) Processes and morphology of continental slope between Santa Monica and Dume submarine canyons, Southern California. *Mar. Geol.*, **28**, 77–87.

Heezen, B.C. and Drake, C.L. (1964) Grand Banks slump. *Bull. Am. Assoc. Petrol. Geol.*, **48**, 44–108.

Heezen, B.C. and Ewing, M. (1952) Turbidity currents and submarine slumps and the 1929 Grand Banks Earthquake. *Am. J. Sci.*, **250**, 849–873.

Hein, F.J. and Gorsline, D.S. (1981) Geotechnical aspects of fine-grained mass flow deposits: California Continental Borderland, *Geomar. Lett.*, **1**, 1–5.

Hemphill, T., Campos, W. and Pilehvari, A. (1993) Yield-power law model more accurately predicts mud rheology. *Oil & Gas Journal*. **91**, 45–50.

Henkel, D.J. (1970) The role of waves causing submarine landslides. *Géotechnique*, **20**, 75–80.

Hiscott, R.N. and James, N.P. (1985) Carbonate debris flows, Cow Head group, western Newfoundland. *J. Sediment. Petrol.*, **55**, 735–745.

Holcomb, R.T. and Searle, R.C. (1991) Large landslides from oceanic volcanoes. *Mar. Geotech.*, **10**, 19–32.

Hutchinson, J.N. (1986) A sliding-consolidation model for flow slides. *Can. Geotech. J.*, **23**, 115–126.

Iverson, R.M. (1995) Can magma-injection and ground-water forces cause massive landslides on Hawaiian volcanoes? *Jour. Volcan. Geother. Res.*, 66, 295–308.

Iverson, R.M. (1997) The physics of debris flow. *Rev. Geophys*, 35, 245–296.

Iverson, R.M. and LaHusen, R.G. (1993) Friction in debris flows: Inferences from large-scale flume experiments. *Hydraulic Engineering '93: Proceedings of the 1993 Conference, San Francisco, California, ASCE*, 1604–1609.

Jibson, R.W., Harp, E.L. and Michael, J.A. (1998) A method for producing digital probabilistic seismic landslide hazard maps: an example from the Los Angeles, California, area. *U.S. Geol. Surv. Open-file Rep.*, **98–113**, 17 pp.

Johnson, A.M. (1970) *Physical Processes in Geology.* Freeman, Cooper and Co., San Francisco.

Kachadoorian, R. (1965) Effects of the earthquake of March 27, 1964 at Whittier, Alaska, *U.S. Geol. Surv. Prof. Pap.*, **542-B**.

Kayen, R.E. and Lee, H.J. (1991) Pleistocene slope instability of gas hydrate-laden sediment on the Beaufort Sea margin. *Mar. Geotech.*, **10**, 125–141.

Knebel, H.J. and Carson, B. (1979) Small-scale slump deposits, Middle Atlantic Continental Slope, off eastern United States. *Mar. Geol.*, **29**, 221–236.

Kostaschuk, R.A. and McCann, S.B. (1983) Observations on delta-forming processes in a fjord-head delta, British Columbia. *Sediment. Geol.*, **36**, 269–288.

Kraft, L.M., Gavin, T.M. and Bruton, J.C. (1992) Submarine flow slide in Puget Sound. *J. Geotech. Eng., ASCE*, **118**, 1577–1591.

Ladd, C.C., Foott, R., Ishihara, K., Schlosser, F. and Poulos, H.G. (1977) Stress-deformation and strength characteristics. *Proceedings, Ninth International Conference on Soil Mechanics and Foundation Engineering*, Tokyo, Japan, Vol. 2, pp. 421–494.

Lambe, T.W. and Whitman, R.V. (1969) *Soil Mechanics.* John Wiley & Sons, New York, 553 pp.

Lee, H.J. (1985) State of the art: Laboratory determination of the strength of marine soils. In: *Strength Testing of Marine Sediments: Laboratory and In situ Measurements* (Eds R.C. Chaney and K.R. Demars), pp. 181–250. STP 883, American Society for Testing and Materials.

Lee, H.J. and Edwards, B.D. (1986) Regional method to assess offshore slope stability. *J. Geotech. Eng., ASCE*, **112**, 489–509.

Lee, H.J., Edwards, B.D. and Field, M.E. (1981) Geotechnical analysis of a submarine slump, Eureka, California. *Proceedings of the Offshore Technology Conference, Houston*, pp. 53–59.

Lee, H.J., Schwab, W.C., Edwards, B.D. and Kayen, R.E. (1991) Quantitative controls on submarine slope failure morphology. *Mar. Geotech.*, **10**, 143–158.

Lee, H.J., Locat, J., Dartnell, P., Israel and Wong, F. (1999) Regional variability of slope stability: application to the Eel margin, California. *Mar. Geol.*, **154**, 305–321.

Lee, H.J., Locat, J., Dartnell, P., Minasian, D. and Wong, F. (2000) A GIS-based regional analysis of the potential for shallow-seated submarine slope failure. *Proceedings, 8th International Symposium on Landslides*, Cardiff, Wales, pp. 917–922.

Lee, H.J., Syvitski, J.P.M., Parker, G., *et al.* (2002) Distinguishing sediment waves from slope failure deposits: field examples, including the 'Humboldt Slide' and modeling results. *Mar. Geol.*, **192**, 79–104.

Lee, K.L. and Focht, J.A. (1976) Strength of clay subjected to cyclic loading. *Mar. Geotech.*, **1**, 165–185.

Lehner, P. (1969) Salt tectonics and Pleistocene stratigraphy on continental slope of northern Gulf of Mexico. *Bull. Am. Assoc. Petrol. Geol.*, **53**, 2431–2479.

Lemke, R.W. (1967) Effects of the earthquake of March 27, 1964, at Seward, Alaska, *U.S. Geol. Surv. Prof. Pap.*, **542-E**.

Leroueil, S., Tavenas, F. and LeBihan (1983) Propriétés caractéristiques des argiles de l'est du Canada. *Can. Geotech. J.*, **20**, 681–705.

Leroueil, S., Vaunat, J., Picarelli, L., Locat, J., Lee, H. and Faure, R. (1996) Geotechnical characterization of slope movements. *Proceedings of the International Symposium on Landslides*, Trondheim.

Locat, J. (1982) *Contribution à l'étude de l'origine de la structuration des argiles sensibles de l'est du Canada. Unpublished PhD thesis,* Department of Civil Engineering, University of Sherbrooke, Sherbrooke, Quebec.

Locat, J. (1997) Normalized rheological behavior of fine muds and their flow properties in a pseudoplastic regime. In: *Debris-flow Hazards Mitigation: Mechanics, Prediction and Assessment*, pp. 260–269. Water Resources Engineering Division, American Society of Civil Engineers.

Locat, J. and Demers, D. (1988) Viscosity, yield stress, remolded shear strength and liquidity index relationships for sensitive clays. *Can. Geotech. J.*, **25**, 799–806.

Locat, J. and Lee, H.J. (2002) Submarine landslides: advances and challenges. *Can. Geotech. J.*, **39**, 193–212.

Locat, J. and Lefebvre, G. (1986) The origin of structuration of the Grande-Baleine marine sediments, Québec Canada. *Q. J. Eng. Geol.*, **19**, 365–374.

Locat, J. and Leroueil, S. (1988) Physicochemical and mechanical characteristics of recent Saguenay Fjord sediments. *Can. Geotech. J.*, **25**, 382–388.

Locat, J., Lee, H.J., Nelson, H.C., Schwab, W.C. and Twichell, D.C. (1996) Analysis of the mobility of far reaching debris flows on the Mississippi Fan, Gulf of Mexico. *Proceedings, 7th International Symposium on Landslides*, pp. 555–560.

Locat, J., Lee, H.J., Kayen, R., *et al.* (2002) Shear strength development with burial in Eel River Margin slope sediments. *Mar. Geotech.*, **20**, 111–135.

Locat, J., Desgagnés, P., Leroueil, S. and Lee, H.J. (2003) Stability analysis of the Hudson Apron slope, off New Jersey, U.S.A. In: *Submarine Mass Movements and their Consequences* (Eds J. Locat and J. Mienert), pp. 267–280. Kluwer, Dordrecht.

Mader, C.L. (1997) Modeling the 1994 Skagway tsunami. *Sci. Tsunami Haz.*, **15**, 41–48.

Major, J.J. (2000) Gravity-driven consolidation of granular slurries – Implications for debris-flow deposition and deposit characteristics. *J. Sediment. Res.*, **70**, 64–83.

Major, J.J. and Pierson, T.C. (1992) Debris flow rheology: Experimental analysis of fine grained slurries. *Water Resources. Res.*, **28**, 841–857.

Malahoff, A., Embley, R.W., Perry, R.B. and Fefe, C. (1980) Submarine mass-wasting of sediments on the continental slope and upper rise south of Baltimore Canyon: *Earth and Plan. Sci, Letters*, **49**, 1–7.

Malouta, D.N., Gorsline, D.S. and Thornton, S.E. (1981) Processes and rates of recent (Holocene) basin filling in an active transform margin: Santa Monica Basin, California Continental Borderland. *J. Sediment. Petrol.*, **51**, 1077–1095.

Marlow, M.S., Scholl, D.W., Buffington, E.C., Boyce, R.E., Alpha, T.R., Smith, P.J. and Shipek, C.J. (1970) Buldir Depression – a late Tertiary graben on the Aleutian Ridge, Alaska. *Mar. Geol.*, **8**, 85–108.

Marshall, N.F. (1978) Large storm-induced sediment slump reopens an unknown Scripps Submarine Canyon tributary. In: *Sedimentation in Submarine Canyons, Fans and Trenches* (Eds D.J. Stanley and G. Kelling, G.), pp. 73–84. Dowden, Hutchinson and Ross, Stroudsburg, PA.

Martin, R.G. and Bouma, A.H. (1982) Active diapirism and slope steepening, northern Gulf of Mexico continental slope. *Mar. Geotech.*, **5**, 63–91

Maslin, M., Owen, M., Day, S. and Long, D. (2004) Linking continental-slope failures and climate change; testing the clathrate gun hypothesis. *Geology*, **32**, 53–56

Maurice, F., Locat, J. and Leroueil, S. (2000) Caractéristiques géotechniques et évolution de la couche de sédiment déposée lors du déluge de 1996, dans la Baie des Ha! Ha! (fjord du Saguenay, Québec). *Proceedings of the 53rd Canadian Geotechnical Conference*, Montreal, Quebec.

McAdoo, B.G., Pratson, L.F. and Orange, D.L. (2000) Submarine landslide geomorphology, US continental slope. *Mar. Geol.*, **169**, 103–136.

McGregor, B.A. (1977) Geophysical assessment of submarine slide northeast of Wilmington Canyon. *Mar. Geotech.*, **2**, 229–244.

McGregor, B.A., Bennett, R.H. and Lambert, D.N. (1979) Bottom processes, morphology and geotechnical properties of the continental slope south of Baltimore Canyon. *Appl. Ocean Res.*, **1**, 177–187.

McGregor, B.A., Rothwell, R.G., Kenyon, N.H. and Twichell, D.C. (1993) Salt tectonics and slope failure in an area of salt domes in the northwestern Gulf of Mexico. In: *Submarine Landslides: Selected Studies in the U.S. EEZ* (Eds W.C. Schwab, H.J. Lee and D.C. Twichell). *U.S. Geol. Surv. Bull.*, **2002**, 92–96.

Milliman, J.D. and Meade, R.H. (1983) World-wide delivery of river sediment to the oceans. *J. Geol.*, **91**, 1–21.

Mitchell, R.J. (1978) On the retrogression of landslides in sensitive muddy sediments, discussion. *Can. Geotech. J.*, **15**, 446–450.

Molnia, B.F., Carlson, P.R. and Bruns, T.R. (1977) Large submarine slide in Kayak Trough, Gulf of Alaska. *Geol. Soc. Am. Rev. Eng. Geol.*, **3**, 137–148.

Moore, J.G. (1964) Giant submarine landslides on the Hawaiian Ridge. *U.S. Geol. Surv. Prof. Pap.*, **501-D**, D95–D98.

Moore, J.G., Clague, D.A., Holcomb, R.T., *et al.* (1989) Prodigious submarine slides on the Hawaiian Ridge. *J. Geophys. Res.*, **94**, 17,465–17,484.

Morgenstern, N.R. (1967) Submarine slumping and the initiation of turbidity currents. In: *Marine Geotechnique* (Ed. A.F. Richards), pp. 189–210. University of Illinois Press, Urbana, IL.

Mucci, A. and Edenborn, H.M. (1992) Influence of an organic poor landslide deposit on the early diagenesis of iron and manganese in a coastal marine sediment. *Geochem. Cosmochim. Acta*, **56**, 3909–3921.

Mulder, T. and Cochonat, P. (1996) Classification of offshore mass movements. *J. Sediment. Res.*, **66**, 43–57.

Mulder, T. and Syvitski, J.P.M. (1995) Turbidity currents generated at river mouths during exceptional discharges to the world oceans. *J. Geol.*, **103**, 285–299.

Mulder, T., Tisot, J.-P., Cochonat, P. and Bourillet, J.-F. (1994) Regional assessment of mass failure events in the Baie des Anges Mediterranean Sea. *Mar. Geol.*, **122**, 29–45.

Mullenbach, B.I. and Nittrouer, C.A. (2000) Rapid deposition of fluvial sediment in the Eel Canyon, northern California. *Cont. Shelf Res.*, **20**, 2191–2212.

Mullenbach, B.I., Nittrouer, C.A., Puig, P. and Orange, D.L. (2004) Sediment deposition in a modern submarine canyon. *Mar. Geol.*, **211**, 101–119.

Nardin, T.R., Edwards, B.D. and Gorsline, D.S. (1979a) Santa Cruz Basin, California Borderland: dominance of slope processes in basin sedimentation. In: *Geology of Continental Slopes* (Eds O.H. Pilkey and L.J. Doyle), pp. 209–221. Special Publication 27, Society of Economic Paleontologists and Mineralogists, Tulsa, OK.

Nardin, T.R., Hein, F.J., Gorsline, D.S. and Edwards, B.D. (1979b) A review of mass movement processes, sediment and acoustic characteristics and contrasts in slope and base-of-slope systems versus canyon-fan-basin floor systems. In: *Geology of Continental Slopes* (Eds O.H. Pilkey and L.J. Doyle), pp. 61–73. Special Publication 27, Society of Economic Paleontologists and Mineralogists, Tulsa, OK.

Nelson, C.H. (1976) Late Pleistocene and Holocene depositional trends, processes and history of Astoria Deep-Sea Fan, northeast Pacific. *Mar. Geol.*, **20**, 129–173.

Norem, H., Locat, J. and Schieldrop, B. (1990) An approach to the physics and the modelling of submarine landslides. *Mar. Geotech.*, **9**, 93–111.

Normark, W.R., Hess, G.R., Stow, D.A.V. and Bowen, A.J. (1980) Sediment waves on the Monterey Fan Levee: a preliminary physical interpretation. *Mar. Geol.*, **37**, 1–18.

Normark, W.R., Moore, J.G. and Torresan, M.E. (1993) Giant volcano-related landslides and the development of the Hawaiian Islands. In: *Submarine Landslides: Selected Studies in the U.S. EEZ* (Eds W.C. Schwab, H.J. Lee and D.C. Twichell). *U.S. Geol. Surv. Bull.*, **2002**, 184–196.

O'Brien, J.S. and Julien, P.Y. (1988) Laboratory analysis of mudflow properties. *J. Hydraul. Eng., ASCE*, **114**, 877–887.

O'Leary, D.W. (1986) The Munson–Nygren slide, a major lower-slope slide off Georges Bank. *Mar. Geol.*, **72**, 101–114.

Orange, D.L. and Breen, N.A. (1992) The effects of fluid escape on accretionary wedges 2. seepage force, slope failure, headless submarine canyons and vents. *J. Geophys. Res.*, **97**, 9277–9295.

Orange, D.L., McAdoo, B.G., Moore, J.C., Tobin, H., Screaton, E., Chezar, H., Lee, H., Reid, M. and Vail, R. (1997) Headless submarine canyons and fluid flow on the toe of the Cascadia accretionary complex. *Basin Res.*, **9**, 303–312.

Orange, D.L., Yun, J., Maher, N., Barry, J. and Greene, G. (2002) Tracking California seafloor seeps with bathymetry, backscatter and ROVs. *Cont. Shelf Res.*, **22**, 2273–2290.

Orange, D.L., Saffer, D., Jeanjean, P., *et al.* (2003) Measurements and modeling of the shallow pore pressure regime at the Sigsbee Escarpment: successful prediction of overpressure and ground-truthing with borehole measurements. *Proceedings, Offshore Technology Conference (CD-ROM)*, 11 pp.

Parsons, J.D., Whipple, K.X. and Simoni, A. (2001) Experimental study of the grain-flow, fluid-mud transition in debris flows. *J. Geol.*, **109**, 427–447.

Perret, D. (1995) *Diagénèse mécanique précoce des sédiments fins du fjord du Saguenay.* Unpublished PhD thesis, Department of Geology and Geological Engineering, Laval University, 412 pp.

Perret, D., Locat, J. and Leroueil, S. (1995) Strength development with burial in fine-grained sediments from the Saguenay Fjord, Québec. *Can. Geotech. J.*, **32**, 247–262.

Piper, D.J.W. and Aksu, A.E. (1987) The source and origin of the 1929 Grand Banks turbidity current inferred from sediment budgets. *Geo-Mar. Lett.*, **7**, 177–182.

Piper, D.J.W. and Normark, W.R. (1982) Acoustic interpretation of Quaternary sedimentation and erosion on the channeled upper Laurentian Fan, Atlantic margin of Canada. *Can. J. Earth Sci.*, **19**, 1974–1984.

Piper, D.J.W., Cochonat, P. and Morrison, M.L. (1999) The sequence of events around the epicentre of the 1929 Grand Banks earthquake: initiation of debris flows and turbidity current inferred from sidescan sonar. *Sedimentology*, **46**, 79–97.

Ploessel, M.R., Crissman, S.C., Rudat, J.H., Son, R., Lee, C.F., Randall, R.G. and Norton, M.P. (1979) Summary of potential hazards and engineering constraints, proposed OCS Lease Sale No. 48, offshore Southern California. *Proceedings, 11th Offshore Technology Conference*, Houston, TX, pp. 355–363.

Poulos, S.G., Castro, G. and France, J.W. (1985) Liquefaction evaluation procedure. *J. Geotech. Eng., ASCE*, **111**, 772–791.

Prior, D.B. (1984) Submarine landslides. *Proceedings of the IV International Symposium on Landslides*, Toronto, Vol. 2, pp. 179–196.

Prior, D.B. and Coleman, J.M. (1978) Disintegrating, retrogressive landslides on very low subaqueous slopes, Mississippi Delta. *Mar. Geotech.*, **3**, 37–60.

Prior, D.B. and Coleman, J.M. (1979) Submarine slope instability. In: *Slope Instability* (Eds D. Brunsden and D.B. Prior), pp. 419–455. John Wiley & Sons, Chichester.

Prior, D.B., Wiseman, W.J. and Gilbert, R. (1981) Submarine slope processes on a fan delta, Howe Sound, British Columbia. *Geo-Mar. Lett.*, **1**, 85–90.

Prior, D.B., Bornhold, B.D., Coleman, J.M. and Bryant, W.R. (1982a) Morphology of a submarine slide, Kitimat Arm, British Columbia. *Geology*, **10**, 588–592.

Prior, D.B., Coleman, J.M. and Bornhold, B.D. (1982b) Results of a known sea-floor instability event. *Geo-Mar. Lett.*, **2**, 117–122.

Prior, D.B., Bornhold, B.D. and Johns, M.W. (1986a) Active sand transport along a fjord-bottom channel, Bute Inlet, British Columbia. *Geology*, **14**, 581–584.

Prior, D.B., Yang, Z.S., Bornhold, B.D., *et al.* (1986b) Active slope failure, sediment collapse and silt flows on the modern subaqueous Huanghe (Yellow River) Delta. *Geo-Mar. Lett.*, **6**, 85–95.

Puig, P., Ogston, A.S., Mullenbach, B.L., Nittrouer, C.A. and Sternberg, R.W. (2003) Shelf-to-canyon sediment-transport processes on the Eel continental margin (northern California). *Mar. Geol.*, **193**, 129–149.

Puig, P., Ogston, A.S., Mullenbach, B.I., *et al.* (2004) Storm-induced sediment gravity flows at the head of the Eel submarine canyon, northern California margin. *J. Geophys. Res.*, **109** CO3019, doi: 10.1029/2003JC001918.

Puig, P., Ogston, A.S., Guillen, J., Fain, A. and Palanques, A. (In press) Sediment transport processes from the topset to the foreset of a crenulated clinoform (Adriatic Sea). *Cont. Shelf Res.*

Rabinovich, A.B., Thomson, R.E., Kulikov, E.A., *et al.* (1999) The landslide-generated tsunami November 3, 1994, in Skagway. *Geophys. Res. Lett.*, **26**, 3009–3012.

Reimnitz, E. (1972) Effects in the Copper River delta. In: *The Great Alaska Earthquake of 1964*, Vol. 6, *Oceanography and Coastal Engineering*, pp. 290–302. National Research Council, National Academy of Sciences, Washington, DC.

Reynolds, S. (1987) A recent turbidity current event, Hueneme Fan, California: reconstruction of flow properties. *Sedimentology*, **34**, 129–137.

Richards, A.F. (1976) Marine geotechnics of the Oslo-fjorden region. In: *Laurits Bjerrum Memorial Volume* (Eds N. Janbu, F. Jorsted and B. Kjaernasli), pp. 41–63. Norwegian Geotechnical Institute, Oslo.

Richardson, M.D., Young, D.K. and Briggs, K.B. (1983) Effects of hydrodynamic and biological processes on sediment geoacoustic properties in Long Island Sound, U.S.A. *Mar. Geol.*, **51**, 201–226.

Richmond, W.C. and Burdick, D.J. (1981) Geologic hazards and constraints of offshore northern and central California. *Proceedings, 13th Offshore Technology Conference*, Houston, TX, Vol. 4, pp. 9–17.

Robb, J.M. (1984) Spring sapping on the lower continental slope, offshore New Jersey. *Geology*, **12**, 278–282.

Saffer, D.M. and Bekins, B.A. (1999) Fluid budgets at convergent plate margins: Implications for the extent and duration of fault-zone dilation. *Geology*, **29**, 1095–1098.

Sangrey, D. (1977) Marine geotechnology – state of the art. *Mar. Geotech.*, **2**, 45–80.

Schwab, W.C. and Lee, H.J. (1983) Geotechnical analyses of submarine landslides in glacial marine sediment, northeast Gulf of Alaska. In: *Glacial Marine Sedimentation* (Ed. B.F. Molnia), pp. 145–184. Plenum press, New York.

Schwab, W.C., Lee, H.J. and Molnia, B.F. (1987) Causes of varied sediment gravity flow types on the Alsek prodelta, northeast Gulf of Alaska. *Mar. Geotech.*, **7**, 317–342.

Schwab, W.C., Lee, H.J., Twichell, D.C., *et al.* (1996) Sediment mass-flow processes on a depositional lobe, outer Mississippi Fan. *J. Sediment. Res.*, **66**, 916–927.

Schwehr, K. and Tauxe, L. (2003) Characterization of soft-sediment deformation; detection of cryptoslumps using magnetic methods. *Geology*, **31**, 203–206.

Schwehr, K., Driscoll, N., Tauxe, L. and Lee, H. (2003) Exploration of the Humboldt Slide using anisotropy of magnetic susceptibility comparison with the Gaviota slide. *Proceedings, American Geophysical Union Fall Meeting*, Abstract OS22A-1150.

Seed, H.B. (1968) Landslides during earthquakes due to soil liquefaction. *J. Soil Mech. Found. Div., ASCE*, **94**, 1055–1122.

Seed, H.B. and I.M. Idriss (1971) Simplified procedure for evaluating soil liquefaction potential. *J. Soil Mech. Found. Eng. Div., ASCE*, **97**, 1249–1273.

Seed, H.B. and Rahman, M.S. (1978) Wave-induced pore pressure in relation to ocean floor stability of cohesionless soils. *Mar. Geotech.*, **3**, 123–150.

Seed, H.B., Seed, R.B., Schlosser, F., Blondeau, F. and Juran, I. (1988) *The Landslide at the Port of Nice on October 16, 1979*. Report No. UCB/EERC-88/10, Earthquake Engineering Research Center, University of California, Berkeley, 68 pp.

Shepard, F.P. (1951) Mass movements in submarine canyon heads. *Eos (Trans. Am. Geophys. Union)*, **32**, 405–418.

Shepard, F.P. (1955) Delta-front valleys bordering the Mississippi distributaries. *Geol. Soc. Am. Bull.*, **66**, 1489–1498.

Shepard, F.P. and Marshall, N.F. (1973) Storm-generated current in La Jolla Submarine Canyon, California. *Mar. Geol.*, **15**, Ml9–M24.

Shi, Y. and Wang, C.Y. (1988) Generation of high pore pressures in accretionary prisms: inferences from the Barbados subduction complex. *J. Geopys. Res.*, **93**, 8893–8910.

Shipboard Scientific Party (1995) Site 930. In: *Proceedings of the Ocean Drilling Program, Initial Reports*, Vol. 155 (Eds R.D. Flood, D.J.W. Piper, A. Klaus, *et al.*), pp. 87–122. College Station, TX.

Silva, A.J. (1974) Marine geomechanics: overview and projections. In: *Deep-Sea Sediments, Physical and Mechanical Properties* (Ed. A.L. Inderbitzen), pp. 45–76. Plenum Press, New York.

Silva, A.J., Baxter, C.D.P., LaRosa, P.T. and Bryant, W.R. (2004) Investigation of mass wasting on the continental slope and rise. *Mar. Geol.*, **203**, 355–366.

Silver, M.L., Chan, C.K., Ladd, R.S., *et al.* (1976) Cyclic triaxial strength of standard test sand. *J. Geotech. Eng. Div., ASCE*, **102**, 511–523.

Skempton, A.W. (1953) Soil mechanics in relation to geology. *Proc. York. Geol. Soc.*, **29**, 33–62.

Sterling, G.H. and Strohbeck, G.E. (1973) The failure of South Pass 70B Platform in Hurricane Camille. *Proceedings, 5th Offshore Technology Conference*, Vol. 1, pp. 123–150.

Suhayda, J.N. and Prior D.B. (1978) Explanation of submarine landslide morphology by stability analysis and rheological models. *Proceedings, 10th Annual Offshore Technology Conference*, Houston, TX.

Sultan, N., Cochonat, P., Bourillet, J.F. and Cayocca, F. (2001) Evaluation of the risk of marine slope instability: a psuedo–3D approach for application to large areas. *Mar. Geores. Geotech.*, **19**, 107–133.

Sultan, N., Cochonat, P., Foucher, J.P., *et al.* (2003) Effect of gas hydrates dissociation on seafloor slope stability. In: *Submarine Mass Movements and their Consequences* (Eds J. Locat and J. Mienert), pp. 103–111. Kluwer, Dordrecht.

Suwa, H. (1988) Focusing mechanism of large boulders to a debris-flow front. *Trans. Jpn. Geomorphol. Union*, **9**, 151–178.

Syvitski, J.P.M. and Farrow, G.E. (1983) Structures and processes in bayhead deltas: Knight and Bute Inlet, British Columbia. *Sediment. Geol.*, **36**, 217–244.

Syvitski, J.P.M., Burrell, D.C. and Skei, J.M. (1987) *Fjords: Processes and Products*. Springer-Verlag, New York, 379 pp.

Tanaka, H. and Locat, J. (1999) A microstructural investigation of Osaka Bay clay: impact of microfossils on its mechanical behavior. *Can. Geotech. J.*, **36**, 493–508.

Tappin, D.R., Matsumoto, T. and Shipboard Scientists (1999) Offshore surveys identify sediment slump as likely cause of devastating Papua New Guinea Tsunami 1998. *Eos (Trans. Am. Geophys. Union)*, **80**, 329, 334, 340.

Tappin, D.R., Watts, P. and Matsumoto, T. (2003) Architecture and failure mechanism of the offshore slump responsible for the 1998 Papua New Guinea Tsunami. In: *Submarine Mass Movements and their Consequences* (Eds J. Locat and J. Mienert), pp. 383–389. Kluwer, Dordrecht.

Terzaghi, K. (1956) Varieties of submarine slope failures. In: *Proceedings, 8th Texas Conference of Soil Mechanics and Foundation Engineering*, pp. 1–41. Special Publication 29, Bureau of Engineering Research, Texas University.

Thornton, S.E. (1986) Origin of mass flow sedimentary structures in hemipelagic basin deposits: Santa Barbara Basin, California Borderland. *Geo-Mar. Lett.*, **6**, 15–19.

Traykovski, P., Geyer, W.R., Irish, J.D. and Lynch, J.F. (2000) The role of wave-induced density driven mud flows for cross-shelf transport on the Eel River continental shelf. *Cont. Shelf Res.*, **20**, 2113–2140.

Van Kessel, T. and Kranenburg, C. (1998) Wave-induced liquefaction and flow of subaqueous mud layers. *Coast. Eng.*, **34**, 109–127

Varnes, D.J. (1958) Landslide types and processes. In: *Landslides and Engineering Practice* (Ed. E.D. Eckel), pp. 20–47. Special Report 29, Highway Research Board, Washington, DC.

Whipple, K.X. (1997) Open-channel flow of Bingham fluids: applications in debris-flow research. *J. Geol.*, **105**, 243–262.

Whipple, K.X. and Dunne, T. (1992) The influence of debris-flow rheology on fan morphology, Owens Valley, California. *Geol. Soc. Am. Bull.*, **104**, 887–900.

Woodbury, H.O. (1977) Movement of sediments of the Gulf of Mexico Continental Slope and Upper Continental Shelf. *Mar. Geotech.*, **2**, 263–274.

Woodcock, N.H. (1979) Sizes of submarine slides and their significance. *J. Struct. Geol.*, **1**, 137–142.

Wynn, R.B., Weaver, P.P.E., Ercilla, G., Stow, D.A.V. and Masson, D.G. (2000) Sedimentary processes in the Selvage sediment-wave field, NE Atlantic; new insights into the formation of sediment waves by turbidity currents. *Sedimentology*, **47**, 1181–1197.

Xiao, H. and Suppe, J. (1992) Origin of rollover. *Bull. Am. Assoc. Petrol. Geol.*, **76**, 509–529.

Young, A.G., Bryant, W.R., Slowey, N.C., Brand, J.R. and Gartner, S. (2003) Age dating of past slope failures of the Sigsbee Escarpment within Atlantis and Mad Dog developments. *Proceedings of the Offshore Technology Conference* (CD-ROM).

Yun, J.W., Orange, D.L. and Field, M.E. (1999) Subsurface gas offshore of northern California and its link to submarine geomorphology. *Mar. Geol.*, **154**, 357–368.

The mechanics of marine sediment gravity flows

JEFFREY D. PARSONS*, CARL T. FRIEDRICHS†, PETER A. TRAYKOVSKI‡,
DAVID MOHRIG§, JASIM IMRAN¶, JAMES P.M. SYVITSKI**, GARY PARKER††,
PERE PUIG‡‡, JAMES L. BUTTLES§ and MARCELO H. GARCÍA††

*School of Oceanography, University of Washington, Seattle, WA 98195, USA (Email: parsons@ocean.washington.edu)
†Virginia Institute of Marine Science, College of William & Mary, Gloucester Point, VA 23062, USA
‡Woods Hole Oceanographic Institution, Woods Hole, MA 02543, USA
§Department of Earth, Atmospheric and Planetary Science, Massachusetts Institute of Technology, Cambridge, MA 02139, USA
¶Department of Civil and Environmental Engineering, University of South Carolina, Columbia, SC 29208, USA
**Institute of Arctic and Alpine Research, University of Colorado, Boulder, CO 80309, USA
††Department of Civil and Environmental Engineering, University of Illinois, Urbana, IL 61801, USA
‡‡Geologia Marina i Oceanografia Fisica, Institut de Ciencies del Mar, Barcelona E-08003, Spain

ABSTRACT

Sediment gravity flows, particularly those in the marine environment, are dynamically interesting because of the non-linear interaction of mixing, sediment entrainment/suspension and water-column stratification. Turbidity currents, which are strongly controlled by mixing at their fronts, are the best understood mode of sediment gravity flows. The type of mixing not only controls flow and deposition near the front, but also changes the dynamics of turbidity currents flowing in self-formed channels. Debris flows, on the other hand, mix little with ambient fluid. In fact, they have been shown to hydroplane, i.e. glide on a thin film of water. Hydroplaning enables marine debris flows to runout much farther than their subaerial equivalents. Some sediment gravity flows require external energy, from sources such as surface waves. When these flows are considered as stratification-limited turbidity currents, models are able to predict observed downslope sediment fluxes. Most marine sediment gravity flows are supercritical and thus controlled by sediment supply to the water column. Therefore, the genesis of the flows is the key to their understanding and prediction. Virtually every subaqueous failure produces a turbidity current, but they engage only a small percentage of the initially failed material. Wave-induced resuspension can produce and sustain sediment gravity flows. Flooding rivers can also do this, but the complex interactions of settling and turbulence need to be better understood and measured to quantify this effect and document its occurrence. Ultimately, only integrative numerical models can connect these related phenomena, and supply realistic predictions of the marine record.

Keywords Gravity flows, turbidity currents, debris flows, hydroplaning, fluid mud, surface waves.

INTRODUCTION

A **sediment gravity flow** is any flow by which sediment moves due to its contribution to the density of the surrounding fluid. Sediment gravity flows are not limited to the oceans, or even Earth. However, Earth's oceans represent one of the best places for observation of this unusual phenomenon. The oceans are particularly prone to sediment grav-

ity flows because the particle (sediment) density is generally of the same order of magnitude as, but still larger than, the interstitial fluid. As a result, sediment gravity-flow deposits are ubiquitous in the marine sediment record. Reconstructing the attributes of the sediment record and tying those to the climate at the time of deposition is fundamental to the study of modern marine geology. In addition to palaeoclimate information contained within the

Table 1 Categorization of different types of sediment gravity flows: *Re* is the relevant Reynolds number describing the overall layer thickness

Type of flow	Duration	Speed (m s^{-1})	Concentration	Coarsest grain size	Rheology	*Re*
Submarine slide	Minutes	> 1?	> 1000 kg m^{-3}	Blocks (> 100 m^3)	Non-Newtonian?	< 1
Debris flow	Minutes to hours	0.1–10	> 1000 kg m^{-3}	Boulders (< 100 m^3)	Non-Newtonian	< 100
Estuarine fluid muds	> Hours	> 0.5	> 10 kg m^{-3}	Silty sand	Non-Newtonian	< 100
Wave-supported sediment gravity flow	Hours	0.05–0.3	> 10 kg m^{-3}	Sand	Non-Newtonian?	1–10^4
Turbidity current	Minutes to days	> 0.3	< 10 kg m^{-3}	Coarse sand	Newtonian	> 10^4

sediment record, marine stratigraphy is an important practical concern as the source for much of the world's remaining petroleum reserves.

Sediment gravity flows can be divided into five broad categories. Each flow type has a range of concentrations, Reynolds numbers, duration and grain size, which are summarized in Table 1. **Submarine slides** are large-scale mass-movement events where particle–particle interactions are dominant and interstitial pore fluids play only a minor role. Slides are primarily a result of tectonic forces and cannot be easily treated with fluid-mechanical (continuum) models. They are most often studied by geophysicists and geotechnical engineers and their characteristics have been summarized by Lee *et al.* (this volume, pp. 213–274). **Debris flows** are fast-moving masses of poorly sorted material where particle–particle interactions are important and rheology is a function of interstitial fluid pressure and internal friction. Debris flows are differentiated from slides by their heightened internal deformation and fluid-like properties. **Turbidity currents** are dilute mass concentrations ($C_m < 10$ kg m^{-3}), fully turbulent (Reynolds numbers $Re > 10^4$) flows of poorly sorted sediment. In addition to these three traditional categories of sediment gravity flows, two have emerged recently. Commonly throughout this volume, these gravity flows are described as 'fluid muds'. However, mechanistically it is easiest to divide them into wave-supported sediment gravity flows and estuarine fluid muds. **Wave-supported sediment gravity flows** (WSSGF) require wave-induced resuspension for transport,

while **estuarine fluid muds** result from a convergence of sediment transport within an estuary or sediment-rich shelf environment. Within this paper, estuarine fluid muds are discussed in a limited manner, but have been summarized in Wright & Friedrichs (2006).

Mention of the term sediment gravity flow, turbidity current or debris flow with regard to marine deposits is virtually absent from the scientific literature prior to 1950. **Flysch deposits**, marine-derived sandstones interbedded with shale, have been studied since the nineteenth century, but their formative mechanisms were not understood until Kuenen & Migliorini (1950) linked turbidity currents to these deposits. Turbidity currents gained further recognition from an analysis of the 1929 Grand Banks slope failure. Seagoing oceanographers sampled the deposit and mapped the bathymetry associated with the slide. They proposed that a seismic event produced a turbidity current that travelled rapidly along the seafloor and broke several 'new' transatlantic communication cables (Heezen & Ewing, 1952). The speed of front could be calculated because the timing of the cable breaks was known precisely. The speed was considerably faster than typical ocean currents (> 10 m s^{-1}), indicating that a new type of current, driven by the negative buoyancy resulting from the sediment itself, was responsible for the flow.

Petroleum geologists originally employed ancient analogues to interpret the sedimentary structure of sandstone reservoirs because little was known about the bottom of the ocean. Later, a

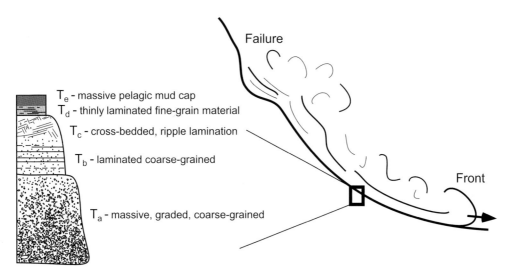

Fig. 1 Schematic of a sediment gravity flow generated from a failure such as the Grand Banks slide. Diagram of the deposit (a **turbidite**) depicts the bedding ($T_a–T_e$) left from such an event (Bouma, 1962; Kneller, 1995).

rich nomenclature was developed to describe these types of deposits. Central to the analyses was the idea of a deposit associated with a failure of poorly sorted material (Bouma, 1962). The **Bouma sequence** consists of a series of layers each formed by a separate phase of a waning, turbulent turbidity current (Fig. 1). In his analysis, Bouma (1962) associated the vertical structure of a single bed (**turbidite**) with the energy of a turbidity current passing over a particular location and waning with time. Although its core principles and relevance have been called into question in recent years (Shanmugam & Moiola, 1995; Shanmugam, 1997), it remains one of the most well accepted and identifiable stratigraphic elements in sedimentology (Kneller, 1995; Kneller & Buckee, 2000). Despite improvements in remote-sensing technology (e.g. seismic tomography: Normark *et al.*, 1993), which made large-basin models possible (i.e. sequence stratigraphy), the connection of outcrop features to basin-scale variability has been difficult to manage.

The results of numerous outcrop studies spawned a small group of researchers to conduct physical experiments of the flow hypothesized in the Bouma sequence. Following advances in experimental techniques for investigating saline gravity currents (Keulegan, 1957a,b), researchers were motivated to perform laboratory experiments of turbidity currents (Kuenen, 1965; Middleton, 1966). Field geologists have benefited greatly from these simple experiments, which supplemented their physical intuition in the analysis of ancient deposits (Kneller & Buckee, 2000).

Engineers and physicists have been drawn to turbidity currents because of their complex interaction of buoyancy, sediment entrainment/deposition and stratified turbulence. One of the first concepts directly attributable to this analysis was ignition. **Ignition**, or **autosuspension**, refers to the ability of a turbidity current to produce enough bed shear (through its motion) to increase its load, and therefore its density, with time. The increase in driving force (i.e. negative buoyancy) propels the flow faster, causing more entrainment, making the flow accelerate, and so on. Suggested first by Bagnold (1962) in a thought experiment, ignition was later rigorously examined in a series of seminal papers (Parker, 1982; Fukushima *et al.*, 1985; Parker *et al.*, 1986). For many years, it was an abstract concept; however, recent laboratory experiments have produced an ignitive flow in the laboratory (Pantin, 2001).

Aside from work regarding ignition, the analytical tools that engineers and physicists brought to bear on turbidity currents have yielded many simple and powerful models describing a variety of sediment gravity flows. Gravity-current-front mechanics have been treated using scaling analysis (Simpson & Britter, 1979; Huppert & Simpson, 1980), while the initial numerical investigations of turbidity-current mechanics came more recently (Bonnecaze *et al.*, 1995; Choi, 1998). In the realm of debris-flow research, where yield-strength models had been used from the earliest predictive work on the subject (Johnson, 1965), large-scale experiments outlined the importance of pore-water pressure and the inherent weakness of these models (Iverson, 1997). Iverson (1997) provided an exhaustive review of the subject, while Iverson & Denlinger (2001) presented a comprehensive numerical debris-flow model based upon these experiments. Advances in the fundamental mechanics of stratified fluids also have paved the way for analysis of dilute, fine-grained sediment transport. Particularly relevant have been the analysis (Howard, 1961) and experiments (Thorpe, 1971, 1973) associated with the development of the Richardson-number criterion in stratified mixing layers.

During the same time that theoretical and experimental advances in the understanding of debris flows and turbidity currents were being made, seagoing researchers addressed the problems of downslope transport, driven by oceanographic variables (i.e. waves and tides), as it pertained to the ultimate burial of terrestrial material. Focus initially was paid to the investigation of sandy turbidity currents (Inman *et al.*, 1976). However, the hazards associated with this effort have been prohibitive in submarine canyons (Inman *et al.*, 1976). Even slow-moving, estuarine fluid muds have consistently made for instrument problems (Kineke & Sternberg, 1992).

Barriers between theoretical, experimental and field-based studies remain today; however, the STRATAFORM (STRATA FORmation on Margins) programme was one of the first programmes to merge the strengths of these disciplines in an attempt to develop a holistic approach to the problem of sediment gravity flows (Nittrouer & Kravitz, 1996). As a result, fundamental discoveries have been made in the mechanics of large sediment gravity flows and their relation to the geological record. The ability to model these natural phenomena has advanced also because of the breakthroughs.

In this paper, recent progress will be outlined with regard to the capability of numerical, physical and analytical models to capture the physics of marine sediment gravity flows, as well as the insight these models lend to the mechanics of continental-margin sediment transport. The paper is organized into three sections representing three dominant modes of transport on the northern California margin: turbidity currents, debris flows and wave-supported sediment gravity flows. Turbidity currents are the best studied and most dilute sediment gravity flow. The other two types of gravity flows represent two end-members common in the ocean: wave-supported sediment gravity flows, which require an external energy source to support the sediment load; and debris flows, where particle–particle interactions are important. Submarine slides, the fourth mode of gravity flow, have been covered in Lee *et al.* (this volume, pp. 213–274), while estuarine fluid muds have been summarized by Wright & Friedrichs (2005). A fourth section describes the interaction of these phenomena, an area of intense research.

TURBIDITY CURRENTS

Turbidity currents are dilute, turbulent flows driven by the horizontal pressure gradient resulting from the increase of hydrostatic pressure due to the addition of particles. They belong to a larger class of flows called **gravity currents**. Gravity currents are any flows where some constituent is added in dilute concentrations to produce a density contrast in a fluid, where that constituent can be anything from salt or temperature to sand. **Conservative gravity currents** do not interact with their boundary and therefore conserve their buoyancy flux as they propagate downslope. Their **runout**, the distance over which the currents travel, is strictly a function of the buoyancy that they supply to the water column. The relation between these two principal forces (gravity and buoyancy) has been exploited in the past to produce relatively simple models that conserve buoyancy flux. However, geologists are primarily interested in flows that deposit or erode along their path. The morphology of the turbidite deposits can be widely varying because of the complex interactions occurring within these currents. A treatment of the dynamics of conservative gravity currents, common to all dilute gravity flows, is necessary to describe the rich phenomena associated with flows that interact strongly with the topography that has been constructed by previous flows.

Basic mechanics

Gravity currents can be divided into conservative and non-conservative flows, and they can also be categorized according to their duration. **Lock-exchange**, or fixed-volume, **gravity currents** are caused by a fixed-volume release of dense material. Failure-induced turbidity currents are a good example of lock-exchange flows (e.g. the 1929 Grand Banks event).

Continuous turbidity currents are possible when the supply of sediment is naturally (in the case of a river mouth) or unnaturally (in the case of a mining operation) continuous. Lock-exhange flows approach continuous flows asymptotically, however (Huppert & Simpson, 1980). Continuous flows are simpler to analyse theoretically and experimentally. As a result, some of the first quantitative work on mixing associated with gravity currents (i.e. Ellison & Turner, 1959) was on this type

of flow. García & Parker (1993) were able to utilize flow steadiness in continuous turbidity currents to examine interaction with the bed in a series of physical experiments. The result was a sediment entrainment model, which serves as the basis for nearly every turbidity-current numerical model proposed to date (Table 2).

In order to understand turbidity-current dynamics, many numerical models have been proposed within the past few years (Table 2). All of these have adapted some version of the shallow-water equations to predict runout and flow characteristics. The **shallow-water equations** are the Reynolds-averaged conservation of momentum and mass equations, with the invocation of the Boussinesq, hydrostatic and boundary-layer approximations (see Box 1). These equations implicitly assume that sediment concentration is a passive tracer (i.e. sediment always travels with the fluid parcel with which it originated). These equations

---

Box 1 The shallow-water equations

Beginning with the incompressible Navier–Stokes equations, which describe the conservation of mass and momentum of a single-phase fluid with density ρ over an arbitrary control volume

$$\frac{D\vec{u}}{Dt} = \frac{\nabla p}{\rho} + \nu \nabla^2 \vec{u} \qquad \text{[momentum]} \qquad \text{(B1.1a)}$$

$$\nabla \cdot \vec{u} = 0 \qquad \text{[mass]} \qquad \text{(B1.1b)}$$

where $D\vec{u}/Dt$ is the 'total derivative', or 'material derivative' described by

$$\frac{D\vec{u}}{Dt} = \frac{\partial \vec{u}}{\partial t} + (\vec{u} \cdot \nabla)\vec{u} \qquad \text{(B1.2)}$$

Here $\vec{u}$ is the velocity field $\vec{u} = u\hat{i} + v\hat{j} + w\hat{k}$, ν is the kinematic viscosity and p is the pressure. By assuming that L_x, $L_y \gg L_z$ and $\partial^n/\partial x^n$, $\partial^n/\partial y^n \ll \partial^n/\partial z^n$ (i.e. the **boundary-layer approximation**), Eqs B1.1a & b simplify to

$$\frac{\partial u}{\partial t} + u\frac{\partial u}{\partial x} + w\frac{\partial u}{\partial z} = -\frac{1}{\rho}\frac{\partial p}{\partial x} + \nu\frac{\partial^2 u}{\partial z^2} \qquad \text{[x-direction momentum]} \qquad \text{(B1.3a)}$$

$$\frac{\partial u}{\partial t} + v\frac{\partial v}{\partial y} + w\frac{\partial v}{\partial z} = -\frac{1}{\rho}\frac{\partial p}{\partial y} + \nu\frac{\partial^2 v}{\partial z^2} \qquad \text{[y-direction momentum]} \qquad \text{(B1.3b)}$$

$$\frac{\partial h}{\partial t} + \frac{\partial uh}{\partial x} + \frac{\partial vh}{\partial y} = 0 \qquad \text{[mass]} \qquad \text{(B1.3c)}$$

Reynolds averaging Eqs B1.3a–c consist of the application of the following rules (of u) to both u and v

$$\overline{\frac{\partial u'}{\partial t}} = \frac{\partial \bar{u}'}{\partial t} = 0, \quad \overline{\frac{\partial u}{\partial x}} = \frac{\partial \bar{u}}{\partial x}, \quad \overline{\frac{\partial u^2}{\partial x}} = \frac{\partial \bar{u}^2}{\partial x} + \frac{\partial \overline{u'^2}}{\partial x}, \quad \overline{\frac{\partial(uv)}{\partial y}} = \frac{\partial}{\partial y}(\overline{uv}) + \frac{\partial}{\partial y}(\overline{u'v'}), \quad \overline{\nabla^2 u} = \nabla^2 \bar{u} \qquad \text{(B1.4)}$$

Folding the Reynolds stress $\overline{\rho u' w'}$ into the bed shear-stress vector $\vec{\tau}_b$

$$\tau_{bx} = \rho\left[-\overline{u'w'} + v\left|\frac{\partial \bar{u}}{\partial z}\right| \right], \ \tau_{by} = \rho\left[-\overline{v'w'} + v\left|\frac{\partial \bar{u}}{\partial z}\right| \right] \tag{B1.5}$$

Invoking the Boussinesq and hydrostatic approximations causes the pressure gradient term (in x, the same is true in y) to become

$$\frac{1}{\rho}\frac{\partial p}{\partial x} = -\frac{1}{\rho}\frac{\partial[(\rho_{sed} - \rho)Cg]}{\partial x} \tag{B1.6}$$

where ρ_{sed} is the density of the sediment in suspension and C is the volumetric concentration of sediment. Defining the submerged specific gravity of the sediment $R = (\rho_{sed} - \rho)/\rho$, and layer-averaged velocities

$$U = \frac{1}{h}\int_0^h \bar{u}\,dz, \ V = \frac{1}{h}\int_0^h \bar{v}\,dz \tag{B1.7}$$

Integrating Eq. B1.3 over the flow thickness h and over topography of height η and substituting in Eqs B1.4–B1.7 yields

$$\underbrace{\frac{\partial U}{\partial t}}_{a} + \underbrace{U\frac{\partial U}{\partial x} + V\frac{\partial U}{\partial y}}_{b} = \underbrace{-\frac{\partial}{\partial x}g'(\eta + h)}_{c} - \underbrace{\frac{\tau_{bx}}{\rho h}}_{d} \qquad \text{[x-direction momentum]} \tag{B1.8a}$$

$$\frac{\partial V}{\partial t} + U\frac{\partial V}{\partial x} + V\frac{\partial V}{\partial y} = -\frac{\partial}{\partial y}g'(\eta + h) - \frac{\tau_{by}}{\rho h} \qquad \text{[y-direction momentum]} \tag{B1.8b}$$

$$\underbrace{\frac{\partial h}{\partial t}}_{e} + \underbrace{\frac{\partial Uh}{\partial x} + \frac{\partial Vh}{\partial y}}_{f} = 0 \qquad \text{[mass]} \tag{B1.8c}$$

where $g' = gR\int_0^h Cy\,dy$ is the layer-averaged reduced gravitational acceleration.

Equations B1.8a–c are the shallow-water equations, the most common governing equations used in numerical turbidity-current models. Term a describes the unsteadiness of the flow. Term b represents the global convective acceleration terms, which describe the change in the velocity caused by the convection (spatial change) of a fluid parcel from one location to another (Granger, 1985). Term c is the driving force, the excess density and resulting pressure gradient supplied by the sediment in suspension. Term d is the dissipation term. Term e is the unsteadiness in mass at a particular location, while the terms in f are the fluxes of mass into and out of that same location. Newer models, which solve for turbulent dissipation (term d) explicitly, typically use Eqs B1.3a–c for their starting point, along with some model of subgrid-scale motions (i.e. a higher-order turbulence-closure model). Direct numeric simulations solve Eqs B1.1a & b directly for all scales of interest (from millimetres to the basin size), although only conservative flows of a few centimetres in height have been performed to date.

are sometimes 'depth-averaged', or integrated, over the thickness of the flow (see Box 1). Newer models using turbulence-closure schemes do not require these approximations (Härtel *et al.*, 2000a,b; Felix, 2001; Choi & García, 2002; Imran *et al.*, 2004). Figure 2 illustrates the parameters involved in these simulations, as well as the variables used for turbidity currents throughout this section.

The models vary in three important respects: (i) sediment entrainment; (ii) turbulent mixing along and within the gravity current; and (iii) the front condition. These are summarized for each model, along with model characteristics, in Table 2. As mentioned previously, the bottom-boundary condition is most often modelled with the entrainment formulation developed by García & Parker

Table 2 Summary of recently proposed turbidity-current models. The columns for mixing, entrainment and front condition list the models used to describe the respective aspects of the flow

Model	Dimension	Solution method	Mixing	Entrainment	Front condition
Bonnecaze et al. (1995)	2.5	Finite-difference	Ellison & Turner (1959)	García & Parker (1993)	Huppert & Simpson (1980)*
Choi (1998)	2.5	Finite-element	Parker et al. (1987)	García (1994)	Huppert & Simpson (1980)
Imran et al. (1998)	2.5	Finite-difference	Parker et al. (1987)	García & Parker (1991)†	Artificial viscosity
Bonnecaze & Lister (1999)	2.5	Finite-difference	Ellison & Turner (1959)	García & Parker (1993)	Huppert & Simpson (1980)*
Bradford & Katopodes (1999a)	2.5	Finite-element	Parker et al. (1987)	García (1994)	Original
Salaheldin et al. (2000)	1.5	Finite-difference	Parker et al. (1987)	García (1994)	Artificial viscosity
Choi & García (2001)	1.5	Finite-element	Parker et al. (1987)	García & Parker (1991)	Huppert & Simpson (1980)
Kassem & Imran (2004)	3	Finite-difference	k-ε‡	?	Original
Felix (2001)	2	Finite-difference	Mellor & Yamada (1974)	García & Parker (1991)	Original
Pratson et al. (2001)	1.5	Finite-difference	Parker et al. (1987)	García (1994)	Huppert & Simpson (1980)
Choi & García (2002)	2	Finite-difference	k-ε	NA	NA – steady-body flow
Imran et al. (2004)	3	Finite-difference	k-ε	García & Parker (1993)	Original
Huang et al. (2005)	3	Finite-difference	k-ε	García & Parker (1993)	Original

NA: not applicable.

*Bonnecaze et al. (1995) and Bonnecaze & Lister (1999) use a modified Huppert & Simpson (1980) front condition, which requires the interstitial fluid within the turbidity current to be somewhat denser than the ambient.

†Imran et al. (1998) modify García & Parker (1991) to account for the effects described in García & Parker (1993) in an original manner.

‡See text for discussion of k-ε turbulence closure scheme.

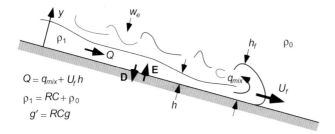

Fig. 2 Schematic of a turbidity current flowing down a fixed slope. See text for descriptions of all terms; definitions are provided in the list of nomenclature.

(1991, 1993). Entrainment rates based upon the shear velocity avoid the controversy over assigning critical shear stresses (Smith & McLean, 1977; Lavelle & Mojfeld, 1987). However, the García & Parker (1991, 1993) formulation is not perfect. It was developed for a particular set of conditions (i.e. steady flow, sandy, well-sorted sediment, and moderately erosive shear stresses). It has been successfully extended to unsteady flows (Admiraal *et al.*, 2000) and turbidity currents where the sediment distribution is non-uniform (García, 1994). However, extension beyond the dimensionless ranges cited in these papers, or the original work, should be made with caution.

Mixing along the interface is another property that must be estimated. Early models were constrained by computation time and therefore used simple empirical relations (Ellison & Turner, 1959; Parker *et al.*, 1987). These relations are typically based upon the **flux Richardson number** $Ri_f = g'h/U^2$, where U is the velocity of the current, g' is the reduced gravitational acceleration induced by the addition of sediment and h is the depth of current. The flux Richardson number Ri_f in a turbidity-current model is a result, rather than an initial estimation. It is possible therefore to calculate the degree of mixing and the effective entrainment velocity w_e along the turbidity current (Fig. 2).

The **drag** associated with bottom (i.e. the bottom-boundary layer) must also be parameterized because the (depth-averaged) shallow-water equations do not account for the turbulence produced there. This is commonly done by invoking a drag coefficient c_D in a formulation for the shear velocity u_* (i.e. $u_*^2 = c_D U^2$). A wide range of drag coefficients are possible (and many simulations choose to vary this parameter); however, a fixed value towards the

lower end of the range ($0.002 < c_D < 0.06$) given by Parker *et al.* (1987) is usually selected.

Recent modelling efforts have been able to avoid using empirical relations for mixing and interaction with the bottom by implementing higher-order turbulence-closure schemes on vertically resolved equations of motion (e.g. Eq. B1.3; Felix, 2001; Choi & García, 2002; Imran *et al.*, 2004; Kassem & Imran, 2004). The most popular of these is the **k-ε turbulence closure scheme**. Felix (2001) used the '2.5-equation' turbulence scheme described by Mellor & Yamada (1982). The **Mellor-Yamada method** incorporates certain assumptions appropriate for flows dominated by stratification. However, both of these models describe situations well only when mixing is not intense ($Ri > 0.25$).

Choi & García (2002) compared higher-order turbulence closure schemes to an established laboratory-derived data set. Their results showed that the assumptions made in layer-averaged models are valid and agree well with the higher-order model. However, when mixing becomes significant, higher-order terms in the conservation of turbulent energy equation become dominant (Choi & García, 2002). In this case, even 2.5-equation turbulence-closure schemes (Mellor & Yamada, 1982) are incapable of accurately describing the velocity field at arbitrary length-scales, owing to the strong influence of fluid-mechanical instabilities (and the importance of variability at small length-scales). In gravity currents, these regions of intense mixing most often occur at the front.

The treatment of the front is the most important of any assumption made in a turbidity current model. The dynamics there play a key role in regulating the overall runout of the flow. As a result, these complicated dynamics, and the recent advances made in understanding them, will be treated in a separate section. However, it is important to consider the various ways the front has been treated in numerical models. The simplest means of treating the front is by dampening it out with the use of an artificial viscosity. Imran *et al.* (1998) adopted this approach in the development of a fast, robust numerical scheme. Imran *et al.* (1998) were able to perform an exhaustive analysis of the turbidity currents and the deposit geometries they generated because of the ease of computation. However, the most popular way to treat the front is to identify the front grid cell and impose a 'front condition'. In this method, the first grid point of the flow is propagated a certain distance given the density and height of the current. Typically, the condition is that of Huppert & Simpson (1980), which uses a Froude number: $Fr_d = U_f / \sqrt{g'h} = 1.19$ for deeply submerged flows ($h/H < 0.075$), or $Fr_d = U_f / \sqrt{g'h} = H^{1/3}/2h^{1/3}$ for shallow flows ($h/H > 0.075$). The **Froude number** is a dimensionless quantity that expresses the relative importance of buoyancy and inertia.

Vertically resolved models capable of describing unsteady flows do not need to model the front specifically (i.e. Felix, 2001; Imran *et al.*, 2004; Kassem & Imran, 2004; Huang *et al.*, 2005). Gross numerical instabilities generated in layer-averaged mode are not a significant problem in vertically resolved simulations. However, the fronts in *k*-*ε* and Mellor-Yamada simulations are not entirely realistic. Many features described in the subsequent section are muted or absent. This is the result of incomplete resolution of all length scales of motion. At the time of this writing, it is uncertain how close these altered fronts are to natural flows; however, they are most likely insignificant for the single-deposit, channelized flows the models were intended to describe.

Frontal dynamics

Front dynamics play a key role in the transport within any gravity current and provide a variable boundary condition to the flow. The **front** also moves more slowly than the body due to enhanced mixing there. At the front, the entire thickness of the gravity current mixes with the ambient fluid, causing the actual transport of material to be somewhat greater than what would be expected from the product of the front velocity, U_f, and the current height, h. Analysis has been slow to develop for front-dominated flows because the details within the front make traditional analytical techniques intractable (Benjamin, 1968). All existing numerical models (at least those described in the previous section) are incapable of resolving the small scales required for the resolution of the fluid-mechanical instabilities at the leading edge of the flow. Physical experiments likewise suffer from the difficulty of following a dynamic front.

As a result, most experiments have focused on uncovering the underlying physical processes and developing empirical relationships for the prediction of frontal mixing based upon bulk current characteristics. Britter & Simpson (1978), and later Simpson & Britter (1979), were the first to quantitatively describe a gravity-current front. They used a laboratory apparatus to arrest a steady conservative (saline) gravity current and measure the flux of material mixed out of the front (Fig. 3). Their experiments indicated that when the flux mixed out of the front, q_{mix} was made dimensionless with the front velocity, U_f, and the reduced gravitational acceleration, g' (for explanations of terms, see Fig. 2); the dimensionless mix rate $g'q_{mix}/U_f^3$ was constant and equal to 0.1. These

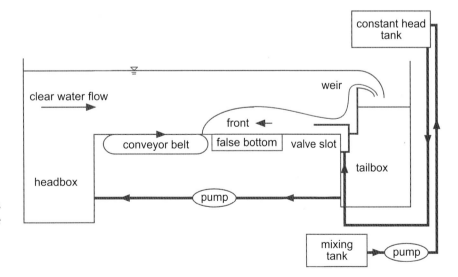

Fig. 3 Schematic of a laboratory flume capable of arresting a gravity current front. The diagram illustrates the general principle upon which the studies of Britter & Simpson (1978) and Parsons & García (1998) were based.

workers drew on earlier qualitative observations to determine that most of the mixing was associated with the **Kelvin–Helmholtz instability**, the result of a vortex-intensification process associated with shear. It is extremely common in natural flows and results in everything from vortex streets behind blunt objects to the raised front seen in most gravity currents.

More recently, Hallworth *et al.* (1996) added a pH-dependent dye to dense fluid that formed a fixed-volume gravity current. The dye illustrated the point at which a gravity current would mix a certain volume of ambient fluid (or a certain amount was mixed out of the current, depending on your perspective). By comparing gravity currents that traversed a fixed bed (no-slip boundary) and a free surface (slip boundary), they ascertained that the lobe-clefts induced approximately half the mixing in bottom-bounded (no-slip) gravity currents. These flows, however, were of different sizes, and no mention of Reynolds (scale) effects was made.

The problem with these early experiments was their extremely small scale. There has been a growing appreciation that frontal dynamics and the processes of turbulent mixing in these flows have been overly simplified (Droegemeier & Wilhelmson, 1987; García & Parsons, 1996; Lingel, 1997; Parsons & García, 1998; Härtel *et al.*, 2000a). Nearly all of these works agree that large flows (i.e. high Reynolds number) mix fluid more efficiently than their smaller counterparts do. As computational speed has advanced, it is now possible to solve directly the Navier–Stokes equations at all relevant scales of motion for small, conservative gravity currents (i.e. Eq. B1.1, where a turbulence-closure scheme is no longer required). Härtel *et al.* (2000a,b) performed a series of **direct-numerical simulation** experiments on conservative (saline) gravity-current fronts. Direct-numerical simulation can solve the Navier–Stokes equations (Eq. B1.1) directly at all the length scales of motion. Computationally, this is extremely expensive and can be done realistically only for laboratory-scale experiments. As a result, the three-dimensional results of Hartel *et al.* (2000a,b) were limited to bulk Reynolds numbers $Re = U_f h_f / v$ of less than 1000, while their two-dimensional results extended to bulk Reynolds numbers in excess of 10^5. The Reynolds numbers express the relative importance of inertia and viscosity.

Härtel *et al.* (2000b) used the three-dimensional results and their companion two-dimensional flows to identify a fluid-mechanical instability that was the result of the no-slip condition at the bed. With a stability analysis of the local region around the nose, they were able to identify a fluid-mechanical instability that arises from the elevation of the nose within a gravity current. The topology of the fronts looked very different than the classic conception of lobe and clefts. The difference was primarily that the dominant mode (wavenumber) of the instability was extremely small compared with the height of the front. Lobes and clefts have traditionally been characterized as an instability that roughly scales with height of the flow. The lobes and clefts of Härtel *et al.* (2000b) also did not extend beyond the nose to the highest portion of the front.

In unpublished experiments performed by John Simpson, even small-Reynolds-number flows evolve from a high-wavenumber instability into the common lobe-and-cleft form. Figure 4 illustrates the evolution of a front that had an instantaneous change in boundary condition. John Simpson performed the experiment about 30 yr ago using the facility described in Fig. 3. To change the boundary condition, a front was arrested on top of the motionless conveyor belt. Then, the belt was quickly started at velocity equal to the mean velocity in the overbearing flow. As can be seen in Fig. 4, the wavenumber of the initial instability decreases as the flow evolves into its final, equilibrium state.

Complementing this analytical work, Parsons & García (1998) used a device similar to, but much larger than, Britter & Simpson (1978). Although their experiments could only realize fully turbulent conditions for a limited number of parameter combinations, flow visualization by **laser-induced fluorescence** allowed them to study the morphology of the secondary structures with the decay of the Kelvin–Helmholtz billow. Two different approaches were taken. In the first set of experiments, Parsons & García (1998) investigated the spanwise development of the Kelvin–Helmholtz billow. They noted finger-like structures behind the billow (within the Kelvin–Helmholtz core), similar to observations of stratified-mixing layers (Sullivan & List, 1994). These structures were found to be responsible for the build-up of the turbulent

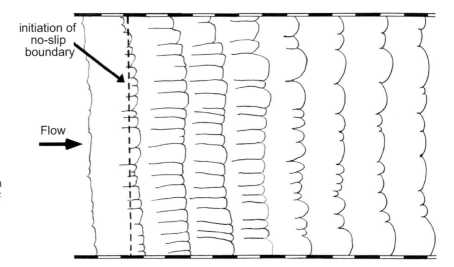

Fig. 4 Unpublished drawings of John Simpson showing the development of lobes and clefts in a gravity current with the floor impulsively arrested. Note that the frequency of lobes and clefts decreases with time.

cascade, the increase in mixing for larger flows and the approach to fully turbulent behaviour (Broadwell & Briedenthal, 1982).

Later experiments sought to understand the physical origin of the finger-like structures (Parsons, 1998). In this set of experiments, the flow was visualized in the spanwise direction (i.e. parallel to the front). Using a stratified-mixing-layer analogue, Parsons (1998) noticed a similarity to the **Klaasen–Peltier instability** observed in stratified-mixing layers (Schowalter *et al.*, 1994). The Klaasen–Peltier instability occurs when dense fluid is forced over light fluid by the Kelvin–Helmholtz instability in a stratified flow (Klaasen & Peltier, 1985). Baroclinic torque causes the thin layer of dense material ejected into the lighter ambient fluid to become unstable in the spanwise direction (Fig. 5). The result is a pair of counter-rotating, streamwise vortices. The vortices will grow to the size of the mixing layer, in this case, roughly the size of the front. Figure 5 shows a series of flow visualization photographs illustrating the breakage of the ejection sheet and the transport of that dense material to the Kelvin–Helmholtz core. Considering that most of the work regarding the Klaasen–Peltier instability requires some initial perturbation, it is reasonable to assume that the instability posed by Härtel *et al.* (2000b) initiates the development of the larger Klaasen–Peltier streamwise vortices. In a separate analysis, however, Parsons (1998) showed that the Klaasen–Peltier instability is responsible for most of the extra mixing produced at large Reynolds numbers.

It is important to note that all of the above results (Parsons & García, 1998; Härtel, 2000a,b) were obtained from experiments of conservative (e.g. saline) gravity-current fronts. Extension of the dynamics to non-conservative flows comprised of sediment is not trivial. A single turbidity-current-front experiment has been run (Parsons, 1998). The results are fraught with potential problems (e.g. the unrealistically low bed shear in the entrance section of the device). Problems aside, the dimensionless mix rate measured was nearly triple the value compared with a conservative flow of similar conditions (i.e. similar Froude and Reynolds number). It is difficult to speculate about the mechanism responsible for the dramatic increase in mixing, or whether the increase is simply an experimental artefact. Only future experimentation, either numerical (direct-numerical simulation) or physical (laboratory), can explain whether there are additional complications to the analysis of turbidity-current fronts.

Turbidity-current fans

It is important to remember that one of the most important applications to the knowledge gained about turbidity currents is the interpretation of turbidites. Although turbidity currents transport material everywhere they travel, these flows only deposit material in areas of reduced bed shear. In these areas, material will be sorted depending on the temporal and spatial variability within the flow itself (Kneller, 1995; Kneller & McCaffrey,

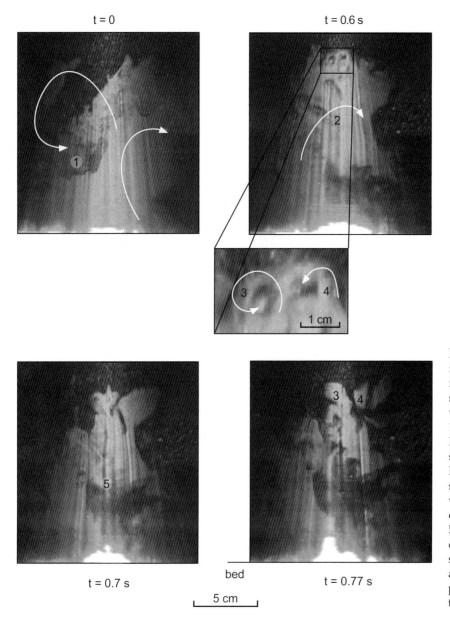

Fig. 5 Streamwise vorticity formation and deposition of filaments from the broken ejection sheet to a Kelvin–Helmholtz core within a gravity current front: 1, filaments or fingers in the core of Kelvin–Helmholtz billow; 2, large streamwise vortex associated with Klaasen–Peltier instability; 3 & 4, smaller Kelvin–Helmholtz-induced vortices rotating in the opposite sense of the large Klaasen–Peltier vortex; 5, source of dense material for downstream filaments. The view shown is a plane dissecting across a gravity-current front. Flow is primarily perpendicular to (out of) the plane of the page.

2003). A number of models have been presented which have attempted to deal with the complexities of the depositional patterns from unconfined (i.e. three dimensional) turbidity currents (Imran *et al.*, 1998; Bradford & Katopodes, 1999b). Unfortuately, there have been few experimental data to support these conclusions. When data have become available, they generally have been in two-dimensional configurations (García, 1990), strongly influenced by scale effects (Kneller & McCaffrey, 1995), or both (Bradford & Katopodes, 1999a).

Despite these limitations, Imran *et al.* (1998) derived a theory about channel formation based upon their innovative numerical model. Using a thought experiment, Imran *et al.* (1998) described the bulk behaviour of an expanding turbidity current from a finite source. A graphical description of their thought experiment appears alongside the deposit from a model run of Imran *et al.* (1998) in Fig. 6. The theory makes use of the non-linear relationship between shear velocity and the entrainment rate. The formulation of García & Parker

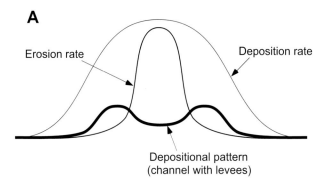

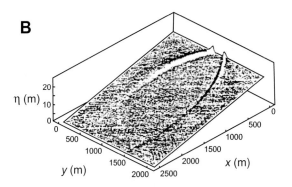

Fig. 6 Turbidite channel formation. (A) An expanding turbidity current will necessarily produce topography that tends to further confine the flow with time. The distribution of deposition rate is nearly always linearly dependent on concentration (and intensity of flow), whereas erosion rate is highly non-linear. (B) Numerical results of Imran *et al.* (1998) that illustrate the deposit from a channelizing flow.

(1993) indicates that the **entrainment rate** $E \propto u_*^5$. **Deposition rate** D is linearly proportional to the volumetric concentration C and the settling velocity of the sediment w_s (i.e. $D \propto w_s C_b$), which is related to the square of the shear velocity u_*. The periphery of the flow will therefore become depositionally dominated, while the core can remain erosional. The result, as shown in Fig. 6, is the construction of levees beside a central channel. A **levee** is a deposit along the periphery of a channel that acts to contain flow within the channel.

A **fan** is made of many deposits that ultimately influence the future locus of deposition, so modelling of the effects of successive flows is extremely difficult. Most of the efforts have been focused on resolving the deposit of a single flow (Luthi, 1981; Imran *et al.*, 1998). The experiments of Luthi (1981) represented the first step towards large-

experimental confirmation of turbidity currents. The turbidite deposits were found to be symmetrical with respect to the centreline of the deposit. The deposits were also dramatically thinner away from the source. The levee features seen in Imran *et al.* (1998), where the currents were dominantly depositional and were not expected to produce erosional features, were not developed.

To rectify this gap in laboratory data, experiments have examined a series of flow deposits from a single pulsed source in the MIT Experimental Geomorphology and Sedimentology Laboratory (Parsons *et al.*, 2002). These experiments were able to observe more realistic fan geometries, although the results must be considered to be preliminary (i.e. only two fans were made). The flows were buoyantly driven and large ($Re > 10^5$). The fronts of these flows exceeded the criterion established for fully turbulent flow by the experiments of Parsons & García (1998), like the earlier single-bed experiments of Luthi (1981). Two fans were produced, with the second fan consisting of more than 30 deposits. These individual deposits were approximately 5 mm near the source, and thinned to a few grain diameters (~100 μm) at the distal edge of the fan surface.

Similar to Luthi (1981), the first few beds were generally symmetrical, as were the flows that produced them. However, after approximately 5–10 event beds, the deposits began to form a depositional mound on the left-hand side of the tank. The mound had a depressed centre, which tended to focus later flows and deposition. Figure 7 illustrates two flows: (A) the first flow in the production of the second fan; (B) a flow representing the later stages in the development of the depositional mound (run 14). It is clear that the lobes and clefts at the front controlled the direction and the intensity of the flow in all of the currents where deposition was concentrated in a depositional mound.

Unlike the first fan, where only one depositional mound formed, the second fan (because it possessed more event beds) contained several lobe switches. After 19 event beds, a particularly coarse run was made. The coarse material deposited quickly within the central depression of the depositional lobe. Consequently, future deposition was steered farther to the left section of the tank, forming a new depositional lobe (Fig. 8). After approximately another six flows, a final lobe was formed on the

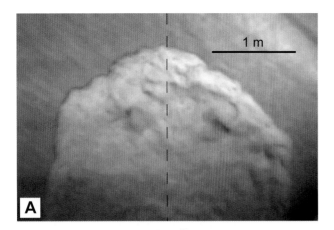

centre line

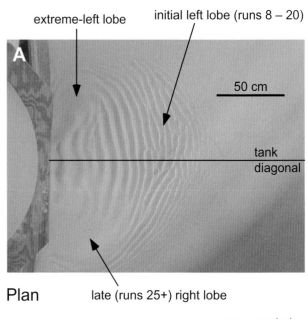

Plan

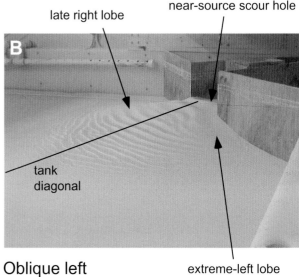

Oblique left

Fig. 7 Development of asymmetric turbidity flows. (A) An initial turbidity current flowing over flat topography. (B) A turbidity current flowing over topography generated from previous events. Clarity of the later experiment (B) is degraded because of low concentrations (~0.001 kg m^{-3}) of suspended sediment from earlier runs. A lobate shape was observed in both currents, although the poor photographic quality of (B) obscures the later current. Velocities were calculated from image analysis of preceding images. Both vectors were obtained at a 25° angle from the source.

Fig. 8 Photographs of the shape of the final deposit formed by Parsons et al. (2002). (A) Plan and (B) oblique left view. The deposit has several mounds corresponding to three distinct depocentres. The transfer from one depocentre to another was usually sudden. Internal structure consistent with three mounds was also observed and documented (Parsons et al., 2002). The origin of the bedforms is uncertain, although Parsons et al. (2002) suggested that they are either 'conventional' unidirectional ripples or miniature sand waves, which scale with the current thickness.

opposite (right) side of the tank. The result, shown in Fig. 8, is a lumpy cone.

Internal structure was also resolved in the second fan. It appeared that the centre depression, at least in the case of the first depositional mound, was similar in many ways to channel forms discussed in Imran et al. (1998). Due to limitations in the run-out of the flow and the strength of the initial flow, the flows were not strongly erosional and there-fore were not efficient at producing topography. Channels represent the dominant form observed in natural turbidite fans and require a separate treatment.

Channelization and channel processes

Submarine channels formed by turbidity currents in deep water are often bounded by natural levees, and tend to divide into **distributaries** in the downslope direction. As discussed in the previous section, turbidity currents force the quasi-periodic formation of depositional lobes and channel avulsions necessary to build up an entire fan surface (Imran *et al.*, 1998). Like fluvial channels, **meandering** is perhaps the most commonly observed planform among submarine channels. Although not common, straight (Klaucke *et al.*, 1998) and braided (Ercilla *et al.*, 1998) channel patterns have also been observed in the submarine environment. The length of submarine channels varies from a few kilometres to several thousand kilometres.

Widely studied submarine channel systems include those of the Bengal, Indus, North Atlantic Mid-Ocean Channel (NAMOC), Mississippi and the Amazon. These channels are thousands of kilometres long and many of them display planform characteristics that are remarkably similar to subaerial meandering channels (Pirmez, 1994). With recent advances in three-dimensional-seismic-imaging techniques, numerous small-scale channels are also being found buried in shallow as well as deep-water settings (Kolla *et al.*, 2001). Like many of the largest subaqueous channels, these buried channels show intricate meandering patterns and simple to very complex architecture. High channel sinuosity, bend cutoffs, point bars, scroll bars, meander belts, chute channels and pools, and crevasse splays are some of the quasi-fluvial features recognized in many subaqueous channels (Hagen *et al.*, 1994; Klaucke & Hesse, 1996; Peakall *et al.*, 2000, Kolla *et al.*, 2001).

There are, however, some significant differences between channels in submarine and subaerial environments. Submarine channels may display levee asymmetry due to the effect of the Coriolis force that is not prevalent in even the largest of subaerial channels (Klaucke *et al.*, 1998). **Flow stripping** occurs in submarine channels when a large part of the upper portion turbidity current is 'stripped' from the main body of the flow and begins to propagate away from the confined channel to other portions of the depositional fan (Piper & Normark, 1983). **Nested-mound formations** are also found on the outsides of channel bends (Clark & Pickering, 1996). These are also thought to be unique to sub-

marine environments. In many muddy submarine fans, channels are surrounded by high natural levees and have beds perched well above the elevation of the adjacent non-channelized regions (Flood *et al.*, 1991). Highly perched channels cannot generally be maintained on subaerial fluvial fans except by artificial means. Continuous mixing between a turbidity current and the ambient water leads to increased flow thickness. Consequently, the dilute upper part of a turbidity current easily spills in the lateral direction, and builds the levee system by depositing dominantly fine-grained sediment. Significant flow stripping can occur at channel bends where centrifugal force causes highly exaggerated **superelevation** of the interface between the turbidity current and the ambient water.

Considerable progress has been made in understanding and quantifying the initiation, deformation and migration of meander bends in rivers (Ikeda *et al.*, 1981, Beck *et al.*, 1983; Johannesson, 1988; Johannesson & Parker, 1989a,b; Seminara & Tubino, 1989; Howard, 1992; Sun *et al.*, 1996). The flow field (and related mechanics) in a meandering subaqueous channel is a complex process, however. Based upon observations of fluvial channel morphology, scientists have developed several conceptual models of submarine channel morphology (Peakall *et al.*, 2000). These conceptual models are speculative since little is known about the mechanics of a turbidity current in a submarine channel, especially the two most important mechanisms: spilling and stripping. Most experimental and numerical studies involving turbidity currents have been conducted in a straight, confined channel configuration. However, submarine channels are rarely straight and the current typically delivers material to adjacent areas. In order to understand the morphology of submarine fans, it is important to understand the underlying fluid mechanics of not only the flow inside the channel, but also the lateral flow that spills into overbank areas. One of the most pressing questions that remain to be addressed is how a turbidity current can maintain its momentum over thousands of kilometres of channel length while it is expected to continuously lose its momentum by water entrainment, and overbanking. Mechanistic models developed by various researchers fail to address this issue, as the loss of sediment due to spilling and stripping cannot be included in these models. The simple force balance derived by

Komar (1969) is still frequently used to estimate the relationship between flow velocity and sediment transport in a subaqueous channel (Hay, 1987; Pirmez, 1994; Klaucke *et al.*, 1998).

Imran *et al.* (1999) developed a two-dimensional model to study depth-averaged primary and lateral velocity and the superelevation of flow thickness in a subaqueous channel with a sinuous planform. A steady well-developed conservative current was considered and it was assumed that the secondary flow had a vertical structure similar to that in a meandering open-channel flow. In order to obtain an analytical solution, the flow was constrained within the channel and it was not allowed lateral overflow. However, to understand the processes of flow stripping, channel migration and variation of flow deposit, it is of foremost importance to resolve the structure of a current that is allowed to vary in time and in all three spatial directions, and is not necessarily forced to remain constrained within a channel.

Kassem & Imran (2004) utilized a robust three-dimensional numerical model (described above, Table 2) to simulate a turbidity current travelling in a sinuous channel within a horizontal domain that is unbounded. The model was applied to a laboratory-scale channel with sine-generated planform and the current was allowed to spill over the banks of the channel on both sides. Initially, the

basin was assumed to be sediment-free. Heavier fluid was then injected at the upstream end. The current proceeded forward until it reached the downstream end where it was allowed to leave the basin. The incoming heavier fluid had a velocity of 161 mm s^{-1} and a density of 1033 kg m^{-3}, while the ambient density was set at 1000 kg m^{-3}. As a result, the inlet densimetric Froude number was 1.22 indicating a mildly supercritical flow. Figure 9 shows a contour map of fluid density at a horizontal plane 60 mm above the channel bottom (slightly above the bank) after 300 s of flow. The current has entrained ambient water from above and at the front, resulting in an increased thickness in the downstream direction. Within a short distance from the inlet, the current thickness has exceeded the bank height and the current has started to spill into the overbank area. Spilling has been symmetrical where the reach is straight, but as soon as the current has approached a channel bend, the flow has begun to react to the curvature. Near the bend apices, the flow has become highly superelevated, and a significant amount of flow stripping has occurred near the outer bank (Fig. 9). Asymmetry of the density distribution has occurred inside the channel as well as in the overbank area.

The computed velocity distribution and flow density in the lateral direction at a cross-section 0.75 m downstream of the channel inlet are shown

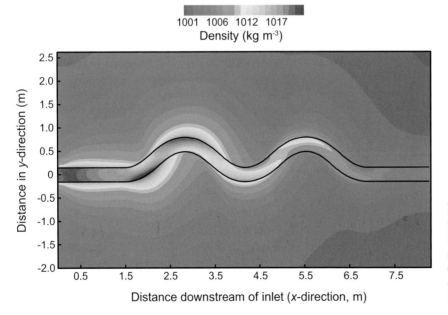

Fig. 9 Modelled density distribution in a horizontal plane slightly above the channel bank of a sinuous submarine channel. The density contour map clearly shows the effect of spilling and stripping processes. (After Kassem & Imran, 2004.)

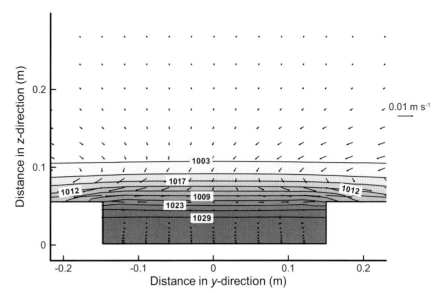

Fig. 10 Distribution of density ($kg\,m^{-3}$) and lateral velocity in the straight reach at a section 0.75 m from the inlet (Fig. 9). The lateral velocity is close to zero inside the channel. Just above the channel bank the lateral velocity increases, leading to enhanced spilling. (After Kassem & Imran, 2004.)

in Fig. 10. The thickness of the turbidity current at this section is larger than the inflow thickness due to entrainment of the ambient fluid. The increased thickness leads to the continuous process of spilling to the overbank area. Figure 10 also illustrates that there is a vertical gradient in the density of the current. A dense lower part and an overlying dilute plume characterize the concentration profile. This is consistent with the sediment-concentration structure observed in laboratory turbidity currents (García, 1990; Lee & Yu, 1997). The magnitude of the lateral velocity inside the channel in a straight reach is not significant compared with the streamwise velocity. However, above the channel, lateral flow associated with overflow is important (Fig. 9). As the flow depth exceeds the bank height, the current begins to spill onto areas outside of the channel. This process is referred in the literature as **overspill** (Hiscott *et al.*, 1997). Overspill results in the formation of continuous levees, including inner-bend levees and inflection points within bends on submarine fans (Peakall *et al.*, 2000).

The driving force of a turbidity current in the lateral direction is the same as that in the streamwise direction (i.e. pressure gradient generated by a density difference). Since there is a density difference between the current of supralevee fluid and the surrounding ambient lower-density water, a pressure difference is developed that forces the current to move out of the channel. The overspill

generates a lateral velocity in the upper part of the density current that increases from zero at the channel centreline to a maximum near the bank. However, the magnitude of the lateral velocity in the overbank area remains small compared with the streamwise velocity inside the channel. This indicates that inertia in the uppermost portion of the current maintains significant streamwise momentum there, whereas a small portion is dragged across the levee due to the lateral velocity. Such an observation agrees with the interpretation made by Hesse (1995) that overspill is a slow gradual process. Development of spatially extensive overspill has important geological implications, because it can create levees with a width that can be an order of magnitude larger than the channel width.

In subaerial channels, the centrifugal force generated by curvature in a bend is balanced by a superelevated air–water interface. While travelling through a channel bend, the interface between a density current and the clear water above is expected to be superelevated in the same fashion. The degree of superelevation of flow around bends is modest in a subaerial channel (i.e. in the order of a few centimetres). However, in the subaqueous case, a superelevation comparable to the flow thickness at channel bends is common (Hay, 1987; Imran *et al.*, 1999). The difference in behaviour is the result of the excess density difference, or reduced-gravitational-acceleration term g', which is equivalent to the gravitational-acceleration term g

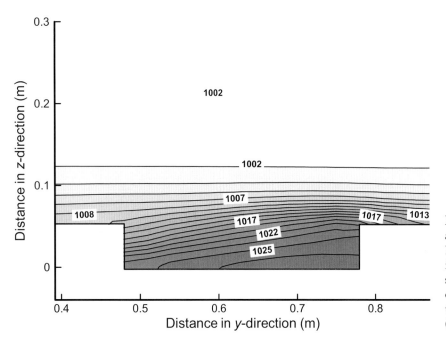

Fig. 11 Density (kg m^{-3}) distribution at the cross-section of a bend apex. Different layers of the stratified flow show varied degree of superelevation. There is a significant difference in density of the current in the outer and inner overbank area. (After Kassem & Imran, 2004.)

(i.e. 9.81 m s^{-2}) in rivers. For turbidity currents, the value is typically ~0.01 g.

The density field at the cross-section of the first bend apex of the channel is shown in Fig. 11 (the plan view is shown in Fig. 9). The different layers of the turbidity current are found to be superelevated to differing degrees. The inclination of the concentration layers increases from the bottom to the middle of the density current and then decreases upward until the top of the current is reached. The fluid density on the outer bank approaches that inside the channel, because of the outward flow and inclination of different flow layers, causing a significant loss in flow density. Flow stripping is thought to be a major submarine-channel process responsible for discrete overbank sedimentation and intrachannel deposition, including nested-mound formation at sharp bends, initiation of channel abandonment and channel plugging (Peakall *et al.*, 2000; Normark & Serra, 2001). Flow stripping at channel bends is more dramatic than the continuous overspill that occurs throughout the channel length. Flow stripping can occur also due to the effect of Coriolis force (Komar, 1969; Klaucke *et al.*, 1998). In a Coriolis-force-dominated flow, levee height should be always larger on one bank. However, if the centrifugal force is the dominant cause of stripping, larger levee height should alternate between the two banks.

In rivers, a relatively small superelevation of flow depth, a parabolic or linear distribution of depth-averaged streamwise velocity between the inner and the outer bank, and a single cell of secondary circulation characterize flow in a meander bend. In submarine meandering channels, centrifugal effects are also expected to cause secondary flow. Significant tilting of the different layers within a turbidity current has already been described in Fig. 11. The inclination of the density and velocity fields generates secondary circulation (Fig. 12). Figure 12 shows the velocity vectors in the lateral-vertical plane at the first bend apex of the channel. The circulation patterns observed here clearly differ from the familiar single-cell helical flow in the cross-section of a sinuous river. The difference between the circulation due to an open-channel (river) flow and a turbidity current can be attributed to the vertical structure of the primary velocity and the dominance of the flow-stripping process. Figure 13 shows the vertical structure of the lateral and the streamwise velocity at the channel centreline. The lateral velocity reverses direction at several locations above the bed, indicating complex circulation patterns. Since the flow is allowed to leave the confines of the channel, a net lateral flow is established towards the outside of the channel. Except for a very small vertical distance near the channel bottom and near the top of the current,

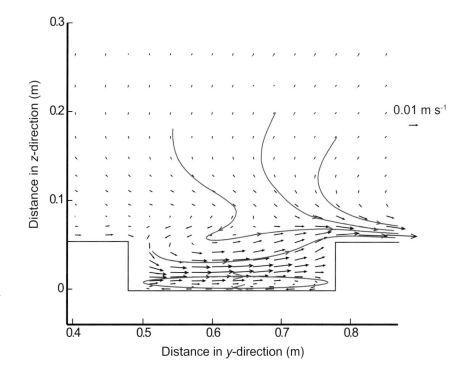

Fig. 12 Distribution of lateral velocity at a bend apex. The streamlines display the formation of a suppressed near-bed circulation cell and strong outward flow. At this cross-section, stripping completely dominates over spill. (After Kassem & Imran, 2004.)

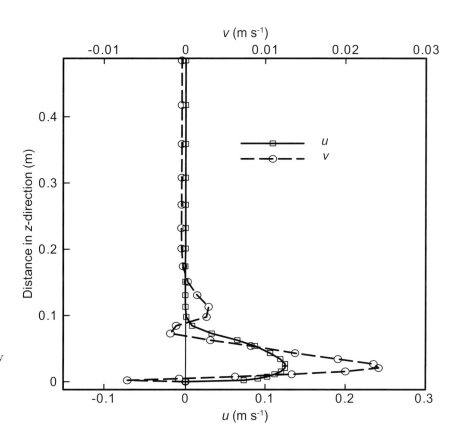

Fig. 13 Vertical structure of streamwise (u) and lateral (v) velocity at a bend apex. Multiple reversals of the lateral velocity clearly indicate a complex circulation field. (After Kassem & Imran, 2004.)

centrifugal force completely dominates over the pressure-gradient force, resulting in the suppression of the channel-confined circulation cell.

The simulations of Kassem & Imran (2004) illustrate the complexity in the flow–bed interactions in a single-event turbidity current. Recent work has begun to focus on the problem over arbitrary (irregular) topography (Huang *et al.*, 2005). The ability to model multiple flows over arbitrary topography has been a major barrier to interpreting recent studies of turbidite sequences. To date, no simulation has been run over undulating topography. Undulations in the seascape are difficult to model because they can produce ponding within an individual turbidity current. Ponding occurs when sediment-laden fluid piles up behind an obstruction. It is extremely common in natural systems and it is an important process in the formation of thick turbidite beds (Lamb *et al.*, 2004b). Several different approaches are being undertaken to overcome this problem. The most straightforward is to simply suppress numerical instabilities within the simulation using extra artificial viscosity, similar to the method that Imran *et al.* (1998) used to stabilize a turbidity-current front. However, this approach sacrifices realism in the vicinity of the bed-height changes, often the place of greatest interest. Other ongoing work is focused on using direct-numerical simulation of turbidity currents (Eckhart Meiburg, personal communication). These simulations are essentially identical to laboratory experiments, so they remain subject to the scale effects of previous laboratory investigations (Parsons *et al.*, 2002).

Observations of turbidity currents

The first serious attempt at observing a turbidity current came in the late 1960s with monitoring of Scripps Canyon offshore of La Jolla, California (Inman *et al.*, 1976). However, the flows observed in this study were so violent that the instrumentation was always lost during the events, making detailed analysis of the flows impossible. The lack of observational data on turbidity currents has led to an undercurrent of skepticism about their importance, particularly on modern continental margins (Shanmugam, 1997). However, within the past few years, a number of novel observations have been made that verify the existence of modern turbidity currents and underscore their power to sculpt continental margins.

Khripounoff *et al.* (2003) was the first investigation to record a turbidity current throughout its lifetime. This study deployed two moorings in 4000 m water depth on the Congo Fan. One of the moorings was inside the Congo Fan Channel. It was equipped with electromagnetic current meters at 30 m and 150 m above the bed (mab) and a sediment trap and an optical backscatter sensor (OBS) at 40 mab. The other mooring was located approximately 5 km away from the fan channel on the broad, flat fan levee. It had sediment traps at 30 mab and 400 mab and current meters 10 m above each trap. At 2200 PST on 8 March 2001, the 150 mab current meter on the channel mooring recorded a velocity of 1.21 m s^{-1} averaged over 1 h. This occurred at exactly the same time as an increase in turbidity observed by the OBS at 40 mab. The lowermost current meter was also broken at this time. The sediment trap at 40 mab simultaneously was filled with terrigenous sediment, including plant debris and wood fragments. Considering that the average currents in the area were typically less than 10 cm s^{-1}, Khripounoff *et al.* (2003) hypothesized that they observed a turbidity current in the channel at that time. Heightened sediment concentrations were observed at the levee station some 3 days later, although no high-velocity front was ever observed.

Monterey Canyon has also been the site of intense turbidity-current observation within the past 5 yr (Johnson *et al.*, 2001; Paull *et al.*, 2003; Xu *et al.*, 2004). Johnson *et al.* (2001) were the first to find a strong connection between flooding of the Salinas River and turbidity-current activity within Monterey Canyon. The modern Salinas River mouth is located just over 1 km from the Monterey Canyon head. Using a single long-term mooring deployed within the canyon, Johnson *et al.* (2001) identified several turbidity currents over several years by a substantial decrease in salinity (≫ 0.1 psu) and an increase in turbidity as observed by a transmissometer at 200 mab. It was noticed that these departures typically occurred just after large flood events on the Salinas River. Paull *et al.* (2003) deployed a bottom tripod in the Monterey Canyon during the winter of 2000–2001. The tripod, replete with a CTD (conductivity-temperature-depth) sensor and an electromagnetic

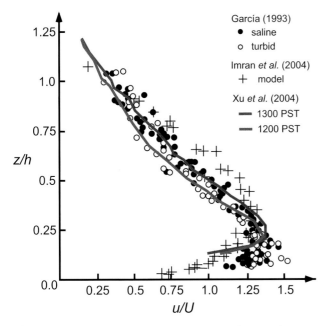

Fig. 14 Comparison of the observations of Xu *et al.* (2004) with the laboratory experiments of García (1993) and the numerical model of Imran *et al.* (2004). The Xu *et al.* (2004) profiles were taken from the mid-depth mooring from the 24 March 2003 event at 1200 and 1300 PST. The inferred U is 0.60 and 0.46 m s^{-1} for the 1200 and 1300 observations, respectively. The inferred H is likewise 51 m and 55 m. The data from García (1993) were obtained near the inlet of the flume on a steep slope where the currents were supercritical. Kassem & Imran (2004) illustrated the agreement of observations with their model for a supercritical current in a straight reach. Note that the poorest agreement between the models and observations is near the bed.

current meter, was found to have moved over 100 m down-canyon and to have been buried by more than 1 m of sediments after a stormy period in Monterey Bay. That deposit was a massive fine sand with a thin organic mud overlying it. It was suggested that this massive bed was a typical Bouma-A unit (see Fig. 1 for details). Xu *et al.* (2004) observed three turbidity-current events over a winter-long deployment in 2003, and measured a turbidity-current velocity profile during a mooring deployment in Monterey Canyon. The origin of two of the events was not addressed (although one occurred on the same day as the annual maximum discharge from the Salinas River), whereas the third (14 March, 2003, shown in Fig. 14) was definitively linked to the release of dredge spoils near the Salinas River mouth.

The novel measurements of Xu *et al.* (2004) provide an unusual opportunity to test the validity of existing laboratory and numerical models. As shown in Fig. 14, the turbidity currents observed by Xu *et al.* (2004) show agreement with the data from García (1993) and the numerical model results of Kassem & Imran (2004). Agreement is particularly good away from the boundary, where transport is dependent mostly upon the fluid motions associated with mixing. Closer to the bed, however, the differences are more significant. Unfortunately, it is the near-bed region that is most crucial to the evolution of a turbidity current. Near-bed shear regulates resuspension and deposition of sediment, which is the primary mechanism responsible for growth in a turbidity current. As a result, modelling errors in this critical area have the potential to produce unrealistic outcomes.

Large destructive turbidity currents have not been the only turbulent sediment gravity flows observed in recent years. As instrumentation improves resolution of energetic wave boundary layers, researchers have been able to resolve transport within the bottom 1 m of the water column. Wright *et al.* (2001), Storlazzi & Jaffe (2002) and Puig *et al.* (2003, 2004) have all found net downslope transport on energetic, sandy continental shelves. Although Storlazzi & Jaffe (2002) suggested that transport is a result of asymmetric wave motions, their results are also consistent with, and indistinguishable from, wave-supported sediment gravity flows. The dynamics of wave-supported sediment gravity flows are significantly different than their unidirectional, buoyancy-dominated counterparts and will be treated in a separate section.

DEBRIS FLOWS

Debris flows have been a popular topic in the scientific literature in recent years, due to the intermittent havoc they wreak in mountainous areas (Costa, 1984). Subaerial debris flows have been the focus of this work, but many of their subaqueous counterparts are significant participants in the marine sediment record (Elverhøi *et al.*, 1997). The force balance within a debris flow is between interstitial fluid pressure and grain–grain interactions, so the boundary conditions imposed by the ambient fluid are usually, but not always (e.g.

hydroplaning), of little consequence. As a result, many of the concepts formulated for subaerial flows can be extended to the marine environment.

Central to any discussion of debris flows is the topic of rheology. **Rheology** is the study of the movement of a material under an applied stress. It is highly dependent on the grain size and mineralogy of the constituent material, the water content of the sample, and a host of other geotechnical (lithological) parameters. A detailed description of subaqueous-debris-flow and slide rheology has been presented in Lee *et al.* (this volume, pp. 213–274). In the present paper, recent advances will be discussed in the resolution of the mechanics for fluids of arbitrary rheology (i.e. non-Newtonian fluids).

Basic mechanics

As mentioned in Lee *et al.* (this volume, pp. 213–274), rheology of debris flow material plays an important role in the stability and mobility of the flow itself. Two models have emerged, which have different underlying assumptions about the relationship between stress and strain in debris flows. The first, oldest, and simplest is the **Herschel–Bulkley model**, which characterizes the relationship between the imposed **stress** τ_x and the **strain** rate $\partial u/\partial z$ through a power law. Mathematically, this relationship can be expressed as

$$\tau_x - \tau_0 = K \left| \frac{\partial u}{\partial z} \right|^n \tag{1}$$

for a two-dimensional flow in the x–z plane, with velocity strictly in the x-direction. Here the yield strength τ_0, K and n are all empirically derived and material specific. These parameters can have a large range of values, which are generally scale dependent (Parsons *et al.*, 2001b). However, for the largest flows (volumes in excess of 0.001 km^3), they tend to approach a fixed value (Dade & Huppert, 1998).

A common simplification of the Herschel–Bulkley model is the **Bingham-plastic model**. A Bingham-plastic model occurs when $n = 1$ in Eq. 1. In this case, μ becomes equivalent to a dynamic viscosity (units: Pa · s = N · s m^{-2}). The dynamic viscosity of most natural debris flows is > 100 Pa s (Whipple

& Dunne, 1992), so Reynolds numbers of debris flows are usually in the order of 1. As a result, turbulence is generally thought to be unimportant. Bingham materials, like paint, also have a tendency to 'freeze' once their thickness becomes less than that required to shear the material. That is, the critical thickness

$$h_{cr} = \frac{\tau_0}{\rho g \tan \beta} \tag{2}$$

where ρ is the density of the debris flow material (usually ~2.2 kg m^{-3}) and β is the bed slope. The Bingham model (viscoelasticity) is particularly relevant when h_{cr} is a significant portion of the flow thickness. This is often the case when either the material is clay-rich or when the flows being studied are small (Parsons *et al.*, 2001b). Viscoelastic effects are generally less important when the flows become large and coarse-grained (Iverson, 1997; Lee *et al.*, this volume, pp. 213–274).

A more sophisticated approach is to assume that there are two distinct phases present within a single flow. Dubbed **mixture theory**, the governing equations being solved are similar to those discussed with regard to the shallow-water equations (see Box 1). For highly concentrated mixtures, complications arise as the density is not fixed, but is a function of the solids content within the flow. The governing equations become increasingly complex with the addition of the second phase, so analytical solutions are limited (Denlinger & Iverson, 2001). Even numerical solutions of mixture theory are computationally demanding, particularly when flow is over irregular (i.e. realistic) topography (Denlinger & Iverson, 2001, 2004).

Considering the potential importance of intergranular friction in the dissipation of energy in large submarine debris flows, future modelling studies are likely to use mixture theory. However, mixture theory has never been adapted for the marine environment. Hydroplaning, which was discovered during the STRATAFORM programme and will be discussed next, presents unusual challenges not typically encountered in subaerial applications. It is uncertain how to incorporate the mechanics of the ambient fluid (important for hydroplaning) into the formulation derived in Box 2. As a result of this uncertainty and the general lack of coarse

Box 2 Mixture theory (Iverson, 1997; Iverson & Denlinger, 2001)

Beginning with the conservation of momentum and mass

$$\rho \frac{D\vec{u}}{Dt} = -\nabla \cdot (\mathbf{T_s} + \mathbf{T_f} + \mathbf{T'}) + \rho\vec{g} \qquad \text{[momentum]} \qquad \text{(B2.1a)}$$

$$\partial\rho/\partial t + \nabla \cdot (\rho\vec{u}) = 0 \qquad \text{[mass]} \qquad \text{(B2.1b)}$$

Unlike in the single-phase shallow-water equation derivation (see Box 1), the density of the fluid ρ is not constant, but dependent on the mixture of the volume of solids (sediment) V_s and fluid V_f, such that

$$\rho = \rho_s V_s + \rho_f V_f \qquad \text{(B2.2)}$$

$\mathbf{T'}$ is the stress tensor associated with the relative motion of the solid and fluid components with respect to the flow as a whole, so

$$\mathbf{T'} = -\rho_s V_s(\vec{u}_s - \vec{u})(\vec{u}_s - \vec{u}) - \rho_f V_f(\vec{u}_f - \vec{u})(\vec{u}_f - \vec{u}) \qquad \text{(B2.3)}$$

where the subscripts f and s denote fluid and solid phase, respectively. The relative stress $\mathbf{T'}$ is assumed negligible and the reference frame is taken with respect to the solid phase only, which simplifies Eqs B2.1 to

$$\rho \frac{D\vec{u}_s}{Dt} = -\nabla \cdot (\mathbf{T_s} + \mathbf{T_f}) + \rho\vec{g} \qquad \text{[momentum]} \qquad \text{(B2.4a)}$$

$$\nabla \cdot \vec{u}_s = 0 \qquad \text{[mass]} \qquad \text{(B2.4b)}$$

Now assuming a rigid boundary and depth averaging the conservation equations (i.e. similar to Eq. B1.8), it is found that

$$\frac{\partial h}{\partial t} + \frac{\partial(Uh)}{\partial x} + \frac{\partial(Vh)}{\partial x} = 0 \qquad \text{(B2.5a)}$$

$$\rho\left[\frac{\partial(Uh)}{\partial t} + \frac{\partial(U^2h)}{\partial x} + \frac{\partial(UVh)}{\partial y}\right] = -\int_0^h \left[\frac{\partial T_{s(xx)}}{\partial x} + \frac{\partial T_{f(xx)}}{\partial x} + \frac{\partial T_{s(yx)}}{\partial y} + \frac{\partial T_{f(yx)}}{\partial y} + \frac{\partial T_{s(zx)}}{\partial z} + \frac{\partial T_{f(zx)}}{\partial z} - \rho_{df}g_x\right]dz \qquad \text{(B2.5b)}$$

where $T_{s(ij)}$ and $T_{f(ij)}$ are the ij component in the stress tensor for the solid and fluid phase, respectively. U and V are the depth-averaged velocities in the x and y direction. Equation B2.5a is the (final) conservation of mass equation and Eq. B2.5b is the x-direction conservation of momentum. Note that the conservation of mass is identical to its shallow-water-equation equivalent (i.e. Eq. B1.8c). The y-direction conservation of momentum is obtained by interchanging x with y and U with V.

Using Leibniz' theorem to evaluate the solid stress portion of the right-hand side of Eq. B2.5b

$$-\int_0^h \left[\frac{\partial T_{s(xx)}}{\partial x} + \frac{\partial T_{s(yx)}}{\partial y} + \frac{\partial T_{s(zx)}}{\partial z}\right]dz = \frac{\partial(h\bar{T}_{s(xx)})}{\partial x} + \frac{\partial(h\bar{T}_{s(yx)})}{\partial y} + \tau_{bx} \qquad \text{(B2.6)}$$

where $\bar{T}_{s(ij)}$ is the ijth component of the depth-averaged stress and $\tau_{bx} = T_{s(zx)}\big|_{z=0}$ is the bed shear stress in the x-direction. Again, the same equation holds for the y-direction when x and y are interchanged. The weight of the slurry is

$$\bar{T}_{s(zz)} + \bar{T}_{f(zz)} = \rho g_z h/2 \qquad \text{(B2.7)}$$

which can be related to the depth-averaged normal stresses (e.g. $\bar{T}_{s(xx)}$). Here, Iverson & Denlinger (2001) used a lateral stress coefficient $k_{act/pass}$ derived from Coulomb theory. By definition

$$\bar{T}_{s(xx)} + \bar{T}_{s(yy)} = k_{act/pass}\bar{T}_{s(zz)} \qquad \text{(B2.8)}$$

The lateral stress coefficient takes a value of 1 if Coulomb failure does not occur within the material. However, the primary case of interest is when failure occurs, such that it can be shown from Coulomb theory that

$$k_{\text{act/pass}} = \frac{1 + \sin^2 \phi_{\text{int}}}{1 - \sin^2 \phi_{\text{int}}} \tag{B2.9}$$

where ϕ_{int} is the internal friction angle of the solid material composing the debris flow. Assuming that the fluid within the debris flow is hydrostatic ($\bar{T}_{\text{s(zz)}} = p_{\text{bed}}/2$), and substituting this into Eqs B2.7 and B2.8, the depth-averaged components of the solid-phase stress tensor become

$$\bar{T}_{\text{s(xx)}} = \bar{T}_{\text{s(yy)}} = k_{\text{act/pass}}[(\rho g_z h - p_{\text{bed}})/2] \tag{B2.10}$$

The normal stresses are geometrically related to the tranverse shear stresses (via 'Mohr's circle') in a Coloumb mixture, such that

$$\bar{T}_{\text{s(yx)}} = \bar{T}_{\text{s(xy)}} = -\text{sgn}(\partial U/\partial y)\{k_{\text{act/pass}}[(\rho g_z h - p_{\text{bed}})/2]\}\sin \phi_{\text{int}} \tag{B2.11}$$

The bed shear stress τ_{bx} is similarly related to the normal stresses (i.e. through Mohr's circle), through the friction angle at the bed $\phi_{\text{bed}} \cdot \phi_{\text{bed}} > \phi_{\text{int}}$, if there is bed roughness. The resulting expression is

$$\bar{T}_{\text{s(zx)}} = -\text{sgn}(U)(\rho g_z h - p_{\text{bed}})\tan \phi_{\text{bed}} \tag{B2.12}$$

The fluid stresses are more straightforward. Here the fluid stresses are expressed as they are on the right-hand side of Eq. B1.3

$$-\int_0^h \left[\frac{\partial T_{\text{f(xx)}}}{\partial x} + \frac{\partial T_{\text{f(yx)}}}{\partial y} + \frac{\partial T_{\text{f(zx)}}}{\partial z} \right]dz = -\int_0^h \left[\frac{\partial p}{\partial x} - V_f \mu \left(\frac{\partial^2 U}{\partial x^2} + \frac{\partial^2 U}{\partial y^2} + \frac{\partial^2 U}{\partial z^2} \right) \right]dz \tag{B2.13}$$

where μ is the dynamic viscosity of interstitial fluid. The first term on the right-hand side is the product of the pressure gradient at the bed and the flow depth h, owing to the hydrostatic approximation. Integration of the next two terms requires an application of Leibniz' theorem, which simplifies to the simple product of the existing terms when a 'slab approximation' ($\partial h/\partial x = 0$) is made. A slab approximation is implicit in the application of the Coulomb equations to the problem at hand. Finally, the application of a no-slip boundary and an assumption of a parabolic vertical velocity profile yields the last term, such that

$$-\int_0^h \left[\frac{\partial T_{\text{f(xx)}}}{\partial x} + \frac{\partial T_{\text{f(yx)}}}{\partial y} + \frac{\partial T_{\text{f(zx)}}}{\partial z} \right]dz = -h\frac{\partial p_{\text{bed}}}{\partial x} + V_f \mu h \frac{\partial^2 U}{\partial x^2} + V_f \mu h \frac{\partial^2 U}{\partial y^2} - 3V_f \mu \frac{U}{h} \tag{B2.14}$$

Combining Eqs B2.10–B2.12 and B2.14 results in the final x-direction conservation of momentum equation

$$\rho \underbrace{\left[\frac{\partial (Uh)}{\partial t} + \frac{\partial (U^2 h)}{\partial x} + \frac{\partial (UVh)}{\partial y} \right]}_{a} = \underbrace{-\text{sgn}(U)(\rho g_z h - p_{\text{bed}})\tan \phi_{\text{bed}} - 3V_f \mu \frac{U}{h}}_{b}$$

$$\underbrace{-hk_{\text{act/pass}} \frac{\partial}{\partial x}(\rho g_z h - p_{\text{bed}}) - h\frac{\partial p_{\text{bed}}}{\partial x} + V_f \mu h \frac{\partial^2 u}{\partial x^2}}_{c}$$

$$\underbrace{-\text{sgn}\left(\frac{\partial U}{\partial y}\right)hk_{\text{act/pass}} \frac{\partial}{\partial y}(\rho g_z h - p_{\text{bed}})\sin \phi_{\text{int}} + V_f \mu h \frac{\partial^2 U}{\partial y^2}}_{d} + \underbrace{\rho g_x h}_{e} \tag{B2.15}$$

Equation B2.5a combined with Eq. B2.15 and its y-direction complement (interchanging x and U with y and V) are the final governing equations. Term a is the material derivative describing debris-flow motion, b is the bed shear stress, c is the streamwise normal stress, d is the transverse stress and e is the driving (gravity induced) stress. For a detailed description of the conservation of mass equation (B2.5a), see Box 1. For more details regarding the assumptions made in the derivation above, consult the primary work of Iverson & Denlinger (2001).

material in the deep-sea environment, all of the analysis performed to date on submarine debris flows has assumed a Bingham or Herschel–Bulkley rheology.

Hydroplaning

Although submarine debris flows share many similarities with their better-studied subaerial counterparts, they differ in striking ways. Recent laboratory comparisons of subaqueous and sub-aerial debris flows and their products have not only highlighted these differences, but have provided physical explanations for the differences as well (Mohrig *et al.*, 1998, 1999; Marr *et al.*, 2001; Toniolo *et al.*, 2004). The physical insight gained in these studies is guiding the development of numerical models for submarine debris flows (Harbitz *et al.*, 2003). The knowledge also aids the interpretation of ancient debris-flow deposits. Since most of the available data on submarine debris flows consist of geometric and compositional information from their deposits, advancing inversion methods are essential. Inverse models and the information gleaned from them are particularly useful in con-straining runout and emplacement processes.

Laboratory experiments have documented the significant role of ambient-fluid density on the beha-viour of subaqueous and subaerial debris flows. Water is about 800 times more dense than air, so the inertial forces associated with a debris flow, accelerating fluid out of its path as it moves down-slope produce substantially larger dynamic pressures at the front of a subaqueous flow (as compared with a subaerial debris flow). Accentuating this effect is the relatively small difference (by about a factor of two) between the density of the ambient fluid (water) and the debris-flow slurry. When pressures develop that are comparable in magnitude to the submerged weight per unit bed area of the flow, it can no longer displace water from its contact area with the bed fast enough and a layer of fluid begins to separate the debris from the bed. Mohrig *et al.* (1998) proposed that the instigation of dynamic **hydroplaning** is suitably character-ized by a **densimetric Froude number** $Fr_d = U_f/\sqrt{g'h\cos\beta}$, where U_f is the head velocity of the flow or slide, g' is the reduced gravitational acceleration (i.e. $(\rho_1 - \rho_0)g/\rho_0$, with g the gravitation accelera-tion, ρ_1 the density of the debris flow and ρ_0 the

ambient density), h is the average thickness of the flow or slide and β denotes the slope of the bed. The densimetric Froude number describes the balance between hydrodynamic pressure and the submerged debris load. It has been used with great success in scaling the effects of the resultant pressure force on a moving body (Brown, 1975). Laboratory flows constrain hydroplaning to cases where Fr_d is greater than 0.4, a value that is con-sistent with fast moving, natural debris flows (Mohrig *et al.*, 1998).

The initial laboratory experiments that deter-mined the conditions for and the consequences of hydroplaning were all conducted in the 'Fish Tank', a glass-walled tank 10 m long, 3 m high and 1 m wide in the St Anthony Falls Laboratory. Suspended within the Fish Tank is a 0.2-m wide channel, sufficiently narrow to ensure conditions for two-dimensional flow. The tank was filled with water or drained for subaqueous and subaerial runs, respectively. The front velocities for most of the observed subaqueous debris flows were greater than their subaerial counterparts. The sub-aqueous runout distances were also larger (Mohrig *et al.*, 1998, 1999). Direct observation revealed that the thin layer of water that penetrated beneath the front of the hydroplaning debris flow acted as a lubricant (Figs 15A & B & 16). These easily sheared lubricating layers of water dramatically reduced the bed resistance acting on the debris, thereby increasing their relative mobility. Average thicknesses for the lubricating layers were small and in all cases < 2% of the thickness of the over-riding debris flow or slide.

The thickness of a lubricating water layer beneath a hydroplaning debris flow was always observed to decrease with distance from the flow front. This spatial change in layer thickness caused a differ-ence in forward velocity between the debris riding on a discrete layer of water and the trailing portion of the flow that was more attached to the bed. The result was a stretching and attenuation of the flow directly behind its front. In some runs, the head **autoacephalates** (i.e. separates from the body), causing a new head to form. The detached blocks slid to the end of the channel in front of the newly formed head of the flow. These isolated **outrunner blocks** (Figs 17 & 18) are commonly observed associated with the deposits of submarine debris flows (Prior *et al.*, 1984; Lipman *et al.*, 1988; Nissen

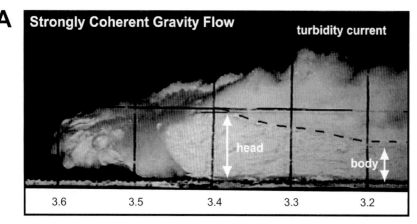

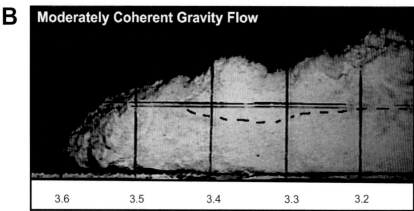

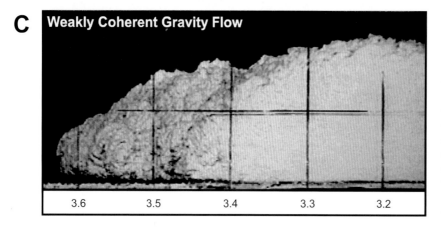

Fig. 15 Sandy gravity flows. (A) Head of a strongly coherent sandy gravity flow. The composition of the slurry by weight was 35% kaolinite, 40% water, 20% 110 μm sand and 5% 500 μm sand. (B) Head of a moderately coherent sandy gravity flow. The composition of the slurry by weight was 25% kaolinite, 40% water, 30% 110 μm sand and 5% 500 μm sand. (C) Head of a weakly coherent sandy gravity flow. The composition of the slurry by weight was 15% kaolinite, 40% water, 40% 110 μm sand and 5% 500 μm sand. (After Marr *et al.*, 2001.)

et al., 1999) and are perhaps the best evidence of hydroplaning in the ocean. Extension at the fronts of submarine flows and slides, a manifestation of the lubricating layer thickening toward the fronts of flows, produces a pattern of deposition quite different from that observed in the subaerial environment, where the fronts and margins of flows are always under compression (Major & Iverson, 1999).

Remobilization and erosion of a substrate by a debris flow are governed by the ability of this flow to overcome resisting stresses of the substrate. Hydroplaning of subaqueous flows has been found to reduce the shear coupling between a flow and its substrate, thereby reducing the degree of substrate remobilization relative to subaerial counterparts (Mohrig *et al.*, 1999; Toniolo *et al.*, 2004). In a number of laboratory runs, a hydroplaning flow was

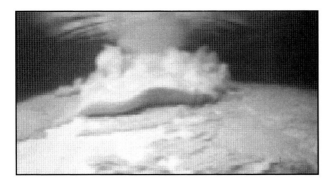

Fig. 16 Hydroplaning head of an unconfined subaqueous debris flow. The photograph was taken from an underwater video camera looking upstream.

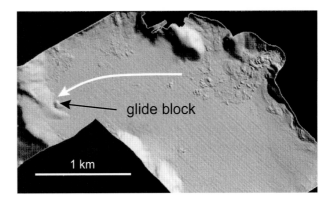

Fig. 18 Multibeam bathymetric image of the deposits from two submarine flows in a Norwegian fjord. The deposits were caused by an adjacent subaerial landslide that flowed into the fjord in 1996. A single glide block (black arrow) hydroplaned to the opposite side of the fjord (path of white arrow). (Image appears courtesy of Anders Elverhøi.)

found to have moved over a very soft substrate without reactivating any measurable amount of the recent deposit. The role of a thin lubricating layer on reducing the transmission of shear stress from an overriding flow into the bed was quantitatively developed by Harbitz *et al.* (2003).

Further experiments on subaqueous debris flows in the Fish Tank have focused on the role of clay content on the dynamics and deposition of sub-aqueous debris flows (Marr *et al.*, 2001). Studies were performed using pre-mixed slurries containing 110 μm sand, clay and water. The experiments indicated that as little as 0.7–5% by weight of bentonite clay or 7–25% by weight of kaolinite clay at water contents ranging from 25 to 40% by weight was required to generate coherent gravity flows

with a substantial basal debris-flow component. At lower clay contents, the flows were transitional to highly concentrated turbidity currents. Increased water content also favoured the formation of turbidity currents.

Examination of the boundary between debris flows and turbidity currents is important for identifying and interpreting sediment gravity-flow deposits, but also for the creation of a turbidity current from failed material. Transition from debris flows to turbidity currents is discussed at length in

Fig. 17 A glide block autoacephalated from the body of a subaqueous debris flow. (A) View looking upstream of a subaqueous debris-flow deposit that passed over an antecedent deposit. Most of the material did not participate in hydroplaning; however, a small mass (an outrunner block) hydroplaned near the front. The path of the outrunner block can be seen in the glide track just downstream of the main flow mass. (B) Plan view of the glide block that autoacephalated and hydroplaned off the main platform (just below the view of camera in A).

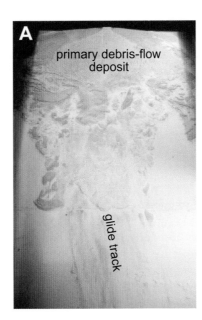

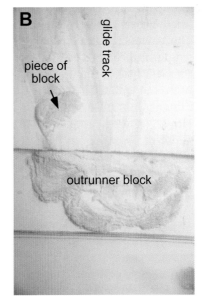

the final section of this paper. The experiments of Marr *et al.* (2001) covered the entire range of sandy gravity flows. These sediment gravity flows were classified in terms of whether or not a coherent, non-turbulent flow head could be discerned. Three types of sandy gravity flows are illustrated in Fig. 15. Figure 15A shows a strongly coherent gravity flow, the head of which is clearly that of a debris flow, with only a weak subsidiary turbidity current peeling off the head. Figure 15B shows a moderately coherent flow, where the debris flow is definable almost to the head of the flow as a whole, but where turbulent entrainment into suspension is intense. Figure 15C shows a weakly coherent flow, for which the head is completely turbulent. This flow can be characterized as a highly concentrated turbidity current, where a definable, non-turbulent debris flow only appears well behind the head.

Toniolo *et al.* (2004) have performed experiments on subaqueous and subaerial debris flows in an unconfined configuration. The experiments have confirmed that lateral confinement is not a necessary condition for hydroplaning. An example of a hydroplaning head for an unconfined subaqueous debris flow is given in Fig. 16. Like their two-dimensional cohorts, these heads often autoacephalate (i.e. detach from their body), forming a glide block that moves out ahead of the main deposit. A glide block that autoacephalated from the body of a subaqueous debris flow is illustrated in Fig. 17. A similar glide block observed on the modern seafloor is shown in Fig. 18. Toniolo *et al.* (2004) also characterized the tendency for unconfined debris flows to rework antecedent deposits. The reworking by a debris flow passing over a single antecedent deposit was found to be suppressed in the subaqueous setting as compared with the subaerial setting, confirming the two-dimensional observations of Mohrig *et al.* (1999). However, when repeated subaqueous flows were allowed to stack and block downslope motion, reworking was significant.

Advances in analytical and numerical solutions

The mathematical simplicity of the Bingham model allows for the precise prediction of flow characteristics throughout the flow, given some knowledge of the rheological parameters. Mei & Liu (1987) first described the governing equations of a one-dimensional Bingham-plastic flow. They took advantage of a long-wave approximation (i.e. the characteristic length of the flow is large with respect to the overall water depth), as well as assuming small-amplitude conditions (i.e. the thickness of flow is much smaller than the flow length). These equations have been used in a number of applications. For example, Jiang & LeBlond (1993) have investigated the effects of a mudflow on a free surface. They used the governing equations presented by Mei & Liu (1987) to numerically solve for the wave characteristics of a tsunami (free-surface wave) produced by a submarine Bingham-plastic mudflow.

These numerical methods are powerful; however, analytical solutions are important for constraining the numerical models and for simplified, exact analysis. As a result, Huang & García (1997) used a matched-perturbation method to analytically solve the equations of motion for one-dimensional transport of material on a fixed slope. In short, Huang & García (1997) adapted the boundary-layer equations for a Bingham plastic material developed by Mei & Liu (1987) and cast them into dimensionless form. The terms in the equations were made dimensionless by two length scales: the initial flow depth h_0 (a vertical length scale) and the initial flow length l (a horizontal length scale). By assuming the characteristic dimensionless depth h_0/l was significantly less than $\sin \beta$ (sine of the bed slope), Huang & García (1997) were able to further approximate the equations of motion to a form equivalent to a kinematic-wave model. These equations are effective at modelling motions well away from the front of the flow. Huang & García (1997) refer to this as the outer solution.

To describe the flow near the front, Huang & García (1997) manipulated the original generalized equations of motion to form several new 'inner variables'. The technique used was patterned after Hunt (1994) who solved the simplified case of a Newtonian fluid (i.e. $\tau_0 = 0$, $n = 1$ in Eq. 1) using the perturbation solution techniques of Nayfeh (1973). The technique consists of taking the limit of $h_0/l \rightarrow 0$ in the dimensionless equations of motion. The continuity equation is integrated, solved and finally substituted into the momentum equation for each layer. The layer depths are then added to form the free-surface profile in the vicinity of the front (i.e. the inner solution). At this point, the profiles

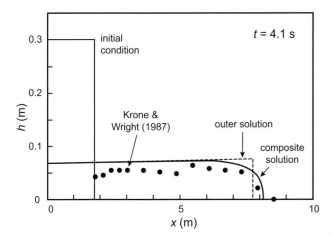

Fig. 19 Comparison of the matched-asymptotic solution of Huang & García (1997) with the experimental results of Krone & Wright (1987). Experimental conditions (other than those shown) are $\sin \beta = 0.06$, $\tau_y = 42.5$ Pa, $\mu = 0.22$ Pa-s, $\rho = 1073$ kg m^{-3}. The slight discrepancy between the height of the flow in the experiment and in the theory is most likely a result of the loss of material due to coating of the sidewalls in the flume.

near the front and the outer solution are redimensionalized. Doing so, Huang & García (1997) found that the characteristic length scales l and h_0 drop out, leaving a solution entirely of known variables. The result is shown in Fig. 19 as the composite solution. It is asymptotically accurate both near the front and along the body. After a number of tests, Huang & García (1997) found that the solution is generally acceptable after the flow runs out three times the initial height.

Huang & García (1998) extended the results of Huang & García (1997) for the more general case of a Herschel–Bulkley fluid (Eq. 1). Due to the addition of another variable (the exponent, n in Eq. 1), the analysis becomes increasingly complicated. Despite the complications, Huang & García (1998) were able to find a solution in a similar manner, thus forming a composite solution capable of characterizing the complex behaviour associated with the more generalized rheological model.

Imran *et al.* (2001a,b) have extended the analytical model of Huang & García (1998) to form 'BING', a one-dimensional integral numerical model of Herschel–Bulkley subaerial and subaqueous muddy debris flows. The bilinear case does not have a closed-form asymptotic solution. Pratson *et al.* (2001) have used it and a shallow-water-equation

model for turbidity currents, 'BANG', to compare the stacked deposits of turbidity currents and submarine debris flows. Hydroplaning is not yet incorporated in BING. Criteria have been developed by Huang & García (1999) to assess the point at which hydroplaning will occur in standard one-dimensional Herschel–Bulkley flows. Resolution of the mechanics once the flow begins to hydroplane requires additional theory, however. In this vein, Harbitz *et al.* (2003) have modified lubrication theory to describe the equilibrium hydroplaning of a thin block over a sloping bed. Their theoretical formulation provides insight into the dynamics of autoacephalated glide blocks.

Another challenge remaining is to extend the numerical models of debris flows into two dimensions. It may seem that numerical models of realistic, two-dimensional debris flows should be attainable, particularly considering the analytical solutions available in the one-dimensional case (and the success of two-dimensional turbidity-current models). However, numerical models of higher-dimensional debris flows have proved to be particularly difficult. Unlike the one-dimensional models discussed above, matching analytical solutions are not easily available for the higher-dimensional case. Further, most finite-difference schemes are particularly sensitive to the moving boundary at the debris-flow front. As a result, researchers are forced to solve the moving-boundary problem with a finite-element approach, similar to that taken by Choi & García (2001) for turbidity currents. Work proceeds to capture the dynamics of debris flows in realistic geometries, but results have yet to be fully tested.

Observations of submarine debris flows

In the search for oil, the petroleum companies and the corporations that service their data-acquisition needs have imaged large portions of the seafloor. These new data have illustrated that submarine-debris-flow deposits are relatively common geological features. Figure 20 provides an interesting contrast between debris-flow deposits on two very different continental margins, both of which are actively being explored for petroleum. These examples depict the dichotomy in debris-flow processes that is embodied in the two different rheological models discussed in this paper.

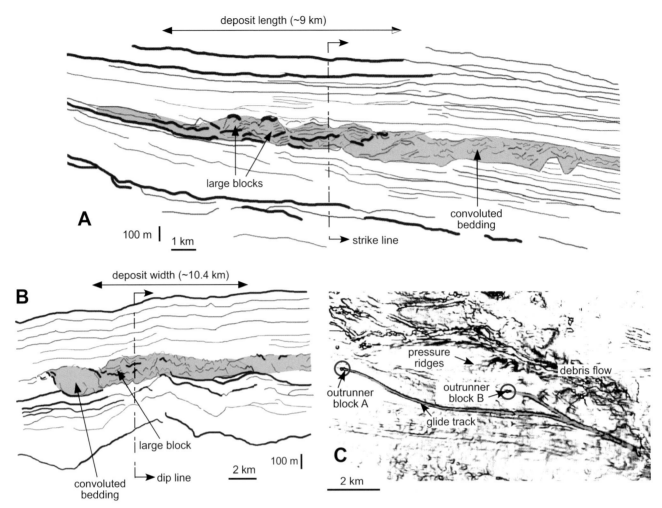

Fig. 20 Contrast of different types of submarine debris flows. (A) Interpreted seismic (dip) cross-section of a blocky submarine debris-flow deposit from the Nova Scotian continental margin. The blue area corresponds to the debris-flow deposit, while orange corresponds to a region of deformed bedding found around the primary deposit. Regions associated with the debris-flow deposit (composed of ancient lithified carbonate-reef fragments) and deformed bedding (siliciclastic mud) were confirmed from several cores. (B) Interpreted strike cross-section from the same Nova Scotia debris flow. (Data appear courtesy of L.G. Kessler.) (C) Depositional signature of a fine-grained, subaqueous debris flow. The image is a 'time' slice obtained from three-dimensional seismic tomography of the Nigerian margin (Nissen *et al.*, 1999). The streaks seen in the figure are qualitatively identical to the smear that connects the outrunner block to its parent in Fig. 17.

Figure 20A & B illustrates two seismic cross-sections of an erosive, blocky debris flow comprised primarily of fractured carbonate on the Nova Scotian margin. The flow that produced the deposit travelled over 10 km down a relatively uniform slope of 9°. As can be seen in the image, the flow was extremely erosive, embedding itself in the underlying stratigraphy. Similar seabed ploughing has been observed in other large submarine debris flows (Prior *et al.*, 1984). Large blocks that appear to retain the stratigraphy of their source

are also common, and have been previously documented in large, carbonate debris-flow deposits (Hine *et al.*, 1992). Finally, the structure of the seabed surrounding the deposit is disturbed, even beyond the point at which the flow stopped. All of these characteristics suggest significant interaction with the bed and substantial energy loss associated with frictional contacts. To properly predict debris-flow runout in this type of setting would require mixture theory (see Box 2). However, it is important to mention that even mixture theory does not include processes likely present in the Nova Scotian case (e.g. fluidization from seismic and acoustic waves; Melosh, 1979).

Figure 20C is a 'time slice' from a seismic cube acquired from the West African margin (Nissen *et al.*, 1999). A time slice represents a single reflector in the seismic cube, so it essentially yields a map of an ancient seafloor. Highlighted in green are glide tracks from outrunner blocks circled in red. These features bear an uncanny resemblance to the glide tracks and outrunner blocks seen in Fig. 18. Also present are 'pressure ridges' commonly found in laboratory experiments of Bingham-like materials (Major & Pierson, 1992; Parsons *et al.*, 2001b). On the muddy, siliciclastic margin of West Africa, a single-phase fluid model (e.g. in Huang & García (1997) and the Fish Tank experiments) appears to encapsulate the most important physical processes.

WAVE-SUPPORTED SEDIMENT GRAVITY FLOWS

Wave-supported sediment gravity flows and estuarine fluid muds, unlike debris flows and turbidity currents, have been studied primarily in the natural environment. They result from a balance between wave, tidal and gravity-driven motions. Unlike the traditional conceptions of a sediment gravity flow, they require some additional energy source (waves, tides or currents) to maintain their integrity. Due to difficulties associated with sampling the large fluxes of sediment close to the bed, progress in understanding their dynamics on continental shelves has come largely from recent multi-investigator studies (e.g. AmasSeds, STRATAFORM). These analyses are discussed below, along with the presentation of a model that is capable of predicting transport based upon basic sedimentological variables (e.g. wave orbital velocity U_w, critical bed shear stress τ_{cr}).

Estuarine fluid muds have been found in many coastal settings where sediment supply exceeds transport capacity (Wells, 1983; Wright *et al.*, 1990; Kineke *et al.*, 1996). However, conditions are not amenable to the formation of estuarine fluid muds on the shelf near the Eel River. Much of the knowledge about the mechanics of estuarine fluid muds stems from the work done on the Amazon shelf (Kineke & Sternberg, 1992, 1995; Trowbridge & Kineke, 1994; Cacchione *et al.*, 1995; Kineke *et al.*,

1996). In particular, Trowbridge & Kineke (1994) set forth a simple model of an impulsively generated, periodic boundary layer, which can be applied to all boundary layers dominated by fine sediment, and will be used below to describe wave-supported sediment gravity flows. Work in other locales has documented that tidally derived fluid muds are common and important for the cross-shelf transport of sediment, as well as the construction of shelf clinoforms in many settings (e.g. Fly River margin, Papua New Guinea; Walsh *et al.*, 2004).

Unlike estuarine fluid muds, wave-supported sediment gravity flows were commonly observed on the northern California margin. As discussed in Hill *et al.* (this volume, pp. 49–99), high concentrations of wave-suspended sediment isolated near the bed are an important cause for downslope transport on the Eel shelf during and after flood events (Fig. 21). Wave-supported sediment gravity flows are also important for creating mid-shelf flood deposits. These flows are fundamentally different from other gravity flows discussed in this paper, because the turbulent energy required to keep the sediment in suspension is supplied primarily by surface waves and not by the flow itself. This was seen in the observations made at a 60-m Eel shelf site (K60) during January 1998 (Fig. 21). The thickness of the highly concentrated near-bed layer is roughly equivalent to the wave-boundary-layer height predicted by

$$\delta_w = \frac{U_w}{\omega} \sqrt{\frac{f_w}{\delta}} \qquad (3)$$

where the wave friction factor f_w is described by Swart (1974), ω is the radial frequency and U_w is the root-mean-square wave orbital velocity (Traykovski *et al.*, 2000). Further support for the role of waves comes from the correlation of wave cessation with the collapse of the gravity-flow **lutocline** (i.e. turbidity discontinuity). As will be shown below, similar behaviour has been seen in laboratory experiments.

Wave-boundary-layer mechanics

The internal mechanics within wave boundary layers, due to their fluid-mechanical complexity (highly sheared, unsteady flows confined within

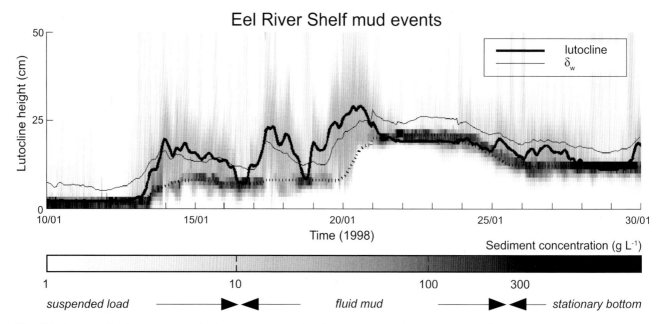

Fig. 21 Acoustic backscatter record of wave-supported sediment gravity flow events (i.e. fluid muds) from the Eel shelf (K60). The lutocline and predicted wave-boundary-layer height are shown relative to the changing bottom elevation. (From Traykovski *et al.*, 2000.)

only a few centimetres above the bed), were only observed and described from first principles in the 1970s. Grant & Madsen (1979) used an eddy-viscosity turbulence closure to solve the equations of motion (see Box 3). Their model has been extremely effective at predicting the onset of sediment motion based upon basic wave variables (i.e. wave-orbital velocity U_w and the velocity of ambient currents V_c; Wiberg *et al.*, 2002). The velocity profile predicted by their theory matches experimental and observational data well, despite the simplistic assumptions present within the eddy-viscosity model (e.g. time-averaged; Madsen & Wikramanayake, 1991; Wiberg, 1995).

A series of recent studies, however, have called into question the predictions of Grant & Madsen (1979) near the bed when sediment transport is significant. For instance, Foda (2003) suggests that **slip** associated with the multiphase nature of the bed is the dominant control on near-bed shear. Laboratory measurements of large breaking waves over sand substantiate this conclusion (Dohmen-Janssen & Hanes, 2002). In these experiments, inclusion of multiphase effects (e.g. permeability of the bed surface) is required to predict observed velocity profiles (Hsu & Hanes, 2004). Even in laboratory experiments of fine-grained materials,

typical of the middle continental shelf, boundary-layer stress fails to predict the sediment entrained by wave motion or the turbulence characteristics near the bed (Lamb *et al.*, 2004a; Lamb & Parsons, 2005).

It is important to mention that none of the above papers examined wave boundary layers that possessed a gravity-flow component. However, it is easy to imagine that the high concentrations studied in all of these experiments could have generated gravity-driven transport, if a slope was present. Until an alternative bed-shear-stress model is developed that incorporates these effects in a general way, Grant & Madsen (1979) will remain the most useful theoretical tool to predict gravitational sediment fluxes associated with wave motions. The lack of data within wave boundary layers also means that most of the discussion contained within this section will be confined to theoretical analysis.

Wave-supported sediment gravity flows and the role of buoyancy

The basic dynamics of wave-supported sediment gravity flows are governed by the balance between the gravitational force acting in the downslope

Box 3 Shear stress and velocity distribution within wave boundary layers (Grant & Madsen, 1979)

Once again beginning with the incompressible, irrotational Navier–Stokes equations, which describe the conservation of mass and momentum over an arbitrary control volume

$$\frac{D\vec{u}}{Dt} = \frac{\nabla p}{\rho} + \nu\nabla^2\vec{u} \tag{B3.1}$$

Defining an eddy viscosity K

$$K = \kappa\left|u_{*\mathrm{cw}}\right|z \tag{B3.2}$$

where $\left|u_{*\mathrm{cw}}\right|$ is the total shear stress both from wave and currents, κ is the von Karman constant. Also defining the wave motions at the outer edge of the wave boundary layer

$$\vec{u}_\infty = \left|\vec{U}_{\mathrm{w}}\right|\exp(i\omega t) \tag{B3.3}$$

where $i = \sqrt{-1}$, $\vec{U}_{\mathrm{w}}$ is the near-bed wave orbital velocity (outside the wave boundary layer) and ω is the wave frequency. Assuming that the convective acceleration terms are encapsulated within the eddy-viscosity model K and the background current is much smaller, Eq. B3.1 becomes

$$\frac{\partial\vec{u}_{\mathrm{w}}}{\partial t} = -\frac{\nabla p_{\mathrm{w}}}{\rho} + \frac{\partial}{\partial z}\left(\kappa\left|\vec{u}_{*\mathrm{cw}}\right|z\frac{\partial\vec{u}_{\mathrm{w}}}{\partial z}\right) \tag{B3.4}$$

If vertical velocities near the bed are assumed negligible, the y-direction of Eq. B3.4 yields

$$\frac{\partial\vec{u}_\infty}{\partial t} = -\frac{\nabla p_{\mathrm{w}}}{\rho} \tag{B3.5}$$

Substituting Eq. B3.5 into Eq. B3.4 and defining $(u_{\mathrm{w}} - u_\infty) = w_0 = \left|w_0\right|\exp(i\omega t)$ yields

$$\frac{\partial}{\partial z}\left(\frac{\kappa\left|\vec{u}_{*\mathrm{cw}}\right|}{\omega}z\frac{\partial\vec{u}_{\mathrm{w}}}{\partial z}\right) - i\left|w_0\right| = 0 \tag{B3.6}$$

Making the change of variables, Eq. B3.6 becomes

$$\frac{\partial}{\partial\zeta}\left(\zeta\frac{\partial\left|w_0\right|}{\partial\zeta}\right) - i^3\left|w_0\right| = 0 \tag{B3.7}$$

Applying the boundary conditions,

near bed $\Rightarrow w_0 \to -u_\infty$ at $\zeta_0 = k_{\mathrm{b}}\omega/30\kappa\left|u_{*\mathrm{cw}}\right|$

at the top of the boundary layer $\Rightarrow w_0 \to 0$ as $\zeta \to \infty$,

the general solution of Eq. B3.7 given these boundary conditions is

$$w_0 = A_1(\mathrm{Ber}\,2\sqrt{\zeta} + i\,\mathrm{Bei}\,2\sqrt{\zeta}) + A_2(\mathrm{Ker}\,2\sqrt{\zeta} + i\,\mathrm{Kei}\,2\sqrt{\zeta}) \tag{B3.8}$$

where Ber, Bei, Ker and Kei are Kelvin functions of zeroth order. $A_1 = 0$ because Ber and Bei become large for $\zeta \to \infty$, which allows for the solution of the second integration constant

$$A_2 = \frac{-u_\infty}{(\mathrm{Ker}\,2\sqrt{\zeta} + i\,\mathrm{Kei}\,2\sqrt{\zeta})} \tag{B3.9}$$

Substituting Eq. B3.9 into Eq. B3.8 and exchanging variables yields the velocity profile

$$u_{\mathrm{w}} = \left(1 - \frac{\mathrm{Ker}\,2\sqrt{\zeta} + i\,\mathrm{Kei}\,2\sqrt{\zeta}}{\mathrm{Ker}\,2\sqrt{\zeta_0} + i\,\mathrm{Kei}\,2\sqrt{\zeta_0}}\right)u_\infty \tag{B3.10}$$

direction and the vertical distribution of shear stress in the water column

$$g' \sin \beta = d\tau/dz \qquad (4)$$

where g' is the reduced gravitational acceleration with respect to the ambient water column, τ is the shear stress and z is the direction perpendicular to the bed. The value of g' is given by $(\rho_1 - \rho_0)g/\rho_0$, where g is the gravitation acceleration, ρ_1 is the density within the gravity flow and ρ_0 is the ambient seawater density. For sediment-driven flows, $g' = RCg$, where C is the volumetric sediment concentration and $R = (\rho_{sed} - \rho_0)/\rho_0$ is the submerged weight of the sediment in water. Integration of Eq. 4 over the wave-boundary-layer thickness yields the familiar Chezy balance between frictional drag on the stationary seafloor τ_b, interfacial drag τ_i from the water above and the gravitational force

$$(\tau_b + \tau_i)/\rho_a = B \sin \beta \qquad (5)$$

where B is the depth-integrated reduced gravitational acceleration or 'buoyancy', given by

$$B = gR \int_0^{\delta_w} C \, dz \qquad (6)$$

Other terms such as Coriolis acceleration, larger scale pressure gradients and fluid accelerations are generally found or assumed to be small and therefore can be neglected (Traykovski *et al.*, 2000).

Determination of the buoyancy B and the distribution of shear in the water column $d\tau/dz$ in terms of measurable sedimentological properties (e.g. wave-orbital velocity, settling velocity) is central to any wave-supported sediment gravity flow model. As part of the STRATAFORM programme, several different approaches to modelling these quantities have been investigated. To calculate the buoyancy B, two different analyses have been used. The first assumes that the wave boundary layer is in equilibrium and well mixed. The result is an approximation akin to the Exner approximation used in classic sediment-transport theory. However, in wave-supported sediment gravity flows, the large values of sediment concentration strongly stratify the water column and suppress turbulence. Therefore, an alternative approach to modelling the buoyancy has been devel-

oped to determine the maximum suspended load possible without suppressing the turbulence below that level.

Regardless of approach, relation of the sediment in suspension to commonly measured wave variables requires a model for the near-bed sediment concentration. The near-bed sediment concentration can often be calculated by equating the upward and downward flux of sediment. Here, the balance will be considered as it applies to gravity flows trapped near the bed. Assuming that the downward flux of sediment occurs through ballistic settling with a settling velocity w_s, the volumetric depositional flux $\mathbf{D} = C_b w_s$, where C_b is the near-bed volumetric concentration and w_s is the settling velocity of the sediment. Upward sediment flux from the seabed is parameterized by an entrainment rate $\mathbf{E} = E w_s$, where E is the dimensionless entrainment rate. From early studies, it was noted that the dimensionless entrainment rate could be alternatively thought of as a concentration. As a result, E is sometimes called the **reference concentration** required for sediment resuspension (Smith & McLean, 1977; Harris & Wiberg, 2001).

Description of the dimensionless entrainment rate in terms of water-column properties has been the focus of innumerable studies over the past 30 yr. Two of these models have emerged as the most popular. García & Parker (1991, 1993), discussed at length in the section on turbidity currents, is most often used by numerical modellers because it prescribes, for all entrainment rates, the distance above the bed at which E is to be observed (i.e. 5% of the total flow depth). The model of Smith & McLean (1977), on the other hand, is constructed such that the distance above the bed corresponding to E is also an empirical function. Prescription of the **reference height** (i.e. the height at which E is obtained) and the dimensionless entrainment rate makes numerical modelling on fixed grids difficult, but it allows for greater flexibility in the formation of the model. It is also well tailored to temporal observations made at a fixed distance above the bed. As a result, the Smith & McLean (1977) formulation has been used primarily to calculate entrainment fluxes from field data, where particle size and behaviour (i.e. flocculated versus unflocculated) may differ significantly from the non-cohesive, sandy flows explored by García & Parker (1991). The Smith & McLean (1977) model has been tested extensively with tripod measurements on many margins (including the Eel) with fine-sand and silty beds (Cacchione *et al.*, 1999; Harris & Wiberg, 2001). The model has the general form

$$E = \frac{(1-n)\gamma_0 S_s}{1 + \gamma_0 S_s} \tag{7}$$

where $S_s = (\tau_b - \tau_{cr})/\tau_{cr}$ is the excess shear stress, τ_b is the bed shear stress, τ_{cr} is the critical shear stress to initiate motion, n is the porosity of the bed and γ_0 is an empirical constant. It also requires a formulation for the reference height. However, it will be shown that the reference height is not a relevant parameter in the calculation of the near-bed buoyancy.

To calculate the concentration distribution in the wave boundary layer, the upward flux of sediment above the reference height must also be known. The flux of sediment due to turbulent diffusion is usually thought to scale with the shear velocity, u_*. If these quantities are assumed to be directly related, the ratio of w_s/u_* determines the shape of the steady-state concentration profile. If $u_* \gg w_s$ (which is the case for fine sediment within the energetic wave boundary layers), then the concentration within the wave boundary layer will not vary significantly with depth. Sediment-induced stratification can, of course, modify this vertical dependence. Modelling results (discussed in subsequent sections) show that stratification dominates at the top of the wave boundary layer. Observations from site K60 on the Eel shelf show that when currents are weak, very little sediment escapes the bottom 0.1 m of the water column. For example, concentrations of approximately 1 kg m^{-3} were typically observed at ~0.1 mab in areas that had concentrations in excess of 10 kg m^{-3} within the wave boundary layer (Fig. 21).

As a result, a model can be constructed that assumes that the net flux into or out of the wave boundary layer is zero (i.e. **E = D**). Finally, when the concentration within the wave boundary layer is assumed to be roughly constant with depth (consistent with a strongly turbulent boundary layer) and equal to the near-bed volumetric concentration C_b, the buoyancy B becomes

$$B = gRE\delta_w \tag{8}$$

while sediment load per unit area of the wave boundary layer L_s is

$$L_s = \frac{\rho_{sed} B}{Rg} = \rho_{sed} E \delta_w \tag{9}$$

Since the dimensionless entrainment rate E scales with the bottom shear stress, which is related to wave velocity squared (from Eq. 7), and the wave-boundary-layer height δ_w scales linearly with the wave velocity, the buoyancy of the wave boundary layer B and the sediment load transported within it (L_s) scales with wave velocity to the third power (Fig. 22).

Vertical distribution of momentum and sediment concentration

An alternative approach to estimating suspended load carried by a wave-supported sediment gravity flow is to calculate the maximum possible sediment concentration that still allows for the generation of shear-induced turbulence. This limit occurs when the **gradient Richardson number** Ri is near its critical value, $Ri_{cr} = 0.25$, where

$$Ri = -\frac{\partial g'/\partial z}{(\partial u/\partial z)^2} \tag{10}$$

where u is the local velocity and g' is the local reduced gravitational acceleration. Qualitatively, a negative-feedback mechanism is thought to maintain the gradient Richardson number near its critical value in highly energetic environments with unlimited supplies of easily suspended fine sediment. If $Ri < Ri_{cr}$, intense turbulence suspends more sediment, increasing stratification $\partial g'/\partial z$, which increases Ri toward Ri_{cr}. If $Ri > Ri_{cr}$, production of turbulence by shear instability is dramatically reduced, sediment settles out of the water column and Ri decreases towards Ri_{cr}. In tidally controlled, sediment-laden shelves and estuaries, Ri has been observed to be close to Ri_{cr} (Geyer & Smith, 1987; Trowbridge & Kineke, 1994; Friedrichs *et al.*, 2000). These effects may also play an important role in controlling stratification in bottom boundary layers in general (Trowbridge, 1992; Trowbridge & Lentz, 1998).

Trowbridge & Kineke (1994) have applied the Ri_{cr} constraint to the momentum-deficit portion of a

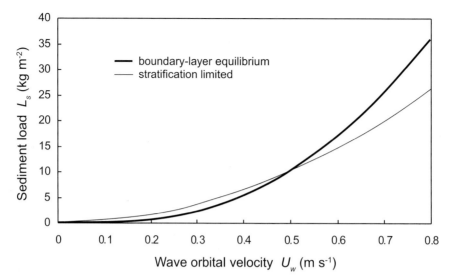

Fig. 22 Wave-velocity dependence on the predicted sediment load within a fluid mud. The boundary-layer-equilibrium solution is derived assuming no net transport into or out of the wave boundary layer (Eq. B3.4). The stratification-limited solution is found using the expression proposed by Wright *et al.* (2001) (Eq. B3.7). The particular boundary-layer-equilibrium solution shown here assumes $f_w = 0.04$, $\gamma_0 = 0.0024$ (after Smith & McLean, 1977), $n = 0.35$ and $\tau_{cr} = 0.1$ Pa, while the stratification-limited solution assumes $Ri_{cr} = 0.25$ (after Howard, 1961). Both solutions assume $\rho_{sed} = 2650$ kg m^{-3}. It should be noted that alternative combinations of the empirical parameters (in particular, f_w and γ_0) are capable of producing significant deviation between the two expressions. However, alteration of these variables does not change the underlying similarity between the two non-linear relationships shown in the figure.

tidal boundary layer containing a fluid mud by approximating the periodic tidal pressure gradient as an impulse. Support for the use of an impulsively started flow to approximate an oscillatory boundary layer is provided by the classic laminar Rayleigh and Stokes solutions, which indicate that a comparison of the two cases is meaningful if carried out when the free-stream speed in the oscillatory flow reaches a relative maximum (Trowbridge & Kineke, 1994).

Here, the scaling of Trowbridge & Kineke (1994) is applied to a wave boundary layer with the additional assumption that the momentum-deficit layer encompasses most of the boundary layer. The momentum-deficit layer is illustrated, along with various models and laboratory data, in Fig. 23. The assumption that the momentum-deficit layer is large with respect to the near-wall layer (where friction from the bed is dominant) is equivalent to the constraint imposed by Wright *et al.* (2001) that the critical Richardson number through most of the wave boundary layer is equal to

$$Ri_{cr} = Ri_f = B/U_w^2 \qquad (11)$$

where U_w is wave orbital velocity at the top of the wave boundary layer (i.e. $z = \delta_w$) and Ri_{cr} is the critical Richardson number approximately equal to 0.25 (Howard, 1961). The quantity B/U_w^2 is sometimes called a **flux Richardson number**. Flux Richardson numbers are usually constructed in terms of bulk-flow parameters as opposed to the local definition implied in the gradient Richardson number (Eq. 10). The relationship described in Eq. 11 is supported by its ability to approximate the sediment load predicted by the boundary-layer-equilibrium model described in the previous section (Fig. 22). The sediment load within the wave boundary layer using Eq. 11 is described by

$$L_s = \frac{\rho_{sed}B}{Rg} = \frac{\rho_{sed}Ri_{cr}U_w^2}{Rg} \qquad (12)$$

Applying Eq. 11 to the solutions of Trowbridge & Kineke (1994), the depth-varying similarity solutions for the vertical distribution of the amplitude of wave motions u_w, buoyancy anomaly $b = RCg/\rho_{sed}$ and the eddy viscosity K are then given by

$$u_w = U_w(z/\delta_w)^2 \qquad (13)$$

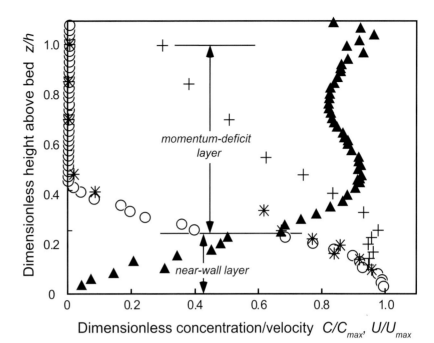

Fig. 23 Illustration of near-wall and momentum-deficit regions within laboratory and natural sediment gravity flows. Solid triangles and open circles represent velocity and sediment-concentration data, respectively, from Trowbridge & Kineke (1994). The maximum value of the velocity in their observations was $U_{max} = 1.8$ m s^{-1}, while the maximum concentration was $C_{max} = 35$ kg m^{-3}. Crosses and asterisks are laboratory-derived velocity and sediment-concentration data from García & Parker (1993). In these experiments, $U_{max} = 0.13$ m s^{-1} and $C_{max} = 10$ kg m^{-3}.

$$b = \frac{4U_w^2[1 - (z/\delta_w)^3]}{3\delta_w} \tag{14}$$

$$K = \frac{\delta_w u_{*w}^2}{\partial U_w}[(\delta_w/z) - (z/\delta_w)^2] \tag{15}$$

where u_{*w} is the shear velocity associated with the waves only. Figure 23 shows data from a tidal boundary layer for which Eqs 12–15 seem to apply reasonably well (Trowbridge & Kineke, 1994). The figure also shows data from analogous laboratory experiments.

Based upon Traykovski *et al.* (2000) and Wright *et al.* (2001), the range of values during periods of rapid flood deposition at 60-m depth off the Eel River are bracketed by $U_w = 0.3$–0.6 m s^{-1} and $\delta_w = 0.05$–0.1 m, which results in depth-averaged concentrations in the order of 100 kg m^{-3} in the wave boundary layer (Fig. 24A & B). Using a wave friction factor $f_w = 0.04$ (discussed previously) and assuming a hindered settling velocity of ~0.1 mm s^{-1} (Mehta, 1989) gives $w_s U_w/u_*^2 \sim 0.01$.

The above formulation (Eqs 13–15) breaks down near the bed where the modelled eddy diffusivity tends to infinity. In reality, very near the bed, gravity-driven motions will begin to dominate the production of shear. This region is extremely small and the gravity-driven component is large in the

case of fine-grained sediments (typical on continental shelves receiving modern sediment), so the law of the wall is assumed to hold, such that $K_{wall} = \kappa u_{*w}z$, where $\kappa = 0.4$ is the von Karman constant. A sensible transition from near-wall to momentum-deficit scaling should then occur where $K = K_{wall}$. Trowbridge (1992) suggested that the transition should occur where the production term in the turbulent-energy balance, $P = K_{wall}(du_w/dz)^2$, predicted by Eqs 13–15 exceeds turbulent production predicted by the law of the wall, namely, $P_{wall} = u_{*w}^3/\kappa z$. It turns out that $K = K_{wall}$ and $P = P_{wall}$ both yield

$$\delta_{wall} = \left(\frac{u_{*w}}{2\kappa U_w}\right)^{1/2} \delta_w \tag{16}$$

For the above conditions on the Eel River shelf, Eq. 16 predicts $\delta_{wall}/\delta_w \approx 0.3$ (Fig. 24C & D). If production in the wall layer is estimated using the Grant & Madsen (1979) solution for u_w instead, then $P/P_{wall} \approx 0.3$ at $\delta_{wall}/\delta_w \approx 0.2$. These values are similar to the dimensionless height at which wall effects become important in unidirectional gravity-current experiments, when those heights are made dimensionless with the gravity-current thickness (Parsons, 1998; Buckee *et al.*, 2001).

The law of the wall produces a velocity profile that is inconsistent with Eq. 13. Superimposed

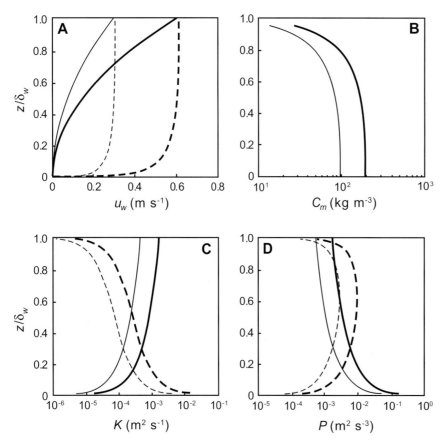

Fig. 24 Comparison of (A) velocity, (B) concentration, (C) eddy-diffusivity and (D) turbulence-production profiles between Grant & Madsen (1979) and the extended theory of Trowbridge & Kineke (1994). In all cases, solid lines correspond to Trowbridge-Kineke, while Grant-Madsen is shown as dashed lines. Cases where $U_w = 0.6$ m s^{-1} are shown in bold, while thinner lines correspond to $U_w = 0.3$ m s^{-1}.

on Fig. 24A are profiles for u_w using the wave-boundary-layer solution of Grant & Madsen (1979). The true solution for u_w in the lower boundary layer is likely to be some intermediate profile, as suggested by the field data displayed in Fig. 25. The

velocity profile close to the bed probably resembles the solution of Grant & Madsen (1979), while velocities nearest the lutocline (the top of the wave boundary layer, $z \sim \delta_w$) resemble the solution of Trowbridge & Kineke (1994). For the discussion of

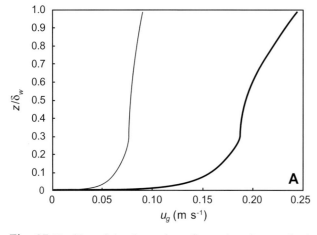

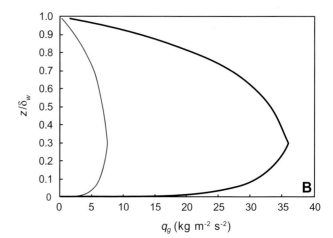

Fig. 25 Profiles of the downslope flux using the matched analysis described in Eq. B4.1. The plot is bracketed by the $U_w = 0.6$ m s^{-1} case in bold, and $U_w = 0.3$ m s^{-1} (thin line). The results shown assume a slope $\beta = 0.5°$. All other parameters correspond to the analysis described in Fig. 22.

wave-supported sediment gravity flows to follow, it is important to recognize that Eq. 15 predicts that shear in the upper portion of the momentum-deficit layer is crucial for regulating the concentration in the lower half of the boundary layer, unlike more conventional, sandy wave boundary layers. In sandy wave boundary layers, the dimensionless entrainment rate E determines concentration throughout the boundary layer because the energy required to mobilize the bed (sand) is significantly greater than the energy required to thoroughly mix the near-bed water column.

The formulation in Box 4 was derived in an attempt to find the simplest possible solutions for wave-supported sediment gravity flows that reasonably represent the underlying physics, including the vertical structure of the mean current. In an attempt to make the problem more tractable analytically, several assumptions were made.

1 A critical Richardson number Ri_{cr} was imposed throughout the momentum-deficit layer (i.e. $\delta_{wall} < z < \delta_w$) and $\delta_{wall} \ll \delta_w$ was assumed.

2 Interfacial drag and entrainment at the top of the boundary layer were neglected, while simultaneously assuming the sediment fall velocity to be small.
3 The structure of the momentum-deficit layer at peak wave orbital velocity was assumed equivalent to an impulsively forced layer and the resulting concentration and eddy diffusivity profiles were assumed to apply under wave-averaged conditions.
4 The eddy diffusivity and turbulent-energy production were matched at the transition from the wall to momentum-deficit layers (i.e. $z = \delta_{wall}$), even though this results in a highly unrealistic discontinuity in the wave-orbital-velocity profile.

The discontinuity in wave-orbital-velocity profile may not critically undermine the predicted structure of the gravity current. The original formulation of Grant & Madsen (1979) for combined-flow boundary layers also possesses similar breaks in the estimation of the eddy diffusivity. These breaks have been shown to be negligible, owing to the relatively minor role that the eddy diffusivity plays in the regulation of the velocity profile, particularly near

---

Box 4 Calculation of the total flux in a wave-supported sediment gravity flow

The above solutions for $K(z > \delta_{wall})$ and $K_{wall}(z < \delta_{wall})$ are assumed to be applicable to the mean conditions represented by Eq. 3 (neglecting interfacial drag at $z = \delta_w$). Since most of the shear in the wall layer (i.e. $z < \delta_{wall}$) is concentrated very near the bed, the solution for the time-averaged downslope velocity (Fig. 25A) between $z_0 = k_r/30$ (assuming rough-wall conditions) and δ_{wall} is approximated by

$$u_g = \frac{Ri_{cr}U_w^2 \sin\beta}{\kappa u_{*w}} \log(z/z_0) \tag{B4.1}$$

For $\delta_{wall} < z < \delta_w$, the time-averaged downslope velocity is given by

$$u_g = u_{gwall}(\delta_{wall}) + Ri_{cr}(U_w^3/u_{*w})\sin\beta[F(Z) - F(\delta_{wall}/\delta_w)] \tag{B4.2}$$

where $Z = \dfrac{z}{\delta_w}$ and $F(Z) = \displaystyle\int_0^z \frac{2(3Z - 4Z^2 + Z^5)}{3(1 - Z^3)}\,dZ$

The depth-dependent, gravitational sediment flux q_g is given by the product of the downslope velocity and the mass of sediment in a given fluid parcel (Fig. 25B). Since shear in the near-wall and momentum-deficit layers is concentrated near the bottom and top of the boundary layer, respectively, the following approximate relationships hold for depth-averaged downslope velocity U_g and depth-integrated downslope transport Q

$$U_g = u_{gwall}(\delta_{wall}) = \frac{Ri_{cr}U_w^2 \sin\beta}{\kappa u_{*w}} \log(\delta_{wall}/z_0) \tag{B4.3}$$

$$Q = \frac{\rho_s U_g B}{Rg} = \frac{\rho_s Ri_{cr}^2 U_w^4 \sin\beta}{Rg\,\kappa u_{*w}} \log(\delta_{wall}/z_0) \tag{B4.4}$$

the bed (Madsen & Wikramanayake, 1991). In addition, because the gravity current is most sensitive to conditions nearest the upper and lower extremes of the boundary layer, the break occurs in a relatively insignificant region.

The Mellor & Yamada (1982) turbulence-closure scheme, as implemented by Reed *et al.* (1999), was used to test the applicability of the assumptions made in the analysis described above. Gravitational forcing was included in the model by adding a downslope pressure-gradient term that is proportional to the sediment concentration and bottom slope. The model accounts for sediment-induced stratification by the modification of stability parameters that are functions of the flux Richardson number B/U_w^2.

The velocity profiles generated by this model show a gravity-forced flow with a maximum downslope (negative) velocity near the top of the wave boundary layer (Fig. 26). The model varied the

empirical parameter γ_0 in the Smith & McLean (1977) entrainment model to identify the value that best fit the current-meter data for the 60-m Eel shelf site (K60). The observed seaward flow was matched with $\gamma_0 \sim 0.005$, resulting in concentrations of 300 kg m^{-3} in the lower half of the wave boundary layer and maximum offshore velocities of 0.3–0.4 m s^{-1}. The empirically derived value of γ_0 is consistent with the range of earlier observations for the Californian margin (Drake & Cacchione, 1985). The concentration profiles show a low-gradient, well-mixed region near the seafloor and a steep-gradient, stratified region above. A small dimensionless entrainment rate E produces a well-mixed region that extends to the top of the wave boundary layer. As E is increased, the stratification extends farther into the wave boundary layer. At extremely large entrainment rates, the stratified region extends to the seafloor. In the limit of vanishing E, the eddy diffusivity is similar to the neutral

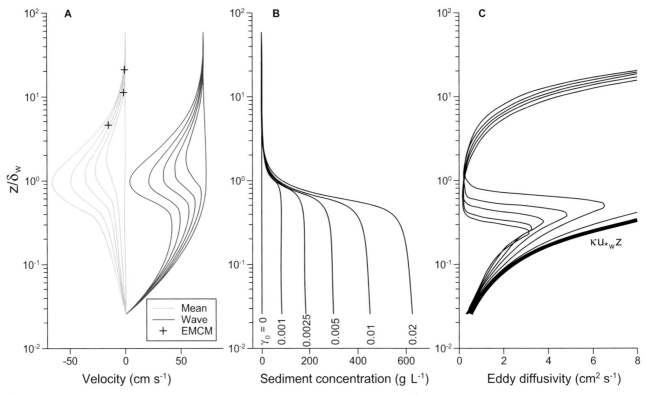

Fig. 26 Results of a k-ε numerical model for fine-grained sediment in a wave boundary layer. (A) Velocity. (B) Concentration. (C) Eddy diffusivity. Elevation above the bed z is non-dimensionalized by the wave boundary thickness δ_w in all plots. The resuspension parameter γ_0 in the Smith & McLean (1977) entrainment formulation was varied to obtain various concentration profiles (B). The gravity-flow velocity increases with increasing γ_0 (i.e. increasing E in C). Eddy diffusivity (thin lines in C) shows a dramatic decrease at the top of the wave boundary layer. A neutral eddy-diffusivity profile ($K = \kappa u_{*w} z$) is also shown for reference.

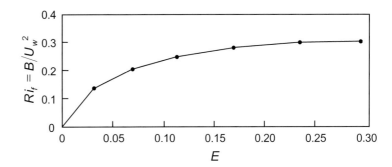

Fig. 27 Flux Richardson number $Ri_f = B/U_w^2$ derived from a k-ε model of fluid mud as a function of the dimensionless entrainment rate E.

solution of $K_e = \kappa u_{*w} z$, consistent with sandy boundary layers (Grant & Madsen, 1979). As sediment load increases, there is only a slight change in the eddy diffusivity within the lower portion of the wave boundary layer, but a drastic decrease in eddy diffusivity occurs at the top of the wave boundary layer due to density stratification. The location of this stratified region becomes closer to the seafloor as the sediment load is increased.

The model was also used to examine the role of density stratification in limiting the maximum sediment load as predicted by Eq. 12. The model predicts that for small dimensionless entrainment rates, the flux Richardson number is an increasing function of the dimensionless entrainment rate E (Fig. 27). However, once the wave boundary layer begins to stratify (i.e. entrainment rates become large and entrained sediment overwhelms the near-bed density), the flux Richardson number tends towards a constant around 0.3.

Finally, to use Eqs 13–16 and B4.1–B4.4 for prediction of downslope sediment fluxes, a way needs to be established to estimate the bed shear stress associated with the various motions. A common method to estimate the wave-averaged shear stress at the base of the wave boundary layer is the time-averaged quadratic formulation given by Grant & Madsen (1979), Feddersen *et al.* (2000) and Wright *et al.* (2001). As seen in Box 5, if a classic quadratic-drag representation is applied to a wave-suspended gravity current, the result is a drag coefficient that is unexpectedly large and inversely proportional to bed slope. However, for Eel shelf conditions, the above analysis yields $c_{Dcr} = 0.014 - 0.017$. These values lie within the large range of drag coefficients ($0.002 < c_D < 0.06$) assembled by Parker *et al.* (1987) for both laboratory and natural flows. Other natural complexities associated with wave-supported sediment gravity flows on large slopes will be treated in the next section.

Observations of wave-supported sediment gravity flows

Wave-supported sediment gravity flows require a substantial sediment supply and strong wave motions. As a result, most observations of these flows have been close to the coast, usually on the relatively flat continental shelf (Traykovski *et al.*, 2000; Ogston *et al.*, 2000). Hill *et al.* (this volume, pp. 49–99) describe wave-supported sediment gravity flows and their transport across the Eel shelf, while their operation is described below for the head of Eel Canyon, a location that meets both conditions for such flows. These conditions make Eel Canyon anomalous among modern slope features, but typical of most canyon systems during lowstands of sea level. As a result of research on the shelf and open slope adjacent to the Eel River, it was realized that a considerable amount of material was making its way into Eel canyon (Mullenbach, 2002; Drexler *et al.*, 2006). As a result, the canyon was instrumented to identify processes associated with sediment transport from the continental shelf to the Eel Canyon. The canyon observations represent a data set that may yield additional insight into the mechanics of wave-supported sediment gravity flows.

Among other instruments, a benthic boundary-layer tripod was deployed from January to April 2000 in the northern thalweg of the Eel Canyon at 120 m depth. Instruments mounted on the tripod included two Marsh McBirney electromagnetic current-meters placed at 0.3 m and 1 m above the bottom (mab), as well as a pressure sensor located at 1.4 mab. These instruments were programmed to sample every hour and collect 450 samples at 1 Hz (7.5 min of data per burst). The tripod also was equipped with a downward-looking video system placed at 1.8 mab that took clips of 7 s every 4 h for seabed-roughness observation. Qualitative analysis

---

Box 5 Assessment of friction

Beginning with a simple force balance to estimate the bed shear stress (Grant & Madsen, 1979)

$$\tau_b = \rho c_D \langle U \rangle \langle \sqrt{U^2 + V^2} \rangle \tag{B5.1}$$

where ρ is the gravity-flow density, c_D is a non-dimensional bottom drag coefficient, $<>$ represents a temporal average, U and V are the velocities in the downslope direction and cross-slope direction at the top of the bottom boundary layer, respectively. If interfacial drag at the top of gravity flow is neglected, assuming $U_g \gg <U>$ and $\rho/\rho_0 \approx 1$, and then substituting Eq. B5.1 into Eq. 5

$$B \sin \beta = c_D U_g U_{max} \tag{B5.2}$$

where $U_{max} = \langle \sqrt{U^2 + V^2} \rangle$ is the characteristic velocity amplitude at the top of the wave boundary layer. Assuming that the waves, current and gravitational flow (i.e. slope) are aligned

$$U_{max} = \sqrt{U_w^2 + U_g^2 + V_c^2} \tag{B5.3}$$

where V_c is the strength of the wave-averaged current at the top of the wave boundary layer.

   If the wave orbital velocity U_w is much larger than either U_g or V_c, then combining Eqs 11, B5.2 and B5.3 provides a compact formula for the velocity of the gravity current (Wright et al., 2001)

$$U_g = \frac{U_w Ri_{cr} \sin \beta}{c_D} \tag{B5.4}$$

Taking the product of Eq. 12 and Eq. B5.4 yields the depth-integrated downslope sediment transport due to the gravity current

$$Q = \frac{\rho_{sed} U_w^3 Ri_{cr}^2 \sin \beta}{R g c_D} \tag{B5.5}$$

Equating Eqs B4.3 and B4.4 with Eqs B5.3 and B5.4 gives

$$c_D = \frac{\kappa u_{*w}}{U_w \log(\delta_{wall}/z_0)} \tag{B5.6}$$

If the balance in Eq. B5.2 is expressed using a classical quadratic drag representation of a turbidity current (Komar, 1977; Traykovski et al., 2000)

$$B \sin \beta = c_{Dcr} U_g^2 \tag{B5.7}$$

Combining Eqs B5.2 and B5.5 yields

$$c_{Dcr} = \frac{c_D^2}{Ri_{cr} \sin \beta} \tag{B5.8}$$

---

of the turbidity of the water, using the opacity of the video images, provided temporal evolution of the suspended-sediment concentration during the entire deployment. **Opacity** units ranged from 100 when the monitor screen was black, due to large amounts of suspended sediment in the water, to 5 when the image of the seabed was perfectly clear. Additionally, an upward-looking 300 kHz acoustic Doppler current-meter profiler (ADCP, RD Instruments) was mounted on the tripod and measured the current components (north, east and vertical) every hour at 1-m bins, profiling from 60–115 m water depth

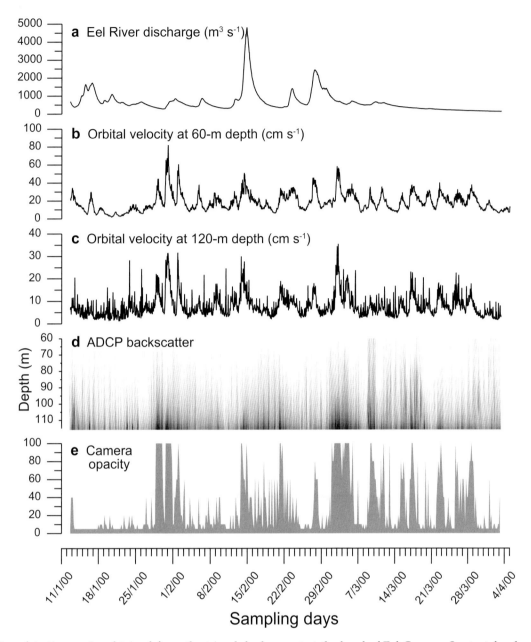

Fig. 28 Complete time series obtained from the tripod deployment at the head of Eel Canyon. See text for description of observations. (From Puig *et al.*, 2004.)

(i.e. from 5 to 60 mab). The mean backscatter signal measured by the four ADCP transducers was also used as an estimate of the suspended-sediment concentration and provided information about the distance above the seabed that particles were suspended in the water column.

The temporal evolution of estimates for near-bottom suspended-sediment concentration reflected a clear link with storm events, but not with the Eel River discharge. Increases in the wave orbital velocity during storms clearly coincided with high values of camera opacity and ADCP backscatter, suggesting a sediment-transport mechanism associated with surface-wave activity (Fig. 28). Puig *et al.* (2003) discussed in detail observational data collected within the Eel Canyon.

Data analysis of large storm events revealed that when camera opacity values reached 100 units

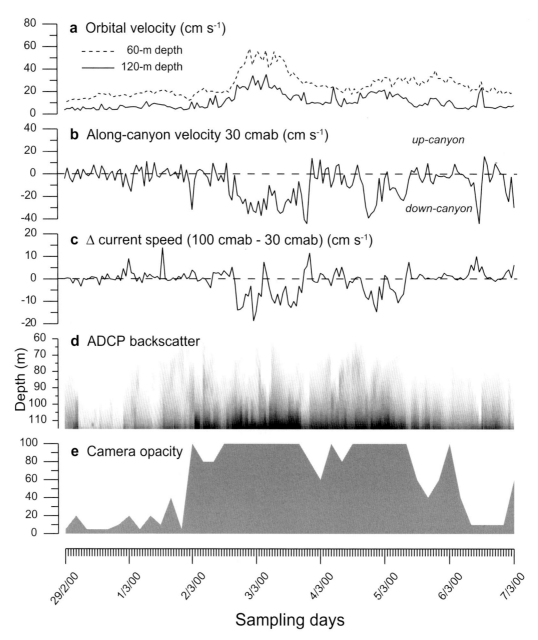

Fig. 29 Detailed time series from a time of high suspended-sediment concentration. Note that the increases in turbidity (camera opacity) and downslope flow come immediately (no time lag) after the wave orbital velocity increases. (From Puig _et al._, 2004.)

('black screen') for several hours, current velocity at 0.3 mab was much higher (~0.15 m s⁻¹) than current velocity at 1 mab and was directed down-canyon (Fig. 29). These near-bottom current profiles combined with the high estimates of suspended-sediment concentration indicate the presence of storm-induced flows driven by negative buoyancy resulting from very high suspended-sediment con-

centrations. These gravity flows transport large amounts of sediment toward deeper parts of the margin.

Coinciding with the occurrence of a gravity flow, the near-bottom current components recorded at the 'burst' time-scale (1 Hz) oscillate at the same periodicity as the pressure and were primarily oriented in the along-canyon direction (Fig. 30). This

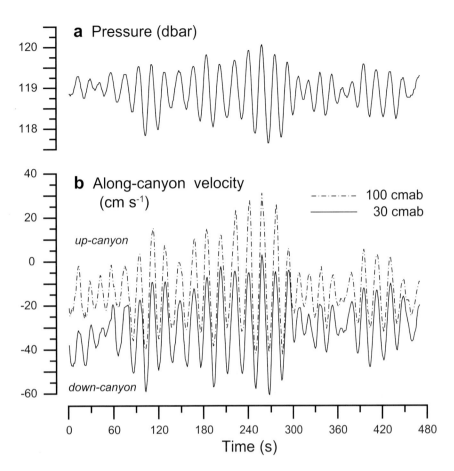

Fig. 30 Burst time series (1 Hz) from the two electromagnetic current meters at 30 cm and 100 cm above the bed. Note that the lowest measurement (30 cm above the bed) has higher downslope velocities. These velocities are strongly modulated by the wave-induced motions, however. (From Puig *et al.*, 2004.)

up- and down-canyon current fluctuation at high frequencies denotes a clear influence of the surface wave activity on the currents at 120-m depth, which may also contribute to maintenance of the gravity flow while it is transported down-canyon.

These gravity flows occur repeatedly through the same stormy period every time that the wave orbital velocity increases, lasting for several hours as long as the wave orbital velocity maintains high values (Figs 29 & 30). This behaviour suggests that the gravity-driven processes are not caused by a catastrophic event removing sediment temporarily deposited at the head of the canyon, but by a continuous process throughout a given storm.

Sediment resuspension by waves appears to be the most probable mechanism for creating high suspended-sediment concentration at the canyon head. However, the rapid formation of the gravity flow, immediately after the increase of the wave orbital velocity, suggests that traditional wave resuspension is probably not the only mechanism contributing to their formation (Puig *et al.*, 2004).

Particle-by-particle entrainment of sediment into suspension requires time to develop high enough near-bottom suspended-sediment concentration to generate fluid-mud suspension, regardless of bed shear stresses (Admiraal *et al.*, 2000). If the entrainment model of García & Parker (1993) is assumed, and the wave characteristics obtained from the tripod are used, the time required to form a wave-supported gravity flow from the onset of strong wave activity is longer (> 1 h) than the observed time of arrival (< 1 h).

Along with generating shear stresses, wave-orbital motions also increase fluid pressure within the sediment bed, which, depending on the resistance of the seabed, can result in **liquefaction** (Clukey *et al.*, 1985, Verbeek & Cornelisse, 1997). During a given storm, infiltration pressures oscillate in wave pulses, while shear-induced excess pore-water pressure builds up progressively. Liquefaction may occur as a direct response to the waves, if the sediment grain structure collapses due to excess pore-water pressure. In addition, the critical shear stress for sediment erosion decreases to almost zero and the volume of the transportable sediment under high wave stresses can increase considerably, due to the disintegration of the bed structure.

Sediment liquefaction can easily occur in unconsolidated cohesive sediments with high water content and containing organic matter, such as those found at the head of the Eel canyon (Mullenbach & Nittrouer, 2000). Therefore, development of excess pore-water pressure may cause potential liquefaction of sediment deposited at the head of the canyon and induce transport of sediment as gravity flows (Puig *et al.*, 2004). Additional shear stresses such as those imposed by elevated slopes at the canyon head may help to initiate sediment transport. Liquefaction may provide an alternative mechanism to resuspension from the seabed for inducing fluid-mud suspensions in high-wave-energy regimes. Therefore, the occurrence of storm-induced gravity flows at the head of submarine canyons could be more frequent than previously assumed. Further bottom-boundary-layer measurements (both in the head of the canyon and in the main channel) will be necessary to fully understand contemporary downslope sediment transport through submarine canyons.

ORIGIN AND TRANSFORMATION OF SEDIMENT GRAVITY FLOWS

A continental margin aggrades and progrades its sedimentary deposits over time as affected by changes to its boundary conditions (sea level, base level, sediment supply, ocean energy). As the margin slowly evolves, different transport processes begin to dominate the shaping of a margin. However, the studies examined so far in this paper have focused primarily on separate processes. The integration of the phenomena is vitally important for understanding and modelling observed stratigraphy and downslope transport of material. The need for modelling interactions in the creation and alteration of sediment gravity flows was understood during the STRATAFORM programme and efforts were made to develop innovative strategies for examining these interactions. This section is a summary of that work, as well as the broader vision of a single, integrated numerical model, SEDFLUX, which is capable of producing realistic stratigraphic architecture.

Failure-induced formation

The 1929 Grand Banks slide long ago motivated workers to understand the episodic processes associated with sediment gravity flows (Lee *et al.*, this volume, pp. 213–274). It also motivated what has been called the 'turbidity-current paradigm' – i.e. most transport beyond the slope was caused by failure-induced debris flows and related turbidity currents (Shanmugam & Moiola, 1995). From the first subaqueous debris flows performed at St Anthony Falls Laboratory, it was clear that most, if not all, debris flows (subaqueous failures) produce a turbidity current on their upper surface. Although mixing is strongly suppressed as a result of the large sediment load, shear on the upper surface of a debris flow is generally enough to entrain the finest material from the debris flow into the ambient fluid. The ambient fluid does not possess the stratification due to sediment load, and therefore readily mixes the material entrained into the water column. This material quickly initiates an efficient, gravity-driven, turbulent flow: i.e. a turbidity current.

Considering that failures along the continental margin are common (Lee *et al.*, this volume, pp. 213–274 for details), many submarine turbidity currents are most likely generated from the collapse of previously stable sedimentary deposits. The production of turbidity currents from these events must involve a significant reduction in sediment concentration, from about 40–60% in the failing deposit, to typically < 10% in the turbidity currents themselves. Processes associated with this transformation are incompletely understood in spite of their importance to defining the initial conditions for the currents. Laboratory studies of subaqueous debris flows have helped to constrain some of the conditions for turbidity-current production from submarine slides and slumps (Mohrig *et al.*, 1998; Marr *et al.*, 2001). These measurements have been directed at answering two basic questions. What volume fraction of sediment from the original dense source is worked into an overriding turbidity current? What percentage of the original flow is diluted through the ingestion of ambient seawater with movement downslope?

The degree to which sediment is exchanged between a subaqueous debris flow and its subsidiary turbidity current is related to the coherence

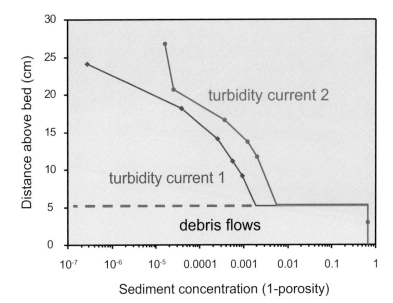

Fig. 31 Sediment concentration profiles for two turbidity currents and their affiliated subaqueous debris flows. A rack of siphons suspended within the water column was used to sample the concentration and composition of suspended sediment in turbidity currents. (From Mohrig & Marr, 2003.)

of the debris flow (i.e. the ability of a slurry to resist breaking apart and becoming fully turbulent under the severe dynamic stresses associated with the head of a debris flow; Marr *et al.*, 2001). **Coherence** describes the extent to which the head of a flow erodes, breaks apart or entrains ambient water for a given dynamic pressure and shear stress. Marr *et al.* (2001) found that the transformation of weakly to moderately coherent debris flows primarily occurs via entrainment of ambient fluid into the slurry, resulting in wholesale dilution. The transformation of moderately to strongly coherent debris flows primarily occurs via grain-by-grain erosion of sediment from the fronts of flows and its subsequent suspension in the overlying water column. While transformation of weakly coherent debris flows is more efficient than that of strongly coherent debris flows, the conversion to turbidity currents was found to be relatively inefficient (i.e. most of the sediment remained in the dense slurry phase). Inefficiency in the transfer of sediment to a subsidiary turbidity current is a consequence of nearly all of the exchange occurring just at the very front of a debris flow. The front is extremely small relative to the total surface area of the flow, but it is the only location on a debris flow that is subjected to significant dynamic stresses. The fluid adjacent to the slurry is accelerated rapidly, reducing stress at other interfacial points, and thus preventing transfer of material from the debris flow to the dilute flow above (Mohrig *et al.*, 1998).

Transformation to turbidity currents by the plucking of individual grains from the heads of strongly coherent debris flows has been studied by Mohrig *et al.* (1998) for an original slurry made up of 34% water, 33% clay and silt, and 33% sand by volume. All grains were quartz; there were no clay minerals. Even for fast-moving subaqueous debris flows travelling a distance ~200 times their average thickness, the fraction of sediment eroded from the debris flow and incorporated in the turbidity current was < 1% of the total volume of sediment in the system. Sediment-concentration profiles for two different turbidity currents and their affiliated debris flows are shown in Fig. 31. These measured profiles show a 100-fold reduction in sediment concentration across the debris-flow–turbidity-current interface. Given that the slurry composition and boundary conditions were not varied, the difference in sediment concentration between the two turbidity currents in Fig. 31 is primarily a function of the debris-flow velocity and hence the front-erosion rate. The turbidity currents were up to approximately six times the average thickness of the associated debris flow. In cases where the debris flow deposited before reaching the end of the flume, the turbidity current would peel off and continue advancing downslope (Fig. 32). At the end of each run, a layer of sediment deposited from the turbidity current was seen mantling the top of the debris flow, as well as any exposed section of the channel bed out in front of the flow. All of these turbidites

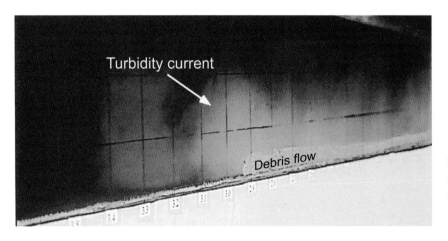

Fig. 32 Turbidity current advancing ahead of the affiliated debris-flow deposit. A 0.1×0.1 m grid is marked on the glass wall of the channel. (From Mohrig & Marr, 2003.)

were enriched in the finer grain sizes relative to the debris flow that was their source (Mohrig *et al.*, 1998; Mohrig & Marr, 2003).

Wave/tide-induced formation

Early examinations of Scripps Canyon, California, illustrated the potential for wave-induced suspension of sand on the inner shelf to trigger turbidity currents where the steep topography of shelf-break canyons dissects the nearshore (Inman *et al.*, 1976). The classic picture of the mangled rebar is an indication that the turbidity-current strength was sufficient to entrain large amounts of material from the seabed and the canyon walls (Fig. 33). These currents were fed from surf-zone resuspension and focusing from infra-gravity waves (i.e. a rip-current cell). Similar canyons along the sandy western margin of southern California and Mexico appear to be formed and supplied in this manner. The upper Eel Canyon and outer portion of the Eel shelf provide examples of deeper shelf settings

where a wave-induced suspension of mud over steep topography may transform into a turbidity current. Model simulations by Scully *et al.* (2002, 2003) suggest that both the seaward and southern boundaries of the Eel shelf flood deposit result from divergent gravity flows that accelerate seaward over increasingly steep bathymetry, possibly transitioning into autosuspending turbidity currents (also see Hill *et al.* (this volume, pp. 49–99) and Syvitski *et al.* (this volume, pp. 459–529)).

On more gently sloping shelves, high-concentration wave- or tide-induced suspensions may accelerate downslope in a manner analogous to an internal bore as strong waves or tidal currents abate (Wright *et al.*, 2001). When waves or tidal currents are strong, mixing near the seabed reduces the velocity at which a gravity current can move downslope (Scully *et al.*, 2002). If ambient currents diminish rapidly, mixing near the seabed may be reduced more quickly than the sediment can settle, and gravity currents may briefly accelerate. On gently sloping shelves such internal bores are short lived. With $\sin \beta < c_D/Ri_{cr}$, shear induced by the gravity current itself cannot compensate for the energy lost from the waves or tides. The inner shelf of the Middle Atlantic Bight provides an example where suspension by wave groups induces short-lived gravity currents of fine sand during storms (Wright *et al.*, 2002). Tripod observations at 12 m depth off Duck, NC, documented periods of accelerated seaward flux for sand 0.1 mab lasting a few tens of seconds during lulls following groups of higher waves. The inner shelf off the mouth of the Huanghe River provides an example where periodic suspension of mud by tidal currents releases metre-thick gravity currents around

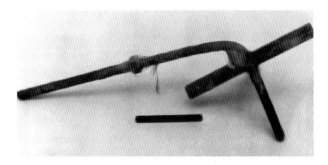

Fig. 33 Bent support pole used in the turbidity-current observations of Inman *et al.* (1976). Measuring stick shown is a standard 12 in (0.3 m) ruler.

slack water that last 2–3 h (Wright *et al.*, 2001). Observations of gravity flows off Duck and the Huanghe River both indicate $Ri > Ri_{cr}$ following slackening of ambient currents, consistent with the rapid collapse of the water column and $\sin \beta < c_D/Ri_{cr}$.

Much work remains in understanding the details of the transformation from a wave-induced suspension to a turbidity current. For example, turbidity currents tend to be coarser and more dilute than critically stratified, wave-supported gravity flows. Liquefaction could also be important in mobilizing material in these flows (Puig *et al.*, 2004). In fact, pore-pressure increase is a common mechanism invoked to explain the production of subaerial debris flows (Iverson, 1997). Work has begun on understanding how a lutocline can form in an energetic environment. Field observations of ancient storm deposits have also given some constraints on the ability of fine-grained flows to generate sediment gravity flows in these settings (Myrow & Hiscott, 1991; Myrow & Southard, 1996). However, no physical experiments have investigated these complicated flows. Observations of active processes in the field have been made (Wheatcroft *et al.*, this volume, pp. 101–155), but it has proven difficult to resolve both the resuspension and the subsequent movement of the resuspended material due to gravity. As a result, physical and numerical experiments will be important for the rectification of not only the flows themselves, but also the deposits that these interesting flows produce.

Direct formation from river loading

River-derived sediment gravity flows, or hyperpycnal plumes, have been hypothesized to occur from the earliest studies of turbidity currents and the deposits they produce (Bell, 1942). **Hyperpycnal plumes** result when the sediment in suspension in a river becomes gravitationally unstable and immediately falls to the seabed due to its collective density being larger than the ambient ocean water. Mulder & Syvitski (1995) first postulated that if concentrations of 40 kg m^{-3} or greater were present in a river, the density of the river water would be greater than the ocean water and a hyperpycnal plume would necessarily result. In this analysis, an extensive data set of the world's rivers was assembled and showed that many small mountainous streams could produce these flows in geological time as a result of the strongly non-linear relationship between sediment entrainment (García & Parker, 1993). Larger rivers, such as the Amazon, could never do so, as the sediment concentrations of their discharges are relatively constant and low (~1 kg m^{-3}).

Recent work on the stability of buoyant layers of sediment-laden fluid has shown that the criterion set forth in Mulder & Syvitski (1995) is conservative, and that sediment-laden surface plumes are generally unstable at lower concentrations (< 1 kg m^{-3}; McCool & Parsons, 2004). The fluid-mechanical process resulting from gravitational instability in a stratified sediment-laden fluid is called **convective sedimentation** (Fig. 34). Maxworthy (1999) was the first to establish that a lock-exchange, fresh, sediment-laden surface plume is unstable when it overrides saline water. In an unrelated study, Hoyal *et al.* (1999) demonstrated that in experiments of sugar-laden water and particles, buoyant surface layers free of sugar, but containing sediment, were universally unstable (with conditions generally being limited to $C_m > 1$ kg m^{-3}). They set forth a stability criterion based upon a critical **Grashof number** (the product of a Reynolds and Froude number). However, neither Hoyal *et al.* (1999) nor Maxworthy (1999) quantified the flux of sediment across the stratified interfaces in their experiments, nor did they identify the reduced critical sediment concentration above which sediment-laden fluid was unstable for typical oceanographic conditions.

As a result, Parsons *et al.* (2001a) attempted to identify the stability field in the absence of turbulence and to compare that to natural riverine-plume processes. Using a device designed to produce an interface without the production of turbulence, Parsons *et al.* (2001a) observed two distinct patterns of convective sedimentation: fingering and leaking. **Fingering** occurred when bulbous plumes emanated from the sediment–saline interface. Fingering was similar to the convective plumes seen by Hoyal *et al.* (1999) and Maxworthy (1999). It was also similar to sedimentary double-diffusion, even though the interfaces examined were stable with respect to this phenomenon (Green, 1987; Parsons & García, 2000). **Leaking** was typified by a series of discrete breaks in the sediment–saline interface, similar to low-Reynolds-number

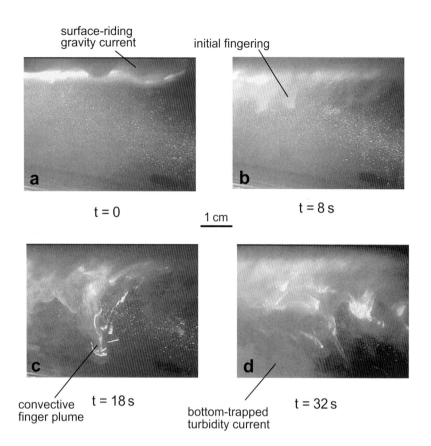

surface-riding
gravity current

initial fingering

a

t = 0

1 cm

b

t = 8 s

c

convective
finger plume

t = 18 s

d

bottom-trapped
turbidity current

t = 32 s

Fig. 34 Sequence of photographs from a strongly scavenged plume. The sequence was obtained from an experiment illustrated in Parsons _et al._ (2001a). The surface-plume sediment concentration $C_m = 12$ kg m^{-3}. Although it is difficult to see in the photographs, a turbidity current moves from the left to right at the bottom of the tank with a velocity in excess of 0.01 m s^{-1}.

multiphase simulations of the upper mantle (Bergantz & Ni, 1999). Leaking was not observed in other vigorously mixed experiments, and therefore Parsons _et al._ (2001a) hypothesized that it would have little application in fully turbulent environments (i.e. natural river plumes). Fingering, on the other hand, was observed in the buoyant turbidity currents of Maxworthy (1999) and Parsons _et al._ (2001a). The criterion for stability with respect to both phenomena was approximately 1 kg m^{-3}, though this limit somewhat depended upon thermal gradients.

Other recent experiments have shown that turbulent mixing intensifies surface-plume instability and reduces the sediment concentration required for the onset of hyperpycnal flow to as little as 0.38 kg m^{-3} in intensely mixed environments. From experimental results, McCool & Parsons (2004) identified the key dimensionless quantity responsible for the vertical flux of sediment associated with convective sedimentation. The removal of sediment from the surface layer was dependent primarily on a single variable, enabling the formulation of a simple predictive relationship

$$\frac{Q}{\rho u} = 1.5 \times 10^5 \frac{C \varepsilon v}{u^4} \tag{17}$$

where Q is the vertical mass flux of sediment associated with convective sedimentation, ρ is the density of the ambient fluid, ε is the viscous dissipation rate, v is the kinematic viscosity of the ambient fluid, C is the volumetric concentration of sediment in the surface layer and u is the characteristic velocity of the surface layer. McCool & Parsons (2004) also used the formulation to predict the removal rates of sediment from Eel River plume. Their results compared favourably with the observations made by Hill _et al._ (2000).

Regardless of the fluid mechanical details of the different experiments, all of these observations indicate that convective-sedimentation processes are robust when concentrations exceed a few grams per litre. The ramifications of this on the dynamics of river plumes are shown schematically in Fig. 35, which marks a striking resemblance to the divergent plumes described by Kineke _et al._ (2000). A **divergent plume** occurs when riverine-derived sediment is simultaneously delivered to a buoy-

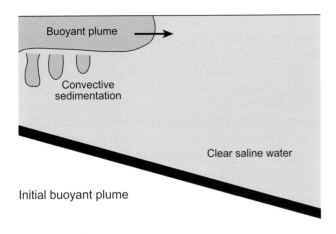

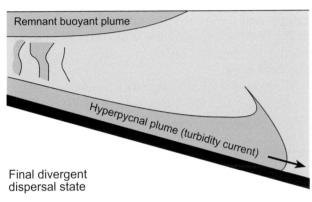

Fig. 35 A schematic diagram of the divergent-plume-formation process observed in the experiments of Parsons *et al.* (2001a) and McCool & Parsons (2004).

ant surface plume and the bottom boundary layer. Kineke *et al.* (2000) based the idea upon field measurements obtained from the Sepik River mouth. The Sepik delivers a large amount of sediment to a steep continental slope in an intensely mixed environment. River-mouth sediment concentrations are typically between 0.1 and 1.0 g m^{-3}, usually exceeding the criterion set forth by McCool & Parsons (2004). Despite the presence of a fresh, sediment-laden surface plume, most of the Sepik sediment delivered to the ocean left within a bottom layer in the form of a sediment gravity flow. The sediment gravity flow was confined to a narrow submarine canyon that intruded into the river mouth. Convective sedimentation provides a reasonable explanation for the divergent nature of the Sepik River mouth.

It is important to note that in the experiments discussed above, particle concentrations could have been locally (i.e. at the interface) enhanced by turbulent processes. **Preferential concentration**

was first suggested by Maxey (1987) to describe the increase in apparent settling velocity in large simulations of particles falling in the presence of isotropic turbulence. He noted that particles falling in a turbulent medium migrate towards areas of diminished vorticity. As a result, they preferentially concentrate, raising the local density, ultimately aiding one another in falling through the fluid. Simulations to date have focused entirely on aerosol dynamics (Hogan & Cuzzi, 2001), but the relevant variable, the Stokes number v_k/w_s, where v_k is the Kolmogorov velocity-scale, is similar in oceanic environments. It remains uncertain whether preferential concentration occurred in the physical experiments mentioned above or whether it is an important mechanism in the natural oceanic environment. If it is, divergent river plumes, and the sediment gravity flows that they produce, may prove to be even more common than currently thought.

Linkages between phenomena

While the laboratory and theoretical (analytical) studies discussed above can isolate the dynamics associated with a few processes simultaneously, it remains difficult for them to deal with all of the sediment transport mechanisms observed in the coastal ocean, particularly when the water-column height is changing. Coupled process–response numerical models provide a solution to this need. This class of models employs a spectrum of sediment-transport mechanisms (fluvial-deltaic deposition, wave and tidal dispersion, alongshore transport, shelf boundary-layer transport, fluid-mud migration, upwelling and downwelling, ocean circulation, hyper- and hypopycnal plume transport, sediment failure, debris-flow and turbidity-current transport, nepheloid transport, internal-wave resuspension, and contour currents). These and other processes are intimately coupled to one another with evolving boundary conditions such as sea-level fluctuations. The resulting deposits therefore range from those formed under the influence of a single dominant process, to those influenced by an admixture of many processes. Integrative models allow scientists to run sensitivity experiments on the multitude of possible realizations in order to cover the natural range of deposit types. These coupled models are therefore the storehouse of sediment-transport knowledge on dynamics, input ranges and conditions, and process interactions. For details on these kinds of numerical models, the reader is referred to Syvitski *et al.* (this volume, pp. 459–529) and Paola (2001).

To understand how integrative models aid understanding of sediment-gravity-flow processes, uncoupled models must first be described, as they are the building blocks to a coupled model. Pratson *et al.* (2000) used single-component numerical models to illustrate the distinction between debris-flows and turbidity-current dynamics and how their contrasting styles of deposition shape a continental margin. The models predict that when begun on a slope that extends onto a basin floor, a debris flow will form a deposit that begins near its point of origin and gradually thicken basinward, ending abruptly at its head. By contrast, deposition from an ignitive turbidity current (i.e. one that causes significant erosion) will largely be restricted to the basin floor and will be separated from its origin on the slope by a zone of no deposition. The modelling effort chose not to worry about a number of phenomena, in order to focus consideration on the comparison of debris-flow-dominated slopes (i.e. arctic margins) to turbidity-current-dominated slopes (i.e. Mediterranean).

In contrast, an integrative model should deal with the phenomena not considered in such an uncoupled modelling approach. First, the sediment source location and characteristics (volume, sediment properties) to be transported by a gravity flow would need to be predicted. Turbidity currents could be introduced either as a hyperpycnal river flood (Imran & Syvitski, 2000; Lee *et al.*, 2002), or from sediment failure. Sediment failure could also create debris flows. Therefore, an integrative model needs to have the facility to simulate river floods that become hyperpycnal and produce excess pore pressures, while simultaneously modelling destabilizing earthquake accelerations, and the evolving properties of the seafloor (bulk density as affected by compaction, grain size, porosity). Second, the integrative model needs to determine the time of occurrence for each sediment failure, and which of the various gravity-flow mechanisms would be appropriate to carry the failed sediment mass. Third, an integrative model should also use routines to reshape existing deposits (e.g. a debris-flow deposit), by currents that flow along the seafloor. Further, an integrative

model should separate individual gravity-flow deposits and hemipelagic deposition, changing both the nature of the seafloor that subsequent gravity flows encounter, and the developing morphology of the margin. Finally, and again for illustrative purposes, the single-component model may use a seafloor-erosion closure scheme based on cohesionless sediment (Pratson *et al.*, 2001), whereas an integrative model should employ more realistic and depth-dependent properties of the seafloor (Skene *et al.*, 1997). For example, Kubo & Nakajima (2002) used the cohesionless sediment-closure scheme of Fukushima *et al.* (1985) when comparing numerical model results with laboratory experiments, and the shear-strength approach of Mulder *et al.* (1998) when using the same turbidity-current model to simulate sediment waves observed at the field scale.

An example of an integrated model is SEDFLUX (Syvitski *et al.*, 1999; Syvitski & Hutton, 2001), which simulates the fill of sedimentary basins (Fig. 36). It can examine the location and attributes of sediment failure on continental margins, and the runout of their associated sediment gravity flows. The model domain interacts with the evolving boundary conditions of sea-level fluctuations, floods, storms, tectonic and other relevant processes to control the rate and size of slope instabilities. By tracking deposit properties (pore pressures, grain size, bulk density and porosity) and their impact by earthquake loading, a finite-slope factor-of-safety analysis of marine deposits examines failure potential. A routine determines whether the failed material will travel downslope as a turbidity current or a debris flow. A full description of SEDFLUX is provided by Syvitski *et al.* (this volume, pp. 459–529).

SEDFLUX has been used to examine how hyperpycnal-flow deposits can interlayer with hemipelagic sedimentation to create the large sediment wave fields located on the margins of many turbidity-current channels, or the wave fields on the slopes off river mouths (Lee *et al.*, 2002). Numerical experiments demonstrate that if a series of turbidity currents flow across a rough seafloor, sediment waves tend to form and migrate upslope. Hemipelagic sedimentation between turbidity current events facilitates this upslope migration of the sediment waves. Morehead *et al.* (2001) used SEDFLUX as a source-to-sink numerical model to evaluate which process dominates the observed variability

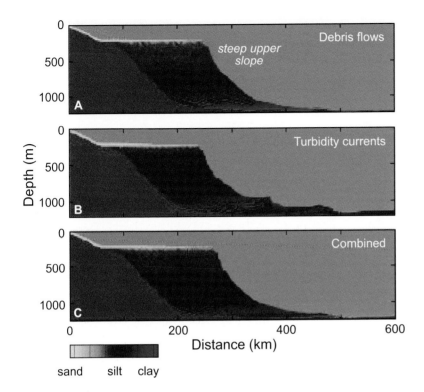

Fig. 36 SEDFLUX realizations of an ideal basin where the sediment supply and sea level were the same in all three runs. Subsidence was not engaged. The time domain of each of the runs was 2 Myr. Sediment failures were allowed to move solely as: (A) debris flows; (B) turbidity currents (with concomitant fan development on the continental rise). (C) The model was allowed to determine which of the gravity-flow transport mechanisms carried the failed material, as determined by the properties of failed sediment mass.

in a sedimentary record in two coastal basins. During the last glacial stage, the Eel River supplied more sediment with a less variable flux to the ocean compared with the modern river, which is dominated by episodic events. Model results show this change in the variability of sediment flux to be as important to the deposit character as is the change in the volume of sediment supply. Due to the complex interaction of flooding events and ocean-storm events, the more episodic flood deposits of recent times are less well preserved than the flood deposits associated with a glacial-stage climate. In Knight Inlet, British Columbia, the evolving boundary conditions (rapidly prograding coastline, secondary transport by gravity flows from sediment failures) are a strong influence on the sedimentary record. Gravity-flow deposits punctuate the sedimentary record otherwise dominated by hemipelagic sedimentation from river plumes. Missing time intervals due to sediment failures take away the advantage of the otherwise amplified lithological record of discharge events, given the enclosed nature of the fjord basin.

O'Grady & Syvitski (2001) used SEDFLUX to evaluate the evolution of the two-dimensional shape of siliciclastic continental slopes by isolating the effect of river plumes, shelf energy, sediment failure, gravity flows, subsidence and sea-level fluctuations on the shape of the slope profile. Simulation results show that hemipelagic sedimentation along with shelf storms produce simple **clinoforms** of varying geometry. Oblique clinoforms are formed in association with low-energy conditions, and sigmoid geometries are associated with more energetic wave conditions. Simulated slope failure steepens the upper continental slope and creates a more textured profile. Topographic smoothing induced by bottom boundary-layer transport enhances the stability of the upper continental slope. Different styles of sediment gravity flows (turbidity currents, debris flows) affect the profile geometry differently (Fig. 36). Debris flows accumulate along the base of the continental slope, leading to slope progradation. Turbidite deposition principally occurs on the basin floor and the continental slope remains a zone of erosion and sediment bypass. Sea-level change and flexural subsidence surprisingly show smaller impacts on profile shape.

CONCLUSIONS

Over the course of the past several years, significant advances have been made in not only the understanding of the mechanics driving sediment gravity flows, but also in the interactions with the many forms these flows can take. Physical experiments are a powerful tool for resolving the processes that transport material under the influence of gravity. From the 'discovery' of hydroplaning to the realization of convective sedimentation processes, new phenomena have been uncovered that have enabled a better understanding of the complexities of the marine sediment record. These results also help us to construct increasingly more accurate and detailed simulations. As illustrated in the previous section, the power of a physics-based, numerical simulation is promising. As computation speeds increase, the ability of these models to produce realistic stratigraphy in arbitrary settings will only improve.

However, there is much work to be done. Many of the experiments described herein have 'discovered' physical processes that are dependent upon variables not traditionally measured in field studies. Commonly these processes are influenced by physics at fundamentally different length and time-scales, which cannot be measured easily or incorporated into a traditional numerical model. For instance, both hydroplaning debris flows and convective sedimentation are regulated by physical processes that have not been observed directly in the field. Developing the models that are capable of incorporating these microscale effects into large-scale numerical models will remain a challenging problem in the years to come. Only when these challenges are met, will numerical models be robust enough to predict stratigraphy in an arbitrary environment.

ACKNOWLEDGEMENTS

A special thanks is given to Joe Kravitz who steadfastly supported much of the research contained within this manuscript. He also provided the unusual opportunity for several of us to concentrate on producing high-quality, scientific theses, through his defence of the now defunct AASERT fellowship programme. Financial support of JDP during the writing of this document was provided by ONR (N00014-03-10138) and NSF (EAR-0309887). Thanks are also due to the many graduate and undergraduate student assistants who loaded sand, mixed sediment and debugged thousands of lines of code during the course of the STRATAFORM programme. The final version of the paper benefited from thorough and insightful reviews by Mike Field, Gail Kineke, Dave Cacchione, Chris Paola, Chuck Nittrouer, Joe Kravitz and an anonymous reviewer.

NOMENCLATURE

Symbol	Definition	Dimensions
A_1, A_2	constants of integration	
B	depth-integrated reduced gravitational acceleration (buoyancy)	$L^2 T^{-2}$
b	local buoyancy anomaly	$L T^{-2}$
C	volumetric concentration	
C_b	near-bed volumetric concentration	
C_m	mass concentration	$M L^{-3}$
C_{max}	maximum volumetric concentration near bed	
c_D	drag coefficient	
D	deposition rate	$L T^{-1}$
E	entrainment rate	$L T^{-1}$
E	dimensionless entrainment rate	
Fr_d	densimetric Froude number	
f_w	wave friction factor	
g	gravitational acceleration	$L T^{-2}$
g'	reduced gravitational acceleration	$L T^{-2}$
h	depth of gravity flow	L
h_0	depth of flow at source	L
h_{cr}	critical height of Bingham material	L
h_f	front height	L
$\hat{i}$	unit vector in x-direction	
$\hat{j}$	unit vector in y-direction	
K	eddy diffusivity	$L^2 T^{-1}$
K_{wall}	eddy diffusivity evaluated in the near-wall region of a wave boundary layer	$L^2 T^{-1}$
$\hat{k}$	unit vector in z-direction	
$k_{act/pass}$	lateral stress coefficient	
k_r	bed roughness	L
L_s	sediment load with a fluid mud	$M L^{-2}$

L_x	length scale in x-direction	L	U	depth-averaged velocity in streamwise direction	L T^{-1}
L_y	length scale in y-direction	L	U_g	velocity associated with gravitational motion in a fluid mud	L T^{-1}
L_z	length scale in z-direction	L	U_f	front velocity	L T^{-1}
l	length of flow at source	L	U_{max}	maximum relevant velocity in a wave-supported gravity current	L T^{-1}
n	exponent in Herschel–Bulkley formulation				
n_p	porosity of bed; volumetric ratio of pores to solids		U_w	root-mean-square velocity at the top of the wave boundary layer	L T^{-1}
P	production of turbulent energy	L^2 T^{-3}	$\vec{u}$	generic velocity vector	
P_{wall}	production of turbulent energy in the near-wall region of a wave boundary layer	L^2 T^{-3}	u_w	root-mean-square velocity (as a function of z) within wave boundary layer	L T^{-1}
p	pressure	M L^{-1} T^{-2}	u	local down-slope velocity	L T^{-1}
Q	mass flux of sediment	M T^{-1}			
q_{mix}	volumetric flux associated with mixing per unit width	L^2 T^{-1}	u'	velocity fluctuations in down-slope velocity	L T^{-1}
q_g	down-slope sediment flux as a function of height above bed z	M L^{-1} T^{-1}	u_∞	velocity at a great distance away from the bed	L T^{-1}
R	submerged specific gravity of sediment		u_*	shear velocity	L T^{-1}
Re	Reynolds number		u_{*cw}	shear velocity associated with both waves and currents	L T^{-1}
Ri	gradient Richardson number		u_{*w}	shear associated with waves only	L T^{-1}
Ri_{cr}	critical gradient Richardson number (usually taken to be $^1/_4$; Howard, 1961)		$\overline{u'w'}$	correlation of streamwise and vertical velocity fluctuations	L^2 T^{-2}
Ri_f	flux Richardson number		V	depth-averaged velocity in flow-perpendicular (spanwise) direction	L T^{-1}
S_s	excess shear stress				
T_{a-e}	Bouma unit a–e		V_c	oceanic current velocity	L T^{-1}
$\mathbf{T}_f$	stress tensor in fluid phase of a debris flow	M L^{-1} T^{-2}	V_f	volume of fluid	L^3
$\mathbf{T}_s$	stress tensor in solid phase of a debris flow	M L^{-1} T^{-2}	V_s	volume of solids	L^3
			v	local spanwise velocity	L T^{-1}
$\mathbf{T}'$	stress tensor associated with relative motions of liquid–solid phase	M L^{-1} T^{-2}	v_k	Kolmogorov velocity scale	L T^{-1}
$T_{s(ij)}$	ijth component of solid-phase stress tensor		w_0	dummy velocity variable used in Grant & Madsen (1979) wave-boundary-layer derivation	
$T_{f(ij)}$	ijth component of fluid-phase stress tensor		w_s	settling velocity of sediment	L T^{-1}
t	time	T	w_e	entrainment velocity	L T^{-1}

x	direction (horizontal) parallel to dominant motion (down-slope, streamwise)	
y	direction (horizontal) perpendicular to dominant motion (spanwise)	
Z	non-dimensional height above the bed (z/δ_w)	
z	direction perpendicular to the seabed	
z_0	roughness height	L
β	bed slope	
γ_0	empirical resuspension parameter in Smith & McLean (1977) entrainment model	
δ_w	thickness of wave boundary layer	L
δ_{wall}	thickness of near-wall region of a fluid mud	L
ε	viscous dissipation rate	$L^2\,T^{-3}$
ϕ_{bed}	friction of bed surface (bed angle of repose)	
ϕ_{int}	angle of internal friction (internal angle of repose)	
η	bed height	m
κ	von Karman constant	
μ	coefficient in Herschel–Bulkley model; dynamic viscosity in Bingham model	$M\,L^{-1}\,T^{-1}$
ν	kinematic viscosity	$L^2\,T^{-1}$
ρ	local density	$M\,L^{-3}$
ρ_0	ambient water density	$M\,L^{-3}$ (1.026 kg m^{-3} for seawater, 1 kg m^{-3} for freshwater)
ρ_1	gravity-flow density	$M\,L^{-3}$
ρ_{sed}	density of sediment particles	$M\,L^{-3}$ (where relevant, assumed 2650 kg m^{-3})
ρ_s	density of solid phase within a debris flow	$M\,L^{-3}$
ρ_f	density of fluid phase within a debris flow	$M\,L^{-3}$
τ_0	yield strength	$M\,L^{-1}\,T^{-2}$

τ_b	bed shear stress	$M\,L^{-1}\,T^{-2}$
τ_{cr}	critical shear stress	$M\,L^{-1}\,T^{-2}$
τ_i	interfacial shear stress on upper surface of gravity flow	$M\,L^{-1}\,T^{-2}$
ζ	dummy variable used in Grant & Madsen (1979) wave-boundary-layer derivation	

REFERENCES

Admiraal, D.M., García, M.H. and Rodriguez, J.F. (2000) Entrainment response of bed sediment to time-varying flows. *Water Resour. Res.*, **36**, 335–348.

Bagnold, R.A. (1962) Auto-suspension of transported sediment; turbidity currents. *Proc. Roy. Soc. London*, **A1322**, 315–319.

Beck, S., Melfi, D. and Yalamanchili, K. (1983) Lateral migration of the Genesee river, New York. In: *River Meanders, Proceedings of the ASCE Conference on Rivers*, pp. 510–517. American Society of Civil Engineers, New York.

Bell, H.S. (1942) Density currents as agents for transporting sediment. *J. Geol.*, **50**, 512–547.

Benjamin, T.B. (1968) Gravity currents and related phenomena. *J. Fluid Mech.*, **31**, 209–248.

Bergantz, G.W. and Ni, J. (1999) A numerical study of sedimentation by dripping instabilities in viscous fluids. *Int. J. Multiphase Flow*, **25**, 307–320.

Bonnecaze, R.T. and Lister, J.R. (1999) Particle-driven gravity currents down planar slopes. *J. Fluid Mech.*, **390**, 75–91.

Bonnecaze, R.T., Hallworth, M.A., Huppert, H.E. and Lister, J.R. (1995) Axisymmetrical particle-driven gravity currents. *J. Fluid Mech.*, **294**, 93–121.

Bouma, A.H. (1962) *Sedimentology of some Flysch Deposits: a Graphic Approach to Facies Interpretation*. Elsevier, Amsterdam.

Bradford, S.F. and Katopodes, N.D. (1999a) Hydrodynamics of turbid underflows. I: Formulation and numerical analysis. *J. Hydraul. Eng.*, **125**, 1006–1015.

Bradford, S.F. and Katopodes, N.D. (1999b) Hydrodynamics of turbid underflows. II: Aggradation, avulsion and channelization. *J. Hydraul. Eng.*, **125**, 1016–1028.

Britter, R.E. and Simpson, J.E. (1978) Experiments on the dynamics of gravity current head. *J. Fluid Mech.*, **88**, 223–240.

Broadwell, J.E. and Breidenthal, R.E. (1982) A simple model of mixing and chemical reaction in a turbulent shear layer. *J. Fluid Mech.*, **125**, 397–410.

Brown, A.L. (1975) Mathematical analysis for pneumatic tire hydroplaning. *Am. Soc. Test. Mater. Spec. Tech. Publ.*, **793**, 75–94.

Buckee, C., Kneller, B. and Peakall, J. (2001) Turbulence structure in steady, solute-driven gravity currents. In: *Particulate Gravity Currents* (Eds W.D. McCaffrey, B.C. Kneller and J. Peakall), pp. 173–188. Special Publication 31, International Association of Sedimentologists. Blackwell Science, Oxford.

Cacchione, D.A., Drake, D.E., Kayen, R.W., *et al.* (1995) Measurements in the bottom boundary layer on the Amazon subaqueous delta. *Mar. Geol.*, **125**, 235–257.

Cacchione, D.A., Wiberg, P.L. Lynch, J. Irish, J. and Traykovski, P. (1999) Estimates of suspended-sediment flux and bedform activity on the inner portion of the Eel continental shelf. *Mar. Geol.*, **154**, 83–97.

Choi, S.-U. (1998) Layer-averaged modeling of two-dimensional turbidity currents with a dissipative-Galerkin finite element method, Part 1: formulation and application example. *J. Hydraul. Res.*, **36**, 339–362.

Choi, S.-U. and García, M.H. (2001) Spreading of gravity plumes on an incline. *Coast. Eng. Jpn*, **43**, 221–237.

Choi, S.-U. and García, M.H. (2002) *k-ε* turbulence modeling of density currents developing two dimensional on a slope. *J. Hydraul. Eng.*, **128**, 55–62.

Clark, J.D. and Pickering, K.T. (1996) Architectural elements and growth patterns of submarine channels: applications to hydrocarbon exploration. *Bull. Am. Assoc. Petrol. Geol.*, **80**, 194–221.

Clukey, E.C., Kulhawy, F.H., Liu, P.L.F. and Tate, G.B. (1985) The impact of wave loads and pore-water pressure generation on initiation of sediment transport. *Geo-Mar. Lett.*, **5**, 177–183.

Costa, J.E. (1984) Physical geomorphology of debris flows. In: *Developments and Applications of Geomorphology* (Eds J.E. Costa and P.J. Fleisher), pp. 269–317. Springer-Verlag, Berlin.

Dade, W.B. and Huppert, H.E. (1998) Long-runout rockfalls. *Geology*, **26**, 803–806.

Denlinger, R.P. and Iverson, R.M. (2001) Flow of variably fluidized granular masses across three-dimensional terrain 2. Numerical predictions and experimental tests. *J. Geophys. Res.*, **106**, 553–566.

Denlinger, R.P. and Iverson R.M. (2004) Granular avalanches across irregular three-dimensional terrain: 1. Theory and computation. *J. Geophys. Res.*, **109**, Art. No. F01014.

Dohmen-Janssen, C.M. and Hanes, D.M. (2002) Sheet flow dynamics under monochromatic nonbreaking waves. *J. Geophys. Res.*, **107**, Art. No. 3149.

Drake, D.E. and Cacchione, D.A. (1985) Seasonal variation in sediment transport on the Russian River shelf, California. *Cont. Shelf Res.*, **4**, 495–514.

Drexler, T.M., Nittrouer, C.A. and Mullenbach, B.L. (2006) Impact of local morphology on sedimentation in a submarine canyon, ROV studies in Eel Canyon. *J. Sediment. Res.*, **76**, 839–853.

Droegemeier, K.K. and Wilhelmson, R.B. (1987) Numerical simulation of thunderstorm outflow dynamics, Part 1: Outflow sensitivity experiments and turbulence dynamics. *J. Atmos. Sci.*, **44**, 1180–1210.

Ellison, T.H. and Turner, J.S. (1959) Turbulent entrainment in stratified flows. *J. Fluid Mech.*, **6**, 423–448.

Elverhøi, A., Norem, H., Andersen, E.S., *et al.* (1997) On the origin and flow behavior of submarine slides on deep-sea fans along the Norwegian Barents Sea continental margin. *Geo-Mar. Lett.*, **17**, 119–125.

Ercilla, G., Alonso, B., Baraza, J., *et al.* (1998) New high-resolution acoustic data from the 'Braided System' of the Orinoco deep sea fan. *Mar. Geol.*, **146**, 243–250.

Feddersen, F., Guza, R.T., Elgar, S. and Herbers, T.H.C. (2000) Velocity moments in along-bottom stress parameterizations. *J. Geophys. Res.*, **105**, 8673–8686.

Felix, M. (2001) A two-dimensional numerical model for a turbidity current. In: *Particulate Gravity Currents* (Eds W.D. McCaffrey, B.C. Kneller and J. Peakall), pp. 71–81. Special Publication 31, International Association of Sedimentologists. Blackwell Science, Oxford.

Flood, R.D., Manley, P.L., Kowsmann, R.O., Appi, C.J. and Pirmez, C. (1991) Seismic facies and late Quaternary growth of Amazon submarine fan. In: *Seismic Facies and Sedimentary Processes of Modern and Ancient Submarine Fans* (Eds P. Weimer and M.H. Link), pp. 415–433. Spring-Verlag, New York.

Foda, M.A. (2003) Role of wave pressure in bedload sediment transport. *J.Waterw. Port Coast. Ocean Eng.*, **129**, 243–249.

Friedrichs, C.T., Wright, L.D., Hepworth, D.A. and Kim, S.C. (2000) Bottom-boundary-layer processes associated with fine sediment accumulation in coastal seas and bays. *Cont. Shelf Res.*, **20**, 807–841.

Fukushima, Y., Parker, G. and Pantin, H.M. (1985) Prediction of ignitive turbidity currents in Scripps Submarine Canyon. *Mar. Geol.*, **67**, 55–81.

García, M.H. (1990) *Depositing and Eroding Turbidity Sediment-driven Flows: Turbidity Currents*. Project Report No. 306, St Anthony Falls Hydraulics Laboratory, University of Minnesota, Minneapolis, 179 pp.

García, M.H. (1993) Hydraulic jumps in sediment-driven bottom currents. *J. Hydraul. Eng.*, **119**, 1094–1117.

García, M.H. (1994) Depositional turbidity currents laden with poorly sorted sediment. *J. Hydraul. Eng.*, **120**, 1240–1263.

García, M. and Parker, G. (1991) Entrainment of bed sediment into suspension. *J. Hydraul. Eng.*, **117**, 414–435.

García, M. and Parker, G. (1993) Experiments on the entrainment of sediment into suspension by a dense bottom current. *J. Geophys. Res.*, **98**, 4793–4807.

García, M.H. and Parsons, J.D. (1996) Mixing at the front of gravity currents. *Dyn. Atmos. Oceans*, **24**, 197–205

Geyer, W.R. and Smith, J.D. (1987) Shear instability in a highly stratified estuary. *J. Phys. Oceanogr.*, **17**, 1668–1679.

Granger, R.A. (1985) *Fluid Mechanics*. CBS College Publishing, New York.

Grant, W.D. and Madsen, O.S. (1979) Combined wave and current interaction with a rough bottom. *J. Geophys. Res.*, **84**, 1797–1808.

Green, T. (1987) The importance of double diffusion to the settling of suspended material. *Sedimentology*, **34**, 319–331.

Hagen, R.A., Bergersen, D., Moberly, R. and Coulbourn, W.T. (1994) Morphology of a large meandering submarine canyon system the Peru–Chile forearc. *Mar. Geol.*, **119**, 7–38.

Hallworth, M.A., Huppert, H.E., Phillips, J.C. and Sparks, R.S.J. (1996) Entrainment into two-dimensional and axisymmetric turbulent gravity currents. *J. Fluid Mech.*, **308**, 289–311.

Harbitz, C., Parker, G., Elverhøi, A., Mohrig, D. and Harff, P. (2003) Hydroplaning of debris glide blocks: analytical solutions and discussion. *J. Geophys. Res. (Solid Earth)*, **108**, Art. No. 2349.

Harris, C.K. and Wiberg, P.L. (2001) A two-dimensional, time-depedent model of suspended sediment transport and bed reworking for continental shelves. *Comput. Geosci.*, **27**, 675–690.

Härtel, C., Meiburg, E. and Necker, F. (2000a) Analysis and direct numerical simulation of the flow at a gravity-current head, Part 1. Flow topology and front speed for slip and no-slip boundaries. *J. Fluid Mech.*, **418**, 189–212.

Härtel, C., Carlsson, F. and Thunblom, M. (2000b) Analysis and direct numerical simulation of the flow at a gravity-current head, Part 2. The lobe-and-cleft instability. *J. Fluid Mech.*, **418**, 213–229.

Hay, A.E. (1987) Turbidity currents and submarine channel formation in Rupert Inlet, British Columbia, Part 2: the roles of continuous and surge type flow. *J. Geophys. Res.*, **92**, 2883–2900.

Heezen, B.C. and Ewing, W.M. (1952) Turbidity currents and submarine slumps and the 1929 Grand Banks earthquake. *Am. J. Sci.*, **250**, 849–873.

Hesse, R. (1995) Long-distance correlation of spill-over turbidites on the western levee of the Northwest Atlantic Mid-Ocean Channel (NAMOC), Labrador Sea. In: *Atlas of Deep Water Environments: Architectural Style in Turbidite Systems* (Eds K.T. Pickering, R.N. Hiscott, N.H. Kenyon, F. Ricci Lucchi and R.D. Smith), pp. 276–281. Chapman and Hall, London.

Hill, P.S., Milligan, T.G. and Geyer, W.R. (2000) Controls on effective settling velocity of suspended sediment in the Eel River flood plume. *Cont. Shelf Res.*, **20**, 2095–2111.

Hine, A.C., Locker, S.D., Tedesco, L.P., Mullins, H.T., Hallock, P., Belknap, D.F., Gonzales, J.L., Neumann, A.C. and Snyder, S.W. (1992) Megabreccia shedding from modern low-relief carbonate platforms, Nicaraguan Rise. *Geol. Soc. Am. Bull.*, **104**, 928–943.

Hiscott, R.N., Hall, F.R. and Pirmez, C. (1997) Turbidity-current overspill from the Amazon channel: Texture of the silt/sand load, pale flow from anisotropy of magnetic susceptibility and implications for flow processes. *Proc. ODP Sci. Res.*, **155**, 53–78.

Hogan, R.C. and Cuzzi, J.N. (2001) Stokes and Reynolds number dependence of preferential concentration in simulated three-dimensional turbulence. *Phys. Fluids*, **13**, 2938–2945.

Howard, A.D. (1992) Modeling channel migration and floodplain sedimentation in meandering streams. In: *Lowland Floodplain Rivers, Geomorphological Perspectives* (Eds P.A. Carling and G.E. Petts), pp. 1–41. John Wiley & Sons, New York.

Howard, L.N. (1961) Note on a paper from John W. Miles. *J. Fluid Mech.*, **10**, 509–512.

Hoyal, D.C.J.D., Bursik, M.I. and Atkinson, J.F. (1999) Settling-driven convection: a mechanism of sedimentation from stratified fluids. *J. Geophys. Res.*, **104**, 7953–7966.

Hsu, T.J. and Hanes, D.M. (2004) Effects of wave shape on sheet flow sediment transport. *J. Geophys. Res.*, **109**, Art. No. C05025.

Huang, H.Q., Imran, J. and Pirmez, C. (2005) Numerical model of turbidity currents with a deforming bottom boundary. *J. Hydraul. Eng.*, **131**, 283–293.

Huang, X. and García, M.H. (1997) Perturbation solution for Bingham-plastic mudflows. *J. Hydraul. Eng.*, **123**, 986–994.

Huang, X. and García, M.H. (1998) A Herschel–Bulkley model for mud flow down a slope. *J. Fluid Mech.*, **374**, 305–333.

Huang, X. and García, M.H. (1999) Modeling of non-hydroplaning mudflows on continental slopes. *Mar. Geol.*, **154**, 131–142.

Hunt, B. (1994) Newtoninan fluid mechanics treatment of debris flows and avalanches. *J. Hydraul. Eng.*, **120**, 1350–1363.

Huppert, H.E. and Simpson, J.E. (1980) Slumping of gravity currents. *J. Fluid Mech.*, **99**, 785–799.

Ikeda, S., Parker, G. and Sawai, K. (1981) Bend theory of river meanders, Part 1: linear development. *J. Fluid Mech.*, **112**, 363–377.

Imran, J. and Syvitski, J.P.M. (2000) Impact of extreme river events on the coastal ocean. *Oceanography*, **13**, 85–92.

Imran, J., Parker, G. and Katapodes, N. (1998) A numerical model of channel inception on submarine fans. *J. Geophys. Res.*, **103**, 1219–1238.

Imran, J., Parker, G. and Pirmez, C. (1999) A nonlinear model of flow in meandering submarine and subaerial channels. *J. Fluid Mech.*, **400**, 295–331.

Imran, J., Harff, P. and Parker, G. (2001a) A numerical model of submarine debris flows with graphical user interface. *Comput. Geosci.*, **27**, 717–729.

Imran, J., Parker, G., Locat, J. and Lee, H. (2001b) A 1-D numerical model of muddy subaqueous and sub-aerial debris flows. *J. Hydraul. Eng.*, **127**, 959–968.

Imran, J., Kassem, A. and Khan, S.M. (2004) Three-dimensional modeling of density current. I. Flow in straight confined and unconfined channels. *J. Hydraul. Res.*, **42**, 578–590.

Inman, D.L., Nordstrum, C.E. and Flick, R.E. (1976) Currents in submarine canyons: an air–sea–land interaction. *Ann. Rev. Fluid Mech.*, **8**, 275–310.

Iverson, R.M. (1997) The physics of debris flows. *Rev. Geophys.*, **35**, 245–296.

Iverson, R.M. and Denlinger, R.P. (2001) Flow of variably fluidized granular masses across three-dimensional terrain 1. Coulomb mixture theory. *J. Geophys. Res.*, **106**, 537–552.

Jiang, L. and LeBlond, P.H. (1993) Numerical modeling of an underwater Bingham plastic mudslide and the wave which it generates. *J. Geophys. Res.*, **98**, 10303–10317.

Johannesson, H. (1988) *Theory of river meanders*. Unpublished PhD thesis, Department of Civil Engineering, University of Minnesota.

Johannesson, H. and Parker, G. (1989a) Linear theory of river meanders. In: *River Meandering* (Eds S. Ikeda and G. Parker), pp. 181–213. American Geophysical Union, Washington, DC.

Johanneson, H. and Parker, G. (1989b) Secondary flow in mildly sinuous channel. *J. Hydraul. Eng.*, **115**, 289–308.

Johnson, A.M. (1965) *A model for debris flow*. Unpublished PhD thesis, Pennsylvania State University, University Park.

Johnson, K.S., Paull, C.K., Barry, J.P. and Chavez, F.P. (2001) A decadal record of underflows from a coastal river into the deep sea. *Geology*, **29**, 1019–1022.

Kassem, A. and Imran, J. (2001) Simulation of turbid underflow generated by the plunging of a river. *Geology*, **29**, 655–658.

Kassem, A. and Imran, J. (2004) Three-dimensional modeling of density current. II. Flow in sinuous confined and unconfined channels. *J. Hydraul. Res.*, **42**, 591–602.

Keulegan, G.H. (1957a) *An Experimental Study of the Motion of Saline Water from Locks into Fresh Water Channels*. Report 5168, US National Bureau of Standards, Washington, DC.

Keulegan, G.H. (1957b) *Form Characteristics of Arrested Saline Wedges*. Report 5482, US National Bureau of Standards, Washington, DC.

Khripounoff, A., Vangriesheim, A., Babonneau, N., *et al.* (2003) Direct observation of intense turbidity current activity in the Zaire submarine valley at 4000 m water depth. *Mar. Geol.*, **194**, 151–158.

Kineke, G.C. and Sternberg, R.W. (1992) Measurements of high concentration suspended sediments using the optical backscatterance sensor. *Mar. Geol.*, **108**, 253–268.

Kineke, G.C. and Sternberg, R.W. (1995) Distribution of fluid muds on the Amazon continental shelf. *Mar. Geol.*, **125**, 193–233.

Kineke, G.C., Sternberg, R.W., Trowbridge, J.H. and Geyer, W.R. (1996) Fluid-mud processes on the Amazon continental shelf. *Cont. Shelf Res.*, **16**, 667–696.

Kineke, G.C., Woolfe, K.J., Kuehl, S.A., *et al.* (2000) Sediment export from the Sepik River, Papua New Guinea: Evidence for a divergent dispersal system. *Cont. Shelf Res.*, **20**, 2239–2266.

Klaasen G.P. and Peltier, W.R. (1985) The onset of turbulence in finite amplitude Kelvin-Helmholtz billows. *J. Fluid Mech.*, **155**, 1–35.

Klaucke, I. and Hesse, R. (1996) Fluvial features in the deep-sea: new insights from the glacigenic submarine drainage system of the Northwest Atlantic Mid-Ocean Channel in the Labrador Sea. *Sediment. Geol.*, **106**, 223–234.

Klaucke, I., Hesse, R. and Ryan, W.B.F. (1998) Morphology and structure of a distal submarine trunk channel: the North-West Atlantic Mid-Ocean Channel between lat 53°N and 44°30′N. *Geol. Soc. Am. Bull.*, **110**, 22–34.

Kneller, B. (1995) Beyond the turbidite paradigm: Physical models for deposition of turbidites and their implications for reservoir prediction. In: *Characterization of Deep Marine Clastic Systems* (Eds A.J. Hartley and D.J. Prosser), pp. 31–49. Special Publication 94, Geological Society Publishing House, Bath.

Kneller, B. and Buckee, C. (2000) The structure and fluid mechanics of turbidity currents: a review of some recent studies and their geological implications. *Sedimentology*, **47**, 62–94.

Kneller, B. and McCaffrey, B. (1995) Modelling the effects of salt-induced topography on deposition from turbidity currents. In: *GCSSEPM Foundation 16th Annual Research Conference Proceedings*, pp. 137–145.

Kneller, B.C. and McCaffrey, W.D. (2003) The interpretation of vertical sequences in turbidite beds: the influence of longitudinal flow structure. *J. Sediment. Res.*, 73, 706–713.

Kolla, V., Bourges, P., Urruty, J. and Safa, P. (2001) Evolution of deep-water tertiary sinuous channels offshore Angola (West Africa) and implications for reservoir architecture. *Bull. Am. Assoc. Petrol. Geol.*, 85, 1373–1405.

Komar, P.D. (1969) The channelized flow of turbidity currents with application to Monterey deep-sea fan channel. *J. Geophys. Res.*, 74, 4544–4557.

Krone, R.B. and Wright, V. (1987) *Laboratory and Numerical Study of Mud and Debris Flow*. Report of the Department of Civil Engineering, University of California, Davis, pp. 1–2.

Kubo, Y. and Nakajima, T. (2002) Laboratory experiments and numerical simulation of sediment wave formation by turbidity currents. *Mar. Geol.*, 192, 105–121.

Kuenen, Ph.H. (1965) Experiments in connection with turbidity currents and clay suspensions. In: *Submarine Geology and Geophysics* (Eds W.F. Whittard and R. Bradshaw), pp. 47–74. Butterworths, London.

Kuenen, Ph.H. and Migliorini, C.I. (1950) Turbidity currents as a cause of graded bedding. *J. Geol.*, 58, 91–127.

Lamb, M.P. and Parsons, J.D. (2005) High-density suspensions formed under waves. *J. Sediment. Res.*, 75, 386–397.

Lamb, M.P., D'Asaro E. and Parsons, J.D. (2004a) Turbulent structure of high-density suspensions formed under waves. *J. Geophys. Res.*, 109, Art. No. C12026.

Lamb, M.P., Hickson, T., Marr, J.G., et al. (2004b) Surging versus continuous turbidity currents: Flow dynamics and deposits in an experimental intraslope minibasin. *J. Sediment. Res.*, 74, 148–155.

Lavelle, J.W. and Mofjeld, H.O. (1987) Do critical stresses for incipient motion and erosion really exist? *J. Hydraul. Eng.*, 113, 370–385.

Lee, H.Y. and Yu, W.S. (1997) Experimental study of reservoir turbidity current. *J. Hydraul. Eng.*, 123, 520–528.

Lee, H.J., Syvitski, J.P.M., Parker, G., et al. (2002) Turbidity-current generated sediment waves: modeling and field examples. *Mar. Geol.*, 192, 79–104.

Lingel, S.L. (1997) *Scaling effects on the mixing process of lock-exchange gravity currents*. Unpublished PhD thesis, University of Washington, Seattle.

Lipman, P.W., Normark, W.R., Moore, J.G., Wilson, J.B. and Gutmacher, C.E. (1988) The giant submarine Alika Debris Slide: Mauna Loa, Hawaii. *J. Geophys. Res.*, 93, 4279–4299.

Luthi, S. (1981) Experiments on non-channelized turbidity currents and their deposits. *Mar. Geol.*, 40, M59-M68.

Madsen, O.S. and Wikramanayake, P.N. (1991) *Simple Models for Turbulent Wave-currents Bottom Boundary Layer Flow*. Contract Report DRP-91-1, Coastal Engineering Research Center, US Army Corps of Engineers, Vicksberg, MS.

Major, J.J. and Iverson, R.M. (1999) Debris-flow deposition: Effects of pore-fluid pressure and friction concentrated at flow margins. *Geol. Soc. Am. Bull.*, 111, 1424–1434.

Major, J.J. and Pierson, T.C. (1992) Debris flow rheology: Experimental analysis of fine grained slurries. *Water Resour. Res.*, 28, 841–857.

Marr, J.G., Harff, P.A., Shanmugam, G. and Parker, G. (2001) Experiments on subaqueous sandy gravity flows: the role of clay and water content in flow dynamics and depositional structures. *Geol. Soc. Am. Bull.*, 113, 1377–1386.

Maxey, M.R. (1987) The gravitational settling of aerosol particles in homogeneous turbulence and random flow fields. *J. Fluid Mech.*, 174, 442–465.

Maxworthy, T. (1999) The dynamics of sedimenting surface gravity currents. *J. Fluid Mech.*, 392, 27–44.

McCool, W.W. and Parsons, J.D. (2004) Sedimentation from buoyant fine-grained suspensions. *Cont. Shelf Res.*, 24, 1129–1142.

Mehta, A.J. (1989) On estuarine cohesive sediment suspension behavior. *J. Geophys. Res.*, 94, 14303–14314.

Mei, C.C. and Liu, K.F. (1987) A Bingham-plastic model for a muddy seabed under long waves. *J. Geophys. Res.*, 92, 14581–14594.

Mellor, G.L. and Yamada, T. (1974) Hierarchy of turbulence closure models for planetary boundary layers. *J. Atmos. Sci.*, 31, 1791–1806.

Mellor, G.L. and Yamada, T. (1982) Development of a turbulence closure model for geophysical fluid problems. *Rev. Geophys., Space Phys.*, 20, 851–875.

Melosh, H.J. (1979) Acoustic fluidization – new geologic process. *J. Geophys. Res.*, 84, 7513–7520.

Middleton, G.V. (1966) Experiments on density and turbidity currents: Motion of the head. *Can. J. Earth Sci.*, 3, 523–546.

Mohrig, D. and Marr, J.G. (2003) Constraining the efficiency of turbidity current generation from submarine debris flows and slides using laboratory experiments. *Mar. Petrol. Geol.*, 20, 883–899.

Mohrig, D., Whipple, K.X., Hondzo, M., Ellis, C. and Parker, G. (1998) Hydroplaning of subaqueous debris flows. *Geol. Soc. Am. Bull.*, 110, 387–394.

Mohrig, D., Elverhøi, A. and Parker, G. (1999) Experiments on the relative mobility of muddy subaqueous and subaerial debris flows and their capacity to remobilize antecedent deposits. *Mar. Geol.*, **154**, 117–129.

Morehead, M., Syvitski, J.P. and Hutton, E.W.H. (2001) The link between abrupt climate change and basin stratigraphy: a numerical approach. *Global Planet. Change*, **28**, 115–135.

Mulder, T. and Syvitski, J.P.M. (1995) Turbidity currents generated at river mouths during exceptional discharges to the world oceans. *J. Geol.*, **103**, 285–299.

Mulder, T., Syvitski, J.P.M. and Skene, K. (1998) Modeling of erosion and deposition by sediment gravity flows generated at river mouths. *J. Sediment. Res.*, **67**, 124–137.

Mullenbach, B.L. (2002) *Characterization of modern off-shelf sediment export on the Eel margin, northern California.* Unpublished PhD thesis, University of Washington, Seattle.

Mullenbach, B.L. and Nittrouer, C.A. (2000) Rapid deposition of fluvial sediment in the Eel Canyon, northern California. *Cont. Shelf Res.*, **20**, 2191–2212.

Myrow, P.M. and Hiscott, R.N. (1991) Shallow-water gravity-flow deposits, Chapel Island Formation, southeast Newfoundland, Canada. *Sedimentology*, **38**, 935–959.

Myrow, P.M. and Southard, J.B. (1996) Tempestite deposition. *J. Sediment. Res.*, **66**, 875–887.

Nayfeh, A.H. (1973) *Perturbation Methods.* John Wiley & Sons, New York.

Nissen, S.E., Haskell, N.L., Steiner, C.T. and Coterill, K.L. (1999) Debris flow outrunner blocks, glide tracks and pressure ridges identified on the Nigerian continental slope using 3-D seismic coherency. *Leading Edge*, **18**, 595–599

Nittrouer, C.A. and Kravitz, J.H. (1996) STRATAFORM: a program to study the creation and interpretation of sedimentary strata on continental margins. *Oceanography*, **9**, 146–152.

Normark, W.R. and Serra, F. (2001) Vertical tectonics in northern Escanaba Trough as recorded by thick late Quaternary turbidites. *J. Geophys. Res.*, **106**, 13793–13802.

Normark, W.R., Posamentier, H. and Mutti, E. (1993) Turbidite systems: State of the art and future directions. *Rev. Geophys.*, 31, 91–116.

Ogston, A.S., Cacchione, D.A., Sternberg, R.W. and Kineke, G.C. (2000) Observations of storm and river flood-driven sediment transport on the northern California continental shelf. *Cont. Shelf Res.*, **20**, 2141–2162.

O'Grady, D.B. and Syvitski, J.P.M. (2001) Predicting profile geometry of continental slopes with a multi-process sedimentation model. In: *Geological Modeling and Simulation: Sedimentary Systems* (Eds D.F. Merriam and J.C. Davis), pp. 99–117. Kluwer Academic/Plenum Publishers, New York.

Pantin, H.M. (2001) Experimental evidence for autosuspension. In: *Particulate Gravity Currents* (Eds W.D. McCaffrey, B.C. Kneller and J. Peakall), pp. 189–205. Special Publication 31, International Association of Sedimentologists. Blackwell Science, Oxford.

Paola, C. (2001) Quantitative models of sedimentary basin filling. *Sedimentology*, **47**, 121–178.

Parker, G. (1982) Conditions for the ignition of catastrophically erosive turbidity currents. *Mar. Geol.*, **46**, 307–327.

Parker, G., Fukushima, Y. and Pantin, H.M. (1986) Self-accelerating turbidity currents. *J. Fluid Mech.*, **171**, 145–181.

Parker, G., García, M., Fukushima, Y. and Yu, W. (1987) Experiments on turbidity currents over an erodible bed. *J. Hydraul. Res.*, **25**, 123–147.

Parsons, J.D. (1998) *Mixing mechanisms in density intrusions.* Unpublished PhD thesis, University of Illinois, Urbana-Champaign.

Parsons, J.D. and García, M.H. (1998) Similarity of gravity current fronts. *Phys. Fluids*, **10**, 3209–3213.

Parsons, J.D. and García, M.H. (2000) Enhanced sediment scavenging due to double-diffusive convection. *J. Sediment. Res.*, **70**, 47–52.

Parsons, J.D., Bush, J.W.M. and Syvitski, J.P.M. (2001a) Hyperpycnal plumes with small sediment concentrations. *Sedimentology*, **48**, 465–478.

Parsons, J.D., Whipple, K.X. and Simoni, A. (2001b) Laboratory experiments of the grain-flow, fluid-mud transition in well-graded debris flows. *J. Geol.*, **109**, 427–447.

Parsons, J.D., Schweller, W.J., Stelting, C.W., et al. (2002) A preliminary experimental study of turbidite fans. *J. Sediment. Res.*, **72**, 619–628.

Paull, C.K, Ussler, W., Greene, H.G., et al. (2003) Caught in the act: the 20 December 2001 gravity flow event in Monterey Canyon. *Geo-Mar. Lett.*, **22**, 227–232.

Peakall, J., McCaffrey, B. and Kneller, B. (2000) A process model for the evolution and architecture of sinuous submarine channels. *J. Sediment. Res.*, 70, 434–448.

Piper, D.J.W. and Normark, W.R. (1983) Turbidite depositional patterns and flow characteristics, Navy submarine fan, California Borderland. *Sedimentology*, **30**, 681–694.

Pirmez, C. (1994) *Growth of a submarine meandering channel-levee system on the Amazon Fan.* Unpublished PhD thesis, Columbia University, New York.

Pratson, L., Imran, J., Parker, G., Syvitski, J.P. and Hutton, E. (2000) Debris flow versus turbidity currents: a modeling comparison of their dynamics and deposits. In: *Fine-Grained Turbidite Systems* (Eds A.H. Bouma and C.G. Stone), pp. 57–71. Memoir 72, American Association of Petroleum Geologists; Special Publication 68, Society for Sedimentary Geology, Tulsa, OK.

Pratson, L.F., Imran, J., Hutton, E., Parker, G. and Syvitski, J.P.M. (2001) BANG1D: a one-dimensional Lagrangian model of turbidity current mechanics. *Comput. Geosci.*, **27**, 701–716.

Prior, D.B., Bornhold, B.D. and Johns, M.W. (1984) Depositional characteristics of a submarine debris flow. *J. Geol.*, **92**, 707–727.

Puig, P., Ogston, A.S., Mullenbach, B.L., Nittrouer, C.A. and Sternberg, R.W. (2003) Shelf-to-canyon sediment-transport processes on the Eel continental margin (northern California). *Mar. Geol.*, **193**, 129–149.

Puig, P., Ogston, A.S., Mullenbach, B.L., *et al.* (2004) Storm-induced density-driven currents at the head of the Eel submarine canyon. *J. Geophys. Res.*, **109**, Art. No. C03019

Reed, C.W., Niedoroda, A.W. and Swift, D.J.P. (1999) Modeling sediment entrainment and transport processes limited by bed armoring. *Mar. Geol.*, **154**, 143–154.

Salaheldin, T.M., Imran, J., Chaudhry, M.H. and Reed, C. (2000) Role of fine-grained sediment in turbidity current flow dynamics and resulting deposits. *Mar. Geol.*, **171**, 21–38.

Schowalter, D.G., Van Atta, C.W. and Lasheras, J.C. (1994) A study of streamwise vortex structure in a stratified shear layer. *J. Fluid Mech.*, **281**, 247–281.

Scully, M.E., Friedrichs, C.T. and Wright, L.D. (2002) Application of an analytical model of critically stratified gravity-driven sediment transport and deposition to observations from the Eel River continental shelf, northern California. *Cont. Shelf Res.*, **22**, 1951–1974.

Scully, M.E., Friedrichs, C.T. and Wright, L.D. (2003) Numerical modeling of gravity-driven sediment transport and deposition on an energetic continental shelf: Eel River, northern California. *J. Geophys. Res.*, **108**, Art. No. 3120.

Seminara, G. and Tubino, M. (1989) Alternate bars and meandering: free, forced and mixed interaction. In: *River Meandering* (Eds S. Ikeda and G. Parker), pp. 267–320, American Geophysical Union, Washington, DC.

Shanmugam, G. (1997) The Bouma Sequence and the turbidite mind set. *Earth-Sci. Rev.*, **42**, 201–229.

Shanmugam, G. and Moiola, R.J. (1995) Reinterpretation of depositional processes in a classic flysch sequence (Pennsylvanian Jackfork Group), Ouachita mountains, Arkansas. *Bull. Am. Assoc. Petrol. Geol.*, **79**, 672–695.

Simpson, J.E. and Britter, R.E. (1979) The dynamics of the head of a gravity current advancing over a horizontal surface. *J. Fluid Mech.*, **94**, 477–495.

Skene, K., Mulder, T. and Syvitski, J.P.M. (1997) INFLO1: a model predicting the behaviour of turbidity currents generated at a river mouth. *Comput. Geosci.*, **23**, 975–991.

Smith, J.D. and McLean, S.R. (1977) Spatially averaged flow over a wavy surface. *J. Geophys. Res.*, **82**, 1735–1746.

Storlazzi, C.D. and Jaffe, B.E. (2002) Flow and sediment suspension events on the inner shelf of central California. *Mar. Geol.*, **181**, 195–213.

Sullivan, G.D. and List, E.J. (1994) On mixing and transport at a sheared density interface. *J. Fluid Mech.*, **273**, 213–239.

Sun, T., Meakin, P. and Jøssang, T. (1996) A simulation model for meandering rivers. *Water Resour. Res.*, **32**, 2937–2954.

Syvitski, J.P.M. and Hutton, E.H. (2001) 2D SEDFLUX 1.0C: an advanced process-response numerical model for the fill of marine sedimentary basins. *Comput. Geosci.*, **27**, 731–754.

Syvitski, J.P.M., Pratson, L. and O'Grady, D. (1999) Stratigraphic Predictions of Continental Margins for the Navy. In: *Numerical Experiments in Stratigraphy: Recent Advances in Stratigraphic and Computer Simulations* (Eds J. Harbaugh, L. Watney, G. Rankey, *et al.*), pp. 219–236. Memoir 62, Society of Economic Paleontologists and Mineralogists, Tulsa, OK.

Thorpe, S.A. (1971) Experiments on the instability of stratified shear flows: miscible fluids. *J. Fluid Mech.*, **46**, 299–319.

Thorpe, S.A. (1973) Experiments on instability and turbulence in a stratified shear flow. *J. Fluid Mech.*, **61**, 731–751.

Toniolo, H., Harff, P., Marr, J., Paola, C. and Parker, G. (2004) Experiments on reworking by successive unconfined subaqueous and subaerial muddy debris flows, *J. Hydraul. Eng.*, **130**, 38–48.

Traykovski, P., Geyer, W.R., Irish, J.D. and Lynch, J.F. (2000) The role of wave-induced density-driven fluid mud flows for cross-shelf transport on the Eel River continental shelf. *Cont. Shelf Res.*, **20**, 2113–2140.

Trowbridge, J.H. (1992) A simple description of the deepening and structure of a stably stratified flow driven by a surface stress. *J. Geophys. Res.*, **97**, 15529–15543.

Trowbridge, J.H. and Kineke, G.C. (1994) Structure and dynamics of fluid muds on the Amazon continental shelf. *J. Geophys. Res.*, **99**, 865–874.

Trowbridge, J.H. and Lentz, S.J. (1998) Dynamics of the bottom boundary layer on the northern California shelf. *J. Phys. Oceanogr.*, **28**, 2075–2093.

Verbeek, H. and Cornelisse, J.M. (1997) Erosion and liquefaction of natural mud under surface waves. In: *Cohesive Sediments* (Eds N. Burt, R. Parker and J. Watts), pp. 353–363. John Wiley & Sons, New York.

Walsh, J.P., Nittrouer, C.A., Palinkas, C.M., *et al.* (2004) Clinoform mechanics in the Gulf of Papua, New Guinea, *Cont. Shelf Res.*, **24**, 2487–2510.

Wells, J.T. (1983) Dynamics of coastal fluid muds in low-, moderate-, high-tide-range environments. *Can. J. Fish. Aquat. Sci.*, **40**, 130–142.

Whipple, K.X. and Dunne, T. (1992) The influence of debris-flow rheology on fan morphology, Owens Valley, California. *Geol. Soc. Am. Bull.*, **104**, 887–900.

Wiberg, P.L. (1995) A theoretical investigation of boundary-layer flow and bottom shear-stress for smooth, transitional and rough flow under waves. *J. Geophys. Res.*, **100**, 22667–22679.

Wiberg, P.L., Drake, D.E., Harris, C.K. and Noble, M. (2002) Sediment transport on the Palos Verdes shelf over seasonal to decadal time scales. *Cont. Shelf Res.*, **22**, 987–1004.

Wright, L.D. and Friedrichs, C.T. (2006) Gravity driven sediment transport on continental shelves: a status report. *Cont. Shelf Res.*, **26**, 2092–2107.

Wright, L.D., Wiseman, W.J.Jr., Yang, Z.-S., *et al.* (1990) Processes of marine dispersal and deposition of suspended silts off the modern mouth of the Huanghe (Yellow) River. *Cont. Shelf Res.*, **10**, 1–40.

Wright, L.D., Friedrichs, C.T., Kim, S.C. and Scully, M.E. (2001) Effects of ambient currents and waves on gravity-driven sediment transport on continental shelves. *Mar. Geol.*, **175**, 25–45.

Wright, L.D., Friedrichs, C.T. and Scully, M.E. (2002) Pulsational gravity-driven sediment transport on two energetic shelves. *Cont. Shelf Res.*, **22**, 2443–2460.

Xu, J.P., Noble, M.A. and Rosenfeld, L.K. (2004) *In-situ* measurements of velocity structure within turbidity currents. *Geophys. Res. Lett.*, **31**, Art. No. L09311.

Seascape evolution on clastic continental shelves and slopes

LINCOLN F. PRATSON*, CHARLES A. NITTROUER†, PATRICIA L. WIBERG‡,
MICHAEL S. STECKLER§, JOHN B. SWENSON¶, DAVID A. CACCHIONE**,
JEFFERY A. KARSON*, A. BRADLEY MURRAY*, MATTHEW A. WOLINSKY*,
THOMAS P. GERBER*, BETH L. MULLENBACH††, GLENN A. SPINELLI‡‡,
CRAIG S. FULTHORPE§§, DAMIAN B. O'GRADY¶¶, GARY PARKER***, NEAL W.
DRISCOLL†††, ROBERT L. BURGER‡‡‡, CHRISTOPHER PAOLA§§§, DANIEL L. ORANGE¶¶¶,
MICHAEL E. FIELD****, CARL T. FRIEDRICHS†††† *and* JUAN J. FEDELE§§§

*Earth & Ocean Sciences, Nicholas School of the Environment and Earth Sciences, Duke University, Durham, NC 27708, USA
(Email: lincoln.pratson@duke.edu)
†School of Oceanography, University of Washington, Seattle, WA 98195, USA
‡Department of Environmental Sciences, University of Virginia, Charlottesville, VA 22904, USA
§Lamont-Doherty Earth Observatory, Columbia University, Palisades, NY 10964, USA
¶Department of Geological Sciences, University of Minnesota Duluth, Duluth, MN 55812, USA
**Coastal & Marine Environments, Redwood City, CA 94065, USA
††Department of Oceanography, Texas A&M University, College Station, TX 77843, USA
‡‡Department of Geological Sciences, University of Missouri, Columbia, MO 65211, USA
§§Institute for Geophysics, University of Texas, Austin, TX 78759, USA
¶¶ExxonMobil Development Co., Houston, TX 77252, USA
***Department of Civil Engineering, University of Illinois, Urbana-Champagne, IL 61801 USA
†††Geosciences Research Division, Scripps Institution of Oceanography, La Jolla, CA 92093, USA
‡‡‡Yale University, New Haven, CT 06520, USA
§§§St. Anthony Falls Laboratory, University of Minnesota, Minneapolis, MN 55455, USA
¶¶¶AOA Geophysics, Mass Landing, CA 95039, USA
****Coastal & Marine Geology Program, US Geological Survey, Santa Cruz, CA 95060, USA
††††Virginia Institute of Marine Science, Gloucester Point, VA 23062, USA

ABSTRACT

The morphology of clastic continental margins directly reflects their formative processes. These include interactions between plate movements and isostasy, which establish the characteristic stair-step shape of margins. Other factors are thermal and loading-induced subsidence, compaction and faulting/folding, which create and/or destroy accommodation space for sediment supplied by rivers and glaciers. These processes are primary controls on margin size and shape. Rivers and glaciers can also directly sculpt the margin surface when it is subaerially exposed by sea-level lowstands. Otherwise, they deposit their sediment load at or near the shoreline. Whether this deposition builds a delta depends on sea level and the energy of the ocean waves and currents. Delta formation will be prevented when sea level is rising faster than sediment supply can build the shoreline. Vigorous wave and current activity can slow or even arrest subaerial delta development by moving sediments seaward to form a subaqueous delta. This sediment movement is accomplished in part by wave-supported sediment gravity flows. Over the continental slope, turbidity currents are driven by gravity and, in combination with slides, cut submarine canyons and gullies. However, turbidity currents also deposit sediment across the continental slope. The average angle of continental slopes (~4°) lies near the threshold angle above which turbidity currents will erode the seafloor and below which they will deposit their sediment load. Therefore, turbidity currents may help regulate the dip of the continental slope. Internal waves exert a maximum shear on the continental-slope surface at about the same angle, and may be another controlling factor.

Keywords Seascape, plate tectonics, deltas, clinoforms, submarine canyons, internal waves, turbidity currents.

INTRODUCTION

From bathymetry to seascape evolution

A great deal of what is known about the evolution of continental shelves and slopes is based on bathymetry. Shelf and slope bathymetry provide key information about underlying structures (Pratson & Haxby, 1997), the distributions of sediment facies (Posamentier *et al.*, 1988), past geological events (Shepard, 1934), and processes that have and may continue to affect a continental margin (Driscoll *et al.*, 2000). Advances in seafloor mapping (e.g. global positioning system, satellite altimetry, multibeam bathymetry and sidescan sonar imagery) and extensive surveys (e.g. the US Geological

Survey and NOAA surveys of the US Exclusive Economic Zone) achieved over the past several decades are especially noteworthy in this regard. These accomplishments are providing a perspective of seafloor morphology that is finally starting to compare in detail and scope with that which has long been available for subaerial and even some extraterrestrial planetary landscapes (Fig. 1) (Pratson & Edwards, 1996). Correspondingly, marine geologists are increasingly taking a geomorphologist's viewpoint of the seafloor, trying to understand seafloor evolution in the same terms as landscape evolution by surface processes acting over geological time.

Inherent in the process–product viewpoint of strata formation is the recognition that **seascapes**

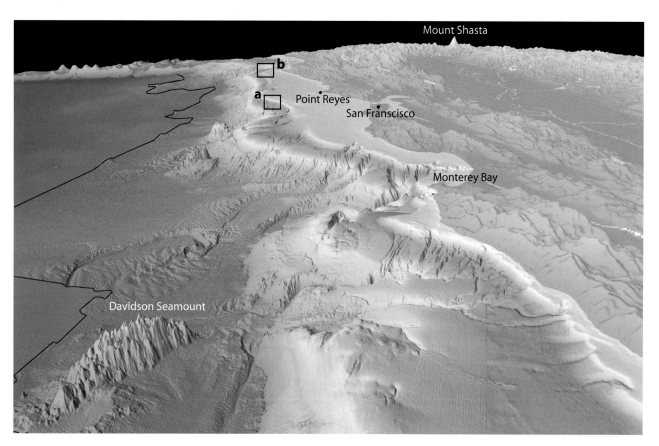

Fig. 1 Computer-generated image of the landscape and seascape along the western US continental margin. Bathymetric data of the seafloor are approaching the detail and scope of topographic data for land. The image looks northward over the northern half of California toward Oregon. Elevations are exaggerated fivefold relative to distances, with the straight-line distance between San Francisco and Point Reyes being ~70 km. Land elevations are from the USGS 30-arc-second digital elevation model (DEM). Seafloor elevations extending from the shoreline out to the black line are from bathymetry collected by the National Oceanic and Atmospheric Administration (NOAA) and gridded by Pratson & Haxby (1996). Bathymetry beyond the black line are from the 2-min bathymetric grid created by Smith & Sandwell (1997). Box (a) is the approximate location of the bathymetry shown in Fig. 16A, while box (b) is the location of the bathymetry shown in Fig. 19A. (Image modified from Pratson & Haxby, 1997.)

do not passively evolve with the build-up of strata, but strongly influence where deposition and erosion occur. Landscapes exert a comparable influence in subaerial settings. They also become shelves when sea level rises, while shelves and even the uppermost slope can become part of the landscape when sea level falls. Thus, in some ways it is not surprising that landscapes and shelf-slope seascapes often look similar. In other ways though, it is surprising, because parts of the shelf and just about the entire slope are purely of marine origin. While the shelf and slope often share the dominantly erosional appearance of landscapes, these provinces contain the thickest accumulations of sediment in the world (Kennett, 1982).

Scope of paper

This paper is about how the overall shapes and dimensions of shelves and slopes are created by the major processes that operate across clastic continental margins in temperate settings, particularly those bordering the continental USA. Not addressed by this paper are the formation of shelves and slopes along carbonate (e.g. Florida, Bahama and north-eastern Australian) or tropical (e.g. Papua New Guinea and Central America) margins, or those that occur in epeiric seas such as the Adriatic. Margins that have been glaciated (e.g. Alaska, Iceland and Antarctica) and those deformed by salt or mud tectonics are also largely omitted (e.g. the USA Gulf of Mexico, Brazilian and Angolan), although impacts of these processes on shelf and slope evolution are considered.

Similarly, this paper does not attempt to review the formation of shelf and slope morphologies across all spatial scales, especially leaving fine scales for other papers in this volume (e.g. the Humboldt 'slide' discussed in Lee *et al.* (this volume, pp. 213–274). The full range of features is too broad and the literature addressing them too substantial to accomplish a complete synthesis here. Instead, the goal of the paper is simply to summarize the basic ways that ubiquitous processes ranging from plate tectonics to internal waves contribute to the large-scale evolution of shelves and slopes over geological time. The impacts are illustrated using conceptual, often geometrically based models that greatly simplify, but also hopefully clarify, the processes that contribute to shelf and slope development.

BACKGROUND

Physiographical definitions

The continental shelf and slope are components of continental margins. Slopes are the transition zones between the continents and ocean basins, and merge with continental rises in a seaward direction (Fig. 2). As with the rise, the shelf and slope are sufficiently distinct from other regions and from one another to be delineated largely on the basis of water depth and average seafloor gradient.

The continental shelf extends seaward from the shoreline to the shelf break, where the seafloor gradient begins to increase markedly (Fig. 2). On average, the shelf break occurs ~80 km offshore at a water depth ~130 m (Kennett, 1982), but the range about these averages is significant. The shelf break is as shallow as several tens of metres seaward of the Ganges-Brahmaputra River, and as deep as 350 m at points along the Antarctic margin. Shelf width varies from being non-existent seaward of the Markham River in Papua New Guinea, to more than 300 km wide in the East China Sea. In all cases though, the shelf is a relatively flat surface characterized by a very gentle overall seafloor gradient of ~0.05° (1:1000; Kennett, 1982). As will be discussed, there are distinct features on the shelf with steeper local gradients. The most important of these is the **shoreface**, the sloping descent from the shoreline to water depths in the order of 10 m that is formed by waves as they impinge upon the shore.

The continental slope extends from the shelf break to water depths of 1.5–3.5 km, although depths are even greater for slopes that descend into oceanic trenches. Seafloor gradients on the slope are similarly variable, ranging from < 1° to > 25°, with the overall average being ~4° (Kennett, 1982). Consequently, the relief and steepness of continental slopes rival those of many mountain ranges.

Historical interest in and importance of the continental shelf and slope

Interest in the bathymetry of shelves and slopes primarily began with concern for charting shoals upon which ships could run aground (Vogt & Tucholke, 1986). Starting at least as early as 85 BC, bathymetric measurements were collected using a weight attached to a measured line. This technique

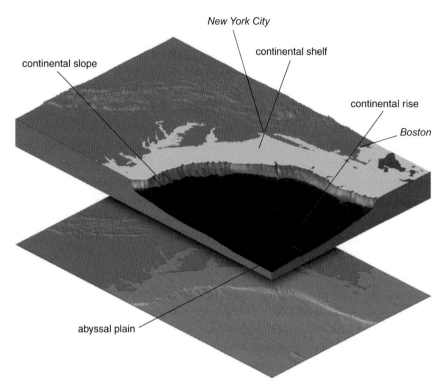

Fig. 2 Continental margin of the north-eastern USA divided into the continental shelf, the continental slope, the continental rise and the abyssal plain, based on water depth and seafloor gradient. Substantial vertical exaggeration is still required to recognize these zones (top). Imaging the same area without such distortion (bottom) reveals that the actual topography of the seafloor is quite modest. The steepest inclines, found on the continental slope, are typically just a few degrees. (After Cacchione & Pratson, 2004. Reprinted with permission of *American Scientist*, magazine of Sigma Xi, The Scientific Research Society.)

was used through the 1800s when the submarine telegraph was introduced, and bathymetric surveys were needed to map routes for laying the cables (Vogt & Tucholke, 1986).

Since the advent of the echosounder in the early 1900s, knowledge of the continental shelf and slope has expanded in both breadth and detail. This in turn has led to the incorporation of shelf and slope physiography in the formulation of many revolutionary geological concepts. For example, prior to the mid-1900s, shelf and slope morphology was used to support theories for the existence of continents and ocean basins, and the formation of mountains (Kay, 1951). Before any sediment-transport observations were made, studies of seafloor morphology led to the realization that many features on shelves and slopes had to be shaped by submarine processes and could not be explained solely by subaerial processes acting during periods of lowered sea level (Daly, 1936; Stetson, 1936; Stetson & Smith, 1938). Shelf and slope morphology has gone on to be explained in such unifying theories as **plate tectonics** (Dewey, 1969; Bird & Dewey, 1970) and the underpinnings of **sequence stratigraphy** (Vail & Mitchum, 1977). It has also become important for:

1 managing offshore resources, including deepwater (i.e. > 300 m) oil and gas reserves, and marine fisheries (Cook & Carleton, 2000);
2 identifying regions subject to marine geohazards such as submarine landslides that can trigger tsunamis (Coulter & Migliaccio, 1966) and damage offshore infrastructure (Bea, 1971);
3 mapping potential pathways and sinks for contaminants introduced offshore (Poppe & Polloni, 1998; Lee *et al.*, 2003);
4 reconstructing sea-level change over the Holocene, and its implications for future changes resulting from global warming (Fedje & Josenhans, 2000).

PROCESSES GOVERNING SHELF WIDTH AND SLOPE RELIEF

Plate tectonics and the stair-step shape of continental margins

The 'stair-step' shape of continental margins is first and foremost the result of plate tectonics. All margins begin where **lithospheric** plates (that make up the rigid outer shell of the Earth) interact with one another as they move over the plastically deforming **asthenosphere** in the upper mantle (Dewey,

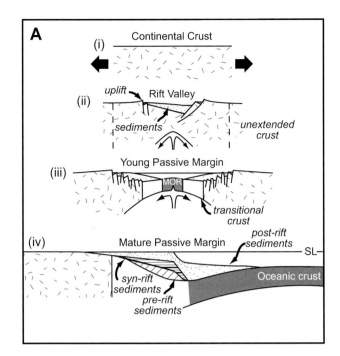

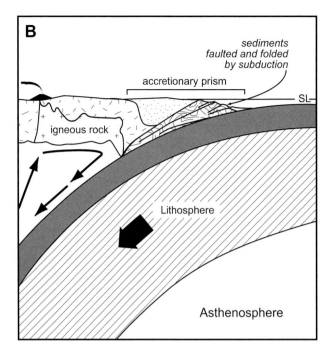

Fig. 3 Continental margin classification as a (A) passive or (B) active margin. (A) Passive margins form where continental crust is rifted apart (i–iv). The shelf and slope are built from any sediments accumulated before rifting (pre-rift, i), as well as during (syn-rift, ii) and after rifting (post-rift, iii–iv). (Modified from Busby & Ingersoll, 1995.) Passive margins take tens of millions of years to form (Klitgord *et al.*, 1988), while active margins form as soon as one lithospheric plate begins grinding past or subducting another. The first type of active margin, a transform margin, is rare. (B) Convergent margins are the only type of active margin considered here. Along these margins, the shelf and slope are generally founded upon the sediments being folded and faulted as they are scraped off the downgoing plate and subducted into the asthenosphere: SL, sea level; MOR, mid-ocean ridge. (Modified from Allen & Allen, 2005.)

1972; Fig. 3). These interactions have generated the granitic and andesitic crust of the continents, and the thinner, denser, basaltic crust that floors the oceans (Fig. 3). Just by virtue of this difference in thickness and density, granitic and basaltic crust form a step where they meet along the edges of continents. However, the shape of this step and its subsequent evolution are strongly influenced by the regional tectonic environment.

Broadly speaking, plate tectonics leads to two types of continental margins. **Passive margins** are those that ride 'passively' within the interior of a lithospheric plate (Kennett, 1982) (Fig. 3A). They first form when the plate and continental crust are rifted. During this process, the continental crust is stretched and fractured (i–ii, Fig. 3A). Basaltic magma derived from partial melting of the underlying asthenosphere rises to fill the fractures. The magma supply eventually becomes focused along a mid-ocean ridge that follows the axis of the old rift (iii, Fig. 3A). Along this ridge, new basaltic crust

forms to floor an expanding and deepening ocean basin. Thus the underlying structure of passive margins, created during the rifting process, is a faulted, several kilometre descent from unextended continental lithosphere down to oceanic lithosphere (iv, Fig. 3A).

Active margins lie along boundaries between plates and are shaken by the plate movements (Kennett, 1982). Active margins form along:

1 convergent plate boundaries where continental lithosphere overrides or **subducts** oceanic lithosphere (Fig. 3B);
2 transform plate boundaries where continental and oceanic lithosphere slide past one another.

The latter type of margin is relatively rare, so here the focus will be on convergent margins. Beneath these, partial melting of the subducted oceanic lithosphere generates andesitic to granitic magmas that penetrate and add to the overriding continental lithosphere (Fig. 3B). The convergent margin itself, however, tends to be located seaward of the igneous terrain over an **accretionary prism**. This is a faulted wedge of sediments that the overriding plate has bulldozed off the downgoing plate (Fig. 3B).

First-order effects of thermal subsidence and tectonic uplift

Sedimentary shelves and slopes along convergent margins are not only built of debris scraped off the downgoing plate, but are partly comprised of sediments eroded off the continent. The continents are also the primary source of clastic sediments to shelves and slopes along passive margins. In general, the greater the accumulation of sediments along a margin, the farther the shelf reaches out into the ocean basin and the greater the water depth to

which the slope extends (Jervey, 1988). An equally important control on shelf width and slope relief, however, is the **accommodation space** available along a continental margin for storing the sediments (Van Wagoner *et al.*, 1988). Abrupt water depth increases with distance offshore cause narrow shelves, and allow for taller slopes to be built from a given volume of sediments (Fig. 4A).

The first-order controls on accommodation space along continental margins are thermal subsidence and tectonic uplift (Hays & Pitman, 1973; Pitman, 1978). Thermal subsidence is geomorphically most important along passive margins, and arises from the cooling of the adjacent oceanic lithosphere following its formation at the mid-ocean ridge (iii–iv, Fig. 3A). As it cools, the oceanic lithosphere thickens and becomes denser, causing it to subside at a rate that tapers off according to $1/\sqrt{age}$ (Parsons & Sclater, 1977). The rate decreases from 40–100 m

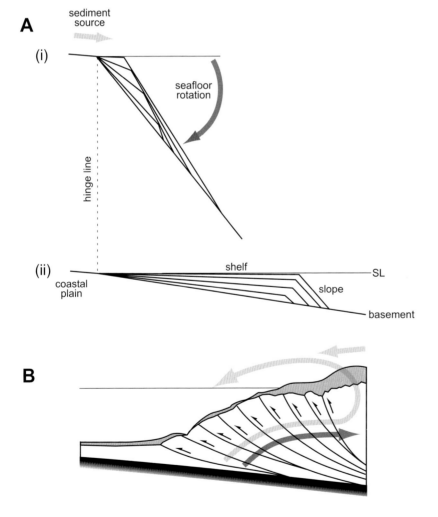

Fig. 4 Crustal controls on continental-margin morphology. (A) The steeper the descent of transitional crust across a continental margin, the more space there is for accommodating sediments eroded from the continent, and thus the narrower the resulting shelf and the greater the relief of the slope (i). Slower subsidence leads to broader shelves and smaller slopes (ii). This relationship holds true to zero subsidence, after which uplift reverses the trend. The faster the uplift, the quicker the slope and shelf are destroyed as they are raised up and incorporated into the continent, leading to narrower shelves and deeper slopes. (Modified from Reynolds *et al.*, 1991.) (B) As a result of uplift, shelf and slope sediments along active margins are recycled, and add to the sediments being eroded from the continent and bulldozed off the subducting plate. (Modified from Kulm & Fowler, 1974.)

Myr^{-1} along a young margin, such as the Gulf of Lions (Burrus, 1989), to < 3 m Myr^{-1} along an old margin, such as that off New Jersey (Steckler *et al.*, 1999). The rate also decreases landward across the margin to zero at the **hinge line** or boundary between extended and unextended continental lithosphere at the edge of the initial rift (Pitman, 1978; dashed line in ii, Fig. 3A).

The changing rate of thermal subsidence with time and with distance offshore adds complexity to the evolution of shelf width and slope relief. If the rate of subsidence on the outer shelf exceeds the rate at which sediments are supplied to the margin, more space is created than can be filled by the sediments, and sediment accumulation shifts landward (Reynolds *et al.*, 1991; Fig. 4A). This scenario is most likely to occur along a young, fast-subsiding margin, and when it does occur, the shelf narrows and builds upward or **aggrades** over time. The relief of the slope increases as well. As a margin ages, however, subsidence rates decrease. When the maximum subsidence rate on the shelf is exceeded by the sediment supply rate, infilling occurs faster than new space is created. Under these conditions, sediment accumulation is forced out into the deeper waters over the slope, which builds seaward or **progrades** into deeper water. As a result, the shelf widens and the slope lengthens (Reynolds *et al.*, 1991; Fig. 4A).

Along convergent margins, tectonic uplift is the most important control on accommodation space. The uplift stems from the overriding continental lithosphere being thickened by:

1 buckling and folding as it ploughs over the downgoing oceanic lithosphere;
2 intrusions and extrusions of magma generated by the subduction;
3 the build-up of accretionary sediments 'bulldozed' off the subducting lithosphere (Allen & Allen, 2005).

Importantly, uplift can also occur along passive margins when the upper lithosphere is stretched less than the lower lithosphere during rifting (Allen & Allen, 2005) (ii, Fig. 3A). The non-uniform stretching exposes the trailing edge of the upper lithosphere to greater heating by the asthenosphere, leading to thermally driven uplift of the margin. This uplift tends to be geologically short lived, however (Fig. 3A), and so it simply sets back the

development of accommodation space by thermal subsidence.

The tectonically driven uplift along convergent margins rotates the shelf and slope strata upwards (Kulm & Fowler, 1974; Kulm *et al.*, 1975) (Fig. 4B). The uplift destroys accommodation space on the shelf and forces the greatest sediment accumulation to occur on the slope, which is where tectonic accretion from below is also greatest. The end result is both aggradation and progradation of the margin. However, the uplift also causes the slope strata to eventually become shelf strata and ultimately part of the continental land mass. Here the strata are once again subject to erosion and form a source of sediments for new shelf and slope strata. As a result, there is a continuous recycling of sediments along convergent margins, as well as the addition of materials by subduction and igneous rock formation (Fig. 4B).

Second-order effects of isostasy, compaction and faulting

If the New Jersey margin is representative, ~40% of the space that accommodates sediments on mature passive margins is created by thermal subsidence (Steckler *et al.*, 1988), about 40% is produced by isostatic subsidence and 20% by compaction. Consequently, the subsidence generated by isostatic and compaction processes has a significant influence on shelf width and slope relief, but it has a less uniform influence. This is because thermal subsidence is an externally driven process that affects the entire breadth of a continental margin. Isostatic subsidence and compaction are more localized and occur in response to sediment loading.

As sediments accumulate on continental margins, they cause isostatic subsidence of the lithosphere and they compact. **Isostatic subsidence** is the sinking of the lithosphere lower into the asthenosphere to reach a new level of buoyant equilibrium under the sediment load. The amount of subsidence depends on the mass of the load and the rigidity of the lithosphere (Watts & Ryan, 1976). If the lithosphere has no rigidity, it will only subside directly beneath the load to a level at which the buoyancy force produced by the displaced asthenosphere equals the mass of the load. This case is known as **Airy isostasy** (Fig. 5A). However, if the lithosphere

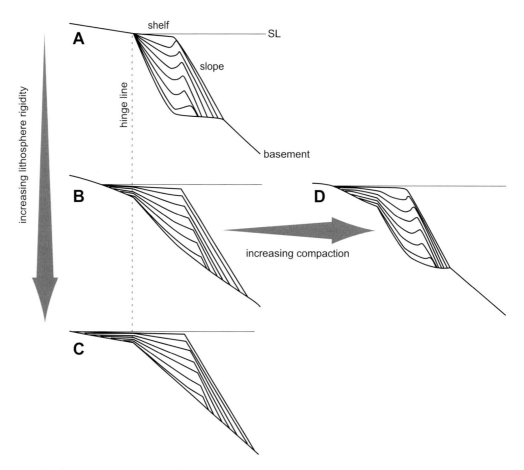

Fig. 5 Isostatic subsidence due to sediment loading and compaction. (A–C) The amount of isostatic subsidence that occurs along a passive margin, as sediments accumulate, depends on the rigidity of the lithosphere; the stronger the lithosphere (i.e. more rigid), the deeper the base of the slope. Even more significantly, rigid lithospheres cause broader shelves, with sediment accumulating landward of the hinge line (compare A–C). Not illustrated here is isostatic subsidence due to sediment loading along active margins, which can be even more significant due to the fact that the subducting plate is not connected to the overriding plate, and so is even less rigid (Allen & Allen, 2005). When the lithosphere has zero rigidity (A), the isostatic subsidence due to sediment loading is referred to as Airy isostasy. (D) Similar subsidence is caused by compaction (compare A and D). (Modified from Reynolds *et al.*, 1991.)

has rigidity, then it behaves like an elastic beam. The load is distributed over a broader region such that beneath the load the subsidence is less but it extends laterally beyond the load. This is known as **flexural isostasy** (Fig. 5B & C).

Sediment compaction is similar to Airy isostasy in that it occurs only directly beneath the region of sediment loading (compare Fig. 5A & D). This loading forces the sediment grains to assume tighter packing, resulting in a reduction of the intervening pore space. Compaction is greatest within about the first kilometre of the seabed, and is completed by a burial depth of ~4–5 km (Sclater & Christie, 1980). Over this range, porosities of sand, silt and clay decrease exponentially from ~0.5,

0.6 and 0.6 to ~0.3, 0.2 and 0.2, respectively (Bahr *et al.*, 2001).

Isostatic subsidence commonly lags the loading, due to the sluggish, viscous response of the displaced asthenosphere, and the thermal readjustment of the lithosphere as it moves to its new level (Le Meur & Hindmarsh, 2000; Watts & Zhong, 2000). Measurements of post-glacial rebound indicate that following loading/unloading, subsidence/rebound occurs relatively rapidly for a period of 10 kyr (Watts *et al.*, 1982). Over the next 300–400 kyr, the isostatic readjustment decays exponentially to zero assuming the load remains unchanged.

Compaction will also lag loading, if the loading rate exceeds the rate at which water between the grains can evacuate the pore space (Suppe, 1985). This is because water is incompressible at near-surface temperatures and pressures, and so will resist compaction if it has nowhere to go. This resistance, which can be great enough to bear the complete weight of the overlying sediment load and thus prevent any reduction in pore space, will slow compaction until enough water escapes the pores that the grain framework can again assume the submerged weight of the load. The lower the **permeability** of the sediments or connectivity between pores, the slower the rate of dewatering and the longer it takes the sediments to compact (Bredehoeft & Hanshaw, 1968).

Being load dependent, isostatic subsidence and compaction generate the greatest accommodation space where the thickest accumulation of sediments occurs, which on most continental margins is beneath the shelf break (Reynolds *et al.*, 1991; Fig. 5). The space created by these processes sets up a positive feedback by promoting additional sediment accumulation here, and so acts to stabilize the position of the shelf break. This influence diminishes, however, when the rate of lithospheric subsidence and compaction falls below the sediment-supply rate. When this happens, the margin can prograde even on lithosphere with zero rigidity. However, the subsidence generated by this loading will then continue long after the input of sediments slows or even stops.

While the impacts of isostatic subsidence and compaction on shelf width and slope relief share major similarities, there are also some important differences. Flexural isostasy affects a broader region than compaction (Reynolds *et al.*, 1991). As the lithosphere flexes beneath a margin, it creates accommodation space to either side of existing strata. When incoming sediments then fill this space, the shelf is extended landward and the slope seaward even if the shelf break does not move (Fig. 5A–C). Thus lithosphere with higher rigidity facilitates the development of a broader margin.

Compaction, on the other hand, can cause smaller-scale subsidence than isostasy, as a result of shorter temporal and smaller spatial variations in sediment accumulation. This subsidence generally develops along the coast and inner shelf. Where rapid sediment supply from a large river has been greatly reduced or even cut off, deflation of the river deposits by compaction can lead to inundation of the coast and landward retreat of the

shoreline (Milliman & Haq, 1996; Fig. 6). Where the subsidence becomes great enough, it may ultimately alter coastal topography to the point that river drainage to the area is recaptured and it once again becomes a major shelf depocentre. This process promotes deltaic **lobe switching**.

Added effects of faulting

Many factors (including sediment loading) can cause faulting, which affects far more than just the dimensions of the shelf and slope. Shallow faulting in sedimentary strata can lead to the slumps and slides that scar many continental slopes and even some continental shelves. Deeper faulting, including faults that extend into underlying or adjacent continental crust, can act as important structural controls on the course of rivers and submarine canyons (Kelling & Stanley, 1970; Dolgoff, 1998; McHugh *et al.*, 1998). However, among the most important morphological impacts of faults on continental margins is their effect on accommodation space. Extensional (**normal**) faulting of continental crust during rifting creates the rugged descent to oceanic crust upon which the sedimentary shelves and slopes of passive margins develop (iv, Fig. 3A). Compressional (high-angle **reverse** and lower-angle **thrust**) faulting in accretionary prisms accommodates the addition of new wedges of scraped-off material to the base of the prism and causes the corresponding uplift and back-tilting of debris within the prism (Figs 3B & 4B).

A third important example of faulting that affects accommodation space is growth faulting. **Growth faults** are common in regions where sediment accumulation has been rapid (Suppe, 1985). These are **listric** or curved faults, in which movement along the fault is gradual rather than catastrophic (Fig. 7). The movement is driven by sediment loading on the basinward side of the fault, which acts somewhat like a spring. As it is loaded with sediment, this side slides downward along the fault, capturing the sediment while rotating the underlying strata (Fig. 7). Early fault movement occurs relatively easily, and a lot of sediment can be trapped. However, the accumulation of strata (toward the base of the fault) slows the faulting and reduces the sediment trapping. Eventually the fault freezes, and no further sediments are trapped.

Where common, growth faults can be significant in sequestering sediments on the shelf and

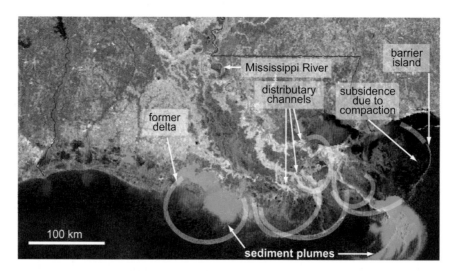

Fig. 6 Colour-composite image of NOAA AVHRR satellite data of the Mississippi Delta. The sediment-laden Mississippi River shows up as the meandering band of green entering the image at the centre top (it follows the Mississippi state line shown in red). As the river nears the coast, it subdivides into a number of distributary channels that fan out and feed sediment to different regions of the delta. Engineering to stabilize the main river channel for shipping has caused the part of the delta in the lower right-hand corner of the image to grow fastest. This is the youngest lobe of the delta, which is seen as an amalgamation of sediment plumes emanating from the distributary channels into the blue Gulf of Mexico and over the delta foreset. A plume is also seen spilling out of Atchafalaya Bay (farther west), which is probably where the Mississippi River would naturally shift its course if its channel were not stabilized. Previous avulsions of the river (i.e. lobe switches) created earlier deltas along much of the USA Gulf of Mexico coast, helping build it seaward. The former shorelines of six of these deltas are outlined in the image, their extent in some cases still demarcated by what are now barrier islands. Much of the outer regions of the delta have subsided due to compaction and are now inundated. This is an ongoing process, and the major reason is that sea level is rising at ~1 m 100 yr^{-1} along the Louisiana coast (Penland *et al.*, 1990). Former delta lobe locations from Elliott (1979). (Modified version of satellite image prepared by R. Sterner, Johns Hopkins University.)

inhibiting margin progradation (Mitchum *et al.*, 1990; Fig. 7). Large growth faults near the shelf break also can lower the slope gradient (Emery & Uchupi, 1984). In these instances, the seaward sides of the growth faults include parts of the slope, which are lowered when the strata are back-tilted as they are rotated by the curved form of the faults.

PROCESSES THAT FORM THE SHELF PROFILE

Rivers, deltas and growth of the coastal plain

Rivers deliver some 84% of the total sediment load that reaches the oceans (Milliman & Meade, 1983) and are the dominant means by which clastic sediments are transported to continental margins. Glaciers come in second, supplying 9% of the sediment load to the oceans, or more than twice the contribution of any remaining subaerial transport process (Scholle, 1996). However, temperate glaciers (e.g. south-east Alaska and west side

of the New Zealand south island) have the largest natural sediment yields (i.e. mass/area) of any runoff in the world (Milliman & Syvitski, 1992). The sediment supply from glaciers during the last glacial maximum may have been many times larger than it is today. The Pleistocene glaciers bulldozed sediments onto continental shelves and even to the slope. As they retreated, the glaciers left **terminal moraines** that still rise above the seafloor and even above sea level (e.g. Long Island and Cape Cod; Shor & McClennen, 1988), and they released catastrophic ice-dam breaks (**jökulhlaups**) that deposited thick sediment wedges on the shelf (e.g. the Hudson Apron; Davies *et al.*, 1992; Milliman *et al.*, 1996; Uchupi *et al.*, 2001).

Even during the glacial epochs of the Pleistocene, however, the majority of the world's continental margins were more affected by rivers than by glaciers. Prior to the Pleistocene, the only glaciers affecting a continental margin still likely to be in existence today were those covering Antarctica (Kennett, 1982). Thus, while glaciers have had

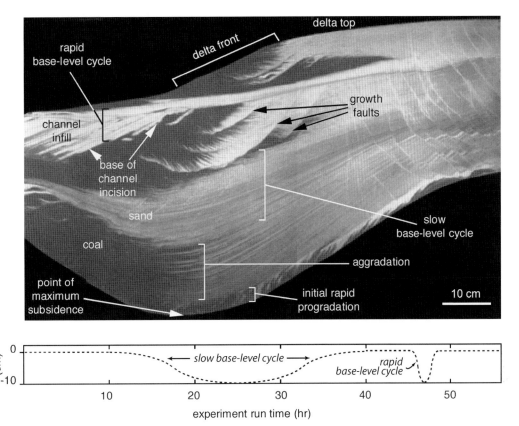

Fig. 7 Sediment geometries in a dip section through experimental strata, formed as subsidence and sediment supply were held constant. Growth faults inhibit progradation of the shelf break by trapping sediments that otherwise would be transported onto the upper slope. The deposit varies in thickness because the rate of subsidence varied basinward across the tank in which the strata formed, but these rates did not change over the course of the experiment. A mixture of two sediments was used: quartz sand and coal. The latter served as a visibly distinct proxy for clays because its specific gravity is half that of quartz. The approximate position of the 'shoreline' during the experiment is demarcated by the abrupt boundary in sediment type between sand (light) and coal (dark). Note that during the stages in the experiment in which growth faults formed, 'shoreline' progradation was significantly impaired, with the sands being captured and rotated downward along the faults. Mass is conserved, so the 'slope' must still advance when the shelf break is held stationary. This occurs through the basinward movement of the base of 'slope', which is pushed outward by the toes of the growth faults, and results in an overall lowering of the 'slope' gradient. The terms 'shoreline' and 'slope' are in quotes because the features in the experimental strata do not scale up to continental margins. However, large growth faults are found on the Brazil and Niger continental slopes, and show clear indication of having contributed to their markedly low gradients relative to other margins (Emery & Uchupi, 1972). (Figure after Paola *et al.*, 2001.)

a major effect on the present-day morphology of continental margins as low as 40° latitude, the effects of rivers have been even greater and so are focused on here.

Rivers have been active in delivering sediments to continental margins since their inception, but individual rivers are ephemeral on geological time-scales. Tectonics, glaciations and stream pirating among other processes alter river courses and have created new rivers from old ones. So, while fluvial input to margins is continuous, the cast

of rivers responsible for this input is subject to change (Bridge, 2003).

Large individual rivers, such as the Amazon, Indus, Yangtze and Mississippi, often act as a single **point source** of sediment to a continental margin. Where multiple rivers feed a margin, they can combine to form a **line source** (Jaeger & Nittrouer, 2000). Sediment emanates from the mouth of each river, but on the shelf these contributions are smeared together by waves and currents, making it impossible to define where the input of one river ends and that of another begins. Broad lateral dispersal of river sediments is also caused by **avulsions** or wholesale shifts in the location of a river mouth (Bridge, 2003). These occur when the river channel upstream of its mouth is sufficiently breached during one or more flooding events that the river follows a new route to the ocean and abandons the old one (i.e. lobe switching). Using the Mississippi River as an example, avulsions ultimately can spread deposition across $\sim 10^3$ km of coastline, with major shifts in the river mouth location occurring roughly once every 1000 yr (Frazier, 1967; Fig. 6).

In settings where the descent of the river valley continues steeply below sea level, much of the fluvial sediment load may continue to be moved directly toward the deep ocean via slides, debris flows, turbidity currents or some other form of gravity-driven transport (e.g. the Markam River in Papua New Guinea; Krause *et al.*, 1970). In these settings, little if any continental shelf is present.

Along more gradual continental margins, the sediments will come to rest close to shore, and if sea level is constant, the river will begin to build a delta (Elliott, 1979). Deltas are partly subaerial, partly submarine deposits that emanate from the river mouth in a fan shape (Fig. 6). They are characterized by a **clinoform** geometry in cross-section that can be subdivided into three parts (Fig. 8). The clinoform **topset** is a gentle seaward-dipping surface that on a delta corresponds to the outer **coastal plain** leading to the shore. It connects to a steeper inclined **foreset** (also known as the delta front), which on the delta extends below water level reaching another gradually dipping surface, the **bottomset** (also known as the **prodelta**).

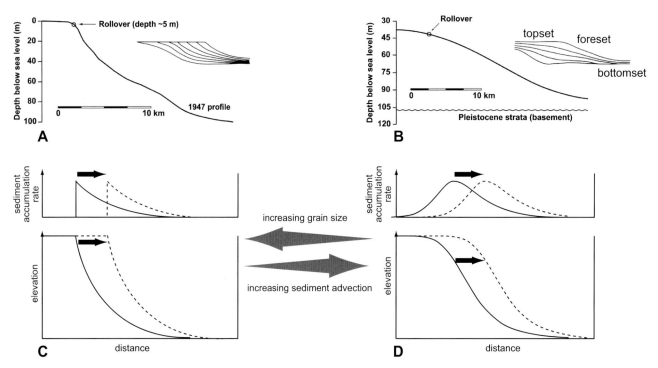

Fig. 8 Factors controlling clinoform geometry in deltas. (A) Mississippi River delta and (B) Ganges–Brahmaputra subaqueous delta bathymetric profiles are examples of oblique and sigmoidal clinoform shapes (after Swenson *et al.*, 2005), respectively, which are also shown schematically to the right of each profile (after Mitchum *et al.*, 1977). These shapes result from variations in the rate of sediment accumulation across the clinoform foreset. (C) Oblique clinoforms are created when sediment accumulation is greatest at the top of the foreset. (D) Sigmoidal clinoforms result when accumulation is greatest farther down on the foreset. The accumulation rate is tied to the foreset because flow divergence here leads to a loss of velocity and carrying capacity. The faster a flow loses its velocity over the foreset and/or the coarser the sediments it is carrying, the more abrupt the decay in accumulation rates down the foreset and the more oblique the clinoform that is created. Note that on subaerial deltas like the Mississippi, the clinoform rollover lies near sea level. On subaqueous deltas, it is often located tens of metres below sea level.

The clinoform is seen in sedimentary bodies ranging from bedforms to entire continental margins, where the shelf corresponds to the topset, the slope to the foreset and the rise to the bottomset (Thorne, 1995). The clinoform is the reflection of a basinward change in sediment flux, in which sediment accumulation is greatest over the foreset. Consequently, the clinoform is the fundamental shape of prograding strata. However, significant sediment accumulation also can occur across the topset and bottomset regions of a clinoform. In fact, aggradation of the coastal-plain topset is an integral part of delta progradation. As the shoreline moves seaward with the delta, the distance the river must travel to reach the ocean increases. The river responds by depositing sediments along its path across the coastal plain to steepen its profile and give the river enough momentum to carry its remaining sediment load to the ocean (Swenson *et al.*, 2005). This profile is characteristically exponential in form. The curvature results from the increase in river discharge toward the coast as tributary inflow is collected, which in turn lowers the gradient needed to drive the river toward the shore (Ritter, 1986).

Deltas and their coastal plains are important to margin evolution for several reasons. As rivers build deltas toward the shelf break, the width of the shelf is reduced (Fig. 9). When sea level rises, the deltas are drowned and their coastal plains become the foundation for a new, broader shelf (Fig. 9). This then becomes the platform upon which new deltas develop. How this development unfolds during sea-level change is the subject of a later section. The main point here is that fluctuations in sea level over geological time have led to numerous advances and retreats of deltas and their coastal plains. In turn, these have contributed significantly to the upward and outward building of the shelf and slope, with many shelves being the product of stacked delta/coastal-plain sequences. Examples include the ancient Miocene deltas underlying the current New Jersey and Angolan continental shelves (Steckler *et al.*, 1999; Lavier *et al.*, 2001).

Bedload deposition, sediment plumes and clinoforms

In general, about 90% of the sediments moved by rivers are carried in the water column as **suspended load** (Meade *et al.*, 1990). The remainder is dragged and bounced along the riverbed as **bedload**. Both types of loads contribute to delta formation. On entering the ocean or an estuary, the discharge from a river spreads, loses velocity and thus loses its competence to move the bedload any farther, depositing it at the river mouth. In the absence of any other flow (e.g. tides), the bedload builds up to form a river-mouth bar. The river is then forced to flow around the bar leading to channel bifurcation. Through this process and also through the breaking of levees (i.e. **crevasses**) during floods, deltas develop distributary-channel networks (Elliott, 1979).

The suspended load of the river largely passes through the distributary-channel network and continues out into the ocean or estuary. Generally the mixture of fine sediments and freshwater is less dense and so spreads out as a surface sediment plume (Fig. 6). Under rare circumstances though, the suspended load may become so concentrated and/or the temperature of the water carrying the sediments may be so cold that the mixture temporarily exceeds the density of seawater and plunges beneath the surface as a **hyperpycnal flow** (Mulder & Syvitski, 1995; Mulder *et al.*, 2003). Being denser than seawater at all depths, hyperpycnal flows move along the bed, as a gravity-driven flow, a mechanism that will be discussed later.

Sediment plumes carry the suspended sediments over the submerged delta foreset. The plumes rarely extend directly offshore. Instead, they are turned parallel to the coast by alongshore winds, currents (e.g. the Amazon shelf; Curtin, 1986; Geyer *et al.*, 1991), or even the Coriolis force if the plume is large enough (Hill *et al.*, this volume, pp. 49–99). Due to lateral spreading and mixing with the ambient seawater, the speed of the plume drops, although ocean currents will continue to move it even after it has lost its river-imparted momentum (Geyer *et al.*, 2000). Sediments suspended within the plume settle. These sediments are typically clays, but under rare circumstances silts and fine sands also have been observed (Syvitski *et al.*, 1985). Consequently, plumes supply sediment to the submarine elements of the delta, principally the foreset and bottomset.

The curvature of the clinoform **rollover** between the topset and foreset of a delta is dictated by the change in sediment flux across this boundary; the more abruptly the flux drops, the steeper the

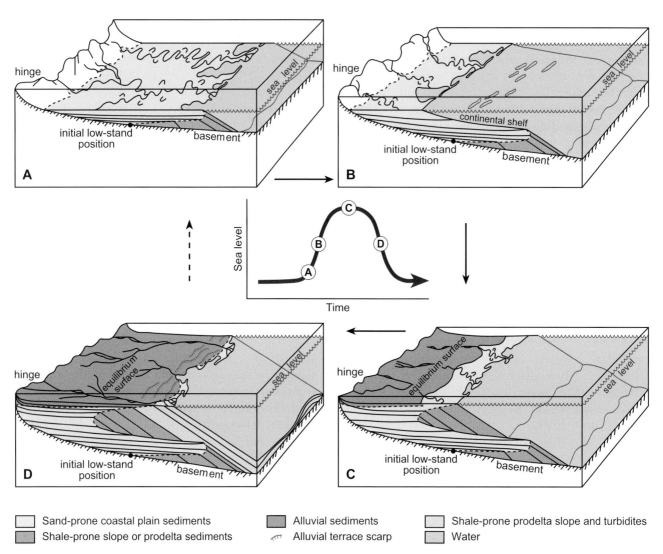

Fig. 9 Response of continental-margin sedimentation to sea-level change. The majority of sediments delivered by rivers to the ocean accumulate along the coast. (A) In the absence of any change in sea level, the coastal plain will grow toward the shelf break and the shelf will narrow. (B) During a sea-level rise this pattern is reversed, if the rate of rise exceeds the rate of sediment supply from the rivers. When this occurs, the coastal plain is flooded and river valleys are turned into estuaries with the inundated regions becoming part of a broadening shelf. (C) When sea-level rise slows, and when it falls back below the sediment supply rate, the latter builds the coastal plain once again. (D) If the shoreline reaches the shelf break, the shelf disappears and the coastal plain links directly to the slope. Continental shelves and slopes evolve through the repetition of this cycle over the numerous rises and falls in sea level that occur during their existence. (Modified from Jervey, 1988.)

break in slope. Rivers carrying a high fraction of their sediment load as bedload, or those that enter a relatively protected margin in which wave and current energy is low, characteristically produce deltas with an abrupt rollover or **oblique clinoform** shape (Mitchum _et al._, 1977; Pirmez _et al._, 1998; Fig. 8A & C). The rapid deposition of sediment near the shore can build a foreset that becomes too steep and fails. As sediments at the top of the

foreset cascade to its base, they lower its overall gradient. Accumulation at the top of the foreset then resumes, the process repeats itself and the foreset progrades by avalanching in a manner similar to a ripple or dune.

A different clinoform shape tends to result if the sediment load of a river is kept in suspension by waves and currents in shallow water on the shelf. In this case, most of the sediments bypass the topset and are carried to the foreset. Sediment accumulation gradually rises to a maximum over the upper foreset (Fig. 8D) then falls gradually over the remainder of the foreset and bottomset. The broader accumulation produces a low-angle foreset and a gentle clinoform rollover, known as a **sigmoidal clinoform** shape (Mitchum *et al.*, 1977; Pirmez *et al.*, 1998) (Fig. 8B & D). The enhanced accumulation of sediments below the top of the foreset elevates the bed. Ultimately this region becomes part of the topset, but as it is raised into the more energetic surface waters, deposition is increasingly inhibited and sedimentation is displaced into deeper waters over the evolving foreset (Pirmez *et al.*, 1998). While oblique and sigmoidal clinoforms are different, they represent end members in a continuum of deltaic shapes. As the operative marine processes change, the clinoform shape can evolve toward one end member or the other.

The impacts of waves and currents on the shelf and shoreface profile

While rivers shape the profile of the coastal plain, waves and currents act to modify the profile of the shelf by redistributing the sediments deposited there (Wright, 1995). Waves and currents exert their most obvious influence on continental margins near the shore. A number of excellent texts treat the subject of coastal morphodynamics far more thoroughly than can be covered here (Komar, 1976; Bascom, 1980; Dyer, 1986; Wright, 1995; Short, 1999). The discussion here is limited to the general role of waves and currents in creating the shoreface and to a lesser extent the shelf profile seaward of it.

Neither waves nor currents are very effective at causing sediment movement on their own (see Hill *et al.*, this volume, pp. 49–99). In the case of ocean currents, the bottom shear stresses they generate tend to be too weak to resuspend even the most moveable sediments (Grant & Madsen, 1979). There are exceptions of course. Strong tidal currents on the shelf can generate enough bottom shear in some settings to elongate and segment river deltas (Wright, 1977). The currents also can form **flood** and **ebb tidal** deltas at the mouths of inlets that connect estuaries and lagoons to the open ocean (Wright, 1995). Along the western margin of the North Atlantic, geostrophic currents can reach speeds strong enough to erode the continental

slope (Poag & Mountain, 1987) and deposit large sediment drifts on the continental rise (Tucholke & Laine, 1982; Mountain & Tucholke, 1985). Generally speaking though, the flow divergence promoted by the large dimensions of the ocean successfully mitigates the generation of bottom currents vigorous enough to cause substantial erosion of the seafloor.

In contrast, shoaling waves can often generate bottom shear stresses strong enough to move sediments, particularly in shallow water (Grant & Madsen, 1979). The term **shoaling** is used here to refer to those waves that generate water motion that extends all the way to the seafloor (i.e. **transitional** and **shallow-water waves**; Fig. 10). The orbital water motion induced by waves is largest near the surface and decreases exponentially with depth (Knauss, 1997). When a water parcel is beneath a wave crest, it moves in the same direction as the waves. When the water parcel is beneath a wave trough, it moves in the opposite direction to the waves. In deep water, wave orbital motion is circular, and becomes negligible before reaching the bed (i.e. **deep-water waves**; Fig. 10). In shallower water, this orbital motion becomes elliptical due to interaction with the bottom (Knauss, 1997; Fig. 10). As water depth decreases, the bed-parallel diameter of the near-bed orbital motion increases, leading to greater near-bed wave orbital velocities.

Near-bed wave orbital motion generates shear stresses on the seafloor. These stresses are proportional to the square of near-bed wave orbital velocity, and therefore increase with decreasing water depth for a given surface-wave condition. When large enough, these shear stresses are able to move the bottom sediments, transporting them either as bedload or in suspension. The sediments are carried forward with the wave as the crest passes and are moved offshore beneath the following wave trough. If the sediment is relatively coarse, it may stop moving or settle to the bed between the forward and backward motion of the wave. Finer sediment can stay suspended throughout the wave cycle.

During storms, wave-generated orbital motion can affect the seafloor over much or all of the shelf. During fair-weather conditions, the influence of waves on the seabed is limited to shallow water depths. The region persistently affected by waves, the shoreface, typically extends from the outer

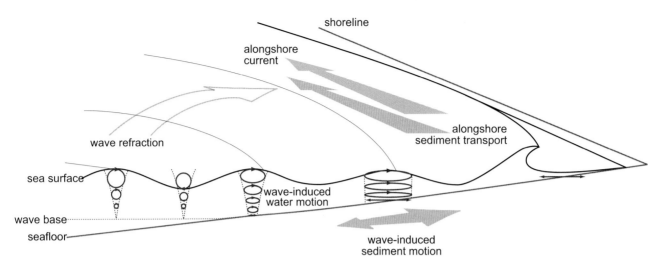

Fig. 10 Water movement and currents approaching a shoreline. Shoaling waves cause both water and sediment to undergo a back-and-forth motion. For water, the motion is a circular orbit, if the depth is large. In this case, the bottom does not affect the wave motion, which decays exponentially with depth and becomes negligible at a level referred to as **wave base**. The orbital path becomes increasingly flattened with decreasing water depth as the waves and bottom interact. The shear at the bottom mobilizes sediments, which are moved shoreward by the water, as it passes through the top of its orbit. The sediments also can be moved along the shore, if the waves are travelling toward the beach at an angle, and generate an **alongshore current**. These currents are significant in carrying sediments from rivers to beaches lacking other significant forms of terrestrial input, and so promote a more uniform distribution of sediments along the coast.

surf zone to the shoreward portions of the inner shelf. As waves cross the shoreface, they are modified by frictional interactions with the bed. As a result, onshore flow beneath wave crests becomes faster than the offshore flow beneath wave troughs, and there is a net shoreward movement of sediment transported by the waves (Komar, 1976). Net sediment transport rates are greatest, however, when waves and currents occur simultaneously (Grant & Madsen, 1979). Even if the currents alone are too weak to mobilize the sediments, the currents will still impose direction of transport when particles are resuspended by the waves. Thus the combined action of waves and currents gradually transports sediments across the shelf.

Waves and currents shift sediment grains until they are moved to a water depth where they no longer undergo net transport. This depth is referred to as **wave base**, or the depth at which 'wave action ceases to stir the sediments' (Baker *et al.*, 1966; Fig. 11). Wave base has been equated to the shoreward limit of deep-water waves (Davies, 1980), but here we take the broader view of wave base also being linked to sediment grain size and reflecting a long-term equilibrium profile (Gulliver,

1899). In this context, wave base is a theoretical level more than an easily defined surface for two reasons. The first is that the water depth at which a wave begins to have an impact on the seafloor depends on its deep-ocean height and wavelength, and both vary with time and distance along a margin. For example, waves can become so large during major storms that they touch bottom over the shelf break. It is during such storms that waves should cause the greatest resuspension of sediments. However, waves during less severe but more frequent storms may cause more sediment movement over geological time, so the wave activity that has the greatest influence on the shelf profile is hard to quantify.

Second, in addition to varying with wave properties, wave base varies with sediment grain-size and density, as well as other sediment properties (e.g. cohesion). Larger near-bed shear stresses are required to move coarse sediments, and their transport is generally limited to the relatively shallow depths of the shoreface. Net shoreward transport under shoaling waves tends to accumulate sediment on the shoreface, steepening its slope relative to the remainder of the shelf. The end result then is an equilibrium shoreface profile that

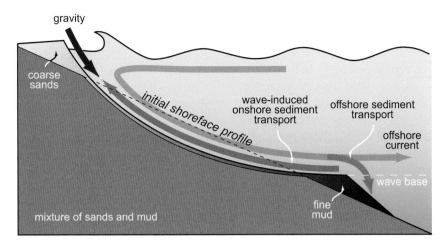

Fig. 11 Wave-induced modification of a shoreline profile. Wave base depends on more than just water depth. It also depends on the erodibility and thus the grain size of the sediments covering the shelf. This is because the water motion caused by waves decreases with depth, so the depth at which a wave will no longer move a grain of sand is shallower than the depth at which the same wave will no longer move a grain of clay. Given enough time, waves will, in theory, rework an initially planar shelf consisting of a sediment mixture into a concave surface across which grain size decreases from sand to mud (silts and clays) in the offshore direction. The sands are carried shoreward by the waves and create a shoreface that steepens until their onshore movement is counteracted by the downslope pull of gravity. The fines stay in suspension and ultimately are moved by return flows to deeper water, where they are no longer reworked by waves. In unusual cases (e.g. river floods), more mud can be supplied to the seabed above wave base than transport processes can remove. Wave climate also can vary, such that, during unusually calm periods, mud can temporarily deposit above normal wave base. Temporal variability of sediment supply and wave conditions can make the interpretation of wave base complex.

shoals, coarsens and steepens landward. Deeper exposures of predominantly sandy sediment are likely to be relict beds, last active during a period of lower sea level.

Finer sediment is more easily resuspended, and will remain in suspension until shear stresses decrease enough to allow it to settle. These conditions are most likely to occur in deeper water, where wave stresses are relatively low. There is a tendency for net seaward transport of fine sediment (Harris & Wiberg, 2002). Consolidation of fine-grained deposits increases their resistance to erosion, so that fine sediment deposition followed by adequate time for consolidation will produce beds with a low probability of remobilization. These conditions are likely to describe fine-grained deposits that form below wave base for typical storm conditions. However, observations on the Eel shelf and elsewhere demonstrate that muddy sediment can also accumulate at depths on the shelf characterized by storm-generated shear stresses that exceed critical values for erosion and deposition. Three key processes appear to be responsible for the development of these deposits:

1 flocculation of sediment suspended in the water column, which increases particle settling velocity (Hill *et al.*, this volume, pp. 49–99);
2 consolidation of muddy deposits on time-scales of weeks to months (Wheatcroft *et al.*, this volume, pp. 101–155);
3 deposition from wave-supported sediment gravity flows of high-concentration, fine-grained suspensions (Parsons *et al.*, this volume, pp. 275–337).

The long-term effect of deposition on the shelf is to fill available accommodation space, provided there is an adequate supply of sediment to the system from river or coastal sources. The shoreface is the region of greatest storm-generated erosion and deposition (Zhang *et al.*, 1999; Wiberg, 2000; Harris & Wiberg, 2002), which has a large effect on the overall shape of the shelf profile. However, recent observations and modelling suggest that deposition from wave-supported sediment gravity flows that can develop in association with a supply of easily eroded sediment (e.g. after a river flood) may play an important role in mid-shelf strata development.

Subaqueous deltas and wave-supported sediment gravity flows

Mid-shelf clinoforms are increasingly being recognized on energetic shelves that also receive significant sediment flux from point or line sources of fluvial sediment. In these regions, sediment plumes deposit sediments across the inner shelf. In rare circumstances, the sediments may also be delivered by hyperpycnal flows. Waves and currents then rework the finer fraction of these deposits, which move into deeper, calmer waters where they accumulate below wave base and build up to it. Although this is the fate of most muddy sediment, some unusual conditions (e.g. river floods) can supply more mud than transport processes can remove from the inner shelf. In general, however, fine sediment moves seaward, and with large systems a submerged clinoform structure develops,

referred to as a **subaqueous delta** (Fig. 12B). Examples are found associated with the Amazon River in Brazil (Nittrouer *et al.*, 1986), the Huanghe (Yellow) River in China (Alexander *et al.*, 1991), the Ganges–Brahmaputra River in Bangladesh (Kuehl *et al.*, 1997), and the Fly River in Papua New Guinea (Walsh *et al.*, 2004). Depositional bulges seaward of smaller rivers with high sediment yields are also suggestive of subaqueous deltas, but are not as well developed (e.g. the inner shelf seaward of the Eel River mapped by Goff *et al.*, 1996). Like their subaerial counterparts, true subaqueous deltas build strata as a result of variable sediment accumulation across their surfaces. The topset aggrades upward, depending on the erosive character of ambient waves and currents. The foreset receives the bulk of sediment and progrades seaward over the small amount of fine sediment deposited as bottomset (Nittrouer *et al.*, 1986).

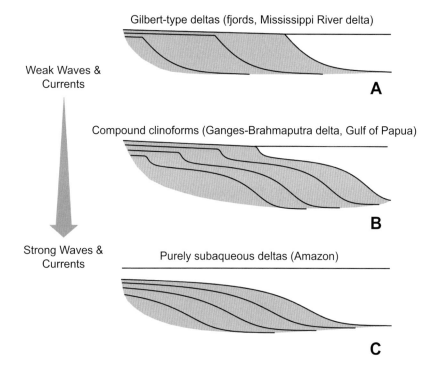

Fig. 12 Wave and current energy effects on delta morphology. (A–C) Subaqueous delta formation is linked to subaerial delta formation, as two ends of a spectrum for river-to-shelf profiles that will evolve depending on how fast sediments are delivered to the ocean by a river versus how fast they are then moved into deeper waters by waves and currents. (A) River-shelf profiles with a subaerial topset are at one end of this spectrum. These form where wave and current activity is low, such that the majority of the riverborne sediments accumulate close to shore. (C) River-shelf profiles with only a subaqueous delta are at the other end of the spectrum. In these settings, waves and currents are strong enough to carry all of the river sediments beyond its mouth and into water tens of metres deep. (B) The more common river-shelf profile where subaqueous deltas are found is one that contains both a subaerial and subaqueous delta. In these settings, the sediment supply from the river apparently overwhelms the capacity of the waves and currents to move all sediment farther seaward, but these processes still transport the majority of the sediments onto the shelf.

The combination of processes that move sediments across the topset of subaqueous deltas is more complex than simple wave or current motion. When the supply of freshwater and fine sediments to such a delta is great, convergent estuarine transport can concentrate suspended sediment and form a **fluid mud** (Kineke, 1993). This is a layer of suspended sediments with concentrations in excess of 10 g L^{-1}. Once developed, fluid muds can move as gravity-driven currents, if the seabed is sloping. More recently, wave-supported fluid muds have been recognized where the suspended sediment is concentrated within a wave boundary layer (Traykovski *et al.*, 2000; Hill *et al.*, this volume, pp. 49–99; Parsons *et al.*, this volume, pp. 275–337).

Turbidity currents are typically thought to form on the continental slope. These are mixtures of sediment and water that entrain additional water, deposit sediment and, if moving fast enough, erode the seabed (Parsons *et al.*, this volume, pp. 275–337). On the slope and rise, the sediment suspensions in turbidity currents are maintained by turbulence that causes erosion as the current moves down an inclined seafloor. In general, the shelf is too flat for turbidity currents to reach erosional speeds. However, waves on the shelf can supply the additional turbulence needed to keep sediments in suspension once the fluid muds begin to flow on a topset (Friedrichs & Wright, 2004). Predominantly wave-supported flows then give way to purely gravity-driven flows when the fluid muds spill below wave base down the relatively steep surface of a subaqueous-delta foreset. Gravity flows have been observed on several subaqueous deltas (Wright *et al.*, 1988; Kineke & Sternberg, 1995; Walsh & Nittrouer, 2004), and these have foresets that steepen downslope at a rate that appears to compensate for the decrease in bottom shear from waves, with increasing water depth (Friedrichs & Wright, 2004).

Independent movements of the shoreline and shelf break

As sediments are issued across the shoreline or shelf break, they pass from one process regime into another. The shoreline separates terrestrial sediment transport processes from their marine counterparts. The shelf break spans a more gradual transition from the predominance of waves and currents in pushing sediments across the shelf, to the ascen-

dancy of gravity-driven mass movements on the slope. What also commonly changes across these boundaries is the sediment flux. This latter change causes deposition or erosion, which in turn leads to movement of the shoreline or shelf break. For example, if rivers deliver more sediment to the shoreline than waves and currents remove, the shoreline builds seaward.

Changes in sediment flux across the shoreline often differ from those that occur across the shelf break, so these two boundaries commonly move at different rates and at times in different directions (Swenson *et al.*, 2005). Two end-member scenarios exist. One is the case in which all terrestrial sediments are trapped on a delta and none are transported seaward across the shelf, which can occur if waves and currents are minimal (Figs 12A & 13A). The shelf narrows as progradation of the delta moves the shoreline seaward toward the sediment-starved and thus immobile shelf break (Fig. 13). Eventually the two boundaries merge; the shelf is eliminated and the coastal plain leads directly to the slope.

The other end-member scenario is when waves and currents are so energetic and thus so effective in transporting sediments across the shelf that all sediments reaching the shore are transported over the shelf break. Under these circumstances, the shore remains fixed or even retreats landward while the slope and thus shelf break progrades seaward (Fig. 13C). The result is an ever-widening shelf.

The evolution of most continental shelves undoubtedly lies between these two end members. It appears that a significant fraction of the fluvial sediment flux to the shore is trapped there, either in an estuary or in a subaerial delta. Of the remaining sediment, some accumulates on the shelf and the balance is transported beyond the shelf break and accumulates on the slope. As in the formation of clinoforms, the partitioning of sediments among the shore, shelf and slope is a function of sediment grain size and the energy of the shelf processes (Swenson *et al.*, 2005). The coarser the sediment supplied and/or the less energetic the waves/currents, the greater the amount of sediment that accumulates nearshore (Fig. 13A). Likewise, the finer grained the sediments and/or the stronger the seaward transport, the farther the sediments can be transported toward the shelf break (Fig. 13B & C).

Where subaqueous deltas occur, a third boundary becomes important, the clinoform rollover. On

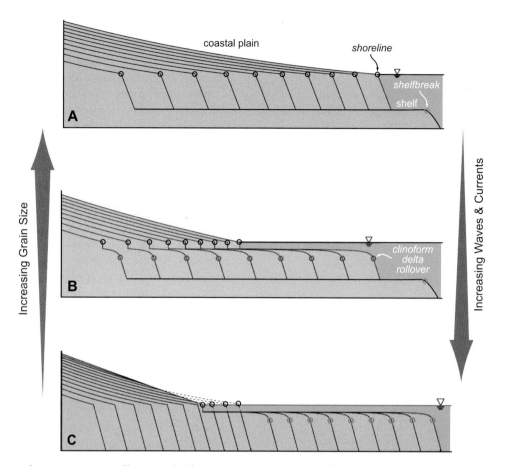

Fig. 13 Wave and current energy effects on shelf morphology. Three critical boundaries can exist on continental shelves: the shoreline (black circle), the shelf break (green circle), and, if a subaqueous delta is present, the clinoform rollover (red circle). Changes in the positions of these boundaries over time produce corresponding changes in the width of the shelf and the shape of its surface. Deposition and erosion rates commonly differ across these boundaries, so they move independently of each other. Commonly, the shelf break moves the slowest because the slope requires such a vast volume of sediment to build seaward. (A) Sediments accumulating along the shoreface cause it to advance upon the shelf break, reducing the shelf width. (B) If the majority of sediment arriving at the shore is moved seaward, a subaqueous delta will develop, transforming the shelf surface into an upper and lower terrace. (C) In the extreme, waves and currents can be so energetic that they erode the shoreline, while simultaneously moving the sediments across the shelf and onto the slope to prograde the shelf break. Under this circumstance, the shelf will widen. (Modifed from Swenson *et al.*, 2005.)

these shelves, most sediments from the river are trapped on the subaqueous delta foreset and do not make it to the shelf break (Fig. 12B & C). This sequestration of sediments leads to progradation of the subaqueous delta, so the position of its clinoform rollover with respect to the shoreline and shelf break can vary depending on the change in sediment flux across all three boundaries (Swenson *et al.*, 2005) (Fig. 13B).

Of the three boundaries, the shelf break will most likely be prograding the slowest. For the shelf break to prograde, the slope must prograde, and, given its large accommodation space, the

slope requires far more sediment to prograde than either the shoreface or the foreset of a subaqueous delta. In addition, the slope is likely to receive the least amount of sediment because it is depositionally 'downstream' of the shore and subaqueous delta. Thus with enough time, the subaqueous delta and/or the shore will reach and merge with the shelf break (Fig. 13).

The existence of an active subaqueous delta indicates that sediment is escaping to the shelf (Nittrouer & Wright, 1994). Additionally, unless there is some change in the sediment flux bypassing the shore or in the wave and current activity moving these sediments to the foreset, the subaqueous delta will continue to outpace the shore and should merge with the shelf break first (Fig. 13B). The shore will then follow, as the subaerial delta encroaches on the more slowly prograding slope and shelf break.

Shelf evolution during sea-level change

As alluded to in the previous section, the evolution of the shelf and slope is subject to change as environmental conditions change. Possibly the most important of these in terms of direct impacts is change in sea level, which has at least three fundamental and simultaneous effects on shelf and slope morphology. The changes:

1 create or destroy space for accommodating sediments on the margin;
2 move the supply of terrigenous sediments from rivers and coasts away from or toward the shelf break;
3 force a corresponding translation back and forth across continental margins of all processes except those that are driven by plate tectonics.

Sea level is defined in terms of either eustasy or relative sea level. **Eustasy** is global sea level and is measured with respect to a fixed datum such as the centre of the Earth (Van Wagoner *et al.*, 1988) (Fig. 14). The two most important controls on eustasy are tectonics and climate (Pitman, 1978). Tectonic events affect eustasy by changing the volume of the ocean basins. These events include the growth of the continents and, most importantly, changes in seafloor spreading rates. The latter alter the volume of mid-ocean ridges and the average age, temperature, isostatic buoyancy and thus depth of the seafloor. Climate affects eustasy principally through global cooling and warming spells. During global cooling, water is transferred from the oceans to glaciers, resulting in a drawdown of eustasy, while during global warming, the transfer is reversed and eustasy is raised.

When measured with respect to a more local datum, such as depth to basement, sea level is referred to as **relative sea level** (Van Wagoner *et al.*, 1988) (Fig. 14). Relative sea level differs from eustasy in that it also depends on local subsidence/uplift, which can cause the water depth at a location to deepen or shoal without any change in eustasy (Fig. 14). In this paper, the term sea level is used to mean relative sea level.

The geological record indicates that changes in sea level have rarely exceeded a few hundred metres (see summary in Allen & Allen, 2005).

Fig. 14 Sea level defined in terms of either eustasy or relative sea level. Eustasy is sea level referenced to a fixed datum such as the centre of the Earth. Relative sea level is referenced to a local datum, often the Earth's crust. Note that both types of sea level differ from water depth in that they pertain to the elevation of the sea surface at a location and not the amount of water there. Eustatic changes in sea level are caused by global variations in the total volume of ocean water and/or the total volume of the ocean basins. Relative changes in sea level also include any local subsidence or uplift. Even if eustatic sea level is constant, relative sea level will rise if the region subsides or it will fall if the region is uplifted. In this paper, the term sea level is used to mean relative sea level. (Modified from Emery & Myers, 1996.)

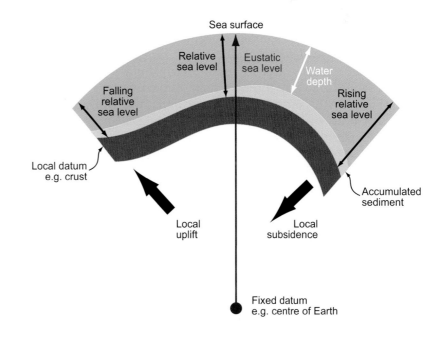

Exceptions include the Mediterranean (Ryan *et al.*, 1973) and Black Seas (Ryan & Pitman, 2000). During different eustatic falls in the Cenozoic, each was cut off from communication with the open ocean by a bedrock sill located at its mouth. While isolated, these Seas underwent significant evaporation, leaving not only their shelves but also their slopes exposed to subaerial erosion. When sea level eventually rose and topped the rock sills, flooding of the seas was catastrophic (sea level rose ~1900 m in the Mediterranean during the Messinian event, which caused a corresponding drop in eustasy of ~10 m; Ryan *et al.*, 1973). The old shelves and slopes worn by subaerial erosion were drowned and became the slate upon which the current shelves and slopes have since formed. Along other margins, only the largest sea-level falls ever expose the uppermost slope. Consequently, it is the width and depth of the shelf that undergo the most dramatic transformation during sea-level change.

Maximum shelf depth is primarily controlled by the amount of sea-level change. Shelf width is also controlled by the pre-existing surface gradient. The lower the gradient, the greater the increase/decrease in shelf width for a given rise/fall in sea level (respectively), and thus the greater the change in accommodation space on the shelf.

In the absence of any other processes, falling sea level moves the shoreline seaward in a **regression**, while rising sea level moves the shoreline landward as a **transgression** (Fig. 9). Where sediments are simultaneously being supplied to the coast by rivers, movement of the shoreline begins to fall out of phase with sea-level change. This is because the sediment supply acts to continuously build the shoreline seaward through delta progradation (Fig. 13A). Consequently, regressions will occur not only whenever sea level is falling, but also whenever the rate of sediment supply to the shoreface exceeds the rate of sea-level rise. Transgressions will only occur when the sediment-supply rate to the shoreface falls below the rate of a sea-level rise.

Consider a cycle in which sea level falls and then returns to its original elevation (Fig. 9). During the fall, delta progradation will accelerate the shoreline regression (Posamentier *et al.*, 1992). In general, the delta will continue prograding as long as the surface of the continental shelf being exposed by falling sea level has a lower gradient than the coastal plain created by deposition from the lengthening river feeding the delta (Blum & Tornqvist, 2000). However, where the sea-level fall exposes a surface with a greater dip than the coastal plain, the river will accelerate and potentially begin to incise the surface. Such incision appears to be common when a large fall in sea level exposes the steeper upper slope, or when a rapid fall exposes the shoreface or delta foreset (Vail *et al.*, 1991). When it occurs, the incision propagates upstream along the river channel, smoothing the gradient increase while deepening and steepening its profile (Posamentier & Vail, 1988).

As sea level then rises, shoreline regression slows and eventually reverses into a transgression, when the rate of rise exceeds the sediment supply (Jervey, 1988; Fig. 9). The rise drowns the delta and coastal plain, and floods the river valleys, transforming their reaches in the vicinity of the new shoreline into semi-enclosed estuaries (Fig. 9). The estuaries develop until the supply of sediments from the river once again exceeds the rate of sea-level rise or sea level stabilizes. At this point, the rivers develop deltas, shoreline regression is renewed, and the cycle is repeated (Fig. 9).

The degree to which waves and currents have an impact on shelf morphology during sea-level change depends on how quickly they can reshape the shelf profile. If sea-level change is too rapid and/or the shelf sediments are too hard to erode, waves and currents will have little effect on the profile. Deltas, beaches, islands and the coastal plains will be drowned by the rise and become relict features on the new shelf that may be gradually smoothed by wave and current activity. For relatively slow changes in sea level across shelves with erodible sediments, waves and currents will continuously be reshaping the profile toward one that is in equilibrium with and thus follows wave base (Fig. 15).

Waves and currents probably cause their greatest erosion during a sea-level fall, because it also lowers wave base (Fig. 15A). Shelf depths in the zone between the shoreline and wave base are increasingly eroded by waves. Where they are eroding, the waves cut down into finer sediments initially deposited in deeper water under calmer conditions. These sediments are moved seaward, increasing accumulation rates farther seaward on the shelf or even the slope (Fig. 15A). At the same

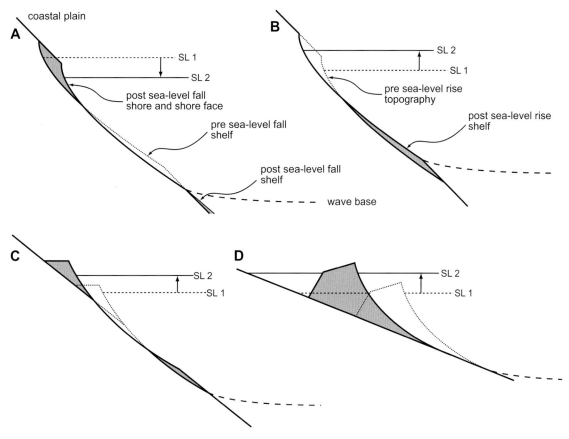

Fig. 15 Shoreline responses to sea-level change. (A) As sea level falls, so does wave base resulting in the remobilization of shelf sediments that lay below the action of waves. The finer fraction of these sediments are picked up and moved closer to the shelf break, but the coarser-grained sediments are moved in the opposite direction toward the shore. The result is that even as falling sea level exposes the old shoreface, waves are building a new one, contributing to the shoreline regression (A). When sea level rises, wave base rises as well. If the drowning coastal plain is too steep for the waves to build a beach, the waves will erode into the coastal plain and all of the mobilized sediments will accumulate farther offshore (B). If the coastal plain has a gentle gradient, however, and the waves can build a beach (C) or a barrier island (D), then these sediments will be transported landward as sea level rises. (Modified from Cowell *et al.*, 1995.)

time, the asymmetric wave motions saltate coarser-grained sediments landward, helping build the beach seaward and promoting shoreline regression (Fig. 15A).

During a sea-level rise, the motions are reversed. Waves erode the flooding coastal plain, with erosion being greatest when a portion of the coastal plain is first submerged. The erosion diminishes as water depth increases and wave base rises across the new shelf surface. If wave base eventually rises above the seafloor at a location, erosion may switch to deposition as fine-grained sediments are swept offshore by wave and current activity. In areas away from rivers and thus receiving no sedimenta-

tion, the waves erode and rework coastal-plain sediments laid down during or prior to the sea-level fall (Fig. 15B). In areas where river deposition is occurring as sea level rises, the waves erode these more recent deposits and older coastal-plain sediments are preserved.

Coastal-plain preservation is enhanced even more where waves build coarse-grained beaches and barrier islands along the shoreline (Fig. 15C & D). As sea level rises, the waves have to rework these deposits before they can erode into coastal-plain sediments. In the process, the waves move the beaches or barrier islands landward. In fact, where these deposits have been built up to significant enough relief, all of the reworking caused by waves can be expended in back-rolling the beaches/barrier islands such that there is no erosion of the former coastal-plain deposits (Cowell *et al.*, 1995, 1999; Fig. 15C).

Again, the above scenarios assume wave reworking happens faster than sea-level change. This may occur on some shelves but probably not all. In fact, variations in the degree to which shelves are reworked to be in equilibrium with the wave climate may be one of the reasons shelf profiles are so varied.

PROCESSES THAT ACT TO LIMIT THE SLOPE OF THE CONTINENTAL SLOPE

Seafloor failure and submarine groundwater flow

Most continental slopes exhibit one of three types of profiles: (i) an oblique clinoform profile, (ii) a sigmoidal clinoform profile or (iii) a simple linear profile (Adams & Schlager, 2000). All three have been interpreted to result in part (the sigmoidal profile) or wholly (the oblique and linear profiles) from sediment mass movements caused by seafloor oversteepening and failure. If correct, then continental slopes should dip at an angle that approaches the threshold for sediment failure. Slope sediments are predominantly muds, which have a maximum angle of repose ~25–35° when dry. When saturated, the buoyancy force exerted by the pore waters in these sediments reduces this maximum angle by about half to 12.5–17.5° (Allen, 1985). This range is still about three to five times steeper than the ~4° dips that characterize most continental slopes, so factors other than just gravity must be at play in limiting slope gradients.

One of these factors may be earthquakes (Pratson & Haxby, 1996). Earthquakes are geologically frequent even on passive margins. For example along the USA east coast, the recurrence time of a magnitude 8 earthquake on the Richter scale is only 2000 yr (Seeber & Armbruster, 1988). The seismic shaking that results from an earthquake can cause a slope to fail at much lower angles than its sediments would support in a static environment. The threshold for failure may be lowered even further if the seismic shaking induces a cyclic loading on the sediments, which can increase sediment pore

pressures and thus decrease the resistance of slope sediments to shearing under the downslope pull of gravity (Hampton *et al.*, 1996; Lee *et al.*, this volume, pp. 213–274). Anecdotal support for the impact of earthquakes is seen in the average slope gradients along USA active margins, which are roughly half those of USA passive margins (Pratson & Haxby, 1996). The one exception to this is the passive margin in the USA Gulf of Mexico. Ongoing deformation and flow of deeply buried salt have so disrupted the slope that its average gradient is < 0.5°. The slopes offshore of Brazil and Nigeria have been similarly lowered by deformation of deeply buried muds (O'Grady *et al.*, 2000).

The force driving the deformation of substrates along continental margins is the weight of rapidly deposited sediments. In fact, a correlation exists globally between gentle slopes, high sediment input and unstable substrates, making rapid sediment loading a second potentially important cause for the low average angle of continental slopes (O'Grady *et al.*, 2000). Rapid sediment loading can destablize slopes in a number of ways, the most important being the generation of pore pressure in excess of **hydrostatic pressure,** or that due to the weight of the overlying water column. Excess pore pressure results when sediments accumulate faster than they dewater (Suppe, 1985). Hydrostatic pressure reduces the maximum angle of repose for sediments by about half. Excess pore pressure can reduce the angle for failure even more, dropping it all the way to 0° if the pressure reaches **lithostatic** levels, i.e. it completely supports the weight of the overlying sediment column.

Under static conditions, failure of a mud slope at 2–4° requires a pore pressure approaching ~90% of lithostatic pressure (Prior & Suhayda, 1982). This may occur on a local basis and it may be an important cause for certain seafloor failures, but most continental-slope sediments have much lower pore pressures. Clear evidence of this are the dips of 10° to > 15° on the walls of submarine canyons, which could never be achieved if the slope strata were so overpressured.

If pore waters are flowing, then pore pressures do not need to be significantly greater than hydrostatic to cause low seafloor gradients. This is because flowing pore waters exert a drag or seepage force on the sediments, and if a component of this force is directed downslope it adds to the pull of gravity, lowering the threshold angle for seafloor failure (Iverson & Major, 1986). Seepage force appears to be an important factor in causing submarine slides and slumps along active margins (Orange & Breen, 1992; Fig. 16A–C). The force results from the expulsion of pore waters being driven out of the slope sediments as they are tectonically accreted to the margin. Along passive margins, seepage forces arise from differential sediment loading. Borehole measurements of anomalously high porosities in strata from the New Jersey (Dugan & Flemings, 2000; Fig. 16D) and Louisiana (Gordon & Flemings, 1998) continental slopes suggest pore waters are moving out of the lower slope in response to excessive sediment loading on the upper slope (Fig. 16E). In the Gulf of Mexico, permeability barriers to this

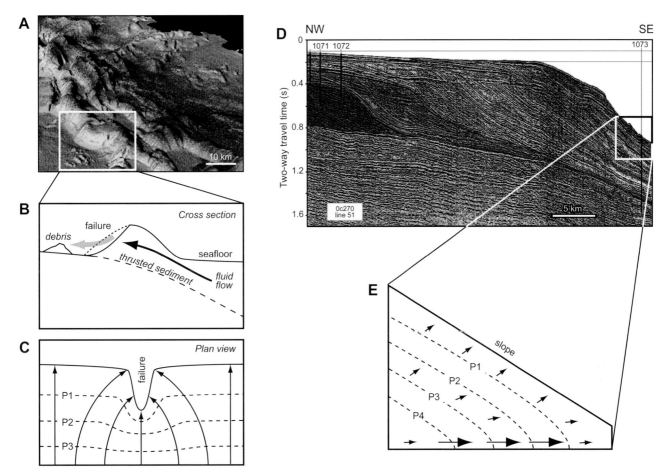

Fig. 16 Seafloor failures along continental slopes associated with submarine groundwater flow. Fluid flow through sediments exerts a drag or seepage force. This force can combine with the downslope pull of gravity and any reduction in sediment shear strength (due to elevated pore pressures) to cause the sediments to fail at significantly lower gradients than they would if the pore waters were not flowing. (A–C) Along active margins, the compression, folding and faulting of sediments in the accretionary prism expel pore waters, inducing a surfaceward flow that in turn can trigger failure. (A & B) This process appears to have played an important role in causing the numerous failures found along the seaward flanks of folded strata on the Oregon slope. (C) The process also may be promoting additional failures at these sites, because the pore-pressure field within the strata should be directed into the chutes and canyons excavated by the failures. (D & E) Fluid flow can occur beneath continental slopes along passive margins. (D) The driving force in these settings is enhanced sediment accumulation on the upper slope, which increases loading and compaction, and (E) drives pore waters toward the surface of the lower slope. Modelling suggests that even without considering the destabilizing effect of this fluid flow, the process can raise pore pressures to near lithostatic levels. (A, modified from Pratson & Haxby, 1996; B, modified from Orange & Breen, 1992; C, modified from Orange *et al.*, 1994; D & E, modified from Dugan & Flemings, 2000.)

flow have produced lithostatic pore pressures. Oil companies have suffered considerable financial setbacks when drilling these regions, because the boreholes create an outlet for the overpressured pore waters, which cause the borehole walls to collapse.

Bottom shear from internal waves

Another important mechanism that may be limiting the steepness of the continental slope is **internal waves** (Fig. 17). These are waves that form within the ocean interior due to perturbations along density interfaces. Triggering mechanisms include the convergence and then divergence of ocean currents and tides as they flow over rugged seafloor topography, and the rise and collapse of sea-surface surges caused by the passage of low-pressure atmospheric storms (Knauss, 1997).

Internal waves form across small changes in density, so the gravitational restoring force damp-

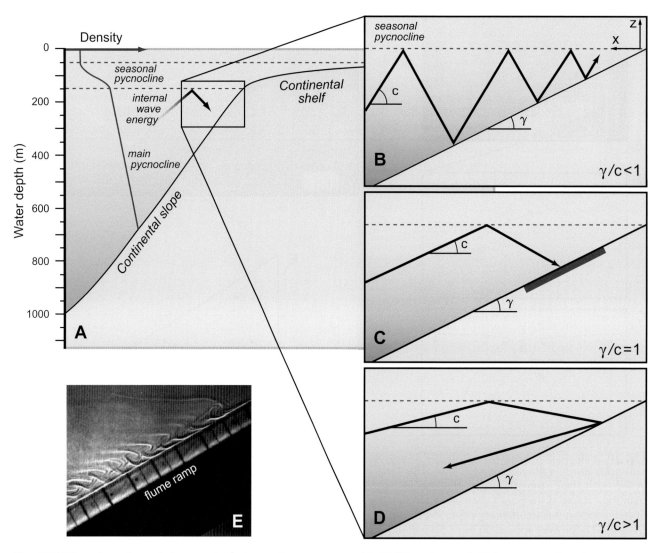

Fig. 17 Effect of continental slope angle on internal wave energy. (A–D) When internal tidal energy strikes the continental slope, the result varies. (B) If the characteristic angle of incidence is steeper than the slope, internal tidal energy is transmitted landward as it bounces between the seafloor and the base of the mixed layer. (D) If the characteristic angle is less than the slope, the energy is reflected back toward the deep ocean. (C) When these two angles are equal, the energy is trapped near the seafloor, where it may be sufficient to erode the sediments or at least to prevent suspended sediments from deposition. (E) Wave-tank experiments confirm the importance of geometry. Adjusting the inclined portion of the tank to match the characteristic angle of the internal waves causes a great deal of energy to be dissipated in turbulent movements along the bottom. (After Cacchione & Pratson, 2004.)

ening the waves is much less than at the sea surface; consequently, internal waves can be much higher and longer than their surface counterparts (Knauss, 1997). Internal waves can have heights reaching tens to hundreds of metres, periods that range from minutes to about 12 h (the internal tide), and at the longer periods, wavelengths that can span tens to hundreds of kilometres. Unlike surface waves, internal waves are not limited to propagating horizontally. The propagation angle, c, depends on latitude and the density structure of the water column, and can vary between being horizontal ($c = 0°$) to vertical ($c = 90°$; Knauss, 1997).

When internal waves encounter the continental shelf and slope, they interact with the seafloor in one of three ways, depending on the magnitude of c relative to the bottom slope angle γ (Cacchione et al., 2002; Cacchione & Pratson, 2004). If $c > \gamma$, then the internal waves are reflected shoreward (Fig. 17B). This is largely the case on the continental shelf where the bottom gradients are relatively small. If $c < \gamma$, then the waves are reflected back out to sea (Fig. 17D). Such reflection is common on the steeper continental slope. The third case is when $c = \gamma$ (Fig. 17C). In this case, the internal-wave energy becomes trapped along the slope and the wave itself develops a bore-like form that propagates up the slope surface.

In all three instances, internal waves exert a shear stress on the seabed, but the stress is greatest when γ is within ± 20% of c. Under such conditions, bottom shear velocities can exceed 1 cm s^{-1}, which is enough to inhibit sediment deposition, generate intermediate and bottom nepheloid layers, produce bedforms, and possibly even erode the seafloor (Cacchione & Southard, 1974).

In the open ocean, the mean angle of the continental slope is often comparable to the propagation angle of the internal tide (Cacchione et al., 2002). This is the longest of internal waves and the most dominant form in terms of energy content. The propagation angle for the internal tide changes with water-column properties, and so varies across the ocean. The mean angle of the continental slope also varies, changing with geological setting. Consequently, the correspondence between these two angles from one margin to another is not only surprising, but probably more than just coincidence. The internal tide may exert a major influence on the gradient of the continental slope by increasingly inhibiting deposition on the slope as it steepens

toward the propagation angle of the internal tide and as bottom shear stresses increase. The end result would be continental slopes that are in equilibrium with the internal tide worldwide (Cacchione et al., 2002).

This hypothesis is attractive because it would help explain why the mean angle of continental slopes is so much lower than the maximum angle of repose for the sediments that compose them. Unlike other processes that can affect the gradient of the continental slope, the internal tide is a global process that has been operating continuously since the existence of the oceans. The idea that internal tides control the gradient of the continental slope is not without problems. There are many places in the world where the mean gradient of the slope is greater than c. In these regions, the internal tide will not act to lower the gradient because its bottom shear stress decreases as γ increases relative to c. Clearly other processes that affect the morphology of the continental slope are also playing a role in keeping the gradient low (Cacchione et al., 2002; Cacchione & Pratson, 2004).

Turbidity-current erosion and deposition

Turbidity currents also may be contributing to the characteristically gentle gradient of continental slopes. In addition to other formative mechanisms, turbidity currents can evolve from seafloor failures (Lee et al., this volume, pp. 213–274; Parsons et al., this volume, pp. 275–337). In fact, this is the way that a number of the best-known turbidity-current events have been generated (Heezen & Ewing, 1952; Malinverno et al., 1988). There is a strong association between eroded continental slopes scarred by numerous seafloor failures, and vast accumulations of turbidite deposits extending from the base of these slopes onto the continental rise, commonly in the form of submarine fans (Ericson et al., 1961; Walker, 1978). This combination has contributed to the mistaken viewpoint that the slope is a region of erosion and sediment bypassing, largely caused by turbidity currents. However when the thickness of marine sediments is tallied globally, a different picture of the slope emerges. While underlying only 6% of the world ocean surface area, slopes contain 41% of ocean sediments, the largest volume for any marine physiographical province (Kennett, 1982). Thus continental slopes are sites of sediment accumulation far more than they are regions of erosion and sediment bypassing (Pratson, 2001).

Turbidity currents may be an important cause for this. One reason is that when large turbidity currents pass downslope through submarine canyons, they can overtop the canyon walls and spread sediments into intercanyon areas. This overbank deposition decreases with distance away from the canyon in a fashion analogous to floodplain deposition adjacent to rivers. As a result, the process creates levees along the walls of a canyon, increasing its relief (Rona, 1970). For canyons that have been subject to many large flows, the overbank deposition commonly shows evidence of having been deflected by the Coriolis force, for one side of these canyons (the right side in the northern hemisphere) is significantly higher than the other (e.g. Hudson Canyon; Shor & McClennen, 1988).

Turbidity currents may also build up intercanyon areas on the slope by directly flowing down them. As previously discussed, turbidity currents can be initiated by numerous triggering mechanisms (e.g. earthquakes, waves; Parsons *et al.*, this volume, pp. 275–337), which cause resuspension of seafloor sediments on a sloping bottom. As these sediments settle back onto the seafloor, they are also acted on by the downslope pull of gravity. If concentrated enough, the sediments may begin to move as a turbidity current rather than simply coming to rest (Traykovski *et al.*, 2000). Whether or not this turbidity current causes net deposition hinges on the amount of sediment it gains through erosion of the seabed versus the amount it simultaneously loses to settling. When erosion exceeds deposition in a turbidity current, it gains mass and thus accelerates, causing more erosion in a positive feedback termed **ignition** (Parker *et al.*, 1986). If deposition exceeds erosion, then the turbidity current gradually loses mass and along with it the excess density driving its movement. Under these circumstances, the current slowly dies, as deposition falls off exponentially with distance downslope (Parker *et al.*, 1986).

One of the most important variables determining whether a turbidity current ignites or dies is the existing seafloor gradient. Theoretical calculations supported by experimental modelling show that if the gradient of the slope is steep enough to cause ignition, repeated turbidity currents will eventually reduce the gradient until the mean drag force caused by the flows combined with the downslope pull of gravity is exactly balanced by the shear strength of the slope sediments (Kostic *et al.*, 2002). If the slope gradient is so low as to cause a turbidity current to die, the seaward-thinning deposits from repeated turbidity currents will eventually steepen the gradient, and the currents will no longer deposit any sediments on the slope but instead bypass it (Gerber *et al.*, 2004). Interestingly, the equilibrium slope in both cases is projected to be within several degrees of the mean angles characteristic of continental slopes (~4°). Consequently, if the slope is more or less steep, repeated turbidity currents should bring it back toward the angle of observed slopes (Fig. 18). Thus, turbidity currents join other mechanisms (earthquakes, pore-water flow and internal waves) that may be keeping the mean gradient of the continental slope far below the internal angle of repose for slope sediments. Turbidity currents are the only one of the mechanisms that will both build and erode the slope to maintain the observed gradients.

PROCESSES THAT CREATE SUBMARINE CANYONS AND SLOPE GULLIES

Turbidity currents versus seafloor failure in forming submarine canyons

Of the many facets to continental slope morphology, the one that has received the most study is **submarine canyons**. These are steep-sided, deep valleys that are often linear and largely directed downslope. However, they may also be meandering (von der Borch *et al.*, 1985), or have sharp turns caused by structures such as faults (Song *et al.*, 2000). In general, submarine canyons vary from hundreds of metres to kilometres in width, and from tens to hundreds of metres in relief. The relief is greatest on the upper to middle slope and diminishes toward both the continental shelf and rise.

Essentially all submarine canyons extend to the base of the continental slope. Where canyons begin on the slope, though, appears to relate to their width, relief and overall form. The narrowest and shallowest submarine canyons begin on the middle to lower continental slope, and are commonly U-shaped in cross-section (Fig. 19A). Canyons that begin on the upper slope are often larger, and can have V-shaped as well as U-shaped cross-sections

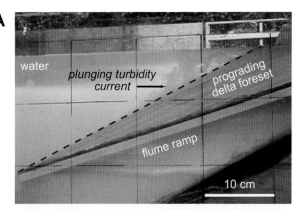

Fig. 18 Relationship between turbidity currents and slope angle. A turbidity current will either erode, bypass or deposit on the slope depending upon the seafloor gradient and the character of the current, including the amount of sediment it is carrying. The effect of these two variables is illustrated in a sequence of three frames from a videotape of a laboratory experiment in which a continuous turbidity current flowed down a ramp. (A) The initial incline of the ramp (18°) was too low to sustain the small current so it deposited its sediment load on the ramp, building up a prograding clinoform. The foreset of this clinoform eventually steepened to an angle (red dashed line) such that sediment was bypassed (i.e. not deposited). (B) The initial sediment concentration of the turbidity current was then reduced by half. This decreased its excess density relative to the water in the tank, immediately eliminating the ability for sediment bypassing. Once again, the turbidity current began to deposit its sediment load, but this time it built an even steeper foreset (green dashed line). (C) The initial concentration of the turbidity current was increased back to the level used at the beginning of the experiment. This increased the density of the turbidity current, which increased its speed, caused erosion and reduced its gradient (blue dashed line) back toward the angle at which bypassing first occurred. These experiments show that if the dip of the seafloor is too low to sustain a turbidity current, the current will increase the seafloor gradient through deposition toward an angle at which the current will bypass the region. Conversely, if the seafloor is too steep for the turbidity current, it will erode the seafloor, lowering its gradient toward the same bypassing angle. The sediments used in this experiment were coarser than most continental-slope sediments, therefore the equilibrium angles are greater than those of natural continental slopes.

(Mountain, 1987; Fig. 19A). The largest submarine canyons are those that incise the shelf break, particularly those that are or at one time were directly connected to a river (Twichell & Roberts, 1982; Farre *et al.*, 1983). These canyons are predominantly V-shaped in cross-section.

Submarine canyons are morphologically similar to large subaerial canyons cut by rivers. Both share similar cross-sectional profiles (e.g. Scripps Submarine Canyon and the Grand Canyon of the Colorado River; Shepard, 1977), and concave-upward profiles along their channel axes. Like rivers, submarine canyons commonly link to form

tributary drainage networks that gather sediments from across large reaches of a continental margin and funnel them onto the continental rise (Pratson, 1993). The striking similarity between submarine and subaerial canyons implies potential parallels in their evolution. This attraction has engendered significant investigation into the genesis of submarine canyons, a subject that continues to be debated.

Active submarine canyons occurring along narrow continental shelves suggest that there is a clear linkage between submarine-canyon formation and the supply of sediments from rivers and/or

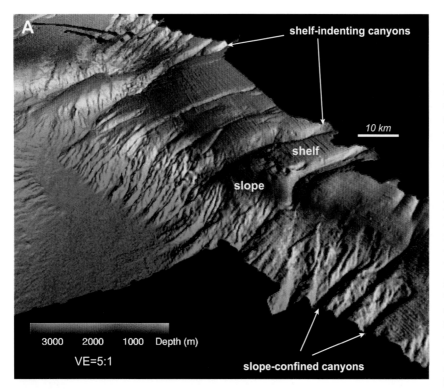

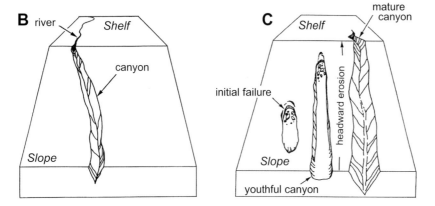

Fig. 19 Submarine canyons on continental margins. (A) Some submarine canyons are confined to the slope, and others incise through the shelf break and connect to channels on the shelf, such as is seen in this perspective image of the continental slope in the vicinity of Medicino, California (box b in Fig. 1). Many processes can contribute to the formation of submarine canyons, and the history of each is undoubtedly unique. (B–C) Submarine canyons are generally thought to originate in one of two ways. (B) Canyons may be carved by turbidity currents with a sediment source from or near rivers, when sea level is low. At that time, the shelf is narrow and the rivers are close to the shelf break. (C) Submarine canyons may evolve from slope failures. Repeated collapsing of the headwall results in the development of a chute that migrates upslope toward the shelf break. If this chute reaches the shelf break, it can then capture sediment drainage across the shelf and evolve into a canyon similar to the first model. However in this second case, the canyon can evolve independently of sediment supply from the shelf and thus independently even of sea level. (B & C modified from Pratson *et al.*, 1994.)

the continental shelf. In a few locations, rivers can be traced flowing directly into the heads of these canyons (Fig. 19B), but often the two are separated and even laterally offset from one another by an intervening shelf. In these latter cases, the sediments are supplied to the canyons by shelf currents, some of which flow in well-defined circulation cells (Gorsline, 1970; Felix & Gorsline, 1971). Advecting grains resuspended by waves, these currents gradually transport the sediments into the heads of the canyons where they indent the shelf break. The sediments are then periodically flushed down the remainder of a canyon, often when the build-

up fails under cyclic loading by large waves from storms (Dill, 1964). The down-canyon movement of sediments can occur as a debris flow, but many submarine canyons incise older slope strata, and such erosion is more likely the result of turbidity currents. The abundance of turbidites that emanate from the mouths of canyons onto the continental rise suggests that downcutting by turbidity currents is one of the main ways that submarine canyons form (Ericson *et al.*, 1961).

Submarine canyons currently separated from rivers and the nearshore by broad continental shelves commonly appear to be relict features, last active when shorelines and thus sediment supply were near the shelf break during sea-level lowerings (Emery & Uchupi, 1972). However, a significant fraction of canyons have heads that occur well below the shelf break and show little if any morphological evidence of having been cut by turbidity currents (Twichell & Roberts, 1982; Fig. 19A). Instead, the scars of submarine slides often form the head and sidewalls of these canyons. The scars suggest that in addition to turbidity currents, seafloor failures can form submarine canyons, initiating them on the middle to lower slope (Fig. 19C). Such canyons then appear to grow upslope through repeated failures of their headwall (Farre *et al.*, 1983). Where these canyons breach the shelf break, they then tap into the supply of sediments stored on the shelf and become subject to turbidity-current erosion (Fig. 19C).

Slope failures need not correlate with times of high sediment input to the slope and low sea level, therefore canyon formation by this process can be similarly independent (Farre *et al.*, 1983). However, some mechanism is required to cause continued failure of a canyon's head wall and its upslope growth. One of the most viable is **spring sapping** (Robb, 1984, 1988; Orange & Breen, 1992; Orange *et al.*, 1994). In this process, groundwater discharging from the face of a sedimented slope exerts a seepage force on the sediments, which in combination with the pull of gravity offsets the frictional forces pinning the sediments in place and can cause them to fail (Fig. 16). Where seepage failure is recurring and the material is carried away, a pit forms. The pit can then grow upslope as the seepage failure undercuts the pit headwall and triggers **retrogressive slides** (headward-eroding) (Fig. 19C).

One of the attractions of spring sapping as a canyon-forming process is that it provides a possible explanation for the seeming regularity to the spacing of submarine canyons along certain continental slopes (Orange *et al.*, 1994). Creation of a canyon dramatically increases the drop in pore pressure between the seafloor and the subsurface in regions where slope sediments are overpressured. This drives pore fluids to flow toward the canyon head, increasing the seepage force there and facilitating headward growth of the canyon through retrogressive failure (Fig. 16C). The convergence of flow at the canyon head also leads to

a divergence of flow between canyons, reducing the likelihood of failure in these regions. Thus a self-organized feedback between hydrological and geomorphological processes develops, which limits the spacing between canyons. The factors that govern this spacing are the bathymetric gradient of the slope, the pore-pressure gradients within the slope sediments, and their material strength. Growth rate and thus size of the canyons is also important, as larger canyons can capture the hydrological discharge of smaller canyons, terminating their development.

However, not all canyons that appear to be created by retrogressive slope failure truly have that origin (Pratson *et al.*, 1994). Some **slope-confined canyons** have completely buried, upslope extensions that may once have indented the shelf break. When followed downslope, these buried canyons connect with modern canyons, usually where the sediment cover has thinned and a seafloor trough has developed (Mountain *et al.*, this volume, pp. 381–458). Seaward, the modern canyons have exhumed the buried canyons and follow the older paths to the base of the slope. These canyons have captured turbidity currents initiated along the upper slope and shelf break, and confined them to the former paths of the buried canyons. Downcutting by the turbidity currents then oversteepened the canyon walls, triggering failures and giving the appearance of having been formed entirely by slope failure (Pratson *et al.*, 1994).

A greater understanding of submarine-canyon formation would come from knowing their age, but here too there are uncertainties and disagreements. The timing of canyon formation has rarely been established (Miller *et al.*, 1987) because submarine canyons are erosive features that cut through the sediments that mark the onset of their formation. The problem is complicated even more by the likelihood that submarine canyons are formed by multiple processes (Shepard, 1981). Some canyons are long-lived and/or have complicated histories due to repeated episodes of downcutting, infilling and rejuvenation (Ryan *et al.*, 1978; Goodwin & Prior, 1989; Mountain *et al.*, 1996). Even the relationship to sea level is complicated, because sediment can be transported into the canyons at times other than lowstands (Carson *et al.*, 1986; Kuehl *et al.*, 1989; Weber *et al.*, 1997; Droz *et al.*, 2001; Mulder *et al.*, 2001).

Consequently, existing theories for the timing of canyon formation fall between two end-member viewpoints. One is that submarine canyons are most active and probably form during lowstands in relative sea level, when rivers deliver sediment to a shoreline that is near or below the shelf break (Daly, 1936). At these times, there is no space left on the shelf to store sediments arriving from the continent and they are transported to the deep sea by turbidity currents, which play a major role in cutting the canyons (Kolla & Macurda, 1988; Rasmussen, 1994). This is supported by the common occurrence of abandoned (Emery & Uchupi, 1972) and buried shelf channels (McMaster & Ashraf, 1973; Berryhill et al., 1986; Burger et al., 2001) that lead toward canyon heads.

The other line of thinking is that submarine canyons initiated by slope failure may or may not be related to sea-level change. Mechanisms that can trigger failures independent of sea level and sediment supply include earthquakes (Garfield et al., 1994) and spring sapping (Robb, 1984). Another is the dissolution of sediments or sedimentary rock beneath the slope surface (Paull & Neumann, 1987), such as that occurring in carbonate rocks being infiltrated by groundwater. Two others are subduction (Orange & Breen, 1992) and diagenesis (McHugh et al., 1993), both of which can expel fluids through sediments with such force as to cause repeated slope failure. However, even spring sapping can be in-phase with sea level. Lowstands can allow the boundary between fresh and saline groundwaters to be pushed beyond the continental slope, causing freshwater to escape due to aquifer pressures (Johnson, 1939).

Multiple buried canyons incising regional unconformities are interpreted to have formed and infilled over the course of a single fall and rise in sea level (Fulthorpe et al., 2000). However, other nearby buried canyons preserved between sequence boundaries indicate that canyon formation is sufficiently variable in geological history and location on the slope as to occur at times other than just lowstands (Fulthorpe et al., 2000).

Turbidity currents versus seafloor failure in forming slope gullies

Smaller counterparts to submarine canyons on the continental slope are **gullies** (Fig. 20). These are channel-like features that can have as little as 1 m of relief. Slope gullies have not engendered the same level of interest as submarine canyons, but they are starting to receive greater attention because many recent studies have discovered them throughout the world.

Slope gullies can be separated into two general categories: erosional and aggradational. Erosional gullies can be further subdived into submarine rills and dendritic gullies. **Submarine rills** are downslope-trending gullies that converge with one another or a nearby submarine canyon at low angles, in a fashion analogous to subaerial rills formed on terrestrial landscapes by runoff (Pratson et al., 1994; Fig. 20A). They range from narrow furrows 5–10 m wide and 1–2 m deep, to broader features 50–300 m wide and 10–40 m deep. **Dendritic gullies** on the other hand intersect submarine canyons at high angles to give the latter a leaf-like, dendritic erosional pattern (Farre et al., 1983; McGregor et al., 1983; Fig. 20A & B). They are relatively short (0.5–5 km long), range from 75 to 250 m wide, and are ~10–20 m deep.

Dendritic gullies and rills are probably formed by different processes. Dendritic gullies occur on the steep slopes of submarine canyon walls (~14°–20°) and their sharply defined, irregular headwalls show clear evidence of having been excavated by slumps and slides (Fig. 20A & B). These failures are likely to be caused when the walls of the canyon

Fig. 20 (*opposite*) Submarine gullies on continental margins. Submarine gullies are smaller than submarine canyons but appear to be intimately related to them. (A & B) Dendritic gullies, which incise the walls of submarine canyons. Submarine rills are narrow, relatively straight gullies that extend down open-slope areas. (C–E) Aggradational gullies appear to originate from submarine rills, but persist even as the slope surface aggrades. Dendritic gullies are an integral part of submarine canyons. Submarine rills and aggradational gullies (A & D) are located on open slopes between canyons. However, on the New Jersey margin, submarine canyons have captured the drainage of submarine rills (left side, A), suggesting the two have grown from a common sediment source. On the California margin, aggradational gullies (C & D) feed into Trinity Canyon, forming an extensive tributary network (E). (A, modified from Farre et al., 1987; B, modified from Ryan, 1982; C, modified from Spinelli & Field, 2000; D & E, modified from Orange et al., 1994.)

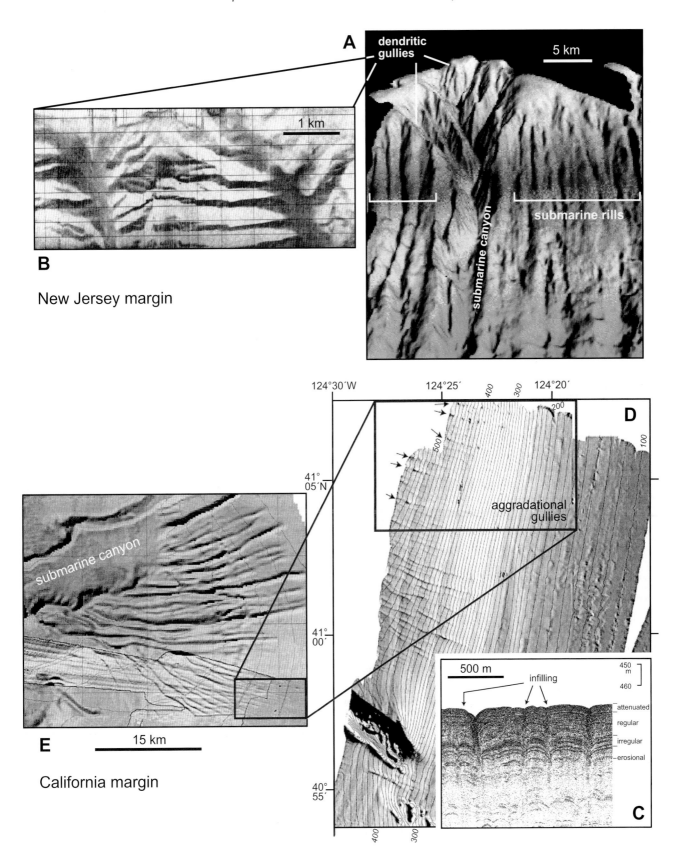

New Jersey margin

California margin

are undercut by turbidity currents flowing through the canyon's thalweg (McGregor *et al.*, 1983). By comparison, the narrower rills are found on much-lower-gradient open-slope areas between canyons (Pratson *et al.*, 1994) (Fig. 20A). This is also where aggradational gullies are found (Spinelli & Field, 2001) (Fig. 20C). Both appear to be created by turbidity currents. Whereas rills are cut by turbidity-current erosion, **aggradational gullies** grow by differential sediment accumulation, with less sediment accumulating on gully floors than on inter-gully areas (Spinelli & Field, 2001).

How turbidity currents generate rills and aggradational gullies on an open slope is an interesting problem. In this setting, turbidity currents have a tendency to diverge and form a **sheet flow**. The gullies, however, are an expression of flow convergence. In the absence of any pre-existing depressions, such convergence could develop if the head of the spreading turbidity current becomes destabilized and starts to **finger** (i.e. create isolated extensions of the turbidity-current head; Parsons *et al.*, this volume, pp. 275–337). Such fingering could arise from small spatial and temporal variations in flow concentration and thus speed. Destabilization may also result from the increase in the centrifugal force acting on the turbidity current when the current passes over a curved break in slope. Fingering of a turbidity-current head will lead to variations in deposition and erosion that, if significant enough, could begin to channelize the flow by subsequent turbidity currents.

Submarine rills should form where the channelizing turbidity currents cause net erosion, but why the aggradational gullies develop is less clear. Analyses of seismic-reflection data indicate that aggradational gullies first develop where erosional rills have been buried by uniform and gradual sedimentation across large stretches of the continental slope (Burger *et al.*, 2001; Spinelli & Field, 2001). This switch from an erosional to depositional regime is interpreted to occur following a lowstand, as sea level rises and the supply of sediments to the slope is reduced. While the rills are buried, their topographic expression is preserved even as the seafloor aggrades. Thus the rills are transformed into aggradational gullies. Complete infilling of the gullies is prevented, because turbidity currents and their high shear stresses are focused on gully floors.

When viewed from a more regional perspective, submarine rills and aggradational gullies are often found to feed into a large submarine canyon, and thus are part of a more extensive submarine drainage system (Orange, 1999; Fig. 20D). This relationship suggests that the rills and gullies could play a direct role in the evolution of submarine canyons. One idea in this regard is that submarine rills in an open-slope area grow and coalesce like their subaerial counterparts (Pratson & Coakley, 1996). Due to spatial variability in the frequency and size of the turbidity currents spilling down the slope, some rills will deepen and widen faster than their neighbours, altering the local slope, and capturing drainage in a fashion analogous to stream piracy (Fig. 20A). Eventually, the larger rills become deep enough and their walls steep enough that failure occurs. The chutes excavated by these failures then evolve into headward-eroding canyons that advance upslope following the path of the rills. This advance is caused by retrogressive failure of the canyon head wall each time it is undercut by turbidity currents that flow down the rill.

The above idea invokes the need for both turbidity-current erosion and retrogressive mass wasting in submarine-canyon formation. It also links the evolution of slope-confined canyons to those that breach the shelf break, and provides an alternative mechanism to spring sapping for the continued episodic failure of the canyon headwall (Pratson & Coakley, 1996). Finally, it provides a mechanism for creating submarine canyons with multiple heads, such as Hudson Canyon on the New Jersey slope and Eel Canyon in northern California. Sediment flows that enter a submarine canyon downslope of its existing head can trigger retrogressive failures, which then excavate an additional canyon head along the path of the flows, back to their shelf-break source.

FUTURE RESEARCH

This paper has attempted to synthesize the fundamental ways that major processes operating along shelves and slopes affect the morphological evolution of seabed surfaces. The information has come from work in a variety of geological disciplines. It draws upon findings from plate tectonics, geomorphology and hydrology, physical oceanography, geophysics and geological engineering, and, of course, sedimentology and stratigraphy. In many respects, the last discipline, stratigraphy (i.e. sequence stratigraphy), has provided the best context for understanding the evolution of seabed surfaces in shelf and slope settings.

As discussed in Mountain *et al.* (this volume, pp. 381–458), sequence stratigraphy is founded upon a cogent theory for why strata should be grouped into sequences and systems tracts. More importantly, this theory has engendered a number of landmark studies in a broad range of fields that have refined its fundamental tenets. These include seminal papers on such topics as: the role of thermal and isostatic subsidence in sequence formation (Pitman, 1978; Watts & Steckler, 1979); the synchronicity of global changes in sea level and their implications for past climates (Vail & Mitchum, 1977; Haq *et al.*, 1988); and the power of numerical modelling for gaining insights to sequence evolution (Burton *et al.*, 1987; Jervey, 1988).

The main focus of sequence stratigraphy has been to clarify why certain sediments are found where they are in relation to other sediments. However, development of the underlying theory also has been utilized to develop linkages between shelves and slopes, and the sediments accumulated in both regions. Therefore, much of the recent research described in this paper reinforces ideas envisioned by early pioneers in sequence stratigraphy. However, this is only partially true.

As originally conceived, sequence stratigraphy addressed strata formation under a well-defined, but limited, set of environmental drivers. These were principally changes in eustasy, subsidence and sediment supply. The causes and consequences of changes in eustasy and subsidence are now better understood. By comparison, the role of sediment supply in shelf and slope evolution remains poorly constrained. This is because sediment supply is far more variable in both space and time than either eustasy or subsidence. In addition to sea level, sediment supply to a margin is determined by the three-dimensional geometry of both the ocean basin and the drainage systems feeding into it. These in turn are subject to internal reorganization by the processes transporting sediments to the basin, including river avulsion, delta lobe switching, and large-scale failures on land and beneath the sea. Changes in lithology can also have a significant

effect, contributing to lags in the response of sediment supply to environmental change. Then there is the basic question of how and how fast sediment processes actually modify the shelf and slope, and ultimately contribute to the development of their stratigraphy.

The latter is the question initially addressed by the STRATAFORM programme. Continued work needs to be done in documenting the tie between what happens on the seabed, and what is preserved in the shelf-slope stratigraphic record. However, STRATAFORM has demonstrated the importance of viewing strata formation from the dual perspectives of processes and products.

ACKNOWLEDGEMENTS

The authors offer their sincere thanks to the reviewers of this paper, the initial version of which was an attempt to say everything that could possibly be said about shelf and slope evolution. These patient souls include Courtney Harris, John Goff, Carl Amos and most especially Jamie Austin. The lead author particularly thanks Jamie for his thorough, thoughtful and incisive comments. Hopefully, this final version of the manuscript is an improvement. The authors also thank the editors of this book, with special thanks to Charles Nittrouer. Finally, we are indebted to Joe Kravitz for his contributions to marine geology and geophysics, as the programme manager.

REFERENCES

Adams, E.W. and Schlager, W. (2000) Basic types of submarine slope curvature. *J. Sediment. Res.*, **70**, 814–828.

Alexander, C.R., DeMaster, D.J. and Nittrouer, C.A. (1991) Sediment accumulation in a modern epicontinental-shelf setting; the Yellow Sea. *Mar. Geol.*, **98**, 51–72.

Allen, J.R.L. (1985) *Principles of Physical Sedimentology.* Allen & Unwin, Boston, 272 pp.

Allen, P.A. and Allen, J.R. (2005) *Basin Analysis: Principles and Applications.* Blackwell Publications, Oxford, 549 pp.

Bahr, D.B., Hutton, E.W.H., Syvitski, J.P.M. and Pratson, L.F. (2001) Exponential approximations to compacted sediment porosity profiles. *Comput. Geosci.*, **27**, 691–700.

Baker Jr., B.B., Deebel, W.R. and Geisenderfer, R.D. (Eds) (1966) *Glossary of Oceanographic Terms.* Special Publication, US Naval Oceanographic Research Office, Washington, DC, 204 pp.

Bascom, W. (1980) *Waves and Beaches; the Dynamics of the Ocean Surface.* Anchor Press/Doubleday, Garden City, 267 pp.

Bea, R.G. (1971) How sea-floor slides affect offshore structures. *Oil Gas J.,* **69,** 88–92.

Berryhill, H.L., Jr., Suter, J.R. and Hardin, N.S. (1986) *Late Quaternary Facies and Structure, Northern Gulf of Mexico; Interpretations from Seismic Data.* Studies in Geology 23, American Association of Petroleum Geologists, Tulsa, OK, 289 pp.

Bird, J.M. and Dewey, J.F. (1970) Lithosphere plate-continental margin tectonics and the evolution of the Appalachian orogen. *Geol. Soc. Am. Bull.,* **81,** 1031–1059.

Blum, M.D. and Tornqvist, T.E. (2000) Fluvial responses to climate and sea-level change; a review and look forward. *Sedimentology,* **47**(Suppl. 1), 2–48.

Bredehoeft, J.D. and Hanshaw, B.B. (1968) On the maintenance of anomalous fluid pressures; 1, Thick sedimentary sequences. *Geol. Soc. Am. Bull.,* **79,** 1097–1106.

Bridge, J.S. (2003) *Rivers and Floodplains: Forms, Processes and Sedimentary Record.* Blackwell Publishers, Oxford, 512 pp.

Burger, R.L., Fulthorpe, C.S. and Austin, J.A., Jr. (2001) Late Pleistocene channel incisions in the southern Eel River basin, Northern California; implications for tectonic vs. eustatic influences on shelf sedimentation patterns. *Mar. Geol.,* **177,** 317–330.

Burrus, J. (1989) Review of geodynamic models for extensional basins; the paradox of stretching in the Gulf of Lions (Northwest Mediterranean). *Bull. Soc. Geol. France, Huitiéme Ser.,* **5,** 377–393.

Burton, R., Kendall, C.G.S.C. and Lerche, I. (1987) Out of our depth; on the impossibility of fathoming eustasy from the stratigraphic record. *Earth-Sci. Rev.,* **24,** 237–277.

Busby, C.J. and Ingersoll, R.V. (1995) *Tectonics of Sedimentary Basins.* Blackwell Science, Oxford.

Cacchione, D.A. and Pratson, L.F. (2004) Internal tides and the continental slope. *Am. Sci.,* **92,** 130–137.

Cacchione, D.A. and Southard, J.B. (1974) Incipient sediment movement by shoaling internal gravity waves. *J. Geophys. Res.,* **79,** 2237–2242.

Cacchione, D.A., Pratson, L.F. and Ogston, A.S. (2002) The shaping of continental slopes by internal tides. *Science,* **296,** 724–727.

Carson, B., Baker, E.T., Hickey, N.M., *et al.* (1986) Modern sediment dispersal and accumulation in Quinault Submarine Canyon – a summary. *Mar. Geol.,* **71,** 1–13.

Cook, P.J. and Carleton, C.M. (2000) *Continental Shelf Limits; the Scientific and Legal Interface.* Oxford University Press, New York, 363 pp.

Coulter, H.W. and Migliaccio, R.R. (1966) Effects of the earthquake of March 27, 1964 at Valdez, Alaska. *U.S. Geol. Surv. Prof. Pap.,* **542-C,** 36 pp.

Cowell, P.J., Roy, P.S. and Jones, R.A. (1995) Simulation of large-scale coastal change using a morphological behaviour model. *Mar. Geol.,* **126,** 45–61.

Cowell, P.J., Hanslow, D.J. and Meleo, J.F. (1999) The shoreface. In: *Handbook of Beach and Shoreface Morphodynamics* (Ed. A.D. Short), pp. 41–71. John Wiley & Sons, Chichester.

Curtin, T.B. (1986) Physical observations in the plume region of the Amazon River during peak discharge; II, water masses. *Cont. Shelf Res.,* **6,** 53–71.

Daly, R.A. (1936) Origin of submarine canyons. *Am. J. Sci.,* **31,** 401–420.

Davies, J.L. (1980) *Geographical Variation in Coastal Development,* 2nd edn. Longman, London, 212 pp.

Davies, T.A., Austin, J.A., Jr., Lagoe, M.B. and Milliman, J.D. (1992) Late Quaternary sedimentation off New Jersey; new results using 3-D seismic profiles and cores. *Mar. Geol.,* **108,** 323–343.

Dewey, J.F. (1969) Continental margins; a model for conversion of Atlantic type to Andean type. *Earth Planet. Sci. Lett.,* **6,** 189–197.

Dewey, J.F. (1972) Plate tectonics. *Sci. Am.,* **226,** 56–68.

Dill, R.F. (1964) Sedimentation and erosion in Scripps Submarine Canyon head. In: *Papers in Marine Geology – Shepard Commemorative Volume* (Ed. R.L. Miller), pp. 23–41. Macmillan, New York.

Dolgoff, A. (1998) *Physical Geology.* Houghton Mifflin, Boston, 638 pp.

Driscoll, N.W., Weissel, J.K. and Goff, J.A. (2000) Potential for large-scale submarine slope failure and tsunami generation along the U.S. Mid-Atlantic coast. *Geology,* **28,** 407–410.

Droz, L., Kergoat, R., Cochonat, P. and Berne, S. (2001) Recent sedimentary events in the western Gulf of Lions (western Mediterranean). *Mar. Geol.,* **176,** 23–37.

Dugan, B. and Flemings, P.B. (2000) Overpressure and fluid flow in the New Jersey continental slope; implications for slope failure and cold seeps. *Science,* **289,** 288–291.

Dyer, K.R. (1986) *Coastal and Estuarine Sediment Dynamics.* J. Wiley & Sons, New York, 358 pp.

Elliott, T. (1979) Deltas. In: *Sedimentary Environments and Facies* (Ed. H.G. Reading), pp. 97–142. Elsevier, New York.

Emery, D. and Myers, K. (1996) *Sequence Stratigraphy.* Blackwell Science, Oxford, 297 pp.

Emery, K.O. and Uchupi, E. (1972) *Western North Atlantic Ocean; Topography, Rocks, Structure, Water,*

Life and Sediments. Memoir 17, American Association of Petroleum Geologists, Tulsa, OK, 532 pp.

Emery, K.O. and Uchupi, E. (1984) *The Geology of the Atlantic Ocean*. Springer-Verlag, New York, 1050 pp.

Ericson, D.B., Ewing, M., Wollin, G. and Heezen, B.C. (1961) Atlantic deep-sea sediment cores. *Geol. Soc. Am. Bull.*, **72**, 193–285.

Farre, J.A., McGregor, B.A., Ryan, W.B.F. and Robb, J.M. (1983) Breaching the shelf break; passage from youthful to mature phase in submarine canyon evolution. In: *The Shelf Break; Critical Interface on Continental Margins* (Eds D.J. Stanley and G.T. Moore), pp. 25–39. Special Publication 33, Society of Economic Paleontologists and Mineralogists, Tulsa, OK.

Fedje, D.W. and Josenhans, H. (2000) Drowned forests and archaeology on the continental shelf of British Columbia, Canada. *Geology*, **28**, 99–102.

Felix, D.W. and Gorsline, D.S. (1971) Newport submarine canyon, California; an example of the effects of shifting loci of sand supply upon canyon position. *Mar. Geol.*, **10**, 177–198.

Frazier, D.E. (1967) Recent deltaic deposits of the Mississippi River; their development and chronology. In: *Symposium on the Geological History of the Gulf of Mexico, Antillean-Caribbean Region. Trans. Gulf Coast Assoc. Geol. Soc.*, **17**, 287–315.

Friedrichs, C.T. and Wright, L.D. (2004) Gravity-driven sediment transport on the continental shelf: implications for equilibrium profiles near river mouths. *Coast. Eng.*, **51**, 795–811.

Fulthorpe, C.S., Austin, J.A., Jr. and Mountain, G.S. (2000) Morphology and distribution of Miocene slope incisions off New Jersey; are they diagnostic of sequence boundaries? *Geol. Soc. Am. Bull.*, **112**, 817–828.

Garfield, N., Rago, T.A., Schnebele, K.J. and Collins, C.A. (1994) Evidence of a turbidity current in Monterey submarine canyon associated with the 1989 Loma Prieta earthquake. *Cont. Shelf Res.*, **14**, 673–686.

Gerber, T.P., Pratson, L.F., Wolinsky, M.A., *et al.* (2004) Autocyclic behavior of experimental turbidity currents. *Eos (Trans. Am. Geophys. Union)*, **85**(47), abstract OS23C-1328.

Geyer, W.R., Beardsley, R.C., Candela, J., *et al.* (1991) The physical oceanography of the Amazon outflow. *Oceanography*, **4**, 8–14.

Geyer, W.R., Hill, P.S., Milligan, T. and Traykovski, P. (2000) The structure of the Eel River plume during floods. *Cont. Shelf Res.*, **20**, 2067–2093.

Goff, J., Mayer, L., Hughes-Clarke, J. and Pratson, L.F. (1996) Swath mapping on the continental shelf and slope: the Eel River basin, Northern California. *Oceanography*, **9**, 178–182.

Goodwin, R.H. and Prior, D.B. (1989) Geometry and depositional sequences of the Mississippi Canyon, Gulf of Mexico. *J. Sediment. Petrol.*, **59**, 318–329.

Gordon, D.S. and Flemings, P.B. (1998) Generation of overpressure and compaction-driven fluid flow in a Plio-Pleistocene growth-faulted basin, Eugene Island 330, offshore Louisiana. *Basin Res.*, **10**, 177–196.

Gorsline, D.S. (1970) Submarine canyons; an introduction. *Mar. Geol.*, **8**, 183–186.

Grant, W.D. and Madsen, O.S. (1979) Combined wave and current interaction with a rough bottom. *J. Geophys. Res.*, **84**, 1797–1808.

Gulliver, F. (1899) Shoreline topography. *Proc. Am. Acad. Arts Sci.*, **34**, 151–258.

Hampton, M.A., Lee, H.J. and Locat, J. (1996) Submarine landslides. *Rev. Geophys.*, **34**, 33–59.

Haq, B.U., Hardenbol, J. and Vail, P.R. (1988) Mesozoic and Cenozoic chronostratigraphy and cycles of sea-level change. In: *Sea-Level Changes; an Integrated Approach* (Eds C.K. Wilgus, B.S. Hastings, C.A. Ross, *et al.*), pp. 72–108. Special Publication 42, Society of Economic Paleontologists and Mineralogists, Tulsa, OK.

Harris, C.K. and Wiberg, P.L. (2002) Across-shelf sediment transport: interactions between suspended sediment and bed sediment. *J. Geophys. Res.*, **107**(C1), 10.1029/2000JC000634.

Hays, J.D. and Pitman, W.C., III. (1973) Lithospheric plate motion, sea level changes and climatic and ecological consequences. *Nature*, **246**, 18–22.

Heezen, B.C. and Ewing, W.M. (1952) Turbidity currents and submarine slumps and the 1929 Grand Banks [Newfoundland] earthquake. *Am. J. Sci.*, **250**, 849–873.

Iverson, R.M. and Major, J.J. (1986) Groundwater seepage vectors and the potential for hillslope failure and debris flow mobilization. *Water Resour. Res.*, **22**, 1543–1548.

Jaeger, J.M. and Nittrouer, C.A. (2000) The formation of point- and multiple-source deposits on continental shelves. In: *Coastal Ocean Processes (CoOP): Transport and Transformation Processes over Continental Shelves with Substantial Freshwater Inflows* (Eds S. Henrichs, N. Bond, R. Garvine, G. Kineke and S. Lohrenz), pp. 78–89. Technical Report TS-237-00, Center for Environmental Science, University of Maryland, Cambridge, MD.

Jervey, M.T. (1988) Quantitative geological modeling of siliciclastic rock sequences and their seismic expression. In: *Sea-level Changes; an Integrated Approach* (Eds C.K. Wilgus, B.S. Hastings, C.A. Ross, *et al.*), pp. 47–69. Special Publication 42, Society of Economic Paleontologists and Mineralogists, Tulsa, OK.

Johnson, D. (1939) *The Origin of Submarine Canyons; a Critical Review of Hypotheses.* Columbia University Press, New York, 114 pp.

Kay, G.M. (1951) North American geosynclines. *Geol. Soc. Am. Mem.,* **48**, 143 pp.

Kelling, G. and Stanley, D.J. (1970) Morphology and structure of Wilmington and Baltimore submarine canyons, eastern United States. *J. Geol.,* **78**, 637–660.

Kennett, J.P. (1982) *Marine Geology.* Prentice-Hall, Englewood Cliffs, NJ, 813 pp.

Kineke, G.C. (1993) *Fluid muds on the Amazon continental shelf.* Unpublished PhD Dissertation, University of Washington, Seattle, 372 pp.

Kineke, G.C. and Sternberg, R.W. (1995) Distribution of fluid muds on the Amazon continental shelf. *Mar. Geol.,* **125**, 193–233.

Klitgord, K.D., Hutchinson, D.R. and Schouten, H. (1988) U.S. Atlantic continental margin; structural and tectonic framework. In: *The Geology of North America,* Vol. I-2, *The Atlantic Continental Margin* (Eds R.E. Sheridan and J.A. Grow), pp. 19–55. Geological Society of America, Boulder, CO.

Knauss, J.A. (1997) *Introduction to Physical Oceanography,* 2nd edn. Prentice Hall, Upper Saddle River, NJ, 309 pp.

Kolla, V. and Macurda, D.B., Jr. (1988) Sea-level changes and timing of turbidity-current events in deep-sea fan systems. In: *Sea-level Changes; an Integrated Approach* (Eds C.K. Wilgus, B.S. Hastings, C.A. Ross, *et al.*), pp. 381–392. Special Publication 42, Society of Economic Paleontologists and Mineralogists, Tulsa, OK.

Komar, P.D. (1976) *Beach Processes and Sedimentation,* 2nd edn. Prentice-Hall, Englewood Cliffs, NJ, 544 pp.

Kostic, S., Parker, G. and Marr, J.G. (2002) Role of turbidity currents in setting the foreset slope of clinoforms prograding into standing fresh water. *J. Sediment. Res.,* **72**, 353–362.

Krause, D.C., White, W.C., Piper, D.J.W. and Heezen, B.C. (1970) Turbidity currents and cable breaks in the western New Britain trench. *Geol. Soc. Am. Bull.,* **81**, 2153–2160.

Kuehl, S.A., Hariu, T.M. and Moore, W.S. (1989) Shelf sedimentation off the Ganges–Brahmaputra river system; evidence for sediment bypassing to the Bengal Fan. *Geology,* **17**, 1132–1135.

Kuehl, S.A., Levy, B.M., Moore, W.S. and Allison, M.A. (1997) Subaqueous delta of the Ganges-Brahmaputra river system. *Mar. Geol.,* **144**, 81–96.

Kulm, L.D. and Fowler, G.A. (1974) Oregon continental margin structure and stratigraphy; a test of the imbricate thrust model. In: *The Geology of Continental Margins* (Eds C.A. Burk and C.L. Drake), pp. 261–233. Springer-Verlag, New York.

Kulm, L.D., Roush, R.C., Harlett, J.C., *et al.* (1975) Oregon continental shelf sedimentation; interrelationships of facies distribution and sedimentary processes. *J. Geol.,* **83**, 145–175.

Lavier, L., Steckler, M.S. and Brigaud, F. (2001) Climatic and tectonic controls on the Cenozoic evolution of the West African continental margin. *Mar. Geol.,* **178**, 63–80.

Le Meur, E. and Hindmarsh, R.C.A. (2000) A comparison of two spectral approaches for computing the Earth response to surface loads. *Geophys. J. Int.,* **141**, 282–298.

Lee, H.J., Noble, M.A. and Xu, J. (2003) Sediment transport and deposition processes near ocean outfalls in Southern California. In: *Contaminated Sediments; Characterization, Evaluation, Mitigation/Restoration and Management Strategy Performance* (Eds J. Locat, C.R. Galvez, R. Chaney and K. Demars), pp. 253–265. American Society for Testing and Materials, Philadelphia.

Malinverno, A., Ryan, W.B.F., Auffret, G.A. and Pautot, G. (1988) Sonar images of the path of recent failure events on the continental margin off Nice, France. In: *Sedimentologic Consequences of Convulsive Geologic Events* (Ed. H.E. Clifton), pp. 59–75. Special Paper 229, Geological Society of America, Boulder, CO.

McGregor, B.A., Stubblefield, W.L., Ryan, W.B.F. and Twichell, D.C. (1983) Wilmington submarine canyon area; a marine fluvial-like system. In: *Environmental Geologic Studies on the United States Mid- and North Atlantic Outer Continental Shelf Area 1980–1982,* Vol. I, *Executive Summary* (Ed. B.A. McGregor). *U.S. Geol. Surv. Open File Rep.,* **83–824**, 14–15.

McHugh, C.M., Ryan, W.B.F. and Schreiber, B.C. (1993) The role of diagenesis in exfoliation of submarine canyons. *Bull. Am. Assoc. Petrol. Geol.,* **77**, 145–172.

McHugh, C.M.G., Ryan, W.B.F., Eittreim, S.L. and Reed, D. (1998) The influence of the San Gregorio Fault on the morphology of Monterey Canyon. *Mar. Geol.,* **146**, 63–91.

McMaster, R.L. and Ashraf, A. (1973) Drowned and buried valleys on the southern New England continental shelf. *Mar. Geol.,* **15**, 249–268.

Meade, R.H., Yuzyk, T.R. and Day, T.J. (1990) Movement and storage of sediment in rivers of the United States and Canada. In: *The Geology of North America,* Vol. O-1, *Surface Water Hydrology* (Eds M.G. Wolman and H.C. Riggs), pp. 255–280. Geological Society of America, Boulder, CO.

Miller, K.G., Melillo, A.J., Mountain, G.S., Farre, J.A. and Poag, C.W. (1987) Middle to late Miocene canyon cutting on the New Jersey continental slope; biostratigraphic and seismic stratigraphic evidence. *Geology,* **15**, 509–512.

Milliman, J.D. and Haq, B.U. (1996) *Sea-level Rise and Coastal Subsidence; Causes, Consequences and Strategies*. Kluwer Academic Publishers, Dordrecht, 384 pp.

Milliman, J.D. and Meade, R.H. (1983) World-wide delivery of river sediment to the oceans. *J. Geol.*, **91**, 1–21.

Milliman, J.D. and Syvitski, J.P.M. (1992) Geomorphic/ tectonic control of sediment discharge to the ocean; the importance of small mountainous rivers. *J. Geol.*, **100**, 525–544.

Milliman, J.D., Snow, J., Jaeger, J. and Nittrouer, C.A. (1996) Catastrophic discharge of fluvial sediment to the ocean; evidence of jokulhlaups events in the Alsek Sea valley, Southeast Alaska (USA). In: *Proceedings of an International Symposium on Erosion and Sediment Yield; Global and Regional Perspectives* (Eds D.E. Walling and B.W. Webb), pp. 367–379. International Association of Hydrological Sciences, Wallingford.

Mitchum, R.M., Jr., Vail, P.R. and Thompson, S., III. (1977) Seismic stratigraphy and global changes of sea level; Part 2. The depositional sequence as a basic unit for stratigraphic analysis. In: *Seismic Stratigraphy; Applications to Hydrocarbon Exploration* (Ed. C.E. Payton), pp. 53–62. Memoir 26, American Association of Petroleum Geologists, Tulsa, OK.

Mitchum, R.M., Jr., Sangree, J.B., Vail, P.R. and Wornhardt, W.W. (1990) Sequence stratigraphy in late Cenozoic expanded sections, Gulf of Mexico. *GCSSEPM Foundation 11th Annual Research Conference*, Houston, TX, pp. 237–256.

Mountain, G. (1987) Cenozoic margin construction and destruction offshore New Jersey. In: *Timing and Depositional History of Eustatic Sequences; Constraints on Seismic Stratigraphy* (Eds C.A. Ross and D. Haman), pp. 57–83. Cushman Foundation for Foraminiferal Research, Ithaca, NY.

Mountain, G.S. and Tucholke, B.E. (1985) Mesozoic and Cenozoic geology of the U.S. Atlantic continental slope and rise. In: *Geologic Evolution of the United States Atlantic Margin* (Ed. C.W. Poag), pp. 293–341. Van Nostrand Reinhold, New York.

Mountain, G.S., Damuth, J.E., McHugh, C.M.G., Lorenzo, J.M. and Fulthorpe, C.S. (1996) Origin, reburial and significance of a middle Miocene canyon, New Jersey continental slope. In: *Proceedings of the Ocean Drilling Program; Scientific results, New Jersey Continental Slope and Rise; Covering Leg 150 of the Cruises of the Drilling Vessel JOIDES Resolution, Lisbon, Portugal, to St. John's, Newfoundland, Sites 902–906, 25 May–24 July 1993* (Eds G.S. Mountain, K.G. Miller, P. Blum, et al.), pp. 283–292. Texas A & M University, Ocean Drilling Program, College Station, TX.

Mulder, T. and Syvitski, J.P.M. (1995) Turbidity currents generated at river mouths during exceptional discharges to the world oceans. *J. Geol.*, **103**, 285–299.

Mulder, T., Migeon, S., Savoye, B. and Jouanneau, J.M. (2001) Twentieth century floods recorded in the deep Mediterranean sediments. *Geology*, **29**, 1011–1014.

Mulder, T., Syvitski, J.P.M., Migeon, S., Faugeres, J.C. and Savoye, B. (2003) Marine hyperpycnal flows; initiation, behavior and related deposits; a review. *Mar. Petrol. Geol.*, **20**, 861–882.

Nittrouer, C.A. and Wright, L.D. (1994) Transport of particles across continental shelves. *Rev. Geophys.*, **32**, 85–113.

Nittrouer, C.A., Kuehl, S.A., DeMaster, D.J. and Kowsmann, R.O. (1986) The deltaic nature of Amazon shelf sedimentation. *Geol. Soc. Am. Bull.*, **97**, 444–458.

O'Grady, D.B., Syvitski, J.P.M., Pratson, L.F. and Sarg, J.F. (2000) Categorizing the morphologic variability of siliciclastic passive continental margins. *Geology*, **28**, 207–210.

Orange, D.L. (1999) Tectonics, sedimentation and erosion in Northern California; submarine geomorphology and sediment preservation potential as a result of three competing processes. *Mar. Geol.*, **154**, 369–382.

Orange, D.L. and Breen, N.A. (1992) The effects of fluid escape on accretionary wedges; 2, Seepage force, slope failure, headless submarine canyons and vents. *J. Geophys. Res., B, Solid Earth and Planets*, **97**, 9277–9295.

Orange, D.L., Anderson, R.S. and Breen, N.A. (1994) Regular canyon spacing in the submarine environment; the link between hydrology and geomorphology. *GSA Today*, **4**, 29, 36–39.

Paola, C., Mullin, J., Ellis, C., *et al.* (2001) Experimental stratigraphy. *GSA Today*, **11**, 4–9.

Parker, G., Fukushima, Y. and Pantin, H. (1986) Self accelerating turbidity currents. *J. Fluid Mech.*, **171**, 145–181.

Parsons, B. and Sclater, J.G. (1977) An analysis of the variation of ocean floor bathymetry and heat flow with age. *J. Geophys. Res.*, **82**, 803–827.

Paull, C.K. and Neumann, A.C. (1987) Continental margin brine seeps: their geologic consequences. *Geology*, **15**, 545–548.

Penland, S., Roberts, H.H., Williams, S.J., *et al.* (1990) Coastal land loss in Louisiana. *Trans. Gulf Coast Assoc. Geol. Soc.*, **40**, 685–699.

Pirmez, C., Pratson, L.F. and Steckler, M.S. (1998) Clinoform development by advection-diffusion of suspended sediment; modeling and comparison to natural systems. *J. Geophys. Res., B, Solid Earth Planets*, **103**, 24141–24157.

Pitman, W.C., III. (1978) Relationship between eustacy and stratigraphic sequences of passive margins. *Geol. Soc. Am. Bull.*, **89**, 1389–1403.

Poag, C.W. and Mountain, G.S. (1987) Late Cretaceous and Cenozoic evolution of the New Jersey continental slope and upper rise; an integration of borehole data with seismic reflection profiles. In: *Initial Reports of the Deep Sea Drilling Project Covering Leg 95 of the Cruises of the Drilling Vessel Glomar Challenger, St. John's, Newfoundland, to Ft. Lauderdale, Florida, August–September 1983* (Eds C.W. Poag, A.B. Watts, M. Cousin, *et al.*), pp. 673–724. Texas A & M University, Ocean Drilling Program, College Station, TX.

Poppe, L.J. and Polloni, C.F. (1998) *Long Island Sound Environmental Studies. U.S. Geol. Surv. Open File Rep.*, (CD-ROM).

Posamentier, H.W. and Vail, P.R. (1988) Eustatic controls on clastic deposition; II, Sequence and systems tract models. In: *Sea-level Changes; an Integrated Approach* (Eds C.K. Wilgus, B.S. Hastings, C.A. Ross, *et al.*), pp. 125–154. Special Publication 42, Society of Economic Paleontologists and Mineralogists, Tulsa, OK.

Posamentier, H.W., Jervey, M.T. and Vail, P.R. (1988) Eustatic controls on clastic deposition; I, Conceptual framework. In: *Sea-level Changes; an Integrated Approach* (Eds C.K. Wilgus, B.S. Hastings, C.A. Ross, *et al.*), pp. 109–124. Special Publication 42, Society of Economic Paleontologists and Mineralogists, Tulsa, OK.

Posamentier, H.W., Allen, G.P., James, D.P. and Tesson, M. (1992) Forced regressions in a sequence stratigraphic framework; concepts, examples and exploration significance. *Bull. Am. Assoc. Petrol. Geol.*, **76**, 1687–1709.

Pratson, L.F. (1993) *Morphologic studies of submarine sediment drainage*. PhD Dissertation, Columbia University, New York.

Pratson, L.F. (2001) A perspective on what is known and not known about seafloor instability in the context of continental margin evolution. *Mar. Petrol. Geol.*, **18**, 499–501.

Pratson, L.F. and Coakley, B.J. (1996) A model for the headward erosion of submarine canyons induced by downslope-eroding sediment flows. *Geol. Soc. Am. Bull.*, **108**, 225–234.

Pratson, L.F. and Edwards, M.H. (1996) Introduction to advances in seafloor mapping using sidescan sonar and multibeam bathymetry data. *Mar. Geophys. Res.*, **18**, 601–605.

Pratson, L.F. and Haxby, W.F. (1996) What is the slope of the U.S. continental slope? *Geology*, **24**, 3–6.

Pratson, L.F. and Haxby, W.F. (1997) Panoramas of the seafloor. *Sci. Am.*, **276**, 82–87.

Pratson, L.F., Ryan, W.B.F., Mountain, G.S. and Twichell, D.C. (1994) Submarine canyon initiation by downslope-eroding sediment flows; evidence in late Cenozoic strata on the New Jersey continental slope. *Geol. Soc. Am. Bull.*, **106**, 395–412.

Prior, D.B. and Suhayda, J.N. (1982) Application of infinite slope analysis to subaqueous sediment instability, Mississippi Delta. In: *Offshore Geologic Hazards; a Short Course Presented at Rice University, May 2–3, 1981 for the Offshore Technology Conference* (Eds A.H. Bouma, D.A. Sangrey, J. Coleman, *et al.*), pp. 5.76–5.92. Continuing Course Note Series 18, American Association of Petroleum Geologists, Tulsa, OK.

Rasmussen, E.S. (1994) The relationship between submarine canyon fill and sea-level change; an example from middle Miocene offshore Gabon, West Africa. *Sediment. Geol.*, **90**, 61–75.

Reynolds, D.J., Steckler, M.S. and Coakley, B.J. (1991) The role of the sediment load in sequence stratigraphy; the influence of flexural isostasy and compaction. *J. Geophys. Res., B, Solid Earth and Planets*, **96**, 6931–6949.

Ritter, D.F. (1986) *Process Geomorphology*, 2nd edn. William C. Brown, Dubuque, 546 pp.

Robb, J.M. (1984) Spring sapping on the lower continental slope, offshore New Jersey. *Geology*, **12**, 278–282.

Rona, P.A. (1970) Submarine canyon origin on upper continental slope off Cape Hatteras. *J. Geol.*, **78**, 141–152.

Ryan, W.B.F. (1982) Imaging of submarine landslides with wide-swath sonar. In: *Marine Slides and other Mass Movements* (Eds S. Saxov and J.K. Nieuwenhuis), pp. 175–188. NATO Conference Series, Vol. 6, Plenum Press, New York.

Ryan, W.B.F. and Pitman, W.C., III. (2000) *Noah's Flood: the New Scientific Discoveries about the Event that Changed History*. Simon & Schuster, New York, 320 pp.

Ryan, W.B.F., Cita, M.B., Miller, E.L., *et al.* (1978) Bedrock geology in New England submarine canyons. *Ocean. Acta*, **1**, 233–254.

Ryan, W.B.F., Hsue, K.J., Cita, M.B., *et al.* (1973) *Initial Reports of the Deep Sea Drilling Project, covering Leg 13 of the Cruises of the Drilling Vessel Glomar Challenger Lisbon, Portugal to Lisbon, Portugal, August–October 1970*. Texas A & M University, Ocean Drilling Program, College Station, TX.

Scholle, P.A. (1996) *Oceanography 1: Origins, History, Structure and Plate Tectonics, Margins, Basins and Sediments*. Society of Economic Paleontologists and Mineralogists, Tulsa, OK, Photo CD-3.

Sclater, J.G. and Christie, P.A.F. (1980) Continental stretching; an explanation of the post-Mid-Cretaceous subsidence of the central North Sea basin. *J. Geophys. Res.*, **85**, 3711–3739.

Seeber, L. and Armbruster, J.G. (1988) Seismicity along the Atlantic seaboard of the U.S.; intraplate neotectonics and earthquake hazards. In: *The Geology of North America*, Vol. I-2, *The Atlantic Continental Margin*

(Eds R.E. Sheridan and J.A. Grow), pp. 565–582. Geological Society of America, Boulder, CO.

Shepard, F.P. (1934) Canyons off the New England coast. *Am. J. Sci.*, **27**, 24–36.

Shepard, F.P. (1977) *Geological Oceanography; Evolution of Coasts, Continental Margins and the Deep-Sea Floor*, Crane, Russak, New York, 214 pp.

Shepard, F.P. (1981) Submarine canyons; multiple causes and long-time persistence. *Bull. Am. Assoc. Petrol. Geol.*, **65**, 1062–1077.

Shor, A.N. and McClennen, C.E. (1988) Marine physiography of the U.S. Atlantic margin. In: *The Geology of North America*, Vol. I-2, *The Atlantic Continental Margin* (Eds R.E. Sheridan and J.A. Grow), pp. 9–18. Geological Society of America, Boulder, CO.

Short, A.D. (1999) *Handbook of Beach and Shoreface Morphodynamics*. J. Wiley & Sons, New York, 392 pp.

Smith, W.H.F. and Sandwell, D.T. (1997) Global sea floor topography from satellite altimetry and ship depth soundings. *Science*, **277**, 1956–1962.

Song, G.S., Ma, C.P. and Yu, H.S. (2000) Fault-controlled genesis of the Chilung sea valley (northern Taiwan) revealed by topographic lineaments. *Mar. Geol.*, **169**, 305–325.

Spinelli, G.A. and Field, M.E. (2001) Evolution of continental slope gullies on the northern California margin. *J. Sediment. Res.*, **71**, 237–245.

Steckler, M.S., Watts, A.B. and Thorne, J.A. (1988) Subsidence and basin modeling at the U.S. Atlantic passive margin. In: *The Geology of North America*, Vol. I-2, *The Atlantic Continental Margin* (Eds R.E. Sheridan and J.A. Grow), pp. 399–416. Geological Society of America, Boulder, CO.

Steckler, M.S., Mountain, G.S., Miller, K.G. and Christie, B.N. (1999) Reconstruction of Tertiary progradation and clinoform development on the New Jersey passive margin by 2-D backstripping. *Mar. Geol.*, **154**, 399–420.

Stetson, H.C. (1936) Dredge samples from the submarine canyons between the Hudson Gorge and Chesapeake Bay. *Trans. Am. Geophys. Union*, **Part 1**, 223–225.

Stetson, H.C. and Smith, J.F., Jr. (1938) Behavior of suspension currents and mud slides on the continental slope. *Am. J. Sci.*, **35**, 1–13.

Suppe, J. (1985) *Principles of Structural Geology*. Prentice-Hall, Englewood Cliffs, NJ, 537 pp.

Swenson, J.B., Paola, C., Pratson, L., Voller, V.R. and Murray, A.B. (2005) Fluvial and marine controls on combined subaerial and subaqueous delta progradation: morphodynamic modeling of compound clinoform development. *J. Geophys. Res.*, **110**, 2013–2029.

Syvitski, J.P.M., Asprey, K.W., Clattenburg, D.A. and Hodge, G.D. (1985) The prodelta environment of a fjord; suspended particle dynamics. *Sedimentology*, **32**, 83–107.

Thorne, J.A. (1995) On the scale independent shape of prograding stratigraphic units: applications to sequence stratigraphy. In: *Fractals in Petroleum Geology and Earth Processes* (Eds C. Barton and P.R. La Pointe), pp. 97–112. Plenum, New York.

Traykovski, P., Geyer, W.R., Irish, J.D. and Lynch, J.F. (2000) The role of wave-induced density-driven fluid mud flows for cross-shelf transport on the Eel River continental shelf. *Cont. Shelf Res.*, **20**, 2113–2140.

Tucholke, B.E. and Laine, E.P. (1982) Neogene and Quaternary development of the lower continental rise off the central U.S. East Coast. In: *Studies in Continental Margin Geology* (Eds J.S. Watkins and C.L. Drake), pp. 295–305. Memoir 34, American Association of Petroleum Geologists, Tulsa, OK.

Twichell, D.C. and Roberts, D.G. (1982) Morphology, distribution and development of submarine canyons on the United States Atlantic continental slope between Hudson and Baltimore canyons. *Geology*, **10**, 408–412.

Uchupi, E., Driscoll, N., Ballard, R.D. and Bolmer, S.T. (2001) Drainage of late Wisconsin glacial lakes and the morphology and late Quaternary stratigraphy of the New Jersey–southern New England continental shelf and slope. *Mar. Geol.*, **172**, 117–145.

Vail, P.R. and Mitchum, R.M., Jr. (1977) Seismic stratigraphy and global changes of sea level; Part 1, Overview. In: *Seismic Stratigraphy – Applications to Hydrocarbon Exploration* (Ed. C.E. Payton), pp. 51–52. Memoir 26, American Association of Petroleum Geologists, Tulsa, OK.

Vail, P.R., Audemard, F., Bowman, S.A., Eisner, P.N. and Perez, C.G. (1991) The stratigraphic signatures of tectonics, eustacy and sedimentology; an overview. In: *Cycles and Events in Stratigraphy* (Eds G. Einsele, W. Ricken and A. Seilacher), pp. 617–659. Springer-Verlag, Berlin.

Van Wagoner, J.C., Posamentier, H.W., Mitchum, R.M., Jr., *et al.* (1988) An overview of the fundamentals of sequence stratigraphy and key definitions. In: *Sea-level Changes; an Integrated Approach* (Eds C.K. Wilgus, B.S. Hastings, C.A. Ross, *et al.*), pp. 39–45. Special Publication 42, Society of Economic Paleontologists and Mineralogists, Tulsa, OK.

Vogt, P.R. and Tucholke, B.E. (1986) Imaging the ocean floor; history and state of the art. In: *Geology of North America*, Vol. M, *The Western North Atlantic Region* (Eds P.R. Vogt and B.E. Tucholke), pp. 19–44. Geological Society of America, Boulder, CO.

Von der Borch, C.C., Grady, A.E., Aldam, R., Miller, D., Neumann, R., Rovira, A. and Eickhoff, K. (1985) A large-scale meandering submarine canyon; outcrop example from the late Proterozoic Adelaide Geosyncline, South Australia. *Sedimentology*, **32**, 507–518.

Walker, R.G. (1978) Deep-water sandstone facies and ancient submarine fans; models for exploration for stratigraphic traps. *Bull. Am. Assoc. Petrol. Geol.*, **62**, 932–966.

Walsh, J.P. and Nittrouer, C.A. (2003) Contrasting styles of off-shelf sediment accumulation in New Guinea. *Mar. Geol.*, **196**, 105–125.

Walsh, J.P., Nittrouer, C.A., Palinkas, C.M., *et al.* (2004) Clinoform mechanics in the Gulf of Papua, New Guinea. *Cont. Shelf Res.*, **24**, 2487–2510.

Watts, A.B. and Ryan, W.B.F. (1976) Flexure of the lithosphere and continental margin basins. *Tectonophysics*, **36**, 25–44.

Watts, A.B. and Steckler, M.S. (1979) Subsidence and eustasy at the continental margin of eastern North America. In: *Deep Drilling Results in the Atlantic Ocean; Continental Margins and Paleoenvironment* (Eds M. Talwani, W. Hay and W.B.F. Ryan), pp. 218–234. Maurice Ewing Series, Vol. 3, American Geophysical Union, Washington, DC.

Watts, A.B. and Zhong, S. (2000) Observations of flexure and the rheology of oceanic lithosphere. *Geophys. J. Int.*, **142**, 855–875.

Watts, A.B., Karner, G.D. and Steckler, M.S. (1982) Lithospheric flexure and the evolution of sedimentary basins. In: *Evolution of Sedimentary Basins* (Eds P. Kent, M.H.P. Botts, D.P. McKenzie and C.A. Williams), pp. 249–281. Royal Society of London, London.

Weber, M.E., Wiedicke, M.H., Kudrass, H.R., Huebscher, C. and Erlenkeuser, H. (1997) Active growth of the Bengal Fan during sea-level rise and highstand. *Geology*, **25**, 315–318.

Wiberg, P.L. (2000) A perfect storm: formation and potential for preservation of storm beds on the continental shelf. *Oceanography*, **13**, 93–99.

Wright, L.D. (1977) Sediment transport and deposition at river mouths; a synthesis. *Geol. Soc. Am. Bull.*, **88**, 857–868.

Wright, L.D. (1995) *Morphodynamics of Inner Continental Shelves*. CRC Press, Boca Raton, 241 pp.

Wright, L.D., Wiseman, W.J., Bornhold, B.D., *et al.* (1988) Marine dispersal and deposition of Yellow River silts by gravity-driven underflows. *Nature*, **332**, 629–632.

Zhang, Y., Swift, D.J.P., Fan, S., Niederoda, A.W. and Reed, C.W. (1999) Two-dimensional numerical modeling of storm deposition on the northern California shelf. *Mar. Geol.*, **154**, 155–167.

The long-term stratigraphic record on continental margins

GREGORY S. MOUNTAIN*, ROBERT L. BURGER†, HEIKE DELIUS‡, CRAIG S. FULTHORPE§,
JAMIE A. AUSTIN§, DAVID S. GOLDBERG¶, MICHAEL S. STECKLER¶, CECILIA M. McHUGH**,
KENNETH G. MILLER*, DONALD H. MONTEVERDE*, DANIEL L. ORANGE†† and
LINCOLN F. PRATSON‡‡

*Rutgers, the State University of New Jersey, Piscataway, NJ 08854, USA (Email: gmtn@rci.rutgers.edu)
†Yale University, New Haven, CT 06520, USA
‡Task Geoscience, Ltd, Aberdeen AB23 8GX, UK
§University of Texas Institute for Geophysics, Austin TX 78759-8500
¶Lamont-Doherty Earth Observatory, Palisades, NY 10964, USA
**Queens College, Flushing, NY 11367, USA
††AOA Geophysics, Inc., Moss Landing, CA 95039, USA
‡‡Duke University, Durham, NC 27708, USA

ABSTRACT

Processes that build continental-margin stratigraphy on time-scales of > 20 kyr have been investigated. Eustatic sea-level exerts a major influence on sedimentation, but the Eel River margin shows that its effects can be interwoven with those of tectonism. Rapid Oligocene subsidence along the Cascadia subduction zone resulted in a foundered forearc basin. Regression and sedimentary reconstruction began in the Pliocene, and up to 1 km of sediment has accumulated since then, with rotating faults, synclines, anticlines and regional uplifts marking plate interactions. Fourteen seismic unconformities along structural highs can be traced into synclines. Many are ravinements formed during rising sea level, and ~70–100 kyr cyclicity suggests a glacio-eustatic signal. Incised channels formed during regressions over the past ~360 kyr, when rivers drained into Eel Canyon. In contrast, the New Jersey margin has long been dormant tectonically, providing clearer access to a eustatic imprint. Lack of Paleogene sediment supply resulted in a carbonate ramp prior to development of Oligocene deltas. With little accommodation space to allow aggradation, clinoforms prograded ~100 km seaward, reaching the shelf break by Late Pleistocene. Coastal-plain drilling recovered ~15 Oligocene and Miocene highstand deposits, which correlate with glacio-eustatic oscillations. Beneath the mid-to-outer shelf, incised valleys have been preserved, and clinoform strata suggest reworking of lobate deposits. Four Late Pleistocene sequences reveal no hiatuses at sequence boundaries, and no correlations between glacio-eustatic oscillations and stratal architecture. Stratal discontinuities are a common feature in margin sediments and provide objective means of interpreting the geological record. Continuous coring is essential to understand the processes that create stratal architecture.

Keywords Eustatic sea level, forearc basin, unconformities, tectonism, accommodation space, shelf valleys, sequence stratigraphy, clinoforms.

INTRODUCTION

This paper describes the production of the long-term stratigraphic record on continental margins. Concepts and analytical techniques are presented first, and then two diverse margins (Eel and New Jersey) are described and contrasted. The goal is to provide a general understanding of the long-term stratigraphic record, reinforced by specific examples of distinctly different continental margins.

Distinguishing time-scales

While there is no generally accepted distinction between long- and short-term sedimentary records, and furthermore, no standardization as to what constitutes the briefest of all – the event-scale record – the following criteria are adopted here. **Events** are processes lasting minutes to days that include (but are not limited to) ash falls, massive storms and floods, abrupt tectonic disturbances, tsunamis and extraterrestrial impacts. Each can leave a distinct imprint on the sedimentary record at a regional and global scale and may provide the ability to establish near-synchroneity to the origin of widely separated stratigraphic features. Many papers in this volume exploit event-scale processes recorded in cores or monitored in the water column above the seabed during observations that span months to years. These processes combine to create the **short-term** sedimentary record, defined here as sediments deposited since the **Last Glacial Maximum** (LGM; ~20 ka).

The **long-term** record pertains to sediments older than ~20 ka, and contains the history of shorter term processes. Typically, events are aperiodic and unpredictable, in contrast to long-term processes that are either the integrated sum of individual events, or occur with measurable periodicities. For example, earthquakes and resultant debris flows constitute events that, if adequately preserved, can provide a correlation marker. Such events are short-lived and unpredictable, but if they occur often enough their net effect is an important control on sediment distribution, and is relevant to analysing the long-term record. Therefore, this paper will touch on some parts of the short-term record that provide a context for better understanding of the long-term record.

The importance of the long-term record

If well-preserved, aperiodic events can provide distinct stratigraphic markers, and the 'short-term' record can yield information on time-scales relevant to human activities; why should the 'long-term' record be considered valuable? There are several answers.

First, the long-term record documents behaviour of the complex Earth system under boundary conditions that may be very different from those of today. Monitoring active processes or consulting the short-term record simply cannot capture the complete range of Earth system activity. For example, to understand fully the role of sea level in the delivery of sediment to a continental shelf and beyond, only the long-term record provides the opportunity to examine this process during glacial maxima, when sea level was as much as 120 m below present. At these times, rivers were downcutting and flushing previously sequestered sediment from floodplains; but how much and by what processes did this sediment remain on an exposed shelf, or how much and by what processes did it bypass the shelf entirely? This cannot be determined by examining either the modern world or the short-term record. As another example, global temperatures during the late middle Eocene were as high as at any time in the past 100 Myr (Savin, 1977; Miller *et al.*, 1987a); what effects on precipitation/weathering/runoff can be ascribed to this extreme state, and what insight does this provide to understanding changes that may be developing today as global warming becomes increasingly significant? To begin answering these questions, the long-term record must be analysed.

Second, consulting the long-term record provides the chance to evaluate numerical models that are intended to duplicate real-world phenomena (see Syvitski *et al.*, this volume, pp. 459–529). This model-checking confirms or refutes that the Earth system works in the manner described by the model. For example, it is known that rifted lithosphere becomes more rigid with time, but is this actually manifested (as the rigid-plate model predicts; e.g. Watts *et al.*, 1982) in a long-term onlap of the craton, measured in hundreds of kilometres over tens of millions of years? Model-checking also provides sensitivity tests concerning the significance of various parameters to the working of the whole system. For example, does a doubling in the rate of sea-level rise double the rate of shoreline retreat?

Third, many fundamental processes that drive the Earth system, or allow us to monitor it, act on time-scales far longer than the short term (~20 kyr). Late Pleistocene glaciers advanced and retreated on 100-kyr cycles (Imbrie *et al.*, 1984); total magnetic-field reversals occur at ~0.1–1 Myr intervals (Lowrie & Kent, 2004); spreading-rate changes at mid-ocean ridges can occur on 10-Myr time-scales with significant impact on global sea level (Hays & Pitman, 1973); even evolutionary change itself, providing one of the most fundamental yardsticks for measuring geological time, proceeds too slowly for the past 20 kyr to detect meaningful change. In sum, only studies based on the long-term record can capture the full range of fundamental processes that characterize and shape continental margins.

Long-term geochronology: dating continental-margin records

Radiometric dating is generally not applicable to the study of long-term continental-margin records because suitable material is not available. Consequently, time equivalency (correlation) is usually established through **biostratigraphy** using fossils, **magnetostratigraphy** using reversals of the Earth's magnetic field, and **chemostratigraphy** using variations in isotope ratios ($^{18}O/^{16}O$, $^{14}C/^{12}C$ and $^{87}Sr/^{86}Sr$). Dates are then assigned by tying the fossil, magnetic and isotopic records to a standard time-scale. Pleistocene sediments older than 40–60 ka (the extent of the radiocarbon technique) are the most difficult to date due to sparse biostratigraphic events as well as to the length of the most recent epoch of normal magnetic polarity (0–790 ka; Lowrie & Kent, 2004). Tertiary sediments are frequently dated by integrating biostratigraphy with Sr-isotopic stratigraphy (Miller et al., 1998a). Age resolution for Tertiary strata varies; it can be as fine as ±0.5 Myr, but in some cases age uncertainties may reach many millions of years.

Planktonic foraminifera, coccolithophores, diatoms and pollen are the taxa most commonly used to establish the biostratigraphy of the continental-margin successions considered in this paper (i.e. 20 ka to 40 Ma). Most studies from the submerged subsurface utilize core samples or drill cuttings, and consequently they depend on microfossils and fragments of mollusc shells. The resolution of dates derived from fossil content depends ultimately on the rate of evolution, the state of fossil preservation and the ability to identify changes in morphology from one stratigraphic level to the next. At any given position in the rock record, the ability of fossils to provide an age determination requires the appearance or disappearance of key index fossils. There are times of rapid evolutionary

change that provide temporal resolution approaching ±0.5 Myr (e.g. the early to middle Eocene) and others in which resolution remains ±1–2 Myr (e.g. the late Oligocene) (Berggren et al., 1995).

The application of magnetogeochronology relies on measuring the direction of remanent magnetization in sediments. Total field reversals appear to be aperiodic, with time between reversals varying from 0.1 to 1 Myr for the interval considered in this paper (e.g. Berggren et al., 1995; Lowrie & Kent, 2004). Consequently, when coupled with other dating tools, the identification of remanent direction can potentially narrow the probable age by a factor of two or more.

Four types of geochemical dating are commonly used in the study of continental-margin sediments: $^{87/86}Sr$, $^{18/16}O$, $^{14/12}C$ and amino acid racemization. The first two depend on an independently derived curve of isotopic change, to which a measured value is compared. The latter two depend on knowing a time-dependent rate of change that needs no independent information; the value of the sample itself provides its age. A variety of post-depositional processes can degrade the reliability of these measurements because they depend on a geochemical system remaining closed since the time of deposition.

ANALYSING THE LONG-TERM RECORD

Basin-wide surfaces and long-term processes

The development of sequence stratigraphy

Despite all that has been learned in the past 200 yr about stratigraphy (Nystuen, 1998), controversy over what controls the long-term sedimentary record still focuses on eustasy versus tectonism. One observation that has endured, however, is that widespread **unconformities** divide the rock record into distinct sedimentary packages. Debate may continue as to what caused these breaks, but their existence is universally accepted.

Lateral changes in facies were described by Walther (1894), when he recognized the critical aspect of time-transgressive deposition that results in horizontal facies changes repeated in vertical successions. Therefore, the physical character of rocks can be unreliable for evaluating time equivalency between widely separated units, and bounding unconformities may be extremely valuable in establishing stratigraphic correlation. However, using stratal discontinuities has serious shortcomings, because local response to a global process can mask the true sequence of events. For example, glacial cycles have long been recognized as a mechanism for the origin of boundaries in sedimentary rocks (Agassiz, 1840; Maclaren, 1842), but large ice sheets can depress the Earth's surface, and, as they melt, the crust progressively rebounds to create a purely local regression (Jamieson, 1865).

The debate regarding the cause for widespread unconformities continued into the mid-20th century (Stille, 1924; Grabau, 1940), but many scientists accepted evidence of ocean water periodically spilling onto the continents to form widespread marine deposits, and receding to form long-period, continent-wide unconformities. For example, six unconformably bounded units of Cambrian to Tertiary age were traced across North America by Sloss (1963). These rock-stratigraphic units were termed **sequences** and their significance lay in the unifying processes invoked to explain such widespread distribution.

The digital revolution of the 1970s dramatically improved the quality of seismic-reflection data, revealing the geometry of buried strata with previously unseen clarity. These improvements, coupled with the search for fossil fuels, accelerated the pace of seismic surveying on continental margins, and soon a community of classically trained stratigraphers began to develop new ways of applying reflection profiles to the interpretation of basin history. The technique, termed **seismic stratigraphy** (Payton, 1977), was quickly accepted, and with some refinements remains the most widely used approach to study the long-term record of continental margins.

Seismic stratigraphy exploits the fact that physical contrasts across bedding planes or between packages of beds lead to acoustic reflections. Consequently, patterns of seismic reflections in the subsurface closely match buried stratal geometry. To be detected, these contrasts must be vertically abrupt at the scale of metres to tens of metres, and must persist laterally for tens to hundreds of metres. Hence, gradual velocity/density changes that coincide with lateral (time-transgressive) facies changes are typically too subtle to be detected acoustically.

Mitchum *et al.* (1977a,b) described patterns of buried geometries revealed by seismic profiles, and demonstrated that strata could be grouped into unconformably bounded units. Vail *et al.* (1977) identified many more unconformities than were known from previous studies of the rock record. For example, the two youngest units described by Sloss (1963) were divided into more than two dozen unconformity-bounded subunits, and these were later revealed to contain still more subunits (Haq *et al.*, 1987). The underlying approach to these studies, whether based on outcrops, boreholes or seismic profiles (e.g. Van Waggoner *et al.*, 1988), held to the same tenet: stratal discontinuities comprise an imprint of global processes on the long-term record of depositional systems.

Basin-wide stratal gaps spanning similar time intervals suggested the possibility of a common cause operating on a 1–2 Myr time-scale associated with changes in global sea level (**eustasy**; Vail *et al.*, 1977; Posamentier *et al.*, 1988). This explanation has generated controversy for a variety of reasons (e.g. Christie-Blick, 1991; Miall, 1991; Karner & Driscoll, 1997). Although the utility of analysing the geological record in terms of unconformably bound stratal units is widely accepted, many scientists are unwilling to accept that synchrony in widely separated basins can be established with sufficient confidence to prove a single cause. Numerical models also indicate that the stratigraphic imprint of eustatic change would be asynchronous in basins having different tectonic histories. Furthermore, variations in deltaic sedimentation, even in the absence of eustatic change, can build stratigraphic architecture that is very difficult to distinguish from those imposed by true changes in global sea-level. Eustatic changes every 1–2 Myr strongly suggest the growth and decay of polar ice sheets (no other mechanism is known to be of sufficient amplitude or frequency, see Harrison, 1990), but there is no geological or isotopic evidence to support their unbroken, cyclic occurrence over the past 200 Myr as Haq *et al.* (1987) proposed.

Current stratigraphic models

Despite the lack of consensus as to cause, there is general agreement that the sedimentary record is discontinuous at a wide range of scales. These pervasive breaks are expressed as abrupt changes

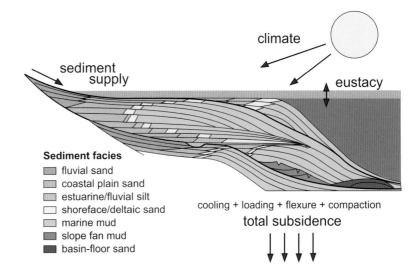

Fig. 1 An idealized clinoform, comprising a fundamental stratigraphic element of detrital sediment deposited adjacent to a point source flowing into standing water. It has been proposed that the influences of sediment supply, total subsidence, climate and eustasy control the distribution of facies as indicated. (After Mitchum, 1977a; Vail *et al.*, 1977.)

in biofacies, lithofacies, and/or the angular termination of stratal surfaces. Seismic profiles can reveal in a single cross-section the arrangement of stratal surfaces that cannot be resolved by outcrops or boreholes. The basic unit of seismic stratigraphic analysis is the **depositional sequence**, defined as a genetically related package of sediments bounded above and below by unconformities and their correlative conformities (Mitchum, 1977a). On continental margins, a prominent type of sequence develops an overall sigmoidal shape (Fig. 1), and is termed a **clinoform**. These are characterized by relatively gently dipping strata at their landward (**topset**) and basinward (**bottomset**) extremes, and more steeply dipping (**foreset**) strata between these locations. The transition between topset and foreset beds is termed the **rollover**.

Clinoforms are a fundamental stratigraphic element of continental margins, and develop wherever detrital sediments are transported from shallow to deeper water (Posamentier *et al.*, 1988). Vertical dimensions range from centimetres to hundreds of metres, and horizontal dimensions from metres to hundreds of kilometres. Although a clinoform resembles the cross-sectional shape of sediments comprising an entire continental margin (Posamentier & Vail, 1988), it is incorrect to equate these structures and conclude that they record the action of similar processes. Bottomsets of a typical sequence lie immediately basinward of foresets of the previous sequence, and, with time, onlap the clinoform and completely bury it. The height of typical continental slopes places any 'bottomset'

strata hundreds to thousands of metres below the 'topset' strata of the shelf. This amount of topographic relief cannot be accounted for within time spans represented by sequences. Clinoform rollovers and continental shelf breaks may coincide spatially, but they are the result of different processes (Steckler *et al.*, 1993).

Clinoforms typically stack one above the other, and despite complexities (see Christie-Blick & Driscoll, 1995), each sequence is distinguished by its bounding unconformities. Three principal types of angular discordance are recognized: **onlap**, **downlap** and **toplap** (Mitchum, 1977b) (Fig. 2). The first describes an angular termination of strata building landward against the basal unconformity of a sequence. The second describes angular terminations building basinward across an underlying unconformity. The third describes stratal terminations caused by an unconformity that defines the top of a sequence, created either by sediment bypassing or erosion. Pre-existing topography, local tectonism, ocean currents and localized sediment sources are some of the many processes and features that lead to complex three-dimensional stratal geometries, and make it possible for one surface to exhibit all three classes of termination in the same basin. Hence, the complexity of a basin dictates how many profiles must be examined before identifying basin-wide sequence boundaries (see Karner & Driscoll, 1997).

An important concept that is still being evaluated is that sequence boundaries possess a time-stratigraphic significance (Vail *et al.*, 1977): all strata

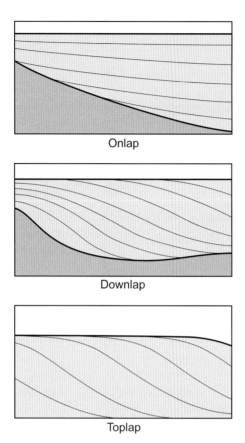

Onlap

Downlap

Toplap

Fig. 2 Three principal types of angular discordance between reflectors (thin lines) and bounding unconformities (bold lines) interpreted as sequence boundaries. (After Mitchum *et al.*, 1977b.)

above a sequence boundary are younger than all strata below. Once identified in this manner, the strata confined to a given sequence can be interpreted in a chronostratigraphic context distinct from that of other sequences in that same basin.

The arrangement of strata within a sequence usually follows one of two patterns, depending on the type of basal sequence boundary. A **Type I** boundary is the result of subaerial erosion across the entire upper surface of the previous sequence (Van Waggoner *et al.*, 1988). Consequently, the strata show evidence of erosion (including valley incision) into the previous sequence, abrupt upward shoaling of facies across the basal sequence boundary, and shallow-water sediments deposited in a deep-water turbidite system basinward of the clinoform. **Type II** boundaries develop without subaerial exposure of marine sediment, submarine-fan deposition, or abrupt upward shoaling

(Van Waggoner *et al.*, 1988). Regardless of the type of basal unconformity, all sequences show an upward progression from onlap to toplap, and the downlap surface that divides these patterns represents a major intrasequence feature termed the **maximum flooding surface** or MFS (Fig. 3). Some workers consider this to be the most readily recognized and useful marker within siliciclastic, continental-margin settings. This approach to basin analysis, termed **genetic stratigraphy** (Galloway, 1989), exploits the ease of identifying a MFS over the occasional difficulty of locating a basin-wide unconformity (sequence boundary) in an up-dip setting where there are many other hiatuses.

Detecting intrasequence geometry attests to sharp changes in physical properties that coincide with bedding or groups of beds within a single unconformably bounded sequence. The most common change in detrital sediments along a continental margin is for relatively coarse-grained sediment to be overlain by finer-grained sediment. In turn, the fine sediments typically coarsen upwards in a gradual manner, without abrupt contrasts that could otherwise generate reflections, until replaced at a sharp contact by finer-grained sediment. These coarsening upward units are termed **parasequences** and are the building blocks of sequences (Van Waggoner *et al.*, 1988). The geometric relationships of parasequences define **system tracts** (Posamentier *et al.*, 1988).

The major influences on stratigraphic succession are sediment supply, global sea level (eustasy) and **total tectonism** (the sum of basin subsidence/uplift, thermal subsidence, sediment loading and sediment compaction). Both physical and numerical models predict the effects of these factors individually (Parker *et al.*, 1986; Reynolds *et al.*, 1991; Steckler *et al.*, 1993; Niedoroda *et al.*, 1995; Syvitski *et al.*, 1999). However, the interactions among factors also exert significant control on the long-term record. For example, the manner in which a large supply of fluvial sediment will disperse and accumulate on the shelf is determined by the space available below wavebase (i.e. **accommodation space**), which in turn is controlled by the balance between eustasy and total subsidence. Choosing among these factors presents a daunting challenge to the geologist attempting to extract the history of events that produce the geometries revealed in seismic profiles.

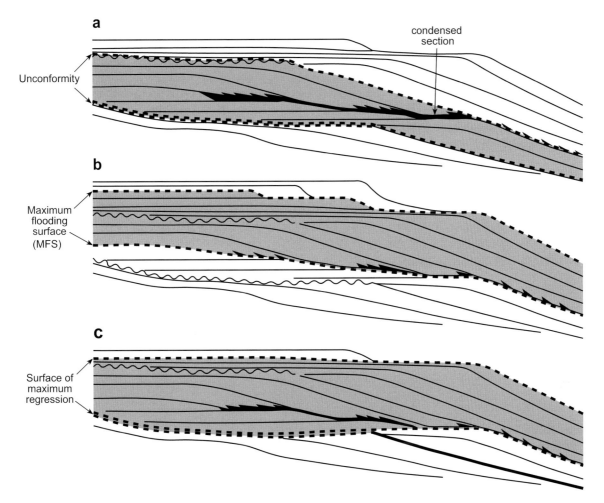

Fig. 3 Three ways of defining sequences based on stratal relationships. (a) Sequences are bounded by unconformities or their correlative conformities. (b) Sequences are bounded by maximum flooding surfaces. (c) Sequences are bounded by surfaces of maximum regression. (From Posamentier & Allen, 1999.) Reproduced with permission from SEPM (Society for Sedimentary Geology).

Tools for accessing the long-term record

Seismic stratigraphy

Seismic reflections are the result of downgoing acoustic energy encountering abrupt vertical changes in sound velocity and/or bulk density beneath the seafloor. The product of velocity and density is defined as **acoustic impedance** and, in general, the greater the vertical change in impedance, the larger the amplitude of the reflected energy. However, sound energy decreases as the square of the distance travelled. Particle-to-particle friction reduces energy further (as a function of frequency, distance, velocity and absorption coefficient of the medium). Low frequencies and slow velocities allow energy to propagate the greatest distance. For this reason, reflected wavelets can be appreciably different from those that enter the subsurface (e.g. Tucker & Yorston, 1973).

The reliability of interpretations based on weak reflections depends on the ability to distinguish true reflected energy from noise. Common sources of noise include rough sea state and poor towing characteristics of the sound receiver (streamer) and sound source. Artefacts that generate unwanted reflections include, but are not limited to: seafloor and internal ('peg-leg') multiples, diffracted arrivals and out-of-plane reflections. Although each can be minimized with appropriate acquisition techniques or specially designed data processing, their occurrence cannot be eliminated.

The practical limits to vertical seismic resolution have been thoroughly studied and shown to depend largely on the frequency of the seismic source (Widess, 1973; Neidell & Poggiagliolmi, 1977; Sheriff, 1977, 1985; Mahradi, 1983). High frequencies resolve more closely spaced reflectors than do low frequencies, but they are unable to penetrate as far into the seabed. Hence, the desirable seismic source is as broadband as practical; the higher frequencies provide optimum resolving power in the shallow part of the section, and lower frequencies penetrate to greater depth and return reflected energy (but with reduced resolution). Experiments and numerical models show that reflections from two surfaces can be clearly resolved when separated by a distance of at least one wavelength of the downgoing acoustic pulse. As the vertical separation narrows, the two reflections begin to interfere, but can be resolved down to separations of one-eighth wavelength. At still narrower distances, the contribution of reflected energy from the two surfaces can no longer be distinguished, despite the fact that some sort of surface or group of surfaces can be detected (Widess, 1973).

Acoustic energy travels as spherically spreading waves, and each impedance change acts as a secondary source of these waves. As a result, reflections that appear to come from directly beneath the streamer may originate at a different location. The distance ahead, behind or to either side of the shortest path to a point on a buried reflector gives rise to the concept of a **Fresnel Zone** (Sheriff, 1977), or acoustic footprint. The smaller the footprint, the better one can distinguish features separated by a narrow horizontal distance (see Sheriff (1985) for details). The size of the footprint is governed by velocity and frequency of the seismic pulse, and the distance to the reflector.

Backstripping

Seismic profiles provide information about how continental-margin strata are arranged in space (i.e. through conversion from travel-time), but identifying the effects of tectonism, climate change, sea-level change and sediment supply in building that record presents a significant challenge (see additional discussion in Syvitski *et al.*, this volume, pp. 459–529). Backstripping is frequently used to estimate tectonic subsidence by accounting for and removing the effects of other causes of subsidence, such as loading due to the weight of the sedimentary column. However, where tectonic subsidence can be estimated, as is the case for the New Jersey margin, the backstripping approach can be used instead to calculate the palaeobathymetry (e.g. Kominz *et al.*, 1998) and, from there, to reconstruct the stratigraphy through time (Steckler *et al.*, 1988, 1999). The rate of tectonic subsidence on this old passive margin is low (0–4 m Myr^{-1}), and can be readily estimated from previous tectonic analyses (e.g. Steckler *et al.*, 1988; Keen & Beaumont, 1990). In contrast, the subsidence of active margins such as the Eel River basin is complex (Clarke, 1992; Gulick *et al.*, 2002) and cannot be estimated sufficiently for backstripping.

Reconstructing sedimentation on a continental margin using the backstripping technique comprises several steps (Fig. 4a–f) (Steckler *et al.*, 1993) that sequentially remove the accumulated effects of subsidence or deformation. The first step is to strip off all sediment above the horizon of interest (Fig. 4b). The weight of sediment corresponding to the removed interval is then calculated, and the remaining layers are flexurally unloaded (Fig. 4c). However, the correct value for **flexural rigidity** of continental margins is a matter of debate (Watts, 1988; Fowler & McKenzie, 1989; Karner, 1991; Kooi *et al.*, 1992). To estimate its value on the New Jersey margin, for example, a parameterization has been used that accounts for the influence of large variations in sediment and crustal thickness at the margin, as well as other factors (Lavier & Steckler, 1997). This model predicts that the Miocene elastic thickness on the New Jersey margin varies between 23 and 30 km, and this range has been used to compute the **isostatic rebound** of the underlying layers (Steckler *et al.*, 1999).

Backstripping is an iterative process, and the next step is to correct for the **compaction** that occurred due to the weight of overlying sediments removed in the previous step (Fig. 4d). For this decompaction at the New Jersey margin, an exponential decrease in porosity with depth (e.g. Sclater & Christie, 1980) has been assigned, based on lithologies known from wells drilled in the region. The layers have then been decompacted to their

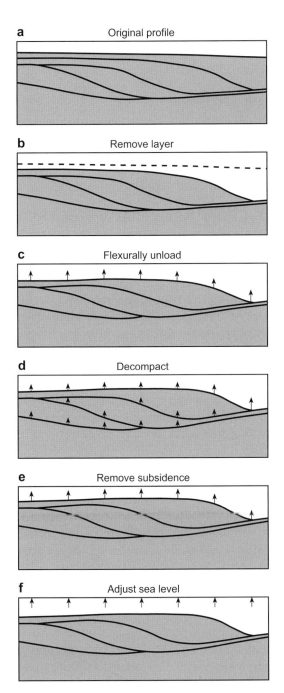

a Original profile

b Remove layer

c Flexurally unload

d Decompact

e Remove subsidence

f Adjust sea level

Fig. 4 Diagram illustrating steps used in reconstructing stratigraphy and palaeobathymetry using backstripping. See text for details. (From Steckler *et al.*, 1999.)

original porosities and their depths adjusted. In a margin as old as the USA Atlantic margin, the observed tectonic subsidence during the Tertiary follows a smoothly decaying exponential curve. Knowledge of the sedimentary and lithospheric

structure (Diebold *et al.*, 1988; Grow *et al.*, 1988; Sheridan *et al.*, 1988; Holbrook & Kelemen, 1993; Klitgord *et al.*, 1994) provides an estimate of the overall tectonic subsidence (Fig. 4e; Watts & Thorne, 1984; Steckler *et al.*, 1988; Keen & Beaumont, 1990). Thermal subsidence has been calculated for the interval between the Oligocene and the present, using a two-dimensional thermal model for the New Jersey margin similar to that of Steckler *et al.* (1988). The last step in backstripping a profile is adjusting for the change in sea level since the time of the reconstruction (Fig. 4f). The effect of water loading is removed using the long-term estimates of Kominz (1984) for the sea-level history, which does not include high-frequency glacio-eustatic fluctuation. A different sea-level curve would cause vertical shifts of the palaeodepths across the entire reconstruction. The final result is a reconstruction of the palaeobathymetry and underlying horizons across a continental-margin profile for the sequential time intervals of the reconstruction.

Uncertainties in the reconstructions can be assessed by varying the flexure and compaction used for the reconstruction. Compaction errors arise from uncertainties in the lithology and the **compaction coefficients** of those lithologies, and are generally proportional to the thicknesses of sediments removed. Consequently, backstripping errors on a continental margin decrease in the landward direction. Uncertainties of 25% in compaction parameters, for example, can produce errors of up to 65 m for a 1300-m-thick section at the shelf–slope transition. However, similar uncertainties for clinoform rollovers 300–500 m beneath the New Jersey shelf lead to errors of only 5–15 m. The compaction depends upon the local overburden removed, thus it influences the height of the reconstructed clinoforms. Uncertainties from varying the flexural rigidity are up to ~100 m for the deepest reflectors beneath the outer shelf, decrease for younger surfaces, and taper to zero beneath the modern coastal plain. Flexural unloading produces a smooth isostatic response, and thus primarily affects the clinoform depth to ±30 m.

Drilling and logging

Recovering samples of the long-term record buried more than a few tens of metres in continental margins is expensive, technically challenging and, therefore, is rarely attempted by the research community. Although cuttings and sidewall samples are routinely collected in oil or gas wells, commercial interests generally do not take the time and expense to recover continuous drill cores. **Downhole measurements** in commercial wells provide a substitute. Unlike measurements made directly on fragmented and incomplete drill cores, downhole data provide continuous information and sample a larger volume of rock (Goldberg, 1997). There are three general categories of downhole logging devices: electrical, nuclear and acoustic (see Doveton, 1986; Ellis, 1987; Paillet *et al.*, 1992; Goldberg, 1997). In addition, borehole imaging, temperature and various *in situ* properties can be measured with wireline tools. Most devices have a vertical resolution better than 0.5 m (Allen *et al.*, 1989; Tittman, 1991). Wireline logs have much greater vertical resolution than seismic profiles but little lateral resolution, so the combination of the two defines subsurface structures better than either data type can alone. Core-seismic correlations are supplemented by downhole measurements of two types. The first combines density and acoustic velocity logs to produce reflection coefficients versus depth. These can be used to produce a **synthetic seismogram** (Doveton, 1986) that provides a direct comparison between the seismic response of a drill core at various depths (metres) and two-way reflection time (seconds) of a seismic profile passing over the drill site. The second type of correlation is based on a **vertical seismic profile**. A hydrophone is lowered into a drill hole, clamped at a series of discrete depths, and the travel-times are recorded for acoustic pulses generated at the sea surface; the resultant profile provides a direct tie between depth and travel-time.

Sediment instability near the seafloor dictates that drill pipe must remain in the uppermost 40–80 m of the hole while logging, and few wireline tools provide useful measurements through metal pipe. However, two types of non-wireline techniques overcome this limitation: **logging while drilling** and **measuring while drilling** (LWD and MWD, respectively; see Allen *et al.*, 1989; Bonner *et al.*, 1992; Murphy, 1993). Both use specially designed logging tools that are part of the drill string, located a few centimetres to 10 m above the drill bit. Both record data while drilling, and, consequently, sediment properties are measured beginning at the

seafloor in a borehole that was drilled only a few seconds to a few minutes previously.

THE EEL RIVER BASIN

Tectonism – a major control of sediment distribution and preservation

Tectonic setting of the Eel River Basin

The **Eel River Basin** (ERB) of northern coastal California is an active **forearc basin** whose history has been strongly influenced by the proximity of three tectonic plates: Pacific, Juan de Fuca and North America (Fig. 5). These plates meet today at the Mendocino Triple Junction (MTJ), with subduction of the Juan de Fuca Plate beneath North America north of the triple junction, transform motion between the Pacific and North American plates south of the triple junction, and transform motion along the Mendocino Fracture Zone (MFZ) west of the triple junction. Basement beneath the western Eel River Basin and much of its offshore extension is coastal Franciscan terrane (Fig. 6). Overall, this basement is a forearc accretionary complex composed of pre-Neogene trench sediments. Subsidence to form a forearc basin began in Miocene time, generating what is now the Eel River Basin (Fig. 7). By early Miocene, subduction rates increased and subsidence of the forearc region allowed deposition in the Eel River Basin.

The main depositional phase of the Eel River Basin is represented by the regressive Wildcat Group (Fig. 6). This succession is roughly 3000 m of upper Miocene marine mudstone at the base (Pullen Formation) grading upward to mid-Pleistocene non-marine sandstones and conglomerates at the top (Carlotta Formation). Such voluminous sediment accumulation in the forearc (Dickinson *et al.*, 1979; Nilsen & Clarke, 1987) implies that the margin experienced less coupling between the downgoing plate and the overriding North American Plate than previously (Orange, 1999).

A basin-wide unconformity developed in the mid-Pleistocene, indicating a transition from regional subsidence and sediment accumulation to spatially variable deposition and erosion. This transition occurred when the Blanco Fracture Zone

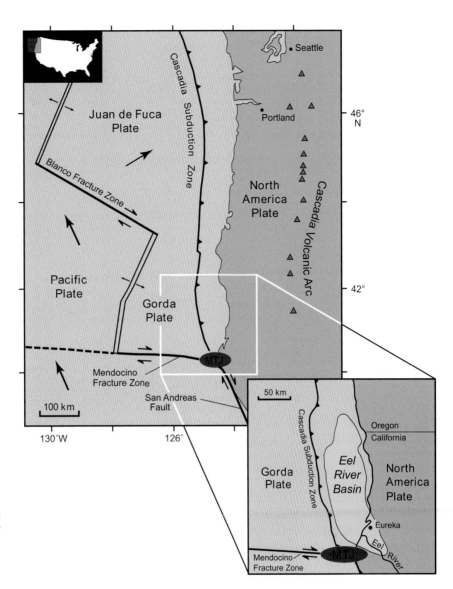

Fig. 5 Regional setting for the Eel River Basin, showing major bounding structural features: large arrows show relative plate movements; MTJ, Mendocino Triple Junction. (After Clarke, 1987; Aalto *et al.*, 1995.)

passed north of the Eel River Basin. This event at 0.7–1.0 Ma marked the change from subducting the older, colder Juan de Fuca crust to the younger, warmer, more buoyant Gorda crust (Carver, 1987). Orange (1999) suggested that the increase in coupling between the warm, young subducting slab and the overriding plate led to uplift and a change from regional deposition to regional erosion, or, at a minimum, to a change in depositional style.

The increase in plate coupling also led to an increase in offscraping of the sedimentary section from the downgoing slab (Carver, 1987). Today, the plate convergence between the Gorda and North American plates is ~3 cm yr^{-1} oriented NE–SW

(Fig. 8; DeMets *et al.*, 1990). Within the southern portion of the Eel River Basin, the late Quaternary arrival of the Mendocino Triple Junction resulted in localized uplift, and a rotation of the folds and thrusts to a more WNW–ESE orientation (Clarke, 1992; Orange, 1999).

Seismic unconformities provide chronology

Fourteen unconformities defined by erosional truncation, downlapping and onlapping relationships have been identified in seismic profiles of the offshore Eel River Basin (Figs 9 & 10; Burger *et al.*, 2002). These surfaces are not to be confused with

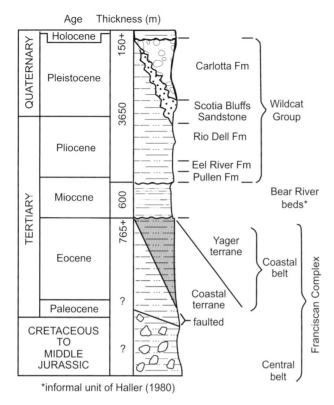

Fig. 6 Composite stratigraphic column for the Eel River Basin, constructed from outcrop studies. (From Clarke, 1992.)

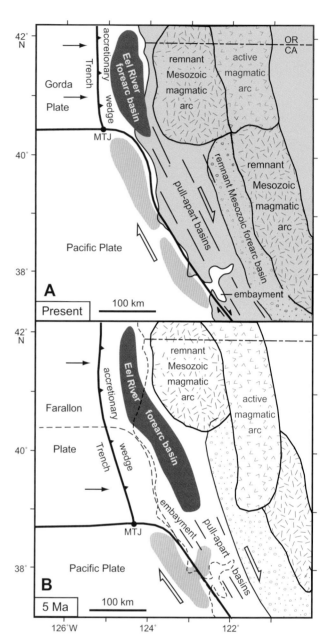

Fig. 7 Palaeotectonic maps showing the evolution of northern California during the late Neogene. Note the shortening of the Eel River Basin in response to northward Mendocino Triple Junction (MTJ) migration. The Gorda Plate is a small fragment of the former Farallon Plate. (After Nilsen & Clarke, 1987.)

the ten V-shaped, localized incised channels of presumed fluvial origin (Figs 10 & 11; Burger *et al.*, 2001). The smoother, more extensive unconformities are most clearly recognized in uplifted areas; they appear to become conformable in adjacent synclines (see Fig. 12). The most extensive, distinctive unconformities are surfaces 13, 9, 7, 5 and 1 (from youngest to oldest). Two of the most prominent unconformities, 1 and 9, have been correlated to features mapped in the onshore Eel River Basin, thereby providing age estimates for the entire package of offshore unconformities.

Surface 1 has been identified by Gulick & Meltzer (2002) as coeval with the onshore Wildcat Unconformity (Woodward-Clyde Consultants, 1980; McCrory, 1995, 1996, 2000). This constrains the seismically imaged sedimentary section offshore, equivalent to the upper ~1.5 s in synclinal areas, to be younger than ~1.0 Ma. Surface 9 has been identified by Gulick & Meltzer (2002) as coeval with the Hookton Datum onshore (McCrory, 1995, 1996). This correlation provides an age of ~500 ka to surface 9.

Burger *et al.* (2002) measured the thicknesses between all 14 surfaces on 6 seismic strike lines, averaged the results, and assigned ages (Fig. 13) based on the assumption of continuous and uniform deposition between surfaces 1 and 9, and

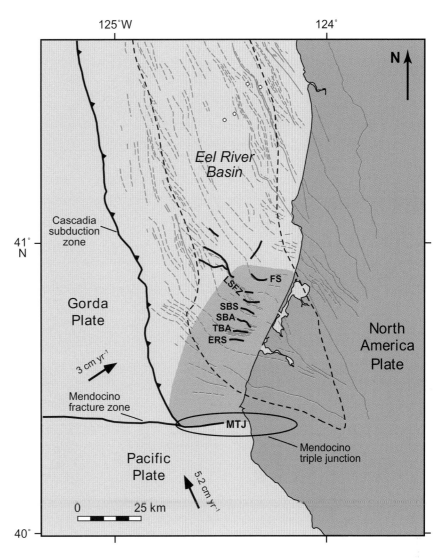

Fig. 8 Structural and tectonic setting of the Eel River Basin. Structural grain of faults (light solid lines; Clarke, 1987), anticlines and synclines (light dashed lines; Clarke, 1987) and plate motions (Gulick *et al.*, 2002), influencing deformation of Eel River Basin sediments are indicated. Dark dashed line denotes extent of the Eel River Basin. Mapped structural trends are indicated by bold lines. Note the change from a general NNW–SSE orientation to NW–SE and then almost E–W orientations nearer the Mendocino Triple Junction (MTJ). Black arrows indicate modern motions (with rates) of the Gorda and eastern Pacific plates with respect to North America (adapted from Gulick *et al.*, 2002). Purple area indicates estimate of the region directly impacted by the MTJ, based on inactive pre-existing folding and apparent counterclockwise rotation of these features. ERS, Eel River Syncline; TBA, Table Bluff Anticline; SBA, South Bay Anticline; SBS, South Bay Syncline; LSFZ, Little Salmon Fault Zone; FS, Freshwater Syncline. (MTJ location after McCrory, 2000; after Hoskins & Griffiths, 1971.)

similarly between 9 and the seafloor. The validity of these assumptions cannot be tested, but given that each was measured in the most conformable section of a synclinal setting on the shelf, these estimates provide useful first attempts. Burger *et al.* (2002) compared the age estimates of these surfaces to the global sea-level proxy curve derived from $\delta^{18}O$ measurements (Mederios *et al.*, 2000), and concluded that glacio-eustasy was very likely an important agent in forming these surfaces.

Offshore deformation since 1 Ma

Based on seismic interpretation, several faults and fold structures are recognized beneath the shelf portion of the Eel River Basin (Fig. 14). While the amount and age of their offsets have not been determined with great accuracy, they allow basic generalizations to be made concerning the role that Gorda Plate subduction and Mendocino Triple Junction migration have played in the structural evolution of the Eel River Basin since 1 Ma.

The northward approach of the Mendocino Triple Junction impinged on the Eel River Basin beginning at ~500 ka (McCrory, 1989). Broad folds (Fig. 14) comprising the South Bay Anticline, South Bay Syncline and Little Salmon Fault Zone (SBA, SBS and LSFZ) all pre-date 1.0 Ma (surface 1, Figs 10 & 12), suggesting that Gorda Plate subduction, not Mendocino Triple Junction migration, caused this

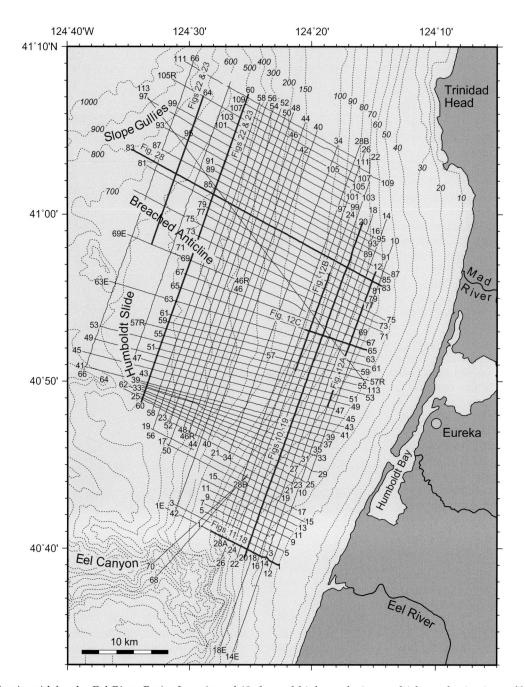

Fig. 9 Seismic grid for the Eel River Basin. Location of 48-channel high-resolution multichannel seismic profiles (numbered 1–113) collected across the basin during *R/V Wecoma* cruise W9605b, using a single 45/45 GI airgun and 600-m streamer, and then stacked 24-fold.

deformation. These older compressional features are currently oriented WNW–ESE (McCrory, 2000; Fig. 8), but when initially formed, pre-1.0-Ma features probably had a more NW–SE orientation.

Compressional folding of Eel River Basin sediments progressively ceased from south to north.

Specifically, deformation of the South Bay Anticline and South Bay Syncline ended at ~1 Ma, then folding of the Little Salmon Fault Zone ended at ~690 ka, and within the Freshwater Syncline at ~600 ka (Burger *et al.*, 2002). Subduction-related folding continues today both on the shelf north

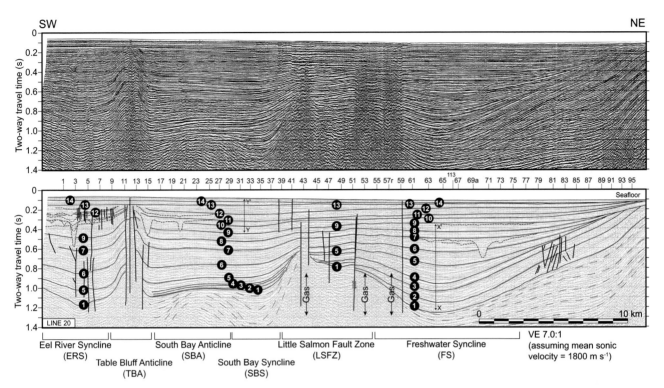

Fig. 10 Uninterpreted and interpreted versions of multichannel seismic profile 20, showing mapped faults, deformational structures and unconformities (numbered circles) in the offshore Eel River Basin. Along-strike section; see Fig. 9 for location. Near-vertical bold lines represent faults; thinner solid lines represent regional unconformities; short-dashed lines represent local incised unconformities; and long-dashed lines represent the trends of folded and truncated reflectors beneath surface 1. Vertical zones of presumed gas wipeout (Yun *et al.*, 1999) are present within the Table Bluff Anticline and the Little Salmon Fault Zone. Onshore structures mapped offshore are indicated at the bottom of the interpreted profile. Locations where travel-times between surfaces were measured to estimate unconformity ages (see Fig. 9) are indicated by vertical lines labelled 'X–X'' and 'Y–Y''. Numbers across the top of the interpreted profile identify locations of crossing profiles. (From Burger *et al.*, 2002.)

of Humboldt Bay (Fig. 14) and beneath the upper slope. The influence of northward motion of the Mendocino Triple Junction appears to be limited to an area ~60–70 km north of the triple junction (Fig. 8).

North–south compression is most noticeable close to the modern triple junction. Uplift indicated by a proliferation of fluvial incisions at the southern end of available seismic coverage (McLaughlin *et al.*, 1994; Aalto *et al.*, 1995; Gulick *et al.*, 2002) occurred sometime since ~330 ka (Figs 10 & 11; Burger *et al.*, 2002). A 95° orientation of the Table Bluff Anticline (TBA) within the past ~330 ka also suggests N–S compression. Counterclockwise rotation of pre-existing features, and right-lateral motion across the Table Bluff Anticline (Fig. 15), all support localized transpression (Gulick & Meltzer, 2002).

Contemporary deformation at the Mendocino Triple Junction is expressed as a zone of localized uplift, measuring 2.5–2.8 mm yr^{-1} of vertical motion (Lajoie *et al.*, 1982; McCrory, 1996). This uplift affects the trend of structures along the Gorda–North American plate boundary, beginning with NE–SW deformation features in the accretionary wedge immediately north of the triple junction. These trends gradually swing toward WNW–ESE farther north, and eventually become NNW–SSE in the Eel River Basin itself (Fig. 8).

Modern tectonic activity within the Eel River Basin results in localized uplift on anticlines and subsidence on synclines. These forearc structures trend across the slope and shelf and project onshore to features such as the Russ Fault, Eel River Syncline, Table Bluff Anticline and Little Salmon Fault (RF, ERS, TBA, LSF). This alignment

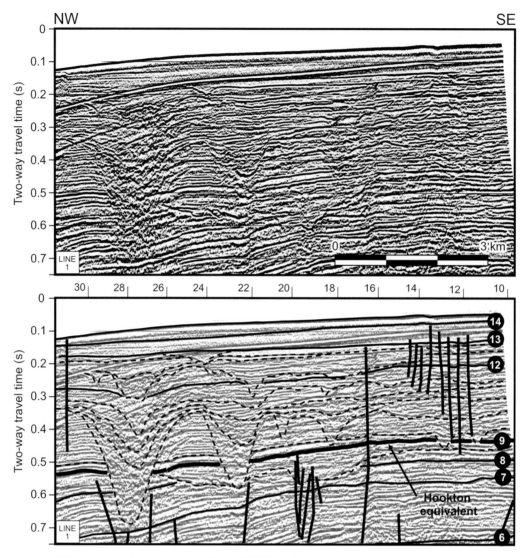

VE 7.1:1 (assuming mean sonic velocity of 1800 m s⁻¹)

Fig. 11 Uninterpreted and interpreted portion of multichannel seismic profile 1 showing regional unconformities (solid, numbered lines). See Fig. 9 for location. Dashed lines indicate local incisions (compare with Fig. 18), which dominate the MCS grid south of the Table Bluff Anticline (see also Fig. 10). Near-vertical bold lines represent faults. The near-horizontal bold line represents surface 9, correlated to the Hookton Datum of McCrory (1995), with an estimated age of ~500 ka. Numbers at the top of the interpreted profile show locations of crossing strike profiles. (After Burger *et al.*, 2002.)

and continuity suggests that the patterns, and perhaps even the rates, of deformation measured onshore can be projected offshore to provide first-order estimates of the tectonic effects on sedimentation (Orange, 1999). Onshore, modern deformation controls the topography, river drainage and configuration of wave-cut terraces. Uplift rates on the subaerial portion of the Little Salmon Fault are in the order of 2.5 mm yr⁻¹, for the Table

Bluff Anticline they are 0.4–0.75 mm yr⁻¹, and for the Mad River fault zone they are 1.3–2.5 mm yr⁻¹; subsidence rates of the Freshwater Syncline are 1.4–3.3 mm yr⁻¹ (see McCrory (1996) for data compilation and references). These structures are prominent despite very high erosion and denudation rates that result from the combination of climate, tectonics and exposure of relatively young sediments.

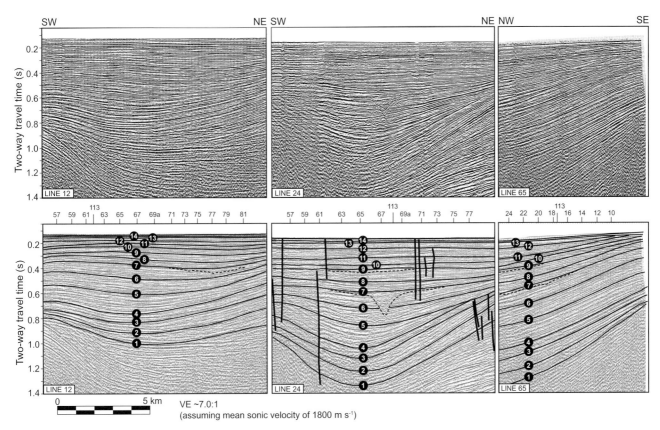

Fig. 12 Uninterpreted and interpreted portions of multichannel seismic strike profiles 12 and 24, and dip profile 65, showing the offshore expression of the Freshwater Syncline (FS). Location indicated in Fig. 9. Line 65 is oriented slightly oblique to the synclinal axis as it crosses the shelf. Bold, near-vertical lines represent faults, and numbers at the top of interpreted profiles indicate locations of crossing profiles. (From Burger *et al.*, 2002.)

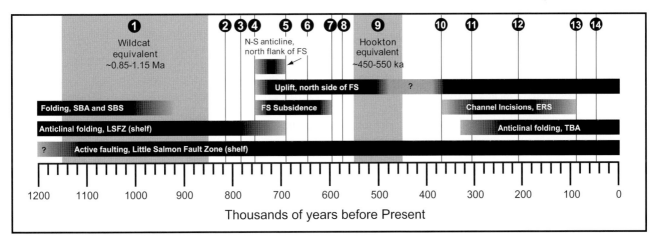

Fig. 13 Timeline summarizing the periods of deformation and channel incision on the Eel River Basin shelf. Age estimates are based on the stratigraphic arrangement of 14 seismic unconformities (numbered vertical lines), two of which are tied to the Wildcat and Hookton surfaces (~1 Ma and ~500 ka, respectively). Assumptions of uniform accumulation rates provide the ages of intervening surfaces (see Fig. 10). The structures abbreviated in this figure are listed in Fig. 10. (From Burger *et al.*, 2002.)

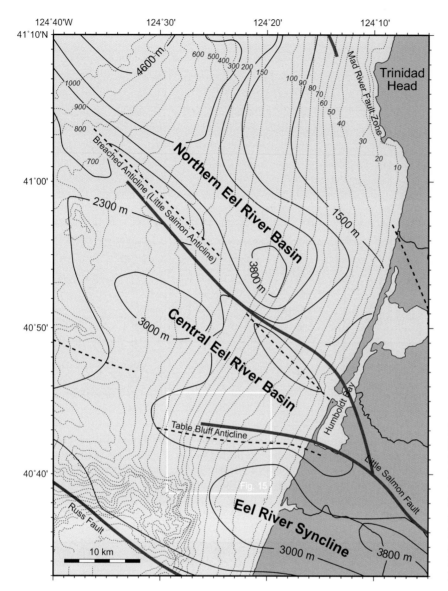

Fig. 14 Structures and sediment isopachs (solid lines) above Franciscan basement in Eel River Basin (after Crouch & Bachman, 1987). Bathymetry (italic numbers) is in metres. Three offshore sub-basins are defined by the Russ and Mad River Fault Zones, plus two antiformal trends offshore that branch from the Little Salmon Fault. Dashed lines are prominent anticlines.

Offshore stratigraphy – local variations of governing processes

The development of shelf sub-basins since 1 Ma

Isochron maps (refer to Figs 16 & 17 for all discussion in this section) show that the region of greatest sediment preservation in the Eel River Basin has gradually shifted from north to south during the past ~1.0 Myr, presumably due to structural controls and changes in the relative importance of sediment sources (Burger *et al.*, 2002). These maps also show that overall sediment accumulation rates on the shelf have decreased since ~500 ka. This

latter change coincides with the northward migration of the Mendocino Triple Junction and related uplift across the southern part of the basin.

Between ~850 ka and 750 ka, an irregular depocentre oriented approximately 055° was located west of Trinidad Head and was supplied by a northern sediment source. Syndepositional uplift of the Little Salmon Fault Zone separated this depocentre from the central Eel River Basin, where little sediment accumulated during this interval (Figs 16A–D & 17A–D).

Folding of the Freshwater Syncline (FS) and relative uplift north of this feature began at ~750 ka; the associated depocentre is currently oriented

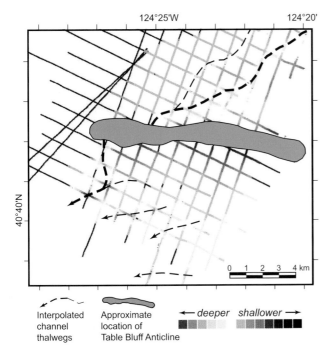

Fig. 15 Right-lateral strike-slip displacement across the Table Bluff Anticline, inferred by the offset of a buried channel (bold dashed line) on both sides of the anticline. The consistent NE–SW gradient across the anticline (relative depths of channel surface are indicated by colour along the seismic grid) strongly suggests that this was once a continuous feature, indicating that its formation pre-dates deformation of the anticline. The observed offset is ~2.0 km, indicating a slip rate of 0.6 cm yr^{-1}. (From Burger *et al.*, 2002.)

~115°, probably due to structural control by the syncline. Sediment continued to be supplied from the north. By ~690 ka, a new depocentre to the south indicates that a southern source, perhaps the ancestral Eel River, has begun to contribute sediment to the basin (Figs 16E & 17E). Less distinct partitioning of depocentres occurs after ~690 ka, when Little Salmon Fault Zone folding and uplift ended. There are depocentres beneath both northern and southern portions of the shelf from ~750 ka to ~550 ka, evidently supplied by sources from both the north and the south.

Between ~450 and 300 ka (Figs 16I–J & 17I–J), offshore sequences appear more continuous, indicating a decrease in the influence of structural deformation and resultant partitioning of sediment accumulation and preservation. Overall thicknesses per unit time also decrease, suggesting either a decrease in sediment supply after ~500 ka

or an increase in sediment bypassing to the slope. Formation of the Eel Canyon after ~500 ka probably increased the amount of sediment bypassing the shelf. Relative sediment thicknesses in the north decreased after ~500 ka, and particularly after ~200 ka, indicating a decrease in sediment contribution from the postulated northern source (Figs 16L–M & 17L–M). However, minor sediment maxima west of the modern Mad River mouth after ~200 ka identify it as a sediment source, at least from then to the present.

The most recent interval considered, ~43 ka–present, reveals modern patterns of shelf sedimentation (Borgeld, 1985; Wheatcroft *et al.*, 1996; Morehead & Syvitski, 1999; Sommerfield *et al.*, this volume, pp. 157–212). A large N–S-oriented depocentre in the south (Figs 16N & 17N) is the result of input from the Eel River. A smaller depocentre to the north suggests continuing, minor input from the Mad River. Modern sediment dispersal patterns (Sommerfield *et al.*, this volume, pp. 157–212) and the dominance of the Eel River as a sediment source have been maintained for at least the last ~43 kyr.

Buried fluvial channels

Seismic profiles reveal numerous V-shaped channels, 10–250 m deep and 0.1–1.0 km wide (Figs 18–20) that incise 10 surfaces as deep as 0.5 s (~450 m) below the shelf of the Eel River Basin. These surfaces (designated with letters A–E in Fig. 18) are distinct from 14 seismic unconformities (designated with numbers; e.g. Fig. 11) also seen in profiles across the shelf (Burger *et al.*, 2002); the latter are smoother, more continuous, and more laterally extensive (Fig. 19). Many unconformities extend across the entire multichannel-seismic (MCS) grid (Fig. 9) and disappear only where they are truncated on the crests of anticlines or eroded by incisions. The origin of these regional unconformities is discussed in a later section of this paper.

An E–W trending structural high, correlative along-strike with the onshore Table Bluff Anticline (TBA; Clarke, 1987) crosses the study area. Most of the incised surfaces are concentrated south of the Table Bluff Anticline (Fig. 20); only one channel cuts into and crosses the Table Bluff Anticline, trending NE–SW. Five incised surfaces have been mapped south of the Table Bluff Anticline (Fig. 20; Burger

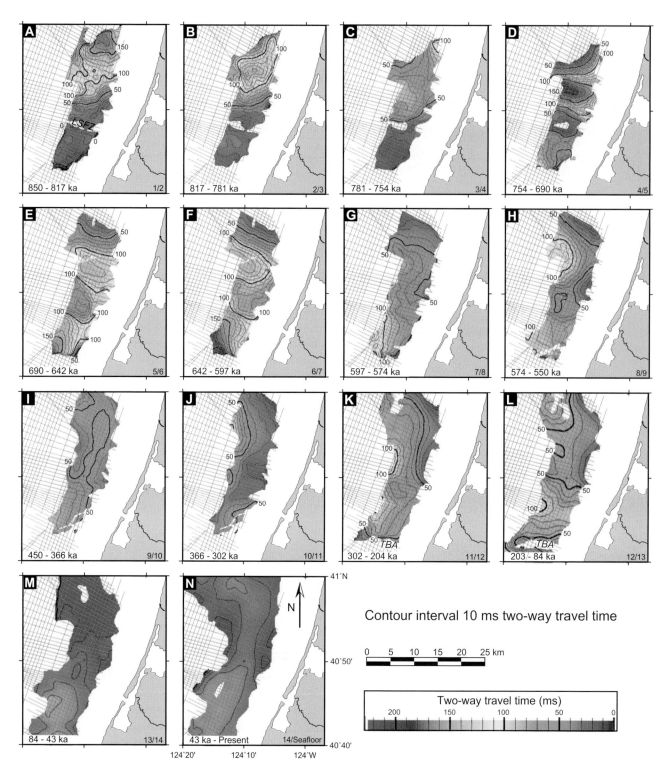

Fig. 16 Isochron maps of sequences bounded by regional seismic unconformities (including the seafloor) beneath the Eel River Basin shelf north of the Table Bluff Anticline. Ages noted on each map (bottom left) are derived from correlation to onshore stratigraphy (see text), and indicate the presumed interval of sediment accumulation for each sequence. Numbers in the lower right corner of each map identify the unconformities that define the interval shown. (From Burger *et al.*, 2002.)

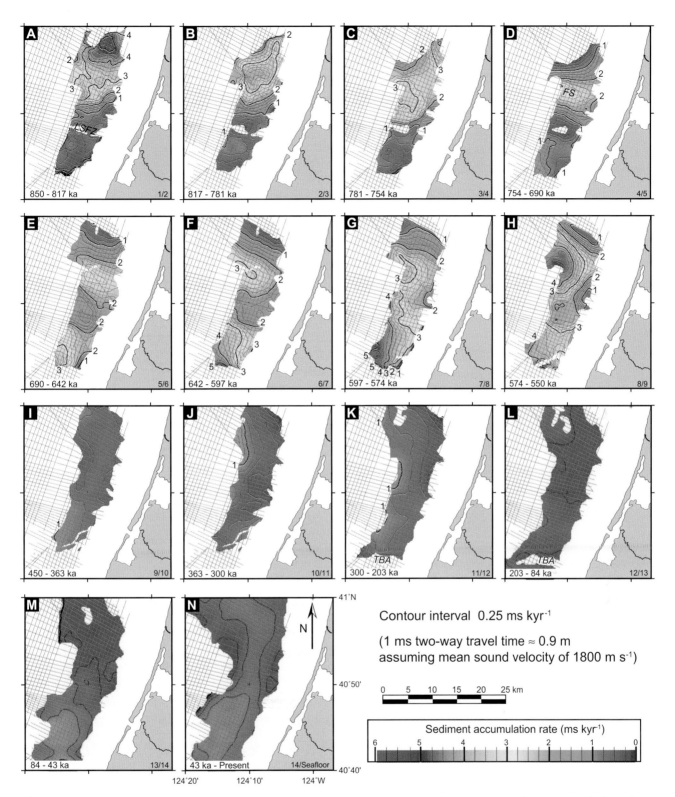

Fig. 17 Maps of sediment accumulation rate for sequences bounded by regional seismic unconformities (including the seafloor) beneath the Eel River Basin shelf north of the Table Bluff Anticline. Travel-time thicknesses from each isochron map (Fig. 16) were converted to sediment accumulation rates based on estimated time over which each sequence was deposited. Inset ages (bottom left) indicate the presumed interval of sediment accumulation for each sequence. Numbers in the lower right corner of each map identify the unconformities that define the interval shown. Note the dramatic decrease in accumulation rates across the margin after ~450 ka, presumably coincident with arrival of uplift and deformation in the southern Eel River Basin related to the Mendocino Triple Junction. (From Burger *et al.*, 2002.)

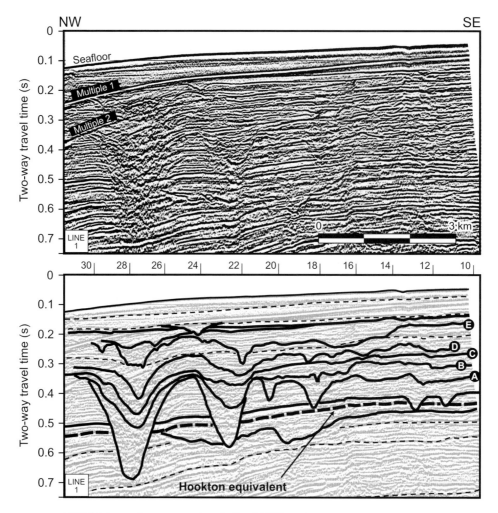

VE 7.1:1 (assuming mean sonic velocity of 1800 m s⁻¹)

Fig. 18 Uninterpreted and interpreted versions of seismic line 1 (see Fig. 9 for location), showing superimposed/ stacked incisions (lettered, solid lines). Labelled incisions are mapped in Fig. 20. Numbered, vertical lines indicate locations of crossing profiles. Thin, dashed lines (see Fig. 11) indicate regional unconformities; bold, dashed line indicates a surface correlated to the Hookton Datum of McCrory (1996). (After Burger *et al.*, 2001.)

et al., 2001). All channels that cut into and define these surfaces trend south-west and deepen toward modern embayments along the north side of Eel Canyon (Fig. 21). North of the Table Bluff Anticline, incisions are observed on only two surfaces, and are prominent on only one. However, like those to the south, these incisions also display southwest orientations and they deepen seaward. The deepest incised channel (surface A, Figs 18 & 19) maintains a nearly constant gradient of ~0.4° both north-east (landward) and south-west (seaward) of the anticline (Fig. 21).

Buried Eel River Basin channels have been mapped in the shallow subsurface at water depths

from ~50 m to 100 m. Their dendritic patterns (Fig. 20) are similar to latest Pleistocene–Holocene drainage features mapped by Duncan *et al.* (2000) on the New Jersey shelf that have been interpreted as either lowstand or early transgressive fluvial systems (Austin *et al.*, 1996; Duncan *et al.*, 2000). A fluvial origin for the Eel River Basin channels also has been proposed (Burger *et al.*, 2001), based on the compelling seismic evidence of dendritic trends, low gradients, cross-shelf extent and location on the shelf where lowstand subaerial exposure is likely to have occurred.

Although Eel River Basin incised surfaces mapped using MCS profiles have not yet been dated

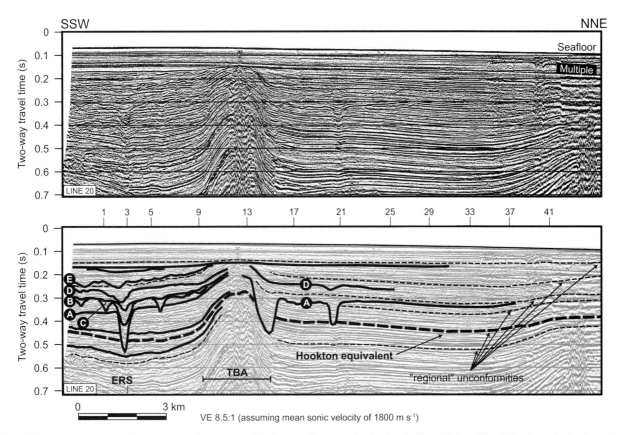

Fig. 19 Uninterpreted and interpreted versions for the southern end of seismic line 20 (see Fig. 9 for location), showing superimposed/stacked incisions. Solid, lettered lines indicate the incised surfaces mapped in Fig. 20. Numbered, vertical lines indicate locations of crossing profiles. Thin, dashed lines indicate regional unconformities that extend across most of the southern shelf; bold, dashed line correlates to the Hookton equivalent (Fig. 18). TBA, Table Bluff Anticline; ERS, Eel River Syncline. Width of TBA deformation as derived from this profile is shown in plan view for each of the shelf strike lines in Fig. 20. (After Burger *et al.*, 2002.)

directly, Burger *et al.* (2001) estimated their ages by correlating a prominent unconformity recognized in the offshore seismic grid to the onshore Hookton Datum (McCrory, 1995). The latter is an angular unconformity correlated across the onshore Eel River Basin that represents a hiatus spanning 450–550 ka ± 35 kyr (McCrory, US Geological Survey, personal communication, 2001). All 10 offshore incised surfaces post-date the offshore Hookton equivalent. Consequently, if these features are the result of fluvial incision into an exposed coastal plain, they could not have originated during glacio-eustatic lowstands with 100 kyr cyclicity. It is likely that some channels were initiated by relative lowstands resulting from local tectonic uplift (Burger *et al.*, 2001).

The consistently NE–SW trend for Surface A incisions on both sides of the Table Bluff Anticline

suggests that the observed drainage represents a single system disrupted by uplift (Fig. 20). The continuity of Surface A (Fig. 21) further suggests that it pre-dates the anticlinal deformation. Continuity of the Hookton equivalent across the trend of the Table Bluff Anticline confirms that uplift must be younger than 500 ka. The incised surfaces are almost exclusively located south-west of the Table Bluff Anticline (Figs 18–20), so the rising Table Bluff Anticline may have been the source for those shelf-drainage systems.

The large number of incised surfaces leading to Eel Canyon indicates that this feature must have existed through numerous oscillations of relative sea level, and strongly controlled off-shelf drainage during lowstands. Based on initiation of incised unconformities at the southern end of the basin after the formation of the Hookton equivalent, it

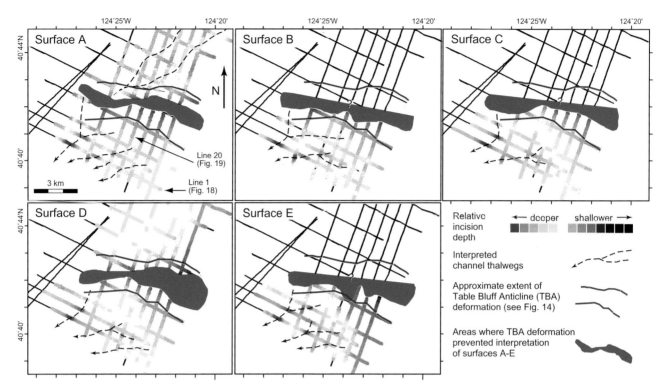

Fig. 20 Maps of the southernmost portion of the seismic grid (Fig. 9), showing the five most deeply incised surfaces (A is oldest, E is youngest; see Figs 18 & 19). Colours of interpreted portions indicate relative incision depth, according to the colour bar (right). The range of the colour scale was normalized by the seismic interpretation software to the maximum and minimum time-depth values for each surface, so the represented colour-scale range is different for each surface. Dashed lines are interpreted orientations of channel thalwegs. The red lines indicate the approximate lateral extent of Table Bluff Anticline (TBA) deformation as defined on each of the shelf strike lines. The blue solid areas indicate where surfaces could not be mapped because of TBA deformation; northern boundaries of these areas are not defined for surfaces B, C and E, because they could not be identified north of the TBA. (From Burger *et al.*, 2001.)

appears that the canyon formed soon after ~360 ka. Burger *et al.* (2002) speculated that sedimentation in the Eel River Basin shifts dramatically in response to variations in relative sea level: during high-stands, some fluvial suspended load can travel northward and seaward; during lowstands, rivers disgorge both suspended and bedload sediments much closer to the shelf edge where they are more readily captured by shelf-indenting canyons.

Processes on the Eel continental slope

The tectonically active Eel River Basin provides sharp along-margin contrasts that help to distinguish processes controlling the accumulation, preservation, and post-depositional deformation of sedimentary strata on the upper continental slope. The combination of large sediment supply from the Eel River (Brown & Ritter, 1971), late Pleistocene

glacio-eustatic sea-level fluctuations (e.g. Ruddiman *et al.*, 1989) and offshore deformation (Orange, 1999; Burger *et al.*, 2001, 2002; Gulick & Meltzer, 2002; Gulick *et al.*, 2002), including frequent earth-quakes (Couch *et al.*, 1974; Field & Barber, 1993), have all contributed to a lengthy history preserved in the slope stratigraphy.

The Little Salmon Fault Zone forms a prominent anticline onshore, but has little to no bathymetric expression on the shelf. Seaward of the shelf break, however, the rates of tectonic uplift exceed the rates of sediment accumulation. As a result, the Little Salmon Fault, and its associated hanging-wall anti-cline (Little Salmon Anticline), crop out on the upper and middle slope in a right-stepping, en échelon pattern (indicated by the **breached anticline** in earlier Figs 9 & 14). This structure forms a drain-age divide on the slope, separating two different slope morphologies and their underlying sediment

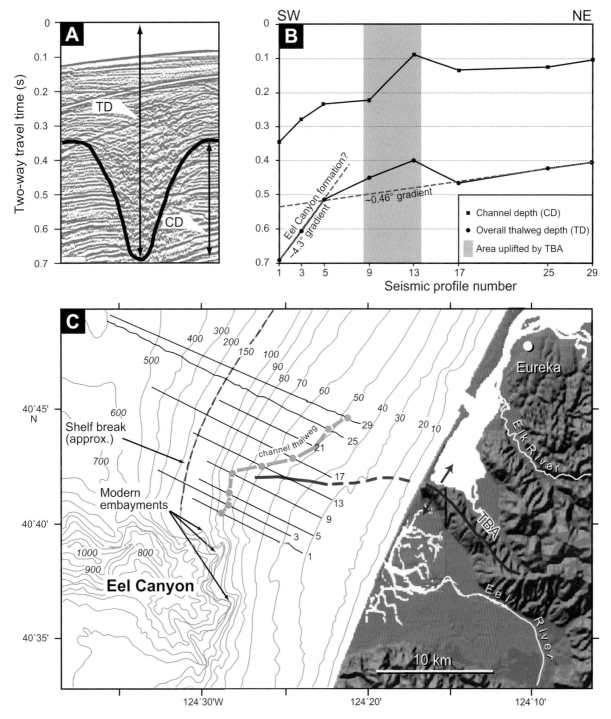

Fig. 21 Influence of Table Bluff Anticline deformation on channel development. (A) Time/depth used to indicate changes in channel gradients and incision depths across the Eel River shelf: CD, channel depth, measured from overbank to thalweg; TD, overall thalweg depth, measured from present sea level to centre of interpreted thalweg. (B) Trends of measurements from north to south for channels in surface A (Figs 18–20). Dashed lines show interpolated channel gradients (calculated using a mean sediment velocity of 1.8 km s^{-1}), suggesting a fluvial source to the north-east. Deviations are attributed to uplift caused by the Table Bluff Anticline (TBA), and deepening as a result of a transition from fluvial to canyon processes. (C) The southern end of the Eel River Basin, showing the spatial relationship between dip seismic profiles, the channel mapped in (B), the TBA (red) and the Eel Canyon. Solid orange circles on the dip profiles represent positions of the channel thalweg. Modern embayments along the flanks of the Eel Canyon are where the mapped channel may have discharged during a preceding relative lowstand. Bathymetry in metres; dashed purple line represents the shelf break, which occurs at ~150-m isobath. (From Burger *et al.*, 2001.)

packages. The slope south of the Little Salmon Anticline is discussed in detail in Lee *et al.* (this volume, pp. 213–274) with reference to the 'Humboldt Slide', and the corresponding region north of the Little Salmon Fault is discussed here.

There is no canyon north of the Little Salmon Anticline that would alter the transport of sediment to the Eel River Basin slope between times of sea-level change. As a result, slope sedimentation to

the north is more uniform through time than it is south of the Little Salmon Anticline. Concordant, parallel reflectors observed in most areas of the northern upper slope suggest hemipelagic settling dominated by fine-grained sediment. Despite this comparatively uniform sedimentation, the slope north of the Little Salmon Anticline is characterized by an abundance of buried channels (Figs 22 & 23). These features incise 11 surfaces that have

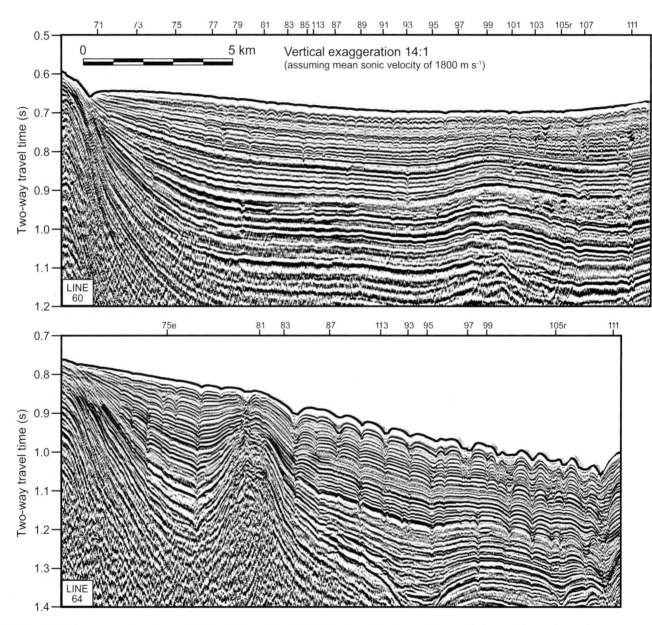

Fig. 22 Uninterpreted portions of seismic lines 60 and 64, showing the downslope evolution of slope channel geometries. Interpretation sections are shown in Fig. 23. Locations of profiles are indicated in Fig. 9. Numbered lines at top indicate locations of crossing profiles. (From Burger *et al.*, 2002.)

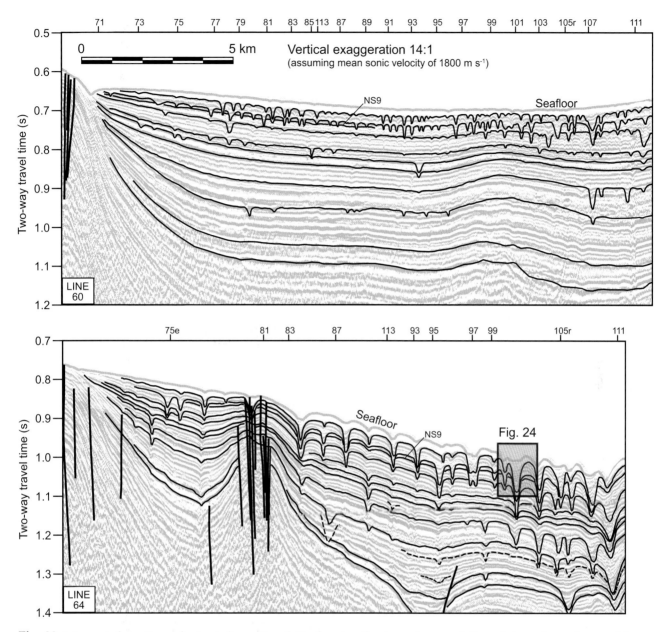

Fig. 23 Interpreted portions of seismic lines 60 and 64, showing the downslope evolution of slope channel geometries. See Fig. 22 for uninterpreted data. Near-vertical bold lines represent faults. Note the expression of larger channels on older surfaces seaward, as well as the increase in numbers of channels on younger surfaces. Channel highlighted on line 64 is shown in greater detail in Fig. 24. (From Burger *et al.*, 2002.)

been traced seaward from the shelf break by Burger *et al.* (2002). The channels trend directly downslope, are 5–110 m deep, 50–1000 m wide, and are generally spaced 1–2 km apart. Older channels are more deeply incised and wider than younger channels. All are V-shaped in cross-section and commonly stacked directly above one another. The density of channels on any given surface increases seaward from the upper slope, and then decreases again as they coalesce farther downslope. Most have a distinctive sedimentary fill consisting of high-amplitude basal reflections that decrease in amplitude upwards and become more draping in character (Fig. 24). Subdued but detectable topography of most channels appears to be the nucleus for the start of the next younger channel.

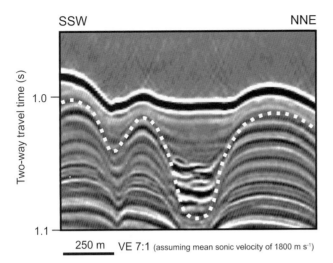

SSW   NNE

Two-way travel time (s)

1.0

1.1

250 m   VE 7:1 (assuming mean sonic velocity of 1800 m s⁻¹)

Fig. 24 Detailed image of a slope channel (white dotted line) from seismic profile 64 (see Fig. 23). Note the high-amplitude basal fill, buried by reflectors of decreasing amplitude, suggesting a fining upward progression of channel-fill sediment. (From Burger *et al.*, 2002.)

Gullies ~1–3 m deep have been observed in high-resolution seismic profiles to ~65 m below the seafloor in the same upper-slope area where channels occur (Field *et al.*, 1999; Spinelli & Field, 2001). These gullies are below the vertical resolution of airgun-based MCS profiling, and are generally more closely spaced (~100–1000 m) than the more deeply buried channels. However, the morphologies of the two types of features are similar and the gullies merge into the larger channels downslope, suggesting they are simply tributaries to the channels (Spinelli & Field, 2001).

Gullies are incised during regressions by downslope transport of coarse-grained sediment

(Spinelli & Field, 2001). Today, they are much closer to Mad River than to Eel River. With evidence that Eel River sediments preferentially debouch into Eel Canyon during lowstands, Mad River is implicated as the source of gully-carving sediments (Fig. 25). If the deeper, larger channels are linked to the same process of formation, then it is likely that the channels are gullies that were either too large or too distal to be totally filled more often than once every 100 kyr. Pratson *et al.* (1994) have shown that mid- to lower-slope canyons off New Jersey can remain clear of sediment through several glacio-eustatic cycles, while the heads of canyons on the upper slope are repeatedly buried and re-excavated. A similar process may have occurred on the Eel slope, making the gullies shorter-lived equivalents of the downslope extensions termed 'channels'.

The density and distribution of channels have changed through time. Younger surfaces display many large channels, distributed fairly evenly across the upper slope. In contrast, older surfaces display a smaller number of large channels concentrated at the northern end of the study area (Figs 22 & 23). Gas-charged sediments that degrade the image quality of seismic profiles are less prevalent north of the Little Salmon Anticline than to the south, and consequently northern surfaces can be correlated to surfaces mapped beneath the shelf.

Gas in the sediment column

Gas is generated in the subseafloor by two mechanisms (Kvenvolden *et al.*, 1993): bacterial degradation of scavenged organic matter; and chemical production of complex hydrocarbons (from organic

Fig. 25 (*opposite*) Interpreted areas of sediment transport, erosion and deposition on the upper slope of Eel River Basin during various stages of sea level. (A) Highstand: suspended sediment from the Eel River bypasses the shelf to the upper slope, depositing fine-grained sediment in the Humboldt Slide area, and hemipelagic sediment on the more distal slope north of the Little Salmon Fault Zone (LSFZ). Slope channels and gullies north of the LSFZ are inactive. Sediment also bypasses the shelf and upper slope through Eel Canyon. (B) Falling stage: increasing amounts of Eel River sediment bypass the shelf and upper slope by way of Eel Canyon. Sediment is eroded, supplying increasing amounts of fine-grained sediment to the upper slope. Sediment discharged from the Mad River deposits progressively closer to the upper slope, increasing slope sediment input north of the LSFZ. Channels north of the LSFZ reactivate along pre-existing trends, and also become bypass conduits. (C) Lowstand: most Eel River sediment bypasses the shelf and upper slope through Eel Canyon. Minimal sediment deposition is expected in the Humboldt Slide area. The Mad River extends to the shelf edge and its sediment causes erosion of slope channels seaward of its mouth. (D) Transgression: bypass of sediment through Eel Canyon decreases, and supply of eroded shelf sediment resumes in the Humboldt Slide area. However, as shelf sediment sources become more distal, slope channels north of the LSFZ begin to infill, first by relatively coarse basal lags, then by finer grained hemipelagic sediment (see Fig. 24). (From Burger *et al.*, 2002.)

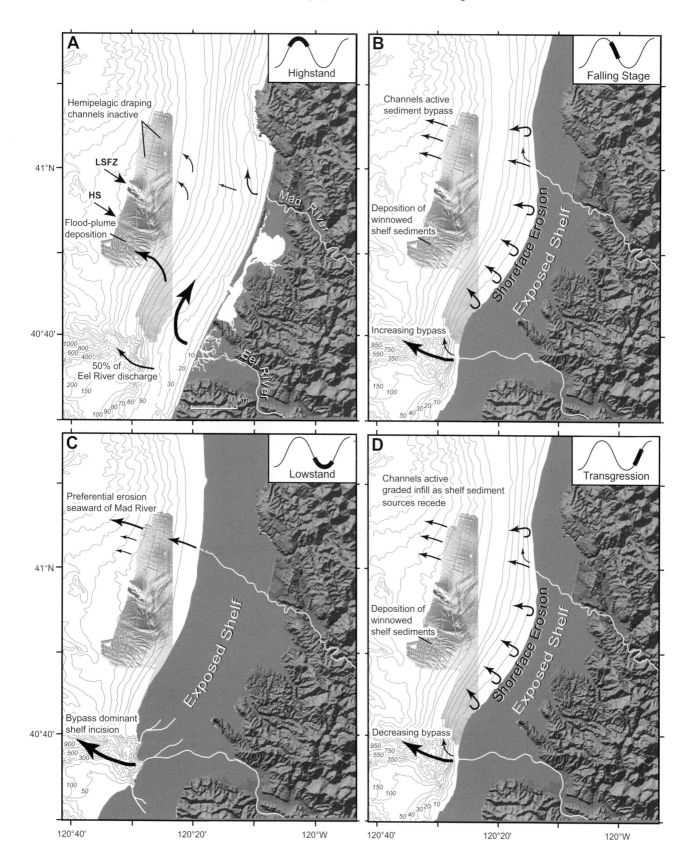

matter and pre-existing oil) under favourable conditions of pressure, temperature and anoxia. The former, **biogenic**, origin predominantly generates methane; the latter, **thermogenic**, origin generates methane at early stages that under proper conditions can lead to ethane, pentane and other long-chain hydrocarbons. Biogenic gas is generated in the upper few hundred metres below the seafloor (the presumed bacterial habitat), but if sealed completely can become buried to greater depths. By contrast, thermogenic processes require conditions

generally found at considerably greater depths, and any gas of this type that is detected a few hundred metres below the seafloor has very likely migrated up from deeper levels of origin.

The Eel River Basin (Fig. 26) has potential for hosting both types of gas (Lorenson *et al.*, 1998). Biogenic gas can be expected to result from the rapid burial of both organic-rich fluvial sediment and marine plankton that flourish in the nutrient-rich upwelling waters of the northern California margin (see Leithold & Blair, 2001). The tectonic

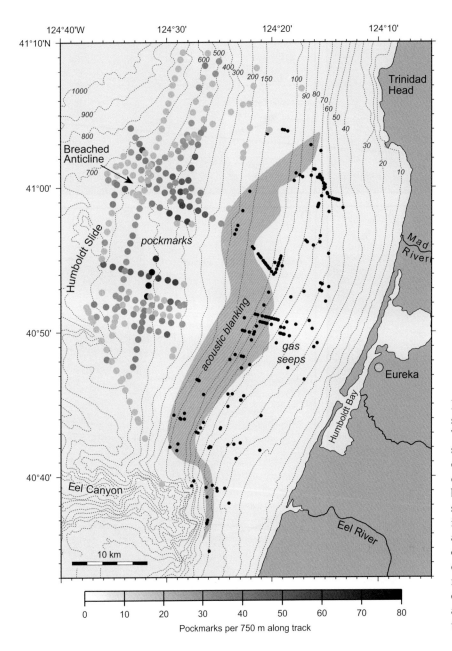

Fig. 26 Indicators of gas in the sediments of the Eel River Basin. The dark blue region along the outer shelf is an area of acoustic blanking of MCS reflections, interpreted to be due to gas in the sediments. Small black circles predominantly on the shelf are water-column plumes identified in 3.5-kHz records and assumed to be gas seeps. Larger circles on the slope are pockmarks detected with deep-towed sidescan; range of purple hues denotes number of pockmarks per 750 m along and across track (i.e. counts ranged from 1 to 80). (After Yun *et al.*, 1999.)

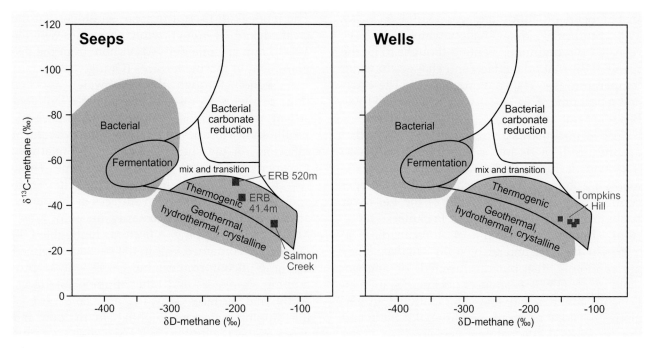

Fig. 27 Isotopic character of gas in the Eel River Basin (ERB) measured in seeps (offshore) and wells (onshore). The $\delta^{13}C$ versus deuterium in methane plot distinguishes between biogenic and thermogenic sources, and suggests that methane in the Eel River Basin is predominantly thermogenic. (From Lorenson *et al.*, 1998.)

setting of this accretionary prism provides post-depositional conditions that lead to thermogenic hydrocarbons as well (Kvenvolden & Field, 1981). Limited geochemical analyses of gas from onshore wells and from seeps both onshore and offshore have determined that thermogenic gas is the more common (Fig. 27; Lorenson *et al.*, 1998).

The abundance of gas in the offshore Eel River Basin has a significant impact on our ability to understand the long-term record of this continental margin. Gas in sufficient quantity impedes the acquisition of an acoustic image, which is usually the most informative stratigraphic approach through remote sensing. Furthermore, the evidence of gas seen venting into the water column or inferred by epifaunal chemosynthetic communities and explosive craters demonstrates that gas has influenced the development of the long-term record.

The loss of reflector strength due to gas-charged formations, **acoustic wipeout**, has proven to be insurmountable in the Eel River Basin. The towed Huntec boomer system (Field *et al.*, 1999; Spinelli & Field, 2001) found large areas of acoustically impenetrable seafloor. A single 45/45-cubic-inch GI gun was also largely unsuccessful in imaging subsurface features in a margin-parallel corridor between 80 m and 200 m water depths (Fig. 26). Amoco Corporation provided reconnaissance profiles acquired by JEBCO Geophysical using a 3380-cubic-inch airgun array that could penetrate to more than 1 km below the seafloor in adjacent regions. Despite this powerful source, the same wipeout zone was encountered on the outer shelf and upper slope. This zone generally parallels isobaths, and is only locally affected by NW–SE structural trends (Yun *et al.*, 1999).

Active gas **seeps** in the Eel River Basin have been detected by water-column reflections in 3.5-kHz records from less than 300 m of water depth (Yun *et al.*, 1999). Side-scan-sonar images show no association with seafloor topography, and MCS profiles indicate no correspondence with underlying structures. Seeps have been the target of camera and water-sampling surveys using the *ROV Ventana* (Monterey Bay Aquarium Research Institute; Orange *et al.*, 1999). Active seeps have been located by gas bubbles seen escaping from the seabed, while dormant seeps have been inferred by the presence of chemosynthetic communities that depend on fluids and gas (Kulm & Suess, 1990). Studies of similar shallow-water seeps have revealed discontinuous venting thought to be the result of diurnal tidal effects and changes in water temperature (see Hagen & Vogt, 1999). The cause and repeat time of active gas venting in the Eel River Basin are not known.

Seabed **pockmarks** are conical depressions that are the result of recent or active fluid expulsion (Hovland & Judd, 1988). Abundant small pockmarks are observed almost exclusively on the upper continental slope (not the shelf) of the Eel River Basin (Yun *et al.*, 1999) (Fig. 26). Pockmarks range from ~10 m (grid-size limit of sidescan data) to > 25 m in diameter, and are typically 3 m deep. They are randomly distributed, clustered, or aligned. Field *et al.* (1999) examined pockmarks in relationship to erosional and aggradational gullies along the slope north of the Humboldt Slide and the breached Little Salmon Anticline, and concluded that there is no compelling evidence linking these morphological features. Yun *et al.* (1999) have counted > 3500 pockmarks along roughly 1500 km of shiptrack with sidescan swath width of 500–750 m, and find that > 95% occur in water depths > 400 m. Pockmarks are most abundant on the immediate north and south sides of the breached Little Salmon Anticline, including the region of the Humboldt Slide (Lee *et al.*, this volume, pp. 213–274).

Stratigraphic modelling

Matching stratigraphy to observations

Backstripping (Syvitski *et al.*, this volume, pp. 459–529) has been used to simulate the Eel River Basin long-term stratigraphic development (Fig. 28), using a sea-level curve for the past 125 kyr imposed on the relatively steep dips, rapid subsidence and high sediment supply that characterize this margin. The complex, multistage sea-level fall from marine Oxygen Isotope Stage 5 (125 ka) to Stage 2 (18 ka) produces a series of progradational packages. Considerable sediment removal during both regressive and transgressive parts of the observed cycles leaves the margin interrupted by numerous erosional surfaces. The large sea-level rise since the Last Glacial Maximum has covered the entire section with a smooth marine transgressive drape.

The model run used a tectonic subsidence at the shelf edge of 2.1 mm yr^{-1} which coincides with the 1.4–3.3 mm yr^{-1} rate derived onshore from the displacement of marine terraces (Orange, 1999).

A section through the Freshwater Syncline (Fig. 28) helps to resolve patterns that are consistent with the model predictions. The seismic profile shows a series of strong, slightly divergent reflectors on the shelf, topped by the relatively transparent Holocene transgressive drape. These reflectors have been highlighted in red (Fig. 28c) to correspond with the erosion surfaces in the model, although not all are necessarily unconformities. More steeply dipping reflectors, highlighted in yellow, are contained within the packages defined by these unconformities, and correspond to prograding shorefaces and other dipping interfaces. Additional surfaces seaward and landward of the dipping reflectors (highlighted maroon and green, respectively; Fig. 28c) have shallower dips subparallel to the major reflectors. Based on their position relative to the assumed shoreface, these surfaces have been correlated to marine and fluvial deposits, respectively. Although these features are at the limits of the resolution of the image, similarity between observations and predictions is encouraging.

Several discrepancies should be noted. The initially rapid progradation of the margin depicted in this model (Fig. 28A & B) is an artefact of the start-up conditions, as is the depositional minimum at the base of the slope; neither feature corresponds to observations from the actual data. The model predicts much less sediment accumulation on the slope than is observed in profiles; sediment-supply and/or slope-failure parameters of the model apparently do not match reality. Other parameters should be adjusted to maintain the correct shelf width. Another limitation is that the model does not accumulate sediment on the uppermost slope, even when the modelled amount of sediment failure is reduced significantly. This appears to be due to the erosion of the shelf edge during modelled transgression, a process that has been reported previously to explain hardgrounds on the uppermost slope (Lee *et al.*, 1999). On the shelf, the model predicts that the strata below the Holocene transgressive drape should be a complex package of nearshore deposits interfingered with marine shales and a limited amount of non-marine strata

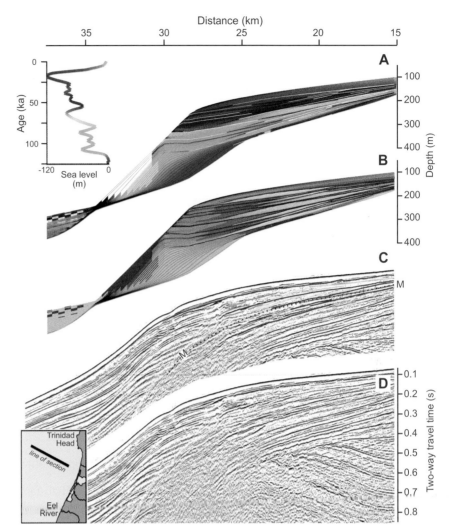

Fig. 28 Model of stratigraphic development in the Eel River Basin compared with seismic stratigraphy. Data shown are from line 83 (Fig. 9). (A) Model at top shows 5-kyr time steps keyed by colour to time and position within sea-level curve shown at left. Sea-level history since 125 ka comes from SPECMAP (Imbrie *et al.*, 1984), assuming sea level was 120 m below present at marine Oxygen Isotope Stage 2. The model fails to account for slow accumulation rates on the lower slope shown in white. (B) Model and (C) seismic profile show the same stratigraphic units, with colours indicating environments of deposition. Red and yellow lines on the profile denote erosion surfaces and prograding shorefaces (see text). (D) The uninterpreted profile plotted against two-way travel time, with a location map (inset).

(Fig. 28A & B). Below the thick transgressive drape, the strata of the Eel River margin do not reflect the seafloor bathymetry; a similar contrast between bathymetry and buried stratal geometry can be seen on the New Jersey margin as well (see below).

There are intriguing and unresolved differences in the stratigraphy between the Eel River margin (ERM) and New Jersey margin (NJM) that have been captured in the respective forward models. The New Jersey margin has extensive, well-developed prograding clinoforms; similar progradation is lacking off California. The Eel River margin exhibits considerable tectonic deformation that includes much greater subsidence than is found off New Jersey, with the result that Eel River margin sediments reveal considerably greater seaward divergence of reflectors beneath the shelf. Over the long term,

the accommodation space created on the Eel River shelf is filled by sediments delivered to the margin. The balance between progradation and subsidence results in a stationary position of the shelf edge (Ulicny *et al.*, 2002). Within sea-level cycles, sediment depocentres transgress and regress across this margin, but each sequence builds to a similar position.

A model of sequence-stratigraphic development

Based on STRATAFORM observations of the Eel River Basin, a conceptual model can explain sequence development that results in syncline filling modulated by sea-level variations (Burger *et al.*, 2002). The Eel River shelf becomes progressively exposed during times of falling sea level (Fig. 29).

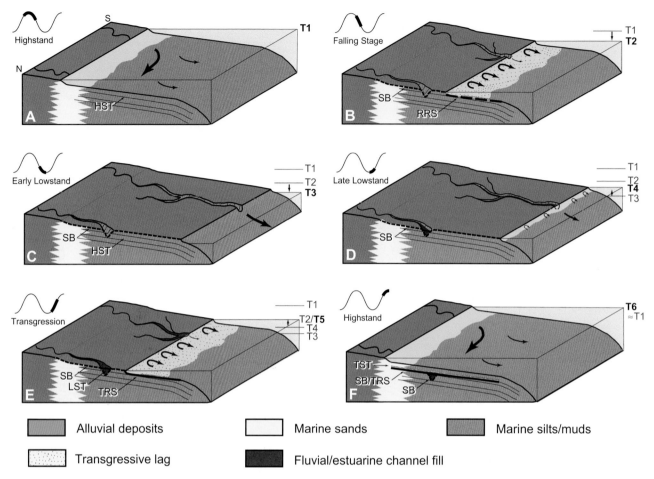

Fig. 29 Proposed sequence-stratigraphic model for the Eel shelf. (A) Highstand: alongshelf-directed, shallow-marine sediment transport and accumulation occurs, as at present. (B) Falling stage: shelf is progressively exposed, and a regressive ravinement forms by shoreface erosion as it crosses the shelf. The exposed shelf represents the lowstand sequence boundary. (C) Early lowstand: sea level approaches the former shelf edge; sediments largely bypass the shelf through fluvial channels. (D) Late lowstand: fluvial gradients decrease, initiating channel infilling seaward; shoreface erosion of the shelf resumes. (E) Transgression: shelf channels are progressively infilled seaward to landward; vigorous erosion forms a prominent transgressive ravinement as the shoreface advances across the shelf. Lowstand sediments and the sequence boundary are eroded in all but deeply incised areas. (F) Highstand: alongshelf accumulation of sediments resumes. Over most of the shelf, highstand deposits directly overlie highstand sediments from the previous sequence. SB, sequence boundary; RRS, regressive ravinement surface; TRS, transgressive ravinement surface; SB/TRS, composite surface; HST, highstand systems tract; LST, lowstand systems tract; TST, transgressive systems tract. (After Burger *et al.*, 2002.)

Rivers incise a widening coastal plain and deliver sediment to a narrowing shelf (Burger *et al.*, 2001). It is likely that a regressive ravinement surface forms by shoreface erosion during these periods. Numerous profiles in the southern Eel River Basin show that the Eel River often has diverted sediment from the shelf by extending close to or into the head of Eel Canyon, through which fluvial sediments

continued farther seaward. Incised channels also have formed in response to uplift of the Table Bluff Anticline (Burger *et al.*, 2001). Farther north, the Mad River shelf valley also has extended to the shelf edge during times of falling sea level; fluvial sediment is then funnelled directly to the slope and the Gorda Fan. Poorly preserved interfluves between shelf valleys, as well as occasionally preserved valley fill, are the only deposits retained on the shelf during times of lowered sea level. As sea level begins to rise, fluvial gradients decrease, and incised shelf channels fill preferentially with an upward-fining sequence of fluvial sands grading into estuarine facies. Accumulation in the areas between channels very likely remains low to absent.

During transgression, the shoreface retreats across a widening shelf (Fig. 29), and erosion removes low-stand sediments in all but the most deeply incised areas and transports these materials seaward. The regressive ravinement and interfluve portions of the sequence boundary formed during the previous sea-level fall are probably completely cannibalized. Erosion then forms a transgressive ravinement, winnowing removes the fine-grained sediment, and a thin, coarse lag deposit forms as a result, immediately above the ravinement unconformity. Where the transgressive ravinement truncates the previous lowstand surface (in most areas), the surface becomes a composite sequence boundary. Highstand silts and muds from the previous sequence underlie similar sediments from the ensuing highstand, separated only by the ravinement and the thin transgressive lag. Lowstand deposits are preserved only as hummocky channel fills, between the sequence boundary and ravinement surface in deeply incised areas.

THE NEW JERSEY MARGIN

Cenozoic sedimentation on a passive margin

Observations

The New Jersey margin (NJM; Fig. 30) is a classic example of a passive margin. Rifting began in Late Triassic times (Grow & Sheridan, 1988), and seafloor spreading commenced by the Callovian (~165 Ma; Middle Jurassic; Sheridan et al., 1983). The subsequent tectonic history has been dominated by simple thermal subsidence, sediment loading and flexure (Watts & Steckler, 1979; Reynolds et al., 1991). The Jurassic section beneath the central to outer shelf is composed of thick (typically 8–12 km), shallow-water limestones and shales. A barrier-reef complex fringed the margin until the Mid-Cretaceous (Jansa, 1981). Accumulation rates were generally low during Late Cretaceous to Paleogene time, when the offshore region was predominantly a starved **carbonate ramp** (Poag, 1985). Carbonate accumulation ended in the late middle Eocene beneath today's coastal plain and shelf, and in earliest Oligocene times beneath the slope, probably in response to global and regional cooling (Miller et al., 1996b).

Sediment accumulation rates increased dramatically on the New Jersey margin in the late Oligocene to Miocene (Poag, 1985; Miller et al., 1997b). Although the cause of this increase is not known with certainty, it has been attributed to tectonics in the hinterland (Poag & Sevon, 1989). Sediments prograded across the margin throughout the Miocene and Pleistocene, and accumulated at rates high enough (~10–100 m Myr⁻¹) to provide detailed seismic resolution of stratal relationships (Poag, 1977; Schlee, 1981; Greenlee et al., 1988, 1992). Tectonic subsidence throughout the Cenozoic has been along the relatively well-defined, nearly linear part of the thermal subsidence curve (Steckler & Watts, 1982). There is little seismic or outcrop evidence to suggest faulting, rotation or other significant disturbances of the Cenozoic section (Poag, 1985), although some differential subsidence may have occurred between the Delmarva Peninsula and New Jersey (Owens & Gohn, 1985). This long, relatively undeformed, and rapidly deposited record has led many workers to conduct studies of continental-margin evolution using seismic, well and outcrop data (e.g. Hathaway et al., 1976; Poag, 1978, 1980, 1985, 1987; Kidwell, 1984, 1988; Olsson & Wise, 1987; Greenlee et al., 1988, 1992).

The earliest drilling along the mid-Atlantic margin comprised the Atlantic Slope Project (ASP) in 1967, conducted by a consortium of oil companies investigating the stratigraphy of the continental slope (Weed et al., 1974; Poag, 1978). Each site was limited to 330 m sub-bottom penetration; all were open-hole drilling from a dynamically positioned ship; no hydrocarbons were reported. The US Geological Survey (USGS) conducted similar operations on the shelf and upper slope in 1976 (the Atlantic Margin Coring Project, or AMCOR; Hathaway et al., 1979). Drilling was open-hole from an anchored vessel, and again was limited to 330 m sub-bottom depth.

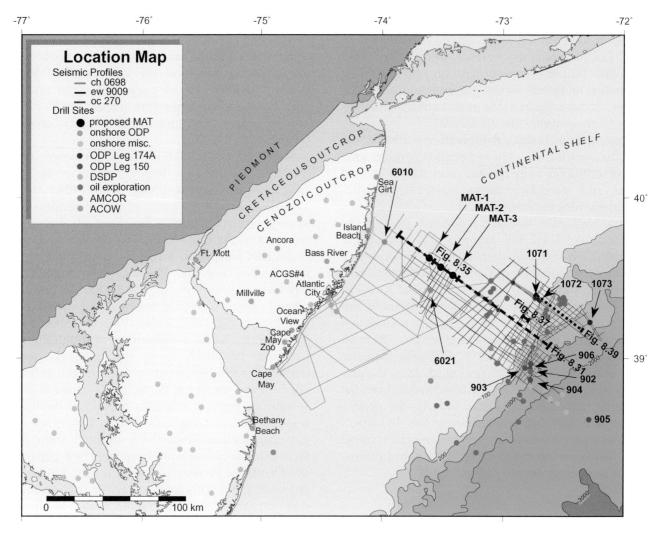

Fig. 30 Location map of New Jersey data discussed in the text. Three grids of multichannel seismic profiles (lines) collected aboard the *R/V Ewing* (Ew), *R/V Oceanus* (Oc), and *R/V Cape Hatteras* (Ch) and many wells (dots) are shown. The locations of seismic profiles shown in Figs 31, 35, 37 and 39 are shown.

Industry exploration continued along the margin in the 1970s and early 1980s (Libby-French, 1982; Prather, 1991). Roughly three dozen wells were drilled, and while some hydrocarbon shows were reported on the outer shelf, none was large enough to be commercially profitable. Some of the survey data (reconnaissance geophysics, wireline logs and petrological analyses of sidewall cores) has proven useful to researchers for baseline studies of the long-term record (e.g. Greenlee *et al.*, 1992; Fulthorpe & Austin, 1998).

Legs 93 and 95 of the Deep Sea Drilling Project (DSDP) were designed as the beginning of

a margin-wide New Jersey Transect (NJT; also known as the Mid-Atlantic Transect) (Figs 30 & 31) to investigate sea-level effects on passive margins (Poag *et al.*, 1987; van Hinte *et al.*, 1987). Ocean Drilling Program (ODP) Leg 150 (Fig. 31) moved closer to land by drilling four sites (902, 903, 904 and 906) between 445 m and 1250 m water depths (Mountain *et al.*, 1994), although still on the slope and more than 140 km from shore.

A land-based drilling programme (Figs 30 & 31) was designed to complement the slope portion of the New Jersey Transect (Miller *et al.*, 1994). Three holes comprising ODP Leg 150X were cored and

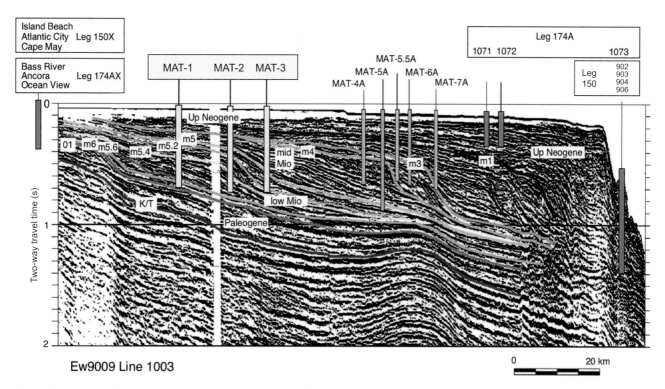

Fig. 31 Interpreted dip-oriented multichannel seismic line Ew1003. Location shown in Fig. 30. As much as 14 km of post-rift sediments lie beneath the outermost shelf, but only the upper 2 s are shown. Major sequence boundaries of post-Eocene age have been traced from the inner shelf to the slope, where they were sampled by ODP Leg 150 (general site locations shown in this profile). Sandy sediments prevented comparable penetration and recovery on the outer shelf during ODP Leg 174 (general site locations shown). Several sites have successfully recovered samples and logged across the equivalent surfaces beneath the coastal plain (Miller *et al.*, 1994, among others). Sites designed to recover lower and middle Miocene clinoform strata have been proposed as ODP MAT sites (also shown). (After Miller & Mountain, 1994.)

logged along the outer edge of the coastal plain (Island Beach, Atlantic City and Cape May; Miller *et al.*, 1997b) to a maximum depth of 457 m and oldest strata of Maastrichtian age. Subsequent drilling by ODP Leg 174AX (Bass River, Ancora, NJ) reached Cenomanian strata at 596 m (Miller *et al.*, 1998b, 1999). These efforts focused on understanding sea-level changes during the Paleogene and Cretaceous, when global climate appears to have been too warm to support large polar ice caps.

A high-resolution multichannel-seismic grid was collected (Oc270; Austin *et al.*, 1996) across the outer shelf and slope prior to more offshore drilling. With improved technology (shallow and short offset GI gun/streamer geometry, 12.5-m shot spacing) these data significantly increased the number of resolvable sequences. Subsequently, ODP Leg 174A

(Austin *et al.*, 1998) attempted to drill into middle to late Miocene clinoform packages on the outer continental margin (Fig. 31). A total of 12 holes at three sites were occupied on the shelf and slope, recovering roughly 1 km of core that ranged in age from late Eocene through to Pleistocene. Slope drilling (Site 1073) recovered a thick Pleistocene section and condensed Pliocene to upper Eocene section with excellent biostratigraphic resolution.

Onshore facies – stacked highstands

Repetitive transgressive–regressive cycles have long been recognized in the New Jersey Coastal Plain (Owens & Sohl, 1969; Owens & Gohn, 1985). The idealized transgressive sequence consists of glauconite sand at the base overlain by a regressive

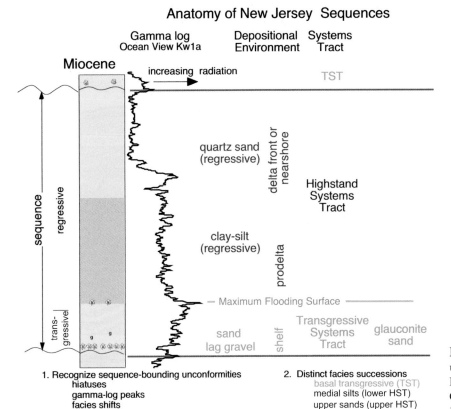

Fig. 32 Characteristics of a typical unconformably bounded sequence in Miocene sediments of the New Jersey Coastal Plain. (After Miller *et al.*, 1997a.)

succession of silt coarsening upwards to quartz sand (Fig. 32) (Owens & Sohl, 1969). The basal glauconite sand (the **condensed section** of Loutit *et al.*, 1988) is equivalent to the upper **transgressive systems tract** (TST) of Posamentier *et al.* (1988), the overlying silt is equivalent to the lower **highstand systems tract** (HST), and the quartz sand corresponds to the upper HST (Sugarman *et al.*, 1993). These patterns result in distinct gamma-ray-log signatures, with high gamma-ray values at the base and low values at the top. Lowstand systems tracts (LSTs) are rarely preserved in the coastal plain and the TSTs are generally thin.

Thinly developed TSTs mean that the maximum flooding surface (MFS) is often difficult to distinguish from an unconformity in the New Jersey Coastal Plain, and both surfaces are typically associated with shell beds. In general, MFSs are recognized by lithofacies successions and benthic foraminiferal changes. The MFS in Miocene sequences of the New Jersey Coastal Plain is usually marked by an upward change from clayey sand to silt.

Studies of New Jersey's onshore sediments, prior to the recognition of these transgressive–regressive cycles, were hampered by a lack of adequate study material due to deeply weathered outcrops and discontinuous coring in wells. Boreholes from ODP Legs 150X and 174A (Fig. 31) led to dramatically improved understanding of regional patterns by providing continuously cored sections with > 85% recovery that enabled facies and age determination in sediments as old as mid-Cretaceous (Miller *et al.*, 1994, 1996a).

Unconformities in core samples from ODP boreholes in the New Jersey Coastal Plain are identified by physical evidence (irregular contacts, reworking, bioturbation, and major facies changes) and well-log characteristics (gamma-ray peaks), and are generally associated with time gaps detected by biostratigraphic and/or chemostratigraphic (Sr-isotopic) breaks. These surfaces are interpreted as sequence boundaries, denoting a rapid fall in **base level**. Hiatuses that are not associated with obvious stratal surfaces or evidence of erosion/non-deposition are interpreted as paraconformities.

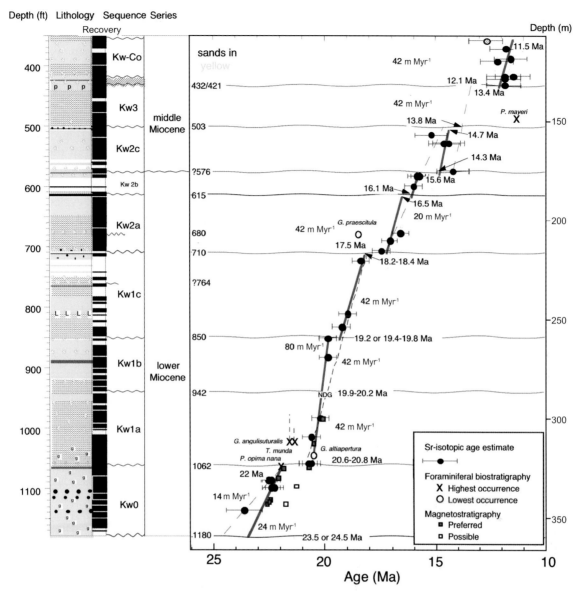

Fig. 33 Age–depth plot of a part of the Miocene section at the Cape May well (Miller *et al.*, 1996b), showing the integration of ages derived from Sr isotopes, magnetic-field reversals and biostratigraphic zonations in determining a geochronology. (After Miller *et al.*, 1997a.)

While hiatuses generally correlate between drill holes, downdip sections are typically more complete than updip correlatives.

The dates of Miocene sequences on the New Jersey Coastal Plain rely primarily on Sr-isotopic ages (Miller *et al.*, 1997a; Sugarman *et al.*, 1997), with a time resolution of ±0.3–0.6 Myr for the earlier part and ±0.9–1.2 Myr for the later part. The chronology of onshore Miocene sequences has been derived from age–depth diagrams in which sediment accumulation rates are linearly inter-

polated between age estimates (Fig. 33). Sequence boundaries are generally associated with hiatuses that occur throughout the coastal plain, and accompanying physical stratigraphy indicating a fall in base level. Specific facies successions vary from the Cretaceous to Miocene in response to differences in palaeodepth, provenance, preservation and deltaic influences (Owens & Gohn, 1985). In general, Miocene sequences follow the generalized pattern of all onshore sequences, and comprise three major lithofacies:

1 thin, shelly, occasionally glauconitic quartz sands of the TST, deposited in neritic (30–100 m palaeo-depth) environments;
2 silty clays of the lower HST, deposited in shallower prodelta environments;
3 upper quartz sands of the upper HST, deposited in nearshore (< 30 m) and delta-front environments.

The TST is thin or absent, so the silty clays and thick sands often stack together as a series of coarsening-upward couplets (Sugarman & Miller, 1997).

Facies patterns within Miocene coastal-plain sequences Kw1a and Kw1b (Fig. 33) reveal variations in both the strike and dip directions that are due to interfingering between marine, transitional marine and deltaic environments (Fig. 34).

Sequences tend to thin updip, although they occasionally thicken along strike. Highstand systems tracts generally become progressively coarser and shallower updip (e.g. the Kw1b between Cape May and Atlantic City), although the Kw1b HST at Atlantic City is finer than at Cape May because of the juxtaposition of prodelta-delta front versus neritic-nearshore environments (Fig. 34). Based on these observations, it appears that the depocentre shifted from near Island Beach during Kw1a to near ACGS#4 (Fig. 30) during Kw1b. Small-scale parasequences (shoaling-upward cycles bounded by flooding surfaces; Van Waggoner *et al.*, 1988) can be detected within several sequences (e.g. within the Kw1b sequence, at Atlantic City, Island Beach and ACGS#4; within Kw1a at Island Beach; Fig. 34).

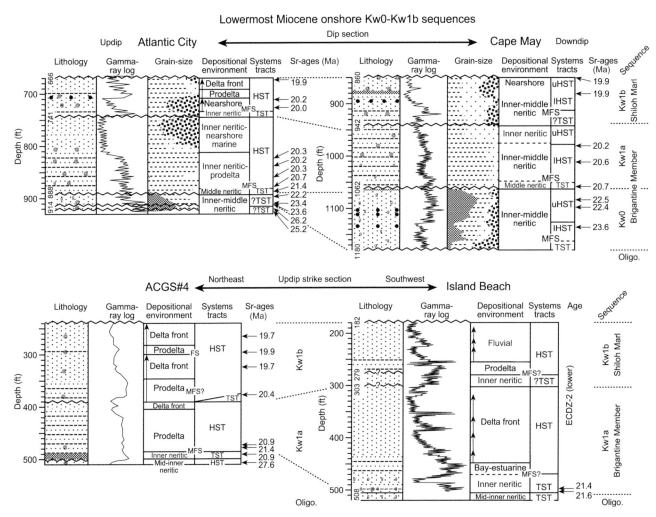

Fig. 34 Comparison of along-strike and down-dip variations of lower Miocene facies in the New Jersey Coastal Plain (see text). (From Miller *et al.*, 1998a.)

Inner-shelf geometries — multiple sediment sources

Using seismic data collected by oil companies, Greenlee & Moore (1988) and Greenlee *et al.* (1988) showed prograding Miocene units beneath the inner and middle shelf. Poag & Sevon (1989) and Poag & Ward (1993) determined first-order patterns in Miocene accumulation. They mapped ancestral Susquehanna/Delaware and Hudson River drainage systems showing south-eastward trends across the margin, and they prepared isopach maps that portrayed a wide but relatively uniform early Miocene margin. Greenlee *et al.* (1992) consolidated seismic data collected to that point and identified eight sequences in well-preserved clinoforms beneath the present inner to middle shelf, thought to span Oligocene through late Miocene age.

The onshore boreholes of ODP Leg 150X were drilled with knowledge of these lower Miocene clinoforms (Miller *et al.*, 1994), but the link of facies to stratal geometry still depended upon a cor-relation across a 30-km gap separating the closest well and the most landward seismic control. High-resolution MCS profiles collected in 1998 (ch0698, Fig. 30; Monteverde *et al.*, 2000) partially filled this gap and defined candidate sequence boundaries (Fig. 35) on the basis of onlap, downlap, toplap and erosional truncation (Mitchum *et al.*, 1977a,b; Vail, 1987; Posamentier & Vail, 1988). Small incised valleys (10–15 m deep; up to 6 km wide) were detected on each sequence boundary landward of the clinoform rollover points. These unconformably bounded packages were recognized on dip lines where clinoform geometries were best defined. Eleven proposed bounding surfaces have since been identified, outlining 10 different sequences (Fig. 36; Monteverde *et al.*, 2000).

There is considerable along-strike variability to these surfaces. Sequences generally thin away from the dip profiles on which they are most easily recognized (Fig. 36), as boundary reflectors are frequently truncated by and/or concatenated

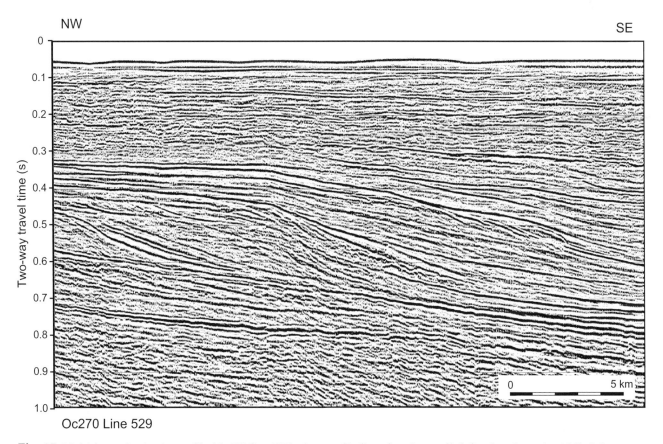

Fig. 35 Multichannel seismic profile (Oc270 line 529) along a dip line showing well-defined early and middle Miocene clinoform geometries beneath the New Jersey middle shelf. Location shown in Fig. 30.

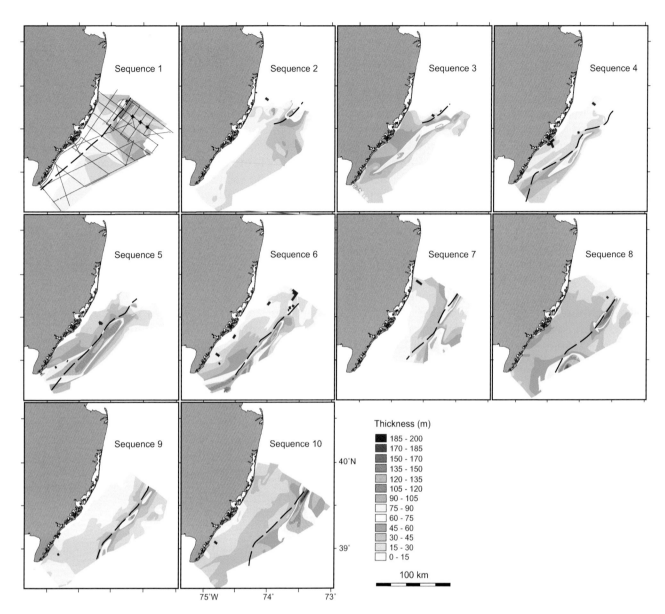

Fig. 36 Isopach maps of ten lower Miocene sequences identified in Ch0698 profiles along the New Jersey inner shelf. Seismic grid shown in Fig. 30. Sequence 1 is the oldest; sequence 10 is the youngest. Note the shift in depocentres from north to south and back to north, while there is a seaward shift in thickness patterns. The dashed line marks the clinoform rollover point at the top of each sequence. The red areas mark probable incised shelf valleys. See text for details.

with over/underlying reflectors in both updip and downdip directions. No single dip line displays all 11 sequence clinoform rollover points.

The arrangement of reflectors within sequences from the northern half of the seismic grid contrasts with those from the southern half. Northern sequences generally are thicker (Fig. 36) and contain more clearly defined internal geometries.

Well-imaged lowstand deposits at the seaward toes of clinoforms are found within most northern sequences. A single shelf-edge delta of limited areal extent rests on a sequence boundary, suggesting development under lowstand conditions. High-stand deposits defined by downlapping reflector truncations prograde seaward, and are thicker and more extensive than transgressive sediments. Onlap is rare away from the preceding rollover due to the very low seaward dip of palaeoshelf sediments landward of each clinoform rollover. Erosional truncation and incised valleys occur on each sequence boundary, and can closely approach the location of the previous clinoform rollover. Erosion and incision predominate along the north-west edge of the seismic grid. Sequences have prograded much less in the northern half of the grid.

By contrast, southern sequences have advanced farther seaward. Sequences are generally thinner and internal geometries are less well defined than those to the north. Lowstand fans and wedges (Posamentier & Vail, 1988) are not detected in the south. Instead, reflectors paralleling the down-dip section of the previous sequence boundaries generally onlap steep sigmoidal clinoform foresets. Similar geometries have been attributed to an early transgressive phase by Posamentier & Allen (1993, 1999). These geometries develop under conditions of high wave energy/low tidal energy that bevel clinoform tops during relative sea-level rise, transporting the eroded sediment seaward of the clinoform rollover. This reflector geometry continues along strike into the northern grid, where it grades into more 'typical' sequence successions of lowstand, transgressive and highstand truncations. As in the northern half of the grid, highstand deposits in the south are much thicker than transgressive sediments. Updip sections are generally thin in both northern and southern regions. Incised valleys are less common in the south than in the north, and enter the seismic grid from the west-southwest.

Along-margin variations in both accommodation space and fluvial sediment supply contributed to these north–south contrasts in sequence development. Greater accommodation space in the north, apparently the result of increased subsidence, led to thicker sequences that did not prograde as far seaward as those in the south. Variable sediment supply is apparent as well. As discussed above, incised valleys are evident along the tops of individual clinoforms, suggesting rivers became entrenched during regressions, and very likely deposited their sediments into a relatively small number of deltaic lobes or wedges. Isopach maps support this speculation. Individual depocentres are relatively margin-parallel bodies with slightly arcuate seaward edges that change geographical location through time (Fig. 36).

Isopachs and the changing location of clinoform rollover points mapped across the grid for each of the 10 early Miocene sequences provide a measure of progradation, and yield insight into margin history. Sediment accumulation began in the north (Sequence 1), supplied by rivers entering the seismic grid from the north-west. Very thin distal accumulation occurred in the south during this time. The next sequence (2) again concentrated accumulation in the northern region. Beginning with Sequence 3, an additional southern sediment source became clear, and by Sequence 4 the thickest accumulation occurred in the central to southern region. The following four cycles (Sequences 5–8) reveal two distinct depocentres that moved progressively seaward with time. Deposition from the southern source either waned or moved seaward of the seismic grid by Sequence 9, and the final lower Miocene sequence (10) formed a linear, margin-parallel centre in the northern region. Onshore borehole data indicate that deltaic sedimentation dominated this interval of shelf progradation (Miller *et al.*, 1997b).

Clinoforms and incised valleys

Clinoforms are observed in ancient continental-margin sediments at a wide range of scales and ages (Vail *et al.*, 1977; Posamentier *et al.*, 1988; Bartek, 1991), but despite this ubiquity the water depths at which they form are unknown and widely debated. One viewpoint claims the gently dipping topsets of clinoforms in any setting are typically exposed during relative falls of sea level ('Type I' events, Van Waggoner *et al.*, 1988), causing the shoreline to move seaward of the rollover point. At these times, base level falls below the top of the clinoform and rivers cut into exposed strata. The contrary view claims the shoreline never passes seaward of the rollover, and the clinoform top is rarely, if ever, exposed to fluvial processes (Cathro *et al.*, 2003).

Incisions into the topsets of six mid-to-late Miocene clinoforms have been reported in profiles along the New Jersey outer shelf (Fig. 37; Fulthorpe *et al.*, 1999). These features are generally 5–12 m deep (at the limit of vertical resolution of these data), with some as deep as 20 m. Widths vary from 50 m (roughly the limit of horizontal detection) to a typical value of 100–400 m. Roughly parallel

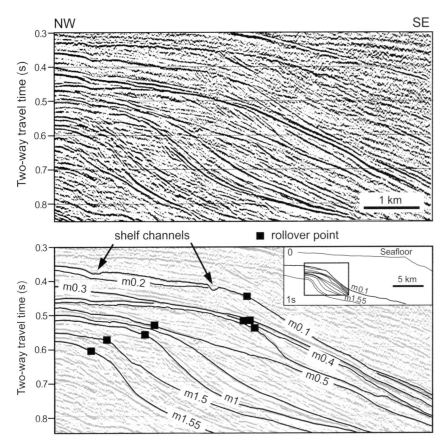

Fig. 37 Uninterpreted and interpreted seismic profile (Oc270 line 35) roughly 10 km from the modern shelf break, showing prograding mid-to-late Miocene clinoforms. See inset and Fig. 30 for location. Black squares mark the rollover points for each highlighted clinoform. Two channels landward of the existing shelf break are incised 8–13 m into the m0.1 palaeoshelf, with apparent widths of 200 m and 430 m. (From Fulthorpe *et al.*, 1999.)

channels appear to be clustered into drainage systems 10–15 km across. Incisions extend seaward of the clinoform rollover on four of 12 mid-to-late Miocene unconformities (Fulthorpe *et al.*, 2000). All are small compared with those of the modern margin (Fig. 38), and consequently most mid-to-late Miocene clinoform foresets are smooth, suggesting that sediment transported downslope was predominantly by non-channelized processes.

The size, grouping, apparent meandering and lack of discernible internal structure suggest that the incisions into clinoform topsets have been formed by rivers that flowed seaward across a gently dipping coastal plain during times of sea-level lowstand (Fulthorpe *et al.*, 1999). All are much smaller and simpler than the composite river systems that appear to have formed the Hudson Shelf Valley or other large depressions crossing modern continental shelves (e.g. Twichell *et al.*, 1977; Knebel *et al.*, 1979). Their interpretation as drainage systems is strongly supported by the recovery of lagoonal facies at ODP Site 1071 (Austin *et al.*, 1998), 3 km

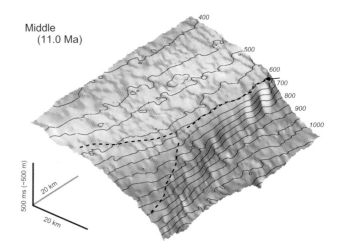

Fig. 38 Isometric view of late Miocene sequence boundary m0.2 beneath the New Jersey outer shelf. Dashed line identifies the break in gradient commonly referred to as the clinoform rollover point; note there are two such abrupt changes in gradient to the south. Evidence for minor incisions into the clinoform foreset is discussed in the text. (From Fulthorpe & Austin, 1998.)

landward of the m0.5 rollover (Fig. 37), and suggests that shorelines may occasionally reach nearly as far seaward as clinoform rollovers. However, the shallow depths of all Miocene incisions and the lack of continuity onto the dipping clinoform foresets strongly suggest that base level never falls more than a few metres below the elevation of the rollover. Early-to-middle Miocene sea-level variations have been estimated from backstripping onshore wells, and fall in the range of 20–30 m (Kominz *et al.*, 1998; Miller *et al.*, 1998a). If magnitudes of sea-level change were roughly the same in the late Miocene, the rollovers under discussion most probably developed in water depths of a few tens of metres at most.

Pleistocene sequences

The Quaternary record on the New Jersey margin is characterized by:

1 a low-relief hinterland that has provided only a modest amount of sediment;
2 sediment reworking across a wide, shallow-gradient shelf;
3 thermal and flexural subsidence that has had little impact on accommodation space (except for the loading effects of intermittent ice sheets).

As a result, the Quaternary section is thin (a few tens of metres) and stacked in complex patterns

across the inner two-thirds of the shelf (Knebel & Spiker, 1977; Knebel *et al.*, 1979; Swift *et al.*, 1980; Ashley *et al.*, 1991; Carey *et al.*, 1998; Sheridan *et al.*, 2000; Uchupi *et al.*, 2001; Schwab *et al.*, 2002). By contrast, large volumes of sediment reached the outer shelf and slope during the mid- to late-Pleistocene (post-750 ka), extending the continental margin several tens of kilometres past the pre-Quaternary shelf edge.

The Hudson Apron is a region of the continental slope largely free of canyons and composed of Hudson Canyon spillover sediment carried southwest by prevailing currents. High-resolution seismic profiles show that it contains four Quaternary sequences: Yellow, Green, Blue and Purple (Figs 30 & 39). The bounding surfaces between each, beginning with the base of Yellow, are p4, p3, p2 and p1, respectively (Fig. 39). These unconformities on the shelf exhibit erosional truncation or toplap. Traced to the slope, these bounding surfaces become more nearly, and in some cases, totally conformable. There is no evidence of marine onlap against these slope unconformities, in contrast to a commonly held model of deposition that buries a sequence boundary after an interval of bypass or erosion (e.g. Vail *et al.*, 1977).

The base of each Quaternary sequence (i.e. surfaces p4, p3, p2 and p1; Fig. 39) exhibits a seaward increase in gradient from gentle (~1:60) to more steeply dipping (~1:20), and the inflection point is referred to as the palaeoshelf break. This buried

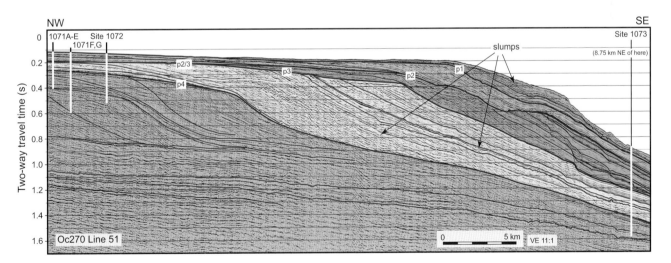

Fig. 39 Seismic profile (from Oc270 line 51) through ODP Sites 1071, 1072 and extending to the continental slope 8.75 km south-west of Site 1073. Pleistocene sequences named Yellow, Green, Blue and Purple are bracketed by sequence boundaries p4, p3, p2 and p1, respectively. See Fig. 30 for location.

physiographical feature at p4 time is 15–18 km landward of the modern shelf break, and each sequence above Yellow steps progressively seaward of the one beneath it. Reflectors within each sequence downlap onto the underlying palaeoshelf and, except for those in sequence Yellow, are conformable with the underlying palaeoslope. The youngest reflectors in each sequence terminate with angular discordance at the overlying palaeoshelf; it is not clear in most cases if this is evidence of sediment bypass and shelf toplap, or actual erosional truncation. By contrast, reflectors at the top of each Quaternary sequence on the palaeoslope are predominantly conformable with the overlying sequence boundary. Many intrasequence reflectors can be traced with ease from the palaeoshelf to the palaeoslope; exceptions occur in apparent slump deposits that are 10–50 m thick, extending continuously from the top of each palaeoslope to the limit of the survey grid at the middle of the modern slope (~1400 m water depth).

Outer-shelf and slope sequences

Three sites were drilled during ODP Leg 174A on the New Jersey margin (Figs 30 & 39); two were on the outer shelf in 88-m and 98-m water depths (Sites 1071 and 1072, respectively), and the other (Site 1073) was on the upper slope at 639 m (Austin *et al.*, 1998). Key objectives were to establish reliable correlations between prominent seismic reflectors and the cored sequence, and evaluate the influence of sea-level change on the observed stratigraphic succession. Unconsolidated sand-rich units, however, were particularly difficult to recover and caused severe drilling-related and hole-stability problems. To tie the complete sequence to seismic reflections, the velocity and density logs as well as a check-shot survey proved to be essential. Figure 40 shows a time–depth comparison using both the sonic-log and check-shot data recorded in Hole 1073A. The difference between the sonic and check-shot profiles is related to frequency differences between the measurements and the presence, or absence, of stratigraphic reflectors in the immediate vicinity of the drill hole.

Where incomplete core recovery during ODP Leg 174A resulted in uncertain relationships between the sedimentary record and reflectors, geophysical data have been used to fill the gaps in the record

(Fig. 41; Delius *et al.*, 2001). The reflectivity coefficient series R have also been calculated from the sonic and density logs at Site 1072, enabling a more confident match between acoustic reflectors and known lithological boundaries (Fig. 41). Key reflectors were found to correspond to erosional surfaces and/or to unrecovered sand-rich units that commonly occurred at sequence boundaries. This example demonstrates that correlation between core, log and seismic data can be essential, and at these locations, in particular, resulted in an improved seismic interpretation of the drilled sequences.

1 *Yellow sequence (p4–p3).* Four notable aspects of margin-building sedimentation occurred during sequence Yellow (Fig. 42D). First, early Pleistocene sediments comprising the base of this sequence prograded several tens of kilometres seaward with no detectable aggradation. Second, at the scale that can be resolved with existing seismic data (~5 m vertical), strata across the palaeoshelf break and on the upper palaeoslope were deposited conformably; there is no evidence of slope bypass and subsequent onlap against p4 or any reflectors within the Yellow sequence (Fig. 39). Third, the palaeoshelf break occurs parallel to and 15–18 km landward of the modern shelf break, at roughly the 110-m isobath. No slope canyons of Yellow age cut into this palaeoshelf break, although several canyons developed exclusively on the slope in the south-west area of study. Fourth, distorted and discontinuous reflectors interpreted as displaced slump sediments comprise ~20% of the Yellow sequence on the Hudson Apron, and can be traced as discrete units from the top of the palaeoslope seaward to the limit of the survey grid. Slumps appear to become more common up-section. However, slope failure occurred in the south-western part of the survey grid sometime after 750 ka; this region has remained a locus of gravity-induced failure since that time, and may have been the nucleus for Hendrickson and other nearby canyons.

2 *Green sequence (p3–p2).* The Green sequence is identified by the same criteria as the other Quaternary units, yet it was partially eroded during emplacement of the overlying p2 sequence boundary, and now exists only beneath the outermost shelf and slope (Fig. 39). Reflectors at the top of the Yellow sequence dip gently seaward, suggesting sediment bypass along the p3 surface; there is no compelling reason to conclude that widespread erosion coincides with the p3

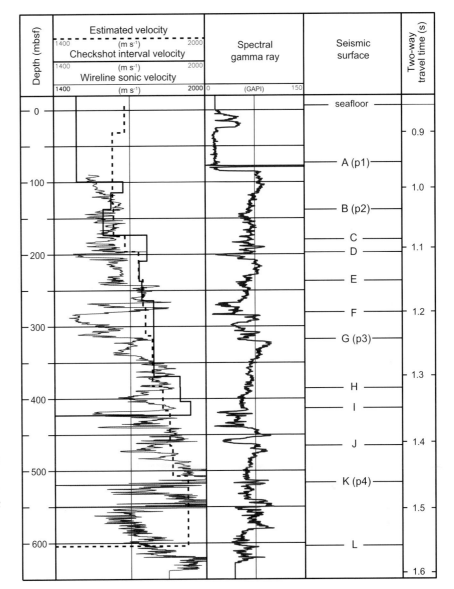

Fig. 40 Velocity-depth plot in ODP Hole 1073A showing wireline sonic velocity (blue) and interval velocities (bold, solid line to 425 m below seafloor, mbsf) calculated from the check-shot survey. Velocity estimates based on previous drilling (dashed line) and the spectral gamma-ray log (red) are also shown. Various seismic reflectors A through L identified on seismic lines are positioned at corresponding depths and the associated two-way travel-time. (From Austin *et al.*, 1998.)

sequence boundary. Strata within the Green sequence accumulated on the p3 surface during steady progradation/downlap across the gently dipping (1:130; 0.4°) palaeoshelf. Thicknesses of palaeoshelf strata (Fig. 42C) now vary from 60–70 m in the northern part of the study area to 10–20 m in the southern part. Basal reflectors on the uppermost palaeoslope, by contrast, are chaotic, suggesting slumping and debris-flow processes (and a gradient of 1:25; 2.3°). Indeed, most of the reflectors within the Green sequence immediately seaward of the palaeoshelf break have a similar chaotic character that gives way downslope to acoustically laminated, well-preserved stratification. The thickest accumulation (~230 m) is found

on the palaeoslope. The palaeoshelf break at p3 time was incised by a slope canyon inherited from the underlying Yellow sequence in the south-western corner of the study area.

3 *Blue sequence (p2–p1).* Erosion at sequence boundary p2 totally removed the Green sequence landward of the modern 125-m isobath; consequently, in this region the Blue sequence rests directly on Yellow (Fig. 39). Elsewhere, the reflectors at the top of Green provide additional evidence of erosion. In contrast to the other three Quaternary sequences, Blue did not build seaward over an abrupt palaeoshelf break; instead, the p2 surface changes very gradually from a shelf gradient of ~1:150 to a slope value

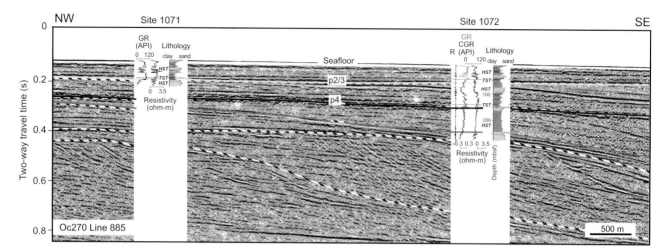

Fig. 41 Composite section showing resistivity and natural gamma-ray logs, synthetic log-lithofacies profiles, and Oc270 seismic line 885 at Sites 1071 and 1072. The reflectivity coefficient series, R, calculated from sonic and density logs allows seismic reflectors to be correlated between sites, revealing erosional surfaces and sand-rich units that were not recovered by coring. (After Austin *et al.*, 1998.)

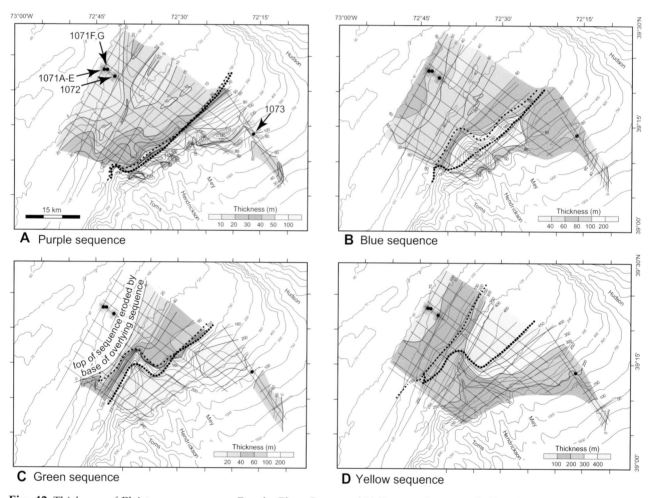

Fig. 42 Thickness of Pleistocene sequences Purple, Blue, Green and Yellow on the outer shelf and upper slope of New Jersey. The light dotted line marks the palaeoshelf break at the base of each sequence; the bold dotted line marks the palaeoshelf break at the top of each sequence. Modern bathymetry (curved grey lines and numerals) is in metres; major slope canyons are labelled; Oc270 profiles used to develop each map are displayed as thin black lines. (After Mountain *et al.*, 2001.)

of 1:20 (0.4° to 2.9°). As with underlying sequences, the only evidence of slope canyons cutting into this shelf break are in the south-west corner of the study area, most probably part of the Toms Canyon system that dominates the modern topography. Roughly 80–100 m of strata (Fig. 42B) prograded across the palaeoshelf in the northern half of the study area; build-up was ~60 m in the southern half. The thickest accumulation (~260 m) is now found immediately seaward of the palaeoshelf break. As in the other Quaternary sequences, chaotic reflections suggest that Blue strata at the top of the palaeoslope were dominated by mass wasting processes, while farther downslope the correlative units are acoustically stratified and draping.

4 *Purple sequence (p1–seafloor).* Strong angular discordance between basal boundary p1 and underlying strata indicate erosion on the outermost palaeoshelf between sequences Blue and Purple (Fig. 39). The p1 surface dividing these units seaward of the 90-m isobath coincides exactly with reflector 'R', first mapped by Milliman *et al.* (1990) and interpreted as the contact between sediments of the Last Glacial Maximum and the subsequent transgression. Davies *et al.* (1992) and Duncan *et al.* (2000) have traced this surface landward and correlated it with the mid-shelf reflector R described by Knebel & Spiker (1979). The palaeoshelf and palaeoslope gradients of p1 are similar to those of older units, and a clear palaeoshelf break is located no more than 2 km landward of the modern shelf break. As in each of the older sequences, chaotic reflectors are gradually replaced by acoustically stratified, draping units in the seaward, downslope direction. Purple is the thinnest of the four Quaternary sequences (Fig. 42A); shelf accumulations are roughly 20 m in the northern part of the study area, and 35–40 m in the southern part. This is opposite to the trend in the other three Pleistocene sequences, and may reflect infilling around the head of Toms Canyon more than any other asymmetry in sediment supply. The thickest Purple accumulation is immediately seaward of the palaeoshelf break, but inexplicably the isopachs do not follow a trend that parallels the palaeoshelf break.

The link to eustasy

Sea level exerts a major influence on the stratal geometry of clinoforms, although the success of linking times and magnitudes of sea-level change to facies distribution continues to be debated (Christie-Blick, 1991; Miall, 1991; Reynolds *et al.*,

1991; Karner & Driscoll, 1997). One supporting argument holds that if sea level falls faster than the subsidence rate beneath a clinoform rollover, the entire clinoform top will become exposed to subaerial erosion, and a prominent unconformity, or 'Type I' sequence boundary, will develop as a result (Posamentier & Vail, 1988). This event constitutes a lowering of base level, meaning the elevation that divides erosion (above) from deposition (below) falls below the top of the clinoform. Hence, marine deposition will be restricted to locations seaward of the rollover. When sea level rises, this model predicts that depocentres will move progressively landward to a location where the rates of sea-level rise, basement subsidence, sediment compaction and sediment accumulation have reached a balance. This will coincide with a time of minimal sedimentation seaward of the clinoform rollover and result in a condensed section at that location (Loutit *et al.*, 1988). Critics of this model note the lack of solid geological confirmation for the link between truly global sea-level change and the facies distribution predicted within a clinoform structure. For many, the uncertainty of the magnitude and timing for past sea-level changes has made it nearly impossible to identify the exact control these changes have had on the stratigraphic record.

At ODP Site 1073 on the New Jersey upper slope, > 500 m of late Pleistocene sediment were continuously cored with 98% recovery (Austin *et al.*, 1998). Global sea-level variations are well defined for this time in the geological record, and the stratigraphy can be compared against an independently established eustatic history to determine controls on stratal geometry. However, Site 1073 is at 639-m water depth on the continental slope, and this is not the setting in which the standard model of clinoform evolution was developed. The majority of sequence-stratigraphic studies have been based on clinoforms found on continental shelves or epieric seas. In these locations, even the sediments that bypass clinoform rollovers during times of rapidly falling sea-level are largely retained on the shelf, where water depths are < 150 m and seafloor gradients are 1:100. The New Jersey Pleistocene clinoforms, by contrast, developed at the edge of the existing continental shelf; the rollovers for these features were also the shelf–slope transition, and the clinoform foresets were > 150 m in height.

With a gradient of roughly 1:40, these foresets continued down to more than 2000-m water depths, before levelling to a grade of less than 1:100 on the continental rise. The interplay between sea-level change, sediment bypass and sediment build-up might follow different histories, depending on whether the clinoform is perched on a shelf or at the edge of one.

Subsidence at Site 1073 was affected by the adjacent Laurentide ice sheet during the late Pleistocene. The ice front extended across northern New Jersey, Staten and Long Island as recently as ~20 ka, roughly 150 km north-west of Site 1073 (Teller, 1987; Stanford *et al.*, 2001). Modern tide gauges backed by models of crustal response to presumed ice load show that New York harbour is currently sinking at roughly 2 mm yr^{-1} due to relaxation of the peripheral bulge that surrounded the former ice sheet (Tushingham & Peltier, 1991; Gornitz *et al.*, 2001). This serves as a warning to researchers that the impact of sea-level change on sequence architecture in this region requires an especially careful assessment of subsidence.

An oxygen isotopic record was constructed from 500 m of Pleistocene sediment recovered at Site 1073 (Fig. 43). The $\delta^{18}O$ record was correlated to the SPECMAP composite curve (Imbrie *et al.*, 1984) to provide a detailed age model extending back to 770 ka (McHugh & Olson, 2002). This correlation was accomplished by visual curve matching constrained in three independent ways. First, datums based on radiocarbon dates, biostratigraphy and magnetostratigraphy anchored several depths to specific ages. Second, planktonic forams were identified as glacial and interglacial assemblages, providing additional confidence in matching peaks and troughs of the measured and composite $\delta^{18}O$ records. Third, surfaces comprising four Pleistocene sequence boundaries were traced to Site 1073, and their depths were correlated to cores with a reliability of roughly ±10 m (Austin *et al.*, 1998; Mountain *et al.*, 2001). The authors assumed that the sedimentation changes expressed at sequence boundaries on the shelf would have a discernable impact on the facies successions at Site 1073, and, in the absence of other information, this helped to choose the depths matching breaks in the $\delta^{18}O$ to SPECMAP correlation.

Sixteen Pleistocene glacial and interglacial fluctuations of global ice volume were identified at Site 1073 on the basis of measured $\delta^{18}O$ variations (Figs 43 & 44): Oxygen Isotope Stages 1 (partial), 2–4, 5 (partial) and 8–18. Stage 6 was a time of major erosion across the shelf and slope (Sheridan *et al.*, 2000), and the resulting unconformity on the upper slope is manifest in profiles by an erosional surface at 79.5 m and in the core samples by the absence of Stages 6 and 7. This hiatus was the only clearly identifiable break in the $\delta^{18}O$ record of Site 1073, despite the compelling evidence of sharp changes in sedimentation, and possible erosion at each of the four correlatives to sequence boundaries p1–p4 traced from the shelf (see Austin *et al.*, 1998). By contrast, the correlations of p2 (145 m) and p3 (325 m; dividing sequences Blue from Green, and Green from Yellow, respectively; see Fig. 39) coincide with the Oxygen Isotope Stage (OIS) 8–9 and 11–12 transitions, respectively. While time may be missing across these two surfaces, this cannot be verified at the resolution of the available $\delta^{18}O$ proxy. The base of the Pleistocene section at Site 1073 (524 mbsf; ~770 ka) rests unconformably on probable Pliocene/Miocene sediments, and corresponds to boundary p4 and the base of sequence Yellow.

These correlations show that sediments deposited in glacial OIS 8 comprise the entire 65-m Blue sequence, supporting the argument that bypass concentrates sediment accumulation seaward of the clinoform rollover during eustatic lowstands. In contrast, the 80 m of sequence Green correspond to OIS 9–11, and the 200 m of sequence Yellow extend without apparent interruption from OIS 12–18. Clearly, there is no consistent, one-to-one correspondence between glacio-eustatic oscillations and the formation of sequence boundaries as they have been identified. Furthermore, sequence boundaries do not even appear to coincide with a consistent stage of glacio-eustasy: p3 matches the transition from glacial to interglacial (OIS 12–11), while p2 does just the opposite (OIS 9–8). Much of the explanation for these observations is probably linked to the location of Site 1073 on the upper continental slope.

There are numerous facies variations (with no detectable hiatuses) confined entirely to individual sequences at Site 1073. These variations suggest oscillations of proximal/distal sediment sources, and may track eustatic changes (McHugh & Olson, 2002). However, the more fundamental structure

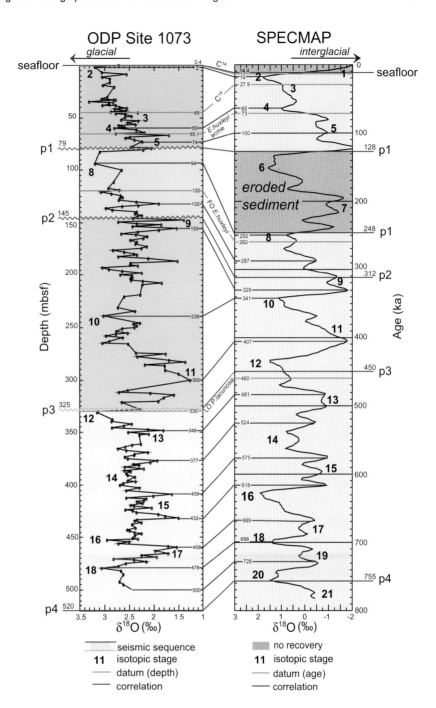

Fig. 43 Measured values of $\delta^{18}O$ at ODP Site 1073 plotted against depth (mbsf, metres below seafloor), compared with the global composite SPECMAP $\delta^{18}O$ curve plotted against age (Imbrie *et al.*, 1984). Green lines mark correlations based on biostratigraphic index fossils; red correlations connect visually matched features of the $\delta^{18}O$ data. Marine Oxygen Isotopic Stages 1 through to 18 are shown, with the exception of Stages 7, 6 and part of 5, which are assumed to have been removed by erosion. The Yellow, Green, Blue and Purple intervals (Fig. 39) at left mark the intervals at Site 1073 bracketed by sequence boundaries p4, p3, p2, p1 and the seafloor, respectively. (After McHugh & Olsson, 2002.)

of stratal architecture is unambiguously defined by the four sequences Yellow, Green, Blue and Purple (Fig. 39). Furthermore, sediment accumulation rates within the three youngest sequences follow an identical pattern, increasing upwards by as much as an order of magnitude (Fig. 44). It remains to be determined just what process is responsible for this pattern, but it does not appear to be global sea

level. The loading/unloading of the craton during advance and retreat of the Laurentide ice sheet, and its consequences on the drainage of rivers and glacial lakes along the adjacent margin, is surely a factor. The timing and magnitude of these effects are not yet known with sufficient precision to evaluate rigorously the impact on Pleistocene sedimentation along the New Jersey shelf and slope.

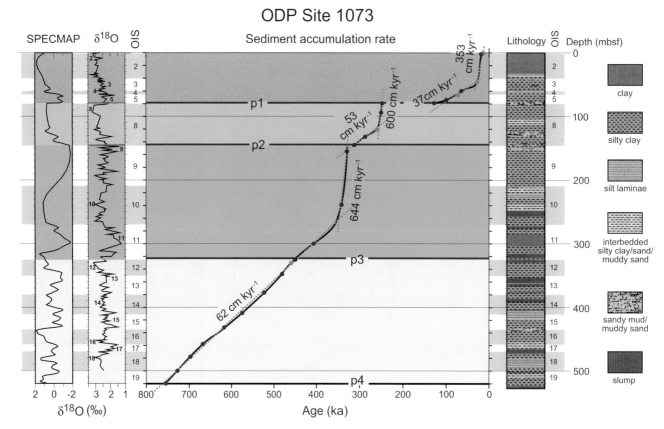

Fig. 44 Age versus depth of Pleistocene sediments at ODP Site 1073 and derived sediment accumulation rates. Ages were obtained by correlation of measured $\delta^{18}O$ values to those of the SPECMAP curve. The SPECMAP curve is plotted versus depth, assuming the age model derived from the correlation shown in Fig. 43. Marine Oxygen Isotope Stages (OIS) 1–19 are labelled, with glacial times highlighted in light blue. Generalized lithofacies are shown at the right. The Yellow, Green, Blue and Purple intervals mark the depths at Site 1073 bracketed by sequence boundaries p4, p3, p2, p1 and the seafloor, respectively. (After Mountain *et al.*, 2001; McHugh & Olsson, 2002.)

The last eustatic cycle and its preserved record

The New Jersey margin is a compelling natural laboratory for understanding the formation and evolution of the seafloor, and for investigating the relationship of shallowly buried subsurface stratigraphy to changing base levels. A range of high-frequency seismic techniques (e.g. sidescan sonar, multibeam bathymetry and backscatter, seismic reflection) and shallow-sampling tools (e.g. grab sampling; push, piston and rotary coring) has been used innovatively to examine the most recent stratigraphic record on the New Jersey shelf and uppermost slope.

Characterizing the seafloor

Surveys of sidescan backscatter and multibeam bathymetry extended from water depths of ~20 m

to the edge of the shelf (Fig. 45; Goff *et al.*, 1999). These data augmented soundings from the National Ocean Survey, and the resultant maps have been used over the past decade to interpret details of seafloor morphology (e.g. sand ridges and dunes, ribbon morphology, iceberg scours) and to relate them to the underlying preserved stratigraphy. The bathymetric data proved that the modern seafloor on this margin bears little resemblance to the geology of the underlying 15–20 m of sedimentary section. Goff *et al.* (1999) showed that swales (floored by sand ribbons) located between clusters of largely relict sand ridges on the middle shelf are erosional in nature, responding to modern south-westward bottom currents. Sinuous furrows on the outer shelf ~100–400 m wide, kilometres long and ~1–4 m deep were identified as iceberg scours preserved in semi-lithified clays of the southern Hudson Apron (Fig. 45). Duncan & Goff

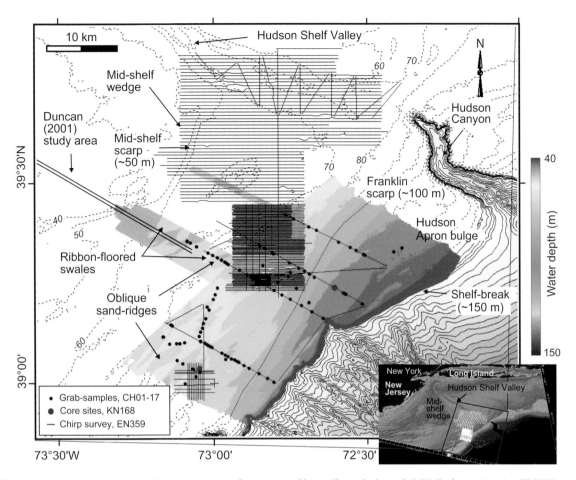

Fig. 45 Locations of deep-towed, Chirp seismic-reflection profiles collected aboard *R/V Endeavor* (cruise EN359), superimposed on multibeam (colour contoured; Goff *et al.*, 1999) and NOAA bathymetry (dashed lines contoured in metres) of the New Jersey middle and outer continental shelf. Inset locates the study area within a three-dimensional rendering of the Mid-Atlantic Bight. Locations of grab samples (CH01–17) are identified.

(2001) mapped these keel marks and interpreted two discrete phases of iceberg rafting, which they related to Laurentide meltwater events at 17 ka and 25 ka.

Goff *et al.* (2000) attempted to correlate back-scatter intensities with grain-size distribution, based upon ~300 grab samples. They identified a prominent linear correlation of backscatter intensity and grain size; coarser grain sizes generally equated to higher backscatter. There was a disproportionate effect on backscatter from the coarsest fraction, in particular shell hash, which proved abundant within older sediments in deeper water. Larger grain sizes also tended to occur on the seaward slopes of dunes oriented transverse to flow. Mid-shelf ridges were generally winnowed of fine-grained material, and muds within swales confirmed their erosional nature. Goff *et al.* (2004) built upon these

earlier results, integrating the bathymetry and grab-sample data with geotechnical measurements. Grain-size distributions were found to be multi-modal. In the absence of shell hash and gravel, backscatter correlated with seismic velocity and the percentage of fine sediments. Mean sand-size and the percentage of fine sediments were the primary controls on compressional-wave velocity of seafloor sediments.

Goff *et al.* (2005) have explored evidence for recent seafloor erosion on the New Jersey shelf, using all available data from the outer shelf (Fig. 45). Timing of erosion is constrained by seafloor truncation of stratigraphic horizons that can be dated. Erosional depths are everywhere ~3 m to > 10 m, and ribbons characterized by alternating areas of high and low backscatter mark eroded shelf strata. High backscatter is associated with shell hash and with concentrations of well-rounded gravels and cobbles, both considered erosional lag deposits. The reworking of the seafloor during transgression is characterized first by sand-ridge evolution to water depths of ~40 m, then by erosional modification caused by selective wave and bottom-current re-suspension of sediments at greater water depths.

Tying stratigraphy to latest Pleistocene–Holocene eustatic base-level changes

The shallow subsurface of the New Jersey shelf represents the composite stratigraphic record of the last regression and transgression (Milliman *et al.*, 1990; Davies *et al.*, 1992; Lagoe *et al.*, 1997; Buck *et al.*, 1999; Duncan *et al.*, 2000). The concept of a single lowstand erosion surface, formed during shelf exposure accompanying the Last Glacial Maximum, has been replaced by a new paradigm involving a complicated surficial stratigraphy evolving con-tinuously throughout the latest eustatic cycle.

Davies *et al.* (1992) imaged a shallowly buried set of incisions beneath the outer shelf. These data illustrated conclusively that the outer shelf repre-sents a complex stratigraphic section completely unrelated to the modern seafloor (Austin, 1996). Duncan *et al.* (2000) mapped three regional seismic stratigraphic horizons on the middle and outer shelf.

1 'R', forming the base of an outer-shelf wedge and earlier identified as the Last Glacial Maximum ero-sion surface (Milliman *et al.*, 1990), was reinterpreted as the product of multiple erosional episodes during the previous regression. Piston coring had previously confirmed that sediments below 'R' are > 45 kyr old (Davies & Austin, 1997).
2 The 'Channels' horizon, initially interpreted as evid-ence for fluvial drainage by Davies *et al.* (1992), was recognized as a widespread surface representing dendritic patterns of fluvial erosion formed during shelf exposure at or near the Last Glacial Maximum. Analyses of a piston core within one of these incisions confirmed that its sedimentary fill records fluctuat-ing marginal-marine and mid-shelf depositional envir-onments during the Holocene transgression, which flooded these fluvial systems ~12.5 ka (Buck *et al.*, 1999).
3 'T', previously unrecognized, was interpreted as a flooding surface or ravinement within the trans-gressive systems tract.

Mapping of multiple dendritic, drowned fluvial drainage systems has since been refined and am-plified. Nordfjord *et al.* (2005), armed with these maps (Fig. 46), have used quantitative geomor-phological analysis to estimate palaeohydrological parameters linking observed channel morphologies to the hydrodynamic setting in which the channels were incised. Large ratios of width to depth, along with low sinuosities and slopes, suggested modern braided streams – but no such braiding was appar-ent. These buried systems are likely to be immature, having never reached equilibrium prior to drown-ing by the Holocene transgression. Furthermore, palaeoflow estimates for these fluvial systems were too high for non-tidal creeks, given the low hydraulic gradients typical of such a coastal-plain setting. Therefore, flows that carved these drain-ages must have been initially high, consistent with the occurrence of meltwater pulses crossing a coastal plain after the Last Glacial Maximum (Nordfjord *et al.*, 2005).

To the south-west, Fulthorpe & Austin (2004) identified and mapped a prominent seismic facies boundary 0–20 m below the seafloor (Fig. 47). This boundary, separating seismically transparent sedi-ments from an underlying stratified facies, defined two populations of apparent incisions trending north-east and east-northeast. Stratified blocks within the transparent facies, lying occasionally within incisions with steep flanks, indicated that the transparent facies was formed by catastrophic disruption of the underlying stratified material (Fig. 48). These incisions may have formed from meltwater discharges associated with the multi-ple breaching of glacial-lake dams to the north ~19–12 ka (Uchupi *et al.*, 2001), lending support to the hypothesis that local forcing can produce regionally significant erosional unconformities on periglacial shelves even in the absence of base-level changes.

Tracking sediment movement across the margin

The regionally significant 'R' horizon (Fig. 48) matches the p1 reflector that defines the base of the Purple sequence mapped with MCS data along the outermost shelf (Figs 42–44). As discussed further earlier, p1/R correlates to 79 mbsf at ODP Site 1073 where Oxygen Isotope Stages 6 and 7 are miss-ing (McHugh & Olson, 2002). This age range points

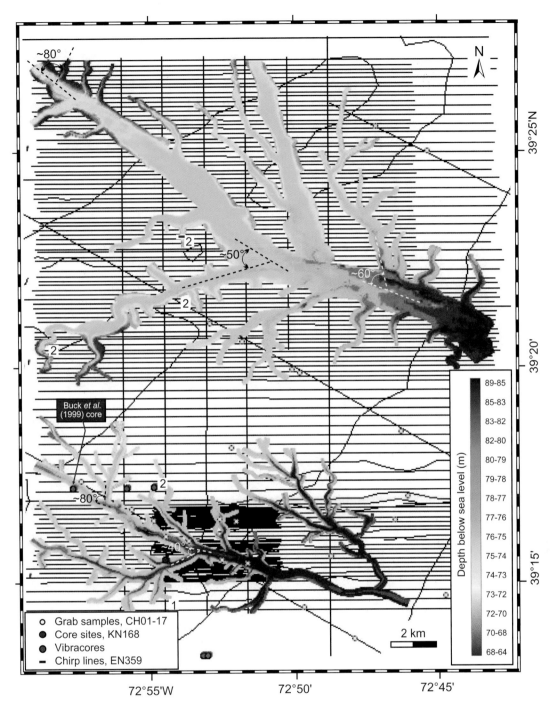

Fig. 46 Interpolated, shallowly buried fluvial drainage systems mapped beneath the New Jersey outer continental shelf. These drainage systems exhibit dendritic patterns. Depths are in metres below ambient sea level. Location of the Buck *et al.* (1999) vibracore is also shown.

toward glacially induced sea-level lowstand as the cause of the hiatus, and suggests that where this surface has been mapped landward to the middle shelf (Knebel *et al.*, 1977; Milliman *et al.*, 1990;

Davies & Austin, 1997; Duncan *et al.*, 2000) it may represent a longer interval of missing record.

Subtle topographic changes during the regression marked by the p1/R surface may have

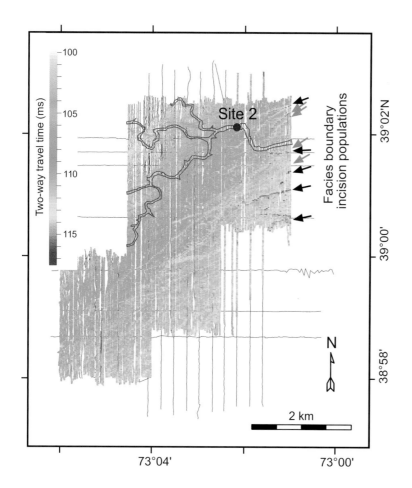

Fig. 47 Map of seismic facies boundary made by interpreting densely distributed Chirp profiles. Facies boundary irregularities align to form two populations of apparent incisions: one trending north-east (orange arrows), and the second with a more easterly trend (black arrows). The facies boundary dips south-east across most of the mapped area, except in the south-east, where it onlaps horizon 'R'. Incisions in both populations meander less than the overlying channel system (in grey), which is perhaps coeval with the 'channels' horizon mapped to the north-east (Nordfjord *et al.*, 2005). These incisions may all be caused by periodic floods crossing the shelf from breached glacial lakes to the north. Depths are in milliseconds, two-way travel-time. (From Fulthorpe & Austin, 2004.)

modulated cross-shelf sedimentation, building the shelf outward as distinct, toplapping wedges above 'R'; in places, these wedges look much like delta fronts (Gulick *et al.*, 2005). They may be important regression markers in clastic-dominated outer-shelf settings along passive margins. Furthermore, seismic-facies data indicated that these wedges represent interbedded lithologies, and in places are sandy, suggesting progressive winnowing of shelf sediments seaward to the shelf-break. Fulthorpe *et al.* (1999) have postulated a similar cross-shelf sediment-transport mechanism for the New Jersey margin during Miocene lowstands; i.e. a line-source delivery of fluvial sediments by small rivers to the outer shelf. From there, both channelized and non-channelized downslope transport processes have moved sediments into deeper water. The presence or absence of canyon incisions on the New Jersey slope has been dictated not only by fluctuations in base level, but by local conditions, including rates of sediment supply, grain-size dis-

tribution and possible slope collapses related to fluid escape (Fulthorpe *et al.*, 2000).

New Jersey submarine canyons

Studies of submarine canyons and their role in continental-margin evolution are based on seismic-reflection data, cores, outcrops and, to a lesser degree, high-resolution bathymetric data. The morphology of modern submarine canyons and theories for their formation are thoroughly discussed in Pratson *et al.* (this volume, pp. 339–380). In this paper, submarine canyons are considered in the context of the stratigraphic record that fills them.

In studying the character of Miocene canyons now buried within the New Jersey slope (Fig. 49), Mountain (1987) interpreted V-shaped canyons to have been cut by turbidity currents and U-shaped canyons to have been carved by mass wasting (see also Twichell & Roberts, 1982; Farre *et al.*, 1983). Buried canyons are typically ~50–100 m deep and

Fig. 48 Seismic facies above horizon 'R'. (A) Relationship between seismically transparent (above) and parallel, continuous/stratified facies (below) within the mapped area and above horizon 'R'. Seismically transparent facies, incised by meandering channels near the seafloor, is up to ~20 m thick. Irregular boundary between these facies resembles steep-sided (commonly 50°–90°) incisions, which truncate underlying reflections. (B) Stratified blocks (circled) occur within the transparent material and within incisions defined by the facies boundary. The presence of these blocks suggests that the parallel, continuous/stratified facies has been catastrophically eroded and then quickly redeposited in topographic lows, along with the remainder of the transparent-facies material. Labels: mbsf, metres below seafloor, VE, vertical exaggeration. (From Fulthorpe & Austin, 2004.)

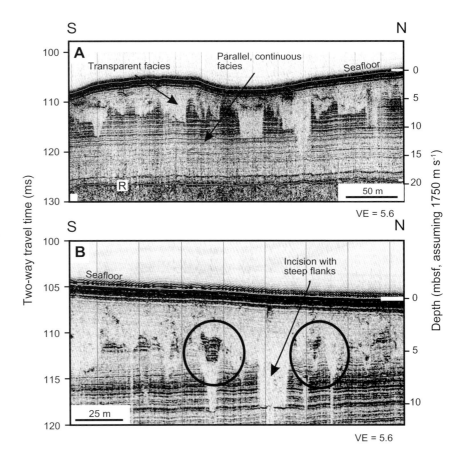

1–2 km wide (Fig. 50; Mountain, 1987; Miller *et al.*, 1987b; Pratson *et al.*, 1994; Mountain *et al.*, 1996). In addition, STRATAFORM studies off New Jersey and in Eel River Basin have identified many smaller, buried gullies in the vicinity of canyons. These features are typically 1–10 m deep, 100–200 m wide, and are directed downslope (Field *et al.*, 1999; Fulthorpe *et al.*, 2000; Spinelli & Field, 2001). Buried gullies tend to be V-shaped, and many may have been tributaries leading into nearby buried canyons (Pratson *et al.*, this volume, pp. 339–380).

The paths of buried canyons and gullies are difficult to trace between seismic lines in typical two-dimensional surveys. Modern gullies are similar in dimension, and they coalesce with increasing regularity downslope to produce a tributary-style drainage pattern. For example, 11 gullies along the New Jersey upper slope merge into just four gullies on the lower slope (Pratson *et al.*, 1994). The buried gullies may have a similar distribution, but the profile coverage is too sparse to determine this. To date, available data suggest that gullies may abruptly start and stop on the slope with

few intersections (Fulthorpe *et al.*, 2000; Spinelli & Field, 2001).

The drainage pattern of buried canyons also appears to be linear and largely directed downslope, though a few may meander (von der Borch *et al.*, 1985) or have sharp turns due to faults and other structures (Kelling & Stanley, 1970; Song *et al.*, 2000). Furthermore, they may branch off into tributaries near the palaeoshelf break (Martin & Emery, 1967; McGregor, 1981). Buried canyons have been found on the New Jersey upper slope where they primarily occur between modern canyons (Mountain, 1987; Pratson *et al.*, 1994; Mountain *et al.*, 1996; Fulthorpe *et al.*, 2000). Farther downslope, all the buried canyons merge into and have been re-excavated by a modern canyon (Pratson *et al.*, 1994). This re-use of the buried canyons appears to be due to their diminishing depth of burial downslope.

Sediments that fill buried canyons are a mixture of mass-wasting debris, turbidites and hemipelagic drape. Although these deposits are commonly interbedded with one another, sedimentological studies of buried canyons have detected a common

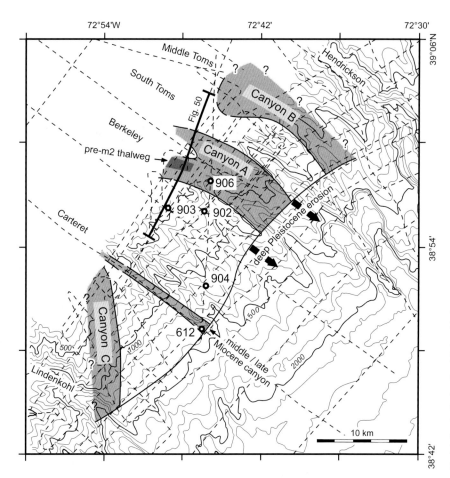

Fig. 49 Locations of drill sites (ODP Leg 150; DSDP Leg 95) and seismic profiles (Ew9009 MCS; ODP Leg 150) used to locate three middle Miocene canyons (grey shaded areas labelled A, B and C) and one younger canyon, all now buried on the New Jersey upper slope. Bathymetry is based on SeaBeam (Pratson *et al.*, 1994). Modern canyons are labelled. (After Mountain *et al.*, 1996.)

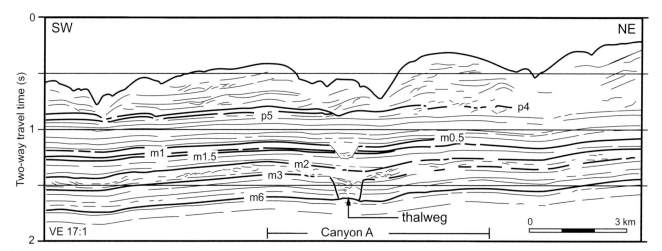

Fig. 50 Line drawing interpretation of seismic profile Ew9009 MCS line 1026 along the New Jersey slope. Several sequence boundaries are traced from the shelf, as noted. At this location (Fig. 49) Canyon A is completely filled below reflector m2. (After Mountain *et al.*, 1996.)

sequence to canyon fill (Stanley & Kelling, 1978; May *et al.*, 1983; Goodwin & Prior, 1989; Bruhn & Walker, 1995; Mountain *et al.*, 1996; Wonham *et al.*, 2000). The lowermost units tend to consist of coarse-grained sediments that are a combination of lag deposits and mass flows derived from either debris flows or turbidity currents (Fig. 51). These typically decline in frequency upwards, where finer-

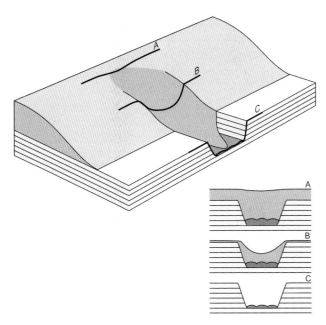

Fig. 51 An isometric view of the New Jersey slope as it may have looked at the time of reflector m2, crossed by hypothetical profiles A, B and C. Drilling at ODP Site 906 showed that the oldest deposits on the canyon floor are debris from the widening and headward erosion of the original slump scar, and are overlain by shallow-water turbidites which demonstrate that the canyon connected with a source of shelf sands. The major volume of infilling sediment is laminated claystone between these sands and reflector m2. (From Mountain *et al.*, 1996.)

grained turbidites become more common. The turbidites are likely to have been deposited by decelerating turbidity currents, and are often capped by hemipelagic units, suggesting periods of even more gradual filling. ODP Site 906 (Mountain *et al.*, 1994) drilled directly through the thalweg of a mid-Miocene slope canyon and determined this three-part succession with clarity. The basal fill is composed predominantly of a conglomerate formed by semi-indurated mid-Miocene slope sediments. These demonstrate that this canyon widened through wall collapse, and some or all of the initial debris was not removed from the canyon floor. Shallow-water turbidite sands rest with angular unconformity on this conglomerate at Site 906, and provide the majority of the fill. These sands thin seaward, and at the 1500-m isobath no fill is observed. Hemipelagic slope sediments overlie the turbidite sands.

According to Posamentier *et al.* (1988), canyon filling should occur during periods of relative

sea-level rise. This conclusion is supported by observations of other margins (Ruch *et al.*, 1993), and by recent laboratory experiments simulating strata formation during changes in base level (Wood *et al.*, 1993; Heller *et al.*, 2001). However, many submarine canyons around the world that were active during the Last Glacial Maximum have remained exposed and unfilled during the subsequent eustatic rise.

Sediment failure and headward erosion can be important precursors to shelf-break sands finding a ready conduit to the deep sea. Sea-level fall is very likely to be a secondary process in the formation of slope canyons. The data at Site 906 has narrowed the critical events to the following (all dates have a standard error of ±0.5 Myr): initial slope failure occurred at 13.5 Ma; continued mass wasting widened and lengthened the canyon, and a source of shallow-water sands began spilling into the canyon at roughly 13.0 Ma. This phase was replaced by a fill of laminated silts very likely as a result of transgression of the shelf. By 12.4 Ma, the history of this canyon was complete. This and presumably other New Jersey canyons were buried on the upper slope before burial on the lower slope; as a result, canyon piracy was more common in the latter setting. Consequently, the stratigraphic record of lower slope canyons can be especially complex.

Sedimentation in open-slope regions (i.e. between canyons) is more uniform and continuous, and occurs as:

1 settling of fine-grained hemipelagic sediment carried off-shelf in dispersed plumes;
2 overflow of turbidity currents from neighbouring canyons;
3 transportation along-slope by prevailing bottom currents.

As canyon thalwegs deepen by erosion, canyon interfluves build, with the net result that canyon morphology becomes more accentuated.

Stratigraphic modelling

Reconstructing the margin

The New Jersey margin developed its present-day character of stacked clinoforms beginning in Oligocene times. Since then, siliciclastic clinoforms

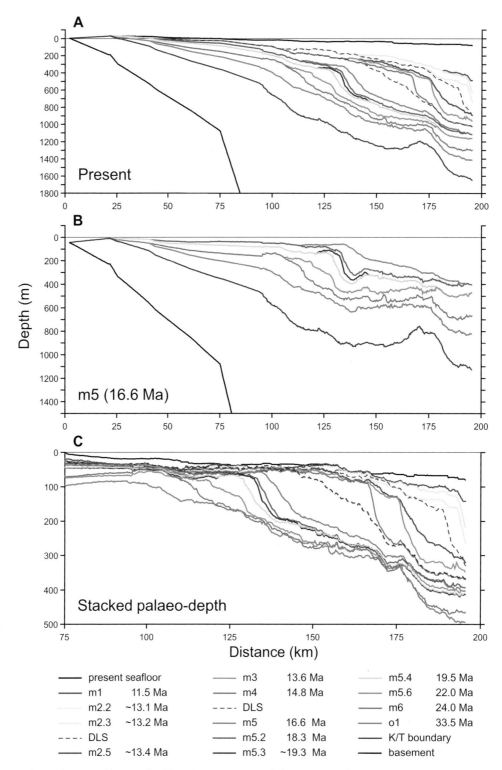

Fig. 52 Sequence boundary profiles on the New Jersey margin. (A) Tracings of 13 post-Eocene sequence boundaries, two widespread downlap surfaces, and the Cretaceous–Tertiary unconformity along profile Ew1003 (see Fig. 31). Names and ages of each are displayed in the colour code at the bottom. Reflection times have been converted to depth using a regional function of sound velocity derived from local seismic stacking velocities. (B) The same depth profiles with all sediment younger than sequence boundary m5 removed; sediment decompaction plus isostatic and flexural rebound have been calculated. (C) The depths of all sequence boundaries at the time of their formation with the added correction for estimated palaeowater depths. Note the changes in both horizontal and vertical scale. (After Steckler *et al.*, 1999.)

have advanced nearly 100 km across the underlying carbonate ramp, and now provide a relatively complete record of late Oligocene to Holocene margin evolution. Clinoforms such as these are a basic building block of many continental margins and have been the subject of considerable debate as to their imprint of eustatic change (Vail *et al.*, 1977; Haq *et al.*, 1987; Posamentier *et al.*, 1988).

However, discussions of Tertiary evolution for the New Jersey margin must recognize that the present geometry of preserved strata is not an accurate indicator of original morphology. Subsidence, compaction, sediment loading and tectonic activity have all deformed these sediments. Backstripping removes sediments layer by layer to restore the morphology of the surface (assumed to be a former seafloor), as well as the geometry of underlying surfaces (Steckler *et al.*, 1988, 1993). The shape and gradient of the past continental shelf can then be estimated, and, in particular, the height of clinoforms can be determined. Furthermore, the depth of any given clinoform rollover and its position relative to the adjacent shoreline can be evaluated.

Numerical experiments have been conducted with initial conditions based on backstripping reconstructions of margin geometries (Steckler *et al.*, 1999; Syvitski *et al.*, this volume, pp. 459–529). Sea-level variations were modelled to fit a sinusoidal cycle with 20-m amplitudes and 2-Myr periods, the latter being the average length of early Miocene cycles in New Jersey. These oscillations were superimposed on a long-term, post-Eocene eustatic rise of 10 m. Doubling the modelled amplitudes to 40 m causes breakdown in the smoothly varying switch from progradation to aggradation within each cycle. However, even 40 m is considerably smaller than previous estimates of 100 m and more, which were derived from sequence geometries and assumptions of lithofacies (Haq *et al.*, 1987). The more modest estimates used in this Steckler *et al.* (1999) modelling effort are in accord with subsequent δ^{18}O and backstripping data based on studies from coastal-plain wells (Kominz & Pekar, 2001). Furthermore, the modelling results yield clinoforms that steepen as they prograde into deeper water, consistent with both observations.

The modern arrangement of 13 post-Eocene sequence boundaries, two widespread downlap surfaces and the Cretaceous–Tertiary unconformity

are displayed in depth in Fig. 52A. Figure 52B shows a depth profile as it would have looked in the early Miocene at 16.6 Ma; all sediment younger than sequence boundary m5 has been removed, and sediment compaction, isostatic and flexural rebound have been accounted for. The true height of the m5 clinoform at 16.6 Ma is predicted in this manner to have been ~110 m. Furthermore, the m5 rollover is predicted to have been in ~60 m of water. Similar backstripped geometric models of other post-Eocene sequence boundaries (Fig. 52C) show that the relief of most clinoforms was at least this large when formed. Indeed, several of the middle Miocene sequence boundaries outline clinoform heights twice as large. This modelling illustrates that the rollover from flat-lying topsets to seaward dipping foresets typically occurs many tens of metres below sea level. This observation challenges the assumption that strata showing coastal onlap against the seaward face of a dipping clinoform are necessarily shallow-water facies.

The backstripped surfaces in Fig. 52 also show that foresets vary in steepness in an organized manner. The pattern of sigmoidal clinoforms appears to change gradually from progradation to aggradation over about six clinoforms. This corresponds as well to a gradual increase in the steepness of the foreset surfaces. If valid, this may indicate either: (i) a higher order (5–15 Myr) record of sea-level variation not yet accounted for; or (ii) a response to varying amounts of sediment supplied to the New Jersey shelf. Figure 53 shows the amount of sediment deposited on the New Jersey margin through time, based on volume calculations of the backstripped models. Note the relatively short-lived peak of sediment deposition centred at ~13 Ma and 1.5 Ma. Both of these spikes lag the times of rapid increase in global ice volume by 1–2 Ma, as indicated from increases in global δ^{18}O (Fig. 53), and suggest a link to climate-induced changes in terrestrial erosion and/or sediment delivery to the continental shelf.

Profiles show that oblique clinoforms prograded roughly 25 km across the New Jersey outer shelf in Pleistocene time, in close agreement with model simulations. Some of the sediments recovered in ODP Leg 174A cores reveal a history of deepening upwards within these deposits (Austin *et al.*, 1998), consistent with the model results that these strata accumulated during a time of transgression.

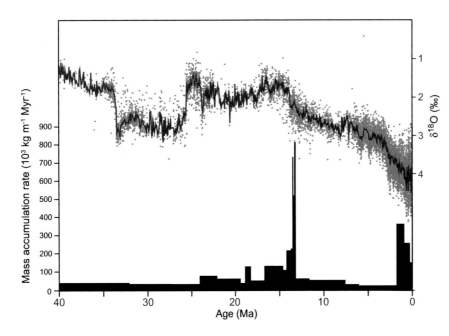

Fig. 53 Mass accumulation rate (black columns along bottom; scale at left) of sediments deposited on the New Jersey margin through time, superimposed with global values of $\delta^{18}O$ (red, right-hand scale; Zachos et al., 2001). Sediment volumes were derived from calculations based on backstripped models (see Fig. 52).

This was a major surprise of ODP Leg 174A, because downstepping of eroded surfaces had suggested that these packages were formed during sea-level fall. However, modelling fails to produce the volume of oblique strata observed in two-dimensional profiles. Increasing the amplitudes of modelled sea-level fluctuations leads to more extensive sequence-boundary unconformities and, with sufficiently large fluctuations, these unconformities eventually reach the shelf edge. Most importantly, if sea-level amplitude is increased to 90 m, models predict extensive oblique clinoforms and non-marine strata overlying the preceding sequence boundary, in a position where ODP Site 1071 recovered possible lagoonal sediments of late Miocene age (Austin et al., 1998). Glacio-eustatic changes implied by variations in global δO^{18} (Miller et al., 1987a) are too small to support a 90-m late Miocene fall, but the drilling evidence is irrefutable: sea level was within a few metres of a clinoform top, only ~3 km from the matching rollover. Future modelling must reconcile this drilling evidence with subsidence caused by loading from sediments, water and possibly continental ice.

Oxygen isotope record

The onshore stratigraphic record can be linked to offshore sequence boundaries and to the global record of $\delta^{18}O$ (Miller & Mountain, 1994; Miller et al., 1996a). A majority (nearly 30 in all) of the regionally correlated Oligocene–Neogene hiatuses documented by onshore drilling separate transgressive–regressive successions (Sugarman et al., 1993; Miller et al., 1997a). Basal quartz sands, shells, glauconite and organic matter deposited in 30–100 m water depths grade upward to middle neritic silts and finally to quartz sands in which biofacies indicate palaeodepths < 30 m. The succession usually ends abruptly, occasionally with evidence of erosion and subaerial exposure. These characteristics point to oscillations in sea level as a major control on coastal-plain facies successions. Geochronology based on biostratigraphic, isotopic (largely $^{87/86}Sr$ ratios) and magnetostratigraphic integration has determined the ages of roughly 10 hiatuses from latest Oligocene to mid-Miocene, and has shown they closely match the ages of inflections in global marine $\delta^{18}O$ records (i.e. inferred glacio-eustatic lowstands, Fig. 54; Sugarman et al., 1997). Thus, the case for a global cause to mid-Tertiary gaps in the onshore record of New Jersey margin is very strong.

The implication that sea-level change is the cause of the Oligocene and mid-Miocene sequence boundaries (defined by facies successions) has been extended to the New Jersey slope (Miller et al., 1996b). Geochronology based on biostratigraphic, isotopic and magnetostratigraphic integra-

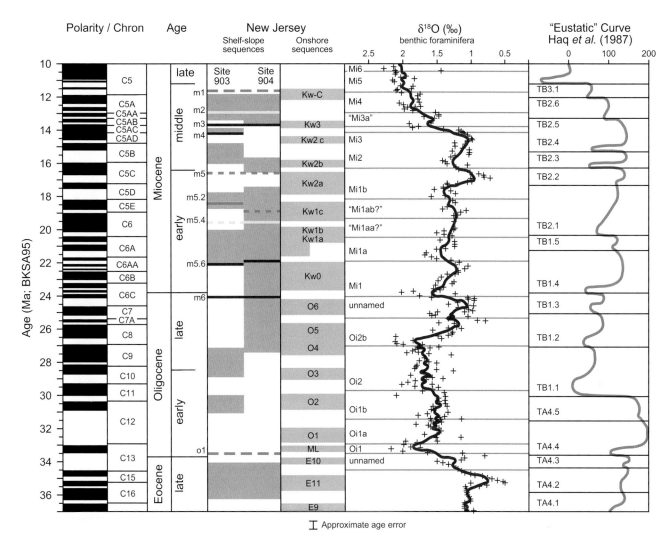

Fig. 54 Oligocene–Miocene sequences for the New Jersey margin (onshore and offshore) correlated to a composite of global $\delta^{18}O$ values and the eustatic curve of Haq *et al.* (1987). Reflector geometries tied to chronologies sampled on the slope define the offshore sequences: facies successions and erosional gaps define the onshore sequences. To a large degree, these correlate in time. The key surfaces and gaps show a close match to $\delta^{18}O$ increases, indicating growth in global ice volume. (From Miller *et al.*, 1998a.)

tion strongly suggests that these surfaces formed during the time that unconformable gaps were developing onshore, and by inference the global marine record of $\delta^{18}O$ was experiencing marked changes as well. Despite the suggestion that similar numbers and ages of events onshore, offshore, and in the $\delta^{18}O$ record point to a common origin, the age comparisons implicated in Fig. 54 do not prove a causal relationship.

The ice-volume changes implied by studies on the New Jersey margin average 1.2×10^6 yr between maxima, but exhibit no clear periodicity. These events are probably composites of Milankovitch-scale (10^4–10^5-yr scale) climate cycles that have merged to yield long-term changes (Zachos *et al.*, 1994). This is illustrated by a moderately high-resolution (~10^4 yr sampling; Flower & Kennett, 1995) $\delta^{18}O$ record showing that the major slope reflectors on the scale of 10^6 yr (m2 through to m5) correlate with major $\delta^{18}O$ increases, although there is higher-order variability contained in both records. Further study of New Jersey sections may continue to detect additional, smaller-scale sequences, such as some of those found onshore at the Cape May drillsite (e.g. subdivisions of Kw2a and Kw3 by de Verteuil, 1997).

SUMMARY

The long-term record: its challenges and rewards

This paper has focused on processes that build continental margins and operate on time-scales of ~20 kyr and greater. In contrast to the impact of processes that operate on shorter periods (see the other papers in this volume), long-term processes:

1 provide the opportunity to examine behaviour of the Earth system under boundary conditions very different from those of today;
2 evaluate a variety of numerical models designed to duplicate natural phenomena;
3 capture the full range of processes that have built the geological record.

Choosing to study the marine record of modern margins provides obvious advantages: sediments are less disrupted and chemically altered than are most sediments seen at outcrop; the modern depositional setting can provide analogues to the ancient environment in which the long-term record accumulated; and the logistics, cost and quality of seismic-reflection profiles are superior to those collected over land.

Gathering samples to document long-term processes inferred from seismic-reflection profiles is nonetheless a challenging task. Rotary drilling is the only technique that can reach > 100 m below seafloor, as needed to sample the long-term record. On continental shelves, these sediments are commonly sand-rich, unlithified and difficult to recover with their primary fabric still intact. Few research-based efforts have attempted to drill to substantial depths in continental margins, and where they have, the recovery has been inadequate for answering many fundamental questions concerning processes that shape the long-term record (Austin *et al.*, 1998). Borehole logging provides information to fill gaps caused by incomplete core recovery, and helps to correlate two or more borehole records (Goldberg, 1997). This is accomplished by sensors lowered into an existing well or installed within the drilling assembly to measure electrical, acoustic or radiometric properties of the sediments and pore fluids. However, logging does not return a physical sample for dating or for more elaborate compositional analysis. Even with complete core recovery, the precision of absolute dating is limited by low accumulation rates, a lack of biostratigraphic markers, and poor preservation of geochemical and palaeomagnetic indexes.

Despite sampling and dating challenges, one of the primary aims in studying continental margins is to understand the causes of depositional gaps that have long been recognized throughout the geological record. Few researchers deny the existence of these widely distributed breaks, typically expressed as abrupt changes in biofacies, lithofacies and/or the angular termination of stratal surfaces. However, agreement as to their cause is far from unanimous. Debate sharpened in the 1970s when the digital revolution dramatically improved the acquisition, processing and display quality of seismic-reflection data. These images of the subsurface revealed stratal patterns in basins and continental margins throughout the world with previously unseen clarity, and classically trained stratigraphers with access to these global, high-quality subsurface data developed new ways of applying seismic-reflection profiles to the study of basin history and the search for fossil fuels. The result was an entirely new approach to interpreting the geological record – termed **seismic sequence stratigraphy**, which relies on the recognition of angular reflector terminations against basin-wide unconformities (Payton, 1977). Proponents claim this approach reveals the distribution of facies sufficiently for recognizing downward shifts in base level that are presumed to track local falls in sea level. Some researchers have taken this one step further to conclude that their records prove the existence of global, i.e. eustatic, changes in sea level every 1–2 Myr for the past 200 Ma (Vail *et al.*, 1977).

Critics have pointed to the absence of a single eustatic mechanism explaining this long record (Christie-Blick, 1991), and have argued that:

1 geochronology can rarely provide the precision needed to prove this level of synchroneity;
2 numerical models show that flexural response of the **lithosphere** to sedimentary loads will generate a hiatus of distinctly different ages in different tectonic settings;
3 numerous basins show that changes in sediment supply, deltaic lobe switching, tectonism and other local processes can lead to reflector geometry that mimics the imprint of eustatic change.

Numerical modelling further demonstrates that reflector geometry, facies successions and the ultimate extraction of a eustatic signal are very sensitive to the sediment compaction history and rate of tectonic response to water and sedimentary load, and also to the regional tectonic adjustments of the underlying crust. Consequently, some researchers have claimed that the record of eustatic change in the range of 10^1 m over 10^5 yr can be overwhelmed by the magnitudes and rates of relative change produced by these other factors (Karner & Driscoll, 1997). This last argument does not deny that eustatic change affects the stratigraphic evolution of continental margins, but clearly places it among other important factors.

The Eel River Basin: difficulties in recognizing eustatic control

Studies of the Eel River Basin confirm the importance of tectonic control on margin evolution. The region surrounding the Eel River Basin has evolved over the past 80 Ma from a transcurrent to a convergent plate boundary (Atwater, 1970). The comparatively narrow shelf (defined locally as seafloor landward of the 150-m isobath) is supplied with sediment derived predominantly from the Eel River. The continental slope descends to the Cascadia Trench. An active magmatic arc plus a region of accreted terrain landward of the shoreline provide sediment to the margin. The modern Eel River margin is a classic forearc basin.

Basement subsided so rapidly in Oligocene to early Miocene time, when the Cascadia subduction zone developed off northern California, that the margin foundered and entered a lengthy period of transgression. Regression and significant sedimentary reconstruction did not begin until Pliocene time. Those processes, too, were altered substantially during 0.7–1.0 Ma, when uplift and a widespread hiatus accompanied the northward migration of the Blanco Fracture Zone along the North American–Gorda plate boundary (Orange, 1999). Since then, as much as 1 km of sediment has accumulated within the Eel River Basin, though the region is faulted and divided into deep synclines and sharp anticlines that attest to convergent plate motion (Clarke, 1987). The regional stress field has been changing since ~0.5 Ma due to the northward approach of the Mendocino Triple Junction, as expressed by rotating

fault trends and basin reorientation throughout this interval (McCrory, 1989). The relatively close line spacing (5 km or less) of the STRATAFORM seismic grid provides good spatial coverage of this post-1-Ma section, but does not have the acoustic penetration to provide insight into older/deeper sedimentary units.

Complications from gas (biogenic and thermogenic) constitute a substantial challenge to documenting the long-term history of the Eel River Basin. Together with the absence of drill holes in the study area, it is apparent why detailed knowledge of controls on the long-term record in the Eel River Basin is currently limited. In particular, the age of basin-wide unconformities in the basin is not well constrained. Fourteen unconformities defined by erosional truncation, downlapping and onlapping reflectors are recognized in uplifted areas; all become conformable in adjacent synclines (Burger *et al.*, 2002). Many or all of these surfaces probably mark transgressive ravinements that formed during shoreface retreat as sea level rose to flood a previously exposed or very shallow (less than storm wavebase) shelf. An observed cyclicity (~70–100 kyr) suggests that there may be a glacio-eustatic signal, but definitive age control is needed (Burger *et al.*, 2001).

Despite the current lack of long-term samples, the following model of sequence-stratigraphic development has been proposed (Burger *et al.*, 2002). Shelf sedimentation during highstand dominates the record and is comprised of fine-grained marine silts and muds. During regression, interfluves between rivers are areas of non-deposition and palaeosol formation; off-shelf sediment transport along the northern Eel River Basin is focused in shelf-break gullies that led into slope channels. In general, very little sediment from times of regression is likely to be found on the continental shelf. During times of trangression, incised channels are filled first, followed by a transgressive lag deposit that in interfluve areas now rests unconformably on the highstand silts and muds of the previous sequence. The recognition of ten additional surfaces incised by sharply defined channels (10–250 m deep and 100–1000 m wide) shows that local tectonism has a major influence on margin development (Burger *et al.*, 2001). Each channel trends south-west from near the mouth of the Eel River and deepens towards embayments along the north side of Eel

Canyon; locations and gradients suggest that they are former river channels that cut into the shelf during shoreline regression, and have been directed toward Eel Canyon since its presumed inception at ~360 ka.

Initial attempts to model sequence development of the Eel River Basin have produced encouraging results. Depocentres transgress and regress across the shelf, but each sequence builds out to approximately the same position in each cycle, as determined by the admittedly oversimplified, uniform rates of sediment supply and margin subsidence. The largest departure between the model and observations occurs on the upper portions of the slope, where the model builds only a thin sedimentary record, and thick deposits are observed.

The New Jersey margin: eustatic imprint, with complications

Factors controlling the long-term stratigraphic development of the New Jersey Margin (NJM) are better documented than they are for the Eel River Basin, largely because they include deep sediment samples of the New Jersey margin record. The margin has evolved under a uniform and more easily documented tectonic setting, with the result that basin history can be more easily determined from seismic-sequence stratigraphy there than it can be in the Eel River Basin. The opening of the Atlantic created a rifted, passive margin along the eastern USA in Middle Jurassic time, and this plate setting has not changed to the present (Sheridan *et al.*, 1988). In contrast to the post-Oligocene Eel River Basin forearc basin, rapid, shallow-water sedimentation on the New Jersey margin largely exceeded basin subsidence throughout the first 65 Myr. This resulted in a continental shelf that continued to widen as flexural loading rolled the shoreline landward (Watts & Steckler, 1979). By mid-Cretaceous time, the shelf was > 150 km wide, with its seaward limit delineated by a fringing reef that extended along the entire eastern North American continent (Jansa, 1981). The reef was overrun by siliciclastic sediment in Barremian time, and has remained buried since then. Sediment supply apparently decreased substantially during the Paleogene, and a wide, seaward-dipping ramp developed as a result. Numerical models

backed by palaeowater depths of microfossils show that the seafloor above the buried mid-Cretaceous reef in Eocene time was roughly 1500 m below sea level (Steckler *et al.*, 1993).

The New Jersey margin entered its present phase of margin evolution in the mid-Oligocene, when one or more rivers began building a deltaic wedge of sediment seaward across the drowned ramp (Poag, 1985). By this time, 130 Myr after rifting, thermal subsidence had diminished to a slow rate, and consequently, significant aggradation could occur only due to loading, flexure and eustatic rise. Sediment supply greatly exceeded the accommodation space provided by these three processes, and from then until Pleistocene time clinoforms prograded ~100 km across the margin (Steckler *et al.*, 1999). Toes of these clinoforms reached the approximate location of the underlying Mesozoic reef in mid-Pleistocene times, and from then until the present sedimentation on the New Jersey margin has been focused on the outer shelf and upper slope. This has resulted in a substantial thickness of post-mid-Pleistocene sediment that has extended the shelf seaward ~25 km in the past 800 kyr.

Onshore drilling into marine sediments of the New Jersey Coastal Plain has documented at least 15 Oligocene to middle Miocene sequences (Miller *et al.*, 1997b). Most begin with shelly, glauconitic, transgressive quartz sands deposited in 30–100 m water depths (Sugarman *et al.*, 1993). These are frequently overlain by regressive, prodelta silts that coarsen upwards to the quartz sands of a nearshore or delta-front setting, followed by an apparent hiatus and then by transgressive sands of the next sequence. Shell fragments provide $^{87}Sr/^{86}Sr$ ages which demonstrate that gaps between sequences match the times of global ice increases independently inferred from the marine $\delta^{18}O$ record, thereby providing strong support for the primary role of glacio-eustasy in the arrangement of sequences on this and possibly other passive continental margins. Furthermore, seismic-sequence boundaries defined on the basis of toplap/onlap/downlap have been traced across the middle to outer shelf and tied to Oligocene–Miocene samples from the continental slope. These boundaries also correlate with the global $\delta^{18}O$ increases. In the future, in order to validate any causal link between eustasy, seismic-sequence development, and facies distribution, samples from successions with as close to 100% recovery as possible must be collected from carefully selected locations within several sequences known to coincide with times of well-established ice-volume change. Only with this confluence of data can the seismic-sequence stratigraphic model linking stratal geometry, facies, water depth and age be properly evaluated. At present, studies from the New Jersey margin have well-documented facies successions in coastal-plain boreholes, but there are no accompanying seismic-reflection profiles to provide stratal geometry. Conversely, stratal geometries are revealed in offshore profiles, but complete samples have not yet been recovered from the crucial topset, foreset or bottomset strata that could prove or disprove the link between facies, stratal geometry and oscillations in global sea level.

Profiles across lower-to-middle Miocene clinoforms on the inner-to-middle New Jersey shelf demonstrate along-strike variability that underscores the importance of acquiring ground-truth samples (Monteverde *et al.*, 2000). Isopach maps of 10 sequences, each estimated to be ~1.5-Myr duration, show lobate accumulations in some intervals, and elongated margin-parallel patterns in others. These differences are most probably due to local variations in both basement response to loading and to rates of sediment supply, but cannot be substantiated without samples that evaluate compaction history and facies. Furthermore, incised valleys across the tops of clinoforms identify ancient coastal plains, but the location of shoreline positions as well as water depths at each of the clinoform rollovers cannot be determined without samples. In contrast to margins with substantially faster rates of thermal subsidence, transgressive ravinement on the New Jersey margin has removed most seismically detectable evidence of shoreface deposits.

The variety of lower Miocene clinoform geometries raises concern that stratigraphic evolution can vary along strike to such an extent that inferences based solely on geometry lead to different conclusions depending on location within a basin. A fundamental lesson to be learned from studies of both margins is that clinoforms do not necessarily build directly seaward during times of margin growth; along-strike variations in sediment supply and lithosphere response to cooling, loading and flexure can lead to local variations in

sequence development that demand a basin-wide analysis.

In direct contrast to variation along the inner New Jersey margin, clinoform development in middle-to-upper Miocene sediments of the middle-to-outer shelf was comparatively uniform. The latter deposits have been mapped using a reasonably dense (~10–15-km line spacing) grid of both academic and commercial seismic data of moderate resolution (10–20 m; Fulthorpe *et al.*, 1999). Along a margin-parallel distance of several tens of kilometres, the clinoform rollover positions for mid-to-late Miocene sequences show no lobate development. This uniformity is undeniable and is made more surprising by the occurrence of incised valleys cutting into the tops of at least six mid-to-late Miocene clinoforms. These are now 5–12 m deep and 50–400 m wide, and indicate that during times of sea-level lowstand fluvial sediment traversed a coastal plain to reach the New Jersey shoreline. Either the margin was fed uniformly by other river systems without one dominating any of the others, or the mid-to-late Miocene was a time of strongly energetic shelf processes that distributed sediment along the margin so effectively that any imprint of localized sediment build-up was removed.

Progradation continued to build the New Jersey margin 25 km seaward of its position in late Miocene time. Four mid-to-late Pleistocene sequences, defined by downlap and toplap stratal geometries with no evidence of steeply dipping shoreface deposits, dominate a seismic grid (55 km × 35 km) across the modern shelf break immediately southwest of Hudson Canyon. This region was chosen for study because the scarcity of slope canyons has made shelf-to-slope correlations possible, but for this same reason it may not be typical of passive margins in general. In contrast to all preceding clinoforms described from the New Jersey margin, these developed above the shelf break established by a Mesozoic reef. Furthermore, these four Pleistocene clinoforms are distinct from all those before them in one very significant way: their foresets coincided with the continental slope that existed at that time. Previously, all Oligocene and Miocene clinoforms developed above a gently dipping ramp inherited from Eocene time. Sediment that bypassed the clinoform rollover still accumulated on the gentle grade of the Eocene ramp, and subsequently built upward to lap onto the clinoform foresets. This was not possible once the clinoform foresets reached the true continental slope in the Pleistocene and encountered seaward gradients of 4° or more that continued to the continental rise at > 2200-m water depths. When an angle of repose was established that could retain sediment on these amalgamated clinoform foresets and continental slope, the absence of slope-cutting canyons assured that the Pleistocene upper slope was maintained as a continuously depositional setting. Consequently, a nearly uninterrupted mid-to-late Pleistocene record can now be found on the upper slope of the Hudson Apron.

All four outer-shelf Pleistocene sequences were drilled at ODP Site 1073 on the Hudson Apron in 639 m of water (Austin *et al.*, 1998), and 520 m of mid-to-late Pleistocene sediment was cored with 98% recovery. A total of 16 glacial and interglacial fluctuations of global ice volume were identified on the basis of changes in δO^{18}, yielding a geochronological proxy back to roughly 755 ka. The only time breaks that could be determined were at the very top of the section (seafloor sediments are 14.4 ka based on ^{14}C dating) and spanning Oxygen Isotope Stages 6 and 7; the latter occurred at the boundary between the shallowest two seismic sequences (McHugh & Olson, 2002). No clear hiatus could be determined from available data at the other Pleistocene sequence boundaries. Furthermore, no consistent, one-to-one correspondence was found between glacio-eustatic oscillations inferred from δO^{18} and any aspect of sequences or their boundaries. Numerous successions from mudstone to siltstone to medium sand (but with no detectable hiatus) were found within each of the four seismic sequences, suggesting oscillations of proximal/distal sediment sources without the development of stratal patterns commonly cited as indicators of sea-level change. It remains to be determined what process was responsible for the four seismic sequences; it does not appear to have been simply changes in global sea level. A likely candidate is abrupt and large changes in sediment supplied to the outer shelf. Loading and unloading effects of the Laurentide ice sheet could have amplified rates of erosion and river discharge to such an extent that sequences developed in roughly 100 kyr packages, overwhelming and encompassing any higher-order cyclicity caused by changes in the global ice budget. It remains to be seen if modelling of Pleistocene sedimentation that includes ice-sheet tectonism can duplicate the observations off New Jersey.

The long-term record – where next?

The STRATAFORM case studies of the long-term record in the Eel River Basin and New Jersey margin underscore the acute need for coring deep into continental-margin sediments. These much-needed samples would document the long-term behaviour of the Earth system in general, and would determine the evolution of these two continental margins in particular. The optimal sample locations would require three-dimensional seismic information.

Stratal discontinuities are a common feature in margin sediments and provide objective and valuable divisions of the geological record. While it is widely accepted that these discontinuities are caused by rapid changes in eustatic sea level, tectonism, sediment supply and climate, the scarcity of samples from modern shelves leaves poorly constrained the explanation of how these processes operate and interact on long time-scales. In particular, the water depth at which basin-wide unconformities develop is fundamental to understanding this complex system, and it can be established only by sampling. Morphometric, numerical and physical models of the long-term record help to identify the processes that most influence shallow-water records, but their power to isolate and explain processes is severely weakened by our limited access to long-term, shallow-water records. The Integrated Ocean Drilling Program (IODP) will continue to target the New Jersey margin in coming years. It remains to be seen if the large technological challenges to continuous core recovery can be surmounted. There is near certainty, however, that any amount of deep samples and logs from this effort will improve understanding of how the long-term stratigraphic record is built.

ACKNOWLEDGEMENTS

Most of the high-resolution reflection profiles used in this study were collected under grants N00014-95-1-0200 and N00014-96-1-0377; the vision and long-standing support of Dr Joseph Kravitz in the Office of Naval Research, in making this work possible, is gratefully acknowledged. Samples of the long-term record offshore New Jersey have been collected by the Ocean Drilling Program. Drilling onshore New Jersey has been supported by cooperative efforts between the ODP, the New Jersey Geological Survey, the US Geological Survey, and the Continental Dynamics Program, Earth Sciences Division of the National Science Foundation. High-resolution profiles in support of ODP drilling were collected under NSF grants OCE89-11810 and OCE97-26273 from the Marine Geology and Geophysics Program, Division of Ocean Sciences. Support for acquiring detailed seismic grids across potential offshore drill sites of the New Jersey margin was provided by Joint Oceanographic Institutions, Inc. The authors thank Michael Field, Roger Flood, James Gardner, James Syvitski, and above all Charles Nittrouer for their very helpful reviews.

REFERENCES

Aalto, K.R, McLaughlin, R.J., Carver, G.A., *et al.* (1995) Uplifted Neogene margin, southernmost Cascadia–Mendocino triple junction region, California. *Tectonics*, **14**(5), 1104–1116.

Agassiz, L. (1840) *Études sur les Glaciers*. Privately published, Neuchatel, 346 pp.

Allen, D., Bergt, D., Best, D., *et al.* (1989) Logging while drilling. *Oilfield Rev.*, **1**, 4–17.

Ashley, G.M., Wellner, R.W., Esker, D. and Sheridan, R.E. (1991) Clastic sequences developed during late Quaternary glacio-eustatic sea level fluctuations on a passive margin: example from the inner continental shelf near Barnegat Inlet, New Jersey. *Geol. Soc. Am. Bull.*, **103**, 1607–1621.

Atwater, T. (1970) Implications of plate tectonics for the Cenozoic tectonic evolution of Western North America. *Geol. Soc. Am. Bull.*, **81**, 3513–3536.

Austin, J.A., Fulthorpe, C.S., Mountain, G.S., Orange, D.L. and Field, M.E. (1996) Continental-margin seismic stratigraphy: Assessing the preservation potential of heterogeneous geological processes operating on continental shelves and slopes. *Oceanography*, **9**, 173–177.

Austin, J.A., Christie-Blick, N., Malone, M. and the Leg 174A Shipboard Party (1998) *Proceedings of the Ocean Drilling Program, Initial Reports*, Vol. 174A. Texas A & M University, Ocean Drilling Program, College Station, TX.

Bartek, L.R., Vail, P.R., Anderson, J.B., Emmet, P.A. and Wu, S. (1991) Effect of Cenozoic ice sheet fluctuations in Antarctica on the stratigraphic signature of the Neogene. *J. Geophys. Res.*, **96**, 6753–6778.

Berggren, W.A., Kent, D.V., Swisher, C.C. and Aubry, M.-P. (1995) A revised Cenozoic geochronology and chronostratigraphy. In: *Geochronology, Time Scales and Global Stratigraphic Correlation* (Eds W.A. Berggren, D.V. Kent, M.-P. Aubry and J. Hardenbol), pp. 129–212. Special Publication 54, Society of Economic Paleontologists and Mineralogists, Tulsa, OK.

Bonner, S., Clark, B., Holenka, J., *et al.* (1992) Logging while drilling – a three-year perspective. *Oilfield Rev.*, **4**, 4–21.

Borgeld, J.C. (1985) *Holocene stratigraphy and sedimentation on the northern California continental shelf*. Unpublished PhD dissertation, University of Washington, Seattle, 177 pp.

Brown, W.M., III and Ritter, J.R. (1971) Sediment transport and turbidity in the Eel River Basin, California. *U.S. Geol. Surv. Wat. Supply Pap.*, **1986**, 70 pp.

Bruhn, C.H.L. and Walker, R.G. (1995) High-resolution stratigraphy and sedimentary evolution of coarse-grained canyon-filling turbidites from the Upper Cretaceous transgressive megasequence, Campos Basin, offshore Brazil. *J. Sediment. Res., Sect. B: Strat. Global Stud.*, **65**, 426–442.

Buck, K.F., Olson, H.C. and Austin, J.A., Jr. (1999) Paleoenvironmental evidence for latest Pleistocene sea-level fluctuations on the New Jersey outer continental shelf: combining high-resolution sequence stratigraphy and foraminiferal analysis. *Mar. Geol.*, **154**, 287–304.

Burger, R.L., Fulthorpe, C.S. and Austin, J.A., Jr. (2001) Late Pleistocene channel incisions in the southern Eel River Basin, northern California: implications for tectonic *vs.* eustatic influences on shelf sedimentation patterns. *Mar. Geol.*, **177**, 317–330.

Burger, R.L., Fulthorpe, C.S., Austin, J.A., Jr. and Gulick, S.P.S. (2002) Lower Pleistocene to Present structural deformation and sequence stratigraphy of the continental shelf, offshore Eel River Basin, northern California. *Mar. Geol.*, **185**(3–4), 249–281.

Carey, J.S., Sheridan, R.E. and Ashley, G.M. (1998) Late Quaternary sequence stratigraphy of a slowly subsiding passive margin: New Jersey Continental Shelf. *Bull. Am. Asssoc. Petrol. Geol.*, **82**, 773–791.

Carver, G.A. (1987) Late Cenozoic tectonics of the Eel River Basin region, coastal northern California. In: *Tectonics, Sedimentation and Evolution of the Eel River and other Coastal Basins of Northern California* (Eds H. Schymiczek and R. Suchland), pp. 61–72. Miscellaneous Publications 37, San Joaquin Geological Society.

Cathro, D.L., Austin, J.A., Jr. and Moss, G.D. (2003) Progradations along a deeply submerged Oligocene-Miocene heterozoan carbonate shelf: how sensitive are clinoforms to sea-level variations? *Bull. Am. Assoc. Petrol. Geol.*, **87**, 1547–1574.

Christie-Blick, N. (1991) Onlap, offlap and the origin of unconformity-bounded depositional sequences. *Mar. Geol.*, **97**, 35–56.

Christie-Blick, N. and Driscoll, N.W. (1995) Sequence stratigraphy. *Ann. Rev. Earth Planet. Sci.*, **23**, 451–478.

Clarke, S.H., Jr. (1987) Late Cenozoic Geology and Structure of the Onshore-Offshore Eel River Basin, Northern California. In: *Tectonics, Sedimentation and Evolution of the Eel River and Associated Coastal Basins of Northern California* (Eds H. Schymiczek and R. Suchsland), pp. 31–40. Miscellaneous Publication 37, San Joaquin Geological Society.

Clarke, S.H., Jr. (1992) Geology of the Eel River Basin and Adjacent Region: Implications for Late Cenozoic Tectonics of the Southern Cascadia Subduction Zone and Mendocino Triple Junction. *Bull. Am. Assoc. Petrol. Geol.*, **76**, 199–224.

Couch, R.W., Victor, L.P. and Keeling, K.M. (1974) *Coastal and Offshore Earthquakes of the Pacific Northwest between 39° and 49° Latitude and 123° and 131° Longitude*. School of Oceanography, Oregon State University Press, Corvallis, OR, 67 pp.

Crouch, J.K. and S.B. Bachman (1987) Exploration potential, offshore Point Arena and Eel River basins. In: *Tectonics, Sedimentation and Evolution of the Eel River and Associated Coastal Basins of Northern California* (Eds H. Schymiczek and R. Suchsland), pp. 99–111. Miscellaneous Publication 37, San Joaquin Geological Society.

Davies, T.A. and Austin, J.A., Jr. (1997) High-resolution 3D seismic reflection and coring techniques applied to late Quaternary deposits on the New Jersey shelf. *Mar. Geol.*, **143**, 137–149.

Davies, T.A., Austin, J.A., Lagoe, M.B. and Milliman, J.D. (1992) Late Quaternary sedimentation off New Jersey: new results using 3-D seismic profiles and cores. *Mar. Geol.*, **108**, 323–343.

De Verteuil, L. (1997) Miocene dinocyst stratigraphy of the Cape May and Atlantic City boreholes. In: *Proceedings of the Ocean Drilling Program, Scientific Results 150X* (Eds K.G. Miller and S.W. Snyder), pp. 129–146. College Station, TX.

Delius, H., Kaupp, A., Muller, A. and Wohlenberg, J. (2001) Stratigraphic correlation of Miocene to Plio-/Pleistocene sequences on the New Jersey shelf based on petrophysical measurements from ODP 174A. *Mar. Geol.*, **175**, 149–165.

DeMets, C., Gordon, R.G., Argus, D.F. and Stein, S. (1990) Current plate motions. *Geophys. J. Int.*, **101**, 425–478.

Dickinson, W.R., Ingersoll, R.V. and Graham, S.A. (1979) Paleogene sediment dispersal and paleotectonics in northern California. *Geol. Soc. Am. Bull.*, **90**, 145–1528.

Diebold, J.B., Stoffa, P.L. and the LASE study group (1988) A large aperture seismic experiment in the Baltimore Canyon Trough. In: *The Geology of North America, Vol. I-2, The Atlantic Continental Margin* (Eds R.E. Sheridan and J.A. Grow), pp. 387–398. Geological Society of America, Boulder, CO.

Doveton, J.H. (1986) *Log Analysis of Subsurface Geology: Concepts and Computer Methods*. Wiley, New York.

Duncan, C.S. and Goff, J.A. (2001) Relict iceberg keel marks on the New Jersey outer shelf, southern Hudson Apron. *Geology,* **29**, 5, 411–414.

Duncan, C.S., Goff, J.A., Austin, J.A., Jr. and Fulthorpe, C.S. (2000) Tracking the last sea-level cycle: seafloor morphology and shallow stratigraphy of the latest Quaternary New Jersey middle continental shelf. *Mar. Geol.*, **170**, 395–421.

Ellis, D.V. (1987) *Well Logging for Earth Scientists*. Elsevier, New York, 532 pp.

Fairbanks, R.G. (1989) A 17,000-year glacio-eustatic sea level record; influence of glacial melting rates on the Younger Dryas event and deep-ocean circulation. *Nature*, **342**, 637–642.

Farre, J.A., McGregor, B.A., Ryan, W.B.F. and Robb, J.M. (1983) Breaching the shelfbreak; passage from youthful to mature phase in submarine canyon evolution. In: *The Shelfbreak; Critical Interface on Continental Margins* (Eds D.J. Stanley and G.T. Moore), pp. 25–39. Special Publication 33, Society of Economic Paleontologists and Mineralogists, Tulsa, OK.

Field, M.E. and Barber, J.H. (1993) A submarine landslide associated with shallow seafloor gas and gas hydrates off northern California. In: *Submarine Landslides: Selected Studies in the U.S. Exclusive Economic Zone* (Eds W.C. Schwab, H.J. Lee and D.C. Twichell). *U.S. Geol. Surv. Bull.*, **2002**, 151–157.

Field, M.E., Gardner, J.V. and Prior, D.B. (1999) Geometry and significance of stacked gullies on the northern California slope. *Mar. Geol.*, **154**, 271–286.

Flower, B.P. and Kennett, J.P. (1995) The middle Miocene climatic transition: East Antarctic ice sheet development, deep ocean circulation and global carbon cycling. *Palaeogeogr. Paleoclimatol. Palaeoecol.*, **108**, 537–555.

Fowler, S. and McKenzie, D.P. (1989) Gravity studies of the Rockall and Exmouth Plateau using SEASAT altimetry. *Basin Res.*, **2**, 27–34.

Fulthorpe, C.S. and Austin, J.A., Jr. (1998) Anatomy of rapid margin progradation: three-dimensional geometries of Miocene clinoforms, New Jersey margin. *Bull. Am. Assoc. Petrol. Geol.*, **82**, 251–273.

Fulthorpe, C.S. and Austin, J.A., Jr. (2004) Shallowly buried, enigmatic seismic stratigraphy on the New Jersey outer shelf: evidence for latest Pleistocene catastrophic erosion? *Geology*, **32**(1), 1013–1016.

Fulthorpe, C.S., Austin, J.A., Jr. and Mountain, G.S. (1999) Buried fluvial channels off New Jersey: did sea-level lowstands expose the entire shelf during the Miocene? *Geology*, **27**(3), 203–206.

Fulthorpe, C.S., Austin, J.A., Jr. and Mountain, G.S. (2000) Morphology and distribution of Miocene slope incisions off New Jersey: are they diagnostic of sequence boundaries? *Geol. Soc. Am. Bull.*, **112**, 817–828.

Galloway, W.E. (1989) Genetic stratigraphic sequences in basin analysis I: Architecture and genesis of flooding-surface bounded depositional units. *Bull. Am. Assoc. Petrol. Geol.*, **73**, 125–142.

Goff, J.A., Swift, D.J.P., Duncan, C.S., Mayer, L.A. and Hughes-Clarke, J. (1999) High-resolution swath sonar investigation of sand ridge, dune and ribbon morphology in the offshore environment of the New Jersey margin. *Mar. Geol.*, **161**, 307–337.

Goff, J.A., Olson, H.C. and Duncan, C.S. (2000) Correlation of side-scan backscatter intensity with grain-size distribution of shelf sediments, New Jersey margin. *Geo-Mar. Lett.*, **20**, 43–49.

Goff, J.A., Kraft, B.J., Mayer, L.A., *et al.* (2004) Seabed characterization on the new Jersey middle and outer shelf: correlatability and spatial variability of sea-floor sediment properties. *Mar. Geol.*, **209**, 147–172.

Goff, J.A., Austin, J.A., Jr., Gulick, S., *et al.* (2005) Recent and modern erosion on the New Jersey outer shelf. *Mar. Geol.*, **216**, 275–296.

Goldberg, D. (1997) The role of downhole measurements in marine geology and geophysics. *Rev. Geophys.*, **35**(3), 315–342.

Goodwin, R.H. and Prior, D.B. (1989) Geometry and depositional sequences of the Mississippi Canyon, Gulf of Mexico. *J. Sediment. Petrol.*, **59**, 318–329.

Gornitz, V., Couch, S. and Hartig, E.K. (2001) Impacts of sea level rise in the New York metropolitan area. *Global Planet. Change*, **32**, 61–88.

Grabau, A.W. (1940) *The Rhythm of the Ages.* Henri Vetch, Peking, 561 pp.

Greenlee, S.M. and Moore, T.C. (1988) Recognition and interpretation of depositional sequences and calculations of sea-level changes from stratigraphic data – offshore New Jersey and Alabama Tertiary. In: *Sea-level Changes; an Integrated Approach* (Eds C.K. Wilgus, B.S. Hastings, C.A. Ross, *et al.*), pp. 329–353. Special Publication 42, Society of Economic Paleontologists and Mineralogists, Tulsa, OK.

Greenlee, S.M., Schroeder, F.W. and Vail, P.R. (1988) Seismic stratigraphic and geohistory analysis of Tertiary strata from the continental shelf off New Jersey

– calculation of eustatic fluctuations from stratigraphic data. In: *The Geology of North America,* Vol. I-2, *The Atlantic Continental Margin* (Eds R.E. Sheridan and J.A. Grow), pp. 437–444. Geological Society of America, Boulder, CO.

Greenlee, S.M., Devlin, W.J., Miller, K.G., Mountain, G.S. and Flemings, P.B. (1992) Integrated sequence stratigraphy of Neogene deposits, New Jersey continental shelf and slope: comparison with the Exxon model. *Geol. Soc. Am. Bull.*, **104**, 1403–1411.

Grow, J.A. and Sheridan, R.E. (1988) U.S. Atlantic continental margin; a typical Atlantic-type or passive continental margin. In: *The Geology of North America,* Vol. I-2, *The Atlantic Continental Margin* (Eds R.E. Sheridan and J.A. Grow), pp. 1–8. Geological Society of America, Boulder, CO.

Grow, J.A., Klitgord, K.D. and Schlee, J.S. (1988) Structure and evolution of Baltimore Canyon Trough. In: *The Geology of North America,* Vol. I-2, *The Atlantic Continental Margin* (Eds R.E. Sheridan and J.A. Grow), pp. 269–290. Geological Society of America, Boulder, CO.

Gulick, S.P. and Meltzer, A.M. (2002) Effect of the northward migrating Mendocino Triple Junction on the Eel River forearc basin, California: structural evolution. *Geol. Soc. Am. Bull.*, **114**(12), 1505–1519.

Gulick, S.P., Meltzer, A.M. and Clarke, S.H., Jr. (2002) Effect of the northward migrating Mendocino Triple Junction on the Eel River forearc basin, California: stratigraphic development. *Geol. Soc. Am. Bull.*, **114**(2), 178–191.

Gulick, S.P.S., Goff, J.A., Austin, J.A., Jr., *et al.* (2005) Basal inflection-controlled shelf-edge wedges off New Jersey track sea-level fall. *Geology*, **33**, 5, 429–432.

Hagen, R.A. and P.R. Vogt (1999) Seasonal variability of shallow biogenic gas in Chesapeake Bay. *Mar. Geol.*, **158**, 75–88.

Haller, C.R. (1980) Pliocene biostratigraphy of California. *Am. Assoc. Petrol. Geol. Stud. Geol.*, **11**, 183–341.

Haq, B.U., Hardenbol, J. and Vail, P.R. (1987) Chronology of fluctuating sea levels since the Triassic (250 million years ago to Present). *Science*, **235**, 1156–1167.

Harrison, C.G.A. (1990) Long-term eustasy and epeirogeny in continents. In: *National Research Council Studies in Geophysics: Sea-level Change*, pp. 141–158. National Academy of Sciences, Washington, DC.

Hathaway, J.C., Schlee, J., Poag, C.W., *et al.* (1976) Preliminary study of the 1976 Atlantic Margin Coring Project of the U.S. Geological Survey. *U.S. Geol. Surv. Open File Rep.*, **76–844**, 217 pp.

Hathaway, J.C., Poag, C.W., Valentine, P.C., *et al.* (1979) U.S. Geological Survey Core Drilling on the Atlantic Shelf. *Science*, **206**, 515–527.

Hays, J.D. and Pitman III, W.C. (1973) Lithospheric plate motion, sea-level changes and climatic and ecological consequences. *Nature*, **246**, 18–22.

Heller, P.L., Paola, C., Hwang, I.-G., John, B. and Steel, R. (2001) Geomorphology and sequence stratigraphy due to slow and rapid base-level changes in an experimental subsiding basin (XES 96-1). *Bull. Am. Assoc. Petrol. Geol.*, **85**, 817–838.

Holbrook, W.S. and Kelemen, P.B. (1993) Large igneous province on the U.S. Atlantic margin and implications for magmatism during continental breakup. *Nature*, **364**, 433–436

Hoskins, E.G. and Griffiths, J.R. (1971) Hydrocarbon potential of northern and central California offshore. In: *Future Petroleum Provinces of the United States, their Geology and Potential* (Ed. K.H. Kram), pp. 212–228. Memoir 15, American Association of Petroleum Geologists, Tulsa, OK.

Hovland, M. and Judd, A. (1988) *Seabed Pockmarks and Seepages: Impact on Geology, Biology and the Marine Environment*. Graham and Trotman, London, 293 pp.

Imbrie, J., Hays, J.D., Martinson, D.G., *et al.* (1984) The orbital theory of Pleistocene climate: Support from a revised chronology of the marine $\delta^{18}O$ record. In: *Milankovitch and Climate* (Eds A.L. Berger, J. Imbrie, J. Hays, G. Kukla and B. Saltzman), pp. 269–305. NATO ASI Series, Series C, Mathematical and Physical Sciences, 126, Part 1. D. Reidel, Dordrecht.

Jamieson, T.E. (1865) On the history of the last geological changes in Scotland. *Quat. J. Geol. Soc. London*, **21**, 161–203.

Jansa, L.F. (1981) Mesozoic carbonate platforms and banks of the eastern North American margin. *Mar. Geol.*, **44**, 97–117.

Karner, G.D. (1991) Sediment blanketing and the flexural strength of extended continental lithosphere. *Basin Res.*, **3**, 177–185.

Karner, G.D. and N.W. Driscoll (1997) Three-dimensional interplay of advective and diffusive processes in the generation of sequence boundaries. *J. Geol. Soc. London*, **154**, 443–449.

Keen, C.E. and Beaumont, C. (1990) Geodynamics of rifted continental margins. In: *Geology of the Continental Margin of Eastern Canada* (Eds M.J. Keen and G.L. Williams), pp. 391–472. Decade of North America Geology, Vol. I-1. Geological Survey of Canada and Geological Society of America.

Kelling, G. and Stanley, D.J. (1970) Morphology and structure of Wilmington and Baltimore submarine canyons, Eastern United States. *J. Geol.*, **78**, 637–660.

Kidwell, S.M. (1984) Outcrop features and origin of basin margin unconformities in the lower Chesapeake Group (Miocene), Atlantic coastal plain. *Am. Assoc. Petr. Geol. Mem.*, **37**, 37–57.

Kidwell, S.M. (1988) Reciprocal sedimentation and non-correlative hiatuses in marine-paralic siliciclastics: Miocene outcrop evidence. *Geology*, **16**, 609–612.

Klitgord, K.D., Poag, C.W., Schneider, C.M. and North, L. (1994) Geophysical database of the east coast of the United States Northern Atlantic margin: cross sections and gridded database (Georges Bank basin, Long Island platform and Baltimore Canyon trough). *U.S. Geol. Surv. Open-File Rep.*, **94–637**, 189 pp.

Knebel, H.J. and Spiker, E. (1977) Thickness and age of surficial sand sheet, Baltimore Canyon trough area. *Bull. Am. Assoc. Petrol. Geol.*, **61**(6), 861–871.

Knebel, H.J., Wood, S.A. and Spiker, E.C. (1979) Hudson River: evidence for extensive migration on the exposed continental shelf during Pleistocene time. *Geology*, **7**, 254–258.

Kominz, M.A. (1984) Oceanic ridge volumes and sea-level change – an error analysis. In: *Interregional Unconformities and Hydrocarbon Accumulation* (Ed. J.S. Schlee), pp. 37–58. Memoir 36, American Association of Petroleum Geologists, Tulsa, OK.

Kominz, M.A. and Pekar, S.F. (2001) Oligocene eustasy from two-dimensional sequence stratigraphic backstripping. *Geol. Soc. Am. Bull.*, **113**, 291–304.

Kominz, M.A., Miller, K.G. and Browning, J.V. (1998) Long-term and short-term global Cenozoic sea-level estimates. *Geology*, **26**, 311–314.

Kooi, H., S. Cloetingh and J. Burrus (1992) Lithospheric necking and regional isostasy at extensional basins, 1, subsidence and gravity modeling with an application to the Gulf of Lions margin (SE France). *J. Geophys. Res.*, **97**, 17,553–17,572.

Kulm, L.D. and Suess, E. (1990) Relationship of carbonate deposits and fluid venting: Oregon accretionary prism. *J. Geophys. Res.*, **95**, 8899–8916.

Kvenvolden, K.A. and Field, M.E. (1981) Thermogenic hydrocarbons in unconsolidated sediments of the Eel River Basin, offshore northern California. *Bull. Am. Assoc. Petrol. Geol.*, **65**, 1642–1646.

Kvenvolden, K.A., G.D. Ginsburg and V.A. Soloviev (1993) Worldwide distribution of subaquatic gas hydrates. *Geo-Mar. Lett.*, **13**, 32–40.

Lagoe, M.B., Davies, T.A., Austin, J.A., Jr. and Olson, H.C. (1997) Foraminiferal constraints on very high-resolution seismic stratigraphy and late Quaternary glacial history, New Jersey continental shelf. *Palaios*, **12**, 249–266.

Lajoie, K.R., Sarna-Wojcicki, A.M. and Ota, Yuko (1982) Emergent Holocene marine terraces at Ventura and Cape Mendocino, California-indicators of high tectonic uplift rates (abstract). *Geol. Soc. Am. Abstr. Progr.*, **14**(4), 178.

Lavier, L. and Steckler, M.S. (1997) Flexural strength of the continental lithosphere: Sediment cover as the last piece of the puzzle. *Nature*, **389**, 476–479.

Lee, H.J., Locat, J., Dartnell, P., Israel, K. and Wong, F. (1999) Regional variability of slope stability: application to the Eel margin, California. *Mar. Geol.*, **154**, 305–321.

Leithold, E.L. and Blair, N.E. (2001) Watershed control on the carbon loading of marine sedimentary particles. *Geochim. Cosmochim. Acta*, **65**, 2231–2240.

Libby-French, J. (1984) Stratigraphic framework and petroleum potential of northeastern Baltimore Canyon Trough, mid-Atlantic outer continental shelf. *Bull. Am. Assoc. Petrol. Geol.*, **68**(1), 50–73.

Lorenson, T.D., McLaughlin, R.J., Kvenvolden, K.A., *et al.* (1998) Comparison of offshore and onshore gas occurrences, Eel River basin, Northern California. *U.S. Geol. Surv. Open-file Rep.*, **98**, 781.

Loutit, T.S., Hardenbol, J., Vail, P.R. and Baum, G.R. (1988) Condensed section: the key to age determination and correlation of continental margin sequences. In: *Sea Level Changes: an Integrated Approach* (Eds C.K. Wilgus, B.S. Hastings, C.G. St.C. Kendall, *et al.*), pp. 183–213, Special Publication 42. Society of Economic Paleontologists and Mineralogists, Tulsa, OK.

Lowrie, W. and Kent, D.V. (2004) Geomagnetic polarity timescale and reversal frequency regimes. In: *Timescales of the Internal Geomagnetic Field* (Eds J.E.T. Channell, D.V. Kent, W. Lowrie and J. Meert), pp. 287–298. Geophysical Monograph 145, American Geophysical Union, Washington, DC.

MacLaren, C. (1842) The glacial theory of Professor Agassiz. *Am. J. Sci.*, **42**, 346–365.

Mahradi, M. (1983) *Physical modeling studies of thin beds.* Unpublished MSc. thesis, University of Houston, 100 pp.

Martin, B.D. and Emery, K.O. (1967) Geology of Monterey Canyon, California. *Bull. Am. Assoc. Petrol. Geol.*, **51**, 2281–2304.

May, J.A., Warme, J.E. and Slater, R.A. (1983) Role of submarine canyons on shelfbreak erosion and sedimentation; modern and ancient examples. In: *The Shelfbreak; Critical Interface on Continental Margins* (Eds D.J. Stanley and G.T. Moore), pp. 315–332. Special Publication 33, Society of Economic Paleontologists and Mineralogists, Tulsa, OK.

McCrory, P.A. (1989) Late Neogene geohistory analysis of the Humboldt Basin and its relationship to convergence of the Juan de Fuca Plate. *J. Geophys. Res.*, **94**(B3), 3126–3138.

McCrory, P.A. (1995) Evolution of a trench-slope basin within the Cascadia subduction margin: the Neogene Humboldt Basin, California. *Sedimentology*, **43**, 223–656.

McCrory, P.A. (1996) Evaluation of fault hazards, northern coastal California. *U.S. Geol. Surv. Open File Rep.*, **96–656**, 87 pp.

McCrory, P.A. (2000) Upper plate contraction north of the migrating Mendocino Triple Junction, northern California: Implications for partitioning of strain. *Tectonics*, **19**(6), 1144–1160.

McGregor, B.A. (1981) Ancestral head of Wilmington Canyon. *Geology*, **9**, 254–257.

McHugh, C.M.G. and Olson, H.C. (2002) Pleistocene chronology of continental margin sedimentation: New insights into traditional models, New Jersey. *Mar. Geol.*, **185**, 389–411.

McLaughlin, R.J., Sliter, W.V., Frederiksen, N.O., Harbert, W.P. and McCulloch, D.S. (1994) Plate motions recorded in tectonostratigraphic terranes of the Franciscan Complex and evolution of the Mendocino Triple Junction, northwestern California. *U.S. Geol. Surv. Bull.*, **1997**, 60 pp.

Mederios, B.P., Karner, D.B., Muller, R.A. and Levine, J. (2000) The global ice volume record as viewed through a benthic $\delta^{18}O$ stack. *Eos (Trans. Am. Geophys. Union)* (suppl.), **81**(48), F597.

Miall, A.D. (1991) Stratigraphic sequences and their chronostratigraphic correlation, *J. Sediment. Petrol*, 61, 497–505.

Miller, K.G. and Mountain, G.S. (1994) Global sea-level change and the New Jersey margin. In: *Proceedings of the Ocean Drilling Program, Initial Reports*, Vol. 150 (G.S. Mountain, K.G. Miller, P. Blum, *et al.*), pp. 11–19. College Station, TX.

Miller, K.G., Fairbanks, R.G. and Mountain, G.S. (1987a) Tertiary oxygen isotope synthesis, sea-level history and continental margin erosion. *Paleoceanography*, **2**, 1–19.

Miller, K.G., Melillo, A.J., Mountain, G.S., Farre, J.A. and Poag, C.W. (1987b) Middle to late Miocene canyon cutting on the New Jersey continental slope; biostratigraphic and seismic stratigraphic evidence. *Geology*, **15**, 509–512.

Miller, K.G., Browning, J.V., Liu, C., *et al.* (1994) Atlantic City site report. In: *Proceedings of the Ocean Drilling Program, Initial Reports*, Vol. 150X (K.G. Miller, *et al.*), pp. 35–55. College Station, TX.

Miller, K.G., Liu, C. and Feigenson, M.D. (1996a) Oligocene to middle Miocene Sr-isotopic stratigraphy of the New Jersey continental slope. In: *Proceedings of the Ocean Drilling Program, Scientific Results*, Vol. 150 (Eds G.S. Mountain, K.G. Miller, P. Blum, *et al.*) pp. 97–114. College Station, TX.

Miller, K.G., Mountain, G.S., the Leg 150 Shipboard Party and Members of the New Jersey Coastal Plain Drilling Project (1996b) Drilling and dating New Jersey Oligocene–Miocene sequences: ice volume, global sea level and Exxon records, *Science*, **271**, 1092–1094.

Miller, K., Rufolo, S., Sugarman, P., Pekar, S., Browning, J. and Gwynn, D. (1997a) Early to middle Miocene sequences, systems tracts and benthic foraminiferal biofacies. In: *Proceedings of the Ocean Drilling Program, Scientific Results*, Vol. 150X (Eds K.G. Miller and S.W. Snyder). College Station, TX.

Miller, K.G., Browning, J.V., Pekar, S.F. and Sugarman, P.J. (1997b) Cenozoic evolution of the New Jersey coastal plain: changes in sea level, tectonics and sediment supply. In: *Proceedings of the Ocean Drilling Program, Scientific Results*, Vol. 150X (Eds K.G. Miller and S.W. Snyder), pp. 361–373. College Station, TX.

Miller, K.G., G.S. Mountain, J.S. Browning, *et al.* (1998a) Cenozoic global sea level, sequences and the New Jersey transect: results from coastal plain and continental slope drilling. *Rev. Geophys.*, **36**(4), 569–601.

Miller, K.G., P.J. Sugarman, J.V. Browning, *et al.* (1998b) Bass River site. In: *Proceedings of the Ocean Drilling Program, Initial Reports*, Vol. 174AX (K.G. Miller, P.J. Sugarman, J.V. Browning, *et al.*), pp. 5–43. College Station, TX.

Miller, K.G., Sugarman, P.J., Browning, J.V., *et al.* (1999) Ancora Site. In: *Proceedings of the Ocean Drilling Program, Initial Reports*, Vol. 174AX (Suppl.) (K.G. Miller, P.J. Sugarman, J.V. Browning, *et al.*), pp. 1–65. College Station, TX.

Milliman, J.D., Zhou, J., Li, A.C. and Ewing, J.I. (1990) Late Quaternary sedimentation on the outer and middle New Jersey continental shelf: result of two local deglaciations? *J. Geol.*, **98**, 966–976.

Mitchum, R.M., Vail, P.R. and Thompson, S., III (1977a) Seismic stratigraphy and global changes of sea level, Part 2: the depositional sequence as a basic unit for stratigraphic analysis. In: *Seismic Stratigraphy – Applications to Hydrocarbon Exploration* (Ed. C.E. Payton), pp. 53–62. Memoir 26, American Association of Petroleum Geologists, Tulsa, OK.

Mitchum, R.M., Vail, P.R. and Sangree, J.B. (1977b) Seismic stratigraphy and global changes of sea level, Part 6: stratigraphic interpretation of seismic reflection patterns in depositional sequences. In: *Seismic Stratigraphy – Applications to Hydrocarbon Exploration* (Ed. C.E. Payton), pp. 117–133. Memoir 26, American Association of Petroleum Geologists, Tulsa, OK.

Monteverde, D., Miller, K. and Mountain, G. (2000) Correlation of offshore seismic profiles with onshore New Jersey Miocene sediments. *Sediment. Geol.*, **134**, 111–127.

Morehead, M.D. and Syvitski, J.P. (1999) River-plume Sedimentation Modeling for Sequence Stratigraphy: Application to the Eel Margin, Northern California. *Mar. Geol.*, **154**, 29–42.

Mountain, G.S. (1987) Cenozoic margin construction and destruction offshore New Jersey. In: *Timing and Depositional History of Eustatic Sequences: Constraints on Seismic Stratigraphy* (Eds C. Ross and D. Haman), pp. 57–83. Special Publication 24, Cushman Foundation for Foraminiferal Research, Ithaca, NY.

Mountain, G.S., Miller, K.G., Blum, P. and the Leg 150 Shipboard Party (1994) *Proc. Ocean Drill. Program, Init. Rep.*, **150**, 885 pp.

Mountain, G.S., Damuth, J.E., McHugh, C.M.G., Lorenzo, J.M. and Fulthorpe, C.S. (1996) Origin, reburial and significance of a middle Miocene canyon, New Jersey continental slope. In: *Proceedings of the Ocean Drilling Program, Scientific Results*, Vol. 150 (Eds G.S. Mountain, K.G. Miller, P. Blum, *et al.*), pp. 283–292. College Station, TX.

Mountain, G., McHugh, C., Olson, H. and Monteverde, D. (2001) Glacioeustasy ain't what it used to be: the search for controls on stratigraphic architecture. *Eos (Trans. Am. Geophys. Union)*, **82**(47), F749.

Murphy, D.P. (1993) What's new in MWD and formation evaluation. *World Oil*, **214**, 47–52.

Neidell, N.S. and Poggiagliolmi, E. (1977) Stratigraphic modeling and interpretation – geophysical principles and techniques. In: *Seismic Stratigraphy – Applications to Hydrocarbon Exploration* (Ed. C.E. Payton), pp. 417–438. Memoir 26, American Association of Petroleum Geologists, Tulsa, OK.

Niedoroda, A.W., Reed, C.W. and Swift, D.J.P., Arato, A. and Hoyanagi, K. (1995) Modeling shore-normal large-scale coastal evolution. *Mar. Geol.*, **126**, 180–200.

Nilsen, T.H. and Clarke, S.H., Jr. (1987) Geological evolution of the late Cenozoic basins of northern California. In: *Tectonics, Sedimentation and Evolution of the Eel River and other Coastal Basins of Northern California* (Eds H. Schymiczek and R. Suchland), pp. 15–29. Miscellaneous Publications 37, San Joaquin Geological Society.

Nordfjord, S., Goff, J.A., Austin, J.A., Jr. and Sommerfield, C.K. (2005) Seismic geomorphology of buried channel systems on the New Jersey outer shelf: Assessing past environmental conditions. *Mar. Geol.*, **214**, 339–364.

Nystuen, J.P. (1998) History and development of sequence stratigraphy. In: *Sequence Stratigraphy – Concepts and Applications* (Eds F.M. Gradstein, K.O. Sundvik and N.J. Milton), pp. 31–116. Special Publication 8, Norwegian Petroleum Society, Oslo.

Olsson, R.K. and Wise, S.W. (1987) Upper Paleocene to middle Eocene depositional sequences and hiatuses in the New Jersey Atlantic Margin. In: *Timing and Depositional History of Eustatic Sequences: Constraints on Seismic Stratigraphy* (Eds C. Ross and D. Haman), pp. 99–112. Special Publication 24, Cushman Foundation for Foraminiferal Research, Ithaca, NY.

Orange, D.L. (1999) Tectonics, sedimentation and erosion in northern California: submarine geomorphology and sediment preservation potential as a result of three competing processes. *Mar. Geol.*, **154**, 369–382.

Orange, D.L., Angell, M.M. and Lapp, D. (1999) Using seafloor mapping (bathymetry and backscatter) and high resolution sub-bottom profiling for both exploration and production: detecting seeps, mapping geohazards and managing data overload with GIS. *Proceedings, 30th Offshore Technical Conference*, Houston, TX.

Owens, J.P. and Gohn, G.S. (1985) Depositional history of the Cretaceous series in the U.S. coastal plain: stratigraphy, paleoenvironments and tectonic controls of sedimentation. In: *Geologic Evolution of the United States Atlantic Margin* (Ed. C.W. Poag), pp. 25–86. Van Nostrand Reinhold, New York.

Owens, J.P. and Sohl, N. (1969) Shelf and deltaic paleoenvironments in the Cretaceous–Tertiary formations of the New Jersey Coastal Plain. In: *Geology of Selected Areas in New Jersey and Eastern Pennsylvania and Guidebook of Excursions* (Ed. S. Subitsky), pp. 235–278. Rutgers University Press, New Brunswick, NJ.

Paillet, F.L., Cheng, C.H. and Pennington, W.D. (1992) Acoustic waveform logging – advances in theory and application. *Log Anal.*, **33**, 239–258.

Parker, G., Fukushima, Y. and Pantin, H.M. (1986) Self-accelerating turbidity currents. *J. Fluid Mech.*, **171**, 145–181.

Payton, C.E. (Ed.) (1977) *Seismic Stratigraphy – Applications to Hydrocarbon Exploration*. Memoir 26, American Association of Petroleum Geologists, Tulsa, OK, 516 pp.

Poag, C.W. (1977) Foraminiferal biostratigraphy. In: *Geological Studies on the COST No. B-2 well, U.S. Mid-Atlantic Outer Continental Shelf Area* (Ed. P.A. Scholle). *U.S. Geol. Circ.*, **750**, 35–36.

Poag, C.W. (1978) Stratigraphy of the Atlantic continental shelf and slope of the United States. *Ann. Rev. Earth Planet. Sci.*, **6**, 251–280.

Poag, C.W. (1980) Foraminiferal stratigraphy, paleoenvironments and depositional cycles in the outer Baltimore Canyon Trough. In: *Geological Studies on the COST No. B-3 well, U.S. Mid-Atlantic Outer Continental Shelf Area* (Ed. P.A. Scholle). *U.S. Geol. Circ.*, **833**, 44–65.

Poag, C.W. (1985) Depositional history and stratigraphic reference section for central Baltimore Canyon trough. In: *Geologic Evolution of the United States Atlantic Margin* (Ed. C.W. Poag), pp. 217–263. Van Nostrand Reinhold, New York.

Poag, C.W. (1987) The New Jersey Transect: stratigraphic framework and depositional history of a sediment-rich passive margin. In: *Initial Reports of the Deep Sea Drilling Project*, Vol. 95 (Eds C.W. Poag, A.B. Watts and the Leg 95 Shipboard Party), pp. 763–817. US Government. Printing Office, Washington, DC.

Poag, C.W. and Sevon, W.D. (1989) A record of Appalachian denudation in postrift Mesozoic and Cenozoic sedimentary deposits of the U.S. middle Atlantic continental margin. *Geomorphology*, **2**, 119–157.

Poag, C.W., Watts, A.B. and the Leg 95 Shipboard Party. (1987) *Initial Reports of the Deep Sea Drilling Project*, Vol. 95. US Government. Printing Office, Washington, DC, 817 pp.

Poag, C.W. and Ward, L.W. (1993) Allostratigraphy of the U.S. middle Atlantic continental margin – characteristics, distribution and depositional history of principal unconformity-bounded Upper Cretaceous and Cenozoic sedimentary units. *U.S. Geol. Surv. Prof. Pap.*, **1542**.

Posamentier, H.W. and Allen, G.P. (1993) Variability of the sequence stratigraphic model: effects of local basin factors, *Sediment. Geol.*, **86**, 91–109.

Posamentier, H.W. and Allen, G.P. (1999) *Siliciclastic Sequence Stratigraphy – Concepts and Applications*. Concepts in Sedimentology and Paleontology No. 7, Society of Sedimentary Geology, Tulsa, OK, 210 pp.

Posamentier, H.W. and Vail, P.R. (1988) Eustatic controls on clastic deposition II – sequence and systems tract models. In: *Sea Level Changes: an Integrated Approach* (Eds C.K. Wilgus, B.S. Hastings, C.G. St.C. Kendall, *et al.*), pp. 125–154. Special Publication 42, Society of Economic Paleontologists and Mineralogists, Tulsa, OK.

Posamentier, H.W., Jervey, M.T. and Vail, P.R. (1988) Eustatic controls on clastic deposition I – conceptual framework. In: *Sea Level Changes: an Integrated Approach* (Eds C.K. Wilgus, B.S. Hastings, C.G. St.C. Kendall, *et al.*), pp. 109–124. Special Publication 42, Society of Economic Paleontologists and Mineralogists, Tulsa, OK.

Prather, B.E. (1991) Petroleum geology of the Upper Jurassic and Lower Cretaceous, Baltimore Canyon Trough, western North Atlantic Ocean. *Bull. Am. Assoc. Petrol. Geol.*, **75**(2), 258–277.

Pratson, L.F., Ryan, W.B.F., Mountain, G.S. and Twichell, D.C. (1994) Submarine canyon initiation by downslope-eroding sediment flows; evidence in late Cenozoic strata on the New Jersey continental slope. *Geol. Soc. Am. Bull.*, **106**, 395–412.

Reynolds, D.J., M.S. Steckler and B.J. Coakley (1991) The role of the sediment load in sequence stratigraphy: the influence of flexural isostasy and compaction, *J. Geophys. Res.*, **96**, 6931–6949.

Ruch, P., Mirmand, M., Jouanneau, J.M. and Latouche, C. (1993) Sediment budget and transfer of suspended sediment from the Gironde Estuary to Cap Ferret Canyon. *Mar. Geol.*, **111**, 109–119.

Ruddiman, W.F., Raymo, M.E., Martinson, D.G., Clement, B.M. and Backman, J. (1989) Pleistocene evolution: northern hemisphere ice sheets and north Atlantic Ocean. *Paleoceanography*, **4**(4), 353–412.

Savin, S. (1977) The history of the Earth's surface temperature during the past 100 million years, *Ann. Rev. Earth Planet. Sci.*, **5**, 319–355.

Schlee, J.S. (1981) Seismic stratigraphy of the Baltimore Canyon Trough. *Bull. Am. Assoc. Petrol. Geol.*, **65**, 26–53.

Schwab, W.C., Denny, J.F., Foster, D.S., *et al.* (2002) High-Resolution Quaternary Seismic Stratigraphy of the New York Bight Continental Shelf. *U.S. Geol. Soc. Open-File Rep.*, **02-152**.

Sclater, J.G. and P.A. Christie (1980) Continental stretching: an explanation of the post Mid-Cretaceous subsidence of the Central North Sea basin. *J. Geophys. Res.*, **85**, 3711–3739.

Sheridan, R.E., Gradstein, F. and the Leg 76 Shipboard Party (1983) *Initial Reports of the Deep Sea Drilling Project*, Vol. 76. US Government. Printing Office, Washington, DC, 947 pp.

Sheridan, R.E., Grow, J.A. and Klitgord, K.C. (1988) Geophysical data. In: *The Geology of North America*, Vol. I-2, *The Atlantic Continental Margin* (Eds R.E. Sheridan and J.A. Grow), pp. 177–195. Geological Society of America, Boulder, CO.

Sheridan, R.E., Ashley, G.M., Miller, K.G., *et al.* (2000) Offshore-onshore correlation of upper Pleistocene strata, New Jersey coastal Plain to continental shelf and slope. *Sediment. Geol.*, **134**, 197–207.

Sheriff, R.E. (1977) Limitations on resolution of seismic reflections and geologic detail derivable from them. In: *Seismic Stratigraphy – Applications to Hydrocarbon Exploration* (Ed. C.E. Payton), pp. 3–14. Memoir 26, American Association of Petroleum Geologists, Tulsa, OK.

Sheriff, R.E. (1985) Aspects of seismic resolution. In: *Seismic Stratigraphy II – an Integrated Approach to Hydrocarbon Exploration* (Eds O.R. Berg and E.G. Woolverton), pp. 1–10. Memoir 39, American Association of Petroleum Geologists, Tulsa, OK.

Sloss, L.L. (1963) Sequences in the cratonic interior of North America, *Geol. Soc. Am. Bull.*, 74, 93–114.

Song, G.S., Ma, C.P. and Yu, H.S. (2000) Fault-controlled genesis of the Chilung sea valley (northern Taiwan) revealed by topographic lineaments. *Mar. Geol.*, **169**, 305–325.

Spinelli, G.A. and Field, M.E. (2001) Evolution of continental slope gullies on the northern California margin, *J. Sediment. Res.*, **71**(2), 237–245.

Stanford, S.D., Ashley, G.M. and Brenner, G.J. (2001) Late Cenozoic fluvial stratigraphy of the New Jersey Piedmont: a record of glacioeustasy, planation and incision on a low-relief passive margin. *J. Geol.*, 109, 265–276.

Stanley, D.J. and Kelling, G. (1978) *Sedimentation in Submarine Canyons, Fans and Trenches*. Dowden, Hutchinson and Ross, Stroudsburg, PA, 382 pp.

Steckler, M.S. and Watts, A.B. (1982) Subsidence history and tectonic evolution of Atlantic-type continental margins. In: *Dynamics of Passive Margins* (Ed. R.A. Scrutton), pp. 184–196. Geodynamic Series, No. 6, American Geophysical Union, Washington, DC.

Steckler, M.S., A.B. Watts and J.A. Thorne (1988) Subsidence and basin modeling at the U.S. Atlantic passive margin. In: *The Geology of North America*, Vol. I-2, *The Atlantic Continental Margin* (Eds R.E. Sheridan and J.A. Grow), pp. 399–416. Geological Society of America, Boulder, CO.

Steckler, M.S., D.J. Reynolds, B.J. Coakley, B.A. Swift and R.D. Jarrard (1993) Modeling passive margin sequence stratigraphy. In: *Sequence Stratigraphy and Facies Associations* (Eds H.W. Posamentier, C.P. Summerhayes, B.U. Haq and G.P. Allen), pp. 19–41. Special Publication 18, International Association of Sedimentologists. Blackwell Scientific Publications, Oxford.

Steckler, M.S., G.S. Mountain, K.G. Miller and N. Christie-Blick (1999) Reconstruction of Tertiary progadation and clinoform development on the New Jersey passive margin by 2-D backstripping. *Mar. Geol.*, **154**, 399–420.

Stille, H. (1924) *Grundfragen der vergleichenden Tektonik*. Borntraeger, Berlin, 443 pp.

Sugarman, P.J. and Miller, K.G. (1997) Correlation of Miocene sequences and hydrogeologic units, New Jersey Coastal Plain, *Sediment. Geol.*, **108**, 3–18.

Sugarman, P.J., Miller, K.G., Owens, J.P. and Feigenson, M.D. (1993) Strontium isotope and sequence stratigraphy of the Miocene Kirkwood Formation, Southern New Jersey, *Geol. Soc. Am. Bull.*, **105**, 423–436.

Sugarman, P., McCartan, L., Miller, K., *et al.* (1997) Strontium-isotopic correlation of Oligocene to Miocene sequences, New Jersey and Florida. In: *Proceedings of the Ocean Drilling Program, Scientific Results*, Vol. 150X (Eds K.G. Miller and S.W. Snyder), pp. 147–159. College Station, TX.

Swift, D.J.P., Moir, R. and Freeland, G.L. (1980) Quaternary rivers on the New Jersey shelf: relation of seafloor to buried valleys. *Geology*, **8**, 276–280.

Syvitski, J.P.M., Pratson, L. and O'Grady, D. (1999) Stratigraphic predictions of continentnal margins for the US Navy. In: *Numerical Experiments in Stratigraphy: Recent Advances in Stratigraphic and Sedimentologic Computer Simulations* (Eds J.W. Harbaugh, W.L. Watney, E.C. Rankey, *et al.*), pp. 219–236. Special Publication 62, Society of Economic Paleontologists and Mineralogists, Tulsa, OK.

Teller, J.T. (1987) Proglacial lakes and the southern margin of the Laurentide ice sheet. In: *North America and Adjacent Oceans during the Last Deglaciation* (Eds W.F. Ruddiman and H.E. Wright), pp. 39–69. Geological Society of America, Boulder, CO.

Tittman, J. (1991) Vertical resolution of well logs – recent developments. *Oilfield Rev.*, **3**, 24–28.

Tucker, P. and Yorston, H. (1973) *Pitfalls in Seismic Interpretation.* Monograph Series, No. 2, Geophysicists Society of Exploration, 50 pp.

Tushingham, A.M. and Peltier, W.R. (1991) ICE–3G: a new global model of late Pleistocene deglaciation based upon geophysical predictions of post glacial relative sea level change, *J. Geophys. Res.*, **96**, 4497–4523.

Twichell, D.C. and Roberts, D.G. (1982) Morphology, distribution and development of submarine canyons on the United States Atlantic continental slope between Hudson and Baltimore canyons. *Geology*, **10**, 408–412.

Twichell, D.C., Knebel, H.J. and Folger, D.W. (1977) Delaware River: evidence for its former extension to Wilmington Submarine Canyon. *Science*, **195**, 483–485.

Uchupi, E., Driscoll, N., Ballard, R.D. and Bolmer, S.T. (2001) Drainage of late Wisconsin glacial lakes and the morphology and late Quaternary stratigraphy of the New Jersey – southern New England continental shelf and slope. *Mar. Geol.*, **172**, 117–145.

Ulicny, D., Nichols, G. and Waltham, D. (2002) Role of initial depth at basin margins in sequence architecture: field examples and computer models. *Basin Res.*, **14**(3), 347–360.

Vail, P.R. (1987) Seismic stratigraphy interpretation using sequence stratigraphy. Part 1: seismic stratigraphy interpretation procedure. In: *Atlas of Seismic Stratigraphy*, Vol. 1 (Ed. A.W. Bally), pp. 1–10. Studies in Geology, No. 27, American Association of Petroleum Geologists, Tulsa, OK.

Vail, P.R., Mitchum, Jr. R.M. and Thompson, S. III, (1977) Seismic stratigraphy and global changes of sea level, Part 4: Global cycles of relative changes of sea level. In: *Seismic Stratigraphy – Applications to Hydrocarbon Exploration* (Ed. C.E. Payton), pp. 83–98. Memoir 26, American Association of Petroleum Geologists, Tulsa, OK.

Van Hinte, J.A., Wise, S.W. and the Leg 93 Shipboard Party (1987) *Initial Reports of the Deep Sea Drilling Project*, Vol. 93. US Government. Printing Office, Washington, DC, 1423 pp.

Van Waggoner, J.C., Posamentier, H.W., Mitchum, R.M., *et al.* (1988) An overview of the fundamentals of sequence stratigraphy and key definitions. In: *Sea-level Changes; an Integrated Approach* (Eds C.K. Wilgus, B.S. Hastings, C.A. Ross, *et al.*), pp. 39–45. Special Publication 42, Society of Economic Paleontologists and Mineralogists, Tulsa, OK.

Von der Borch, C.C., Grady, A.E., Aldam, R., *et al.* (1985) A large-scale meandering submarine canyon; outcrop example from the late Proterozoic Adelaide Geosyncline, South Australia. *Sedimentology*, **32**, 507–518.

Walther, J. (1894) *Einleitung in die Geologie als Historische Wissenschaft, Bd. 3, Lithogenesis der Gegenwart*, pp. 535–1055. G. Fischer Verlag, Jena.

Watts, A.B. (1988) Gravity anomalies, crustal structure and flexure of the lithosphere at the Baltimore Canyon Trough. *Earth Planet. Sci. Lett.*, **89**, 221–138.

Watts, A. and Steckler, M.S. (1979) Subsidence and eustasy at the continental margins of eastern North America. In: *The Continental Margins and Paleoenvironments: Deep Drilling Results in the Atlantic Ocean* (Eds W. Talwani and W. Ryan), pp. 273–310. Maurice Ewing Symposium 3, American Geophysical Union, Washington, DC.

Watts, A.B. and Thorne, J.A. (1984) Tectonics, global changes in sea-level and their relationship to stratigraphic sequences at the U.S. Atlantic continental margin. *Marine Petrol. Geol.*, **1**, 319–339.

Watts, A.B., Karner, G.D. and Steckler, M.S. (1982) Lithospheric flexure and the evolution of sedimentary basins. In: *The Evolution of Sedimentary Basins* (Eds Sr. P. Kent, M.H.P. Bott, D.P. McKenzie and C.A. Williams). *Philos. Trans. Roy. Soc. London*, **305A**, 249–281.

Weed, E.G.A., Minard, J.P., Perry, W.J., Rhodehamel, E.C. and Robbins, E.I. (1974) *Generalized pre-Pleistocene geologic map of the northern United States Atlantic continental margin*, Map I–861 (scale 1:1,000,000). Miscellaneous Geological Investigation Series, US Geological Survey.

Wheatcroft, R.A., Borgeld, J.C., Born, R.S., *et al.* (1996) The anatomy of an oceanic flood deposit. *Oceanography*, **3**(9), 158–162.

Widess, M.B. (1973) How thin is a thin bed? *Geophysics*, **38**(6), 1176–1180.

Wonham, J.P., Jayr, S., Mougamba, R. and Chuilon, P. (2000) 3D sedimentary evolution of a canyon fill (lower Miocene-age) from the Mandorove Formation, offshore Gabon. *Mar. Petrol. Geol.*, **17**, 175–197.

Wood, L.J., Ethridge, F.G. and Schumm, S.A. (1993) The effects of rate of base-level fluctuation on coastalplain, shelf and slope depositional systems; an experimental approach. In: *Sequence Stratigraphy and Facies Associations* (Eds H.W. Posamentier, C.P. Summerhayes, B.U. Haq and G.P. Allen), pp. 43–53. Special Publication 18, International Association of Sedimentologists. Blackwell Scientific Publications, Oxford.

Woodward-Clyde Consultants (1980) *Evaluation of the Potential for Resolving the Geologic and Seismic Issues at the Humboldt Bay Power Plant Unit No. 3.* Unpublished Final Report Prepared for Pacific Gas and Electric Company, 309 pp.

Yun, J.W., Orange, D.L. and Field, M.E. (1999) Subsurface gas offshore of northern California and its link to submarine geomorphology. *Mar. Geol.*, **154**, 357–368.

Zachos, J.C., Stott, L.D. and Lohmann, K.C. (1994) Evolution of early Cenozoic marine temperatures. *Paleoceanography*, **9**, 353–387.

Zachos, J.C., Pagani, M., Sloan, L., Thomas, E. and Billups, K. (2001) Trends, rythms and aberrations in global climate 65 Ma to Present. *Science*, **292**, 686.

Prediction of margin stratigraphy

JAMES P.M. SYVITSKI*, LINCOLN F. PRATSON†, PATRICIA L. WIBERG‡, MICHAEL S.
STECKLER§, MARCELO H. GARCÍA¶, W. ROCKWELL GEYER**, COURTNEY K. HARRIS††,
ERIC W.H. HUTTON*, JASIM IMRAN‡‡, HOMA J. LEE§§, MARK D. MOREHEAD¶¶
and GARY PARKER***

*Environmental Computation and Imaging Facility, INSTAAR, University of Colorado, Boulder, CO 80309-0450, USA
(Email: james.syvitski@Colorado.edu)
†Earth and Ocean Sciences, Duke University, Box 90230, Durham, NC 27708, USA
‡Department of Environmental Sciences, University of Virginia, Charlottesville, VA 22904, USA
§Lamont-Doherty Earth Observatory of Columbia University, Palisades, NY 10964, USA
¶Ven Te Chow Hydrosystems Laboratory, University of Illinois, Urbana, IL 61801, USA
**Department of Applied Ocean Physics and Engineering, WHOI, Woods Hole, MA 02543, USA
††Department of Physical Sciences, Virginia Institute of Marine Sciences, Gloucester Point, VA 23062, USA
‡‡Department of Civil and Environmental Engineering, University of South Carolina, Columbia, SC 29208, USA
§§US Geological Survey, Menlo Park, CA 94025, USA
¶¶Department of Civil Engineering, University of Idaho, Boise, ID 83702, USA
***Department of Civil and Environmental Engineering, University of Illinois, Urbana, IL 61801, USA

ABSTRACT

A new generation of predictive, process–response models provides insight about how sediment-transport processes work to form and destroy strata, and to influence the developing architecture along continental margins. The spectrum of models considered in this paper includes short-term sedimentary processes (river discharge, surface plumes, hyperpycnal plumes, wave-current inter-actions, subaqueous debris flows, turbidity currents), the filling of geological basins where tectonics and subsidence are important controls on sediment dispersal (slope stability, compaction, tectonics, sea-level fluctuations, subsidence), and acoustic models for comparison to seismic images. Recent efforts have coordinated individual modelling studies and catalysed Earth-surface research by:

1 empowering scientists with computing tools and knowledge from interlinked fields;
2 streamlining the process of hypothesis testing through linked surface dynamics models;
3 creating models tailored to specific settings, scientific problems and time-scales.

The extreme ranges of space- and time-scales that define Earth history demand an array of approaches, including model nesting, rather than a single monolithic modelling structure. Numerical models that simulate the development of landscapes and sedimentary architecture are the repositories of our understanding about basic physics and thermodynamics underlying the field of sedimentology.

Keywords Numerical modelling, compaction, acoustic properties, river discharge, gravity flows, clinoforms, river plumes, sequence stratigraphy.

INTRODUCTION

This paper highlights a new generation of predictive process–response models that have been developed by the STRATAFORM programme. These simulators provide insights on how sediment-transport processes work to form and destroy strata, and influence the developing architecture along the world's continental margins. The paper provides an overview of these models, written for

Response Model

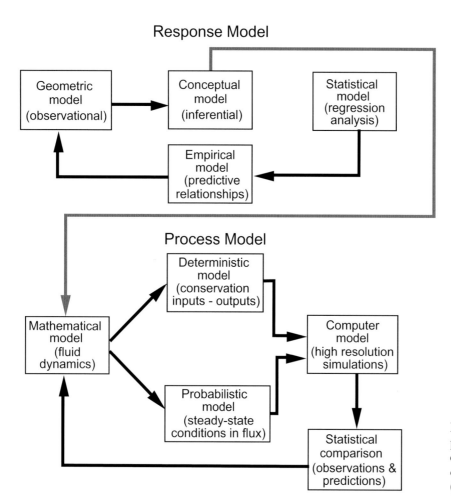

Process Model

Fig. 1 Organization of the process–response modelling environment in the field of quantitative dynamic stratigraphy. (After Syvitski, 1989.)

the non-expert and those not desiring a comprehensive mathematical treatment. The models cover the spectrum of both short time-scale sedimentary processes and those relating to the filling of geological basins where tectonics and subsidence are important controls on sediment dispersal. Students and researchers who are interested in pursuing this subject are encouraged to read the overviews of Slingerland *et al.* (1994), Watney *et al.* (1999), Paola (2000) and Tetzlaff & Priddy (2001).

There are two main approaches to modelling the fill of basins (Fig. 1). The **response model** begins with: (i) observing the stratigraphic record to determine the temporal and spatial distribution of the properties of sedimentary deposits; followed by (ii) constructing a schema of how the sedimentary system works; (iii) developing statistical relationships from these or other observations; (iv) generating an empirical model from the observational statistics; and finally (v) applying the model to new

geographical regions for validation and verification. Markov analysis of the transitions between sedimentary facies is an example where statistically defined relationships are used to develop conceptual models (Sonu & James, 1973). The applicability of response models is usually limited to environments with similar conditions (Fox, 1978).

The **process model** relies on the rudiments of fluid–sediment interactions and the processes that govern those interactions. Key equations are considered to be mathematical approximations of reality, which are linked to describe the physical system. The equations describe the conservation of mass, momentum and sometimes energy. Equations describing the conservation of mass are called **continuity equations**, and, for water, describe the rate of change in water mass as a function of spatial gradients of water discharge. The continuity equation for sediment in transport is called the **Exner equation** and describes the rate of change in bed elevation as a function of spatial gradients in sediment flux. The **conservation of momentum** refers to the spatial balance of forces (change in momentum over time) on a fluid system (gravity and pressure balanced by friction), although secondary terms relating to acceleration may be included. Momentum equations evaluate the operating forces at a given location and provide information on the velocity and boundary shear stresses in sediment transport models. The **conservation of energy** tracks the conversion of turbulence to work done as a function of boundary friction and other actions, for example energy expended on particle suspension, fluid entrainment and bed erosion. Forces that are either steady or vary with time drive the conservation equations. In some models a stochastic element is introduced to simulate the seemingly random nature of Earth-surface processes (earthquakes, floods, storms, and so on).

Computer models are the numerical schemes used to solve these mathematical constructs (Fig. 1). They may be designed using finite-element or finite-difference schemes, and solved on a variety of numerical grids according to the dynamics being simulated or the complexity of the domain boundaries. Solutions may involve the full domain (Eulerian view), or a subset of the domain through the tracking of individual fluid particles (Lagrangian view).

Four classes of models are available to the earth scientist, depending on the objective of the investigation. These include:

1 models whose predictions are at the resolution of modern sampling rates that are used to hindcast and forecast real time and space events (transport or realistic-domain models);
2 models that predict the net product of a particular set of processes at a time or space resolution beyond normal sampling schemes (depositional models);
3 models that principally describe one class of processes and their geological response (single-process models);
4 models that describe the complete sedimentary environment (whole-margin models).

Combinations of these four classes of models are possible and common.

Engineers, oceanographers and environmental scientists use **transport models** (Jewell *et al.*, 1993; Mulder *et al.*, 1997; Lou *et al.*, 2000; Kassem &

Imran, 2001) because they usually predict the consequences of sediment erosion or deposition on very short time-scales. For example, an engineer might want to predict the advent of the next turbidity current and whether it is capable of breaking communication cables or breaching a pipeline. By contrast, geologists who relate modern processes to the stratigraphic record often use **depositional models** that combine a steady-state condition with rare but important geological events (Mulder *et al.*, 1998a). Such models provide simulations of geological environments encompassing thousands of years. The interest here is in the properties of a sedimentary deposit, not the physics of sediment transport, even though the laws of physics are employed to predict the properties of the final deposit.

Single-process models simulate a single transport/deposition process for a particular sedimentary environment. For example, delta progradation can be modelled with sedimentation from river plumes as the dominant process (Bonham-Carter & Sutherland, 1967; Wang & Wei, 1986), or through bulk-transport processes (Kenyon & Turcotte, 1985). Both processes may duplicate how bathymetric profiles of a prograding delta can vary with time. This similarity of result (equifinality) occurs through the judicious use of empirical coefficients. In a **whole-margin model**, these two processes would be linked together with other processes to simulate multiple sediment transfer mechanisms (Syvitski *et al.*, 1988).

In 1988, there were few whole-margin models available (Syvitski, 1989), examples being two-dimensional, dynamic models for the transport, deposition, erosion and compaction of clastic sediments (Bitzer & Harbaugh, 1987; Syvitski *et al.*, 1988), and a diffusion-based model that moved sediment across key boundaries (Jervey, 1988). Many more whole-margin models have since become available in a variety of flavours: (i) geometric models for the study of sequence stratigraphy; (ii) analytical models; (iii) mechanistic models; (iv) hybrid models that combine these elements; and (v) inverse models.

Geometric models build sequences using knowledge of the geometry and timing of deposits (Cant, 1991; Reynolds *et al.*, 1991; Mackey & Bridge, 1995; Prather, 2000). **Analytical models** apply general equations that define the long-term nature of stratal evolution. The analytical equations take the form of diffusion, advection or advective-diffusion formulations (Kaufman *et al.*, 1991; Niedoroda *et al.*, 1995; Flemings & Grotzinger, 1996; Pirmez *et al.*, 1998; Ganjeon & Joseph, 1999; Kendall & Sen, 1999; Swenson *et al.*, 2000; Storms *et al.*, 2002). **Mechanistic models** apply physical laws to model the important sedimentary processes (Tetzlaff & Harbaugh, 1989; Martinez & Harbaugh, 1993; Haupt & Stattegger, 1999; Syvitski & Hutton, 2001). **Hybrid models** combine a variety of approaches and acknowledge that some processes are known well and can employ physical laws while other processes do not yet have accepted physics (Loseth, 1999; Ritchie *et al.*, 1999; Syvitski *et al.*, 1999). **Inverse models** simultaneously solve for the values of multiple process parameters using bounded unique solutions constrained by field observations (Cross & Lessenger, 1999).

Whole-margin models are useful for predicting the time-varying impact of sedimentary processes, and the distribution of lithostratigraphic properties away from points of control (Franseen *et al.*, 1991). Whole-margin models also prove valuable to those working on modern environmental applications (Martinez & Harbaugh, 1993), global warming scenarios (Syvitski & Andrews, 1994; Schäfer-Neth & Stattegger, 1999), natural disaster mitigation efforts (Caruccio, 1989), reservoir characterization (Shelton & Cross, 1989; Pratson *et al.*, 2000a), oil exploration (Tetzlaff & Harbaugh, 1989) and national security (Syvitski *et al.*, 1999).

This paper will first review individual process models that emphasize important margin environments, and sediment transport or seafloor dynamics. A source-to-sink modelling approach is used, first discussing how a river's discharge of water and sediment may be simulated, followed by its dispersal across the shelf as surface (hypopycnal) plumes, or subsurface (hyperpycnal) plumes, and through wave-current interactions within the bottom boundary-layer. Next, the stability of seafloor deposits is discussed, as failures may remobilize material as either debris flows or turbidity currents. Compaction is discussed as a separate modifier of seafloor deposits. It is then considered how these separate modules can be united into whole-margin models, able to examine the role of sea-level fluctuations, subsidence, isostasy and tectonics on the evolving sedimentary deposits, and how whole-margin models can be coupled to acoustic models, for comparison with seismic images. The paper concludes with a sense of the future, the development of a **Community Surface-Dynamics Modelling System**.

COMPONENT SED-STRAT MODULES

River flux

Most sediment is delivered to the ocean by rivers, and this accounts for approximately 95% of the global flux from land to sea (Syvitski, 2003; Syvitski *et al.*, 2003). Most of the flux is as suspended sediment load. Dissolved load and bedload are nearly an order-of-magnitude less than the suspended load. Thus the first step in modelling the evolution of a continental margin is to simulate a time series for the fluxes of water and sediment reaching the sedimentary basin. The temporal variations in a river's discharge and sediment flux to the ocean can strongly influence the creation and preservation of stratigraphic layers (Steckler *et al.*, 1996; Morehead *et al.*, 2001). The temporal variability of river discharge and sediment flux to the ocean ranges from daily cycles due to diurnal snow melt events or diurnal rains, to seasonal moisture and temperature cycles, to annual weather patterns, up to long-term climatic shifts. Modelling of the spatial and temporal variability of river discharge and its associated sediment flux to the ocean is the focus of this section.

The suspended-sediment concentration (c_s) in a river is commonly related to the discharge (Q) of the river and can be approximated by a power law:

$$c_s = aQ^b \tag{1}$$

This is exemplified by work on the Mackenzie River in Canada. Within the cloud of data points depicted in Fig. 2A, well-defined patterns exist that relate to various events occurring in the river basin. For the Mackenzie River, the rising limb of the flood has lower sediment concentrations than the falling flood limb, producing what is known as a **hysteresis loop** (Fig. 2B). Depending on the river basin, its source of discharge variability and its sediment sources, hysteresis loops can proceed clockwise, counterclockwise, or have complex patterns. Hysteresis loops occur on many time-scales, from daily loops associated with snow or glacial melt

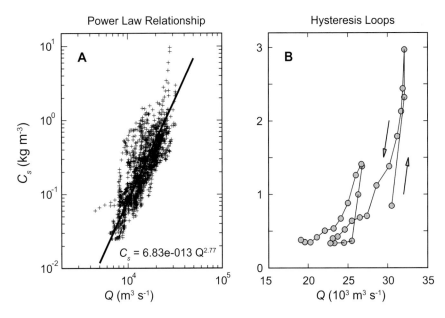

Fig. 2 The complex relationship between discharge (Q) and suspended-sediment concentration (C_s) on the Mackenzie River. (A) The relationship between Q and C_s can be approximated by a power law with the rating coefficient a and exponent b. (B) A counterclockwise hysteresis loop exists on the Mackenzie River, with high concentrations later in the season due to increasing glacial melt water (after Morehead *et al.*, 2003). Hysteresis loops account for part of the suspended-sediment variability at a given discharge. (Data are from Environment Canada, Ottawa.)

(Richards, 1984), to flood-wave loops occurring over several days to weeks (Brown & Ritter, 1971), to annual patterns occurring on larger river systems (Kostaschuk *et al.*, 1989; Morehead *et al.*, 2003).

Floods on rivers larger than creeks ($> 25 \, \text{m}^3 \, \text{s}^{-1}$ mean discharge) typically last for days to weeks. In order to adequately resolve this type of flood in a **hydrological-basin model**, it is necessary to simulate the basin processes on daily time-steps. With the present technology, it is feasible to model sedimentary and hydrological basin processes on a daily time-step for intervals from individual events up to tens of thousands of years (Syvitski & Morehead, 1999). Models designed for longer time-scales typically use bulk properties for the river flux that parameterize the character of the river without explicitly modelling individual events (Steckler, 1999).

In order to simulate the temporal variability and magnitude of the discharge and sediment flux from a river into a sedimentary basin, the main processes controlling runoff must be accounted for in a model (Beschta, 1987). **River discharge** is controlled by many factors including: river-basin characteristics (soil properties, vegetation, relief and size); weather (precipitation rate, timing and type, temperature and lapse rate); glaciated extent; and the hydrogeology (groundwater effects) of the basin. Each of these factors contributes its own signal to the temporal variability of the discharge. A successful hydrological basin model will produce

time-variable output with realistic distribution and timing of events. In order to simulate past climates, the model must either have a weather simulator embedded in it or be able to accept weather generated by another model, such as an atmospheric **global-circulation model** (GCM).

The first part of a hydrological basin model is the inclusion of meteorological information. Observations from meteorological stations can be fed directly into the model. However, direct observations are typically sparse and overrepresent lowland valleys, while underrepresenting higher mountains where much of the precipitation falls. A number of methods have been developed to distribute station data across a landscape to provide more realistic forcing fields. The PRISM climate maps are one example of an expert-weighted regression analysis that provides spatially distributed forcing fields for hydrological modelling (Daly *et al.*, 1994). Other models such as ISNOBAL, a distributed snow energy-balance model (Marks *et al.*, 2001), use similar regression methods to distribute atmospheric forcing fields over a watershed. Both the PRISM and ISNOBAL distribution routines were developed to produce forcing fields using measured data.

HYDROTREND is an example of a two-dimensional (elevation and upstream distance) **water-balance model** used for simulations at the dynamic level (daily) of discharge (Syvitski & Alcott, 1995b; Syvitski *et al.*, 1998a). The model resides within a geographical information system (GIS) where climatic environmental data are fused with basin characteristics to create a synthetic river-discharge time series. HYDROTREND accepts direct input from meteorological-station data, GCM output, or past global-change proxies of climate. The model can either use the forcing data directly or it can generate daily temperature and precipitation using Monte Carlo techniques (Syvitski & Morehead, 1999) based on input climate statistics. The observed or simulated meteorological data are distributed across the model basin using a geographically determined (by latitude) lapse rate and the basin morphology (area, maximum elevation, hypsometry, glacial equilibrium-line altitude). The HYDROTREND climate simulator can generate weather scenarios for modern, past and changing climate.

A second component of any hydrological model is a tracking algorithm for the various water reservoirs and flow pathways (e.g. TOPMODEL; Beven et al., 1995). HYDROTREND also tracks the various water pools (surface water, groundwater, snow, glaciers and overland flow) and accounts for additions (precipitation, infiltration) and reductions (evapotranspiration, runoff, melt) from each pool. Simple **energy-balance models** are incorporated into HYDROTREND to account for snow and glacial accumulation and melt. Each of these pools feeds into a simplified routing routine to simulate daily discharge (velocity, width, depth) at the river mouth.

Once the velocity, width and depth at the river mouth are determined, HYDROTREND utilizes predicted **power-law rating** equations (see Box 1) for the calculation of suspended load (Syvitski et al.,

---

Box 1 Stochastic modelling of a river's sediment load

Morehead et al. (2003) proposed a stochastic model for the simulation of the sediment discharge of ungauged rivers

$$Qs/\overline{Qs} = \psi F(Q/\bar{Q})^C \qquad \text{(B1.1)}$$

where Qs is daily sediment discharge, ψ is a lognormal random variable designed to capture measurement error and flood dynamics, F is a constant of mass flux, $\overline{Qs}$ is power-law average of the long-term average of Qs, Q is daily discharge at the river mouth, $\bar{Q}$ is power-law average of the long-term average of Q, and C is a normal random variable, otherwise known as the rating coefficient. The following relationships define these terms

$$E(\psi) = 1 \qquad \text{(B1.2)}$$

$$\sigma(\psi) = \delta(10^{-\phi\bar{Q}}) \qquad \text{(B1.3)}$$

$$E(C) = a_1 - a_2\bar{T} + a_3H + a_4\ln(\overline{Qs}) \qquad \text{(B1.4)}$$

$$\sigma(C) = b + m\bar{Q} \qquad \text{(B1.5)}$$

$$Q = Q_r + Q_{ice} + Q_n - Q_{evap} \pm Q_g \qquad \text{(B1.6)}$$

where E and σ represent mean and standard deviation of the stated variable, δ, ϕ, a_i, b and m are empirical constants, Q_g is ground-water discharge, Q_r is discharge from rainfall as surface runoff, Q_{ice} is icemelt discharge, Q_n is snowmelt (nival) discharge, and Q_{evap} is discharge component lost via evaporation. Q is thus a function of precipitation, basin location, temperature, elevation and soil properties. Large rivers suppress small perturbations in time-varying sediment and water sources and have relatively small $\sigma(\psi)$ (Syvitski et al., 2000). Small rivers are inherently noisy in terms of their discharge signal with larger values of $\sigma(\psi)$. C varies annually, dependent on $\bar{Q}$. Smaller rivers have more invariant values of C. Rivers with large discharge have a greater range in annual C, depending on the importance of their various tributaries to the overall discharge within a given year. $\overline{Qs}$ is

$$\overline{Qs} = \alpha_1 H^{\alpha_2} A^{\alpha_3} e^{k\bar{T}} \qquad \text{(B1.7)}$$

where H is river-basin relief, A is river-basin area, $\bar{T}$ is mean surface temperature of the drainage basin, and α_i and k are dimensionless constants that vary with the global climate belts.

Table 1 Comparison of HYDROTREND modelled output with US Geological Survey observations for the Eel River (After Morehead *et al.*, 2001)

Parameter	Observation	Model
Total annual load (t yr^{-1})	4×10^7 to 2×10^8	2×10^7 to 1×10^8
Mean discharge (m^3 s^{-1})	240	230
Mean concentration (g m^{-3})	2170	2120
Rating coefficient, a (g m^{-3})	0.35	0.14
Rating exponent, b	1.14	1.06

2000, 2003; Morehead *et al.*, 2003), and Bagnold's (1966) transport equation for the calculation of bedload. By predicting the daily discharge hydraulics at a river's mouth, the model output allows for direct input to shelf circulation and sedimentation models. HYDROTREND has simulated the relatively small icemelt-dominated Klinaklini River (Syvitski *et al.*, 1998b) in British Columbia, the rain-dominated Eel River in northern California (Syvitski & Morehead, 1999), the snowmelt-dominated Liard River in northern Canada (Syvitski, 2002), cyclone-dominated Lanyang, Taiwan (Syvitski *et al.*, 2005a), and the reservoir-influenced Po, Italy (Syvitski *et al.*, 2005b). Model validation of HYDROTREND on the Eel River is based on (Morehead *et al.*, 2001; Table 1): (i) daily discharge hydrographs including flood frequency based on US Geological Survey observations; and (ii) daily and annual sediment-load variability based on 16 yr of US Geological Survey observations.

Hydrological, hydraulic and water quality models are classified as deterministic, stochastic, or some combination of these two types. Processes that are too complex or poorly understood to be modelled deterministically are represented by statistical characteristics, while many statistical models also employ simple process-type mechanisms. Below are some commonly used models (Syvitski *et al.*, 2004):

RORB RAFTS – rainfall-runoff and stream-flow routing model
HEC – surface runoff model suite
IDRO – rainfall-runoff and storm-forecasting model
IRIS – Interactive River System Simulation program
WQRRS – Water Quality for River-Reservoir System
TOPMODEL – hillslope hydrology simulator
HYDROTREND – climate-driven sediment discharge simulator
WEPP – Water Erosion Prediction Project model
ANSWERS – Areal Non-point Source Watershed Environment Response Simulation
FHANTM – Field Hydrological And Nutrient Transport Model
MIKE 11 – river flow simulation model with data assimilation
WATFLOOD – integrated models to forecast watershed flows

Once a river model is shown capable of simulating an appropriately wide range of conditions in the modern environment, it can be used to represent past environmental conditions, assuming the input climatic and basin parameters can be estimated. The ability to simulate the full distribution of discharge events, including the rare, extremely large events likely to be preserved in deposits, is crucial for a hydrological model being used to supply sediment to a stratigraphic model. Over the Holocene period, sea-level fluctuations have greatly impacted the drainage areas of rivers (Mulder & Syvitski, 1996). Although basin area increases for all non-glacial rivers during glacial periods, around former ice sheets, isostatic depression would have caused a local rise in sea level and thus a decrease in a river's drainage area. In most cases, the glacial to interglacial drainage areas of rivers vary by < 30%. Following the global lowering (~120 m) of sea level, small rivers fronted by a wide continental shelf may become tributaries of larger rivers, while large rivers may merge to form enormous rivers (Mulder & Syvitski, 1996).

When modelling the hydrological changes between glacial conditions and Holocene temperate conditions, changes in climate can widely alter the discharge of water and sediment. Many modern temperate rivers may have been under polar conditions where snow or even icemelt would have dominated the discharge to the ocean (Morehead *et al.*, 2001). It is illustrative to compare the modern Eel River to its glacial-stage counterpart. In today's climate, the Eel River Basin is strongly affected by the southern excursion of the Jet Stream in the winter, which propagates storm fronts onto the coast from the Pacific Ocean, producing intense precipitation. The Eel River Basin receives as much as 80% of its annual precipitation in the winter. Sometimes a winter month may experience 20 to 30 days of nearly continuous rainfall, reaching inputs of 0.5 m in a month. As a result of the high discharges and the power-law relation of sediment load to discharge, the Eel River transports six orders-of-magnitude more sediment in the winter than during the drier summer. Most years have a winter month where the river transports between 10^6 and 10^7 t of sediment. The month of December

in 1964 is significant in that the sediment load exceeded 10^8 t. In the three peak days of this December flood, the river may have transported more sediment than during the previous 8 yr of river flow combined (Syvitski & Morehead, 1999). The Eel River Basin is comprised of the loosely consolidated Franciscan and Rio Dell Formations that are easy to erode and land slides are generated on a regular basis (Brown & Ritter, 1971). This makes the Eel River one of the dirtiest rivers in North America (Milliman & Syvitski, 1992).

Climate proxies for the western USA (Hostetler *et al.*, 1994; Benson *et al.*, 1996, 1997a,b) were used to set the climate statistics in HYDROTREND in order to simulate the last glacial period (about 18 kyr BP). These proxies indicate that the Eel River Basin was wetter and colder during the Last Glacial Maximum (LGM) than at present. Figure 3A depicts the climate (averaged weather) trends used as input to HYDROTREND to simulate the transition from a glacial interval to the modern interglacial. Random annual, monthly and daily weather variations were imposed upon these climate trends to create the precipitation and temperature

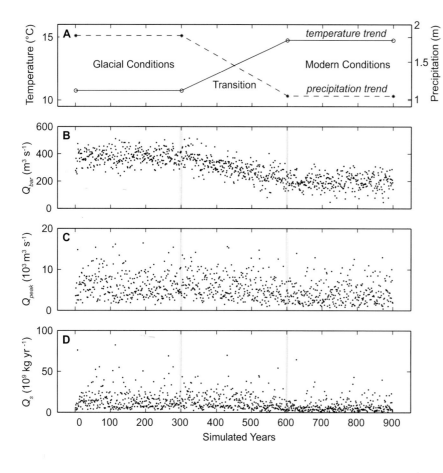

Fig. 3 A HYDROTREND simulation indicating profound changes in the total flux of sediment from the Eel River for an idealized climate-change scenario depicting a 300-yr transition from the last glacial (first 300 yr) to the modern climate (last 300 yr; after Morehead *et al.*, 2001). (A) Modelled climatic values of temperature and precipitation for the Eel River Basin taken from climatic proxies. (B) Mean annual discharge under modern climate is about half that of the glacial scenario. (C) The largest floods in the two scenarios are of similar magnitude, but the average annual floods are smaller in the modern period. (D) The mean annual sediment flux of the glacial period was about twice the modern flux. The modelled modern flux of sediment is more temporally variable.

Table 2 Characteristics of the Eel River, California, as simulated by HYDROTREND (After Syvitski & Morehead, 1999)

Parameter	Glacial	Modern
Total annual load (Mt yr^{-1})	15.6	7.8
Mean discharge (m^3 s^{-1})	380	200
Mean annual peak (m^3 s^{-1})	6160	4240

time-series used to generate river discharge in HYDROTREND. During the LGM, the Jet Stream may have been confined to the central California region causing increased storm frequency, colder temperatures and increased precipitation in the Eel River Basin (the first 300 yr of the simulation experiment; Morehead *et al.*, 2001). As the North American continental glaciers retreated, the Jet Stream moved progressively northward during the summer months. The climate transition is thought to have occurred rapidly as the glaciers retreated (the second 300 yr of the simulation). During modern climate (the last 300 yr of the simulation), the Jet Stream is north of northern California during the summer and only moves over the region during the winter months.

The increased precipitation and cooler climate of the glacial period yielded approximately double the annual discharge and sediment loads of the modern period (Fig. 3B & D; Table 2) according to HYDROTREND (Syvitski & Morehead, 1999). The flow conditions were more constant during the glacial period, due to the basin being snowmelt dominated, consistent with modern river systems (Pitlick, 1994). The mean of the annual peak discharges for each of the periods shows a marked decrease during the modern period (Fig. 3C; Table 2). The largest floods, however, are of similar magnitudes during both periods. The largest modern floods are due to rain events on top of widespread low-elevation snow, which provides more water to the river for runoff than either a pure rain event or a pure snowmelt event. Presumably the largest floods during the last glacial period would have been caused by the same mechanism. Assuming the model is correct, the present-day Eel River has much more interannual variability than during the last glacial period. A change in the total flux of sediment out of the river, along with a change in the variability of the flux, probably occurred between the two periods, and would have had marked effects on the type of sedimentary layers that are accumulated on the continental margin.

Modelling river basins over longer time-scales has revealed a number of important characteristics that affect how rivers behave. A simulation on the Klinaklini River, British Columbia, Canada, demonstrated how a river can switch discharge states rather abruptly during a slow and continuous climate-change scenario (Morehead *et al.*, 2001). During a brief period, as the river switched from a glacial-melt-dominated system to a mixed-glacial-and-snow system, larger than expected fluctuations in discharge were simulated. During this portion of the simulation, certain summers were not warm enough to melt the entire snow pack, leading to lower than normal peak discharge for those years. On subsequent warmer years, accumulated snow would melt entirely, yielding higher than normal peak discharges for those years.

In summary, climate change produces complex temporal changes in a river basin's discharge and sediment load, and this may have profound effects on the resultant sedimentary deposits (Morehead *et al.*, 2001; Syvitski, 2002; Syvitski *et al.*, 2003). It is interesting to speculate that the dominant flux of sediment to temperate continental margins follows the location of the Jet Stream, as the climate system moves in and out of glacial periods. In contrast, tropical systems are less impacted by fluctuations in temperature and its inherent effects on sediment production and transport (permafrost soils, freeze–thaw intervals, snowmelt and icemelt), but are rather more impacted by the intensity and duration of rainfall. An abundance of proxy records show an enhanced monsoonal regime in the early Holocene (Overpeck, 1996), and the monsoon-driven river systems strongly responded to this enhanced regime (Goodbred & Kuehl, 2000a,b). The Ganges–Brahmaputra system then carried over 2.5 times its present load.

Surface plumes from rivers

Freshwater plumes provide the initial dispersal of terrestrially derived particulate matter in the coastal ocean. The lower density of freshwater compared with seawater generally produces buoyant plumes; however, in rare cases the concentration of suspended sediment may be so great that the density of the inflowing turbid water is greater than the ambient seawater, forming a **hyperpycnal plume** (Mulder & Syvitski, 1996). This section addresses the more common case of buoyant plumes.

The behaviour of a **buoyant plume** depends on the freshwater flow, the geometry of the river mouth and receiving waters, and oceanographic conditions in the coastal ocean (Wright & Coleman, 1971; Garvine, 1974; Geyer *et al.*, 2000, 2004). River plumes that enter shallow receiving waters may remain attached to the bottom for significant distances from the mouth (Wright *et al.*, 1990; Geyer & Kineke, 1995). In contrast, rivers entering deep water or narrow continental margins separate from the bottom at locations close to the mouth (Eisma & Kalf, 1984; Geyer *et al.*, 2004). These two types of plumes contain distinctly different sediment transport processes: **bottom-attached plumes** recycle sediment by resuspension in the bottom boundary layer, whereas **detached plumes** lose sediment to the underlying, ambient seawater due to isolation of the plume from the turbulent boundary layer. Surface plumes are particularly sensitive to wind forcing, due both to the direct forcing of wind stress on the plume and to the influence of the winds on the ambient coastal currents (Chao, 1988; Fong & Geyer, 2001).

The Eel River enters a relatively steep and narrow continental shelf, resulting in a surface plume that detaches from the bottom just seaward of the surf zone (Geyer *et al.*, 2000). As sediment enters the coastal ocean and moves away from the river mouth, it settles out of the surface plume at a rate dependent on its settling characteristics (Hill *et al.*, 2000, this volume, pp. 49–99; Curran *et al.*, 2002). Once removed from the surface plume, sediment is dispersed by oceanographic transport in the underlying water, which may include resuspension in the bottom boundary layer. Predicting the deposition of fluvially derived material in an environment such as the Eel shelf depends on the characterization of resuspension, particle settling velocities, and the structures of both plume and oceanographic waters. These evolve in time, due to changes in discharge and oceanographic forcing conditions. Sediment transport in the coastal ocean also responds to significant variability in the vertical, across-shelf and along-shelf dimensions. Full representation of the behaviour of freshwa-

ter plumes in the coastal ocean requires a three-dimensional, time-dependent model. Some basic aspects of plume behaviour and initial deposition can still be captured by more parameterized representations (Scully *et al.*, 2003).

Two numerical models have been developed as part of the STRATAFORM Project in order to characterize plume dispersal on the Eel shelf, northern California (Morehead & Syvitski, 1999; Harris *et al.*, 2000; Harris & Wiberg, 2002). Both incorporate sediment transport within the plume and settling of sediment into underlying waters, but the models differ in the level of detail included in the dynamics, and in the time-scales over which they can be applied. Morehead & Syvitski (1999) represented the freshwater plume as a turbulent-jet entering the coastal ocean (see Box 2), while Harris *et al.* (2000, 2005) used a **primitive-equation ocean model** (see Blumberg & Mellor, 1987) to predict the structure and velocities of the plume and underlying waters. The primitive-equation model used 45-s time-steps to resolve vertical sediment advection. This, and the challenge of specifying robust boundary forcing, have limited the length of their simulations to time-scales of a few weeks. The turbulent jet model used simplified physics of plume dynamics and sediment transport, and longer time-steps (~1 day). This avoided the need for detailed open-boundary conditions and allowed for simulations that are more suitable to making predictions over geological time-scales.

The PLUME model predicts sediment delivery to the coastal ocean as a function of sediment load, along-shelf currents and, as the dominant factor, the initial momentum of the plume as it enters the ocean (Morehead & Syvitski, 1999). The plume is modelled as a turbulent jet that can be steered by underlying currents, and that interacts with seawater through diffusion of momentum and sediment. Moving away from the river mouth, plume width increases, and velocities and turbulence levels within the plume decrease. Turbulent diffusion of the plume's momentum slows the flow, and enables sediment to settle out of the plume and into the underlying water column. The PLUME model handles transport of sediment below the brackish plume, by assuming that it settles directly to the bed or, alternatively, computes its transport by using a bottom-boundary-layer model (Syvitski & Bahr, 2001). This approach simplifies a number of

---

Box 2 Non-conservative, advection-diffusion model for a steady, two-dimensional surface plume

The sediment inventory of a surface plume from a river's discharge into the ocean can be largely captured by three terms: (i) advection of sediment by the river-induced velocity field (term *a*); (ii) diffusion of the sediment by turbulence (term *c*); and (iii) settling of sediment out of the plume at a rate governed by a first-order removal-rate constant λ for each grain size (term *b*). The governing equation in this case is the two-dimensional advection–diffusion equation

$$\underbrace{\frac{\partial uI}{\partial x} + \frac{\partial vI}{\partial y}}_{a} + \underbrace{\lambda I}_{b} = \underbrace{\frac{\partial}{\partial x}\left(K_s \frac{\partial I}{\partial x}\right) + \frac{\partial}{\partial y}\left(K_s \frac{\partial I}{\partial y}\right)}_{c} \tag{B2.1}$$

where x is longitudinal or axial direction, y is lateral direction, u is longitudinal velocity, v is lateral velocity, I is sediment inventory or mass per unit area of the plume, λ is first-order removal-rate constant for a given grain size, and K_s is sediment diffusivity driven by turbulence.

Syvitski *et al.* (1998a) defined the position of the centreline of a plume as:

$$\frac{x}{b_0} = 1.53 + 0.90 \left(\frac{u_0}{v_0}\right)\left(\frac{y}{b_0}\right)^{0.37} \tag{B2.2}$$

where b_0 is the river-mouth width. The suspended-sediment concentration, c_s, around the centreline is

$$c_s(x,y) = C_0 e^{-\lambda t} \sqrt{\frac{b_0}{\sqrt{\pi C_1 x}}} \cdot e^{-\left(\frac{y}{\sqrt{2}C_1 x}\right)^2} \tag{B2.3}$$

where C_0 and C_1 are empirical constants. The flow field of a plume is influenced by: (i) the Coriolis force (i.e. the latitudinal position determines the deflection of the angular acceleration); (ii) angle of the river mouth in relation to the coastline; (iii) a Kelvin wave effect (causing plumes to hug the coast when the plume flows in the direction of the Coriolis force); (iv) shore winds (Ekman effect) causing upwelling or downwelling along the coast; (v) the hydraulics of the river mouth (i.e. supercritical or subcritical nature of flow); and (vi) the general nature of the ambient coastal circulation.

---

potentially important processes, including bottom friction, wind stress and bottom-boundary-layer transport. As an example, simulation of a large Eel River flood in 1995 predicted sediment delivery from the plume to be highest near the river mouth, and to be confined shoreward of the 50-m isobath (Fig. 4).

The primitive-equation model is able to compute the sediment transport and deposition from the Eel plume, based on sediment settling, advection by ocean currents and resuspension in the bottom boundary layer (Harris *et al.*, 2000, 2005). The model predicts velocities and water-column structure under forcing by winds, momentum input from the Eel River discharge, buoyancy, tides and along-shelf currents. Sediment is treated as a settling tracer that is transported with ambient currents, and can be deposited and resuspended from the seafloor. Sediment delivered during floods is trans-

ported with the floodwater until it settles out of the surface plume with a specified settling velocity. Beneath the plume, sediment is transported with ambient water and continues to settle towards the bed. Wave-current interaction and bottom friction are incorporated to account for a bottom boundary layer (Grant & Madsen, 1979). Sediment is deposited on the bed when the supply exceeds the ability of the flow to carry the material in suspension, but can be resuspended should the flow capacity increase or supply decrease. Applied to a large flood of the Eel River (January 1997), the depositional patterns depended most critically on the settling properties of the sediment and on wind forcing during peak discharge (Harris *et al.*, 2000). Winds for this simulation were strong (15 m s^{-1}) and northward during the first 3 days of the flood, and then relaxed to be weaker and southward. Peak sediment discharge (5 g L^{-1}) at

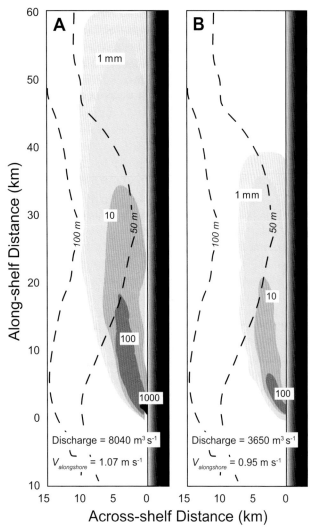

Fig. 4 Snapshots of predicted plume deposition (mm) summed over four grain sizes for the January 1995 flood of the Eel River, CA. (A) The first and largest peak on 8 January. (B) The start of the falling flood limb on 14 January. (Modified from Morehead and Syvitski, 1999.)

the river mouth occurred 3 days into the flood event, and by day six of the simulation most sediment had been delivered to the coastal ocean. Wave energy was held constant, with significant wave heights of 3 m throughout the simulation. Based on field observations of particle aggregation (Hill *et al.*, 2000), 60% of the sediment was flocculated (settling velocity $w_s = 0.1$ cm s^{-1}), and 40% unflocculated ($w_s = 0.01$ cm s^{-1}). Results showed the direction of the plume responds to Earth's rotation, wind forcing and the angle of the river mouth.

These factors deflect the plume to the north, and confine it to the inner shelf during the early stages of the simulation (Fig. 5A). As winds relax and turn southward, the plume thins and spreads out over deeper waters (Fig. 5B).

Dispersal patterns depend critically on settling characteristics. The unflocculated, fine-grained material remains, for the most part, within the surface plume, and its dispersal pattern depends on plume velocity and structure. During strong northward winds, the unflocculated portion of the sediment load is transported beyond Trinidad Head (50 km north of the river mouth), resulting in export of approximately 10–20% of the sediment load (see Fig. 5A). Under southward winds, the unflocculated sediment that remains in the plume is dispersed over a wide area. **Cross-shelf currents** carry some sediment to mid-shelf depths, where a muddy flood deposit is observed. While the location of this mid-shelf deposit matches that of the deposits observed soon after flood events, the thickness of deposition predicted by this mechanism is much smaller (< 1 cm) than the observed deposit (up to 10 cm; Wheatcroft *et al.*, 1997).

Flocculated material is predicted to settle out of the plume within a few kilometres north of the river mouth, and it accounts for the majority of the deposit (outlined in red in Fig. 5A & B). The mass of flocculated sediment delivered by the surface plume accounts for the amount observed in the mid-shelf flood deposit (Fig. 5C), but a cross-shelf transport mechanism is needed to reconcile the predicted inner-shelf delivery with observed mid-shelf deposition. One hypothesis is that resuspension by energetic waves prevents sediment deposition on the energetic inner shelf, and cross-shelf currents transport sediment to the less-energetic mid-shelf depths where deposition occurs (Hill *et al.*, this volume, pp. 49–99). Model calculations indicate that the supply of sediment to the inner-shelf overwhelms resuspension processes there, however, and even simulations that included energetic waves predicted a thick inner-shelf deposit within the timeframe of observable mid-shelf deposition (Fig. 5B).

Field observations suggest that a near-bed fluid-mud layer with significant offshore velocities can develop on the Eel River shelf during times of ample sediment supply and energetic waves (Ogston *et al.*, 2000; Traykovski *et al.*, 2000). The buoyancy

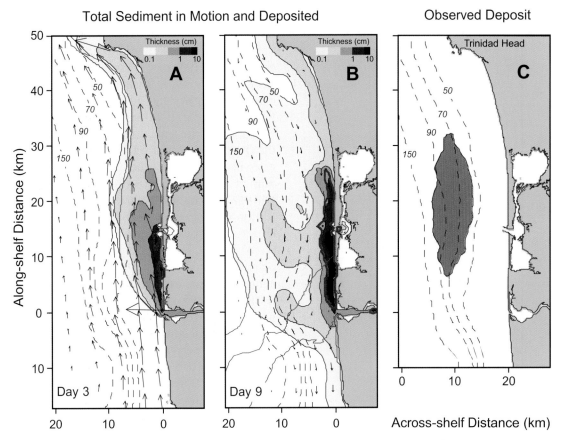

Fig. 5 Plume structure, ocean currents and sediment dispersal simulated for the January 1997 flood of the Eel river, compared with observed deposit (after Harris *et al.*, 2000). Bathymetric contours shown as dashed lines. Snapshots shown at: (A) peak of flood and northward wind forcing; and (B) when discharge and winds are low. (A & B) Colour contours show total suspended and bed sediment for a simulation containing 40% unflocculated and 60% flocculated material. Red line indicates contour of > 1 cm deposited on the bed. Blue lines illustrate the isohaline for a salinity of 30. Arrows indicate depth-averaged velocities. (C) Location of observed deposit following the January 1997 flood. (From Wheatcroft *et al.*, 1997.)

anomaly of the sediment-laden layer could create seaward velocities. This hypothesis has been tested by including a representation of the fluid-mud layer within the three-dimensional model (Harris *et al.*, 2005). Inclusion of a thin near-bed fluid-mud layer resulted in a mid-shelf deposit of appropriate thickness and mass within the time-scale of observed deposition (Fig. 6). The role of gravity-driven fluid mud was parameterized by a two-dimensional ($x - y$) model (see Box 4) which assumes that plume sediment is deposited directly into the wave boundary layer on the inner shelf (Scully *et al.*, 2003). This **fluid-mud model** predicted that 26% of the fine sediment discharge by the Eel River was deposited on the mid-shelf, with a distribution consistent with field observations.

These and other modelling studies indicate that plume dynamics and sediment settling rate determine the along-shelf position of deposition, and the fraction of sediment that remains on the proximal shelf. Even for a bottom-detached plume, such as the Eel River plume, the cross-shelf location of the deposit depends on transport that occurs in the bottom boundary layer after sediment settles from the surface plume.

Hyperpycnal flow from rivers

Turbidity currents are sediment-laden underflows that occur in relatively quiescent bodies of water such as dammed reservoirs, natural lakes, fjords and oceans. The current obtains its driving force

Dispersal with Gravity-driven Flow

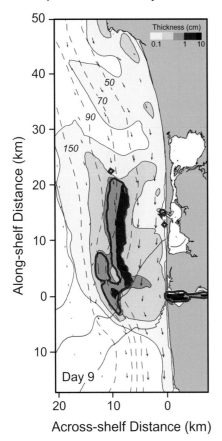

Fig. 6 Simulation of the January 1997 Eel River flood using a model that accounts for a thin fluid-mud layer at the seafloor. Bathymetric contours (m) shown as dashed lines. Simulation represents the end of the flood event, when discharge and winds are low. Colour contours show total suspended and bed sediment for a simulation containing 40% unflocculated, and 60% flocculated material. Red line indicates contour of > 1-cm seabed deposit. Blue lines illustrate the isohaline for a salinity of 30. Arrows indicate depth-averaged velocities. (After Harris *et al.*, 2005.)

from a density difference with the ambient water. A turbidity current can be triggered by one or more of several processes such as wave action, subaqueous slumps induced by seismic or other loading, or plunging of river water as a hyperpycnal plume (see Parsons *et al.* (this volume, pp. 275–337) for further discussion). Among the various mechanisms, a hyperpycnal plume generated at a river mouth is most capable of maintaining a sustained turbidity current event. Even rivers with moderate to small loads of suspended sediment (e.g. < 1.0 kg m^{-3}) can

create hyperpycnal plumes in freshwater lakes (Weirich, 1984). The Rhine plume flowing into Lake Constance, Switzerland, can produce bottom currents of > 100 cm s^{-1} (Lambert, 1982). Concentrations > 0.10 kg m^{-3} in river water are enough to produce underflows in the freshwater Katsurazawa Reservoir, Japan (Chikita, 1989).

Initiation of a sustained turbidity current from the direct delivery of sediment into the ocean by a river is considered to be a rare event, as seawater with a specific gravity of ~1.026 poses a strong density barrier to plunging. The critical concentration of suspended sediment needed to create a hyperpycnal plume in seawater is considered to be 35–45 kg m^{-3}. There are few rivers in the world that discharge into the ocean with such large sediment concentrations, even during flood events (Mulder & Syvitski, 1995). Alternatively, Parsons *et al.* (2001) found that fresh warm water with suspended-sediment concentration as small as 1 kg m^{-3} can produce a hyperpycnal flow in cold brine by means of double diffusive convection. Rare or not, marine hyperpycnal flows associated with large floods can significantly alter the seascape. A case in point is an 11-m-thick deposit in the Saguenay Fjord, eastern Canada, apparently formed from a historical hyperpycnal discharge event (Syvitski & Schafer, 1996; Skene *et al.*, 1997).

Turbidity currents generated at a river mouth consist of three distinct parts: the plunge region, the main body and the leading head. Each part of the current has distinct characteristics and plays an important role in the overall flow and transport processes (Kassem & Imran, 2001). The ambient-water entrainment during plunging affects the remainder of the current. Vertical structures of sediment concentration and velocity in the main body of the current are responsible for sustained scour of, and deposition on, the lake or ocean floor. The mixing at the leading head determines how far the current will travel before it loses its identity. The plunging phenomenon has been widely studied as a practical engineering problem associated with industrial effluent discharge, reservoir sedimentation and stratified flow (Johnson *et al.*, 1987). Various factors, such as the inflow Richardson number, inflow Reynolds number, bottom slope of the receiving water body and confluence divergence angle, influence the plunging process and such factors as plunge depth and location, and the subsequent evolution of the current. Once the current is established the hyperpycnal flow evolves with momentum balance set up between the pull of gravity and fluid pressure with frictional forces that dissipate energy (see Box 3).

Studies on the **plunging phenomenon** have involved laboratory and field observations, and parameterization of different processes in empirical equations (Akiyama & Stefan, 1984; Johnson *et al.*, 1987; Stefan & Johnson, 1987). Such empirical equations address the effect of just one or two variables at a time, and therefore lack universal applicability. A robust numerical model, in contrast, can incor-

porate many of the contributing factors simultaneously. Often empirical relationships are needed to determine upstream boundary conditions when using one-dimensional depth-averaged models of turbidity currents (Skene *et al.*, 1997; Mulder *et al.*, 1998b). Few attempts have been made to model the plunging process and subsequent generation of hyperpycnal plume (Akiyama & Stefan, 1984; Bournet *et al.*, 1999; Kassem & Imran, 2001).

Higher-dimensional numerical models can predict the development of the entire process from the free-surface-flow condition at the upstream end to the formation of the turbidity current, including the

---

Box 3 Layer-averaged hyperpycnal plume in Lagrangian form

The layer-averaged hyperpycnal plume consists of four conservation equations: (i) continuity of water; (ii) continuity of sediment; (iii) conservation of momentum; and (iv) conservation of energy (after Parker *et al.*, 1986; Pratson *et al.*, 2000a).

Continuity of fluid

$$\underbrace{\frac{\partial h}{\partial t} + \frac{\partial Uh}{\partial x}}_{a} = \underbrace{e_{w}U}_{b} \tag{B3.1}$$

where h is current thickness, t is time and x is distance along the flow path, U is layer-averaged flow velocity and e_{w} is a dimensionless entrainment coefficient. Equation (B3.1) reflects the dependency of the height of the current (term a) on the entrainment rate of ambient water (term b).

Continuity of sediment

$$\underbrace{\frac{\partial Ch}{\partial t} + \frac{\partial UCh}{\partial x}}_{a} = \underbrace{w_{s}(E_{s} - r_{o}C)}_{b} \tag{B3.2}$$

where C is the layer-averaged bulk sediment concentration, w_{s} is the settling velocity of particles, E_{s} is dimensionless erosion coefficient that determines the rate that sediment is entrained into the current by erosion, and r_{o} is the ratio of sediment concentrated at the base of the flow. The mass of sediment in the flow (term a) depends on the rate that sediment is either added to the flow or settles out of the flow (term b).

Conservation of momentum

$$\underbrace{\frac{\partial Uh}{\partial t} + \frac{\partial U^{2}h}{\partial x}}_{a} = \underbrace{RgCh\alpha}_{b} - \underbrace{\tfrac{1}{2}Rg\frac{\partial Ch^{2}}{\partial x}}_{c} - \underbrace{u_{*}^{2}}_{d} \tag{B3.3}$$

where R denotes the submerged specific gravity of the sediment, g is gravitational acceleration, α is seafloor slope, and u_{*} is the shear velocity at the base of the current and reflects the loss of momentum to the current flowing across a rough sea floor. Equation (B3.3) describes the balance of momentum in a turbidity current (term a), with the current's downslope acceleration due to gravity (term b), fluid pressure gradients within it (term c), and frictional slowing (term d).

Conservation of energy

$$\frac{\partial Kh}{\partial t} + \frac{\partial UKh}{\partial x} = \underbrace{u_*^2 U + \tfrac{1}{2} U^3 e_w}_{b} - \underbrace{\varepsilon_0 h}_{c} - \underbrace{Rg w_s Ch}_{d} - \underbrace{\tfrac{1}{2} RgChUe_w}_{e} - \underbrace{\tfrac{1}{2} Rghw_s(E_s - r_o C)}_{f} \qquad \text{(B3.4)}$$

where K is the layer-averaged turbulent kinetic energy, and ε_0 is the layer-averaged dissipation rate. The distributed kinetic energy (term a) represents the balance between the rate at which the current produces turbulent kinetic energy (term b) and then consumes it through dissipation by the viscosity of the current (term c), energy consumption needed to keep the sediments in suspension (term d), turbulent kinetic energy converted to potential energy as the flow thickens through entrainment thus raising its centre of gravity (term e), and the expenditure of energy in eroding the bed (term f). Note that if the current deposits sediment, then the current actually gains energy back.

Mixing of freshwater and seawater

Hyperpycnal flows in the marine environment involve freshwater and the sediment it carries mixing with salt water (Skene et al., 1997). River water enters the marine basin with a density of 1000 kg m^{-3}, where it mixes with ocean water having a density typically of 1028 kg m^{-3}. This mixing increases the fluid density of a hyperpycnal current, and alters the value of C and R (above), where

$$\frac{\partial \rho}{\partial x} = \frac{e_w}{h}(\rho_{sw} - \rho), \quad C = \frac{\rho_f - \rho}{\rho_s - \rho}, \quad R = \frac{\rho_s - \rho}{\rho}, \quad \rho_f = (1 - C)\rho + C\rho_s \qquad \text{(B3.5a, b, c, d)}$$

where ρ is density of the fluid in the hyperpycnal current, ρ_{sw} is density of the ambient seawater, ρ_f is density of the hyperpycnal current (sediment and water), and ρ_s is grain density.

stabilization of the plunge point (Fig. 7: Kassem & Imran, 2001). As the sediment-laden water flows in with a dominant dynamic force, the reservoir water is pushed forward and a separation surface becomes pronounced (Fig. 7A). When the pressure force at the bottom becomes significant, it accelerates the flow at the bottom at a rate higher than the movement at the top (Fig. 7B). As the pressure forces continue to grow, the flow plunges to the bottom and begins to move as an underflow (Fig. 7C). At this

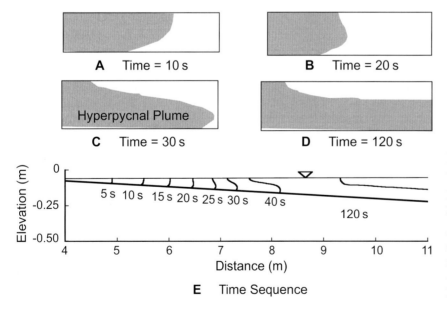

Fig. 7 A numerical simulation that depicts (A) an open channel flow, or river, transforming into a hyperpycnal flow (B through to D) beneath ambient marine water. This simulation is in the dimensions of a similar laboratory experiment where sediment-laden freshwater flowed in a saltwater basin (E) along an inclined plane. (After Kassem & Imran, 2001.)

stage, the velocity at the top surface is still significant enough to move the plunging point forward. When equilibrium is reached, the velocity at the top disappears, a stable plunge point (plunge line in a three-dimensional flow) forms, and the current moves forward with a bulge-shaped head and an elongated body (Fig. 7D). The entire process is chronicled in Fig. 7E.

Of the measured global rivers, only nine can produce hyperpycnal flow on an annual basis (Mulder & Syvitski, 1995). The Eel River in northern California is one of seven rivers located in the USA that can produce hyperpycnal flow during a 100-yr flood event. The Eel experienced large floods in 1995 and 1997. Estimated concentrations obtained using a rating curve (Mulder & Syvitski, 1995) indicated that the 1995 Eel River might have exceeded the critical concentration for plunging over a 2-day period (Imran & Syvitski, 2000). A survey conducted by investigators of the STRATAFORM programme immediately after the 1995 flood revealed that only ~25% of the flood sediment discharged by the Eel River remained on the shelf. While storm-induced waves and currents may play a role in resuspending and transporting the flood sediment, hyperpycnal discharge may be a dominant sediment transport mechanism during the peak flood flow. A hyperpycnal flow may have been directed towards the adjacent Eel Canyon and delivered sediment to the base of the continental slope through the canyon (Puig *et al.*, 2003, 2004; Mullenbach *et al.*, 2004).

Hyperpycnal flows can also be strongly influenced by **along-shelf currents**. The impact of the 1995 Eel hyperpycnal flow was modelled using a two-dimensional depth-averaged formulation that incorporates alongshore currents (Imran *et al.*, 1998; Imran & Syvitski, 2000). To provide the upstream boundary condition, river data were obtained from a nearby gauging station (Scotia, CA; Brown & Ritter, 1971), and the width of the river mouth was estimated to be 1.5 km during peak discharge. A current thickness of 4.4 m, velocity of 1.3 m s^{-1}, sediment concentration of 0.019 by volume, and a reduced specific gravity of 1.58 were specified for the river-mouth boundary condition. Other input parameters were obtained using the empirical formulae developed by Akiyama & Stefan (1984). The bottom slopes were calculated from seafloor bathymetry. Based on wind data (NDBC Buoy

46022), the along-shelf current velocity ranged from 0 to 1.4 m s^{-1} during the flood period (Morehead & Syvitski, 1999). Model results revealed hyperpycnal plumes generated by the Eel River that could flow into the adjacent Eel Canyon, or move northward towards Trinidad Head depending on the strength of along-shelf currents (Imran & Syvitski, 2000). Velocity vectors and contours of plume thickness (Fig. 8A) indicate that, if along-shelf currents do not influence the hyperpycnal plume, it has a tendency to flow towards the canyon along a natural gradient. The presence of an along-shelf current of modest magnitude, directed to the northeast, will turn the the hyperpycnal flow away from the canyon (Fig. 8B).

The above discussion shows the complex nature of hyperpycnal flow as a hydrological event, and its effectiveness in dispersal of sediment delivered by large floods in certain rivers (see global review by Mulder *et al.*, 2003). Hyperpycnal flows form in the marine environment when river discharge enters the ocean with suspended concentrations in excess of 36 kg m^{-3} due to buoyancy considerations, or as little as 1–5 kg m^{-3} when convective instability is considered. They form at a river mouth during floods of small to medium size rivers including extreme events such as jökulhaups, dam breaking and draining, and lahar events. Associated with high suspended-sediment concentration, hyperpycnal flows can transport a considerable volume of sediment to the ocean. The typical deposits or **hyperpycnites** differ from **turbidites** because of their well-developed inversely graded facies and intrasequence erosional contacts (Mulder *et al.*, 2003). Hyperpycnite stacking can locally generate high sediment accumulation rates, in the range of 10–20 cm kyr^{-1}. Hyperpycnites are related to climate through flood frequency and magnitude, so their record should vary with sea level and climate change.

Shelf boundary-layer sediment transport

Flows observed on the continental shelf are typically combinations of unidirectional currents and oscillatory motions due to wind-generated water waves. **Currents**, with variations on time-scales of hours or days, are principally tidal, wind-driven, or related to larger-scale patterns of ocean circulation. **Wave-generated oscillatory flows**, with

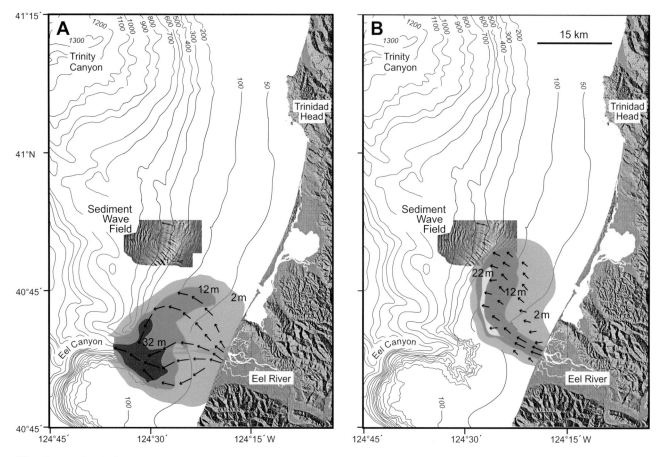

Fig. 8 Simulated thickness of a hyperpycnal current originating from the Eel River, after 8 h of flow. Velocity vectors show the direction and relative magnitude of the current. Base map shows the bathymetry (m) of the margin. (A) Under a no-along-shelf-current scenario, the hyperpycnal flow heads towards the Eel canyon following the natural topography. (B) With an along-shelf current of 35 cm s^{-1}, in a north-east direction, the hyperpycnal flow is directed towards the north-west and the sediment wave field, and away from the canyon.

periods in the order of seconds to tens of seconds, are produced by surface gravity waves, although lower frequency internal waves can also be present. When wave-generated oscillatory fluid motions extend to the seabed, they produce shear stresses on the bed and within the relatively thin (~0.1-m-thick) wave boundary layer. The velocity of near-bed wave orbital motion depends on wave height, period and water depth. In general, the larger the waves and the shallower the water, the larger the near-bed wave orbital velocities and wave-generated shear stresses that could exceed the threshold of grain motion. When both currents and waves contribute to near-bed flow, shear stresses in the wave and the current (bottom) boundary layer are enhanced.

Sediment transport on the continental shelf is strongly affected by the combined presence of waves and currents. **Coastal currents** in many locations are insufficient to mobilize sediment outside of the nearshore zone. In these regions, it is only when moderate to high wave conditions are present that bed shear stresses are sufficient to mobilize sediment at the seabed. Even at depths of 50–100 m, mean winter wave-generated shear velocities can be considerably larger than current shear velocity. Waves, however, are inefficient at transporting sediment owing to their oscillatory motion. Directions and rates of transport depend largely on the currents present during episodes of high wave-generated bed shear stress.

The first and still the most widely used models of shelf sediment transport are one-dimensional models in which vertical profiles of velocity and suspended-sediment concentration are calculated from near-bed wave and current conditions (Smith,

1977; Grant & Madsen, 1979). One-dimensional models assume that flow and sediment conditions are spatially uniform. This yields

$$\frac{\partial c_s}{\partial t} = \frac{\partial}{\partial z}\left(w_s c_s + K_s \frac{\partial c_s}{\partial z}\right) \tag{2}$$

as the governing equation for suspended-sediment concentration, where c_s is sediment concentration, w_s is particle settling velocity and K_s is sediment diffusivity. Additionally, these models often neglect transient effects, instead assuming an instantaneous adjustment of shear stress and suspended-sediment concentration to changes in wave and current velocities. Despite these somewhat limiting assumptions, one-dimensional models of sediment transport on the shelf provide near-bed suspended-sediment concentrations in reasonable agreement with measured concentrations (Li *et al.*, 1996; Cacchione *et al.*, 1999; Wright *et al.*, 1999; Wiberg *et al.*, 2002). This suggests that local resuspension tends to dominate measured near-bed suspended-sediment concentrations. Often these models are applied with at least one adjustable parameter: the resuspension coefficient that scales the bottom boundary condition on suspended-sediment concentrations.

Further development of **shelf sediment-transport models** has followed four directions. First, a number of efforts have been made to improve the representations of near-bed transport processes and model parameterizations. These studies address the parameterization of bottom roughness (Xu & Wright, 1995; Li & Amos, 1998), reference concentration (Li *et al.*, 1996), effects of flocculation (Carr, 2002), near-bed stratification due to suspended sediment (Styles & Glenn, 2000), and bed armouring (Wiberg *et al.*, 1994; Reed *et al.*, 1999). Second, two-dimensional, time-dependent models can resolve effects of advection as well as patterns of net erosion or deposition (Zhang *et al.*, 1999; Harris & Wiberg, 2001). Third, shelf sediment-transport models have been embedded into, or coupled with, models of shelf circulation, allowing spatial patterns of transport, erosion and deposition to be determined (Keen & Slingerland, 1993; Harris *et al.*, 2000). Finally, various approaches are used to represent shelf sediment-transport processes in models predicting the evolution of margin form and stratigraphy (Thorne *et al.*, 1991; Niedoroda *et al.*, 1995) in which time-steps of days to tens of years

are employed, much longer than the time-steps used in most shelf-transport models. Examples from STRATAFORM studies of the Eel River and New Jersey continental shelf are provided below.

The Eel shelf has a very energetic wave environment and yet mud is able to accumulate on the shelf. The Eel River provides a large source of sediment to the shelf, with estimated fine-grained ($< 63\ \mu m$) sediment discharges averaging 15 Mt yr^{-1} (Sommerfield & Nittrouer, 1999). Analysis of modern and ancient shelf deposits indicates that very fine-grained mud (predominantly $< 20\ \mu m$) layers are preserved in the bed (Leithold, 1989; Drake, 1999). Yet analysis of transport on the Eel shelf indicates that wave and current conditions are sufficient to remove at least the very fine-grained mud delivered to the shelf (Harris & Wiberg, 2002). The source of the apparent contradiction appears to lie in the properties of fine-grained deposits that are not adequately accounted for in the most commonly used shelf sediment-transport models, because they were developed for sediment that behaves like sand.

Several processes are important in the transport of mud that generally can be neglected when modelling sand transport: compaction and cohesion, flocculation and near-bed stratification produced by sediment in suspension. Each of these tends to limit the suspended-sediment flux. Compaction and cohesion place a limit on entrainment rates, thereby reducing the volume of sediment in suspension (Wiberg, 2000, Sanford & Maa, 2001). Flocculation of suspended sediment on the Eel shelf increases the settling of sediment from a flood plume to the seabed (Harris *et al.*, 2000; Hill *et al.*, 2000) and is an important control on limiting near-bed suspended-sediment concentrations (Harris & Wiberg, 2001; Carr, 2002). Observation and modelling suggest that near-bed stratification associated with fine-grained suspensions strongly affects the distribution of sediment in the water column (Friedrichs *et al.*, 2000). This is particularly true near the top of the **wave-boundary-layer** when current shear velocities are significantly smaller than wave shear velocities (Reed *et al.*, 1999). Under these conditions, stratification can inhibit the upward diffusion of sediment out of the wave-boundary-layer, resulting in wave-boundary-layer concentrations that under some conditions reach fluid-mud concentrations (Ogston *et al.*, 2000; Traykovski *et al.*, 2000; Scully *et al.*, 2002). The resulting **fluid-mud**

suspensions can travel across the shelf as density currents (Parsons *et al.*, this volume, pp. 275–337). Models of this cross-shelf gravity-driven transport process indicate that it is important in forming the flood deposits found on the Eel shelf (Traykovski *et al.*, 2000; Scully *et al.*, 2003; Harris *et al.*, 2005).

One-dimensional shelf sediment-transport models provide a good framework for investigating the detailed mechanics of transport, but they cannot account for spatial variations in flow and/or bed sediment properties that might lead to divergences in sediment flux. Calculating patterns of net erosion and deposition, as well as changes in bed texture on the shelf, requires two- or three-dimensional modelling. One approach is to extend the equation for conservation of sediment mass to yield a **time-dependent advection-diffusion** equation in two dimensions (vertical and one horizontal dimension)

$$\frac{\partial c_s}{\partial t} = -u\frac{\partial c_s}{\partial x} + \frac{\partial}{\partial x}\left(K_s\frac{\partial c_s}{\partial x}\right) + w_s\frac{\partial c_s}{\partial z} + \frac{\partial}{\partial z}\left(K_s\frac{\partial c_s}{\partial z}\right) \quad (3)$$

and to couple it with an equation for conservation of mass in the bed

$$\frac{\partial \eta}{\partial t} = -\frac{1}{c_{bed}}\left(\frac{\partial q_{sx}}{\partial x} + \frac{\partial V_s}{\partial t}\right) \quad (4)$$

where u is the x-component of flow velocity, η is bed elevation, c_{bed} is the sediment concentration of the bed surface (1-porosity), q_{sx} is the x-component of sediment flux and V_s is the volume of sediment in suspension (Zhang *et al.*, 1999; Harris & Wiberg, 2001, 2002). Applied to a cross-shelf transect, these models capture the effects of sediment advection, bed winnowing, and net erosion and deposition on sediment redistribution and the formation of fine-scale stratigraphy on the shelf. These models show that there is a general tendency for winnowing of fine-grained sediment from the inner shelf with subsequent deposition on the mid-shelf. Resulting depths of net erosion and deposition are in the order of several centimetres for typical storm events on the Eel shelf (Zhang *et al.*, 1999; Harris & Wiberg, 2002).

While useful for developing an understanding of the importance of various processes to cross-shelf sediment redistribution, fully two-dimensional (*x–z*) advection-diffusion models of shelf sediment

transport have several drawbacks that limit their potential for general operational use or direct extension to three dimensions. First, they are computationally intensive because the equations must be solved on a two-dimensional grid at relatively small time-steps to assure numerical stability. Second, the models require input of near-bed wave and current conditions across the shelf. While wave conditions can in some cases be determined from deep-water measurements of surface wave spectra, as in the case of the Eel shelf, no comparable information is available on cross-shelf variations in currents. Currents are typically measured at one or several cross-shore locations, but never at the resolution of a model grid, and rarely for long periods of time. Furthermore, while along-shelf currents tend to be correlated over large distances, across-shelf currents tend to exhibit little or no correlation, even over relatively short distances.

One solution is to use an ocean circulation model to characterize the currents. For example the cross-shelf model SLICE computes boundary-layer currents, sediment transport and bed modification (Reed *et al.*, 1999). It embodies the dynamics of the two-dimensional models described above, while relying only on an imposed wind field and wave and tidal conditions as forcing terms. Spatial wind fields are available globally, which allows the SLICE model to be used in locations where no direct measurements of currents exist. While some shelf systems can be reasonably approximated as a two-dimensional system (e.g. no significant along-shelf variations), many cannot. In these cases, a full **three-dimensional circulation model** must be used to represent the currents on the shelf. Three-dimensional models, such as the Princeton Ocean Model (Blumberg & Mellor, 1987), not only include effects of waves, winds, tides and mean currents, but can also account for freshwater buoyancy, which is important in understanding the delivery of sediment to the shelf during river flood events (Geyer *et al.*, 2000). Application of a three-dimensional circulation model (ECOM-SED) that includes suspended-sediment transport equations has revealed that the volume and location of deposition are critically sensitive to the settling properties of fine-grained flood sediment (Fig. 9; Harris *et al.*, 2000, 2005).

The sediment transport models described above are primarily used to investigate shelf sediment

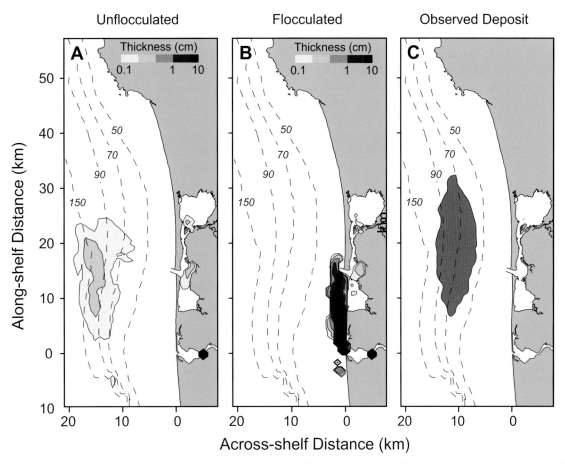

Fig. 9 Flood deposits on the Eel shelf following the January 1997 flood of the Eel River. (A) Calculated when sediment in the flood plume settled as individual grains. (B) Calculated when sediment in the flood plume settled as flocs. (C) The observed flood deposit. The calculations were made using a three-dimensional circulation model with suspended-sediment transport. Although the unflocculated calculations put the flood deposit in approximately the correct position, the thicknesses are too small. Other observations suggest that the calculated flocculated deposit (B) is located in the region of initial plume deposition, and that a cross-shelf gravity-driven fluid-mud flow transferred the sediment from the inner to the mid-shelf shortly after the flood (see Fig. 6). (After Harris *et al.*, 2002.)

transport during storm or flood events for which the forcing conditions (waves and currents or winds) are known. When modelling the long-term evolution of margin morphology or stratigraphy, a different approach is needed. First, there are no direct measurements of forcing conditions over these long time-scales and, second, the small (several hours or less) time-steps used in the models are neither computationally practical nor necessary to capture long-term patterns of shelf sedimentation. There are several ways to represent shelf sediment-transport processes over long time periods. One approach is to force a simple one-dimensional shelf sediment-transport model with conditions representative of a day or a year, rather than an hour or minutes.

In doing so, care must be taken to find statistics of the forcing conditions that, when used as input to the model, yield appropriate average sediment fluxes. For example, characterizing waves in terms of annual average conditions or currents in terms of daily or longer averages will underestimate sediment transport rates owing to the non-linear relationship between transport and wave height (or orbital velocity) and to the neglect of tidal currents, respectively.

Another approach is to develop a time-averaged form of the **conservation-of-mass equation** for suspended-sediment concentration, to yield an equation for long-term average sediment fluxes on the continental shelf. Niedoroda *et al.* (1995), for example, proposed a **two-dimensional, depth-averaged and time-generalized equation** in which the sediment flux on the shelf was related to advection, wave-biased bedload transport, and large-scale diffusion. The diffusive fluxes were specified such that an equilibrium shelf profile is maintained on average (Niedoroda *et al.*, 1995). This model SEQUENCE has been used to represent shelf processes in a model of long-term margin stratigraphy (Steckler *et al.*, 1996), which is described below. Alternatively, available measurements of wave and current conditions can be used to formulate shelf diffusion coefficients representative of specific shelf locations, which can be used in a diffusive model of shelf sediment fluxes (Wiberg, 2001).

A third approach to incorporating shelf sediment-transport processes in **continental-margin models**

is to develop simple analytical representations of the important processes. As an example, a recent **analytical shelf-transport model** based on the importance of wave-supported sediment gravity flows to cross-shelf transport on energetic shelves subject to high loads of fine sediment (Traykovski *et al.*, 2000; Wright *et al.*, 2001; Scully *et al.*, 2002) has been implemented in SEDFLUX. The model is based on two relationships, a linear momentum balance for sediment-laden gravity flows and a critical Richardson number which determines the maximum amount of sediment that can be suspended before stratification effects shut down turbulence production (Box 4). From these, gravity-flow sediment flux, sediment deposition rates and equilibrium shelf profiles can be calculated. Results suggest that margins with large sediment

---

Box 4 Gravity-driven sediment flows from wave resuspension on continental shelves

The maximum sediment load carried in a shelf bottom boundary layer (BBL) is determined by the critical Richardson number, R_{cr}, defined as the ratio of the depth-integrated buoyancy anomaly B to bottom velocity shear U^2_{max} (Trowbridge & Kineke, 1994)

$$Ri_{cr} = \frac{B}{U^2_{max}} = 1/4 \tag{B4.1}$$

and

$$U_{max} = \sqrt{U^2_w + U^2_c + U^2_g} \tag{B4.2}$$

where U_w is near-bed velocity due to the action of waves, U_c is the near-bed velocity due to ambient currents, and U_g is the velocity of the bottom boundary layer (BBL) under the influence of gravity. Shear instabilities occur when $Ri < Ri_{cr}$ and shear instabilities are suppressed for $Ri > Ri_{cr}$. If excess sediment enters the BBL, and Ri increases beyond Ri_{cr}, then turbulence is dampened. In that case, sediment settles out of the boundary layer, stratification in the boundary layer is reduced and Ri returns to Ri_{cr}. If excess sediment settles out of the BBL, or bottom stress increases and Ri decreases beyond Ri_{cr}, then turbulence intensifies. Sediment then will re-enter the base of the boundary layer, stratification will increase in the lower boundary layer and Ri returns to Ri_{cr}.

At high suspended-sediment concentrations in the BBL, the momentum balance for shelf gravity-driven sediment flow is (Wright *et al.*, 2001)

$$\alpha B = C_D \langle |\vec{u}|u \rangle = C_D U_{max} U_g \tag{B4.3}$$

where C_D is the coefficient of drag, $\langle |\vec{u}|u \rangle$ is the wave-averaged, cross-shelf component of the quadratic velocity and U_{max} is the total velocity. The sediment deposition rate D is given by

$$D = \frac{Ri^2_{cr}}{(1 - \phi)C_D g \left(\frac{\rho_{sw}}{\rho_w} - 1 \right)} \frac{d}{dx} (\alpha U^3_{max}) \tag{B4.4}$$

where ϕ is porosity, ρ_s is grain density and ρ_{sw} is the density of the ambient seawater.

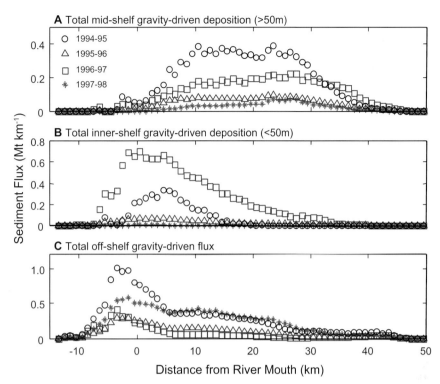

Fig. 10 Modelled along-shelf distribution of cumulative sediment deposition on the Eel shelf. (A) Mid-shelf gravity-driven deposition. (B) Inner-shelf deposition. (C) Off-shelf gravity-driven flux. Predictions based on a two-dimensional model of large-scale gravity-flow deposition run for four flood seasons (1994–1998) on the shelf (Scully *et al.*, 2003). The position of the maximum mid-shelf deposition, roughly 10–30 km north of the river mouth, is consistent with observations of the flood deposit (Fig. 9C). When the river delivers sufficient sediment to critically stratify the wave boundary layer, wave intensity and shelf bathymetry are the main factors controlling the pattern of deposition.

loads and small waves develop narrower shelves than margins with smaller sediment loads and large waves (Fig. 10).

Slope stability analysis

Slope stability analysis determines the likelihood that a sedimentary deposit will fail. Failure may lead to downward translation of some or all of the sediment, as a slide, a slump or a gravity flow (e.g. debris flow, turbidity current). Movement of an underwater sediment mass may also generate a tsunami. The stability of natural slopes is usually evaluated by **limiting equilibrium models**, which can take into account all of the factors that influence the shearing resistance of the sediment in the slope. However, the models do not produce a prediction of how much the slope will deform or how the sediment will behave after failure has occurred. They only predict whether or not the slope will fail (Lee *et al.*, this volume, pp. 213–274).

All limiting equilibrium approaches share a number of elements than can be described in four steps. First, a **slip mechanism** is postulated. The simplest configurations assume failure along planes or circular sliding surfaces. When detailed information about the geometry and properties

of the sediment mass is available, more complex shapes can be assumed.

Second, the **shearing resistance** (shear strength) required for stability of the postulated failure mechanism is calculated. This calculation can commonly be made in a number of different ways depending upon which conditions for equilibrium are required to be satisfied. In fact, the common slope-stability-analysis techniques differ from each other mainly as a result of which conditions are satisfied and which are not. Often, a two-dimensional profile of the slope and assumed failure plane is drawn, and the failing block of sediment is divided into a number of vertical slices (Fig. 11). The forces acting on the base and sides of each slice are calculated so that a force balance can be made, and the shear strength required for stability is determined. As is shown in many soil mechanics textbooks (Lambe & Whitman, 1969), a slope consisting of a series of slices at limiting equilibrium is statistically indeterminant, and the forces cannot be calculated without simplifying assumptions. For example, the 'ordinary method of slices' (Fellenius, 1936) is based on the simplifying assumption that the forces acting between slices are parallel to the failure surface. This results in a straightforward calculation (Nash, 1987) but can lead to possible errors of

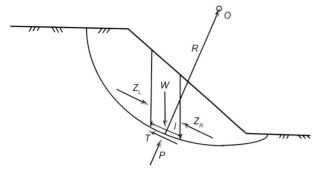

Fig. 11 Depiction of a postulated circular failure surface on a finite slope. The circle has a centre at O and a radius, R. An example slice is shown with weight, W, basal shear force, T, basal normal force, P, and basal length, l. The forces acting on the sides of the slice, Z_R and Z_L, are parallel to the part of the failure surface that is at the bottom of the slice ('ordinary method of slices' after Nash, 1987).

as much as 60% (Whitman & Bailey, 1987). Other methods are more accurate but involve more complex calculations, sometimes including trial and error solutions of equations.

As a third step in evaluating slope stability, the available **shear strength** along the postulated failure surface is calculated. Here, two choices are available: the total-stress approach and the effective-stress approach. Both are based on the Mohr–Coulomb failure criterion:

$$s = c + \sigma \tan \phi \qquad (5)$$

where s is shear strength, c is cohesion, σ is normal stress and ϕ is friction angle.

Empirical studies have shown that the shear strength of saturated sediment is a constant if the volume of the sediment (porosity) stays constant. Such a condition, termed **undrained loading**, occurs when loading is much more rapid than the ability of the sediment to expel or take in water as a result of the loading. An example of this condition might be earthquake or storm-wave loading on relatively fine-grained (low permeability) sediment. For undrained loading the friction angle, ϕ, *appears* equal to zero and the strength *appears* equal to the cohesion, c. When these conditions are satisfied, a total stress approach can be used, in which internal changes in pore pressure can be ignored. This greatly simplifies the analysis.

The **effective stress approach** is more general and applies to slower loading or more permeable sediment. Here, the Mohr–Coulomb failure criterion is generally written as:

$$s = c' + \sigma' \tan \phi' \qquad (6)$$

where σ' is the normal effective stress (equal to the total normal stress minus the pore-water pressure) and c' and ϕ' are the cohesion and friction angle defined in terms of effective stresses. The effective stress approach makes it possible to consider pore pressures generated by rapid sedimentation or groundwater seepage as well as slow loading or fairly pervious sediment. The drawback is that the full distribution of pore-water pressures must be estimated.

In the fourth step, a **factor of safety**, F, is calculated (also see Box 5)

$$F = \frac{\text{shear strength available}}{\text{shear strength required for stability}} \qquad (7)$$

After these first four steps, a different mechanism of failure is assumed (different failure surface, for example), and the above steps are followed again to yield a new factor of safety. The procedure is followed iteratively until a full range of possible failure mechanisms has been explored and a series of factors of safety has been calculated. The failure mechanism that produces the lowest factor of safety is assumed to be the critical one. If the critical factor of safety is ≤ 1, the slope is determined to be unstable.

One popular method is Bishop's simplified method (Bishop, 1955) within which a circular failure surface is assumed, and the forces between slices are taken to be horizontal. Another popular method is that of Janbu *et al.* (1956), which allows for non-circular failure surfaces. Initially it assumes that there are no shear stresses between slices and then provides a correction factor to allow for non-zero shear stresses (Box 5). These slope-stability models are applicable to the seafloor. In many cases, however, they cannot be applied very accurately because the variation of strength properties within the seafloor and the distribution of pore-water pressures are poorly known. As a consequence, scientists investigating seafloor slope stability have often used a less sophisticated approach, namely

---

Box 5 Factor-of-safety analysis of finite surfaces adapted for earthquake accelerations

The geometry and location of failures for finite non-circular slip surfaces can be determined using the Janbu factor-of-safety analysis with the method of slices (Janbu, 1956). The method ignores interslice forces, although a technique is available to correct for these forces (Anderson & Richards, 1987). Interslice forces are invariably small, adjusting the calculated factor of safety by less than 10%, depending on the geometry of the problem as well as the soil condition. The static stability of a possible failure plane is characterized through its factor of safety as (Hutton & Syvitski, 2004)

$$F_T = \frac{\sum_{i=0}^{n}\left[b_i\left(c_i + \left(\dfrac{W_{vi}}{b_i} - u_{e_i}\right)\tan\phi_i\right)\dfrac{\sec\alpha_i}{1+\dfrac{\tan\alpha_i\tan\phi_i}{F_T}}\right]}{\sum_{i=0}^{n}W_{vi}\sin\alpha_i + \sum_{i=0}^{n}W_{hi}\cos\alpha_i} \qquad (B5.1)$$

where F_T is the factor of safety for the entire sediment volume (with iterative convergence to a solution), b_i is the width of the *i*th slice or column, c_i is sediment cohesion; ϕ_i is internal friction angle, W_{vi} is vertical weight of the column, α is the slope of failure plane, W_{hi} is horizontal pull on column, and u_e is excess pore pressure.

The vertical weight of column W_{vi} is then modified by the earthquake load, such that

$$W_{vi} = M(g + a_v) \qquad (B5.2)$$

where M is the mass of sediment column, g is the acceleration due to gravity and a_v is the vertical acceleration due to the earthquake. The horizontal pull on the sediment column W_{hi} is

$$W_{hi} = Ma_h \qquad (B5.3)$$

where a_h is the horizontal acceleration due to the earthquake.

---

the infinite-slope model (Fig. 12). Within this model, the seafloor is represented by a long, wide ramp having a uniform slope steepness, α. Failure is assumed to occur along planes parallel to the slope. The representation is so simple that the factor of safety of the slope can be calculated directly using the **effective-stress approach** (Nash, 1987):

$$F = [c' + (\gamma'z\cos^2\alpha - u)\tan\phi']/(\gamma'z\sin\alpha\cos\alpha) \qquad (8)$$

where z is depth below the seafloor, γ' is submerged density of sediment and u is pore-water pressure. Using the **total-stress approach** (ϕ' appears equal to 0):

$$F = s_u/[\gamma'z\sin\alpha\cos\alpha] \qquad (9)$$

where s_u is the undrained shear strength of the sediment.

The **infinite-slope model** was applied to the seafloor, and modified to allow for seismic loading (Morgenstern, 1967). Later, Morgenstern's infinite-

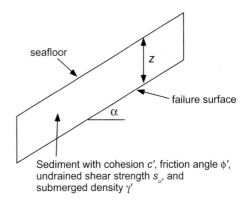

Fig. 12 Illustration of the infinite-slope simplification for seafloor stability analysis.

slope seismic-loading analysis was expanded to include representations of shear-strength profiles using the **normalized-soil-profile** (NSP) **method** (Lee & Edwards, 1986). This approach, in turn, allowed relative slope stability to be calculated at a point, given data for environmental parameters, such as sediment density, slope steepness and regional seismicity patterns. These can be represented as layers within a geographical information system (GIS) that can be manipulated with slope-stability algorithms to yield relative estimates of slope-failure potential (Lee *et al.*, 2000).

Subaqueous debris flows and turbidity currents

Subaqueous debris flows and turbidity currents are the main mechanisms for delivering sediment from shallow water to deep water. They constitute major mechanisms for the sculpting of the morphology and creating the stratigraphy of the continental shelf, continental slope and abyssal plain. Subaqueous debris flows and turbidity currents are cousins (Fig. 13). Both are dense bottom flows that are driven downslope because sediment makes them heavier than the ambient water (Parsons *et al.*, this volume, pp. 275–337, for details).

A dilute sediment-driven underflow, i.e. one for which the layer-averaged concentration of sediment is much smaller than unity, is here termed a turbidity current. **Turbidity currents** are usually, but not universally, turbulent. A very slow-moving laminar turbidity current containing a dilute suspension of clay particles can move long distances before the clay eventually settles out and the current dissipates; the sustained motion is made possible by the very low fall velocity of the particles. It is likely that laminar turbidity currents play a role in moving fine sediment from the distal ends of submarine fans to abyssal plains. Turbulence provides a net upward flux of sediment that counteracts the tendency of grains to settle under their own weight. Turbulence is also a mechanism for the entrainment of bed sediment into the current. Without turbulent resuspension of sand from the bed, turbidity currents could not transport sand over long distances in a single event. If a turbidity current is sufficiently intense, a bedload layer may devolve into a 'sheet flow', i.e. a basal layer of dense flowing grains that is several grain diameters thick (Wilson, 1987; Horikawa, 1988; Fredsoe & Diegaard, 1994; Ribberink & Al-Salem, 1995). This sheet flow can be viewed as a momentum-driven debris flow, i.e. a dense layer moving as a mass flow that is driven by the transfer of momentum from the turbidity current above.

A dense sediment-driven underflow, i.e. one that contains similar volumetric fractions of water

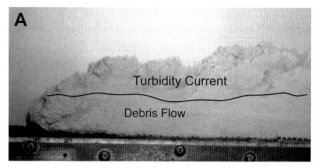

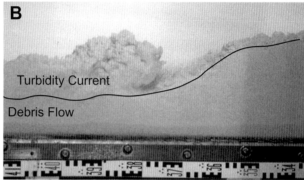

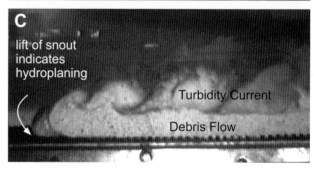

Fig. 13 Snouts of laboratory subaqueous debris flows. (A & B) Secondary turbidity currents occur above the debris flows. (C) Hydroplaning of the debris-flow snout is common. (After Marr, 1999.)

and sediment, is here termed a **debris flow** (Takahashi, 1991). While stony debris flows do occur in the subaqueous environment (e.g. in proglacial environments), flows rich in mud and relatively free of stones are more common. The maintenance of sustained debris flows composed of pure sand presents a problem, in that the excess pore pressures needed to reduce the internal angle of friction should dissipate rather rapidly. Small fractions of clay are sufficient to block the pores and prevent the dissipation of pore pressure (e.g. Fig. 13; Marr *et al.*, 2001). A flow thickness in the order of tens of metres could dissipate pore pressure sufficiently slowly to allow for a sustained debris flow (Iverson, 1997). Subaqueous debris flows may be laminar or turbulent (van Kessel & Kranenburg, 1996). Mechanisms that tend to suppress turbulence relate to mud content. Additionally, dense mixtures of water and sediment become turbulent much less easily than dilute suspensions. Experimental evidence suggests that higher velocity debris flows (tens of metres per second) may become turbulent, especially at the head of the debris flow, while the body remains laminar (Marr *et al.*, 2001; Fig. 13).

Turbidity currents may be either pulse-like or continuous, the latter implying that the current passing a cross-section is sustained for many minutes, hours, or days after the passage of the head. In the case of continuous currents, the body is much longer than the head. Subaqueous debris flows tend to move in pulses associated with the discrete release of material. The pulse will elongate as the flow propagates downslope.

Debris flows and turbidity currents can occur together, i.e. a secondary debris-flow layer under a turbidity current. Debris flows can also generate secondary turbidity currents by the suspension of debris at the head, or along the body. On steep slopes, the debris flow outruns the turbidity current it generates, but as the slope drops the turbidity current can run past the front of the debris flow (Mohrig *et al.*, 1998). The demarcation between the debris flow and the turbidity current is often unclear at the head of the debris flow due to intense turbulence there, but is usually quite distinct behind the head (Fig. 13; Marr, 1999). Flows that are transitional between turbidity currents and debris flows can and do occur, however, transitional flows relatively quickly evolve into one or the other (Marr, 1999; Marr *et al.*, 2001).

Generation

The generation of subaqueous debris flows is largely governed by slope stability (Lee *et al.*, this volume, pp. 213–274). If the failed material simply slides as a block and deposits, it is properly called a **submarine landslide**. If it sufficiently disaggregates to allow for internal deformation, it makes the transition to a debris flow. The process of disaggregation is often incomplete in many submarine debris flows.

The mechanisms generating turbidity currents are far more diverse.

1 If the sediment concentration in a river's discharge leads to water denser than that of the ocean water, the river will plunge and convert directly into a continuous turbidity current referred to as a hyperpycnal flow (see above). Under present-day conditions, hyperpycnal flows into the ocean are rare (Mulder & Syvitski, 1995), but may deliver huge amounts of sediment during flood events.

2 If the head of a submarine canyon is located in shallow water, sediment resuspended by storm waves may move down the canyon as a turbidity current, e.g. Scripps Submarine Canyon, California (Inman *et al.*, 1976) and Eel Canyon, California (Puig *et al.*, 2003, 2004).

3 Failure of relatively well-sorted sediment, delta foresets for example, can lead to the direct formation of turbidity currents without the intermediary of a debris flow (Hay, 1987). Retrogressive failures may help to generate sustained events.

4 Suspension of sediment from the head and forward body of a debris flow can lead to the formation of a turbidity current (Parker, 1982; Norem *et al.*, 1990; Mohrig *et al.*, 1998; Marr *et al.*, 2001). High mud content, however, tends to suppress the disaggregation that allows for easy entrainment of the sediment into suspension.

5 Well-sorted sand may spall off steeply dipping slopes, and the ensuing rain of sediment may convert into a continuous turbidity current (Van den Berg *et al.*, 2002).

6 Breaking internal waves focused by a submarine canyon are here hypothesized to be able to generate turbidity currents that then move down the canyon.

7 Blobs of sediment-laden water that settle to the ocean bottom boundary layer, underneath hypopycnal (surface) river plumes, may coalesce into a turbidity current (Parsons & García, 2000).

Modelling of subaqueous debris flows

An early one-dimensional numerical model depicted the submarine debris flow as a **Bingham fluid**, i.e. a flow with discrete yield strength below which no deformation occurs, and a viscosity relating the strain rate to the excess in shear stress above the yield stress (Jiang & LeBlond, 1993). A slender-flow approximation and a kinematic boundary condition were employed to integrate the governing equations in the upward-normal direction. Such a flow invariably possesses an upper plug layer and a lower deforming shear layer. The one-dimensional integral governing equations consist of one relation for mass balance, one relation for momentum balance in the plug layer, and one relation for mass balance in the shear layer.

Huang & García (1998) generalized the model of Jiang & LeBlond (1993) to a **Herschel–Bulkley rheology**, as advocated by Coussot (1994). The Herschel–Bulkley formulation also possesses a **yield stress**; the excess of the shear stress above the yield stress depends on the strain rate to an arbitrary power. A perturbation method was used to obtain an analytical solution of debris-flow movement, for application to continental-slope dynamics (Huang & García, 1998, 1999). A generalized one-dimensional numerical model for subaqueous debris flows termed BING (see Box 6; Imran *et al.*, 2001) allows the choice of one of three rheologies: Bingham, generalized Herschel–Bulkley and bilinear (Locat, 1997). The **bilinear rheology** replaces an actual yield stress with an 'apparent' yield stress. Well below this yield stress, the fluid deforms as a Newtonian fluid with a very high viscosity; well above it, the fluid deforms as a lower-viscosity Bingham fluid, with a smooth transition over a relatively narrow regime in between.

Box 6 The conservation equations for a debris flow

Debris flows can be modelled using the properties of a Bingham plastic (viscoplastic) fluid, with deformation driven by the excess of stress beyond the yield stress (Pratson *et al.*, 2000a; Imran *et al.*, 2001). Such models neglect the tangential stress acting on the water–mud interface, because the viscosity of water is much smaller than that of the mud, and the basal shear of the mudflow is much greater than the interfacial shear. Often these models allow for a no-slip condition at the slide bottom. The governing equations of the Lagrangian form for the depth-averaged debris flow equations, including both viscous and plug-flow regions, are then

Continuity

$$\underbrace{\frac{\partial h}{\partial t}}_{a} + \underbrace{\frac{\partial}{\partial x}\left[U_p h_p + \tfrac{2}{3} U_p h_s \right]}_{b} = 0 \tag{B6.1}$$

Momentum (shear layer)

$$\underbrace{\tfrac{2}{3}\frac{\partial}{\partial t}(U_p h_s) - U_p\frac{\partial h_s}{\partial t} + \tfrac{8}{15}\frac{\partial}{\partial x}(U_p^2 h_s) - \tfrac{2}{3}U_p\frac{\partial}{\partial x}(U_p h_s)}_{a} = \underbrace{h_s g\left(1 - \frac{\rho_{sw}}{\rho_m}\right)\alpha}_{b} - \underbrace{h_s g\left(1 - \frac{\rho_{sw}}{\rho_m}\right)\frac{\partial h}{\partial x}}_{c} - \underbrace{2\left(\frac{\mu U_p}{\rho_m h_s}\right)}_{d} \tag{B6.2}$$

Momentum (plug-flow layer)

$$\underbrace{\frac{\partial}{\partial t}(U_p h_p) + \frac{\partial}{\partial x}(U_p^2 h_p) + U_p\frac{\partial h_s}{\partial t} + \tfrac{2}{3}U_p\frac{\partial}{\partial x}(U_p h_s)}_{a} = \underbrace{h_p g\frac{\partial}{\partial x}\left(1 - \frac{\rho_{sw}}{\rho_m}\right)\alpha}_{b} - \underbrace{h_p g\left(1 - \frac{\rho_{sw}}{\rho_m}\right)\frac{\partial h}{\partial x}}_{c} - \underbrace{\frac{\tau_r}{\rho_{sw}}}_{d} \tag{B6.3}$$

where h is the total thickness of the debris flow ($h_p + h_s$), h_p and U_p are the thickness and layer-averaged velocity of the upper plug zone, respectively; h_s and U_s are the thickness and layer-averaged velocity of the lower shear layer, respectively; g is acceleration due to gravity; α is slope; ρ_{sw} is density of ocean water; ρ_m is density of the mud flow, τ_y is the yield (or remoulded shear) strength, and μ is the dynamic viscosity.

The thickness of the debris flow (term B6.1*a*) is directly proportional to its velocity (term B6.1*b*). If the flow slows it will thicken, and if it speeds up it will thin. The accelerations of the upper and lower layers of the debris flow (terms B6.2*a* and B6.3*a*, respectively) are directly proportional to the weight of the flow scaled by the seafloor slope (terms B6.2*b* and B6.3*b*, respectively), and fluid pressures produced by lateral variations in flow height (terms B6.2*c* and B6.3*c*, respectively), plus frictional forces (terms B6.2*d* and B6.3*d*, respectively). For the shear layer, the frictional force relates to the viscosity of the flow matrix, whereas for the plug layer it derives from the yield strength of the matrix.

BING employs the Lagrangian formulation of Savage & Hutter (1991), and may be applied to arbitrary slopes in the down-dip direction. Flow is started from the collapse of a parabolically shaped pile of debris of arbitrary length and height, and is allowed to run freely down the slope. In the case of a Bingham or Herschel–Bulkley rheology, the flow eventually comes to a complete stop, due to thinning of the flow, usually abetted by a drop-off of slope in the down-dip direction. BING has been applied to debris flows that are inferred to create stacked deposits in the Norwegian–Barents Sea Margin (Elverhøi *et al.*, 1997) and on the East Greenland margin (Syvitski & Hutton, 2003). Applied to the Bear Island Trough Mouth Fan and the Isfjorden Fan, BING was able to capture the runout of debris flows on two submarine fans that are composed almost entirely of stacked debris-flow deposits, using a Bingham rheology. Yield strengths (1–25 kPa) and dynamic viscosities (30–300 Pa s) used to reproduce the observed patterns of run-out are much larger than those that reproduce the laboratory experiments of Mohrig *et al.* (1999) (~36 Pa and 0.023 Pa s). Numerically generated stacked-debris-flow deposits show a marked thickening in the deposit just downstream of the slope break (Pratson *et al.*, 2000a), similar to stacked-debris-flow deposits at field scale (Laberg & Vorren, 1995). Such deposits are easily distinguished from the morphology and stratigraphy produced by a similar model for turbidity currents (BANG described below).

Existing models of subaqueous debris flows can be improved in a number of important areas: (i) the initiation process; (ii) the gradual disaggregation of consolidated material, with a resulting change in flow characteristics; (iii) flow resistance at the upper interface between debris and water, and thus the formation of a secondary turbidity current from either the head or the body of the debris flow; (iv) variation in pore pressure, of particular importance for describing sandy debris flows; (v) remobilization of antecedent deposits by debris flows; and (vi) the dynamic pressure as the head of the debris flow ploughs through the water in front of it.

A dramatic difference exists between subaerial and subaqueous debris flows; the latter can hydro-plane (Mohrig *et al.*, 1998). The moving front of a subaqueous debris flow creates a dynamic pressure wave in front as it moves water out of the way. The dynamic pressure at the head depends linearly on the density of the ambient flow and as the square of front velocity. If the head velocity exceeds a critical value governed by a densimetric Froude number, a thin layer of water can slip under the head of the flow (Fig. 13C). This thin layer provides lubrication, and results in a dramatic reduction in the basal resistance to flow. The head, and in some cases part of the body of the debris flow, then rides on a cushion of water, allowing them to accelerate and place the body of the flow in tension. If the debris slurry has a tensile strength, thinning of the body due to hydroplaning of the head can lead to '**autoacephalation**', a process by which the debris flow decapitates itself (Fig. 14; Marr *et al.*, 2001). This appears to be the mechanism for the formation of the outrunner 'glide blocks' (Prior, 1984; Nissen *et al.*, 1999). **Hydroplaning** suppresses remobilization of the antecedent sediment over which the flow runs (Mohrig *et al.*, 1999). In experiments on three-dimensional stacking of subaqueous debris flows, however, significant remobilization can occur when autoacephalated glide blocks collide with antecedent glide blocks (Toniolo *et al.*, 2003). An analytical solution to represent hydroplaning in models of submarine debris flows (Harbitz *et al.*, 2003) indicates that once hydroplaning is initiated, the glide block can continue to move downslope at constant speed as long as the bed remains smooth and the slope does not change. The velocity of flow is mediated by an order-one densimetric Froude number.

Subaqueous debris flows usually produce subsidiary turbidity currents. The turbidity current is caused by the shear stress at the interface between the debris flow and the ambient water above. If the debris flow head is sufficiently swift, and the debris material is sufficiently disaggregated into fine-grained particles, suspension at the head can be significant (Mohrig *et al.*, 1998, 1999; Marr *et al.*, 2001). If the head is moving rather slowly, or the slurry contains a highly cohesive element that prevents complete disaggregation, the subsidiary turbidity current may be negligible (Mohrig & Marr, 2003).

Modelling of turbidity currents

Two families of predictive models are used to explore the dynamics of turbidity currents, namely,

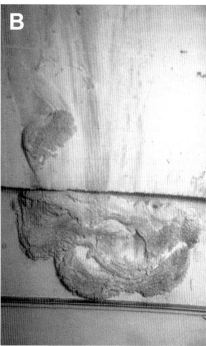

Fig. 14 View of an experimental debris-flow deposit showing (A) autoacephalated heads, and (B) runout blocks that have hydroplaned out in front of the main deposit (shown in A). (After Marr *et al.*, 2001.)

the vertical-structure model and the integral or layer-averaged model. Vertical-structure models employ the **Reynolds equations** in order to obtain flow variables that vary in the direction normal to the flow. **Integral models** are akin to the shallow-water equations for free-surface flows. An integral model for turbidity currents is derived by integrating the vertical structure of a turbidity current with the objective of obtaining a predictive model for the variation of mean flow velocity, current thickness and layer-averaged sediment concentration with distance from a given source.

The origin of **one-dimensional integral models** for turbidity currents can be traced back to the seminal work of Ellison & Turner (1959) on density currents. This model consisted of three layer-integrated conservation equations for flow momentum, flow volume and contaminant mass. The **Ellison–Turner (ET) model** only applies to conservative flows driven by density differences caused by temperature or salinity gradients. However, the ET model can be used as a first approximation to estimate the development in space of a turbidity current laden with fine-grained sediment such as clay and fine silt. The first predictive, one-dimensional integral model for erosive turbidity currents was the **Parker–Fukushima–Pantin (PFP) model**, which accounted for the balance of turbulent kinetic

energy (Parker *et al.*, 1986). The one-dimensional integral PFP model provided a major departure from the simpler Bagnold–Knapp criterion for 'autosuspension' (Knapp, 1938; Bagnold, 1962). The PFP model consists of layer-averaged balance equations for momentum, water volume, sediment mass and turbulent kinetic energy. It is valid only for dilute suspensions having layer-averaged concentrations of suspended sediment that are < 1 or 2% by volume. For its application, the model requires closure relations to estimate entrainment of sediment from the bed into suspension, entrainment of ambient water into the turbidity flow, bed friction and several shape factors. Most of the knowledge needed to close the four-equation model has come from observations of turbidity currents under controlled laboratory conditions (Parker *et al.*, 1987; García & Parker, 1989, 1993; Altinakar *et al.*, 1990, 1996; García, 1993).

Earlier modelling efforts with integral models concentrated on the spatial evolution of steady turbidity flows in water reservoirs (Akiyama & Stefan, 1986) and the ability of swift turbidity currents to scour out submarine canyons (Fukushima *et al.*, 1985). More recently, integral models have been used to study the dynamics of unsteady turbidity currents. An integral, three-equation model together with a **dissipative-Galerkin finite-element method** was used to study the propagation of a turbidity front along a sloping bottom (Choi & García, 1995). The results compared well against the relationship proposed by Britter & Linden (1980) to estimate the speed of density-current fronts and the experimental observations made by Altinakar *et al.* (1990) for weakly-depositional turbidity currents. Satisfactory results have also been obtained for the simulation of internal hydraulic jumps experienced by turbidity currents near a slope break (García & Parker, 1989; García, 1993).

The **method of characteristics** was employed to extract analytical solutions from the Parker–Fukushima–Pantin model (Bradford *et al.*, 1997). For turbid surges flowing into quiescent water, the correct speed of the propagating front could be obtained by either including a turbulent entrainment term in the continuity equation (Choi & García, 1995) or by specifying a finite acceleration of the wave front. The complex nature of turbidity-current fronts is not captured by integral models (Parsons & García, 1998). A one-dimensional layer-averaged model (Mulder *et al.*, 1998b) is still able to capture the general patterns of erosion and deposition by turbidity currents. A one-dimensional PFP model solved in a Lagrangian reference frame with a staggered-grid finite-difference method (BANG1D: Pratson *et al.*, 2000, 2001) is a natural choice for modelling of discontinuous, surge-like turbidity currents resulting from a finite source. As pointed out by Savage & Hutter (1991) for the case of granular flows, a Lagrangian approach helps in overcoming some of the numerical difficulties associated with the more traditional Eulerian framework used by all the integral turbidity-current models described above.

A **moving-boundary formulation** of river deltas, used two moving boundaries for the topset-foreset break and the foreset-basement break (Swenson *et al.*, 2000), with the latter break amended into a foreset-bottomset break by employing a plunging purely depositional turbidity current (Kostic & Parker, 2003a,b). Moving-boundary formulations provide a natural mechanism for connecting zones dominated by turbidity-current processes with zones driven by other mechanisms for sediment transport, such as delta fronts or the continental shelf.

To obtain numerical solutions with higher-dimensional models at continental-margin scales

is computationally very intense and practically impossible to achieve with today's computer power. Imran *et al.* (1998) used a two-dimensional version of the PFP model in conjunction with the Exner equation for bed sediment continuity to study the incipient formation of channels in submarine fans. Rates of sediment entrainment into suspension were obtained from a modified version of the relation proposed by García & Parker (1991) for open-channel suspensions, needed to prevent the oversuspension of fine-grained material.

A **finite-volume method** capable of capturing sharp fronts was developed using a two-dimensional integral model for unsteady, turbid underflows driven by sediment mixtures (Bradford & Katopodes, 1999a,b). The closure relations included the García & Parker (1991) sediment entrainment function for non-uniform bed sediment sizes, as well as an empirical expression to estimate the near-bed concentration of a given size fraction from the corresponding layer-averaged concentration (García, 1993). The model can simulate bed aggradation due to sediment deposition, channel avulsion and channelization in submarine fans.

A two-dimensional laterally spreading turbidity-current model was developed with the help of a dissipative-Galerkin finite-element method (Choi & García, 2002). The spreading of unconfined turbidity currents along a sloping bed could be accomplished for only a rather short period of time before becoming numerically unstable. Laboratory results coupled with dimensional analysis were used to develop a set of spreading laws that can be used to estimate the maximum width of the current at different times as well as the longitudinal spreading rate (Choi & García, 2001).

Vertical-structure models require a turbulence closure scheme in order to compute the vertical distribution of flow velocity and sediment concentration. As an example, a **mixing-length model** can estimate the vertical structure of turbidity currents (Stacey & Bowen, 1988a,b). The conditions for self-accelerating currents have been investigated using a *k-ε* **turbulence model** for turbidity currents (Farrell & Stefan, 1988; Eidsvik & Brors, 1989; Bournet *et al.*, 1999). The vertical structure of turbidity currents can be examined with the help of a **Reynolds stress model** (Brors & Eidsvik, 1992). The shape factors commonly used in the one-dimensional integral models for turbidity currents laden with fine-grained sediment were examined with the k-ε turbulence closure (Choi & García, 2002). Water entrainment rates were found to be in good agreement with the empirical water entrainment commonly used in layer-averaged models (Parker *et al.*, 1987).

A similar second-order **Mellor–Yamada 2.5 turbulence-closure scheme** was able to capture the effect of sediment-induced stratification on the vertical structure of a turbidity current (Felix, 2001). The effect from the presence of the particles on the turbulence other than the buoyancy was incorporated through a drag term leading to an extra dissipation term in the turbulent kinetic-energy equation and in the turbulent-length-scale equation. The equations were solved numerically using a finite-volume method. While the estimated flow magnitudes and thicknesses were of the same order as flows observed in nature, no comparison was made against observations.

Compaction

Compaction reduces the amount of pore space in a deposit as pore water is expelled. This action causes a sediment parcel to become more dense and cohesive and, in turn, less prone to failure. Compaction of a less permeable package of sediment (unable to expel its pore water) will result in an increase in its pore-water pressure, making it more prone to sediment failure. Some applications such as the engineering of foundations, dams and other structures may require a complete numerical compaction model (Das, 1983). However, to answer fundamental questions, such as those concerning the role of sea-level fluctuations on compaction, a simpler model that misses the finer details but captures the essence of the compaction is required. Sediment porosities typically decrease with depth as the weight of the overlying sediment compresses the underlying material. Models of this compaction process have shifted from simple empiricisms to complex numerical models (Bahr *et al.*, 2001). These models, however, do not illuminate the fundamental behaviour underlying this process. A simple exponential is frequently all that is necessary to describe the basic compaction process.

Analytical solution

A variety of fundamental processes affect the **compaction** of sediments, including the rearrangement of sediment grains, crushing of grains, plastic deformation, and the dissolution and re-precipitation of minerals (Waples & Kamata, 1993). For each of these processes, the amount that a parcel of sediment will compact depends on its degree of compactness. A sediment package that is already well packed will not be able to compact as much as one that is poorly packed. This leads to a modified version of **Athy's Law** (Athy, 1930) where the change in porosity due to an applied load is proportional to the porosity rather than burial depth

$$\frac{\mathrm{d}\phi}{\mathrm{d}\sigma} = -c\phi \tag{10}$$

where ϕ is porosity, σ is the applied load and c is an empirical constant that depends on the fundamental compaction processes mentioned above. This modification of Athy's Law is an important distinction, as excess pore-water pressures and vertical variations in sediment density can cause the change in porosity to be independent of burial depth. Thus, sediments that underlie a lighter load will compact less than those under a heavier load.

Sediment is composed of individual grains, so there is a minimum porosity ($\phi_{\min} > 0$) that can be reached when the grains are 'closest packed' (Klein & Hurlbut, 1985) and that cannot be reduced without crushing or deforming the grains. To account for this, Eq. 10 is modified so that there is no compaction for $\phi < \phi_{\min}$. Typical values of $\phi_{\min}$ are 0.3–0.5 for sands and 0.2–0.5 for silt and mud (Coogan & Manus, 1975). Bahr *et al.* (2001) presented an analysis that leads to an analytic solution with $\phi_{\min} = 0$. The result is the exponential solution

$$\phi(z) = \frac{\mathrm{e}^{-cg(\rho_s - \rho_w)z}}{\mathrm{e}^{-cg(\rho_s - \rho_w)z} + k_1} \tag{11}$$

where z is depth of sediment column, g is acceleration due to gravity, ρ_s is grain density, ρ_w is pore-water density and $k_1 = (1 - \phi(0))/\phi(0)$. This solution assumes that both grain density and excess pore-water pressure are constant with depth. The first of these assumptions is reasonable for most cases and the second assumption is only a low-order approximation, but helps to illustrate the underlying structure of the equation. A numerical solution to the more general case of Eq. 11 for $\phi_{\min} > 0$ is presented in the following section.

For large depths, the k_1 term dominates the denominator of Eq. 11. Therefore for large depths, the porosity varies exponentially with depth as

$$\phi(z) \approx e^{-cg(\rho_s - \rho_w)z} \quad (12)$$

In particular, if typical values for ρ_s, c and $\phi(0)$ are used, Eq. 11 becomes exponential for $z \gg 675$ m. To obtain an idea of how the porosity will vary for shallower depths, the Maclaurin expansion of Eq. 11 (with the same typical values) yields

$$\phi(z) \approx 0.6 - 1.43 \times 10^{-4}z - 8.52 \times 10^{-9}z^2 \quad (13)$$

For small depths the porosity is nearly linear with depth, when assuming a constant excess pore pressure.

In addition to the assumptions of constant excess pore pressure and $\phi_{min} = 0$, the analytical solution also assumes that sediment was deposited at a constant porosity of $\phi(0)$. Despite these shortcomings, the analytical solution strongly suggests that the underlying behaviour of compaction is fundamentally exponential in form.

Numerical solution

Bahr *et al.* (2001) devised a numerical method to solve the more general compaction equation (Eq. 11) that eliminates some of the assumptions used in the analytical analysis. Namely, different sediment types should compact at different rates (variable c), and to different minimum porosities (variable ϕ_{min}). For simplicity, the analysis below neglects excess pore-water pressure, but it is easily added. Following Eq. 10, the porosity of a cell of sediment that feels a load σ is

$$\phi = \phi_{min} + (\phi_0 - \phi_{min})e^{-c\sigma} \quad (14)$$

where ϕ_{min} and c are determined by the type of sediment in a cell, and ϕ_0 is the porosity of the sediment under zero loading.

This scheme divides a column of sediment into cells of arbitrary height that can contain any number of sediment types (ϕ_{min} and c values distinguish a sediment type). The assumption here is that when a cell of sediment is compacted, the grains of different sediment types do not interact with one another. Thus, a sediment cell compacts as if the different sediment types were compacted separately. For real sediment mixtures, the different types of grains will interact with one another to produce porosities that will be less than our approximation.

Comparison with empirical data

Bahr *et al.* (2001) used sediment-property data provided by the Deep Sea Drilling Project (DSDP) from a variety of marine cores. The DSDP data are sorted according to sediment type so as to provide aggregate core data on shale, siltstone and sandstone porosities as a function of depth. This separation ensures that the coefficient c, and minimum porosity ϕ_{min}, vary as little as possible from point to point within the data for each sediment type. In addition, these data are from relatively shallow burial depths (< 2000 m) where diagenesis is at a minimum.

The porosity data are apparently non-linear with depth (Fig. 15). However, on a log–linear plot (Fig. 16) the trend appears linear, indicating an exponential trend with depth, as the analytical solution suggests. The exponential fits show consistently better correlation coefficients (0.73 for

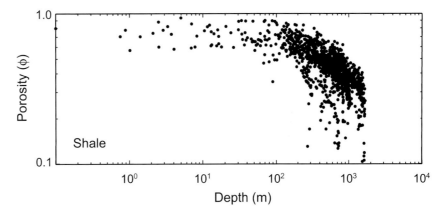

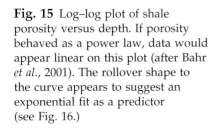

Fig. 15 Log–log plot of shale porosity versus depth. If porosity behaved as a power law, data would appear linear on this plot (after Bahr *et al.*, 2001). The rollover shape to the curve appears to suggest an exponential fit as a predictor (see Fig. 16.)

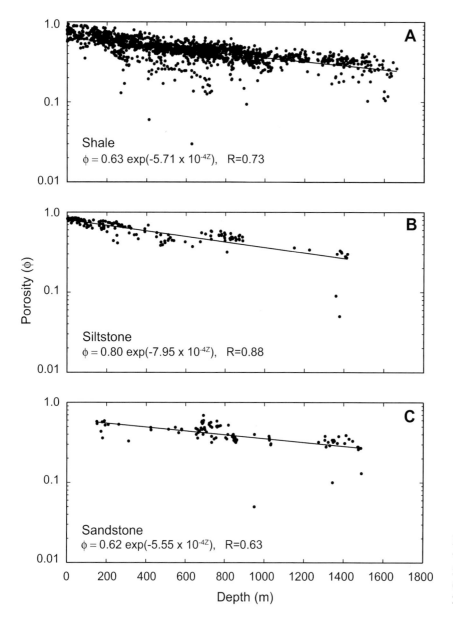

Fig. 16 Log–linear plots of porosity versus depth, along with exponential fits. (A) Shale (clay). (B) Siltstone. (C) Sandstone. (After Bahr *et al.*, 2001.)

shale, 0.88 for silt and 0.63 for sand) than linear or power-law fits. The exponential trend captures the general features of the data despite the scatter introduced by excess pore-water pressures and non-uniform lithologies.

In conclusion, for normally compacting sediments the porosity follows an exponential trend with depth. An analytical evaluation of the compaction equation, and comparison with DSDP marine core data, demonstrate this exponential behaviour. Although complex numerical models are useful in some contexts, empirical exponential curves, like those presented here, provide first-order approximations to the real compaction process. The simple exponential fits do not attempt to reproduce all of the porosity details, but they do model the essential porosity variations with depth.

Effect of compaction on stratigraphy

Compaction only occurs as a reaction to external forcing such as burial associated with sediment accumulation. The effect of compaction can be important over geological time (Reynolds *et al.*,

1991). The empirical data described above (Fig. 16) indicate that the surface sediments have porosities in the range of 0.7–0.8, while sediments at depth drop to porosities of 0.2–0.3. In fact, deeply buried sediments possess porosities of 0.01–0.03, as cementation develops, and for shale burial to 6 km depth, porosities decline to ≈ 0.02. Integrated over this depth range, the average porosity would be about 0.18. That is, compaction is responsible for reducing the height of this sediment column to 30% of its uncompacted thickness. Such an effect is on the same order as subsidence (Reynolds *et al.*, 1991). Furthermore, the large increase in sediment results in an equally large increase in the load felt by the substratum. This, in turn, increases subsidence rates and so again increases **accommodation space** below the sea surface.

INTEGRATED MODELS

Modelling individual sediment-transport processes affecting a continental margin or sedimentary basin has historically been the limit of our predictive capability. Modelling the entire spectrum of processes related to sediment delivery, initial deposition, reworking, dispersion and final burial has been desired but thought overwhelming. Previous multiprocess simulators employing geometric parameterization schemes could provide insight into sequence stratigraphy, but not the properties of the deposit (grain size, bulk density, porosity, bedding planes) formed by the interplay between the various processes during different climatic or sea-level regimes. The difficulty in dynamically modelling sedimentary architecture relates to the numerical task of duplicating inherently complex natural systems and the required computational power (Syvitski & Hutton, 2001). Only recently have advances in computer power and numerical techniques allowed a set of integrated process-based models to be built that evolve sedimentary sequences over short geological time-scales.

SEDFLUX approach

The ONR-sponsored STRATAFORM programme has brought together many investigators to study individual sedimentary processes and to combine individual event models into a fully interactive

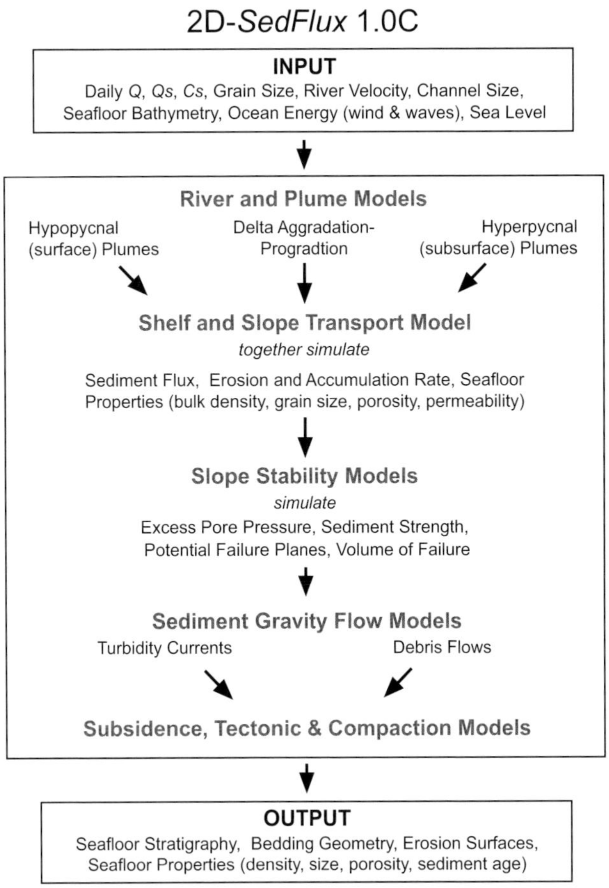

2D-*SedFlux* 1.0C

Fig. 17 Flow chart of 2D-SEDFLUX, showing required model inputs, model components and modelled outputs: Q, water discharge; Qs, sediment discharge; Cs, sediment concentration. (Modified from Syvitski & Hutton, 2003.)

whole-margin model. EARTHWORKS (Syvitski, *et al.*, 1997) allows the event models HYDROTREND and SEDFLUX to communicate with one another. EARTHWORKS takes a source-to-sink approach, modelling the transport, deposition, reworking and burial of sediment, with each process in the chain explicitly resolved. HYDROTREND (Syvitski *et al.*, 1998b; Syvitski & Morehead, 1999) produces a synthetic time series of a river's discharge and sediment load for thousands of years. This is an advance over previous process-response models which were limited in their ability to simulate the natural variability of a river's discharge (Syvitski & Hutton, 2001).

SEDFLUX simulates the lithological character of basin stratigraphy (Fig. 17) by integrating a series of process-based event modules to:

1 spread the fluvial bedload across supratidal and subtidal portions of an evolving delta plain;

2 disperse suspended sediment from a model river through either surface (hypopycnal: Syvitski *et al.*, 1998a) or subsurface (hyperpycnal: Skene *et al.*, 1997) plumes;

3 disperse and sort seafloor sediment by wave-current interactions from ocean storms (Harris & Wiberg, 2001; Wright *et al.*, 2001);

4 check the stability of slopes for failure potential (Syvitski & Alcott, 1995a; Syvitski & Hutton, 2003; Hutton & Syvitski, 2004);

5 redistribute any failed material by turbidity currents (Mulder *et al.*, 1997; Pratson *et al.*, 2001), or debris flows (Pratson *et al.*, 2000a);

6 change the basin's accommodation space due to subsidence, tectonics (Steckler, 1999; Syvitski & Hutton, 2001) and compaction of the deposits (Bahr *et al.*, 2001).

SEDFLUX processes and deposits interact with the ever-evolving boundary conditions (seafloor bathymetry, sea level and coastline position) to create a sedimentary architecture (Skene *et al.*, 1998). The modelled deposits are displayed as plots of sectional sediment properties, synthetic cores, and as synthetic seismic profiles (using acoustic models: Syvitski *et al.*, 1999; Kraft *et al.*, 2006), which together allow comparison of the model results with field or laboratory observations. The modelling environment has been developed by dozens of researchers since 1985 (for details see Syvitski & Hutton, 2001). There are both two-dimensional and three-dimensional versions of the model available (Overeem *et al.*, 2006); only the two-dimensional version is described here.

SEDFLUX *formulation*

2D-SEDFLUX is a two-dimensional basin-filling model able to simulate stratigraphy both in one horizontal dimension and vertically. Some of the process modules that contribute to this stratigraphy are modelled using vertically averaged variables that vary only horizontally (Table 3). Some processes, such as plume deposition, are averaged over a specified basin width. This width is user-defined and is constant in time but can vary in space.

The initial basin geometry defines the initial bathymetry at points along a profile, and has a user-specified resolution (≈ 10 m). The vertical resolution is user-specified (i.e. 1 cm) within which deposited sediment characteristics are averaged. Each component process uses its own unique resolution (both temporal and spatial) independent of the SEDFLUX architecture. This makes SEDFLUX modular. For instance, a debris-flow simulation may require a time-step of one-tenth of a second; a surface plume may produce a deposit only a few millimetres thick. In each case, the final deposit is eventually averaged with contributions from other processes of sediment deposition, into vertical and horizontal bins defined by the overall resolution of the SEDFLUX architecture.

Sediment is supplied to a 2D-SEDFLUX basin through a single river. An input file provides data values that characterize the river-mouth dynamics for every time-step of the SEDFLUX simulation. The model river transports suspended sediment and bedload composed of a user-specified number of distinct grain sizes. The river's suspended-sediment load is distributed either as a hypopycnal plume or as a hyperpycnal flow, depending upon the sediment-laden density of the river water. Fluvial bedload is initially spread across the delta both landward and seaward of the coastline. Seafloor sediment is redistributed in the model domain through a number of processes. Thus, although individual processes do not act directly with one another, the seafloor acts as a medium through which they are able to interact. Unlike some models that track the shoreline as an analytically determined moving boundary, SEDFLUX more naturally handles the position of the shoreline as the position between terrestrial and marine processes, as determined by sea level at any given time. For details on the governing equations of the individual modules, readers are referred to Syvitski & Hutton (2001) and Hutton & Syvitski (2004) and the references therein.

As with any community model, individual modules will see improvements in resolution, speed and sophistication with time. The modelling environment is well suited to test new versions of individual modules, and the way they propagate their physics through sedimentary sequences. Modules can be tested, for example, to examine the trade-offs in computer speed versus numerical resolution (Syvitski *et al.*, 1999). All new modules, however, are first tested independently against field or laboratory data before they are considered for inclusion in the model architecture.

Table 3 Inputs and outputs of SEDFLUX modules. Newer versions of the model contain additional modules, including river-delta dynamics and nearshore breaking-wave dynamics

Module	Inputs	Outputs	Comments
HYDROTREND (see Box 1)	Climate (T, P, L, Ev) Basin (*hyps*, *ELA*, lakes, soil): 2D	Q, Uo, Ho, Bo, Cs, Di, Qb at river mouth	Eulerian, steady-state; TS = daily
PLUME (see Box 2)	**HYDROTREND** Current scalar	Vertical flux in xy, 1 to 10 m	Eulerian, steady-state; TS = daily
LITTORAL	Tidal range **HYDROTREND** Significant Wave H, T	$\partial\eta/\partial x$, $\partial Di/\partial x$, $\partial\rho/\partial x$ 1 m in x	Eulerian, steady-state; TS = daily
INFLO	**HYDROTREND** Internal: $\partial z/\partial x$, $\partial Di/\partial x$, $\partial\rho/\partial xz$	$\partial\eta/\partial x$, $\partial Di/\partial x$, $\partial\rho/\partial x$ 1 m in x	Eulerian, steady-state; TS = 1 s
SHELF BLT (see Box 4)	U_w; T; $U_c\text{-}x$; Internal: $\partial z/\partial x$, $\partial Di/\partial x$, $\partial\rho/\partial x$	$\partial\eta/\partial x$, $\partial Di/\partial x$, $\partial\rho/\partial x$ < 10 m in x	Eulerian, steady-state; TS < daily
FAIL (see Box 5)	Earthquake E pdf Internal: $\partial z/\partial x$, $\partial Di/\partial x$, $\partial\rho/\partial x$, $\partial Pe/\partial x$	ΣDi, 10 to 100 m in x	Eulerian, steady-state; TS = > yr
SAKURA (see Box 3)	Internal: $\partial Di/\partial x$, $\partial\rho/\partial x$, $\partial\eta/\partial x$	$\partial\eta/\partial x$, $\partial Di/\partial x$, $\partial\rho/\partial x$, 1 m in x	Eulerian or Lagrangian, TS = 1 s
BING (see Box 6)	Internal: $\partial Di/\partial x$, $\partial v/\partial x$, $\partial\eta/\partial x$	$\partial\eta/\partial x$, $\partial Di/\partial x$, $\partial\rho/\partial x$ < 10 m in x	Lagrangian, TS = < 0.1 s
COMPACT	Internal: $\partial Di/\partial z$, $\partial Pe/\partial z$, $\partial\rho/\partial z$	$\partial z/\partial x$, $\partial\rho/\partial x$ < 0.1 to 1 m in z	Eulerian, steady-state; TS = > yr
BOUNDARY	Sea level with ∂t Tectonics with ∂t and ∂x Mantle viscosity	$\partial z/\partial x$	Eulerian, steady-state; TS = > daily

T, temperature; P, precipitation; L, lapse rate; Ev, evaporation; *hyps*, hypsometry; *ELA*, equilibrium line altitude; Q, discharge; Uo, velocity at the river mouth; Ho, river mouth depth; Bo, river mouth width; Cs, suspended concentration; Di, grain size of the ith class; Qb, bedload; H, wave height; T, wave period; U_w, wave orbital velocity; U_c, longitudinal current; η, seafloor height above datum; ρ, bulk density; x, y, z, spatial directions in the longitudinal; lateral and vertical, respectively; E, seismic energy; Pe, excess pore pressure; v, viscosity; TS, time-step.

SEDFLUX can be run in a high-fidelity mode where the time-steps are short (e.g. daily: Fig. 18A), or a lower fidelity mode where the time-steps are longer and the important sediment transporting events are evaluated through the use of probability density function for their return interval (Fig. 18B). Often the latter approach is used for long geological simulations (millions of years) and the former is carried out when the hydrology of the input flux is well understood.

An important and recent numerical change to SEDFLUX is the replacement of a number of global-domain parameters with parameterizations determined within the model domain (Fig. 19: Hutton & Syvitski, 2004). These global geotechnical parameters were otherwise set at the beginning of a model run and would be invariant during the run. Replaced parameters include the coefficient of compaction, sediment cohesion, friction angle, shear strength and sediment viscosity. These parameters are presently calculated on a cell-by-cell basis. Cohesion is determined from the sediment load, the friction angle and shear strength. Viscosity is approximated from the void ratio. Shear strength

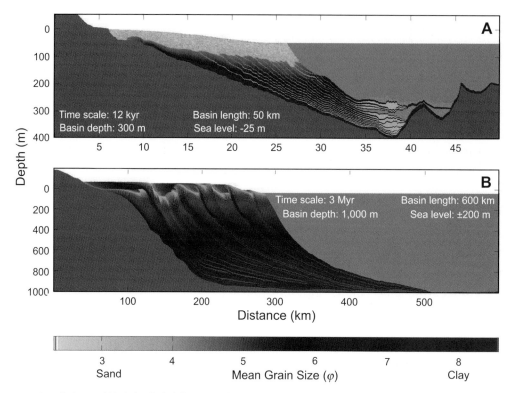

Fig. 18 SEDFLUX simulations. (A) A high-fidelity simulation (time-steps of 1 day) of a small fjord basin. (B) A low-fidelity simulation (time-steps of 100 yr) of a continental margin, where the important sediment-transporting events are evaluated through the use of probability density function for their return intervals. (After Syvitski & Hutton, 2001.)

is calculated from the plasticity index and sediment load. The friction angle is determined from grain-size and bulk-density values. The excess pore pressure is also calculated dynamically as a function of time and sediment properties. Together these changes allow for a reduction in the number of input parameters that a user needs to determine a priori. The changes also allow for a more realistic range in sediment failures, including smaller failures that are produced through retrogressive failure (Fig. 19).

SEDFLUX application to seafloor morphology

Numerical experiments have been widely applied to the study of sedimentary sequences on continental margins (Jervey, 1988; Kendall *et al.*, 1991; Lawrence, 1993; Skene *et al.*, 1998; Steckler, 1999). These studies are often focused on the response of certain geomorphological features (e.g. the shoreline and/or shelf break) to changes in sea level, tectonics and sediment supply. Few numerical experiments have examined the shape and

geomorphological evolution of continental margins (Ross *et al.*, 1994; Pratson & Coakley, 1996). Stratigraphic models are just now reaching a level of sophistication where the myriad sedimentary processes that occur on continental margins can be realistically and efficiently modelled (O'Grady & Syvitski, 2001). Agents controlling sedimentation on margins are numerous and interact on a variety of scales, from geological-scale processes such as thermal and isostatic subsidence, to event-scale phenomena such as underwater landslides (O'Grady *et al.*, 2000).

Numerical experiments conducted with 2D-SEDFLUX provide insight into how continental slopes evolve their shape through time, under varying environmental conditions (O'Grady & Syvitski, 2001). In the experiments, hypothetical passive margins are built on top of a simple initial bathymetric profile (Fig. 20). In each simulation, boundary conditions are switched on, or off. This is done step-wise, starting by having only essential processes active, and then enabling additional processes in subsequent runs. The initial model basin

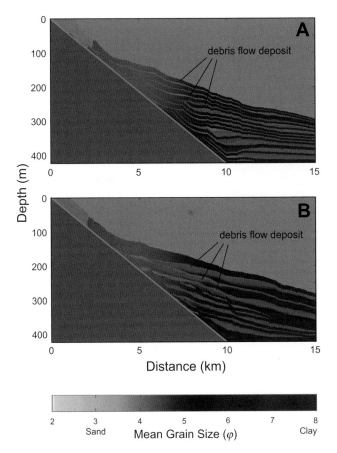

Fig. 19 2D-SEDFLUX simulations. (A) Simulation using time-invariant global-domain parameters (coefficient of compaction, sediment cohesion, friction angle, shear strength and sediment viscosity). (B) For comparison a SEDFLUX simulation where the above parameters are determined locally (cell-by-cell) within the model domain (see text for details). (Modified from Syvitski & Hutton, 2004.)

is 600 km long, with a maximum water depth of 1000 m (Fig. 20). The initial bathymetry resembles that of a small continental margin with a gently dipping shelf 0.01°, a continental slope of approximately 1°, and a very gentle sloping basin floor 0.001°. Most of 2D-SEDFLUX component models require a pseudo third dimension, and in the experiments shown (Fig. 20), the developing sedimentary architecture is tracked across a basin width of 15 km. Each run simulates deposition for 2 Myr, except where noted (e.g. some runs are for 6 Myr). Sediment accumulation is tracked at time-steps of 1 kyr; compaction is operating in all of the experimental runs.

Experiments begin with sediment fed by river plumes in a quiet ocean (Fig. 20A). Next, the ocean is energized through shelf storms (Fig. 20B). Then, seafloor stability is examined and any failed masses moved as turbidity currents (Fig. 20C), or as debris flows (Fig. 20D), or some combination of these flows depending on the characteristics of the failed sediment mass (Fig. 20E). Sea level is then allowed to fluctuate (Fig. 20F). Finally, the weight of the accumulated sediment allows the lithosphere to subside (Fig. 20G). In each case, the shape of the seafloor is tracked throughout each experiment by measuring the seafloor gradient along the simulated margin profiles at regular intervals. 2D-SEDFLUX was able to reproduce realistic margin morphology (within the limits of natural systems), suggesting that the dominant dispersal algorithms are adequately represented in their two-dimensional form.

Hemipelagic sedimentation along with shelf storms produces simple clinoforms of varying geometry. Oblique clinoforms are associated with low-energy conditions and sigmoid geometries are associated with more energetic wave conditions. Slope failure steepens the upper continental slope and creates a more textured profile. Topographic smoothing induced by bottom boundary-layer transport enhances the stability of the upper continental slope. Different styles of sediment gravity flows (turbidity currents, debris flows) affect the profile geometry differently. Debris flows accumulate along the base of the continental slope, leading to slope progradation. Turbidite deposition principally occurs on the basin floor and the continental slope remains a zone of erosion and sediment bypass. Sea level and flexural subsidence surprisingly show smaller impacts on profile shape. Initial basin steepness and water depth have a profound influence on the steepness of the equilibrium profile. When compared with the morphology of modern passive margins, most of the equilibrium profiles compare best with margins under the influence of relatively high sediment input (O'Grady *et al.*, 2000; O'Grady & Syvitski, 2001).

SEDFLUX application to climate change

To use basin stratigraphy for studying past climate change, it is important to understand the influence of evolving boundary conditions (river discharge, sediment loads, initial bathymetry, sea

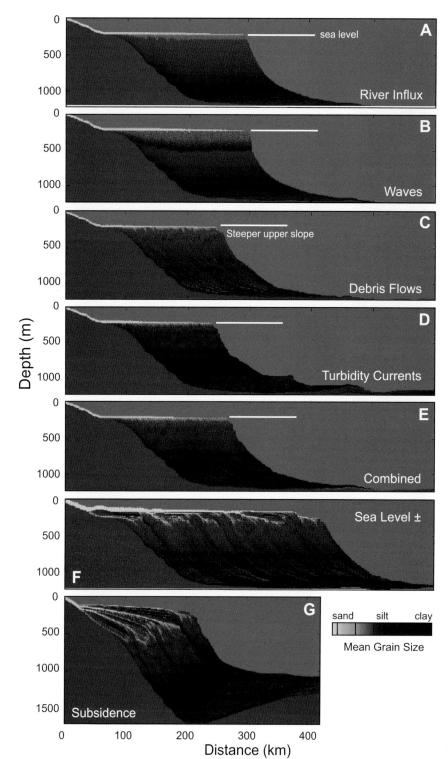

Fig. 20 Basin filling using 2D-SEDFLUX, showing gradational grain-size variations (yellow for medium sand to blue for clay). (A) Quiet ocean, sediment dispersal as surface plumes and bedload dumping. (B) River dispersal of sediment is affected by ocean storms. (C) Same as (B) with the addition of sediment failures and gravity flow transport by debris flows. (D) Same as (B) plus sediment failures and gravity-flow transport by turbidity currents. (E) Same as (B) plus sediment failures and gravity-flow transport by debris flows and turbidity currents. (F) Same as (E) plus sea-level fluctuations. (G) Same as (E) plus isostatic subsidence. (See text for details). Duration of runs is 2 Myr for runs (A) through to (E), and 4 Myr for runs (F) and G). (Modified from O'Grady & Syvitski, 2001.)

level, subsidence) and the complex interplay of the redistribution processes (e.g. plumes, turbidity currents, debris flows). To provide understanding of this complexity, 2D-SEDFLUX was employed as a source-to-sink numerical model to evaluate which process dominates the observed variability in a sedimentary record of two coastal Pacific basins, the Eel Margin of northern California and Knight

Inlet in British Columbia (Morehead *et al.*, 2001). The model is able to simulate the properties of marginal shelf and slope sediments at these test sites (Tables 4 & 5).

The Eel River Basin drains the steep-sloped coastal mountains of northern California. During present climatic conditions, the discharge is rain dominated and during the last glacial period it was rain and snow dominated (Syvitski & Morehead, 1999). The Eel Margin allows an inspection of how an intermittent rain-dominated river may interact with a high-energy coastal environment to create sedimentary sequences and preserve the record of climate. During the last glacial period, the Eel River supplied comparatively more sediment with a less variable flux to the ocean, while today the river is dominated by episodic events (Hill *et al.*,

this volume, pp. 49–99). SEDFLUX results show this change in the variability of sediment flux to be as important to the deposit character as is the change in the volume of sediment supply (Figs 21 & 22). Due to the complex interaction of flooding events and ocean storm events, the episodic flood deposits of recent times are less well preserved than the flood deposits associated with a glacial climate.

The Klinaklini River drains a glacial- and snow-dominated river basin and discharges into Knight Inlet, a fjord in British Columbia (Syvitski *et al.*, 1999). During the Last Glacial Maximum, glaciers covered the Klinaklini basin. Fjords, such as Knight Inlet, limit the lateral spreading of the sediment plume and thus allow for the preservation of an enhanced palaeoclimate signal. Knight Inlet acts as an excellent counter-example to the Eel Margin,

Table 4 Comparison of the SEDFLUX model with observations for the Eel Margin (After Morehead *et al.*, 2001)

Parameter	Distance from river (km)	Observation	Model
Grain size distribution (ϕ)	30	6.2–7.2	6.4–6.9
	20	5.6–6.0	5.6–6.1
	10	4.6–5.4	4.6–5.4
Accumulation rate (cm yr^{-1})	30	0.2	0.1
	20	0.4	0.2
	10	0.7	0.4
Flood deposit (cm)	30	10	12.5
	20	20	15

Table 5 Comparison of the SEDFLUX model with observations for Knight Inlet (for details see Morehead *et al.*, 2001)

Observation	SEDFLUX model
Sandy foresets extend 2 km from river mouth	Sandy foresets extend 2.5 km from river mouth
Thin bedded sandy turbidites (up to 20% sand) mixed with plume muds extend 10 km from river mouth	Thin bedded sandy turbidites (up to 15% sand) mixed with plume muds extend 10 km from river mouth
Thicker bedded sandy turbidites (up to 40% sand) mixed with plume muds extend 10–25 km from river mouth	Thicker bedded sandy turbidites (up to 45% sand) mixed with plume muds extend 10–25 km from river mouth
Hemipelagic (plume) muds (> 40% clay) extend 35–40 km from river mouth	Hemipelagic (plume) muds (> 40% clay) extend 35–40 km from river mouth

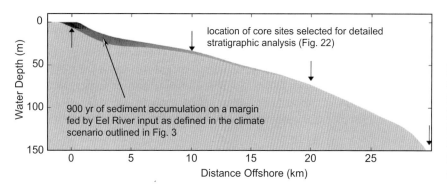

Fig. 21 Simulated (SEDFLUX) deposit thickness for a generic margin similar to the Eel margin (note that the bathymetry is different), using a 900-yr climate scenario. Arrows show the locations of the synthetic cores (Fig. 22). (Modified from Morehead *et al.*, 2001.)

900-yr *SedFlux* Simulation Experiment

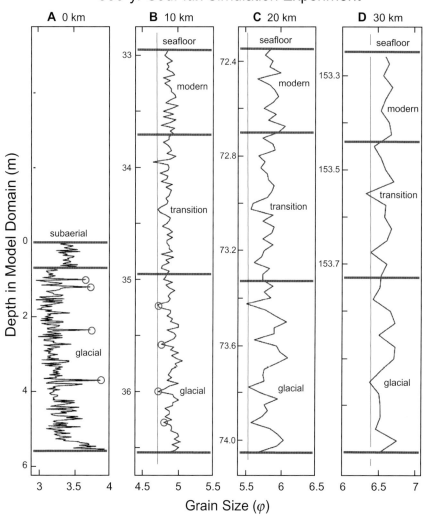

Fig. 22 Grain-size patterns from core sites within the 900-yr SEDFLUX simulation shown in Fig. 21. The horizontal grey lines demark the climatic boundaries: simulation used 300 yr under glacial conditions, 300 yr of modern conditions, separated by a 300 yr transition (see Fig. 3). (Modified from Morehead *et al.*, 2001.)

because the basin is annually subjected to a large number of gravity flow events (Syvitski *et al.*, 1988), which can interrupt the proxy climate record and make interpretations more difficult. In Knight Inlet, the evolving boundary conditions (rapidly prograding coastline, secondary transport by gravity flows from sediment failures) are a strong influence on the sedimentary record (Fig. 18A). The delta and gravity-flow deposits punctuate the sedimentary record formed by hemipelagic sedimentation from river plumes. Missing time intervals due to sediment failures can take away the advantage of the otherwise amplified lithological record of discharge events, given the enclosed nature of the fjord basin (Morehead *et al.*, 2001).

Climate change affects the supply of sediment to a basin both in terms of the volume flux and in the timing and size of events. The interaction of these climate-induced effects with the evolving bathymetry due to seafloor erosion and deposition, and reworking by ocean storms, sediment failures, debris flows and turbidity currents, may provide for a complex sedimentary record. The Eel Margin SEDFLUX simulation shows how changes in sediment flux do not alter the large-scale sedimentary processes, but also shows how fine-scale (cm) bedding can be altered by the frequency of ocean storms that rework the seafloor and river floods that contribute to the sedimentary deposit (Fig. 22).

The Knight Inlet SEDFLUX simulation shows how climate changes within a small river basin can be amplified within a semi-enclosed basin, but that a regional perspective must be taken in order to decipher the complex sedimentary packages (Fig. 18A). Many fjords are subject to the reworking of deposits by sediment failures and subsequent transport by gravity flows.

SEDFLUX summary

SEDFLUX is a basin-fill model able to simulate the delivery and accumulation of sediment over time-scales of tens of thousands of years. SEDFLUX includes the effects of sea-level fluctuations, river floods, ocean storms and other relevant environmental factors (climate trends, random catastrophic events), at a time-step (daily to yearly) that is sensitive to short-term variations in sediment supply. SEDFLUX combines individual process–response models into a single model, delivering a multi-sized sediment load onto and across a continental margin, including sediment redistribution by: (i) river-mouth dynamics, (ii) hypopycnal plumes, (iii) hyperpycnal flows, (iv) ocean storms, (v) slope instabilities, (vi) turbidity currents and (vii) debris flows. The model allows for the deposit to compact, and to experience tectonic processes (faults, uplift) and isostatic subsidence from the sediment load.

Many of these processes work together to sculpt margins, yet their interactions are complex (Pratson *et al.*, this volume, pp. 339–380). In the formation of siliciclastic margins, terrestrial sediment supply and initial bathymetry provide the foundation for overall margin shape and evolution. Hydrodynamic energy, slope failure, mass movement and fluctuations in relative sea level subsequently alter the basic morphology. Overall these processes combine to produce a prograding siliciclastic margin with an equilibrium shape similar to that of some modern passive margins. SEDFLUX can reproduce realistic profile morphology, within the limits of natural systems, suggesting that the dominant dispersal algorithms are adequately represented in their two-dimensional form. SEDFLUX is useful to examine non-linear relationships between variations in external forces (e.g. sediment supply resulting from climate change) and internal dynamics (e.g. episodicity of sediment failure and gravity flows).

Stratigraphic sequences

The stratigraphy of margins is the result of the interplay of accommodation space and its fill by sediment. **Accommodation space** is created by the relative subsidence or uplift of the seafloor, and by changes in the morphology of the seafloor. Sediment fills the space as a function of supply rate and highly coupled transport processes. The net sediment accumulation is controlled by seafloor morphology and the sedimentary deposits in turn modify the seafloor morphology. The fundamental units for subdividing and interpreting the evolution of margin stratigraphy are stratigraphic sequences. These are packages of relatively conformable genetically related strata that are bounded by unconformities or their correlative conformities (Mitchum *et al.*, 1977). Conceptual models have advanced our understanding of the development of stratigraphic sequences and the influence of sea-level fluctuations (Posamentier & Vail, 1988; Posamentier *et al.*, 1988). Simple numerical schemes (Jervey, 1988; Cant, 1991; Nummedal *et al.*, 1993) can help quantify relationships between variables and provide insight into the system. However, to fully understand the influence of different processes and their interactions, more complete numerical models are required.

On continental margins, most transport and accumulation of sediment occurs during infrequent energetic events (i.e. storms and floods). Extrapolating from event scales of days up to geological time-scales of thousands and millions of years is a difficult undertaking. Stratigraphic evolution can be modelled using a scale-integral approach (Carey *et al.*, 1999; Steckler, 1999), rather than modelling individual events. For example, numerical model SEQUENCE is constructed from algorithms that represent the range of processes appropriate to the relevant time and length scales. Here, shelf deposition/erosion is parameterized by a single equation representing the long-term impact of the active processes in their environment instead of employing separate algorithms for each major shelf process. The **scale-integration approach** of SEQUENCE allows for a longer time-step and thus the rapid output of stratigraphic simulations over longer periods of geological time.

Formulation of SEQUENCE

The geophysical framework for the model includes the factors that affect accommodation space: tectonics and eustasy, flexural (regional) isostatic

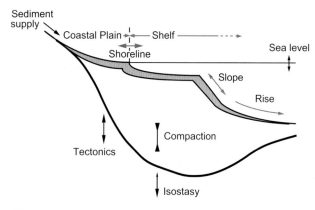

Fig. 23 The SEQUENCE simulation framework estimates accommodation, taking into account tectonic subsidence, sediment compaction, isostasy and fluctuations of sea level as indicated schematically by the black arrows. Sediment filling the available space is supplied from the landward edge of the model, and deposition is calculated using four transport regimes (coastal plain, shelf, slope, rise) whose boundaries are indicated by the grey arrows. The boundary between the non-marine coastal-plain accumulation and the marine shelf accumulation is a moving boundary that tracks the shoreline (see text for details.)

compensation of the sediment and water loads, and compaction (Fig. 23). The sedimentation module uses a four-component moving-boundary formulation for calculating the time-averaged cross-shelf sediment transport. Deposition is calculated using either a geometric scheme in which slope and other morphologicalal parameters are specified (Steckler, 1999) or a differential-equation representation of sediment transport and deposition. To capture the interaction and feedback between accommodation and deposition, calculations are embedded in an iterative loop (Steckler, 1999). Most parameters can be interactively adjusted throughout the model run.

The SEQUENCE framework is a moving-boundary formulation with separate model components for the coastal plain, shelf, upper slope and lower slope/rise (Fig. 23). The model tracks a single grain size. For the differential-equation-based version, the coastal plain is simulated using either a linear or non-linear diffusion that allows for increasing river discharge towards the coast (Paola *et al.*, 1992). Shelf sediment transport is modelled by an advection–diffusion equation dominated by non-linear diffusion (Niedoroda *et al.*, 1995) and by non-linear advection (Pirmez *et al.*, 1998). Both methods show that sediment transport decreases with increasing water depth. The position of the shoreline is calculated iteratively by matching sediment flux across the boundary. Sediment lying on a steep slope is presumed to fail and be transported downslope (Pratson & Coakley, 1996). The sediment is then assumed to deposit as a turbidite using a simplification of the Parker *et al.* (1986) three-equation model yielding an exponential thinning of the deposit.

The model tracks the deposited strata and any subsequent erosion, and includes calculations of tectonic subsidence, sea-level changes, flexural isostatic compensation of water and sediment loads, and compaction of the sediments. In order to accurately calculate vertical motions and resultant stratigraphy at short time-steps (10^3–10^4 yr), isostatic and erosion adjustments to an equilibrium profile are modelled as time-dependent processes.

Application of SEQUENCE to stratigraphy

An archetypical simulation (Fig. 24) illustrates several features of the model. During a sea-level cycle, three distinct unconformity surfaces are devel-

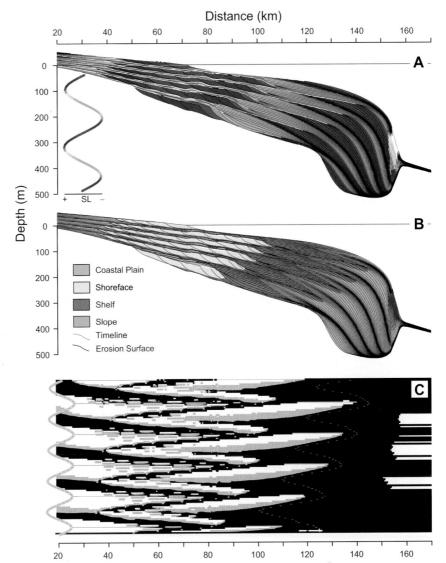

Fig. 24 SEQUENCE stratigraphy simulation using a sea-level curve with a third-order (2 Myr) cyclicity. (A) Stratigraphic section in which each layer is coloured according to its position on the sea-level curve. Timelines are drawn every 100 kyr. Unconformities are indicated by white lines. (B) Stratigraphic section coloured according to the depositional environment. (C) Chronostratigraphic plot, where preserved sediments are black, sediments that were deposited and subsequently eroded are grey, and non-depositional hiatuses are white. The yellow and red marks indicate the positions of the shoreline and clinoform breakpoint, respectively. The sea-level curve (blue) is shown at the left. The sea-level curve had a 50-m amplitude and long-term fall of 2.5 m per cycle.

oped: (i) a subaerial erosion surface, (ii) a marine regressive erosion surface and (iii) a transgressive ravinement surface. These three surfaces are seen in other simulations (Nummedal *et al.*, 1993), and are observed in the field. The **ravinement surface** removes most of the transgressive shoreline deposits; generally, only **regressive shoreface** deposits are preserved. The **sequence boundary** (SB) is the subaerial erosion surface above the prograding shorefaces (van Wagoner *et al.*, 1990; Hunt & Tucker, 1993; van Wagoner, 1995; Plint, 1996) that continues down the front of the shoreface and onto the marine regressive surface across the shelf. In addition, when the clinoform foreset progrades into deeper water, it steepens until it

reaches an angle that produces sediment failures, creating additional erosion surfaces and associated slides and gravity flows.

SEQUENCE produces depositional and facies belts that shift back and forth with sea level. Unlike the classic conceptual model for sequence stratigraphy (Vail, 1987; Posamentier *et al.*, 1988; Van Wagoner, 1990), all of the unconformity surfaces are time-transgressive, developing progressively through time. Furthermore, the timing of different systems tracts in the model differs from the classic conceptual model. **Systems tracts** are a set of contemporaneous depositional systems within a sequence that are characterized by similar internal geometry and facies associations (van Wagoner *et al.*, 1990).

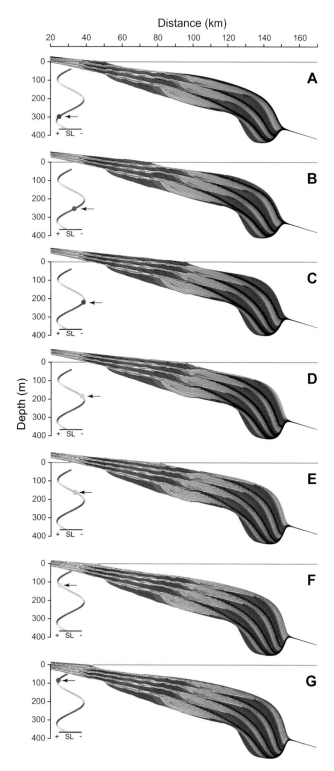

Distance (km)

Depth (m)

Fig. 25 Set of time slices from the SEQUENCE model run in Fig. 24 illustrating the depositional patterns during a sea-level cycle. Arrows indicate the position within a sea-level cycle for each plot. See text for details.

The changing depositional patterns produced by SEQUENCE are best illustrated by example.

Figure 25 shows a series of steps through a sea-level cycle for the archetypical model. In Fig. 25A, sea level has started to fall. As the shoreface grows and steepens, the inner shelf undergoes erosion because of the decreasing water depth. The shore-faces are sharp based; under different conditions, such as greater subsidence, the shoreface bases could be gradational. The base-level lowering has also started to erode the lower coastal plain. Besides the shoreface, deposition is concentrated on the outer shelf but tails off towards the shelf edge. The slope is starved and developing a **condensed section** – a distal deposit of nearly continuous but very low sedimentation.

As sea level reaches its peak rate of fall (Fig. 25B), erosion extends over the entire coastal plain and most of the shelf. Onshore, an erosion surface truncates the tops of the earlier strata, creating an offlapping pattern. The shoreface continues to accumulate sediment due to the elevation step down at the shoreline and the accommodation it creates. The shoreface progrades over the marine erosion surface yielding sharp bases to the deposits, and forms the **falling stage systems tract** (FSST) of Plint & Nummedal (2000). This is the time of sequence-boundary formation according to Vail (1987) and Posamentier *et al.* (1988). However, the model results see no break in the depositional pattern. The shoreface is far from the shelf edge, even though an erosion surface has developed across the shelf. Offshore, the locus of deposition has shifted seaward to the outer shelf. The bathymetry of the outer shelf is also beginning to steepen.

The next image (Fig. 25C) is almost at the time of the sea-level minimum. The coastal plain and shelf erode with the progradational shoreface deposits of the FSST between them. However, with the slow-ing of the sea-level fall, coastal onlap over the previous shoreface has started. The underlying surface is a candidate for the sequence boundary. The marine deposition is now concentrated on the upper slope. Offlapping oblique clinoforms are well developed. The break-in-slope at the depositional shelf edge is sharper and more pronounced.

When sea level starts to rise (Fig. 25D), the de-positional pattern is altered. The shoreface begins to **backstep** along with some deposition on the lower part of the coastal plain. As a result of the backstepping, there is limited accommodation at the shoreface, much of which is erosional. The height of the shoreface is flattened, producing deposition on the inner shelf. This pattern is similar to the 'healing stage' of Posamentier & Allen (1993). With the rise in base level, deposition is slow but continuous across the shelf. Accumulation rates are greatest on the front of the still **prograding clinoform**. An important feature of the model is that the timing of the change at the clinoform front differs from that at the shoreface. Picking the sequence boundary from nearshore geology and marine seismic data does not necessarily yield the same surfaces.

In the following image (Fig. 25E), sea level continues to rise, causing an extensive transgression of the shoreline. There is more deposition on the coastal plain, which onlaps onto the subaerial erosion surface. The ravinement surface planes off the inner-shelf area and removes the limited shoreface deposits. In addition some of the coastal-plain deposits are removed. Offshore, deposition on the shelf smoothes the relief left by the previous lowstand deposits. Slope deposition continues to prograde the clinoform, but at a reduced rate.

As the peak in the sea-level curve is approached (Fig. 25F), the pattern of accumulation continues, i.e the **transgressive systems tract** (TST). Coastal-plain deposition laps onto the edge of the model domain. Previously deposited coastal-plain strata are bevelled by the ravinement surface. As in earlier cycles, overlapping wedges of fluvial and marine deposits are preserved between the shoreface deposits. Depending upon model parameters, these can be entirely fluvial or marine, or there may be stacked shorefaces with no intervening strata. Offshore, deposition has filled the relief on the shelf. Little sediment reaches the slope where the section is condensed. For this simulation, sediment supply was kept constant; if it decreases during transgression, then the condensed section is more extensive and the TST strata on the shelf are reduced.

The last episode (Fig. 25G) is just after the sea-level peak. There is still accumulation on the coastal plain. Progradation of the shoreline has once again started as the sea-level rise slows and sea level begins to fall. Marine deposition is concentrated on the shelf, and the slope remains starved of sediment. Changes in the shape of the clinoforms can lead to

a concentration of slope failure and downstream turbidite deposits, if the angle for stable slopes is exceeded. The variability in the sediment flux going over the shelf edge between high and low sea level will further enhance this effect.

Comparison of SEQUENCE results to the standard model

The numerical model SEQUENCE and the standard sequence model (Fig. 26) contain significant differences as well as similarities. In both models, the shoreline roughly follows sea level. For the numerical model, the shape of the deposits and phase differences from the shoreline curve are small (see chronostratigraphic diagram, Fig. 26B). Differences of the numerical model with the conceptual model are found between the **lowstand systems tract** (LST) architectures following the **Type 1 sequence boundary** (when the shore progrades past the shelf edge). In the numerical model, the clinoform front can develop significant lags or leads from the sea-level curve. In this example, clinoform progradation continues into the early part of the sea-level rise. With sufficient sediment supply, progradation can continue well into the rise and also restart near the end of the rise. This pattern of clinoform motion is closer to the standard sequence model, which is not unexpected, as it is the clinoforms that are generally imaged on seismic data and formed the basis for the initial concepts (Vail *et al.*, 1977). The conceptual model until recently did not differentiate between the depositional shelf edge and shoreface (Vail *et al.*, 1991), which implies that there is no active shelf except during a transgression.

If the stratigraphy produced by the numerical model is de-constructed into systems tracts (Fig. 26), considerable differences appear relative to the standard conceptual model. In the standard model, the sequence boundary occurs mid-way through the relative sea-level fall. To a large extent, this is because in the standard model the shoreline is at the depositional shelf edge. At a ramp margin, where there is not a break in slope near the shoreline, the FSST continues throughout the fall (Plint & Nummedal, 2000). This is what occurs in the numerical model. Without a bathymetric break, one cannot easily separate the regressive deposits as simulated by the model into **highstand systems tract** (HST) and LST; rather one sees a prograding

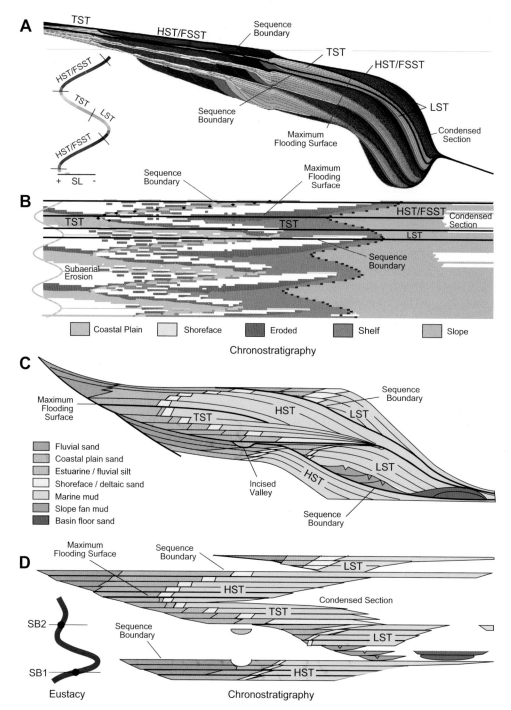

Fig. 26 Comparison of sequence architecture developed in SEQUENCE (A & B) and the standard conceptual model (C & D; Vail, 1987; Posamentier *et al.*, 1988). (A) Results from the numerical simulation divide the evolving stratigraphic architecture into systems tracts using different bounding surfaces than those employed in the standard model (C). As seen in Fig. 24, all unconformities are time transgressive. (B) and (D) display the sedimentary facies in a time–distance domain and are directly equivalent to the depth–distance diagrams of (A) and (C), respectively. TST, transgressive systems tract; FSST, falling-stage systems tract; LST, lowstand systems tract; HST, highstand systems tract. A systems tract defines a package of sediment according to when in the sea-level cycle the material was deposited.

FSST. The **sequence boundary** (SB) is best picked as the top of the subaerial unconformity over the prograding shorefaces, down the front of one of the last shorefaces and across the regressive marine erosion surface. Here the surface below the first coastal onlap is chosen to determine which shoreface is at the SB. This is similar to ramp settings where one may choose the surface below the first preserved onlap as the sequence boundary (Christie-Blick, 1991; Plint, 1996).

While the surface at the base of the shoreface merges into the marine erosion surface, the interpretation of this surface as the sequence boundary (Posamentier *et al.*, 1992; Posamentier & Allen, 1993) results in a **time-transgressive surface** (i.e. a surface with physical continuity, whose parts have developed at different times). The zone of erosion shifts seaward with time to the extent that sediments above the landward parts of the erosion surface can often be younger than sediments below the most seaward parts of the surface. Thus, this interpretation would violate the goal of having the sequence boundary separate older and younger sediment packages.

In the standard model, coastal onlap ceases abruptly at the SB making a clean isochronous surface. This never occurs in the numerical simulation; the end of coastal-plain deposition and the subaerial surface over the shorefaces are distinctly time-transgressive. The subaerial unconformity marches seaward with the shoreface. The non-depositional hiatus is much larger at the landward side of the model.

The LST is a transitional phase between the FSST and the TST. This is often the case in outcrop sequence stratigraphy because the shoreline rarely reaches the shelf edge. It is only during the high-amplitude, high-frequency eustatic fluctuations of the ice age that the shore reaches the shelf edge. The numerical LST starts with the first coastal onlap and, as in the standard model, the shoreline continues to prograde. The shoreline movement reverses with sea level, but these shoreface deposits are not preserved. The end of the LST occurs at the beginning of the marine onlap.

During the TST, marine deposits onlap onto the sequence boundary in the numerical model (Fig. 26). During the TST, with its increasing accommodation, nearly all of the preserved coastal plain develops. This is in contrast with the standard model, which has most of it developed during the

HST. The condensed section in the model peaks at the sea-level peak. This is when deposition is the farthest landward and the least sediment reaches the shelf edge. The turnaround from transgression to regression is very clearly seen in the preserved section (Fig. 26), occurring at or just before the sea-level peak. On the shelf, the start of the HST also marks a transition from sigmoid to oblique clinoforms, because of the erosion on the shelf during the sea-level fall.

Application of SEQUENCE to continental margins

SEQUENCE runs exhibit a number of features similar to those observed on seismic sections from the STRATAFORM field areas. Figure 27 shows a section of the Lower Miocene deposits on the New Jersey margin compared with a simulation performed using the earlier, geometric version of the model. The sequences on the seismic line display sigmoidal clinoforms with a transition from progradation to aggradation. Well control indicates that the shoreline lay ~50 km landward of the clinoform rollovers. The numerical experiments used boundary conditions taken from backstripping reconstructions of the margin (Steckler *et al.*, 1999) and a sinusoidal sea-level cycle of 2 Myr, which is the average length of the early Miocene cycles in New Jersey. The simulation fits the pattern with sea-level cycle amplitude of 20 m combined with a long-term sea-level rise of 10 m. If the sea-level cycle amplitude is increased to 40 m, the progradation to aggradation pattern breaks down. These sea-level results are considerably smaller than previous estimates (Haq *et al.*, 1988), but are in accordance with $\delta^{18}O$ and backstripping data (Pekar *et al.*, 2000; Miller *et al.*, 2005). Model results also show clinoforms that steepened as they prograded into deeper water, consistent with both observations and backstripping.

The Pleistocene section in New Jersey (Mountain *et al.*, this volume, pp. 381–458) shows an extensive progradation of oblique clinoforms that is similar to the results of the archetypical model simulation. Some of the sediments recovered from the ODP Leg 174A boreholes show deepening upwards within the sigmoid deposits (Austin *et al.*, 1997), consistent with the model results that these strata formed during the transgressive systems tract. This build-up of prograding clinoform deposits

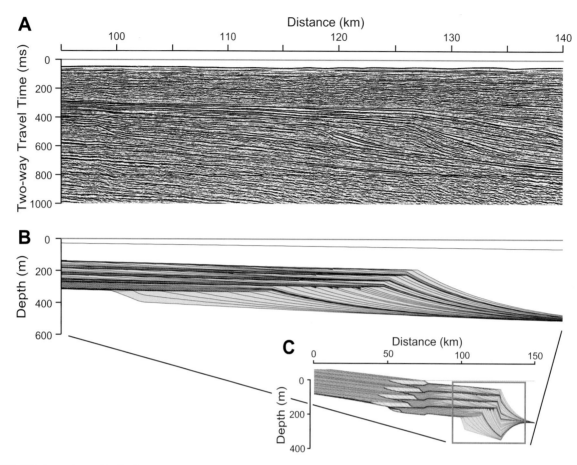

Fig. 27 (A) A section of *R/V Oceanus* seismic line 529 from the New Jersey margin. Early Miocene clinoforms show a progradational to aggradational pattern while growing in height and foreset angle. The palaeoshore for this period was sampled by the Island Beach Well (Miller *et al.*, 1994) and is located near the present shore 35 km landward of the profile. (B) The SEQUENCE simulation (Steckler, 1999) is scaled to fit the seismic line (see text for details). Timelines are in black with erosion surfaces highlighted in red. (C) The entire model-run domain; the box indicates the section shown in (B).

during transgression was one of the surprises of the Leg 174A drilling. The seaward down-stepping of the erosion surface suggests that these packages and the surface were formed during sea-level fall. The simulation (Fig. 24) does not produce as many of these oblique strata as the observations. In the model, increasing the sea-level amplitude causes the sequence boundary unconformity to become more extensive, reaching the shelf edge for high fluctuations. Simulations using the geometric model with the sea-level amplitude increased to ≥ 90 m yielded extensive oblique clinoforms and non-marine strata overlying the sequence boundary in a position where drilling recovered possible estuarine sediments at ODP Site1071.

Figure 28 shows the results of a much higher frequency experiment with a comparison to the

seismic stratigraphy of the Eel River basin. The simulation was run using the relatively steep dips of this margin, rapid subsidence, high sediment supply and a sea-level curve for the past 125 kyr. The complex multistage sea-level fall from Oxygen Isotope Stage 5 (125 ka) to Stage 2 (18 ka) produces a series of progradational packages. These are best seen in the cross-section coloured by time. Considerable erosion during both the regressive and transgressive parts of the cycles leaves a very incomplete section with numerous erosion surfaces. The large sea-level rise since the Last Glacial Maximum has covered the entire section with a smooth marine transgressive drape.

The seismic line shown on Fig. 28 suffers from acoustic masking by gas bubbles in the sediment column, but the high subsidence rate here helps to

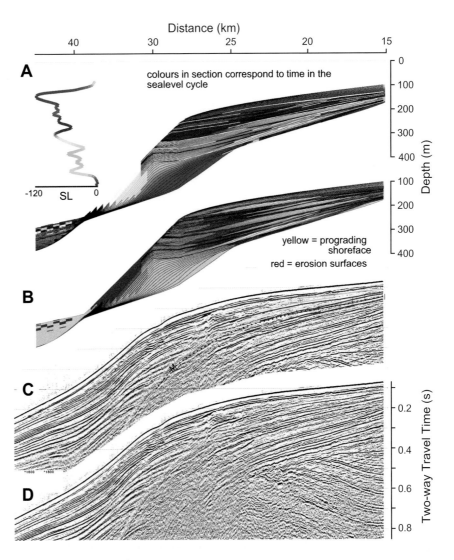

Fig. 28 Simulation of the stratigraphy of the Eel River basin for the past 125 kyr showing (A) the age distribution and (B) the environment of the section scaled to fit the seismic image. (C) Interpreted and (D) observed seismic images from *R/V Wecoma* line 83. See text for discussion.

resolve some patterns that are consistent with the model predictions. The section shows a series of strong, slightly divergent reflectors on the shelf, topped by the relatively transparent Holocene transgressive drape. These have been coloured red (Fig. 28) and correspond to the erosion surfaces in the model, although not all of these reflectors are necessarily unconformities. Within the packages defined by these reflectors are a series of more steeply dipping reflectors (coloured yellow in the interpreted section) corresponding to the prograding shorefaces and other more steeply dipping interfaces in the model. Seaward and landward of the dipping reflectors are additional surfaces with lower dips, subparallel to the major reflectors (coloured green and maroon Fig. 28), and correlating with fluvial and marine deposits.

Sediment accumulation on the slope is not a good match to the seismic profile. Slope failure in the model limits accumulation while the seismic line exhibits a thick drape. In model runs with less slope failure, there is more accumulation. However, non-accumulation on the uppermost slope persists. This is because the final progradation creates a very sharp shelf edge that is smoothed and eroded during the subsequent transgression. It compares favourably with the relative hardground on the uppermost slope (Lee *et al.*, 1999). On the shelf, the model predicts that the strata below the Holocene transgressive drape should be a very complex package of nearshore deposits interfingered with marine shales and a limited amount of non-marine strata. Thus, below the thick transgressive drape, the strata of the Eel River margin do not reflect the seafloor bathymetry, as on the New Jersey margin.

SEQUENCE summary

The stratigraphy of continental margins is constructed of building blocks termed sequences that develop from the interplay of accommodation space and filling of that space by sediments. SEQUENCE numerically represents the time-averaged results of these processes to simulate sequence formation. The model is valid for time spans much longer than the 500–1000 yr numerical time steps. At this scale, the model is able to generate simulations that reproduce a number of details that fit geological observations. The model contributes to our understanding of how sequences develop and to estimates of controlling parameters, such as sea-level amplitudes. However, they conflict with the initial conceptual models of how sequences develop. The early models (Vail *et al.*, 1977; Posamentier *et al.*, 1988) constituted a tremendous advance in our understanding of stratigraphic development, and have been evolving with our understanding of sedimentary processes. Quantitative numerical models highlight areas where further scrutiny is needed.

SEISMIC MODELS

Stratigraphic modelling is used to bridge the gap between observations of sedimentary processes on continental shelves and slopes and the strata these processes produce over geological time. However, much of what is known of continental margins is based on the stratigraphic filter of seismic-reflection data. Therefore, as part of the stratigraphic modelling effort, a capability has been developed for transforming stratigraphic simulations into synthetic seismic reflection data. This capability serves two important purposes (Syvitski *et al.*, 1999):

1 it allows comparison of model predictions to seismic data of the shelf and slope in a more direct way than has generally been accomplished;
2 it offers a means for advancing our understanding of what physical attributes of the rock record are and are not retained when it is imaged acoustically at different frequencies.

This section summarizes methods for generating synthetic seismic data of stratigraphic simulations.

A variety of approaches can be used to simulate the appearance of strata in seismic reflection data, as an aid to understanding how structures and stratal patterns appear in seismic data. There is also a practical interest in how seismic waves behave as they move through strata, in order to enhance the imaging of strata. Examples include:

1 idealized, digital depictions of strata with relatively few layers of simple geometries and uniform properties (Claerbout, 1985; Yilmaz, 1987);
2 more complicated digital depictions of strata that contain a selected mixture of structures and stratal patterns (e.g. the Marmousie model: Versteeg, 1994);
3 pre-defined physical models of strata constructed and seismically imaged in the laboratory (Ebrom & McDonald, 1994; Sherlock & Evans, 2001);
4 stratal patterns mapped from large outcrops and populated with rock properties derived from the outcrops (Stafleu & Schlager, 1993; Bracco & Schlager, 1999).

The use of computer-generated stratigraphic models to simulate synthetic seismic data has been less common, largely because most stratigraphic models have been devoid of the lithological heterogeneity that produces seismic reflection horizons in natural strata. In these models, depositional surfaces or 'time lines' have simply been equated to seismic reflection horizons (Jervey, 1988; Jordan & Flemings, 1991; Reynolds *et al.*, 1991; Flemings & Grotzinger, 1996). However, a few stratigraphic models do predict lithological variability, and the potential for using these models as the input for generating synthetic seismic data has been demonstrated (Shuster & Aigner, 1994; Syvitski *et al.*, 1999). More recently, the use of experimental stratigraphy for generating synthetic seismic data also has been demonstrated (Pratson & Gouveia, 2002). The methodology that follows builds on these initial studies, and is described in detail in Herrick (2001), Pratson & Gouveia (2002) and Pratson *et al.* (2003).

Physical properties

Porosity and bulk density are the minimum physical properties needed to simulate seismic-reflection data. Generally, the **porosity**, ϕ, of each cell in a stratigraphic model is equated to the volume of space in that cell that is not occupied by sediment after compaction, i.e.

$$\phi = 1 - \frac{V_s}{V_t} \tag{15}$$

where V_t is the total volume of the cell and V_s is the volume occupied by sediment.

Once porosity is known, bulk density is given by the porosity-weighted sum of the densities of the fluid (ρ_f) and grains (ρ_g) within the cell. For the simple case in which all the grains have the same density, the bulk density (ρ_b) is

$$\rho_b = (1 - \phi)\rho_g + \phi\rho_f \tag{16}$$

In natural strata, grain packing appears to have a significant impact on porosity and bulk density. The effects are captured in a simple packing model (Fig. 29; Marion *et al.*, 1992), where grain sizes are either sand (i.e. relatively large) or clay (i.e. relatively small). Porosity is then determined from the volumetric fraction of clay, C.

If the fraction of clay is zero, the porosity is that of the sand. At higher fractions, the clay is assumed to partly fill the pore spaces between the sand grains, gradually reducing the porosity. This reduction continues until the clay fraction equals the sand porosity (ϕ_s), i.e.

$$\phi = \phi_s - [C(1 - \phi_c)] \quad 0 \le C \le \phi_s \tag{17}$$

where ϕ_c is the clay porosity. When $C = \phi_s$, porosity is at a minimum, equalling the porosity of the sand minus the space within this porosity that is occupied by clay grains. At higher clay fractions, the clay is assumed to separate the sand grains and thus encase them. The porosity increases towards that for pure clay, as the solid sand grains are replaced by the smaller, porous clay deposits, i.e.

$$\phi = C\phi_c \quad \phi_s \le C \le 1 \tag{18}$$

When the clay fraction reaches unity, the porosity is that of the clays.

Importantly, the fall and rise in porosity with increasing clay content has a reverse effect on bulk density, and consequently on seismic velocity. As the clay fraction increases from zero to the sand porosity, bulk density and velocity rise to their maximum values, but as the clay fraction exceeds the sand porosity and increases further towards unity, the bulk density and velocity decline.

Equations 16 & 17 provide an explanation for the V-shaped pattern in porosity, bulk density and velocity (Fig. 29) that have been observed with increasing clay content in continental-margin strata offshore the USA Gulf of Mexico and Brazil (Marion *et al.*, 1992; Koltermann & Gorelick, 1995;

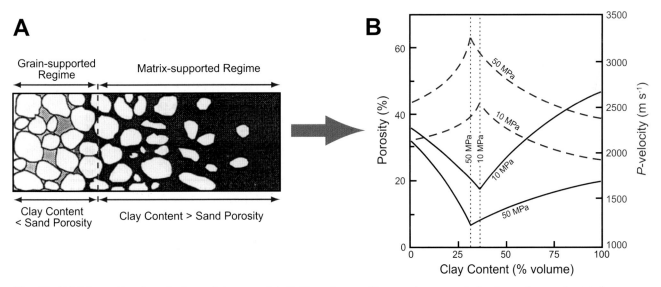

Fig. 29 (A) Schematic of sand–clay mixture model. Sediments are either grain-supported, where clay content is less than sand porosity, or matrix-supported, where clay content is greater than sand porosity. (B) Porosity (blue) and compressional velocity (red) at 10 MPa and 50 MPa overburden pressure as a function of increasing clay content in the sand–clay mixture model (A). Dotted vertical lines represent critical clay contents at the two overburden pressures. (Modified from Marion *et al.*, 1992.)

Flood *et al.*, 1997). However, clay cannot completely fill the pore spaces between sand grains. This limitation is accounted for in modified versions of Eqs 16 & 17 (Koltermann & Gorelick, 1995).

Acoustic properties

The main acoustic property required for simulating seismic data is compressional or **P-wave velocity**, V_p. Empirical relationships offer a way of rapidly estimating *P*-wave velocities from a limited number of physical properties. For example, SEDFLUX uses the following relationship from Hamilton (1980) to obtain *P*-wave velocities solely from porosity (Syvitski *et al.*, 1999)

$$V_p = a_0 + a_1\phi + a_2\phi^2 \qquad (19)$$

In Eq. 19, $a_0 = 1782$, $a_1 = -833$ and $a_2 = 522$.

Other empirical models exist, but many need additional physical properties. For example, the models of Toyasa & Nur (1982) and Kowallis *et al.* (1984) rely on porosity and clay fraction, while the models of Castagna *et al.* (1985) and Eberhardt-Phillips *et al.* (1989) also require confining pressure.

An important limitation in using an empirical model to predict strata velocities is that the model coefficients are based on the data used to develop the model. To predict velocities in new strata, the coefficients may need to be recalculated using measurements from these strata of velocity and the physical properties required to compute it (e.g. porosity, clay content and confining pressure). If such a recalibration is not done, the empirical model can yield unreasonable results. For example, Herrick (2001) found that when applied to borehole data from the Amazon Fan, the empirical model (Eq. 18) typically yielded negative velocities for near-surface sediments and exceptionally high velocities for sediments buried only tens of metres below the seafloor. These problems are compounded by the fact that the measurements needed to tune the model coefficients are often not available.

More robust estimates of velocity can be obtained with far less calibration using the physical theory for how seismic waves propagate in strata. In the simplest form of this theory, seismic waves are waves of elastic deformation that move through the strata, i.e. the strata snap back to their original form after the waves have passed. The velocities of the waves depend on the bulk and shear moduli of the strata, which are elastic constants that describe the resistance of the strata to deformation by compression/dilation and shearing, respectively.

Using this approach then, **bulk** and **shear moduli** must be computed for each cell in the stratigraphic model in order to determine velocities. These moduli are *effective* elastic moduli for they reflect the combined, non-linear response of the moduli for the grains and pore fluids mixed within a cell. Physical bounds on effective moduli can be determined using the approach developed by Hashin & Shtrikman (1963). Parameters needed to compute these bounds (i.e. the Hashin–Shtrikman bounds) are: (i) the volume fractions of the various phases within a volume (e.g. sand, clay, and pore-water); and (ii) the elastic moduli for these phases, which are generally tabulated (Mavko *et al.*, 1998).

To predict the true effective moduli, a third piece of information needs to be provided: the geometric details of how the phases within the volume are arranged relative to each other, i.e. the geometries of the sediment grains and the pore spaces, and how they are distributed. Knowing or assuming this information, the elastic moduli can be estimated using a range of methods (Mavko *et al.*, 1998). The methods employed here (see Herrick, 2001) are those of Berryman (1980a, b) and Xu & White (1995), which are variations on the method developed by Kuster & Toksöz (1974). The effective moduli computed by these latter methods are for dry sediments; for saturated sediments the dry effective moduli in Gassmann's (1951) relations are used.

Finally having obtained the effective bulk and shear moduli for a saturated sediment mixture within the volume, its *P*-wave velocity can be determined and so can the shear wave or *S*-wave velocity, which is an additional advantage of this physics-based approach over empirical models for velocity. The fundamental equations relating the effective bulk and shear moduli to the **P- and S-wave velocities** are

$$V_p = \sqrt{\frac{K + 4/3\mu}{\rho_b}} \qquad (20)$$

$$V_s = \sqrt{\frac{\mu}{\rho_b}} \qquad (21)$$

where V_p and V_s are the P- and S-wave velocities (respectively), and K and μ are the saturated effective bulk and shear moduli (respectively).

Seismic waves are waves of deformation, and as they propagate, they temporarily displace the strata they are moving through. This displacement is described by the wave equation. The one-dimensional wave equation for a P-wave is

$$\frac{\partial^2 u}{\partial t^2} = V_p^2 \frac{\partial^2 u}{\partial x^2} \qquad (22)$$

where u is displacement, t is time and x is distance. This equation gives the acceleration of a point as it is displaced by a P-wave moving past it at a speed or celerity equal to the P-wave velocity of the strata (given by Eq. 19).

For plane waves propagating in homogeneous strata, a solution to Eq. 22 is

$$u(x,t) = u_0 \exp[i(\omega t - kx)] \qquad (23)$$

In this equation, u_0 is the wave amplitude, ω is the angular frequency of the wave and k is the wave number. Equation 23 gives the displacement at any distance in the strata at any time while the wave is propagating through it. It is these displacements that are recorded in seismic reflection data.

An important phenomenon in seismic-reflection data is the **attenuation** of seismic energy with increasing distance from the source. Part of this attenuation is due to the spreading of the finite energy emitted by the source over an increasing area. In terms of displacement, the wave amplitude drops off with increasing distance, R, from the source. Modifying Eq. 23 to reflect this yields

$$u(x,t) = u_0 \frac{R_0}{R} \exp[i(wt - kx)] \qquad (24)$$

where R_0 is a reference distance (normally taken to be 1: Dobrin, 1976).

Seismic energy is also lost to friction. In strata, the friction is of two general forms. One is the friction that results from the grain-to-grain interactions as a seismic wave deforms the sediments it passes through. This friction is known as **intrinsic attenuation**. Its inclusion in the solution to the

wave equation results in the further modification of Eq. 23 to yield

$$u(x,t) = u_0 \frac{R_0}{R} \exp[-\alpha(\omega)x] \exp[i(\omega t - kx)] \qquad (25)$$

The variable α is the intrinsic attenuation coefficient, which is the rate at which the wave amplitude diminishes with increasing distance from the source. Note that α is a function of the angular frequency of the wave ω. Variable α is related to the frequency of the wave, f, via

$$\alpha = \frac{\pi f}{QV} \qquad (26)$$

In this equation, V is the velocity of the strata (either V_p or V_s) and Q is the so-called quality factor, which is a constant of the material and various values of Q for different strata have been tabulated (Clark, 1966). What is important though from the standpoint of attenuation is that α is directly proportional to frequency, meaning that high frequencies attenuate more rapidly than low frequencies. Seismic waves are composed of a range of frequencies, which means that the higher frequency components of the wave are attenuated more quickly than the lower frequency components as the wave moves away from its source. This then is a cause for the blurring and fading in seismic images with increasing travel time.

The other form of friction that attenuates a seismic wave is that which results from the movement of the sediment framework relative to the pore fluid as the seismic wave passes by. This type of energy loss is referred to as fluid attenuation. The theory that describes **fluid attenuation** (Biot, 1956a,b), also known as poroelastic theory, is complex (for details see Stoll, 1989; Wang, 2000). In developing his theory, Biot formulated a pair of coupled equations that describe: (i) the motion of a volume of sediment as a wave passes through it; and (ii) the resulting movement of pore fluid into or out of this volume. The solutions to these equations yield fluid attenuation coefficients that, like the intrinsic attenuation coefficient, are frequency dependent. However, the solutions also yield strata velocities that are frequency dependent. In other words, there is no single P-wave and S-wave velocity for a particular sediment mixture. Instead, seismic waves of different frequency will propagate through the mixture at different speeds.

Having established bulk densities and velocities for each cell in a stratigraphic model, the last rock properties that need to be determined to simulate synthetic seismic data are the **seismic impedances**, **reflectivities** and **transmissivities** of the modelled strata. Impedance, I, is simply

$$I = \rho_b V \tag{27}$$

At a planar interface between two homogeneous volumes, the normal incidence reflectivity for waves travelling from volume 1 into volume 2 is

$$R_{12} = \frac{A_r}{A_i} = \frac{I_2 - I_1}{I_2 + I_1} \tag{28}$$

where R_{12} is the reflectivity, A_R is the amplitude of the reflected wave and A_i is the amplitude of the incident wave.

The normal incident transmissivity is

$$T_{12} = \frac{A_t}{A_i} = \frac{2I_1}{I_2 + I_1} \tag{29}$$

where T_{12} is the transmissivity and A_t is the amplitude of the transmitted wave. Note that continuity at the interface requires that

$$A_i + A_r = A_t \tag{30}$$

Seismic modelling

Once the necessary rock properties exist for a stratigraphic model, synthetic seismic data of the model can be generated. A number of ways exist for producing this synthetic data. Three are highlighted here. From simple and idealistic to sophisticated and realistic, they are convolution, the one-way wave equation and the full two-way wave equation.

Convolution

Seismic-reflection data can be viewed as the result of convolving the wavelet produced by a seismic source with Earth's response. Using this perspective, synthetic seismic traces can be produced by

convolving a source wavelet with each column of reflectivities (R) in the stratigraphic model (i.e. the model of Earth's response)

$$Tr = S(t) \cdot R(t) \tag{31}$$

Here, Tr is the seismic trace and S is the seismic source wavelet. Note that both S and R are functions of time. Thus to carry out this operation, the depth dimension of the stratigraphic model must first be converted to time using the strata velocities in each cell of the model.

The convolution simulates the seismic response of strata as if it were imaged by seismic waves that travel directly down from and back to the seismic source. This is the ultimate imaging objective of seismic data processing. Therefore, the convolutional simulation is theoretically the 'best' data that can be obtained with the seismic source.

Wave equation

Convolution simulates perfectly imaged seismic data in which all seismic events are correctly positioned in time and space. In actual data, seismic events are often compromised by wave-propagation phenomena such as diffractions, interference patterns and misplaced events caused by heterogeneities in the velocity of the strata as the seismic waves propagate through them.

These phenomena can be simulated using the 'one-way' wave equation or **exploding reflector model** (Claerbout & Johnson, 1971; Loewenthal *et al.*, 1976). This version of the wave equation is based on the two-dimensional form of the wave equation for *P*-waves (i.e. the two-dimensional form of Eq. 21). It is referred to as the one-way wave equation because of the way that it is solved. In the approach, every reflectivity in the stratigraphic model is treated as a seismic source, the 'strength' of which is made proportional to the magnitude of the reflectivity. The sources are simultaneously 'detonated', generating a wave field that propagates upward at speeds and in directions that are governed by the velocity structure of the geological model. Solution of the equation simulates unmigrated, zero-offset seismic sections, i.e. sections in which seismic events have not yet been properly positioned in space and time.

Solution of the full 'two-way' wave equation (i.e. for *P*- and *S*-waves) allows the complete propagation path of seismic waves to be simulated, i.e. their departure from the source, their passage through multilayered strata and their arrival at surface receivers (Fei & Larner, 1995). This more realistic form of modelling accounts for such additional effects as geometrical spreading, multiples, refractions and reflections at interfaces. It also provides the added flexibility of simulating multi-channel seismic data, which can then be used to investigate the loss of information due to choices made in processing such data.

Synthetic seismic data generated by both one- and two-way wave equations contain distortions due to spatial mispositioning of seismic events. These can be corrected through time or depth migration. The synthetic multichannel seismic data generated from the two-way wave equation can be further processed to remove other types of seismic artefacts and to estimate a seismic velocity model for the stratigraphy. In theory, such processing should yield seismic images that are the equal of that produced by convolution.

THE WAY FORWARD

Individual efforts to develop numerical stratigraphic and sediment-transport models can become much more effective if better coordinated, and openness is developed between modellers and field-oriented geoscientists. A new initiative is designed to catalyse Earth-surface research over the coming decades by: empowering a broad community of scientists and students with computing tools and knowledge from interlinked fields; streamlining the process of idea generation and hypothesis testing through linked surface dynamics models; and enabling rapid creation and application of models tailored to specific settings, scientific problems and time-scales. The program is called the **Community Surface-Dynamics Modelling System** (CSDMS).

The CSDMS program is based on algorithms that mathematically describe the processes and conditions relevant to sediment transport and deposition, and incorporates the major input and boundary conditions that define a sedimentary system. The effort is being coordinated and funded by government agencies and industry. The CSDMS program adds value to those working on modern environ-

mental applications, future global-warming scenarios, natural-disaster mitigation efforts, natural-hazard efforts, reservoir characterization, oil exploration and national security. It could be argued that the new satellite databases and the large-scale three-dimensional geophysical datasets can only realize their full potential in collaboration with a CSDMS.

The CSDMS initiative in predictive sedimentary dynamics will help us to:

1 improve assessment of risk from natural hazards such as landslides, mudflows, floods and coastal storms;
2 improve predictive capability at all scales of stratal architecture, and consequently improve our ability to explore and exploit energy and mineral source rocks and reservoirs;
3 better manage natural landscapes;
4 understand the role of basin water storage and chemical processing in the hydrological cycle;
5 understand the role of sedimentary basins as incubators of the deep biosphere;
6 understand the manner in which sedimentary basins and erosional landscapes control carbon and other elemental cycles;
7 better interpret the record of global and regional climate change.

Researchers working together can produce a more reliable and more flexible simulation model than can any single agency. The code being developed is free, thus minimizing the rewriting of the same initial algorithms and allowing researchers more time to spend on new advances. A CSDMS architecture also creates honesty in what modellers claim, and allows for faster verification and comparison of different approaches on new datasets. Communication is greatly increased among users and coders; a community is built. If a new component of the model is developed, and the identified community agrees on the substantive improvement, then the new component replaces the old component and a new version of the model is released. Integrated models will allow hypothesis testing, sensitivity experiments on key parameters, and even the identification of new thrusts in the science. The CSDMS effort has strong support from the petroleum industry whose own research laboratories have pioneered initial attempts.

The CSDMS program will provide the modelling framework for understanding and predicting the dynamics of the 'Critical Zone': the heterogeneous, near-surface environment in which complex interactions involving rock, soil, water, air and living organisms regulate the natural habitat and determine the availability of life-sustaining resources. The following fundamental scientific questions form the foundation of CSDMS.

1 What are the fluxes, reservoirs, and flow paths associated with the physical, biological, and chemical transport processes in the Critical Zone? How do these depend on substrate properties like morphology, geology, and ecology, and on human activities?
2 How do material fluxes and surface evolution vary across time and space scales?
3 How are physical and biological processes coupled in surface systems?
4 How is the history of surface evolution recorded in surface morphology and physical, chemical, and biological stratigraphic records?
5 How does coupling occur within the Critical Zone. How do linked surface environments communicate with one another across their dynamic boundaries?
6 How does the Critical Zone couple to the tectosphere, atmosphere, hydrosphere, cryosphere, and biosphere and serve as the dynamic interface among them? How do changes in one part of the global surface system affect other parts?

The extreme range of space and time-scales that define Earth's history demands an array of approaches rather than a single monolithic modelling structure. As its point of departure, CSDMS will use techniques developed in the atmospheric and marine sciences communities, which also deal with a wide range of length and time-scales. A direct approach is simply to develop models at different scales, with length and time-scales generally increasing together. At each scale, processes below the resolution of the model are parameterized, much as ensembles of complex particle dynamics are parameterized by bulk diffusion coefficients. A more sophisticated approach is to use nested models, in which large-scale models using coarse grids interact with finer-scale models that in effect run within subdomains or single grid cells of the larger model. In the simplest version, the large-scale model provides boundary and initial conditions for the smaller-scale model. In more sophisticated schemes, the output of the smaller-scale model feeds back to the larger scales as well.

The modular design of the CSDMS is well suited to accommodate coupling of transport environments across dynamic moving boundaries. Its integrative, multi-environmental nature will make it possible to develop a global view of the highly interconnected surface system. The CSDMS program will include interfaces to allow for flexible connection to comparable modelling systems for other 'spheres' with which the surface interacts (tectosphere, atmosphere, hydrosphere, cryosphere, biosphere).

Numerical models that define the development of landscapes and sedimentary architecture are the repositories of our understanding of basic physics and thermodynamics underlying sedimentology. They force scientists to examine their knowledge level, and the pyramid of assumptions upon which they view the world. The CSDMS program would push scientists to confront nature in terms of whether: (i) one process should be coupled or uncoupled with respect to another; (ii) a particular process is deterministic or stochastic; (iii) processes can be scaled across time and space; (iv) adequate databases for key parameters from field or laboratory measurements are in place.

ACKNOWLEDGEMENTS

Colleagues assigned the complex task of providing the authors with feedback on this paper include Rudy Slingerland, Alan Niedoroda, Pat Wiberg, Chuck Nittrouer, Carl Friedrichs and Gert-Jan Weltje. We are thankful for their kindness and thoroughness, and are deeply indebted. The US Office of Naval Research funded this coordinated effort to develop predictive models of sediment transport and stratigraphy. Particular thanks go to Dr Joseph Kravitz for his oversight of the STRATAFORM programme and Dr Thomas Drake for his oversight of the EuroSTRATAFORM effort.

NOMENCLATURE

Symbol	Definition	Dimensions
a_h	horizontal acceleration due to the earthquake	$L\,T^{-2}$
a_i	empirical constants	
a_v	vertical acceleration due to the earthquake	$L\,T^{-2}$

Symbol	Description	Dimensions
A	river-basin area	L^2
A_i	amplitude of the incident wave	L
A_r	amplitude of the reflected wave	L
b	empirical constant	
b_0	river-mouth width	L
b_i	width of the ith slice or column	L
B	depth-integrated buoyancy anomaly	$M\,L\,T^{-2}$
c	cohesion	$M\,T^{-2}$
c	empirical compaction constant	$L\,T^2\,M^{-1}$
c_{bed}	sediment concentration of the bed surface (1 − porosity)	
c'	cohesion defined in terms of effective stress	$M\,L^{-1}\,T^{-2}$
c_i	sediment cohesion	$M\,T^{-2}$
c_s	suspended sediment concentration	$M\,L^{-3}$
C	layer-averaged bulk sediment concentration	$M\,L^{-3}$
C	rating coefficient	
C	volumetric fraction of clay	
C_0, C_1	empirical constants	
C_D	coefficient of drag	
D	sediment deposition rate	$L\,T^{-1}$
e_w	dimensionless entrainment coefficient	
E	mean (expected) of the stated variable	various
E_s	rate that sediment is entrained into the current by erosion	$M\,L^{-2}\,T^{-1}$
f	frequency of the wave	$L\,T^{-1}$
F	constant of mass flux	
F	factor of safety for a sediment column	
F_T	factor of safety for the entire sediment volume	
g	acceleration due to gravity	$L\,T^{-2}$
h	current or flow thickness	L
h_p	thickness of the upper plug zone in a debris flow	L
h_s	thickness of the lower shear layer in a debris flow	L
H	river-basin relief	L
I	impedance	$M\,L^2\,T^{-1}\,C^{-2}$
I	sediment inventory or mass per unit area of the plume	$M\,L^{-2}$
k	constant representing the global climate belts	
k	wave number	
K	layer-averaged turbulent kinetic energy	$M\,L^2\,T^{-2}$
K	saturated effective bulk modulus	$M\,L^{-1}\,T^{-2}$
K_s	sediment diffusivity	$L^2\,T^{-1}$
m	empirical constant	
M	mass of sediment column	M
Q	discharge of the river	$L^3\,T^{-1}$
Q	the so-called quality factor	
Q_{evap}	discharge component lost via evaporation	$L^3\,T^{-1}$
Q_g	ground-water discharge	$L^3\,T^{-1}$
Q_{ice}	icemelt discharge	$L^3\,T^{-1}$
Q_n	snowmelt (nival) discharge	$L^3\,T^{-1}$
Q_r	discharge from rainfall as surface runoff	$L^3\,T^{-1}$
Qs	daily sediment discharge	$M\,T^{-1}$
$\bar{Q}$	long-term power-law average of discharge	$L^3\,T^{-1}$
$\overline{Qs}$	long-term power-law average of Qs	$M\,T^{-1}$
q_{sx}	x-component of sediment flux	$M\,L^{-1}\,T^{-1}$
r_0	ratio of sediment concentrated at the base of the flow	
R	reflectivity	
R	submerged specific gravity of the sediment	$M\,L^{-3}$
R	distance from acoustic source	L
R_0	reference distance	L
R_{cr}	critical Richardson number	
s	shear strength	$M\,L^{-1}\,T^{-2}$
s_u	undrained shear strength of the sediment	$M\,L^{-1}\,T^{-2}$
S	seismic source wavelet	T
t	time	T
T	normal incident transmissivity	
$\bar{T}$	mean surface temperature of the drainage basin	C
Tr	seismic trace	T
u	displacement	L
u	pore pressure	$M\,L^{-1}\,T^{-2}$

u	x-component of flow velocity (longitudinal)	$L\,T^{-1}$	γ'	submerged density of sediment	$M\,L^{-3}$
u_*	shear velocity at the base of a current	$L\,T^{-1}$	δ	empirical constant	
u_0	wave amplitude	L	ε_0	layer-averaged dissipation rate	$L^2\,T^{-3}$
u_e	excess pore pressure	$M\,L^{-1}\,T^{-2}$	λ	first-order removal-rate constant for a given grain size	T^{-1}
U	layer-averaged flow velocity	$L\,T^{-1}$			
U_{max}^2	bottom velocity shear	$L^2\,T^{-2}$	μ	dynamic viscosity	$M\,L^{-1}\,T^{-1}$
U_c	near-bed velocity due to ambient currents	$L\,T^{-1}$	μ	shear modulus	$M\,L^{-1}\,T^{-2}$
			η	bed elevation	L
U_g	velocity of the bottom boundary layer under the influence of gravity	$L\,T^{-1}$	ρ	density of the fluid in the hyperpycnal current	$M\,L^{-3}$
			ρ_b	bulk density	$M\,L^{-3}$
U_{max}	total velocity	$L\,T^{-1}$	ρ_f	density of the hyperpycnal current (sediment and fluid)	$M\,L^{-3}$
U_p	layer-averaged velocity of the upper plug zone	$L\,T^{-1}$			
U_s	layer-averaged velocity of the lower shear layer	$L\,T^{-1}$	ρ_f	fluid density	$M\,L^{-3}$
			ρ_g	density of grains within the cell	$M\,L^{-3}$
U_w	near-bed velocity due to the action of waves	$L\,T^{-1}$	ρ_m	density of the mudflow	$M\,L^{-3}$
v	y-component of flow velocity (lateral)	$L\,T^{-1}$	ρ_s	grain density	$M\,L^{-3}$
			ρ_{sw}	density of the ambient seawater	$M\,L^{-3}$
V	acoustic velocity through strata (either V_p or V_s)	$L\,T^{-1}$			
			ρ_w	pore-water density	$M\,L^{-3}$
V_p	compressional or P-wave velocity	$L\,T^{-1}$	σ	normal stress, applied load	$M\,L^{-1}\,T^{-2}$
V_s	shear or S-wave velocity	$L\,T^{-1}$	σ	standard deviation of the stated variable	various
V_s	volume in cell occupied by sediment	L^3	σ'	normal effective stress (total normal stress minus the pore-water pressure)	$M\,L^{-1}\,T^{-2}$
V_s	volume of sediment in suspension	L^3			
V_t	total volume of a cell	L^3			
w_s	particle settling velocity	$L\,T^{-1}$	τ_y	yield (or remoulded shear) strength	$M\,L^{-1}\,T^{-2}$
W_{hi}	horizontal pull on sediment column	$L\,T^{-2}$			
W_{vi}	vertical weight of a sediment column	$L\,T^{-2}$	ϕ	empirical constant	
			ϕ	porosity	
x	distance along the flow path, longitudinal or axial direction	L	ϕ	friction angle	
			ϕ'	friction angle defined in terms of effective stress	
			ϕ_c	clay porosity	
y	lateral direction	L	ϕ_i	internal friction angle	
z	depth below the seafloor, depth of sediment column	L	ϕ_{min}	minimum porosity after deep burial	
α	intrinsic attenuation coefficient	T	ϕ_s	sand porosity	
			ψ	lognormal random variable describing measurement error and flood dynamics	
α	slope of failure plane				
α	slope				
α_i	dimensionless constants representing global climate belts		ω	angular frequency of the wave	T^{-1}

REFERENCES

Akiyama, J. and Stefan, H.G. (1984) Plunging flow into reservoir. Theory. *J. Hydraul. Eng.*, **110**, 484–499.

Altinakar, M., Graf, W.H. and Hopfinger, E. (1990) Weakly depositing turbidity current on a small slope. *J. Hydraul. Res.*, **28**, 55–80.

Altinakar, M., Graf, W.H. and Hopfinger, E. (1996) Flow structure in turbidity currents. *J. Hydraul. Res.*, **34**, 713–718.

Anderson, M.G. and Richards, K.S. (1987) *Slope Stability: Geotechnical Engineering and Geomorphology.* John Wiley & Sons, New York.

Athy, L.F. (1930) Density, porosity and compaction of sedimentary rocks. *Bull. Am. Assoc. Petrol. Geol.*, **14**, 1–22.

Austin, J.A., Christie-Blick, N. and Malone, M. (1997) *Ocean Drilling Program Leg 174A Preliminary Report: Continuing the New Jersey Mid-Atlantic Sea-Level Transect.* Ocean Drilling Program, College Station, Texas, 73 pp.

Bagnold, R.A. (1962) Autosuspension of transported sediment: turbidity currents. *Proc. R. Soc. London*, **A265**, 315–319.

Bagnold, R.A. (1966) An approach to the sediment transport problem from general physics. *U.S. Geol. Surv. Prof. Pap.*, **422-I**, 37 pp.

Bahr, D.B., Hutton, E.W.H., Syvitski, J.P.M. and Pratson, L. (2001) Exponential approximation to compacted sediment porosity profiles. *Comput. Goesci.*, **27**, 691–700.

Benson, L.V., Burdett, J., Kashgarian, M., *et al.* (1996) Climatic and hydrologic oscillations in the Owens Lake Basin and adjacent Sierra Nevada, California. *Science*, **274**, 746–749.

Benson, L.V., Burdett, J., Lund, S., Kashgarian, M. and Mensing, S. (1997a) Nearly synchronous climate change in the Northern Hemisphere during the Last glacial termination. *Nature*, **388**, 263–265.

Benson, L.V., Lund, S.P., Burdett, J.W., *et al.* (1997b) Correlation of late-Pleistocene lake-level oscillations in Mono Lake, California, with North Atlantic climate events. *Quat. Rev.*, **17**, 251–260.

Berryman, J.G. (1980a) Long-wavelength propagation in composite elastic media I. Spherical inclusions. *J. Acoust. Soc. Am.*, **68**, 1809–1819.

Berryman, J.G. (1980b) Long-wavelength propagation in composite elastic media II. Ellipsoidal inclusions. *J. Acoust. Soc. Am.*, **68**, 1820–1831.

Beschta, R.L. (1987) Conceptual models of sediment transport in streams. In: *Sediment Transport in Gravel-bed Rivers* (Eds C.R. Thorne, J.C. Bathurst and R.D. Hey), pp. 387–419. John Wiley & Sons, Chichester.

Beven, K.J., Lamb, R., Quinn, P.F., Romanowicz, R. and Freer, J. (1995) TOPMODEL. In: *Computer Models of Watershed Hydrology* (Ed. V.P. Singh), pp. 627–668. Water Resources Publications, Highlands Ranch, CO.

Biot, M.A. (1956a) Theory of propagation of elastic waves in a fluid-saturated porous solid. I. low frequency range. *J. Acoust. Soc. Am.*, **28**, 168–178.

Biot, M.A. (1956b) Theory of propagation of elastic waves in a fluid-saturated porous solid. II. higher frequency range. *J. Acoust. Soc. Am.*, **28**, 178–191.

Bishop, A.W. (1955) The use of the slip circle in the stability analysis of slopes. *Géotechnique*, **5**, 7–17.

Bitzer, K. and Harbaugh, J.W. (1987) DEPOSIM: a Macintosh computer model for two-dimensional simulation of transport, deposition, erosion and compaction of clastic sediments. *Comput. Geosci.*, **13**, 611–637.

Blumberg, A.F. and Mellor, G.L. (1987) A description of a three-dimensional coastal ocean circulation model. In: *Three-dimensional Coastal Ocean Models, Coastal and Estuarine Sciences* (Ed. N.S. Heaps), pp. 1–16. American Geophysical Union, Washington, DC.

Bonham-Carter, G.F. and Sutherland, A.J. (1967) Diffusion and settling of sediments at river mouths: a computer simulation model. *Trans. Gulf Coast Assoc. Geol. Soc.*, **17**, 326–338.

Bournet, P.E., Dartus, D., Tassin, B. and Vincon-Leite, B. (1999) Numerical investigation of plunging density current. *J. Hydraul. Eng.*, **125**, 584–594.

Bracco, G.L.G. and Schlager, W. (1999) Discriminating between onlap and lithologic interfingering in seismic models of outcrops. *Bull. Am. Assoc. Petrol. Geol.*, **83**, 952–971.

Bradford, S.F. and Katopodes, N.D. (1999a) Hydrodynamics of turbid underflows I: formulation and numerical analysis. *J. Hydraul. Eng.*, **125**, 1006–1015.

Bradford, S.F. and Katopodes, N.D. (1999b) Hydrodynamics of turbid underflows II: aggradation, avulsion and channelization. *J. Hydraulic Eng.*, 125, 1016–1028.

Bradford, S.F., Katopodes, N.D. and Parker, G. (1997) Characteristic analysis of turbid underflows. *J. Hydraul. Eng.*, **123**, 420–431.

Britter, R.E. and Linden, P.F. (1980) The motion of the front of gravity current traveling down an incline. *J. Fluid Mech.*, **99**, 532–543.

Brors, B. and Eidsvik, K.J. (1992) Dynamic Reynolds stress modeling of turbidity currents. *J. Geophys. Res.*, **97**, 9645–9652.

Brown, W.M. III and Ritter, J.R. (1971) Sediment transport and turbidity in the Eel River Basin, California. *U.S. Geol. Surv. Prof. Pap.*, **1986**, 70 pp.

Cacchione, D.A., Wiberg, P.L., Lynch, J., Irish, J. and Traykovski, P. (1999) Estimates of suspended-sediment flux and bedform activity on the inner portion of the Eel continental shelf. *Mar. Geol.*, **154**, 83–97.

Cant, D.J. (1991) Geometric modeling of facies successions and local unconformities. *Basin Res.*, **3**, 51–62.

Carey, J.S., Swift, D.J.P., Steckler, M.S., Reed, C. and Niedoroda, A. (1999) High resolution sequence stratigraphic modeling: 2. The influence of sedimentation processes. In: *Numerical Experiments in Stratigraphy: Recent Advances in Stratigraphic and Computer Simulations* (Eds J. Harbaugh, L. Watney, G. Rankey, *et al.*), pp. 151–164. Memoir 62, Society of Economic Paleontologists and Mineralogists, Tulsa, OK.

Carr, J.A. (2002) *The effects of flocculation on contaminant sorption during resuspension events on the continental shelf: a modeling approach.* M.A. thesis, University of Virginia, Charlottesville, VA, 120 pp.

Caruccio, F.T. (1989) Stratigraphic and geochemical controls on the occurrence of acidic mine waters and predictive technologies. In: Quantitative Dynamic Stratigraphy (Ed. T.A. Cross) Prentice-Hall, New York, pp. 581–588.

Castagna, J.P., Batzle, M.L. and Eastwood, R.L. (1985) Relationships between compressional-wave and shear-wave velocities in clastic silicate rocks. *Geophysics*, **50**, 571–581.

Chao, S.-Y. (1988) Wind-driven motion of estuarine plumes. *J. Phys. Ocean.*, **18**, 1144–1166.

Chikita, K. (1989) A field study on turbidity currents initiated from spring runoffs. *Water Resour. Res.*, **25**, 257–271.

Choi, S.-U. and García, M.H. (1995) Modeling of one-dimensional turbidity currents with a dissipative-Galerkin finite element method. *J. Hydraul. Res.*, **33**, 623–648.

Choi, S.-U. and García, M.H. (2001) Spreading of gravity plumes on an incline. *Coast. Eng.*, **43**, 221–237.

Choi, S.-U. and García, M.H. (2002) k-ε turbulence modeling of density currents developing two dimensionally on a slope. *J. Hydraul. Res.*, **128**, 55–63.

Christie-Blick, N. (1991) Onlap, offlap and the origin of unconformity-bounded depositional sequences. *Mar. Geol.*, **97**, 35–56.

Claerbout, J.F. (1985) *Imaging the Earth's Interior*. Blackwell Scientific Publications, Oxford, 385 pp.

Claerbout, J.F. and Johnson, A.G. (1971) Extrapolation of time dependent waveforms along their path of propagation. *Geophys. J. R. Astron. Soc.*, **26**, 285–293.

Clark, S.P., Jr. (1966) Handbook of physical constants. *Geol. Soc. Am. Mem.*, **97**.

Coogan, A.H. and Manus, R.W. (1975) Compaction and diagenesis of carbonate sands. In: *Compaction of Coarse Grained Sediments* (Eds G.V. Chilingarian and K.H. Wolf), pp. 79–166. Elsevier, New York.

Coussot, P. (1994) Steady, laminar flow of concentrated mud suspensions in open channel. *J. Hydraul. Res.*, **32**, 535–560.

Cross, T.A. and Lessenger, M.A. (1999) Construction and application of a stratigraphic inverse model. In: *Numerical Experiments in Stratigraphy: Recent Advances in Stratigraphic and Computer Simulations* (Eds J. Harbaugh, L. Watney, G. Rankey, *et al.*), pp. 69–83. Memoir 62, Society of Economic Paleontologists and Mineralogists, Tulsa, OK.

Curran, K.J., Hill, P.S. and Milligan, T.G. (2002) The role of particle aggregation in size-dependent deposition of drill mud. *Cont. Shelf Res.*, **22**, 405–416.

Daly, C., Neilson, R.P. and Phillips, D.L. (1994) A statistical-topographic model for mapping climatological precipitation over mountainous terrain. *J. Appl. Meteorol.*, **33**, 140–158.

Das, B.M. (1983) *Advanced Soil Mechanics*. McGraw-Hill, New York, 511 pp.

Dobrin, M.B. (1976) *Introduction to Geophysical Prospecting*. McGraw Hill, New York, 630 pp.

Drake, D.E. (1999) Temporal and spatial variability of the sediment grain-size distribution on the Eel shelf: the flood layer of 1995. *Mar. Geol.*, **154**, 169–182.

Eberhardt-Phillips, D., Han, D.-H. and Zoback, M.D. (1989) Empirical relationships among seismic velocity, effective pressure, porosity and clay content in sandstones. *Geophysics*, **54**, 82–89.

Ebrom, D.A. and McDonald, J.A. (1994) *Seismic Physical Modeling*. Society of Exploration Geophyscists, 519 pp.

Eidsvik, K.J. and Brors, B. (1989) Self-accelerated turbidity current prediction based upon k-ε turbulence. *Cont. Shelf Res.*, **9**, 617–627.

Eisma, D. and Kalf, J. (1984) Dispersal of Zaire River suspended matter in the estuary and the Angola Basin. *Neth. J. Sea Res.*, **17**, 385–411.

Ellison, T.H. and Turner, J.S. (1959) turbulent entrainment in stratified flows. *J. Fluid Mech.*, **6**, 423–448.

Elverhøi, A., Norem, H., Andersen, E.S., *et al.* (1997) On the origin and flow behavior of submarine slides on deep-sea fans along the Norwegian-Barents Sea continental margin. *Geo-Mar. Lett.*, **17**, 119–125.

Farrell, G.J. and Stefan, H.G. (1988) Mathematical modeling of plunging reservoir flows. *J. Hydraul. Res.*, **26**, 525–537.

Fei, T. and Larner, K. (1995) Elimination of numerical dispersion in finite-difference modeling and migration by flux-corrected transport. *Geophysics*, **60**, 1830–1842.

Felix, M. (2001) A two-dimensional numerical model for a turbidity current. In: *Particulate Gravity Currents* (Eds W.D. McCaffrey, B.C. Kneller and J. Peakall), pp. 71–81. Special Publication 31, International Association of Sedimentologists. Blackwell Science, Oxford.

Fellenius, W. (1936) Calculation of the stability of earth dams. *Transactions of the 2nd Congress on Large Dams*, Vol. 4, p. 445.

Flemings, P.B. and Grotzinger, J.P. (1996) STRATA: freeware for solving classic stratigraphic problems. *GSA Today*, **6**, 1–7.

Flood, R.D., Pirmez, C. and Yin, H. (1997) The compressional-wave velocity of Amazon Fan sediments: calculation from index properties and variation with clay content. In: *Proceedings of the Ocean Drilling Program, Scientific Results*, Vol. 155. College Station, TX.

Fong, D.A. and Geyer, W.R. (2001) The response of a river plume during an upwelling favorable wind event. *J. Geophys. Res.*, **106**, 1067–1084.

Fox, W.T. (1978) Modeling coastal environments. In: *Coastal Sedimentary Environments* (Ed. R.A. Jr. Davis), pp. 385–413. Springer-Verlag, New York.

Franseen, E.K., Watney, W.L., Kendall, S.G. and Ross, W. (1991) Sedimentary modeling: Computer simulation and methods for improved parameter definition. *Kansas Geol. Surv. Bull.*, **233**, 524 pp.

Fredsoe, J. and Deigaard, R. (1994) *Mechanics of Coastal Sediment Transport*. World Scientific, Singapore, 369 pp.

Friedrichs, C.T., Wright, L.D., Hepworth, D.A. and Kim, S.C. (2000) Bottom-boundary-layer processes associated with fine sediment accumulation in coastal seas and bays. *Cont. Shelf Res.*, **20**, 807–841.

Fukushima, Y., Parker, G. and Pantin, H. (1985) Prediction of igniting turbidity currents in Scripps Submarine Canyon. *Mar. Geol.*, **67**, 55–81.

Ganjeon, D. and Joseph, P. (1999) Concepts and application of a 3-D multiple lithology diffusive model in stratigraphic modeling. In: *Numerical Experiments in Stratigraphy: Recent Advances in Stratigraphic and Computer Simulations* (Eds J. Harbaugh, L. Watney, G. Rankey, *et al.*), pp. 197–210. Memoir 62, Society of Economic Paleontologists and Mineralogists, Tulsa, OK.

García, M.H. (1993) Hydraulic jumps in sediment-laden bottom currents. *J. Hydraul. Eng.*, **199**, 1094–1117.

García, M.H. and Parker, G. (1989) Experiments on hydraulic jumps in turbidity currents near a canyon-fan transition. *Science*, **245**, 393–396.

García, M.H. and Parker, G. (1991) Entrainment of bed sediment into suspension. *J. Hydraul. Eng.*, **117**, 414–435.

García, M.H. and Parker, G. (1993) Experiments on the entrainment of sediment into suspension by a dense bottom current. *J. Geophys. Res.*, **98**, 4793–4807.

Garvine, R.W. (1974) Physical features of the Connecticut River outflow during high discharge. *J. Geophys. Res.*, **79**, 831–846.

Gassmann, F. (1951) Elastic waves through a packing of sphere. *Geophysics*, **16**, 673–685.

Geyer, W.R. and Kineke, G.C. (1995) Observations of currents and water properties in the Amazon frontal zone. *J. Geophys. Res.*, **100**, 2321–2339.

Geyer, W.R., Hill, P.S., Milligan, T.G. and Traykovski, P. (2000) The structure of the Eel River plume during floods. *Continental Shelf Res.*, **20**, 2067–2093.

Geyer, W.R., Hill, P.S. and Kineke, G.C. (2004) The transport and dispersal of sediment by buoyant coastal plumes. *Continental Shelf Res.*, **24**, 927–949.

Goodbred Jr., S.L. and Kuehl, S.A. (2000a) Enormous Ganges–Brahmaputra sediment discharge during strengthened early Holocene monsoon. *Geology*, **28**, 1083–1086.

Goodbred Jr., S.L. and Kuehl, S.A. (2000b) The significance of large sediment supply, active tectonism and eustasy on margin sequence development: late Quaternary stratigraphy and evolution of the Ganges–Brahmaputra delta. *Sediment. Geol.*, **133**, 227–248.

Grant, W.D. and Madsen, O.S. (1979) Combined wave and current interactions with a rough bottom. *J. Geophys. Res.*, **84**, 1797–1808.

Hamilton, E.L. (1980) Geoacoustic modeling of the seafloor. *J. Acoust. Soc. Am.*, **68**, 1313–1340.

Haq, B.U., Hardenbol, J. and Vail, P.R. (1988) Mesozoic and Cenozoic chronostratigraphy and cycles in sea level change. In: *Sea Level Changes: an Integrated Approach* (Eds C.K. Wilgus, B.S. Hastings, C.G. St.C. Kendall, *et al.*), pp. 71–108. Special Publication 42, Society of Economic Paleontologists and Mineralogists, Tulsa, OK.

Harbitz, C.B., Parker, G., Elverhøi, A., *et al.* (2003) Hydroplaning of submarine debris flows and glide blocks: analytical solutions and discussion. *J. Geophys. Res.*, **108**, 2349, doi:10.1029/2001JB001454.

Harris, C.K. and Wiberg, P.L. (2001) A two-dimensional, time-dependent model of suspended sediment transport and bed reworking for continental shelves. *Comput. Goesci.*, **27**, 675–690.

Harris, C.K. and Wiberg, P.L. (2002) Across-shelf sediment transport: interactions between suspended sediment and bed sediment. *J. Geophys. Res.*, **107**: 10.1029/2000JC000634.

Harris, C.K., Geyer, R.W. and Signell, R.P. (2000) Dispersal of flood sediment by oceanographic currents and energetic waves. *Eos (Trans. Am. Geophys. Union)*, **80**(49), OS281.

Harris, C.K., Traykovski, P.A. and Geyer, W.R. (2005) Flood dispersal and deposition by near-bed gravitational sediment flows and oceanographic transport: a numerical modeling study of the Eel River shelf, northern California. *J. Geophys. Res.*, **110**, C09025.

Hashin, Z. and Shtrikman, S. (1963) A variational approach to the theory to the elastic behavior of multiphase materials. *J. Mech. Phys. Solids*, **11**, 127–140.

Haupt, B. and Stattegger, K. (1999) The ocean-sediment system and stratigraphic modeling in large basins. In:

522   J.P.M. Syvitski et al.

Numerical Experiments in Stratigraphy: Recent Advances in Stratigraphic and Computer Simulations (Eds J. Harbaugh, L. Watney, G. Rankey, *et al.*), pp. 313–321. Memoir 62, Society of Economic Paleontologists and Mineralogists, Tulsa, OK.

Hay, A.E. (1987) Turbidity currents and submarine channel formation in Rupert Inlet, British Columbia. II. The roles of continuous and surge-type flow. *J. Geophys. Res.*, **92**, 2883–2900.

Herrick, D. (2001) *A survey of existing velocity models using physical properties derived solely from clay content.* M.S. thesis, Duke University, Durham, NC, 54 pp.

Hill, P.S., Milligan, T.G. and Geyer, W.R. (2000) Controls on effective settling velocity of suspended sediment in the Eel River flood plume. *Cont. Shelf Res.*, **20**, 2095–2111.

Horikawa, K. (1988) *Nearshore Dynamics and Coastal Processes.* Univ. Tokyo Press, 522 pp.

Hostetler, S.W., Giorgi, F., Bates, G.T. and Bartlein, P.J. (1994) Lake-atmosphere feedbacks associated with paleolakes Bonneville and Lahontan. *Science*, **263**, 665–668.

Huang, X. and García, M.H. (1998) A Herschel–Bulkley model for mud flow down a slope. *J. Fluid Mech.*, **374**, 305–333.

Huang, X. and García, M.H. (1999) Modeling of non-hydroplaning mudflows on continental slopes. *Mar. Geol.*, **154**, 131–142.

Hunt, D. and Tucker, M.E. (1993) Sequence stratigraphy of carbonate shelves with an example from the mid-Cretaceous (Urgonian) of southeast France. In: *Sequence Stratigraphy and Facies Associations* (Eds H.W. Posamentier, C.P. Summerhayes, B.U. Haq and G.P. Allen), pp. 307–341. Special Publication 18, International Association of Sedimentologists. Blackwell Scientific Publications, Oxford.

Hutton, E.W.H. and Syvitski, J.P.M. (2004) Advances in the Numerical Modeling of Sediment Failure During the Development of a Continental Margin. *Mar. Geol.*, **203**, 367–380.

Imran, J. and Syvitski, J.P.M. (2000) Impact of extreme river events on coastal oceans. *Oceanography*, **13**, 85–92.

Imran, J., Parker, G. and Katopodes, N.D. (1998) A numerical model of the channel inception on submarine fans. *J. Geophys. Res.*, **103**, 1219–1238.

Imran, J., Harff, P. and Parker, G. (2001) A numerical model of submarine debris flow with graphical user interface. *Comput. Goesci.*, **27**, 717–729.

Inman, D., Nordstrom, C.E. and Flick, R.E. (1976) Currents in submarine canyons: an air-sea-land interaction. *Ann. Rev. Fluid Mech.*, **8**, 275–310.

Iverson, R.M. (1997) The physics of debris flows. *Rev. Geophys.*, **35**, 245–29.

Janbu, N., Bjerrum, L. and Kjaernsli, B. (1956) *Soil Mechanics Applied to some Engineering Problems.* Publication 16, Norwegian Geotechnical Institute, Oslo.

Jervey, M.T. (1988) Quantitative geologic modelling of siliciclastic rock sequences and their seismic expression. In: *Sea Level Changes: an Integrated Approach* (Eds C.K. Wilgus, B.S. Hastings, C.G. St.C. Kendall, *et al.*), pp. 47–69. Special Publication 42, Society of Economic Paleontologists and Mineralogists, Tulsa, OK.

Jewell, P.W., Stallard, R.F. and Mellor, G.L. (1993) Numerical studies of bottom shear stress and sediment distribution on the Amazon continental shelf. *J. Sediment. Petrol.*, **63**, 734–745.

Jiang, L. and LeBlond, P.H. (1993) Numerical modeling of an underwater Bingham plastic mudslide and the waves which it generates. *J. Geophys. Res.*, **98**, 10303–10317.

Johnson, T.R., Ellis, C.R., Farrel, G.J. and Stefan, H.G. (1987) Negatively buoyant flow in a diverging channel. I: flow regimes. *J. Hydraul. Eng.*, **113**, 715–729.

Jordan, T.E. and Flemings, P.B. (1991) Large-scale stratigraphic architecture, eustatic variation and unsteady tectonism: a theoretical evaluation. *J. Geophys. Res.*, **96**, 6681–6699.

Kassem, A. and Imran, J. (2001) Simulation of turbid underflow generated by the plunging of a river. *Geology*, **29**, 655–658.

Kaufman, P., Grotzinger, J.P. and McCormick, D.S. (1991) Depth-dependent diffusion algorithms for simulation of sedimentation in shallow marine depositional systems. In: *Sedimentary Modeling: Computer Simulations and Methods for Improved Parameter Definition* (Eds E.K. Franseen, L.W. Watney, C.G. Kendall and W. Ross). *Kansas State Geol. Surv. Bull.*, **233**, 489–508.

Keen, T.R. and Slingerland, R.L. (1993) A numerical study of sediment transport and event bed genesis during Tropical Storm Delia. *J. Geophys. Res.*, **98**, 4775–4791.

Kendall, C.G. and Sen, A. (1999) Use of sedimentary simulations for dating sequence boundaries and measuring the size of eustatic sea-level changes: an example from the Neogene of the Bahamas. In: *Computerized Modeling of Sedimentary Systems* (Eds J. Harff, W. Lemke and K. Stattegger), pp. 291–307. Springer-Verlag, New York.

Kendall, G., Moore, P., Strobel, J., *et al.* (1991) Simulation of the sedimentary fill of basins. In: *Computer Simulations and Methods for Improved Parameter Definition* (Eds E.K. Franseen, L.W. Watney, C.G. Kendall and W. Ross). *Kansas State Geol. Surv. Bull.*, **233**, 9–30.

Kenyon, P.M. and Turcotte, D.L. (1985) Morphology of a delta prograding by bulk sediment transport. *Geol. Soc. Am. Bull.*, **96**, 1457–1465.

Klein, C. and Hurlbut, C.S. (1985) *Manual of Mineralogy.* Wiley, New York, 596 pp.

Knapp, R.T. (1938) Energy balance in streams carrying suspended load. *Trans. Am. Geophys. Union*, **1**, 501–505.

Koltermann, C.E. and Gorelick, S. (1995) Fractional packing model for hydraulic conductivity derived from sediment mixtures. *Water Resour. Res.*, **31**, 3283–3297.

Kostaschuk, R.A., Luternauer, J.L. and Church, M.A. (1989) Suspended sediment hysteresis in a salt-wedge estuary: Fraser River, Canada. *Mar. Geol.*, **87**, 273–285.

Kostic, S. and Parker, G. (2003a) Progradational sand-mud deltas in lakes and reservoirs. Part 1. Theory and numerical modeling. *J. Hydraul. Res.*, **41**, 127–140.

Kostic, S. and Parker, G. (2003b) Progradational sand-mud deltas in lakes and reservoirs. Part 2. Experiment and numerical simulation. *J. Hydraul. Res.*, **41**, 141–152.

Kowallis, B.J., Jones, L.E.A. and Wang, H.F. (1984) Velocity-porosity-clay content systematics of poorly consolidated sandstones. *J. Geophys. Res.*, **89**, 10355–10364.

Kraft, B.J., Overeem, I., Holland, C.W., Pratson, L.F., Syvitski, J.P.M. and Mayer, L.M. (2006) Stratigraphic model predictions of geoacoustic properties, *IEEE J. Ocean Eng.*, **31**, 266–283.

Kuster, G.T. and Toksöz, M.N. (1974) Velocity and attenuation of seismic waves in two-phase media: Part I. Theoretical Formulations. *Geophysics*, **39**, 587–606.

Laberg, J.S. and Vorren, T.O. (1995) Late Weichselian submarine debris flow deposits on the Bear Island Trough Mouth Fan. *Mar. Geol.*, **127**, 45–72.

Lambe, T.W. and Whitman, R.V. (1969) *Soil Mechanics.* John Wiley & Sons, New York, 553 pp.

Lambert, A. (1982) Trubestrome des Rheins am Grund des Bodensees. *Sonderdruk aus Wasserwirtschaft*, **72**.

Lawrence, D.T. (1993) Evaluation of eustasy, subsidence and sediment input as controls on depositional sequence geometries and the synchroneity of sequence boundaries. In: *Siliciclastic Sequence Stratigraphy: Recent Developments and Applications* (Eds P. Weimer and H. Posamentier), pp. 337–368. Memoir 58, American Association of Petroleum Geologists, Tulsa, OK.

Lee, H.J. and Edwards, B.D. (1986) Regional method to assess offshore slope stability. *J. Geotech. Eng., ASCE*, **112**, 489–509.

Lee, H., Locat, J., Dartnell, P., Israel, K. and Wong, F. (1999) Regional variability of slope stability: application to the Eel margin, California. *Mar. Geol.*, **154**, 305–321.

Lee, H.J., Locat, J., Dartnell, P., Minasian, D. and Wong, F. (2000) A GIS-based regional analysis of the potential for shallow-seated submarine slope failure. *Proceedings of the 8th International Symposium on Landslides*, Cardiff, June, pp. 917–922.

Leithold, E.L. (1989) Depositional processes on an ancient and modern muddy shelf, northern California. *Sedimentology*, **36**, 179–202.

Li, M.Z. and Amos, C.L. (1998) Predicting ripple geometry and bed roughness under combined waves and currents in a continental shelf environment. *Cont. Shelf Res.*, **18**, 941–970.

Li, M.Z., Wright, L.D. and Amos, C.L. (1996) Predicting ripple roughness and sand resuspension under combined flows in a shoreface environment. *Mar. Geol.*, **130**, 139–161.

Locat, J. (1997) Normalized rheological behavior of muds and their flow properties in a pseudoplastic regime. In: *Proceedings of the First International Conference, Debris-flow Hazard Mitigation: Mechanics, Prediction and Assessment*, American Society of Civil Engineers, pp. 260–269.

Loseth, T.M. (1999) *Submarine Massflow Sedimentation: Computer Modelling and Basin Fill Stratigraphy.* Springer Lecture Notes in Earth Science No. 82, 156 pp.

Lou, J., Schwab, D.J., Beletsky, D. and Hawley, N. (2000) A model of sediment resuspension and transport dynamics in southern Lake Michigan. *J. Geophys. Res.*, **105**, 6591–6610.

Lowenthal, D., Lu, L., Roberson, R. and Sherwood, J.W.C. (1976) The wave equation applied to migration. *Geophys. Prospect.*, **24**, 380–399.

Mackey, S.D. and Bridge, J.S. (1995) Three-dimensional model of alluvial stratigraphy: theory and application. *J. Sediment. Res.*, **65**, 7–31.

Marion, D., Nur, A., Yin, H. and Han, D. (1992) Compressional velocity and porosity in sand-clay mixtures. *Geophysics*, **57**, 554–563.

Marks, D., Link, T., Winstral, A. and Garen, D. (2001) Simulating snowmelt processes during rain-on-snow over a semi-arid mountain basin. *Ann. Glaciol.*, **32**, 195–202.

Marr, J.G. (1999) *Experiments on subaqueous sandy gravity flows: flow dynamics and deposit structure.* M.S. thesis, University of Minnesota, 121 pp.

Marr, J.G., Harff, P.A., Shanmugam, G. and Parker, G. (2001) Experiments on subaqueous sandy gravity flows: the role of clay and water content in flow dynamics and depositional structures. *Geol. Soc. Am. Bull.*, **113**, 1377–1386.

Marr, J.G., Elverhøi, A., Harbitz, C., Imran, J. and Harff, P. (2002) Numerical simulation of mud-rich subaqueous debris flows on the glacially active margins of the Svalbard-Barents Sea. *Mar. Geol.*, **188**, 351–364.

Martinez, P.A. and Harbaugh, J.W. (1993) Simulating Nearshore Environments. Computer Methods in Geosciences V. 12, Pergamon Press, 265 pp.

Mavko, G., Mukerji, T. and Dvorkin, J. (1998) The Rock Physics Handbook: Tools for Seismic Analysis in Porous Media. Cambridge, Cambridge University Press, 329 pp.

Miller, K.G., Kominz, M.A., Browning, J.V., et al. (2005) The Phanerozoic Record of Global Sea-Level Change. Science, 310, 1293–1298.

Milliman, J.D. and Syvitski, J.P.M. (1992) Geomorphic/tectonic control of sediment discharge to the ocean: the importance of small mountainous rivers. J. Geol., 100, 525–544.

Mitchum, Jr., R.M., Vail, P.R. and Thompson, III, S. (1977) Part Two: the depositional sequence as a basic unit for stratigraphic analysis. In: Seismic Stratigraphy–Applications to Hydrocarbon Exploration (Ed. C.E. Payton), pp. 53–62. Memoir 26, American Association of Petroleum Geologists, Tulsa, OK.

Mohrig, D. and Marr, J.G. (2003) Constraining the efficiency of turbidity-current generation from submarine slides, slumps and debris flows using laboratory experiments. Mar. Petrol. Geol., 20, 883–899.

Mohrig, D., Whipple, K.X., Hondzo, M., Ellis, C. and Parker, G. (1998) Hydroplaning of subaqueous debris flows. Geol. Soc. Am. Bull., 110, 387–394.

Mohrig, D., Elverhøi, A. and Parker, G. (1999) Experiments on the relative mobility of muddy subaqueous and subaerial debris flows and their capacity to remobilize antecedent deposits. Mar. Geol., 154, 117–129.

Morehead, M.D. and Syvitski, J.P.M. (1999) River-plume sedimentation modeling for sequence stratigraphy: application to the Eel margin, Northern California. Mar. Geol., 154, 29–41.

Morehead, M.D., Syvitski, J.P.M. and Hutton, E.W.H. (2001) The link between abrupt climate change and basin stratigraphy: a numerical approach. Global Planet. Change, 28, 115–135.

Morehead, M.D., Syvitski, J.P.M., Hutton, E.W.H. and Peckham, S.D. (2003) Modeling the inter-annual and intra-annual variability in the flux of sediment in ungauged river basins. Global Planet. Change, 39, 95–110.

Morgenstern, N. (1967) Submarine slumping and the initiation of turbidity currents. In: Marine Geotechnique (Ed. A.F. Richard), pp. 189–210. Univesity of Illinois Press, Urbana.

Mulder, T. and Syvitski, J.P.M. (1995) Turbidity currents generated at river mouths during exceptional discharges to the world oceans. J. Geol., 103, 285–299.

Mulder, T. and Syvitski, J.P.M. (1996) Climatic and morphologic relationships of rivers: Implications of sea level fluctuations on river loads. J. Geol., 104, 509–523.

Mulder, T., Savoye, B. and Syvitski, J.P.M. (1997) Numerical modelling of the sediment budget for a mid-sized gravity flow: the 1979 Nice turbidity current. Sedimentology, 44, 305–326.

Mulder, T., Savoye, B., Syvitski, J.P.M. and Piper, D.J.W. (1998a) The Var submarine sedimentary system: understanding Holocene sediment delivery processes and their importance to the geological record. In: Geological Processes on Continental Margins: Sedimentation, Mass Wasting and Stability (Eds M.S. Stoker, D. Evans, A. Cramp), pp. 145–166. Special Publication 129, Geological Society Publishing House, Bath.

Mulder, T., Syvitski, J.P.M. and Skene, K.I. (1998b) Modeling of erosion and deposition by turbidity currents generated at river mouths. J. Sediment. Res., 68, 124–137.

Mulder, T., Syvitski, J.P.M., Migeon, S., Faugères, J.-C. and Savoye, B. (2003) Marine hyperpycnal flows: initiation, behavior and related deposits: a review. Mar. Petrol. Geol., 20, 861–882.

Mullenbach, B.L., Nittrouer, C.A., Puig, P. and Orange, D.L. (2004) Sediment deposition in a modern submarine canyon: Eel Canyon, northern California. Mar. Geol., 211, 101–119.

Nash, D. (1987) A comparative review of limit equilibrium methods of stability analysis. In: Slope Stability: Geotechnical Engineering and Geomorphology (Eds M.G. Anderson and K.S. Richards), pp. 11–76. John Wiley & Sons, New York.

Niedoroda, A.W., Reed, C.W., Swift, D.J.P., Arato, A. and Hoyanagi, K. (1995) Modeling shore-normal large-scale coastal evolution. Mar. Geol., 126, 180–200.

Nissen, S.E., Haskell, N.L., Steiner, C.T. and Coterill, K.L. (1999) Debris flow outrunner blocks, glide tracks and pressure ridges identified on the Nigerian continental slope using 3-D seismic coherency. The Leading Edge, 18, 595–599.

Norem, H., Locat, J. and Schieldrop, B. (1990) An approach to the physics and modeling of submarine flowslides. Mar. Geotech., 9, 93–111.

Nummedal, D., Riley, G.W. and Templet, P.L. (1993) High-resolution sequence architecture: a chronostratigraphic model based on equilibrium profile studies. In: Sequence Stratigraphy and Facies Associations (Eds H.W. Posamentier, C.P. Summerhayes, B.U. Haq and G.P. Allen), pp. 55–68. Special Publication 18, International Association of Sedimentologists. Blackwell Scientific Publications, Oxford.

O'Grady, D.B. and Syvitski, J.P.M. (2001) Predicting profile geometry of continental slopes with a multiprocess sedimentation model. In: Geological Modeling and Simulation: Sedimentary Systems (Eds D.F. Merriam and J.C. Davis), pp. 99–117. Kluwer Academic/Plenum Publishers, New York.

O'Grady, D.B., Syvitski, J.P.M., Pratson, L.F. and Sarg, J.F. (2000) Categorizing the morphologic variability of siliciclastic passive continental margins. Geology, 28, 207–210.

Ogston, A.S., Cacchione, D.A., Sternberg, R.W. and Kineke, G.C. (2000) Observations of storm and river flood-driven sediment transport on the northern California continental shelf. *Cont. Shelf Res.*, **20**, 2141–2162.

Overeem, I., Syvitski, J.P.M. and Hutton, E.W.H. (2005) Three-dimensional numerical modelling of deltas. In: *River Deltas – Concepts, Models and Examples* (Eds L. Giosan and J.P. Bhattacharya), pp. 13–30. Special Publication 83, Society of Sedimentary Geology, Tulsa, OK.

Overpeck, J.T. (1996) Warm climate surprises. *Science*, **271**, 1820–1821.

Paola, C. (2000) Quantitative models of sedimentary basin filling. *Sedimentology*, **47**, 121–178.

Paola, C., Heller, P.L. and Angevine, C.L. (1992) The large-scale dynamics of grain-size variation in alluvial basins, 1: theory. *Basin Res.*, **4**, 73–90.

Parker, G. (1982) Conditions for the ignition of catastrophically erosive turbidity currents. *Mar. Geol.*, **46**, 307–327.

Parker, G., Fukushima, Y. and Pantin, H.M. (1986) Self-accelerating turbidity currents. *J. Fluid Mech.*, **171**, 145–181.

Parker, G., García, M.H., Fukushima, Y. and Yu, W. (1987) Experiments on turbidity currents on an erodible bed. *J. Hydraul. Res.*, **25**, 123–147.

Parsons, J.D. and García, M.H. (1998) Similarity of gravity current fronts. *Phys. Fluids*, **10**, 3209–3213.

Parsons, J.D. and García, M.H. (2000) Enhanced sediment scavenging due to double-diffusive convection. *J. Sediment. Res.*, **70**, 47–52.

Parsons, J.D., Bush, J.W.M. and Syvitski, J.P.M. (2001) Hyperpycnal plumes with small sediment concentrations. *Sedimentology*, **48**, 465–478.

Pekar, S.F., Miller, K.G. and Kominz, M.A. (2000) Reconstructing the stratal geometry of New Jersey Oligocene sequences: resolving a patchwork distribution into a clear pattern of progradation. *Sediment. Geol.*, **134**, 93–109.

Pirmez, C., Pratson, L.F. and Steckler, M.S. (1998) Clinoform development by advection-diffusion of suspended sediment: Modeling and comparison to natural systems. *J. Geophys. Res.*, **103**, 24141–24157.

Pitlick, J. (1994) Relation between peak flows, precipitation and physiography for five mountainous regions in the western USA. *J. Hydrol.*, **158**, 219–240.

Plint, A.G. (1996) Marine and nonmarine systems tracts in fourth-order sequences in the Early-Middle Cenomanian, Dunvegan Alloformation, northeastern British Columbia, Canada. In: *High Resolution Sequence Stratigraphy: Innovations and Applications* (Eds J.A. Howell and J.F. Aitken), pp. 159–191. Special Publication 104, Geological Society Publishing House, Bath.

Plint, A.G. and Nummedal, D. (2000) The falling stage systems tract: recognition and importance in sequence stratigraphic analysis. In: *Sedimentary Responses to Forced Regressions* (Eds D.R. Hunt and R. Gawthorpe), pp. 1–17. Special Publication, 172, Geological Society Publishing House, Bath.

Posamentier, H.W. and Allen, G.P. (1993) The 'healing phase' – a commonly overlooked component of the transgressive systems tract. *American Association of Petroleum Geologists 1993 Annual Convention Abstracts*, p. 167.

Posamentier, H.W. and Vail, P.R. (1988) Eustatic controls on clastic deposition II – sequence and systems tract models. In: *Sea Level Changes: an Integrated Approach* (Eds C.K. Wilgus, B.S. Hastings, C.G. St.C. Kendall, *et al.*), pp. 125–154. Special Publication 42, Society of Economic Paleontologists and Mineralogists, Tulsa, OK.

Posamentier, H.W., Jervey, M.T. and Vail, P.R. (1988) Eustatic controls on clastic deposition I: conceptual framework. In: *Sea Level Changes: an Integrated Approach* (Eds C.K. Wilgus, B.S. Hastings, C.G. St.C. Kendall, *et al.*), pp. 109–124. Special Publication 42, Society of Economic Paleontologists and Mineralogists, Tulsa, OK.

Posamentier, H.W., Allen, G.P., James, D.P. and Tesson, M. (1992) Forced regressions in a sequence stratigraphic framework: Concepts, examples and exploration significance. *Bull. Am. Assoc. Petrol. Geol.*, **76**, 1687–1709.

Prather, B.E. (2000) Calibration and visualization of depositional process models for above-grade slopes: a case study from the Gulf of Mexico. *Mar. Petrol. Geol.*, **17**, 619–638.

Pratson, L.F. and Coakley, B.J. (1996) A model for the headward erosion of submarine canyons induced by downslope eroding sediment flows. *Geol. Soc. Am. Bull.*, **108**, 225–234.

Pratson, L.F. and Gouveia, W. (2002) Seismic modeling of experimental strata. *Bull. Am. Assoc. Petrol. Geol.*, **86**, 129–144.

Pratson, L., Imran, J., Parker, G., Syvitski, J.P.M. and Hutton, E.W.H. (2000) Debris flow versus turbidity currents: a modeling comparison of their dynamics and deposits. In: *Fine-Grained Turbidite Systems* (Eds A.H. Bouma and C.G. Stone), pp. 57–71. Memoir 72, American Association of Petroleum Geologists; Special Publication 68, Society for Sedimentary Geology, Tulsa, OK.

Pratson, L., Imran, J., Hutton, E.W.H., Parker, G. and Syvitski, J.P.M. (2001) BANG1D: a one-dimensional, Lagrangian model of subaqueous turbid surges. *Comput. Goesci.*, **27**, 701–716.

Pratson, L.F., Stroujkova, A., Herrick, D. and Boadu, F. (2003) Predicting seismic velocity and other rock properties from clay content only. *Geophysics*, **68**, 1847–1856.

Prior, D.D., Bornhold, B.D. and Johns, M.W. (1984) Depositional characteristics of a submarine debris flow. *J. Geol.*, **92**, 707–727.

Puig, P., Ogston, A.S., Mullenbach, B.L., Nittrouer, C.A. and Sternberg, R.W. (2003) Shelf to canyon sediment-transport processes on the Eel continental margin (northern California). *Mar. Geol.*, **193**, 129–149.

Puig, P., Ogston, A.S., Mullenbach, B.L., *et al.* (2004) Storm-induced sediment-gravity flows at the head of the Eel submarine canyon, northern California margin. *J. Geophys. Res.*, **109**, C03019.

Reed, C.W., Niedoroda, A.W. and Swift, D.J.P. (1999) Modeling sediment entrainment and transport processes limited by bed armoring. *Mar. Geol.*, **154**, 143–154.

Reynolds, D.J., Steckler, M.S. and Coakley, B.J. (1991) The role of the sediment load in sequence stratigraphy; the influence of flexural isostasy and compaction. *J. Geophys. Res.*, **96**, 6931–6949.

Ribberink, J.S. and Al-Salem, A.A. (1995) Sheet flow and suspension of sand in oscillatory boundary layers. *Coastal Eng.*, **25**, 205–225.

Richards, K. (1984) Some observations on suspended sediment dynamics in Storbregrova, Jotunheimen. *Earth Surf. Proc. Landf.*, **9**, 101–112.

Ritchie, B.D., Hardy, S. and Gawthorpe, R.L. (1999) Three-dimensional numerical modeling of coarse-grained clastic deposition in sedimentary basins. *J. Geophys. Res.*, **104**, 17759–17780.

Ross, W.C., Halliwell, B.A., May, J.A., Watts, D.E. and Syvitski, J.P.M. (1994) Slope readjustment: a new model for the development of submarine fans and aprons. *Geology*, **22**, 511–514.

Sanford, L.P. and Maa, J.P.-Y. (2001) A unified erosion formulation for fine sediments. *Mar. Geol.*, **179**, 9–23.

Savage, S.V. and Hutter, K. (1991) The dynamics of avalanches of granular materials from initiation to run out. Part 1: analysis. *Acta Mechanica*, **86**, 201–223.

Schäfer-Neth, Ch. and Stattegger, K. (1999) Icebergs in the North Atlantic: Modeling circulation changes and glacio-marine deposition. In: *Computerized Modeling of Sedimentary Systems* (Eds J. Harff, W. Lemke and K. Stattegger), pp. 63–78. Springer-Verlag, New York.

Scully, M.E., Friedrichs, C.T. and Wright, L.D. (2002) Application of an analytical model of critically stratified gravity-driven sediment transport and deposition to observations from the Eel River continental shelf, Northern California. *Cont. Shelf Res.*, **22**, 2443–2460.

Scully, M.E., Friedrichs, C.T. and Wright, L.D. (2003) Numerical modeling results of gravity-driven sedi-

ment transport and deposition on energetic shelves, *J. Geophys. Res.*, **108**, 17.1–17.14.

Shelton, J.L. and Cross, T.A. (1989) The influence of stratigraphy in reservoir simulation. In: *Quantitative Dynamic Stratigraphy* (Ed. T.A. Cross), pp. 589–600. Prentice-Hall, New York.

Sherlock, D.H. and Evans, B.J. (2001) The development of seismic reflection sandbox modeling. *Bull. Am. Assoc. Petrol. Geol.*, **85**, 1645–1659.

Shuster, M.W. and Aigner, T. (1994) Two-dimensional synthetic seismic and log cross sections from stratigraphic forward models. *Bull. Am. Assoc. Petrol. Geol.*, **78**, 409–431.

Skene, K., Mulder, T. and Syvitski, J.P.M. (1997) INFLO1: a model predicting the behaviour of turbidity currents generated at a river mouth. *Comput. Geosci.*, **23**, 975–991.

Skene, K., Piper, D.J.W., Aksu, A.E. and Syvitski, J.P.M. (1998) Evaluation of the global oxygen isotope curve as a proxy for Quaternary sea level by modeling of delta progradation. *J. Sediment. Res.*, **68**, 1077–1092.

Slingerland, R., Harbaugh, J.W. and Furlong, K. (1994) *Simulating Clastic Sedimentary Basins*. PTR Prentice Hall Sedimentary Geology Series, Englewood Cliffs, NJ, 211 pp.

Smith, J.D. (1977) Modeling of sediment transport on continental shelves. In: *The Sea*, Vol. 6 (Eds E.D. Goldberg, I.N. McCave, J.J. O'Brien and J.H. Steele), pp. 539–577. John Wiley & Sons, New York.

Sommerfield, C.K. and Nittrouer, C.A. (1999) Modern accumulation rates and a sediment budget for the Eel shelf: a flood-dominated depositional environment. *Mar. Geol.*, **154**, 227–241.

Sonu, C.J. and James, W.R. (1973) A Markov model for beach profile changes. *J. Geophys. Res.*, **78**, 1462–1471.

Stacey, M.W. and Bowen, A.J. (1988a) The vertical structure of turbidity currents: theory and observations. *J. Geophys. Res.*, **93**, 3528–3542.

Stacey, M.W. and Bowen, A.J. (1988b) The vertical structure of turbidity currents and a necessary condition for self-maintenance. *J. Geophys. Res.*, **93**, 3543–3553.

Stafleu, J. and Schlager, W. (1993) Pseudo-toplap in seismic models of the Schlern-Raibl contact (Sella platform, northern Italy). *Basin Res.*, **5**, 55–65.

Steckler, M.S. (1999) High resolution sequence stratigraphic modeling: 1. The interplay of sedimentation, erosion and subsidence. In: *Numerical Experiments in Stratigraphy: Recent Advances in Stratigraphic and Computer Simulations* (Eds J. Harbaugh, L. Watney, G. Rankey, *et al.*), pp. 139–149. Memoir 62, Society of Economic Paleontologists and Mineralogists, Tulsa, OK.

Steckler, M.S., Syvitski, J.P.M., Swift, D.J.P., Goff, J. and Niedoroda, A.W. (1996) Stratigraphic modeling of continental margins. *Oceanography*, **9**, 193–188.

Steckler, M.S., Mountain, G.S., Miller, K.G. and Christie-Blick, N. (1999) Reconstruction of Tertiary progradation and clinoform development on the New Jersey passive margin by 2-D backstripping. *Mar. Geol.*, **154**, 399–420.

Stefan, H.G. and Johnson, T. (1987) Negatively buoyant flow in a diverging channel III: onset of underflow. *J. Hydraulic Eng.*, **113**, 715–729.

Stoll, R.D. (1989) *Sediment Acoustics*. Springer-Verlag, Berlin, 152 pp.

Storms, J.E.A., Weltje, G.J., van Dijke, J.J., Geel, C.R. and Kroonenberg, S.B. (2002) Process-response modeling of wave-dominated coastal systems: Simulating evolution and stratigraphy on geological timescales. *J. Sediment. Res.*, **72**, 226–239.

Styles, R. and Glenn, S.M. (2000) Modeling stratified wave and current bottom boundary layers on the continental shelf. *J. Geophys. Res.*, **105**, 24119–24139.

Swenson, J.B., Voller, V.R., Paola, C., Parker, G. and Marr, J. (2000) Fluvio-deltaic sedimentation: a generalized Stefan problem. *European J. Applied Math.*, **11**, 433–452.

Syvitski, J.P.M. (1989) The process–response model in quantitative dynamic stratigraphy. In: *Quantitative Dynamic Stratigraphy* (Ed. T.A. Cross), pp. 309–334. Prentice-Hall, New York.

Syvitski, J.P.M. (2002) Sediment transport variability in Arctic rivers: implications for a warmer future. *Polar Res.*, **21**, 323–330.

Syvitski, J.P.M. (2003) Sediment fluxes and rates of sedimentation. In: *Encyclopedia of Sediments and Sedimentary Rocks* (Ed. G.V. Middleton), pp. 600–606. Kluwer Academic Publishers, Dordrecht.

Syvitski, J.P.M. and Alcott, J.M. (1995a). DELTA6: numerical simulation of basin sedimentation affected by slope failure and debris flow runout. In: *Pierre Beghin International Workshop on Rapid Gravitational Mass Movements*, Grenoble, France, pp. 305–312.

Syvitski, J.P.M. and Alcott, J.M. (1995b) RIVER3: Simulation of water and sediment river discharge from climate and drainage basin variables. *Comput. Geosci.*, **21**, 89–151.

Syvitski, J.P.M. and Andrews, J.T. (1994) Climate change: numerical modeling of sedimentation and coastal processes, eastern Canadian Arctic. *Arctic and Alpine Res.*, **26**, 199–212.

Syvitski, J.P.M. and Bahr, D.B. (2001) Numerical models of marine sediment transport and deposition. *Comput. Geosci.*, **27**(6): 617–618.

Syvitski, J.P.M. and Hutton, E.W.H. (2001) 2D-SEDFLUX 1.0C: an advanced process–response numerical model for the fill of marine sedimentary basins. *Comput. Geosci.*, **27**, 731–754.

Syvitski, J.P.M. and Hutton, E.W.H. (2003) Failure of marine deposits and their redistribution by sediment gravity flows. *Pure Applied Geophys.*, **160**, 2053–2069.

Syvitski, J.P.M. and Hutton, E.W.H. (2004) Advances in the numerical modeling of sediment failure during the development of a continental margin. *Mar. Geol.*, **203**, 367–380.

Syvitski, J.P.M. and Morehead, M.D. (1999) Estimating river-sediment discharge to the ocean: application to the Eel Margin, northern California. *Mar. Geol.*, **154**, 13–28.

Syvitski, J.P.M. and Schafer, C.T. (1996) Evidence for an earthquake-triggered basin collapse in Saguenay Fjord, Canada. *Sediment. Geol.*, **104**, 127–153.

Syvitski, J.P.M., Smith, J.N., Boudreau, B. and Calabrese, E.A. (1988) Basin sedimentation and the growth of prograding deltas. *J. Geophys. Res.*, **93**, 6895–6908.

Syvitski, J.P.M., Pratson, L., Perlmutter, M., de Boer, P., Parker, G., García, M., Wiberg, P., Steckler, M., Swift, D. and Lee, H.J. (1997) EARTHWORKS: a large scale and complex numerical model to understand the flux and deposition of sediment over various time scales. In: *Proceedings of the Third Annual Conference of the International Association of Mathmatic Geologists* (Ed. V. Pawlowsky-Glahn), CIMNE-Barcelona, Vol. 3, pp. 29–33.

Syvitski, J.P.M., Nicholson, M., Skene, K. and Morehead, M.D. (1998a) PLUME1.1: deposition of sediment from a fluvial plume. *Comput. Geosci.*, **24**, 159–171.

Syvitski, J.P.M., Morehead, M. and Nicholson, M. (1998b) HYDROTREND: a climate-driven hydrologic-transport model for predicting discharge and sediment to lakes or oceans. *Comput. Geosci.*, **24**, 51–68.

Syvitski, J.P.M., Pratson, L. and O'Grady, D.B. (1999) Stratigraphic predictions of continental margins for the Navy. In: *Numerical Experiments in Stratigraphy: Recent Advances in Stratigraphic and Computer Simulations* (Eds J. Harbaugh, L. Watney, G. Rankey, *et al.*), pp. 219–236. Memoir 62, Society of Economic Paleontologists and Mineralogists, Tulsa, OK.

Syvitski, J.P.M., Morehead, M.D., Bahr, D. and Mulder, T. (2000) Estimating fluvial sediment transport: the rating parameters. *Water Resource Res.*, **36**, 2747–2760.

Syvitski, J.P.M., Peckham, S.D., Hilberman, R.D. and Mulder, T. (2003) Predicting the terrestrial flux of sediment to the global ocean: a planetary perspective. *Sediment. Geol.*, **162**, 5–24.

Syvitski, J.P.M., Tucker, G., Seber, D., *et al.* (2004) *Community Surface Dynamics Modeling System Implementation Plan, A Report to the National Science Foundation*. INSTAAR, University of Colorado, Boulder, 61 pp.

Syvitski, J.P.M., Kettner, A., Peckham, S.D. and Kao, S.-J. (2005a) Predicting the flux of sediment to the coastal zone: application to the Lanyang watershed, northern Taiwan. *J. Coastal Res.*, **21**, 580–587.

Syvitski, J.P.M., Kettner, A.J., Correggiari, A. and Nelson, B.W. (2005b) Distributary channels and their impact on sediment dispersal. *Mar. Geol.*, **222–223**, 75–94.

Takahashi, T. (1991) *Debris Flow*. A.A. Balkema, Rotterdam, 165 pp.

Tetzlaff, D.M. and Harbaugh, J.W. (1989) *Simulating Clastic Sedimentation*. Van Nostrand Reinhold, New York, 202 pp.

Tetzlaff, D.M. and Priddy, G. (2001) Sedimentary process modeling: From academia to industry. In: *Geological Modeling and Simulation: Sedimentary Systems* (Eds D.F. Merriam and J.C. Davis), pp. 45–70. Kluwer Academic/Plenum Publishers, New York.

Thorne, J.A., Grace, E., Swift, D.J.P and Niedoroda, A. (1991) Sedimentation on continental margins, III: the depositional fabric – and analytical approach to stratification and facies identification. In: *Shelf Sand and Sandstone Bodies* (Eds D.J.P. Swift, G.F. Oertel, R.W. Tillman and J.A. Thorne), pp. 59–87. Special Publication 14, International Association of Sedimentologists. Blackwell Scientific Publications, Oxford.

Toniolo, H., Harff, P.A., Marr, J.G., Paola, C. and Parker, G. (2004) Experiments on reworking by successive unconfined subaqueous and subaerial muddy debris flows. *J. Hydraulic Eng.*, **130**, 38–48.

Toyasa, C. and Nur, A. (1982) Effects of diagenesis and clays on compressional velocities in rocks. *Geophys. Res. Lett.*, **9**, 5–8.

Traykovski P., Geyer, W.R., Irish, J.D. and Lynch, J.F. (2000) The role of wave-induced density-driven fluid mud flows for cross-shelf transport on the Eel River continental shelf. *Cont. Shelf Res.*, **20**, 2113–2140.

Trowbridge, J.H. and Kineke, G.C. (1994) Structure and dynamics of fluid muds on the Amazon continental shelf. *J. Geophys. Res.*, **99**, 865–874.

Vail, P.R. (1977) *Seismic Stratigraphy and Global Changes in Sea Level, Parts 1–11*. Memoir 26, American Association of Petroleum Geologists, Tulsa, OK, pp. 51–212.

Vail, P.R. (1987) Seismic stratigraphy interpretation procedure. In: *Atlas of Seismic Stratigraphy* (Ed. A.W. Bally), pp. 1–10. Studies in Geology, No. 27, American Association of Petroleum Geologists, Tulsa, OK.

Vail, P.R., Mitchum, R.M. Jr., Todd, R.G., *et al.* (1977) Seismic stratigraphy and global changes of sea level. In: *Seismic Stratigraphy – Applications to Hydrocarbon Exploration* (Ed. C.E. Payton), pp. 49–212. Memoir 26, American Association of Petroleum Geologists, Tulsa, OK.

Vail, P.R., Audemard, F., Bowman, S.A., Einsele, G. and Perez-Cruz, G. (1991) The stratigraphic signatures of tectonics, eustasy and sedimentation. In: *Cycles and Events in Stratigraphy* (Eds G. Einsele, W. Ricken and A. Seilacher), pp. 617–659. Springer-Verlag, Berlin.

Van den Berg, J.H., van Gelder, A. and Mastbergen, D.R. (2002) The importance of breaching as a mechanism for subaqueous slope failure in fine sand. *Sedimentology*, **49**, 81–95.

Van Kessel, T. and Kranenburg, C. (1996) Gravity current of fluid mud on sloping bed. *J. Hydraulic Eng.*, **122**, 710–717.

Van Wagoner, J.C. (1995) Sequence stratigraphy and marine to nonmarine facies architecture of the Grassy Member, Blackhawk Formation, Book Cliffs, Utah, U.S.A. In: *Sequence Stratigraphy of Foreland Basin Deposits, Outcrop and Subsurface Examples from the Cretaceous of North America* (Eds J.C. Van Wagoner and G.T. Bertram), pp. 137–224. Memoir 64, American Association of Petroleum Geologists, Tulsa, OK.

Van Wagoner, J.C., Mitchum, R.M., Campion, K.M. and Rahmanian, V.D. (1990) Siliciclastic sequence stratigraphy in well logs, cores and outcrops. *Am. Assoc. Petrol. Geol. Meth. Explor. Ser.*, **7**, 55 pp.

Versteeg, R. (1994) The Marmousi experience: velocity model determination on a synthetic complex data set. *The Leading Edge*, **13**, 927–936.

Wang, F.C. and Wei, J.S. (1986) River mouth mechanisms and coastal sediment deposition. In: *River Sedimentation*, Vol. 3 (Eds S.Y. Wang, H.W. Shen and L.Z. Ding), pp. 290–299. University of Mississippi Press.

Wang, H.F. (2000) *Theory of Linear Poroelasticity*. Princeton University Press, Princeton, NJ, 287 pp.

Waples, D.W. and Kamata, H. (1993) Modelling porosity reduction as a series of chemical and physical processes. In: *Basin Modelling: Advances and Applications* (Eds A.G. Dore, J.H. Augustustson, C. Hermanrud, D.J. Stewart and O. Sylta), pp. 303–320. Special Publication 3, Norwegian Petroleum Society, Oslo.

Watney, W.L., Rankey, E.C. and Harbaugh, J. (1999) Perspectives on stratigraphic simulation models: Current approaches and future opportunities. In: *Numerical Experiments in Stratigraphy: Recent Advances in Stratigraphic and Computer Simulations* (Eds J. Harbaugh, L. Watney, G. Rankey, *et al.*), pp. 3–21. Memoir 62, Society of Economic Paleontologists and Mineralogists, Tulsa, OK.

Weirich, F.H. (1984) Turbidity currents: monitoring their occurence and movement with a three-dimensional sensor network. *Science*, **224**, 384–387.

Wheatcroft, R.A., Sommerfield, C.K., Drake, D.E., Borgeld, J.C. and Nittrouer, C.A. (1997) Rapid and widespread dispersal of flood sediment on the northern California continental margin. *Geology*, **25**, 163–166.

Whitman, R.V. and Bailey, W.A. (1967) Use of computers for slope stability analysis. *J. Soil Mech., ASCE*, **93**, 475–498.

Wiberg, P.L. (2000) Coupling of sediment resuspension and post-depositional bed modifications in fine-grained shelf environments. *Eos (Trans. Am. Geophys. Union)*, **81**, OS41M–10.

Wiberg, P.L., Drake, D.E. and Cacchione, D.A. (1994) Sediment resuspension and bed armoring during high bottom stress events on the northern California inner continental shelf: measurements and predictions. *Cont. Shelf Res.*, **14**, 1191–1220.

Wiberg, P.L., Drake, D.E., Harris, C.K. and Noble, M.A. (2002) Sediment transport on the Palos Verdes shelf over seasonal to decadal time scales. *Cont. Shelf Res.*, **22**, 987–1004.

Wilson, K.C. (1987) Analysis of bedload motion at high shear stress. *J. Hydraulic Eng.*, **113**, 97–106.

Wright, L.D. and Coleman, J.M. (1971) Effluent expansion and interfacial mixing in the presence of a salt wedge, Mississippi River Delta. *J. Geophys. Res.*, **76**, 8649–8661.

Wright, L.D., Wiseman Jr., W.J., Yang, Z.-S., *et al.* (1990) Processes of marine dispersal and deposition of suspended silts off the modern mouth of the Huanghe (Yellow River). *Cont. Shelf Res.*, **10**, 1–40.

Wright, L.D., Kim, S.-C. and Friedrichs, C.T. (1999) Across-shelf variations in bed roughness, bed stress and sediment suspension on the northern California shelf. *Mar. Geol.*, **154**, 99–115.

Wright, L.D., Friedrichs, C.T., Kim, S.C. and Scully, M.E. (2001) Effects of ambient currents and waves on gravity-driven sediment transport on continental shelves. *Mar. Geol.*, **175**, 25–45.

Xu, J.P. and Wright, L.D. (1995) Tests of bed roughness models using field data from the Middle Atlantic Bight. *Cont. Shelf Res.*, **15**, 1409–1434.

Xu, S. and White, R.E. (1995) A physical model for shear wave velocity prediction. *Geophys. Prospect.*, **44**, 687–717.

Yilmaz, Ö. (1987) *Seismic Data Processing*. Investigations in Geophysics No. 2, Society of Exploration Geophysicists, Tulsa, OK, 526 pp.

Zhang, Y., Swift, D.J.P., Fan, S., Niedoroda, A.W. and Reed, C.W. (1999) Two-dimensional numerical modeling of storm deposition on the northern California shelf. *Mar. Geol.*, **154**, 155–167.

Index